Microbiological Identification using MALDI-TOF and Tandem Mass Spectrometry

Microbiological Identification using MALDI-TOF and Tandem Mass Spectrometry

Industrial and Environmental Applications

Edited by

Haroun N. Shah
Middlesex University
London, UK

Saheer E. Gharbia
UK Health Security Agency
London, UK

Ajit J. Shah
Middlesex University
London, UK

Erika Y. Tranfield
Bruker UK Limited
Coventry, UK

K. Clive Thompson
ALS, Life Sciences
Rotherham, UK

WILEY

This edition first published 2023

Registered Offices
John Wiley & Sons, Inc., 111 River Street, Hoboken, NJ 07030, USA
John Wiley & Sons Ltd, The Atrium, Southern Gate, Chichester, West Sussex, PO19 8SQ, UK

Editorial Office
John Wiley & Sons Ltd, The Atrium, Southern Gate, Chichester, West Sussex, PO19 8SQ, UK

For details of our global editorial offices, customer services, and more information about Wiley products visit us at www.wiley.com.

Wiley also publishes its books in a variety of electronic formats and by print-on-demand. Some content that appears in standard print versions of this book may not be available in other formats.

Library of Congress Cataloging-in-Publication Data applied for
Hardback ISBN: 9781119814054

Cover Design: Wiley
Cover Image: Courtesy of C.K Fagerquist and O. Sultan; Shiga holotoxin structural image (adapted by Clifton Fagerquist), based on open-access article - Structure of shiga toxin type 2 (Stx2) from Escherichia coli O157:H7

Set in 9.5/12.5pt STIXTwoText by Straive, Pondicherry, India
Printed and bound by CPI Group (UK) Ltd, Croydon, CR0 4YY
C9781119814054_150323

This book is dedicated to the late Professor Franz Hillenkamp who visited and communicated with MISU in the midst of this work and acted as our mentor during the development of MALDI-TOF and tandem mass spectrometry for clinical microbiology

and

Iona, Amaya and Calum.

Contents

List of Contributors

Manuel J. Arroyo
Clover Bioanalytical Software
Granada, Spain

Omar Belgacem
Ascend Diagnostics, Ltd
Manchester, UK

Ken Bruce
School of Cancer and Pharmaceutical Sciences
King's College London
London, UK

Rachel Chalmers
Cryptosporidium Reference Unit
Public Health Wales
Microbiology and Health Protection
Singleton Hospital
Swansea, UK;
Swansea University Medical School
Swansea, UK

Hilary Chan
Global Sterility Assurance and Microbiology
Takeda Pharmaceutical Co. Ltd
Lexington, MA, USA

Michael L. Clawson
United States Department of Agriculture
Agricultural Research Service
U.S. Meat Animal Research Center
Clay Center
NE, USA

Rainer Cramer
Department of Chemistry
University of Reading
Reading, UK

Nadia Darwich
Dŵr Cymru Welsh Water
Glaslyn Laboratory
Newport, UK

Maria E. Dueñas
Laboratory for Biological Mass Spectrometry
Biosciences Institute
Newcastle University
Newcastle upon Tyne, UK

Louise Duncan
School of Cancer and Pharmaceutical Sciences
King's College London
London, UK

Kaveh Emami
FUJIFILM Diosynth Biotechnologies
Billingham, UK

Clifton K. Fagerquist
Produce Safety & Microbiology
Western Regional Research Center
Agricultural Research Service
U.S. Department of Agriculture
CA, USA

Christine E. Farrance
Charles River Laboratories
Newark, DE, USA

Simona Francese
Biomolecular Science Research Centre
Sheffield, UK

Giulia T.M. Getti
School of Science
Faculty of Engineering and Science
University of Greenwich
Kent, UK

Saheer E. Gharbia
Pathogen Genomics, Gastrointestinal
Infection and Food Safety
Clinical Public Health
UK Health Security Agency
London, UK

Victoria Girard
BioMérieux
R&D Microbiologie
France

Radhey S. Gupta
Department of Biochemistry and
Biomedical Sciences
McMaster University
Hamilton, Canada

Alvin Han
Department of Biochemistry and
Biomedical Sciences
McMaster University
Hamilton, Canada

Christoph Hansy
Global Sterility Assurance and Microbiology
Takeda Pharmaceutical Co. Ltd
Vienna, Austria

Jason K. Iles
MAP Sciences Limited
iLab, Priory Park
Bedfordshire, UK

Ray K. Iles
MAP Sciences Limited
iLab, Priory Park
Bedfordshire, UK

Jean-Luc Jany
Université de Brest
INRAE
Laboratoire Universitaire de Biodiversité et
Ecologie Microbienne
France

Lynn Johnson
Central Quality - Microbiology and
Contamination Control
National Resilience, Inc.
USA

Matthew Jones
Dŵr Cymru Welsh Water
Glaslyn Laboratory
Newport, UK

Prasanna D. Khot
Charles River Laboratories
Newark, DE, USA

Gerald Larrouy-Maumus
Faculty of Natural Sciences
Department of Life Sciences
MRC Centre for Molecular Bacteriology &
Infection
Imperial College
London, UK

Sibylle Lebert
Microbiology Group
Matís, Reykjavík, Iceland

Yi Liu
Faculty of Natural Sciences
Department of Life Sciences
MRC Centre for Molecular Bacteriology &
Infection
Imperial College
London, UK

John Dustin Loy
School of Veterinary Medicine and Biomedical Sciences
Institute for Agriculture and Natural Resources
University of Nebraska-Lincoln
Lincoln, NE, USA

Luis Mancera
Clover Bioanalytical Software
Granada, Spain

Viggó Þór Marteinsson
Microbiology Group
Matís, Reykjavík, Iceland;
Faculty of Food Science and Nutrition
University of Iceland, Reykjavík, Iceland;
The Agricultural University of Iceland
Reykjavík, Iceland

Gema Méndez
Clover Bioanalytical Software
Granada, Spain

Valérie Monnin
BioMérieux
R&D Microbiologie
France

Jérôme Mounier
Université de Brest, INRAE
Laboratoire Universitaire de Biodiversité et Ecologie Microbienne
France

Bjorn Nielsen
ALS Environmental
Wakefield, UK

Gudrun Overesch
Institute of Veterinary Bacteriology
Vetsuisse Faculty
University of Bern
Bern, Switzerland

Marina Oviaño
Complejo Hospitalario Universitario de A Coruña
A Coruña, Spain

Cristian Piras
Department of Health Sciences
Magna Graecia University
Catanzaro, Italy

Jade Pizzato
Faculty of Natural Sciences
Department of Life Sciences
MRC Centre for Molecular Bacteriology & Infection
Imperial College
London, UK

Philip L. Poole
School of Science
Faculty of Engineering and Science
University of Greenwich
Kent, UK

Diane Purchase
Department of Natural Sciences
Faculty of Science and Technology
Middlesex University
London, UK

Emmanuel Raptakis
Fasmatech Science and Technology
Athens, Greece

Belén Rodríguez-Sánchez
Instituto de Investigación Sanitaria Gregorio Marañón
Madrid, Spain

Nolwenn Rolland
BioMérieux
R&D Microbiologie
France

Bashudev Rudra
Department of Biochemistry and
Biomedical Sciences
McMaster University
Hamilton, Canada

Cristina Russo
Department of Natural Sciences
Faculty of Science and Technology
Middlesex University
London, UK

Hiroaki Sato
Research Institute for Sustainable
Chemistry
National Institute of Advanced Industrial
Science and Technology (AIST)
Hiroshima, Japan

Ajit J. Shah
Department of Natural Sciences
Middlesex University
London, UK

Haroun N. Shah
Department of Natural Sciences
Middlesex University
London, UK

Laila M.N. Shah
Department of Physical & Theoretical
Chemistry
University of Oxford
Oxford, UK

Joachim Spergser
Department for Pathobiology
Institute of Microbiology
Vetmeduni – University of Veterinary
Medicine Vienna
Vienna, Austria

Hiroto Tamura
Department of Environmental Bioscience
Laboratory of Environmental Microbiology
Meijo University
Nagoya, Japan

K. Clive Thompson
ALS, Life Sciences
Rotherham, UK

Dorina Timofte
Department of Veterinary Anatomy,
Physiology and Pathology
Institute of Infection,
Veterinary and Ecological Sciences
University of Liverpool, Leahurst Campus
Neston, UK

Andrew Tomlin
ALS Laboratories (UK) Ltd
Rotherham, UK

Erika Y. Tranfield
Bruker UK Limited
Coventry, UK

Pauline Vannier
Microbiology Group
Matís, Reykjavík, Iceland

Ruth Walker
Laboratory for Biological Mass
Spectrometry
Biosciences Institute
Newcastle University
Newcastle upon Tyne, UK

Alan Ward
School of Biology
Newcastle University
Newcastle upon Tyne, UK

Malcolm Ward
Department of Natural Sciences
Middlesex University
London, UK

Raminta Zmuidinaite
MAP Sciences Limited
iLab, Priory Park
Bedfordshire, UK

Preface

Clinical symptoms provide clues to the aetiology of infections. However, there is no substitute for accurate identification to confirm disease and enable appropriate treatment to be confidently applied. Consequently, microbiological identification has been the cornerstone of clinical microbiology since Gram introduced his differential staining technique in 1884. Along with morphological characters, phenotypic and metabolic fermentation profiles grew in significance and appeared in the first edition of *Bergey's Manual of Determinative Bacteriology* in 1923. This remained the foundation of diagnostic microbiology and the basis for classification and identification of bacteria up until eighth edition of *Bergey's Manual* in 1974. This framework was incorporated into automated and miniaturized commercial kits (e.g. API® ID strip range by bioMérieux) for clinical and subsequently industrial and environmental laboratories to identify microbial species.

In the late 1970s, a new era of mass spectrometry enhanced microbial identification through higher-resolution analysis of microbial cellular components. Extensive analyses of polar and non-polar lipids revealed immense diversity in the microbial kingdom and specific molecular compositions were found to be unique to taxa. A new phase of chemical structure analysis dominated microbial taxonomy and classification and *Bergey's Manual* transitioned to a systematic approach for classification of the microbial kingdom, reflected in a change of the title of the new first edition to *Bergey's Manual of Systematic Bacteriology* (1984). However, implementation of chemical/bioanalytical methods and, in particular, electrophoresis, chromatography and mass spectrometry into diagnostic laboratories remained limited to specialized research laboratories with access to high-resolution analytical resources. Co-editor HNS's laboratory at the Royal London Hospital Medical College, University of London, was one of the pioneering teams to implement chemotaxonomy for the analysis of atypical and metabolically inert bacteria, including anaerobic pathogens. Wider adoption was found too cumbersome and technically demanding and limited their broader endorsement. However, a breakthrough was achieved by gas chromatography of long-chained fatty acid (LCFA) where analysis was partially automated by MIDI Inc., who developed the first dedicated database to drive analysis of lipid profiles. On the other hand, pyrolysis mass spectrometry failed to establish a wider base and was explored mainly by research laboratories. But lack of interlaboratory reproducibility, high cost and their cumbersome nature curtailed its development.

In 1973, Franz Hillenkamp developed a high-performance laser microprobe mass spectrometer with a spatial resolution of 0.5 μm and sub-attogram limit of detection for lithium atoms. This instrument was commercialized as the LAMMA 500. The more advanced 1000 was one of the first laser desorption mass spectrometers to be used for mass spectrometry imaging of tissues. In 1985, Hillenkamp and Michael Karas used a LAMMA 1000 mass spectrometer to demonstrate for the first time the technique of matrix-assisted laser desorption/ionization (MALDI), which allowed the analysis of large biopolymers. By the mid-1990s several researchers analysed bacterial cells and demonstrated that unique mass spectral profiles could be obtained from different species with MALDI. The concept of exploring the technology for a clinical diagnostic laboratory was not pursued, partly due to a history of failure of mass spectrometry (MS)-based techniques in clinical microbiology.

The UK's Public Health Laboratory Service (PHLS, later PHE and now UK Health Security Agency), a 100-year-old institute that focused on the analysis of pathogens, was structured along the lines of human pathogens/infections; thus a *Staphylococcus* laboratory, enteric or respiratory infections units etc. were led by specialist scientists for principal clinical pathogens. This permitted a high degree of expertise of various pathogens, yet each laboratory still retained a significant level of 'unknown' species in storage that were designated *incertae sedis*. Therefore, in 1997, PHLS established a new laboratory, designated the 'Molecular Identification Services Unit' (MISU), under the directorship of co-editor HNS. The function of MISU was to improve the level of species identification of atypical, rarely isolated and emerging human pathogens through research programmes while providing a more comprehensive diagnostic service function for the organization. Coming from a research background in which 16S rRNA sequence analysis was being developed as a tool for studying microbial phylogenetics, the technique was adopted for bacterial identification and MISU became the first accredited laboratory to implement this approach for human clinical samples. The laboratory also incorporated LCFA profiles as an adjunct to its newly employed 16S rRNA and was therefore well positioned to assess the potential of emerging technologies.

MISU was fortuitously given the opportunity to field-test the first benchtop MALDI-time-of-flight MS (MALDI TOF MS) (Kratos Analytical Inc.) and organized a conference on 27 October 1998 jointly with Kratos Analytical and Manchester Metropolitan University to explore and demonstrate potential applications of MALDI-TOF MS for clinical laboratories. An instrument was placed at the conference lecture theatre and the technique was demonstrated live during the meeting. Its speed of analysis and simplicity had a huge impact on the audience, but the general comment was that it was a new platform for research applications. Undeterred by the negative views, MISU would go on to relentlessly pursue the technology for the next decade for microbial identification of clinical samples and eventually implement it as its frontline approach.

These were interesting times, as concurrently there were two major developments of the technology, one designated surface-enhanced laser desorption/ionization (SELDI)-TOF MS that selectively captured proteins on ProteinChip Arrays prior to MALDI-TOF MS analysis and SEQUENOM's MassArray for genotyping that used reverse transcriptase to produce the more stable RNA for analysis. To our knowledge, MISU was the only laboratory to have implemented the three approaches at the time. The strategy envisioned was that MALDI-TOF MS would be explored for general microbial identification, SELDI-TOF MS for proteotyping of strains, while the MassArray was employed for genotyping of strains by the Genomics, Proteomics and Bioanalytical Laboratory led by co-editor SEG.

Both laboratories jointly organized annual conferences from 1998 to the present to promote and develop new applications of these technologies. These were later supported by co-authors AJS, EYT and KCT until the hiatus caused by the COVID-19 pandemic.

Poor uptake of the ProteinChip and MassArray technologies by microbiologists led to their eventual cessation, while MALDI-TOF MS grappled to gain acceptance. The MS company Micromass (later Waters Inc.) designed the first upright, more compact MS benchtop instrument and was placed at MISU in 2000 to develop the method for clinical microbiology. Because microbial identification was historically based upon patterns of carbohydrate fermentation, the major problem encountered by diagnostic laboratories was the differentiation of non-fermentative species that produced uniformly negative results. However, the introduction of comparative 16S rRNA sequencing permitted insight into the immense diversity of these highly complex microorganisms. Species of the non-fermentative genus *Porphyromonas* that were difficult to distinguish by diagnostic laboratories were used by MISU to establish proof of concept of MALDI-TOF MS for microbial identification. With all 18 species of the genus being unambiguously delineated, work began to standardize protocols and assemble a database using strains from an accredited source, the National Collection of Type Cultures. To assess the potential of MALDI-TOF MS, 16S rRNA and LCFA profiles were included as complementary methods and, by 2004, the first microbial database of 3500 mass spectral profiles were reported and MISU later field-tested the method at the Royal London Hospital. This demonstrated not only the pragmatism of the method but also its value for a clinical laboratory as it was found to be accurate and rapid and offered significantly lower cost.

The annual European Congress of Clinical Microbiology and Infectious Diseases (ECCMID) conferences became the major forum for communications on the development on MALDI-TOF MS. Progress at first was slow and circumspect, with very poor attendance in 2003 in Glasgow to a packed main auditorium capacity in Vienna in 2010 in which data on the success of MALDI-TOF MS to rapidly identify *Clostridium difficile* during an epidemic in UK hospitals were reported by PHE. This congress was a watershed moment for MALDI-TOF MS for several reasons. For example, in the midst of the meeting, bioMérieux entered the field by acquiring the successful diagnostic company AnagnosTec GmbH (Potsdam Golm, Germany), while ThermoFisher announced its engagement with MISU to explore the wider potential of MS for microbial diagnostics. It was also at this meeting that Wiley launched our first book of this series entitled *Mass Spectrometry for Microbial Proteomics* (eds H.N. Shah & S.E. Gharbia), published in 2010. Bruker Microbiology & Diagnostics recognized the impact of dedicated mass spectral profile databases and established a database linked to their specific MALDI-TOF MS for clinical microbiology. They, too, had a significant presence at the Vienna meeting in which there was a dedicated session in MALDI-TOF MS for the first time, entitled ‘MALDI-TOF MS in Clinical Microbiology’. PHE’s presentation at this meeting, entitled ‘MALDI-TOF MS of surface-associated and stable intracellular proteins for identification and resistance profiling of human pathogens’ (Shah, H.N., 10–13 April 2010), led to extensive support to expand clinical applications of the technology.

With the London 2012 Olympics approaching and the UK’s government keen to establish an economic, accurate and rapid diagnostic method in the event of major outbreaks of infection at the games, six Bruker instruments were placed across PHS’s national network of laboratories in October to deliver rapid local identification of potential biological threats and transmission of infections in mass events. Extended applications from clinical to non-clinical samples were pioneered by ASTA, based in Seoul, Korea, developing specialist databases such as Food, Agriculture and Environmental for specific applications using their own ASTA Tinkerbell LT MALDI-TOF MS instrument. Soon, industrial and environmental applications of MALDI-TOF MS were reported using other MS platforms. A common strategy was to utilize 16S rRNA to delineate new diversity, followed by deposition of the mass spectral data of an unknown isolate into existing databases to expand its capability. Our last conference to promote these technologies prior to the COVID-19 pandemic restrictions was held on 21 and 22 June 2018, entitled 'The Impact of Advances in Mass Spectrometry and Analytical Technologies on Detection and Revealing Microbial Behaviour and Interaction with their Environment', and was co-sponsored by a large number of biotechnology companies such as Ascend Diagnostics, bioMérieux, Inc., Bruker Microbiology & Diagnostics, Shimadzu Corp. and ThermoFisher Scientific.

As microbes continue to be exploited for their industrial and environmental properties, profiling the expressed proteome is now essential for developing commercial applications. For example, an extensive collaborative study, entitled 'Feasibility study to assess the potential of electrospray mass spectrometry to provide mass spectral-based identification to the now established MALDI-TOF MS', which utilized a Q-Exactive, was undertaken between 2012 and 2015. This was reported at ECCMID, Denmark, in 2015, under the title 'A global diagnostic approach for microbial identification: accurate characterisation of difficult to differentiate pathogens', by co-author HNS et al. and demonstrated unambiguous delineation of closely aligned pathogens such as *E. coli* and *Shigella sonnei*. We also reported extensive coverage of the proteome of several human pathogens along with unique strain biomarkers using both bottom-up and top-down methods, some of which were reported in the second book *MALDI-TOF and Tandem MS for Clinical Microbiology* (eds H.N. Shah & S.E. Gharbia), published by Wiley in 2017.

The present book focuses on applications of mass spectrometry in industry and the environment and is divided into four overlapping sections. The first (Chapters 1–5) commences with an historical background leading up to MALDI-TOF MS in the clinical laboratory, including recent viral applications and data analysis such as machine learning algorithms that are being championed for strain typing and tandem MS, involving bottom-up and top-down proteomics. New approaches such as liquid atmospheric pressure and advanced applications in imaging and HDX are proposed for future development. Special attention is given to a new instrument, the Omnitrap, which combines several applications of MS. Chapters 6–10 describe environmental applications as MALDI-TOF MS transitions to non-clinical laboratories. Environmental, agricultural, soil and bioremediation research for typing and biopolymer degradation are considered. Veterinary applications of mass spectrometry for diagnostics are at an early stage and have focused almost entirely on MALDI-TOF MS. Its impact has already been substantial, and Chapters 11 and 12 cover both domestic animals and livestock. Industrial use of MALDI-TOF MS is now very advanced and encompasses a diverse range of applications. Chapters 13–19 report applications in the food, water, marine and pharmaceutical industries, including drug discovery, as well as its potential as part of a contamination control strategy for regulated industries.

1

Progress in the Microbiological Applications of Mass Spectrometry: from Electron Impact to Soft Ionization Techniques, MALDI-TOF MS and Beyond

Emmanuel Raptakis[1], Ajit J. Shah[2], Saheer E. Gharbia[3], Laila M.N. Shah[4], Simona Francese[5], Erika Y. Tranfield[6], Louise Duncan[7], and Haroun N. Shah[2]

[1] *Fasmatech Science and Technology, Athens, Greece*
[2] *Department of Natural Sciences, Middlesex University, London, UK*
[3] *Pathogen Genomics, Gastrointestinal Infections and Food Safety, Clinical Public Health, UK Health Security Agency, London, UK*
[4] *Department of Physical & Theoretical Chemistry, University of Oxford, Oxford, UK*
[5] *Biomolecular Science Research Centre, Sheffield, UK*
[6] *Bruker UK Limited, Coventry, UK*
[7] *School of Cancer and Pharmaceutical Sciences, King's College London, London, UK*

1.1 Introduction

Over the past two decades, advances in genomics, proteomics and metabolomics and their key technologies, such as mass spectrometry (MS) and particularly matrix-assisted laser desorption/ionization time-of-flight (MALDI-TOF) MS, have propelled microbiology to the forefront of life sciences, radically altering the workflow of diagnostic laboratories and subsequently expanding into environmental and industrial applications. The road map of microbial classification has been profoundly altered and has progressed from a phenotypic, determinative system to one based on phylogeny as new technologies have been incorporated, modified and applied. This transition has been complex, and hence to illustrate its impact, it is discussed here in the first instance in the context of a single genus, *Bacteroides*, first described in 1898 [1, 2]. This is the dominant taxon of the intestinal tract of humans and animals and therefore plays a pivotal role in health and disease. The nature of this multifaceted ecosystem is central to an understanding of the biology of humans and requires in-depth analysis of the physiology and diversity of its microbiome. MS has been the underlying technology used to probe this ecosystem since studies first commenced over six decades ago (see Drasar and Hill [3]).

1.1.1 Algorithms Based upon Traditional Carbohydrate Fermentation Tests

The Gram stain, together with advances in microscopy, built the foundation for the characterization of microbes, dividing the kingdom into four main domains as Gram-negative

Microbiological Identification using MALDI-TOF and Tandem Mass Spectrometry: Industrial and Environmental Applications, First edition. Edited by Haroun N. Shah, Saheer E. Gharbia, Ajit J. Shah, Erika Y. Tranfield, and K. Clive Thompson.

and Gram-positive rods and cocci [4]. Together with morphological criteria, the capacity to ferment/metabolize individual carbohydrates to produce acid was widely adopted for algorithms to identify species and describe new centres of variation and has been reported in various editions of *Bergey's Manual of Determinative Bacteriology* from its first edition [5]. However, the emphasis from its infancy was strongly biased towards clinical applications. Thus, the first diagnostic compendium, titled *Manual for the Identification of Medical Bacteria* [6], utilized these algorithms based upon carbohydrate fermentation tests to delineate clinical isolates to species level. Although this approach expanded and helped to describe new taxa that were saccharolytic or moderately fermentative, non-fermentative species, which represent a significant component of the microbiome of any habitat, remained poorly circumscribed and overlooked through the years. It is among this latter cluster that MALDI-TOF MS would bring about a paradigm shift in clinical microbiology and its eventual transition to non-clinical sites.

1.1.2 Dynamic Changes in the Chemotaxonomic Era (c. 1970–1985) through the Lens of the Genus *Bacteroides*

Genera such as *Bacteroides* that were described decades earlier using the above algorithms accumulated large numbers of species that only loosely fitted their broad definition. With the introduction of DNA analysis in the 1960s (initially as mol% G+C content), this heterogeneity was reflected in their wide range in DNA base compositions (e.g. Owen et al. [7]). The limit of a genus was then fixed at c. 10–12 mol% G+C and was applied primarily as an exclusionary criterion in systematics. *Bacteroides,* with a 28–61 mol% G+C span, was therefore redefined around the type species *B. fragilis* and related taxa with a reduced base composition of c. 40–50% mol% G+C [8] [9], [10]. Consequently, many taxa were left as *incertae sedis* and were subjected to a range of biochemical and chemical analyses that included protein electrophoretic and lipid analyses [11–14]. Within this more restricted definition of the genus *Bacteroides*, three broad groups of species were clearly discernible based on carbohydrate fermentation tests: (i) saccharolytic species, *B. fragilis* group; (ii) moderately/weakly saccharolytic, *Bacteroides melaninogenicus* cluster; and (iii) non-fermentative species, *Bacteroides asaccharolyticus* group (see Figure 1.1).

This heterogeneity presented enormous difficulties, but it was the third group where the largest clinical and taxonomic problems were encountered because of the paucity of reliable characters to define taxa. Members of this group were uniformly non-fermentative but there were indications of heterogeneity using new techniques (Figure 1.1). Thus, sodium dodecyl sulphate-polyacrylamide gel electrophoresis (SDS-PAGE) of polypeptides and isoelectric focusing (IEF) of cellular proteins revealed profiles that were concordant with the three groups described earlier (see, e.g., [15, 16]). Multilocus enzyme electrophoresis (MLEE) further corroborated these findings, with the *B. fragilis* group being defined by enzymes of the hexose monophosphate shunt/pentose phosphate pathway in addition to malate and glutamate dehydrogenases, whereas the moderately/nonfermentive groups contained only the last two oxidoreductases [17]. These species, in common with other microorganisms were subjected to extensive lipid analyses using hard ionization techniques in MS, including gas chromatography-MS (GC-MS). Subsequently, the arrival of soft ionization methods such as MALDI-TOF MS in the late 1990s were explored for microbial

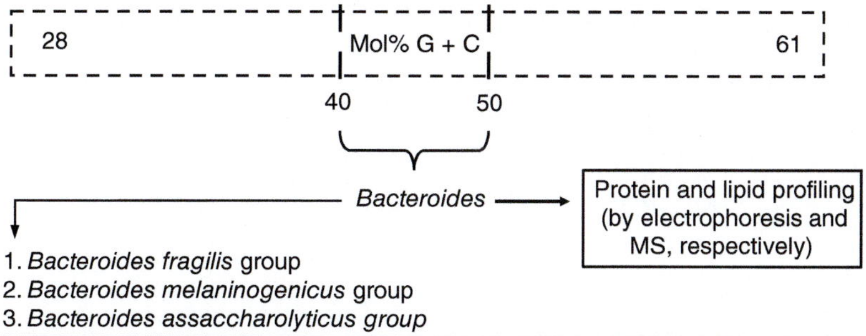

	Glucose	Fructose	Sucrose	Maltose	Lactose	Xylose	Arabinose	Lactose	Raffinose	Sucrose
B. fragilis group	+	+	+	+	+	+	+	+	+	+
B. melaninogenicus	+	+	+	+	+	−	−/+	w	−	−
B. asaccharolyticus	−	−	−	−	−	−	−	−	−	−

MALDI-TOF MS

Figure 1.1 Transition of the genus *Bacteroides* (Mol.% G+C = 28-61%) to more a restricted group of species (40–50 Mol% G+C). Even with this restricted group, carbohydrate fermentation tests, one of the most significant criteria for defining a species, revealed three centres of variation with a prominent non-fermentative cluster. +, acid produced; −, carbohydrate not fermented.

identification, while advances in liquid chromatography with tandem mass spectrometry (LC-MS/MS) were used to elucidate their proteomes particularly for the study of pathogenic mechanisms of variant strains (see reviews in [18], [19]).

1.1.3 Microbial Lipids as Diagnostic Biomarkers; Resurgence of Interest in MALDI-TOF MS with Advances in Lipidomics

Lipids are ubiquitous among living organisms and present as a bewildering array of structures that are considered highly reliable characters for microbial systematics. During the period from the 1970s to the late 1980s, profound changes in microbial taxonomy began to take place in an era regarded as the heyday of the chemotaxonomy. Because lipids ionize particularly well, they are readily amenable to MS (see Chapter 15). Thus, hard ionization techniques such as electron impact, that were available at the time, were widely applied and began to reveal the immense diversity of lipids among microorganisms. (see, e.g., [20, 21]). Cellular lipids may be 'free' or 'bound' to macromolecules to form larger entities such as lipopolysaccharides, polysaccharides or lipoproteins. For bound lipids, the general approach has been extraction via acid or alkaline methanolysis, followed by separation using thin-layer chromatography (TLC) prior to MS. Some of the most common lipid structures found in bacteria, including members of the genus *Bacteroides*, are shown in Figure 1.2.

Gram-negative cells, with their significantly larger wall, were predicated to display the most complex and diverse forms; however, it is among the actinomycetes and related bacteria that some of the most complex lipids were reported (see [21] for a review). During this

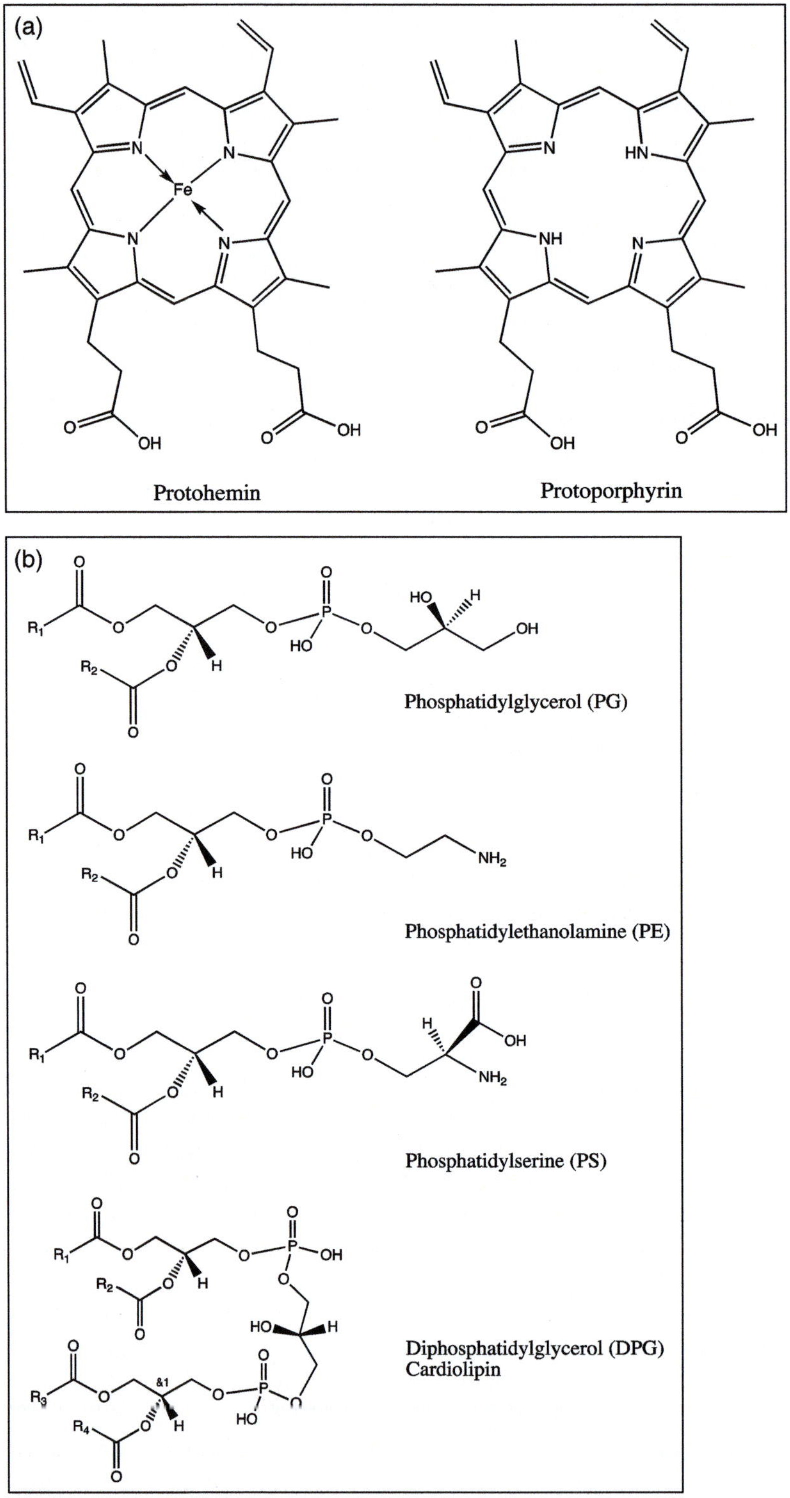

Figure 1.2 Lipids most commonly analysed in *Bacteroides* and other bacteria during the chemotaxonomic era using hard ionization techniques in MS. (a) Porphyrins; (b) phospholipids; (c) long-chain fatty acids; (d) menaquinones. In general, these lipids were extracted/methylated from dried cells by acid methanolysis and separated by TLC prior to mass spectral analysis. More recently, use of these techniques is returning to microbiology, this time employing TLC with MALDI-TOF MS (see text).

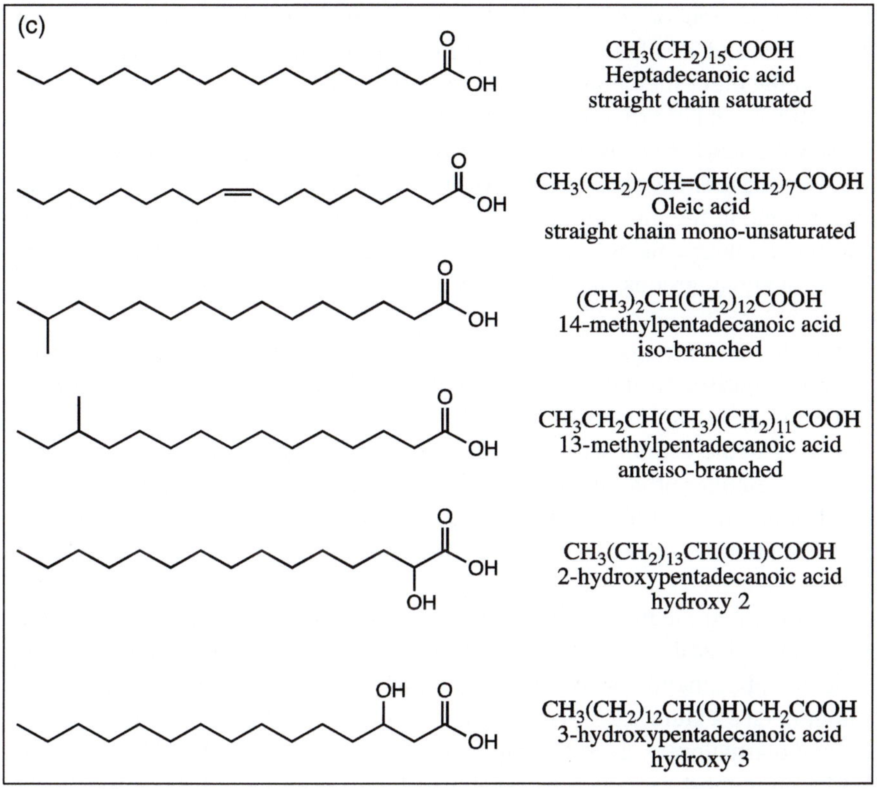

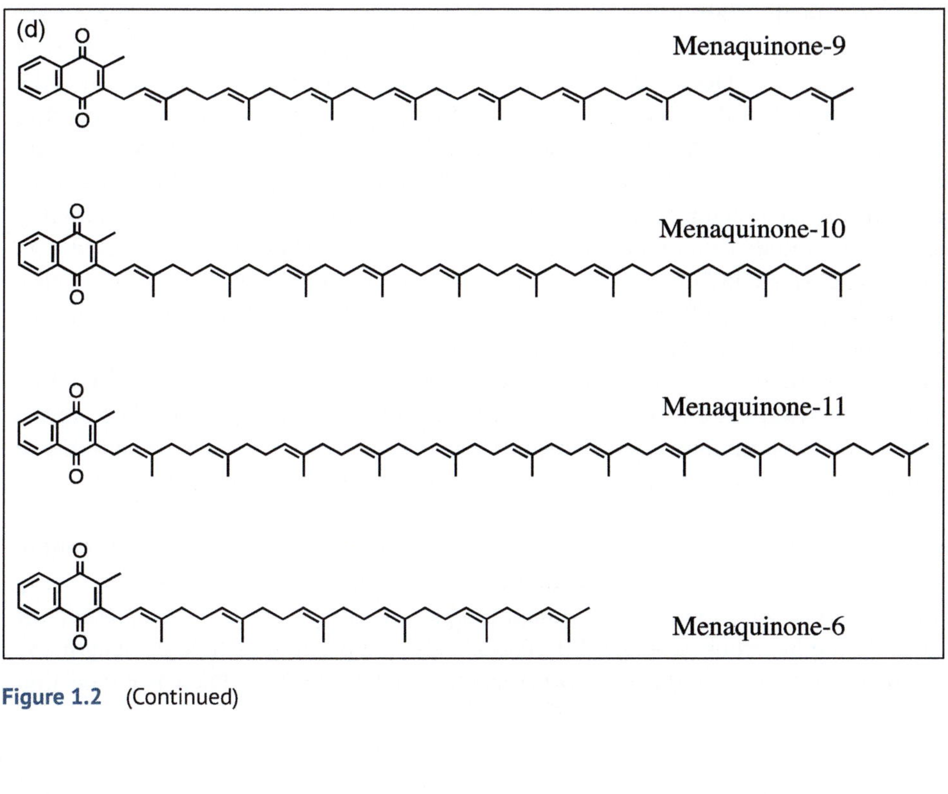

Figure 1.2 (Continued)

period, structural variation of the peptidoglycan was meticulously elucidated among microbes by Schleifer and Kandler [22] and its link to various lipids became apparent. In general, it is from the outer membrane and via long-chain fatty acids (LCFAs) that many lipids are covalently bound but that, when extracted, give rise to species-specific lipid profiles. The latter were used extensively for describing both new and existing taxa while simultaneously serving to elucidate the complex structure of the cell envelope. Mycobacterial mycolic acids are among the most complex reported, ranging between 60 and 90 carbon atoms. Furthermore, the mycolates found in different genera were shown to have carboxy, epoxy, keto and methoxy groups, in addition to its 3-hydroxy units as their functional groups [23].

In common with many groups of bacteria that were analysed, the diversity of lipids revealed by MS led to restructuring of taxa as occurred with *Bacteroides* [14]. Some members designated the '*B. melaninogenicus* group', produced black pigmented colonies on blood agar and exhibited UV fluorescence in the dark. Their nature was highly contentious and attributed to a number of compounds such as melanin, haematin and even ferrous sulphide (see [14] for a review). They were eventually extracted as their dimethyl esters in boron trifluoride/methanol, separated by TLC and shown by MS to be protohaem and protophyrin [24] (see Figure 1.2).

The most common amphipathic polar lipids reported were glycerophospholipids, the structures of which are based upon the glycerol 3-phosphate configuration of phosphatidic acid. These include phosphatidyl glycerol, phosphatidyl ethanolamine, phosphatidyl serine and cardiolipin (see Figure 1.2). Gram-negative taxa such as *Bacteroides* also possess sphingolipids and plasmalogens that were reported many years earlier [25–27].

The LCFAs found in most bacteria are predominantly between 12 and 20 carbon atoms and may be straight-chain saturated and monounsaturated, 2- or 3-hydroxylated, non-hydroxylated, *iso*-methyl and *anteiso*-methyl branched fatty acids (Figure 1.2). This broad spectrum of LCFAs, and its potential for variation and consistency among species led to the creation of commercial databases that were used by diagnostic laboratories (see, e.g., [28]). On the basis of their LCFA composition, *Bacteroides* were shown to comprise two main groups: (i) species that contained major amounts of *iso*- and *anteiso*-methyl branched acids in addition to straight-chain saturated acids; and (ii) taxa which contained predominantly straight-chain saturated and monounsaturated fatty acids. *Bacteroides* that possessed high levels of branched-chain fatty acids generally possessed 12 methyl-tetradecanoic (*anteiso*-$C_{15:0}$) as the major fatty acid. The non-fermentative group (Figure 1.1) possessed 13 methyl-tetradecanoic acid (*iso*-$C_{15:0}$) as their major fatty acid.

Respiratory quinones, crucial to the energy metabolism of cells, have been used to characterize microorganisms during the 1970s and depended almost solely on MS to reveal their detailed structural forms. Aerobic Gram-negative bacteria possess benzoquinone or ubiquinone as their sole respiratory quinone, whereas facultative anaerobes such as *Escherichia coli* or obligate anaerobes such as *Bacteroides* contain naphthoquinones, demethylmenaquinone and menaquinone (MK). The sole respiratory quinones present in *Bacteroides* are menaquinones, 2-methyl-3-polyprenyl-1, 4-naphthoquinone [13]. Variations at the C-2 position of the 1-4 naphthoquinone ring is limited to the presence or absence of a methyl group, whereas differences at the C-3 position include the length and degree of unsaturation of the polyprenyl side-chain (Figure 1.2). Each isoprene unit

(C_5H_8-68 Da) is readily visible in a mass spectrum by their characteristic fragmentation patterns and signature mass ions at 187 and 225 Da [18]. The *B. fragilis–B. melaninogenicus* groups possessed high levels of MK-10 and MK-11, whereas the *B. asaccharolyticus* group comprised species with MK-9 and MK-10. While the menaquinone profiles lacked resolution to the species level, the presence of predominantly MK-9 to MK-11 were used as key characters to redefine the genus [8] [9], [10] so that taxa such as '*B. ochraceus*' which contained MK-6 was reclassified in a new genus [29].

These lipids were used widely during the chemotaxonomic period for microbial classification, but interest declined because they were considered too cumbersome for diagnostic laboratories, they required large biomass, and lipid databases during this period were confined to only LCFAs. Some three decades passed with only intermittent reports of comprehensive lipid analyses of various taxa.

The arrival of MALDI-TOF MS in the mid-1990s witnessed a huge shift towards analysis of proteins which has remained the cornerstone of the technology in microbiology. Its presence has been so dominant that Schiller et al. [30] stated that 'MALDI-TOF MS seemed to be nearly a synonym for protein analysis.' However, this has subsequently prompted interest in exploring MALDI-TOF MS for analysis of lipids (see reviews in [30, 31]). Some studies have reported the use of lipids based on earlier methods in which TLC was coupled with MALDI-TOF MS [31–34]. However, with improvements in direct extraction methods and better understanding of the interaction between the matrix and analyte and the sensitivity of different lipid classes, more precise methods are being developed that are likely to find wide application (see, e.g., [35–41] [42]). More targeted lipids (e.g. membrane phospholipids) are being analysed directly now in bacteria, fungi, and yeasts [43, 44] [45]. This is expanded in Chapter 15.

1.2 The Dawn of MALDI-TOF MS: Establishing Proof of Concept for Diagnostic Microbiology

Of the seven initial papers published on the potential use of MALDI-TOF MS in microbiology, six [46–51] were reported in MS journals and one in *Nature Biotechnology* [52]. These journals are not frequently read by clinical microbiologists and consequently there was little initial interest in the technology. Furthermore, these papers focused mainly on methodology and reported mass spectral profiles between Gram-positive and Gram-negative species and taxa that were already readily distinguishable. Earlier forms of MS (pyrolysis-MS, electron impact, fast atom bombardment; see [18] for a review) did not gain a foothold in diagnostic microbiology and hence MALDI-TOF MS was greeted with considerable scepticism by infectious disease scientists.

Public Health England (PHE), Centre for Infections (formerly Public Health Laboratory Service, Health Protection Agency, and recently UK Health Security Agency) began a programme of restructuring around this period and created a new laboratory, the Molecular Identification Services Unit (MISU), with a remit to focus on the characterization of atypical, rarely isolated and emerging pathogens. One of the authors (Haroun N. Shah) was appointed as its director in late 1997 and immediately opted to use LCFAs to initiate the work of MISU (see Section 1.1.3; Figure 1.2). 16S rRNA sequence analysis, used then for phylogeny, was adopted to confirm species designation by fatty acid analysis as the laboratory became operational.

As genomic methods began to replace traditional algorithms for diagnostics, laboratories shrank in size, and thus housing large pieces of equipment was a major hindrance to progress. One of the most engaging features of the first MALDI-TOF MS (Kratos Kompact Alpha, manufactured by Kratos Analytical, Manchester, UK) was its appeal as a 'benchtop' instrument with sample preparation space being negligible. MISU, being a new laboratory with a new focus, was able to explore several new technologies and soon acquired three different types on MALDI-TOF platforms. In addition to the Kratos Kompact Alpha (on loan from the company), PHE acquired a surface-enhanced laser desorption/ionization (SELDI)-TOF MS from Ciphergen BioSystems for broad-spectrum proteomic analyses using ProteinChip Arrays and a MALDI-TOF MassARRAY System from Sequenom GmbH for DNA sequencing for single nucleotide polymorphisms (SNPs) (Figure 1.3).

To our knowledge PHE was the only establishment to have possessed all three MALDI-TOF mass spectrometers simultaneously, each of which possessed unique features worthy of use today. The emergence of the Kratos/Micromass/Bruker Linear MALDI-TOF type platform for microbiologists may have been due purely to the ease with which it was possible to assemble a preliminary database of mass spectral profiles, and, in the case of PHE,

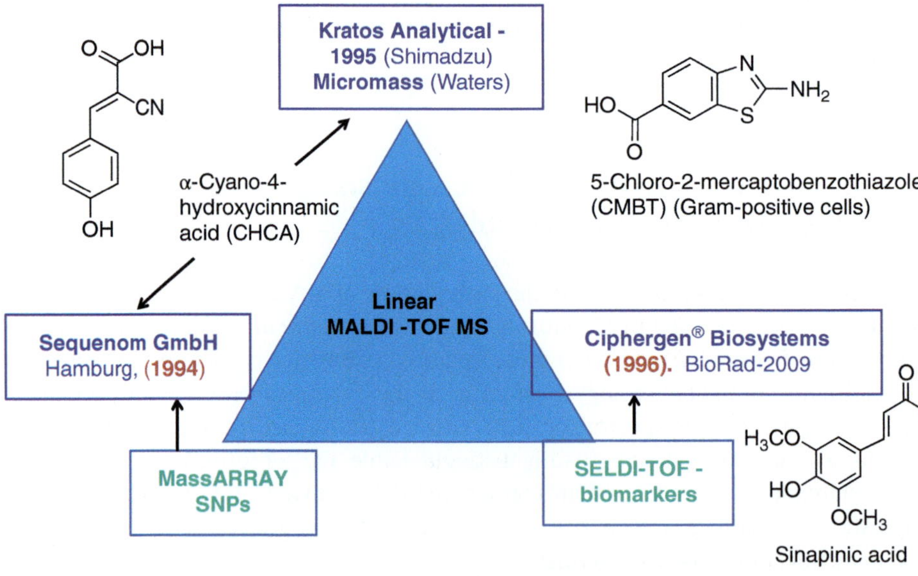

Figure 1.3 Competing MALDI-TOF MS technologies in the mid-1990s and matrices commonly employed for microbial diagnostics. The matrix α-cyano-4-hydroxycinnamic acid (CHCA) was used more generally, whereas 5-chloro-2-mercaptobenzothiazole (CMBT) was often used for Gram-positive cells where their ability to detect glycopeptides and phophopeptides were deemed superior. Sinapinic acid, with its ability to reduce photochemically generated adducts and greatly improve the mass resolution of proteins, was reserved for the SELDI-TOF MS analysis and was prepacked by the company Ciphergen Biosystems.

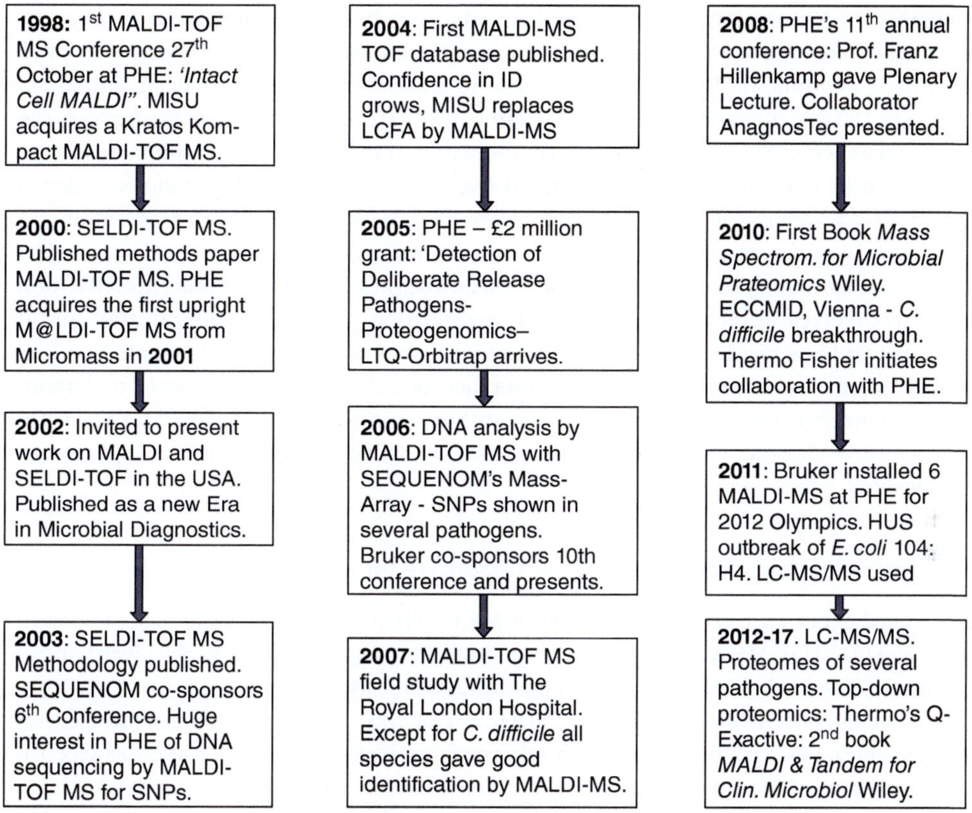

Figure 1.4 Milestones in the development and application of various MALDI-TOF MS platforms and tandem MS/MS in a diagnostic laboratory (1998–2017).

funding was more forthcoming. However, the three technologies were actively pursued over the years at PHE and had a major impact on the service function of the organization (Figure 1.4).

When Bio-Rad acquired the SELDI-TOF MS in 2009 from Ciphergen Biosystems, they introduced an improved platform with updated software and there was evidence to show that the system had the potential as a powerful MS platform for strain typing. The company did not allocate sufficient resources to develop the capabilities of the technology further and it became obsolete, leaving many laboratories with an instrument but no service backup. In a similar manner, Sequenom's MassARRAY was a precise and elegant analytical platform and its versatility was demonstrated for typing of many species through SNPs (see [53] for a review). PHE transferred its traditional serotyping system of Kauffmann-White, which dates back to the 1920s, to MALDI-TOF MS platform using Sequenom's MassArray [54].

Collaborative work with Claydon's group at Manchester Metropolitan University began in October 1998, soon after organizing a one-day conference at PHE on 27 October 1998, titled simply 'Intact Cell MALDI', to assess the potential of MALDI-TOF MS against new and developing technologies. (After this first meeting, conferences in 'Genomics and

Proteomics of Human Pathogens' would be held annually over the next 20 years to follow new developments and these have been instrumental in bringing about changes in proteomics and genomics at PHE and several clinical laboratories. They have been suspended during the COVID-19 pandemic but will resume in 2023). At this inaugural conference, the Kratos Kompact Alpha mass spectrometer was placed at the front of the conference lecture theatre and its versatility was demonstrated live to the audience using various bacterial species. The general view from participants was that MALDI-TOF MS had potential and was a useful tool for research but there were shortcomings for diagnostic applications. For a diagnostic laboratory to report results on clinical samples would necessitate accreditation from a recognized authority, and to obtain certification, rigorous standardization of all methods would be needed, including parameters such as software, database, etc. used on the instrument to generate the data. The manufacturers themselves would need to demonstrate competence and rigour from approved authorities and this alone would take years to achieve. Nevertheless, PHE would pursue its development for diagnostic applications relentlessly.

1.2.1 Development of a MALDI-TOF MS Database for Human Infectious Diseases

Having had experience with earlier forms of MS for chemotaxonomic analyses, the potential of MALDI-TOF MS was immediately apparent and, against all odds, PHE went ahead vigorously with its development for diagnostic applications.

It had already been demonstrated that MALDI-TOF MS was a useful tool for the analysis of intact microbes and, compared with other soft ionization techniques such as electrospray ionization (ESI), it was far more tolerant to salts and medium components which are often present in a sample. Furthermore, since the laser beam is focused on a small portion of the sample, it permits efficient energy transfer without destruction of the analyte molecules, allowing them to be widely separated within the matrix admixture and thus ensured good separation of the molecular ions. However, despite this high degree of tolerance, preliminary work in our laboratory (Molecular Identification Services, PHE) showed that the reproducibility of a mass spectrum was subject to a number of factors, and to assemble a database, these variable parameters would have to be investigated. Even basic considerations such as culture media, the mean generation time of different species, broth versus solid media, temperature, pH, atmospheric culture conditions, their Gram reaction, etc. affected the reproducibility of a mass spectrum from various bacterial species and a rigorous standard operating procedure was needed. It was shown almost immediately that the same bacterial strain grown on blood agar and nutrient agar produced very different spectra, ion suppression being markedly affected on nutrient agar-cultured cells. It was therefore necessary to evaluate the effect of blood from different suppliers before a standard medium could be derived. The method was optimized and evaluated using strains of *B. fragilis* to investigate its diversity using the Kratos Kompact Alpha MS [55]. Mass spectral data of strains run in both positive and negative modes were different, but in all cases spectra in the positive mode were more dense and therefore were used in all further studies.

With a standardized method in use, proof of principle was considered accomplished when the non-fermentative, *B. asaccharolyticus* group of species (Section 1.1.2; Figure 1.5) were shown to be unambiguously delineated. Until the arrival of MALDI-TOF MS, there

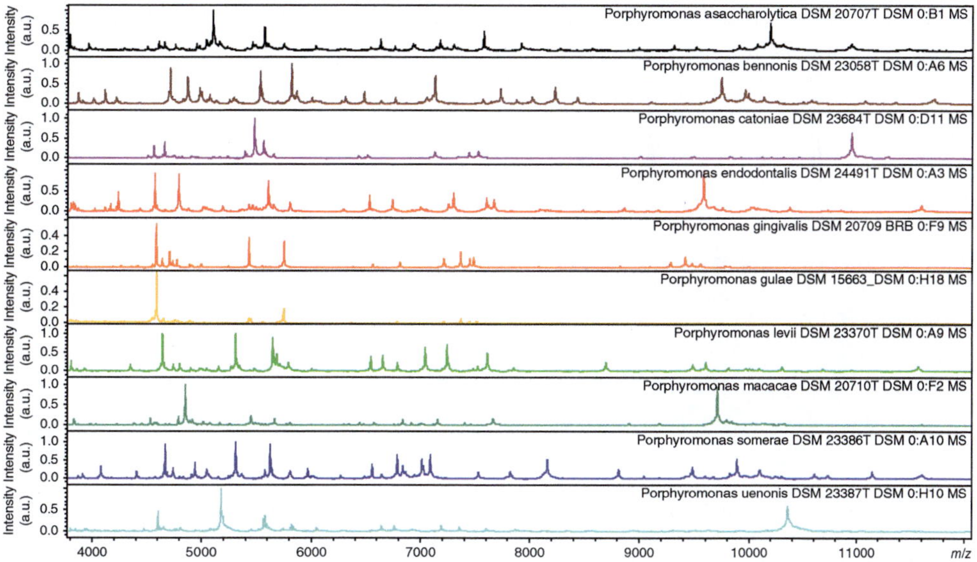

Figure 1.5 Emergence of the genus *Porphyromonas* from the immensely complex genus *Bacteroides* which then contained species between 28 and 61 mol% G+C. Species within this new genus *Porphyromonas* were delineated by 16S rRNA sequence analysis but there were no reliable simple phenotypic characters to differentiate them. This poorly defined genus became the first taxon to be subjected to MALDI-TOF MS analysis in 1998 to establish 'proof of concept' using the first benchtop MALDI-TOF MS instrument manufactured by Kratos Analytical [56]. Here the partial composite spectra of 10 species of *Porphyromonas* (represented by different colours, from Bruker Microbiology & Diagnostics). Strains were from the DSMZ Culture Collection and shows the mass range 4000–12 000 Da where unique species biomarkers are discerned. *Source:* Adapted from Shah et al. [56].

were no phenotypic characters to characterize these poorly defined species and identification was dependent purely on DNA sequence analysis which was very cumbersome at this time [56]. These results were confirmed several years later by Bruker with a larger number of species (Figures 1.5, 1.6; Table 1.1) making it now possible to the investigate the function of these and other non-fermentative species *in vivo*.

The Kratos Kompact Alpha MS, being the first instrument developed and not necessarily designed for microbial analysis, had a number of limitations. Its analytical features were basic (e.g. sample analysis was manual, the target plates had only 20 sample wells, laser energies needed to be manually predetermined, etc.) while data analytics software was in its infancy. However, the experience gained during this period enabled the design of a new instrument in collaboration with the MS company, Micromass, also based in Manchester. A key feature of the new instrument (designated M@ADI-TOF) was the placement of the time-of-flight tube in a vertical position, as opposed to its horizontal location in the Kratos instrument. This narrowed its width, enabled a longer flight tube and increased its resolution. The number of samples on a target plate increased from 20 to 96, while most of its operation was now automated and carried the new analytical software MicrobeLynx. The instrument arrived at PHE on the eve of the fourth annual conference (25–27 June 2001) and its image was used in leaflets to announce the conference and subsequently to promote the diagnostic work of MISU. With PHE's entire National Collection of Type Cultures

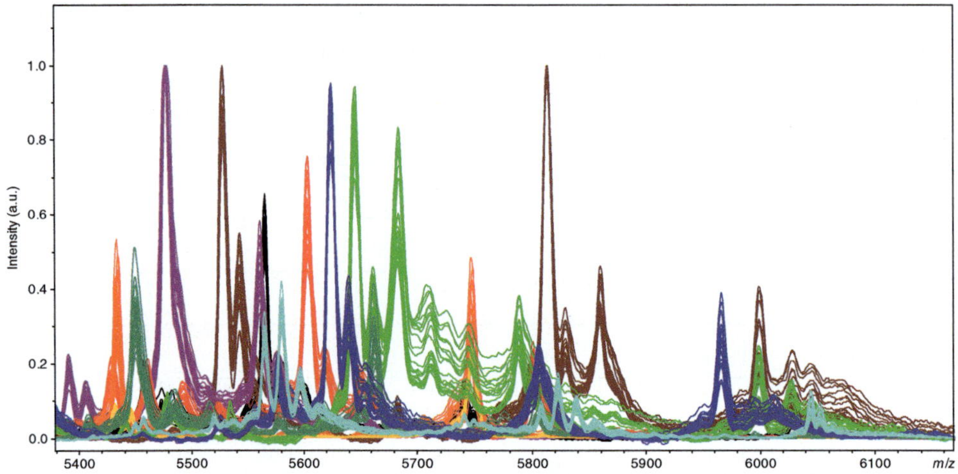

Figure 1.6 A subset of data from Figure 1.5 is expanded within a narrower range of 5400– 6100 Da. Here the partial spectra are overlaid in different colours to vividly highlight differences between species that are otherwise poorly differentiated using other methods.

Table 1.1 MALDI-TOF MS data obtained from Bruker Microbiology & Diagnostics for 11 of the 17 species of the genus *Porphyromonas*. The threshold for a confident identification score is >2 and shows a few isolates below this level. Because of their non-reactive nature to traditional tests and the limited number of reliable criteria, most recent studies have depended entirely on whole-genome sequencing and MALDI-TOF MS.

Results overview

Analyte name	Organism (best match)	Score value	Organism (second best match)	Score Value
Porphyromonas asaccharolytica 29_644 IBS (+++) **(A)**	*Porphyromonas asaccharolytica*	3	Not reliable identification	1.5
Porphyromonas asaccharolytica DSM 20707T DSM (+++) **(A)**	*Porphyromonas asaccharolytica*	3	*Porphyromonas asaccharolytica*	2.47
Porphyromonas asaccharolytica ENR_0216 ENR (+++) **(A)**	*Porphyromonas asaccharolytica*	3	*Porphyromonas asaccharolytica*	2.57
Porphyromonas asaccharolytica ENR-0252 ENR (+++) **(A)**	*Porphyromonas asaccharolytica*	3	*Porphyromonas asaccharolytica*	2.56
Porphyromonas asaccharolytica ENR-0294-2 ENR (+++) **(A)**	*Porphyromonas asaccharolytica*	3	*Porphyromonas asaccharolytica*	2.49
Porphyromonas bennonis DSM 23058T DSM (+++) **(A)**	*Porphyromonas bennonis*	3	*Porphyromonas bennonis*	2.43
Porphyromonas bennonis ENR-0457 ENR (+++) **(A)**	*Porphyromonas bennonis*	3	*Porphyromonas bennonis*	2.43
Porphyromonas catoniae DSM 23684T	*Porphyromonas*	3	Not reliable	0.91

The MALDI-TOF MS results emphasize the paucity of protein biomarkers for some species and the potential to search for other species-specific mass ions such as lipids to delineate such taxa (see Chapter 19).
Source: Bruker Microbiology & Diagnostics.

(NCTC) at its disposal and standard operating procedures now fully established, the laboratory commenced on an ambitious programme to assemble a database of mass spectral profiles of clinically relevant species, initially analysing each strain in triplicate – one at MISU and two at sites in Manchester. Based on earlier data, spectral profiles were collected in the mass range 500–10 000 Da. Ten shots were acquired per spectrum at a laser firing rate of 20 Hz. Fifteen spectra per sample well were collected using the lock mass facility to optimize the quality of the data.

Interlaboratory reproducibility was near 100% and thus after two years the work was confined to two laboratories: Micromass and PHE. The concordance of data between laboratories was so high that a year later each laboratory began to work independently to assemble a quality control database. The group reported the first mass spectral database with 3500 spectra in 2004 [57] after which annual databases were released to the public by Manchester Metropolitan University who funded its development. MISU soon replaced its identification platform based upon LCFAs by MALDI-TOF MS but continued to test each isolate in parallel with 16S rRNA sequence analysis. Recognizing that while NCTC's type and reference strains were accredited (BS EN 9002) and used for whole-genome sequencing databases, they were likely to be variants of their clinical counterparts, and thus a field study was considered necessary. This was undertaken with The Royal London Hospital in collaboration with their Medical Microbiology Department to assess the efficacy of this developing MALDI-TOF MS database, which by then had increased to 5000 spectral profiles. Staff at the hospital were trained in sample preparation and, over a period of 10 months, randomly selected clinical isolates were analysed by MALDI-TOF MS in parallel with traditional diagnostic tests used by the hospital. Excellent correlation was obtained for all species except for putative strains of *Clostridium difficile* in which 90% were misidentified. Initially the hospital's most frequent pathogen, *Staphylococcus aureus* was first selected for further study. Here, putative strains of *S. aureus* identified by the hospital's diagnostic laboratory were also examined by PHE's Staphylococcus Reference Laboratory and analysis was carried out against the MALDI-TOF MS database. The correspondence in the data was high except where samples were mixed. Once purified and re-analysed in tandem with 16S rRNA, confirmation of the species was attained [58].

1.2.2 The Dilemma with *Clostridium difficile*: from Intact Cells to Intracellular Proteins, MALDI-TOF MS Enters a New Phase

The high rate of failure to identify clinical isolates of *C. difficile* in the above study appeared at a period when UK hospitals were experiencing a crisis, with intestinal infections being reported in the tens of thousands with high mortalities due to this species. With MALDI-TOF MS failing to identify the most significant hospital-associated pathogen in Britain at that time, it nearly resulted in termination of a decade of considerable work. A short period of respite was granted to re-examine protocols. A collaborative programme with AnagnosTec GmbH, Potsdam Golm, Germany, led by Dr Wibke Kallow who visited PHE in 2007, focused largely on methodology including data analysis to resolve the problem. Success was eventually achieved by changing the matrix solution to 2,5-dihydroxy benzoic acid in acetronitrile : enthanol : water (1 : 1 : 1) with 0.3% trifluroacetic acid and introduction of formic acid extraction of proteins from cells prior to analysis [18]. These results were reported at the European Conference of Clinical Microbiology and Infectious Diseases (ECCMID) in Vienna at a symposium titled 'MALDI-TOF-MS in Clinical Microbiology', 10–14 April 2010. Through these studies it became evident that the distinctive mass ions in MALDI-TOF MS spectra were mainly ribosomal proteins, deduced from translated genome sequences that matched the theoretical masses of ribosomal subunits proteins (30S and 50S). This new protocol removed the need to standardize growth conditions as done in early methods for development of the database [57]. Ribosomal proteins are constantly present in cells in high copy number irrespective of growth conditions and hence reproducible spectra were obtained even using different instruments. While MISU continued to use its M@LDI-TOF MS from Micromass UK Ltd, AnagnosTec GmbH used a Shimadzu AXIMA platform with their own in-house software (Spectral Archiving and Microbial Identification System, SARAMIS) and comparable spectra were obtained [59]. Many clinical microbiologists consider the Vienna ECCMID conference in 2010 as a watershed, the high point at which MALDI-TOF MS passed its threshold of acceptance and gained the widespread approval of diagnostic laboratories [60]. Its impact was so immense that in the midst of the meeting, bioMerieux acquired AnagnosTec GmbH to the utter dismay of the group of dedicated scientists who started the company at the Max-Planck in Potsdam Golm. It was also during this conference that ThermoFisher Scientific, a leading MS company, signalled its interest to enter the field of clinical microbiology through PHE and in collaboration with its UK laboratories.

With the 2012 Olympics in London on the horizon, MISU responded to the government's call in 2011 for high-throughput, low-cost, simple and accurate technologies for dealing with potential disease outbreaks during the games. Through this programme, PHE successfully established a network of two Bruker Microflex instruments in MISU in October 2011, with 12 additional instruments subsequently in other PHE laboratories in the UK, making it the largest network of MALDI-TOF MS instruments in a single organization. As part of the rationale for the above grant application, data from parallel studies of 16S rRNA and MALDI-TOF MS over a 10-year period were collated and showed that MALDI-TOF MS outperformed 16S rRNA by orders of magnitude [19]. This gained the confidence of its directors such that the method moved from being supportive to the primary means of identifying unknown samples sent into MISU.

Although PHE's responsibility is solely reserved for human infectious disease pathogens, numerous non-clinical samples were often sent in for analysis. These included landfill and environmental sites, contaminated soil, rivers and extreme environments, together with those from companies in which microbes are central to their manufacturing processes such as alcohol, beverages, dairy and, in particular, probiotics. Because the database then comprised mostly clinical isolates, some of the strains were not identified to species level. Contact was made with ASTA Inc., Suwon, Korea which manufactured the MALDI-TOF platform Tinkerbell LT and its accompanying software MicroIDSys [61]. Their core database included a series of focused databases such as 'FOOD DB' (food isolates), 'AGRI DB' (from agricultural sources) and 'ENVIRON DB' (isolates from sea, lakes, rivers etc). In a collaborative project between Middlesex University and ASTA during 2016–2018, ASTA's MALDI-TOF MS, Tinkerbell LT and MicroIDSys system was installed at Middlesex University, and parallel studies were undertaken with the university's Bruker Autoflex MS. There was such high concordance between spectra produced using both instruments

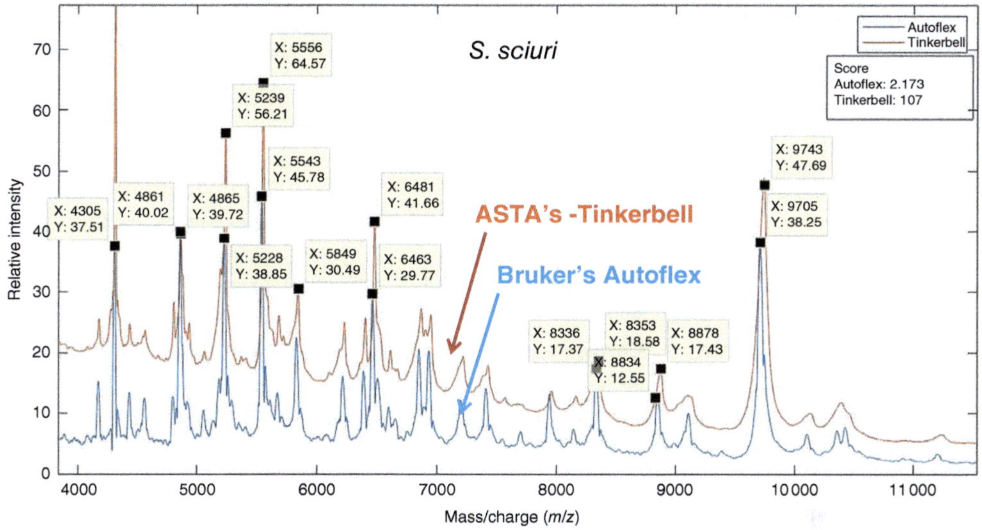

Figure 1.7 Comparison of the MS spectrum of *Staphylococcus sciuri* from the environment analysed on ASTA's MALDI-TOF MS platform, Tinkerbell LT and Bruker's Autoflex III MALDI-TOF MS. Excellent concordance between instruments was found for all strains tested.

(see, e.g., Figure 1.7) that an unknown spectrum on one instrument could be used blindly to interrogate the data from the other instrument.

However, as more environmental samples were analysed and new diversity was found using 16S rRNA or whole-genome sequencing, well-characterized strains were added to existing databases, making it possible to use a single database for identification from non-clinical samples with a high degree of confidence. This has permitted the use of MALDI-TOF MS widely, outside the field of clinical microbiology for which it was initially designed. Table 1.2, obtained from Bruker Microbiology & Diagnostics, shows the wide range of non-clinical applications of Bruker's MALDI-TOF MS platforms and demonstrates how universal the technology has become today and this is expanded upon in various chapters of this book.

1.3 Linear/Reflectron MALDI-TOF MS to Tandem Mass Spectrometry

As described earlier, successful identification and characterization of a microorganism using MALDI-TOF MS require a mass spectrum of the unknown to be matched to one in a predetermined database. An alternative is to use tandem mass spectrometry (MS/MS). Here, fragmentation of the protein in MS/MS produces sequence-specific peptide ions which can be identified through searches of various public protein databases to ascertain its possible biological origin [62]. One drawback to direct analysis is that it may lead to ion suppression and thus the number of proteins detected may be significantly reduced, which may have an impact on the identification of the microorganism. To overcome this, high-performance liquid chromatography (HPLC) or ultra-performance liquid chromatography

Table 1.2 A cross-section of the non-clinical areas in which MALDI-TOF MS is currently being used.

Food safety inspection	Pharmaceutical laboratories	Drinking water supplier/waste water management/energy suppliers
Private service/contract laboratoriess	Veterinary inspection and disease control	Government laboratories – monitoring, e.g. agriculture, surface water, environment
Food manufacturers	Animal health industry (livestock, seafood and pets)	Diagnostics industry: manufacturers of detection kits
Beverage and mineral water manufacturers	Cosmetics industry	Biobanks and culture collections
Manufacturers of probiotics, and yeast and bacteria for starter cultures	Paper and building materials industry	Research and universities, national reference centres: food, agriculture, veterinary

The table was kindly provided by Bruker's marketing department. Various chapters in this book provide detailed examples of some of these applications. *Source:* Bruker.

(UPLC) can be used to separate the proteins prior to fragmentation and subsequent detection. However, it is essential to extract the proteins from the organism to prevent insoluble components being loaded onto the HPLC/UPLC system. Furthermore, fragmentation of proteins in a top-down approach generates large peptide ions. Thus, tandem MS instruments can fragment large proteins and detect the resulting peptide ions with high mass accuracy for identification. An alternative to this strategy is to use bottom-up proteomics where proteins are extracted from the organism and digested by chemicals or enzymatic means prior to LC-MS/MS analysis [63]. The MS/MS spectra generated from the peptides are matched to a database to determine their identity. Although top-down and bottom-up approaches increase both the number of sample preparation steps and analysis time, they do offer higher selectivity and the potential to identify biomarkers such as proteins associated with antibiotic resistance or pathogenic determinants. Higher selectivity also allows mixed cultures to be identified and closely related species to be distinguished [64]. In the case of the latter, it may be feasible to use protein markers for identification of organisms to sub-species level using tandem MALDI-TOF/TOF.

A conventional mass spectrometer consists of three key components: an ion source, a mass analysers(s) and a detector. For detection and characterization of microorganisms, the analysis is typically based on peptides and proteins. These must be converted to gas-phase ions before they enter the analyser. MALDI and ESI are soft ionization techniques that are used for detection and characterization of microorganisms based on peptides and proteins as they result in minimal degradation of these thermally labile molecules. The ions formed can be separated using either a single or multistage analyser. In MALDI the ionization is typically carried out under vacuum although it can be performed under atmospheric conditions [65]. There are numerous single and multistage analysers that are commercially available. Mass analysers that are currently used are Fourier transform ion cyclotron resonance (FTICR), ion trap, quadrupole, Orbitrap and TOF.

The ionization process in MALDI is complex but it predominantly leads to the formation of singly charged ions. The principle of the TOF analyser is to produce discrete pocket of ions with different *m/z* from a pulsed ion source which are then accelerated into the TOF analyser. Ions of different *m/z* drift down a field-free path of a known distance. Lighter ions arrive at the detector first followed by heavier ones. TOF analysers are well suited for methods based on laser desorption/ionization that is used in MALDI. In linear MALDI-TOF instruments, the sample on a target plate is irradiated by a pulsed laser beam. At the same time an acceleration voltage is applied between the target plate and ground counter electrode. During laser pulses, ions are formed and desorbed from the target plate. These are continuously extracted and accelerated to the same kinetic energy. The ions pass into the drift field-free zone towards the detector which is typically a multichannel plate (MCP). TOF analysers can transmit ions of exceptionally high mass. However, they are limited in this respect by poor ionization of high-molecular-weight molecules and poor detectability of these ions by MCP because they possess low kinetic energy and thus produce a small change in currents at the detector surface. Detection and identification of microorganisms are carried out primarily using linear MALDI-TOF MS. Thus, the resulting mass spectra are simpler than those observed with ESI.

MALDI-TOF MS analysis is largely restricted to analysis of microorganisms to genus and species levels. However, microorganisms like bacteria can be genetically highly heterogenous and the typing of bacterial strains with specific phenotypic characteristics such as pathogenicity and antibiotic resistance is also significant. MALDI-TOF MS has been used for typing some bacterial strains (see, e.g., [66, 67]). MALDI-TOF has had mixed success in the detection of antibiotic resistance. It has been used successfully for the detection of intact β-lactamase [68]. These authors analysed bacterial extracts using MALDI-TOF MS and obtained a signal around *m/z* 29 000 that they attributed to β-lactamase. This was confirmed by separation of the protein using SDS and peptide mass fingerprinting.

The resolution of a linear TOF analyser is limited by the energy spread of laser-desorbed ions. This can be improved two- to four-fold using a reflectron TOF analyser. The reflectron comprises a series of ring-shaped electrodes set at increasing potential of opposing polarity to extraction voltage and it is located behind the field-free region. These act as an ion mirror that slow down the ions and turn them around and send them back to a second detector. Thus, it focuses ions of different energies in time. Leading to better resolution. MALDI-TOF in reflectron mode has been used in a new approach for detection of organisms producing β-lactamase (citation). In this methodology, the organism is incubated with β-lactam antibiotic, and hydrolysis products resulting from enzyme activity are detected.

1.3.1 Tandem MALDI-TOF Mass Spectrometry

To obtain sequence information of peptide/protein, tandem MS is used for detecting fragments. However, matrix-assisted laser desorption mass spectrometry (MALDI-MS) suffers from a low degree of fragmentation during desorption/ionization. One approach that was commonly used for enhancing metastable fragmentation of ions produced by MALDI is post-source decay (PSD) [69]. Here, high laser intensity is used to form a small proportion of ions with sufficient internal energy to fragment in the metastable drift region between the ion source and entrance to the reflector. The velocity of precursor and fragment ions is similar, but the mass and kinetic energy of fragment ions are lower and this allows them to reach the reflector detector first. With the typical acceleration voltage used with MALDI-TOF, not all ions can be focused onto the detector at the same time. Thus, the different *m/z* ranges are separated and detected by decreasing the reflecting lens potentials. The mass spectra recorded for each voltage interval are combined to generate a PSD spectrum. This approach is not widely used at present for characterization of proteins and peptides because it is a tedious and time-consuming stepwise process. The development of MALDI-TOF/TOF instruments has allowed peptide and protein characterization to be carried out using laser-induced dissociation (LID), which is the basic process underlying PSD and collision-induced dissociation (CID). In the latter approach, a timed ion selector using a series of voltages allows a designated ion with a specific *m/z* to enter a collision cell. All other ions are deflected. Inside the collision cell, the ions are bombarded by an inert gas such as nitrogen or argon and undergo CID. The fragment ions are accelerated into the second TOF analyser where they are separated and subsequently detected.

1.3.2 Electrospray-based Mass Analysers

Other MS-based approaches that are used to study microorganisms incorporate a LC step prior to detection to improve selectivity and the number of proteins that can be detected. In comparison to MALDI molecules, those in electrospray are ionized out of solution at atmospheric pressure by applying a potential difference of 3–6 kV to a liquid flowing through a capillary into an ion source. This causes a build-up of charge at the liquid surface sited at the end of the capillary and results in the formation of charged droplets. An inert gas introduced coaxially allows droplets to be sprayed as a fine aerosol into the ion source. The high temperature in the ion source and the gas causes the solvent within the droplets to evaporate, enabling them to shrink and increase their charge density. The charged droplets continue to lose solvent and once the electric field on the surface grows to be large enough, ions are desorbed from the surface and are attracted towards the orifice of the mass analyser. Small molecules < 1000 Da mainly produce singly charged ions. However, molecules such as proteins and peptides that have multiple ionizable groups result in the formation of multiple charge states [70]. This makes interpretation of the mass spectrum a little more complex.

One of the first mass analysers that was coupled to ESI to study microorganisms was a quadrupole. Since then, it has been successfully interfaced to ion trap, FTICR, Orbitrap and TOF. Initial studies that were carried out to characterize microorganisms were based on analysis of lipids using a quadrupole instrument [71, 72] [73]. The method involved pyrolysis of lyophilized bacteria and subsequent analysis by MS. Since then, MALDI-TOF MS and tandem MS-based approaches have dominated the way microorganisms are characterized and antibiotic resistance and microbial pathogenic determinants are detected. In comparison to MALDI-TOF MS, LC-MS/MS usually provides higher mass accuracy, resolution, dynamic sensitivity and limit of detection.

1.3.3 Tandem Mass Spectrometry

Tandem MS involves two stages of *m/z* analysis. Ions of interest are mass-selected and subjected to fragmentation followed by second mass spectrometric analysis. There are two categories into which MS/MS instruments fall. One of these is tandem MS in space. These instruments comprise two mass analysers installed in tandem. Here, precursor/parent ions are selected by the first analyser and then fragmented in an intermediate region and the resulting fragments are transmitted via a second analyser to the detector. Mass spectrometers that are beam-transmitting, such as TOF/TOF, triple quadrupole (QqQ) and quadrupole TOF (QqTOF), fall into this route of MS/MS analysis. The second approach is tandem MS in time. In these instruments, a single m/z analyser, such as an FTICR linear ion trap (LIT), quadrupole ion trap (QIT), quadrupole Orbitrap (Q-Orbitrap), LIT-Orbitrap, linear ion trap quadrupole Orbitrap (LTQ-Orbitrap) and Q-Orbitrap-tandem LIT, may be used to carry out different stages of ion selection, activation and analysis of fragment ion in the same place but successively in time. Each mass spectrometer has a different mass accuracy, mass resolution, sensitivity or dynamic range. Many of these instruments have been used in microbial proteomic studies and for measurement of specific microbial biomarkers.

1.3.4 Mass Spectrometry-based Proteomics

Over the last two decades the application of MS to proteomic-based studies has grown significantly. These approaches have been utilized to investigate the microbial proteome [74], post-translational modifications (PTMs) [66], antibiotic resistance [75, 76] and identification of biomarkers [77]. Proteomic studies can be divided into two categories: bottom-up and top-down. In both approaches, it may be necessary to fractionate the sample. Even with advances in MS and LC, the characterization of the proteome is limited by the wide dynamic range of protein concentrations, PTMs and lack of methods used routinely in genomics such as the polymerase chain reaction (PCR) to amplify proteins. The wide dynamic range of protein expression can be addressed by fractionating the sample. This can be achieved using either one- or two-dimensional SDS-PAGE or anion/cation exchange chromatography. In the case where proteins are separated using SDS-PAGE, protein bands of interest are excised and these are subjected to in-gel digestion of protein using either chemical or enzyme (s). Samples fractionated using chromatography are desalted, concentrated and then reconstituted in a suitable buffer before digestion. The resulting peptides are analysed using LC-MS/MS. The MS approach used in bottom-up proteomics can be categorized into discovery and targeted proteomics.

1.3.5 Case Study: LC-MS/MS of Biothreat Agents, Proteomes of Pathogens and Strain-level Tying Using Bottom-up and Top-down Proteomics

Government intelligence agencies in the USA and Europe had placed several cities on high alert from biothreat agents in 2005 and the UK's Home Office put out calls for research proposals to rapidly detect biothreat agents in the environment. PHE's application was based on DNA sequencing approaches by MALDI-TOF MS and LC-MS/MS to obtain in-depth analysis of the proteome to search for novel biomarkers in real time. A Sequenom MALDI-TOF MS MassARRAY system (Figure 1.3) and a ThermoFisher LTQ Orbitrap Classic with a Shimadzu LC System were acquired to undertake SNP and in-depth proteome analysis, respectively. Considerable time was spent optimizing sample extraction by physical methods such as sonication to the use of a French press cell disrupter and the use of several chaotropic agents for cell lysis, prior to trypsin digestion and LC-MS/MS analysis (see, e.g., [78]). These bottom-up approaches yielded protein reference maps of several pathogens, such as *Streptococcus pneumonia*, *Salmonella enterica* and *C. difficile*, were constructed against a background of their whole-genome sequences [79–83].

Studies then progressed to biothreat agents which required cell extracts to be prepared in class 3 cabinets and viability checks to be undertaken on samples prior to mass spectral analysis in an open laboratory. Data obtained from such studies were complex, and eventually a bioinformatics pipeline was developed that correlated the immense mass spectral data with genomics which enabled the discovery of peptide signatures to species and strain level for all pathogens [84]. Soon all target biothreat agents were analysed and over 100 peptide biomarkers could be identified per strain (see, e.g., [85]). The system produced such high-resolution data that when, in the summer of 2011, the *E. coli* O104:H4 outbreak of haemolytic-uraemic syndrome (HUS) occurred in Germany, PHE's genomics and proteomic laboratories were positioned to assist the UK's National Health Service in the event of the outbreak reaching its shores. By May 2011, there were more than 4000 reported cases of severe infection and 50 deaths in Germany. HUS is a devastating and rare disease characterized by disintegration of red blood cells, acute kidney failure and impaired ability to clot blood, and thus bloody diarrhoea is a common symptom. This outbreak remains an exceptional example for PHE as it simultaneously incorporated whole-genome sequencing and in-depth proteomic analyses to effectively reveal the topography of this complex 'hybrid' category 3 pathogen in real time. Within days of arrival of the first case of infection in the UK in June 2011, PHE's Genomics Laboratory sequenced its full genome and deposited its sequence in the National Center for Biotechnology Information library, providing a blueprint for the proteomic work that followed, which has been described in detail [19]. In brief the outcome was that it enabled pathotypes of pathogenic *E. coli* to be delineated through analysis of their proteomes and the data could be used to advise hospitals on treatment and expected outcomes.

As a simple workflow, PHE's method was to culture pathogens overnight in a class 3 cabinet in LB broth and on nutrient agar plates and pelleted cells lysed using a chaotropic agent. Before samples were taken out of the cabinet, all extracts were subcultured into enriched broths and retested for loss of viability by another technical staff to ensure rigour. As only one batch of samples from broth and agar plates was provided, every step was taken to ensure maximum efficiency. The lysates were divided into two aliquots and one batch was run directly on SDS-PAGE after which 1-cm gel slices were digested with trypsin and peptides analysed using nano-LC–MS/MS. With the second batch, the entire cell lysate was digested directly with trypsin and the peptides were injected onto two LC-MS/MS systems – the LTQ Orbitrap Classic at PHE and an LTQ Orbitrap Velos (ThermoFisher, Hemel Hempstead, UK) – each with a front-end Ultimate 3000 Dionex nano/capillary LC system.

Peptide MS/MS spectra were aligned to both protein and in silico genome-translated databases to identify the expressed proteins of strains while the bioinformatics pipeline described earlier [84] was used to acquire unique peptide signatures at the species/strain levels while pathogenic determinants were obtained from Blast and Scarfold. As an example of the data, for *E. coli* O104:H4 strain 280, some 2500 proteins were identified from the outbreak isolates. Interestingly, 68 peptide signatures were unique to the outbreak *E. coli* isolates that were not shared by enteroaggregative *Escherichia coli* (EAEC) or enterohemorrhagic *E. coli* (EHEC), delineating the outbreak strain from other closely related Enterobacteriaceae. Species-level peptide signatures were also detected, including those for the AggR transcription factor, haemolysin protein, Aaf fimbriae protein and Iha adhesion protein. In total, 3031 peptides were identified as unique to the outbreak strains when compared against control isolates.

These results supported the genomic data that the background genome of the outbreak strain came from an EAEC progenitor that acquired plasmids and prophages, and exchanged chromosomal loci leading to the emergence of an aggressive strain with a unique profile. All strains shared 89% of the expressed proteins. The two large plasmids encoded 31 proteins. Peptide signatures for adhesion and multidrug resistance (including β-lactamase, CTX-M extended spectrum β-lactamase and metallo-β-lactamase enzymes) were revealed.

The nano-LC-MS/MS platform used proved a highly sensitive and reliable system as the peptide lists identified proteins that covered a significant percentage of the predicted open reading frames of the sequenced outbreak strain's genome. Interestingly, pathogenic markers were identified without a requirement for enrichment, selective media or antibiotic incorporation that can otherwise delay analysis. The protein signatures detected provided definitive characterization at the genus, species and often strain level, as well as detection of expressed pathogenic determinants and antibiotic resistance mechanisms. The level of resolution achieved in this study could not be done using a linear MALDI-TOF MS and paved the way for MS/MS-based analysis for simultaneous microbial strain typing and pathotyping during outbreak investigations.

A drawback of the system was its requirement for highly skilled technical staff, its inherent complexity in sample processing and the extended time for a clinical laboratory. For these reasons, ThermoFisher set out to design a tandem MS platform comparable in ease of operation to a linear MALDI-TOF MS that would overcome these shortcomings. Their engineers developed a 'hands-free' instrument in which the only work required was the addition of a loopful of cells to a microtube and inserted into the Q-Exactive MS in top-down mode (see Section 1.3.8). Lysis, followed by centrifugation and direct sample injection, was done automatically in approximately five minutes. Initial data with such an instrument enabled separation between *Shigella sonnei* and *E. coli,* two virtually indistinguishable species, and therefore hold much promise. These initial data were presented at ECCMID 2015 in Denmark [86].

The examples described here covered cases at the extreme ends of mass spectrometry viz. work using a simple linear MALDI-TOR MS to the most sophisticated end using a ThermoFisher Q-Exactive Orbitrap with Top-down facilities. However, between these opposite ends of the spectrum, there is a family of MS instruments with different capabilities that may be used to maximize the quality of the data as shown in this section.

1.3.6 Discovery Proteomics

The objective of discovery proteomics is to obtain a comprehensive coverage of the proteome. To do this the mass spectrometer is operated in a data-dependent acquisition (DDA) mode. In this mode, the MS is set to detect the most abundant precursor ion from a survey scan. These peptide ions are then isolated, fragmented and analysed. The peptides selected are within a small m/z range signal intensity. Typically between 10 and 15 peptides per cycle are selected and data acquisition follows sequentially for each peptide. Karlsson et al. [87] reported a bottom-up proteomic study using nano-LC coupled to Q-Orbitrap or Q-Orbitrap-tandem LIT operated in DDA mode with a bioinformatics pipeline and an in-house genome sequence database to identify species-specific peptides as probable biomarkers for *S. aureus*, *Moraxella catarrhalis*, *Haemophilus influenzae* and *Streptococcus pneumoniae*. In another study reported by Hayoun et al. [88], a DDA-based MS method was used for high-throughput proteotyping of a range of bacteria. The authors used SP3 paramagnetic beads for cell lysis and extraction of proteins and were able to show whether isolates were single isolates or a mixture of strains and to identify organisms to a more resolved taxonomic level. DDA has also been used for detection of bacterial toxins. Cheng et al. [77] used DDA for the detection of Shiga toxins (Chapter 3). In their study, several reference strains of *E. coli* were cultured in the presence of mitomycin C, which is known to stimulate production of Shiga toxins. The toxins were enriched using affinity purification, subsequently digested with trypsin and analysed using LC-MS/MS. A curated *E. coli* toxin database was used to classify the toxins. LC-MS/MS methods based on DDA are simple to set up and run, and analysis of data is fast. For these reasons DDA is the most commonly used approach in shotgun proteomics. Limitations of DDA are its inherent lower precision, reproducibility and under-sampling [89].

1.3.7 Targeted Proteomics

In targeted proteomics, the MS may be operated in data-independent acquisition (DIA) mode. In contrast to DDA where MS/MS spectra are only obtained for most abundant precursor ions with DIA, both MS1 and MS2 data are obtained for all precursor ions within the isolation window with no prior selection of the precursor ion [90]. The MS2 spectra are generated by fragmenting all precursor ions that pass into the mass spectrometer at a given time or by sequentially concentrating on a small *m/z* range of precursor ions and fragmenting these. The MS/MS data are collected in parallel across peptides. Peptides are normally identified by comparing mass spectra to a spectral library of known peptides. To determine abundance, a range of different DIA strategies have been developed such as shotgun CID [91], DIA [92] and Sequential Windowed Acquisition of all Theoretical fragment ions Mass Spectrometry (SWATH-MS) [93] but not all of these approaches have been utilized for microbial proteomics. The authors demonstrated that the SWATH-MS approach can be used for qualitative and quantitative analysis of peptides in a single LC-MS/MS injection. The authors used a high-resolution qQTOF to record MS/MS spectra and demonstrated that these could be used to mine the dataset for a combination of fragment ions to identify specific peptides over a linear range of four orders of magnitude. In another study by Curto et al. [94], DIA/SWATH-MS was used to demonstrate a difference in proteome profiles between pathogenic and non-pathogenic *Rickettsia* inside macrophages. To carry out LC-MS/MS analysis in DIA mode, no previous knowledge of the protein content of sample is necessary. It offers better temporal resolution, and it can be used to measure the level of proteins present over a wide dynamic range. Furthermore, it offers greater precision and reproducibility than DDA. However, there are several key drawbacks to using the DIA approach. Analysis of the data generated can be challenging. The limit of detection is lower than DDA because a complete spectrum must be recorded which reduces the acquisition time for each data point and it is more expensive than DDA. Although DIA/SWATH-MS

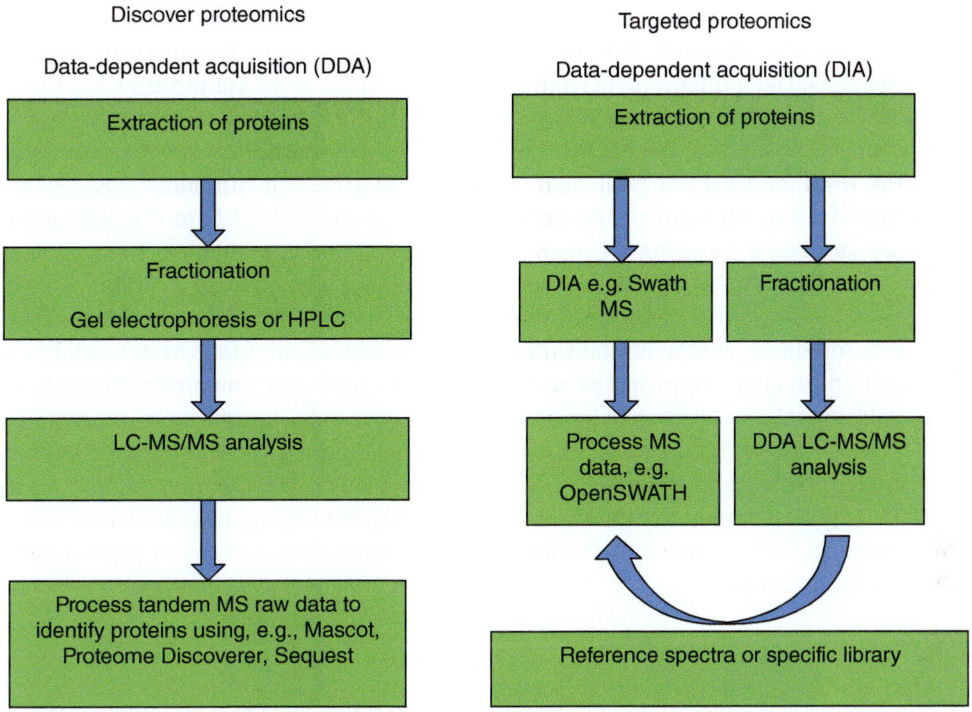

Figure 1.8 DDA-MS and DIA-MS for discover and targeted proteomics workflows.

has been used for quantitative proteomic studies, multiple reaction monitoring (MRM) or selected reaction monitoring (SRM) and parallel reaction monitoring (PRM) are used for quantification of proteins (see Figure 1.8)

1.3.8 Top-down Proteomics

Top-down proteomics involves analysis of intact proteins. This approach can be utilized with MALDI and LC-MS/MS. The focus of the content presented here will be on LC-MS/MS. To increase the dynamic range of proteins that are detected using LC-MS/MS, it is necessary to separate the proteins before MS analysis. This can be achieved by using gel electrophoresis. However, this is time-consuming and the throughput is low. In comparison, methods based on LC separation are more efficient and thus preferred. In this approach, intact proteins eluting from the LC column are converted to the gas phase and ionized in the source. The first step in their analysis is high accuracy mass measurement of the intact protein and the second step is fragmentation of these. The first step under ESI conditions leads to detection of multiple charge states of each protein. To reduce the complexity of these data, the signals throughout the entire chromatogram have to be deconvoluted to a single mass for the neutral intact protein to give a protein expression profile of the sample. The fragmentation spectra generated in the second step in the top-down approach is comparable to the enzymatic digestion step utilized in the bottom-up approach. In comparison to bottom-up proteomics, the top-down approach gives complete sequence coverage of the protein and complete characterization of proteoforms arising from genetic variation, alternative splicing and post-translational modification. Top-down proteomic studies are carried out primarily using ESI combined with FTICR or Orbitrap mass analysers which have high mass accuracy and resolution.

In top-down proteomics, to characterize proteins it is essential to fragment them and detect the resulting products with high resolution. In FTICR instruments this can be accomplished using CID and nozzle-skimmer dissociation (NSD). These fragmentation techniques are based on using an inert gas to break the peptide bond to form b- and y-ions [95]. Alternative to these modes of fragmentation of a protein, is electron capture dissociation (ECD). In this method electrons are used to cleave the N-Cα bond to form c- and z-ions. Top-down proteomic studies have also been carried out using Orbitrap analyser using CID and higher-energy collisional dissociation (HCD). In this approach, higher-energy collisions allow detection of low *m/z* fragment ions and generates more informative fragmentation spectra. Mott et al. [96] used a top-down approach to discriminate pathogenic and non-pathogenic strains of *E. coli*. In another study, McFarland et al. [97] used LC-ESI-MS and top-down LC-ESI-MS/MS proteomics to identify proteins that arise from serovar-specific non-synonymous SNPs and used combinations of these to discriminate bacteria at the strain level.

1.3.9 Targeted Protein Quantitation

Several targeted studies have been reported on the use of QqQ in SRM mode for characterization and analysis of bacteria. In SRM mode the precursor ion is selected by the first quadrupole (Q1). These are subjected to fragmentation by CID usually using an inert gas in the second quadrupole (Q2) and the selected daughter ion(s) are transmitted via the third quadrupole (Q3) to the detector. In comparison to conventional MS and tandem MS where the precursor and daughter ion, respectively, are measured, in QqQ operated in SRM mode the transition from precursor to daughter ion is measured. The peak area of each of the transmitted ions is used for quantitation of the precursor. The SRM targeted approach leads to better reproducibility, selectivity and limits of detection and quantification. Karlsson et al. [98] reported the application of QqQ in SRM mode for selective detection of group A streptococcus strains using a bottom-up proteomic approach. The authors demonstrated that the SRM approach could be used to unravel the group A *Streptococcus* proteome and that it could be applied to other pathogens. Charretier et al. [99] reported a SRM assay for rapid characterization of microorganism, detection of antibiotic resistance and evaluation of virulence by analysis of *S. aureus*. The SRM approach was used by Schmidt et al. [100] to quantitate 41 glycolytic proteins in *E. coli* cultured under 22 different experimental conditions. The authors determined absolute copy numbers for more than 2300 proteins covering the full dynamic range from c. 1 to more than 100 000 copies per cell across the 22 different conditions. The development of hybrid instruments such as Q-IT, Q-TOF and Q-Orbitrap has led to the development of targeted quantification using PRM. In this approach, complete MS/MS spectra are recorded within a *m/z* range for selected precursor ions, and thus all product ions are detected at once [101]. Hence, unlike SRM in PRM there is no need to select and optimize the parameters for each transition. Also, the use of high-resolution instruments with PRM-based methods gives better selectivity than SRM, which gives better separation of target peptides from other peptides in the sample. The measurement of peptides is typically carried out using stable isotope dilution (SID). In this technique, a known concentration of labelled peptide is added to the sample and the level of endogenous peptide is determined based on its signal intensity relative to the internal standard peptide. Dupré et al. [67] developed an immunoextraction LC-MS/MS assay based on PRM for the quantitation of ricin, staphylococcal enterotoxin B and epsilon toxin, in complex human biofluids and food matrices. Quantitation was carried out using SID and a lower limit of 1.0 ng/ml was achieved.

The applications described here in bottom-up and top-down proteomics and targeted protein quantitation demonstrate that these approaches are becoming widely used for characterizing and identifying microorganisms. However, one technique that is an alternative and gives untargeted and multitudinous imaging data is MS imaging.

1.4 The Application of MALDI-MS Profiling and Imaging in Microbial Forensics: Perspectives

MALDI-MS is a soft ionization technique permitting the generation of intact ions in the gas phase over a wide m/z range (50 Da to > 500 kDa). The ability to profile molecules, macromolecules, biomolecules and supramolecular complexes belonging to different molecular classes, such as metabolites, drugs, peptides, proteins, nucleic acids and polymers, confers this technique with high versatility, which in turn has led to the development of a wide range of life science applications spanning from clinical diagnostics to plant biology and pharmaceutical analyses.

Since the landmark paper by Holland et al. [47] on whole cell bacteria, the application of MALDI-MS profiling (MALDI-MSP) to detect and identify viruses, fungi and bacteria [102] in clinical diagnostics, environmental analysis and food monitoring has yielded a substantial number of peer-reviewed papers over the last decade and, consequently, a considerable body of knowledge [103]. The vast majority of microbial profiling by MALDI-MSP is conducted by detecting peptides and proteins, with substantially fewer reports exploiting microorganisms' small molecules, metabolites and lipids (see Chapter 19). In particular, the detection of bacteria has had a significant impact in clinical diagnostic microbiology [104, 105], with two leading examples of implementation in hospital settings provided by the Biotyper® (Bruker Daltonics) and VITEK® MS Plus (bioMérieux) for the rapid and accurate identification of bacteria and bacterial strains. This new capability has: (i) greatly reduced the diagnosis turnaround times and costs; (ii) increased the diagnostic confidence by increased selectivity; and (iii) opened additional avenues for determination of antibiotic resistance.

The introduction of the imaging modality [MALDI-MS imaging (MALDI-MSI)] towards the end of 1990s [106] permits visualization of molecules and biomolecules in a variety of specimens, biological tissues and tissue cultures. Furthermore, it has boosted the use of this technique, including microbial analysis and the development of new applications, particularly in forensic science. The latter can be broken down further into several branches such as forensic toxicology, fingermarks, hair and body fluid analysis, forensic botany, food forensics and microbial forensics. Some applications are more developed than others with the technique showing a high level of technology readiness for hair and fingermark analysis [107]. Conversely, microbial forensics is one of the areas in which the uptake of MALDI-MS is perceived as slow, with fewer published papers compared with other areas.

Microbial forensics is invested with the identification of microorganisms and their products to provide ancillary intelligence for assisting investigations of biocrimes and bioterrorism (Homeland Security). Budowle et al. and Robinson et al. [108, 109] have offered a comprehensive and up-to-date outlook of all the intelligence that microbial forensics can contribute in a non-exhaustive list that includes geo-location, personal identification, manner and cause of death and post-mortem interval (PMI). For the most part, genetic techniques, such as the recently introduced massive parallel sequencing [110], are used to connect a crime (or an accidental release) to an individual or groups of individuals.

Mass spectrometry is among the non-genetic techniques that has been reported to have the potential to assist forensic investigations. Foundational work in this context has been published in 2015 and established a 3D molecular cartography of human skin with respect to the different types of microorganisms colonizing different parts of the body [111]. In this work, MALDI-MSP provided complementary information through the detection of the microorganism peptide/protein spectral profiles. This work was expanded by Kapono et al. [112] who demonstrated the potential to assist in the reconstruction of crime scene dynamics, specifically with regard to human–environment interaction.; They combined 3D molecular cartography with MS and microbial libraries to obtain individual-specific chemical and microbial signatures enabling the identification of the individuals, specifically with regard to touching given objects in an office environment (e.g. computers, cell and desk phone). However, in this case MALDI-MSP was not employed.

1.4.1 MALDI-MSP of Microorganisms and their Products

MALDI-MSP has been widely reported in the literature for the detection and quantification of bacterial toxins [113, 114] such as from *Clostridium botulinum* ([115] [116]) and *Bacillus anthracis* [117] or for the direct identification of microorganisms, such as *Coxiella burnetii* [118], through their peptide/protein spectral profiles which can assist in bioterrorism and biocrime investigations. An emerging and interesting area of work concerns microbial identification to infer PMI [119]. In a recent report [107], it was suggested that MALDI-MSP and MSI could be used to detect and image post-mortem tissue sections such as from skin (where the epinecrotic microbial consortia proliferate) and internal organs (the so-called thanatomicrobiome) to inform on the manner of death and even determine PMI. A more recent study assessed the potential of MALDI-MSP as an alternative method to identify species associated with the thanatomicrobiota and epinecrotic communities [120]. Although the study was conducted on murine cadavers, it did demonstrate the possibility to determine PMI from the determination of the changes over time of the post-mortem microbiota through their protein spectral profiles. Zhang et al. have also recently reported on the opportunity to predict PMI based on the microbial community succession using genetic techniques in burial human cadavers [121], complementing previously published knowledge on the use of changing microbiota in exposed cadavers. The combined know-how from murine models with the knowledge that changing microbiota in human cadavers can act as PMI indicators poses the basis for a promising microbial application of MALDI-MSP for the estimation of PMI in forensics. With respect to the manner of death, Lee et al. [122] investigated the possibility of distinguishing between victims who were drowned and those whose body was disposed of in water post-mortem using murine models. The findings showed that this differentiation could be made based on the differential presence (i) of the microbial communities and (ii) of the pulmonary surfactant protein A (SP-A) and that MALDI-MSP was able to identify culturable acquatic microorganisms.

Another interesting report investigated the death of porpoises caused by grey seal predatory attack [123]. Although in this report only animal models were investigated, the application of MALDI-MSP for the detection and identification of *Neisseria animaloris* infection in the wounds of the porpoises enabled the 'perpetrator' to be linked to the 'victim'. This is a concept that could well be used in violent and fatal assaults of individuals, especially when bite marks are recovered, and thus contribute to the elucidation of the manner of death; MALDI-MSI could be used in addition to, or instead of, MALDI-MSP to localize the microbial community colonizing the bite mark area.

In contrast to MALDI-MSP, MALDI-MSI has not been used to address forensic queries. This may in part reflect the initial technical challenges in preparing cultures/biofilms (including matrix application) for imaging purposes [124] and may explain the lower number of reports in the literature. Some technical MALDI-MSI know-how has been generated for the understanding of the metabolism of microorganisms [125, 126], response to a drug [127] and chemical interactions between intra- and inter-microorganism communities ([128], [129]) as well as with animal and plant tissues [130, 131]. Some of these studies have been reviewed by Dunham et al. [132] and examples shown in their review and reported in Figure 1.9.

Although these reports provide sound understanding and prove the feasibility and usefulness of the application of MALDI-MSI in microbiology, there has been no uptake in microbial forensics and, as such, it currently remains an unexplored field of application for MALDI-MSI.

1.5 Hydrogen/Deuterium Exchange Mass Spectrometry in Microbiology

A rapidly emerging technique used to study the dynamics of proteins is hydrogen/deuterium exchange mass spectrometry (HDX-MS). The technique was originally combined with nuclear magnetic resonance (NMR) and used by Linderstrøm-Lang to study hydrogen bonds involved in the secondary structure formation, but was later shown to correspond to conformational changes [134, 135]. Until the 1980s, NMR was used as there was no method to transfer macromolecules into the gas phase but this was overtaken by soft ionization methods such as the use of a matrix or proton donating chemicals (e.g. α-cyano-4-hydroxycinnamic acid used in MALDI-TOF MS) and subsequently ESI as described earlier (Sections 1.2, 1.3.2). HDX-MS overtook HDX-NMR due to the fact that it has basically no size limitation and usually requires much less of the sample. However, NMR still provides higher resolution, being able to pinpoint changes to the individual atom [136]. HDX-MS is able to provide dynamic information using minute samples of proteins without chemical

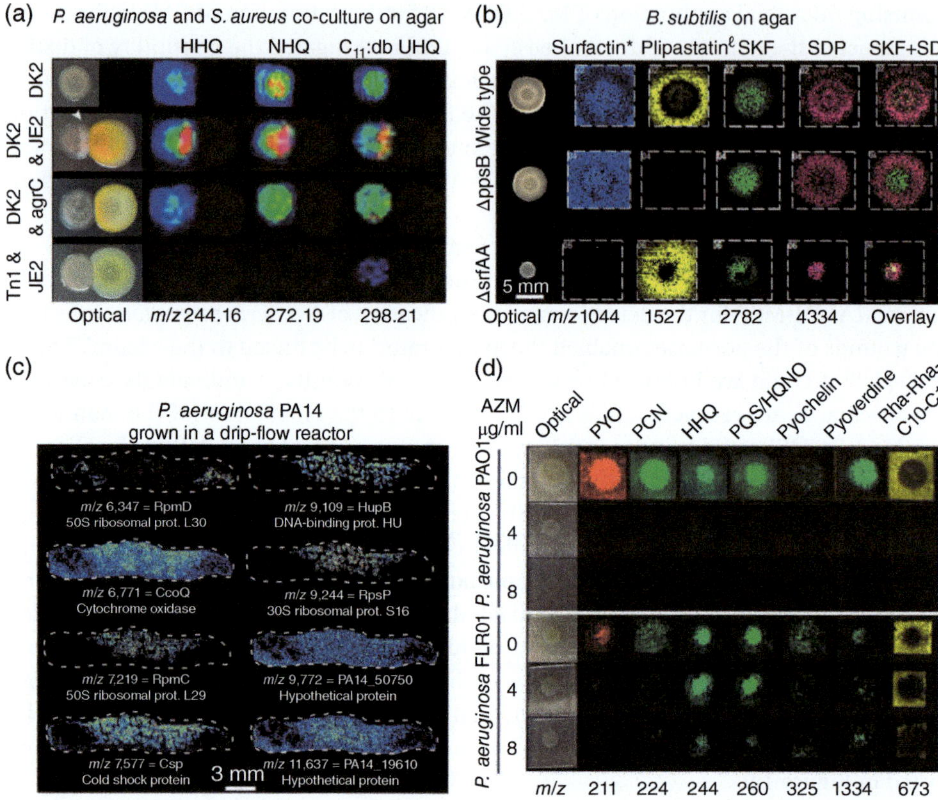

Figure 1.9 Examples of MALDI MS imaging in microbiology. (a) Alkyl quinolines produced by *Pseudomonas aeruginosa* in the presence of *Staphylococcus aureus* [133]. (b) Surfactants and peptides produced by colony biofilms of *Bacillus subtilis* [125]. (c) Nutritionally dependent *P. aeruginosa* proteins from a heterogeneous biofilm grown in a drip-flow reactor [126]. (d) Chemical response of two strains of *P. aeruginosa* in the presence of the antibiotic azithromycin [127]. See original references for additional details on the specific bacterial strains used and the identities of all of the ions. *; Surfactin-C14; Plipastatin-C17-Val. *Source:* Dunham et al. [132]/reproduced with permission from ACS.

labelling which may affect their structures [137]. An experiment usually consists of four steps as described in Figure 1.10 [140].

Recently, HDX-MS has been used to study mechanisms in order to assess inhibitors and further elucidate dynamics of proteins involved in antibiotic resistance [141, 142]. Reading et al. [142] studied the inner membrane transporter AcrB – a subcomplex within the AcrAB-TolC efflux pump in *E. coli* to investigate interactions with efflux pump inhibitor Phe-Arg β-naphthylamide (PaβN) and licensed antibiotic ciprofloxacin. In the presence of PaβN, there were reduced rates of HDX throughout the drug-binding pockets and connecting loop, potentially demonstrating structural stabilization of the drug-binding pocket entrances. However, the study revealed a lack of extensive HDX in the transmembrane region, highlighting a major issue with studying membrane proteins successfully – the coverage of the protein that can be obtained. With transmembrane proteins such as the AcrB

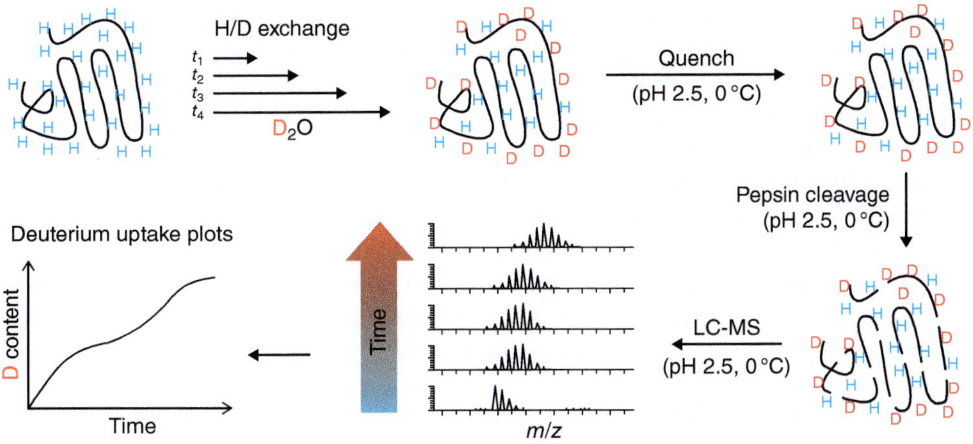

Figure 1.10 (i) The protein is incubated in D_2O for different time periods, enabling disordered regions with weak hydrogen bonding (and, to a lesser extent, ease of solvent access) to exchange via acid, base or water catalysis. This happens at any site that contains a labile hydrogen, but exchange of backbone amides of the peptide is by far the most commonly measured. This is because they are evenly distributed across the polypeptide chain (excluding proline) and form important hydrogen bonding networks across the secondary structure [138]. (ii) The reaction is then quenched by reducing the pH to 2.4–2.6 and the temperature to 0°C, minimizing the loss of deuterium (back-exchange) [139]. (iii) The protein is then digested into peptides which are later computationally reconstructed to give information on the original polypeptide. (iv) The peptides are injected into the LC-MS and the generated mass spectra are analysed computationally to assess the confidence of the presence of different peptides and map them onto the sequence. *Source:* Weis [140]/Reproduced with permission of John Wiley & Sons.

complex, this value can be low due to the inability of the protease to access transmembrane areas protected by the detergent micelle which stops solvent access and therefore deuteration. However, there are certain methods for increasing this coverage, including the use of alternative proteases and quench buffer additives. For example, Möller et al. [143] reported that the addition of 3 M urea increased LeuT transporter coverage by 11%.

Similarly, Huang et al. [141] used HDX-MS to investigate antibiotic resistance, this time in conjunction with molecular dynamics (MD) simulations. The binding interactions and flexibility of β-lactamase inhibitory protein (BLIP) produced by *Streptomyces clavuligerus* with several types of class A β-lactamases was demonstrated. Huang et al. [141] suggested that this ability could offer an alternative approach to studying β-lactamase-mediated antibiotic resistance. The HDX data collected provided novel information on the dynamics of BLIP, highlighting residues involved in its flexibility and therefore important in its inhibitory binding with class A β-lactamases.

1.6 The Omnitrap, a Novel MS Instrument that Combines Many Applications of Mass Spectrometry

Ion trapping is an extremely powerful technique that gives instrument designers the opportunity to maintain ions in known position and energy states for a long time and to manipulate ion clouds in various ways to ultimately reveal the identity and structure of the sample constituents. A very successful ion trap was introduced in 1953 by W. Paul and H. Steinwedel [144]. The Paul trap trapped ions between two hyperbolic endcap electrodes with the aid of a ring electrode between them. A high-frequency and high-voltage sinusoidal waveform is typically applied to the ring electrode, while the two endcap electrodes are near-ground potential [hence such traps are also called radio frequency (RF) traps or 3D traps]. Ions found near the centre between the endcap electrodes are typically forced by the alternating electrical field to maintain stable trajectories around the centre, depending on their mass-to-charge ratio and following mathematical principles described by Paul. Several commercial instruments based on this design were launched in the years since its discovery, with the latest being the MALDImini-1 by Shimadzu [145]. By applying carefully selected waveforms to the endcap electrodes, ions of specific m/z can be preferentially excited (resonance excitation) and made to oscillate away from the centre of the trap, acquiring higher energies. Taking advantage of this effect, a Paul trap can be used to isolate a single m/z from an ion mixture (by exciting all other ions until they are lost at the endcap electrodes), to fragment selected ions via collisional activation or even to measure the m/z spectrum by sequentially scanning ions out of a small hole on the centre of the endcaps towards an ion detector situated outside the trap. Combining these effects, a 3D Paul trap can be used to produce high-order tandem spectra by sequentially isolating fragment ions for reactivation/fragmentation (MS^n). A review of early instrumentation and applications of quadrupole ion traps was published by the inventor Wolfgang Paul [146], who was awarded the Nobel Prize in Physics in 1989.

A variation of the Paul trap is the linear ion trap, which uses four spatially shaped quadrupole rod electrodes to trap ions radially; various arrangements of electrodes at both ends of the rods are designed to keep ions within the confines of the linear trap, while allowing ions to enter or exit axially when required. In a typical implementation, an opposed pair of rods have complementary high-frequency/high-voltage sinusoidal waveforms applied to them, while the other pair is reserved for excitation waveforms and exit slots for ion measurement. Linear ion traps resolve a fundamental limitation of 3D ion traps, space charge: when a high ion population is stored at the centre of a 3D trap, their combined charge changes the trap characteristics and interferes with the manipulation of target ions, while lowering the resolution and mass accuracy of the device. This has a detrimental effect on the practical dynamic range of 3D traps. Linear traps allow the ion cloud to spread across the geometrical axis between the rods, multiplying the charge capacity of the device before adverse effects are observed. Several linear trap instruments have been commercialized, one of which is the ThermoFisher LTQ described by Schwartz et al. [147]. This technology still plays an important role in some of the most sophisticated mass spectrometers for proteomics research, including the ThermoFisher Orbitrap Tribrid mass spectrometer. In this example an advanced, dual-pressure version of the linear ion trap is combined with an Orbitrap mass analyser to take advantage of the high-resolution capabilities of the platform.

Despite the success of linear ion traps and their extensive use in academic and pharmaceutical research, there are significant drawbacks, including the limited array of activation methods employed in commercial devices and the low mass accuracy and resolution (when the traps are used as mass analysers). The Omnitrap® Platform [148] was proposed and developed by Fasmatech Science and Technology (Athens, Greece) to overcome these limitations. The Omnitrap platform is a novel linear ion trap that combines a number of innovations: the trapping rods are cut in a number of segments, each segment being optimally designed to perform a specific set of functions: a fast-changing controllable floating potential can be applied independently to the rods of each segment; pressure in the trap can be dynamically controlled; and the trapping high-voltage and high-frequency waveform is not sinusoidal but it is a bipolar rectangular wave, driven by fast, high-power transistor arrays.

Separating the trapping rods into independent segments gave Fasmatech engineers the opportunity to optimize the design of each segment for a specific set of tasks, while maintaining the trapping waveforms applied throughout the device (Figure 1.11). For example, segment Q2 is longer than other segments to spread the ion cloud across a larger area and minimize space charge. This is essential because segment Q2 is designed to perform high-resolution ion selection and collisional activation fragmentation, and thus space charge is an important consideration. Segments Q5 and Q8 are shorter and have small apertures drilled at opposing ends of the side rod segments. This is to allow externally generated electrons or ions/radicals (Q5) and laser light (Q8) to enter from either side of the segment, while ions are compressed axially to form a small-sized cloud in the centre of the segments to maximize interaction.

Being able to apply a fast-changing independent floating voltage to each segment (Figure 1.12) allows the segments to create electrical potential slopes and wells across the length of the Omnitrap platform and therefore to push ions to any segment at any time of

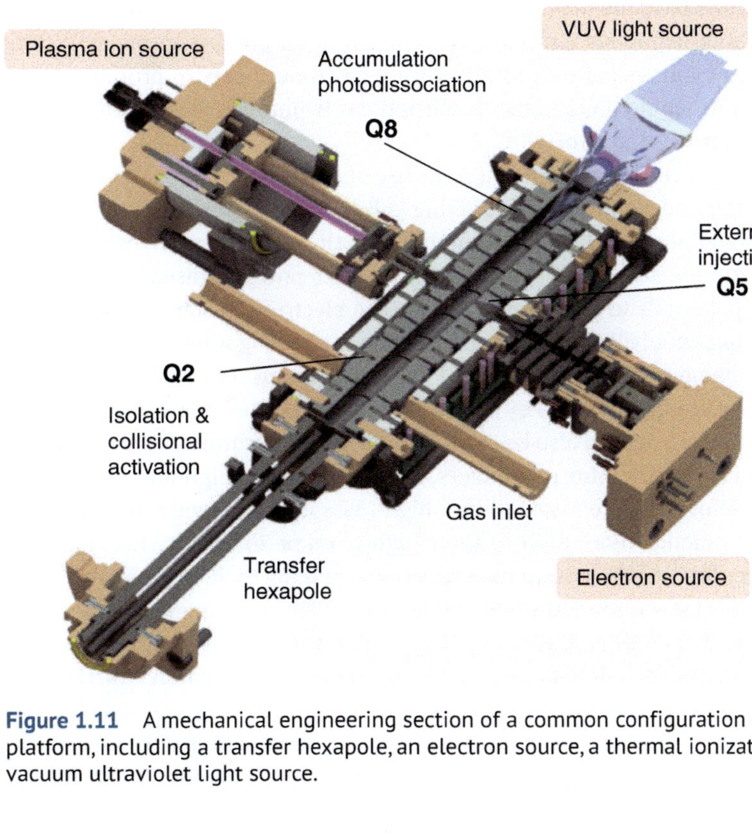

Figure 1.11 A mechanical engineering section of a common configuration of the Omnitrap platform, including a transfer hexapole, an electron source, a thermal ionization cavity source and a vacuum ultraviolet light source.

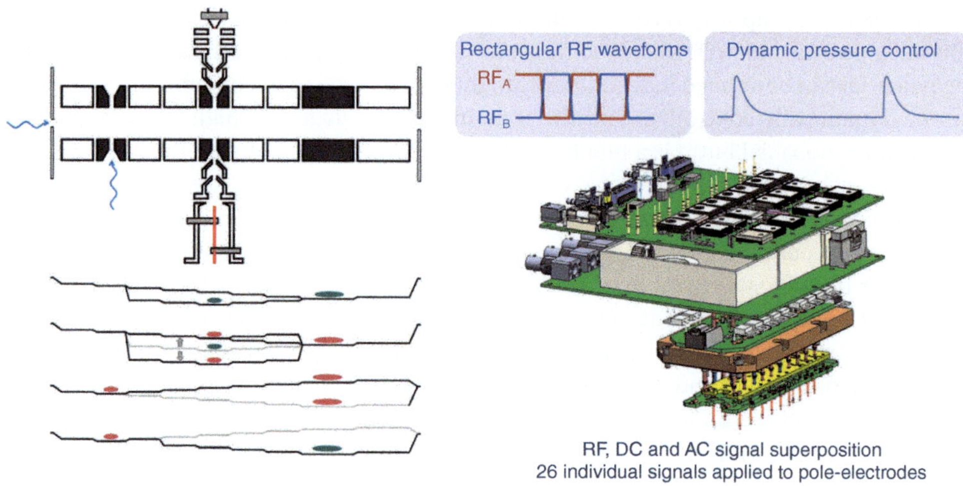

Figure 1.12 Schematic and practical implementation of the Omnitrap platform. Each segment can float at an independently controlled potential, creating slopes and wells that can transport ions to the appropriate segment for each experimental step.

an experimental sequence. Ion transport is virtually lossless, as ion population can be trapped in the device for several seconds without significant ion population loss. Operating the trap at a relatively low baseline pressure of helium buffer gas helps with the speed of ion transfer and results in negligible ion loss. Partial pressure of an appropriate collision gas (e.g. argon) is raised by means of pulsed valves only when necessary to promote energetic collisions for CID in segment Q2. Excess collision gas is pumped away within a few milliseconds (dynamic pressure control).

The key innovation of the Omnitrap platform is the use of rectangular pulses for trapping. This technique, also described by Shimadzu's Li Ding [149], provides similar trapping characteristics to the sinusoidal ion trap through a modified mathematical framework. Instead an LC circuit to generate the sinusoidal waveform, it requires power transistors to switch positive and negative potentials to the trapping rods. The electrical design of sinusoidal-driven ion traps is fundamentally limited to single-frequency trapping waveforms and engineers must rely on analogue amplitude control for trap tuning. On the other hand, rectangular pulses can be readily triggered by digital circuitry at a wide range of frequencies, while in principle the rail voltages can also be modified, allowing instrumentation engineers higher flexibility in optimizing trapping parameters. For example, very high-amplitude (several kV) sinusoidal waveforms are required to trap high-mass ions efficiently in typical RF ion traps, which poses practical limitation to their high-mass performance. Rectangular-wave traps require lower frequencies to trap high mass ions, making their ability to trap very high m/z species far superior. Additionally, ions undergoing CID in rectangular-wave traps can be driven to reach higher collisional energies (higher-energy collision-induced dissociation, heCID), or be set to undergo slow-heating collision-induced dissociation (shCID), the latter resulting in fragmentation patterns typical of conventional traps.

A more subtle but very important characteristic is that rectangular-pulses oscillate between two distinct electrical field states: they occupy each static state for almost 50% of the experimental time (transitions between states are very rapid). By contrast, the electric field of sinusoidal-driven ion traps is constantly changing. Having a well-defined 'electrostatic' state for half the experimental time allows the opportunity to launch externally generated electron packets with highly controlled and well characterized energies to interact with the ion cloud. The pulsed electron source designed for the Omnitrap platform can generate electrons that reach the ion cloud with energies from very low eVs up to thousands of eVs, allowing several fine-tuned electron-induced phenomena to be carried out in a single device. These include low-energy ECD, hot-ECD (using higher-energy electrons), electron-induced dissociation (EID), using high-energy electrons and electron meta-ionization (EmI), when electron-generated radical ions are subsequently fragmented with slow-heating CID. All these techniques can provide substantial new information that can aid the structural characterization of complex species (proteins, peptides, glycans, lipids, non-tryptic peptides, etc.).

The Omnitrap platform can also employ a thermal ionization cavity (TIC) (Figure 1.13) to drive ions or radical species to collide with the ion cloud, creating a hydrogen atom attachment (HAA) effect. Finally, a wide variety of photo-dissociation methods can be applied in the Omnitrap, including Excimer Laser ultraviolet photodissociation (UVPD), continuous light source vacuum ultraviolet photodissociation (VUVPD) or infrared multiphoton dissociation (IRMPD) using a focused infrared laser beam (e.g. a CO_2 laser). The light beams can be introduced radially into the trap through either aperture of section Q8 or, in some implementations, axially from the far end of the device. In the latter case, the light beam can be made to interact with the ion cloud in multiple segments. Prototypes of all the above are currently under advanced development.

The extreme analytical power of the Omnitrap platform comes with its ability to employ any one of the available activation methods (Figure 1.14) in sequential tandem MS steps and interrogate structures with the most appropriate fragmentation method for each investigative step. The most common implementation of the instrument to date is in a hybrid

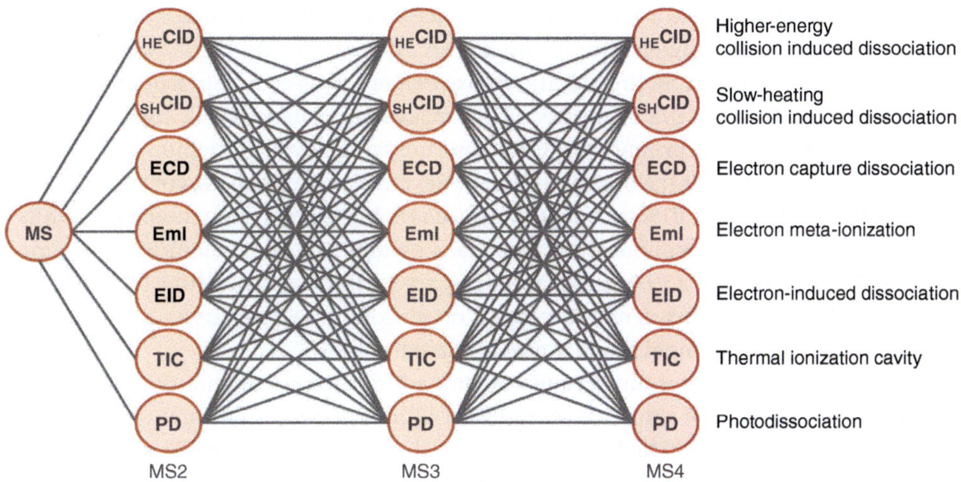

Figure 1.13 The ion activation network accessible for multiple fragmentation steps with an Omnitrap platform.

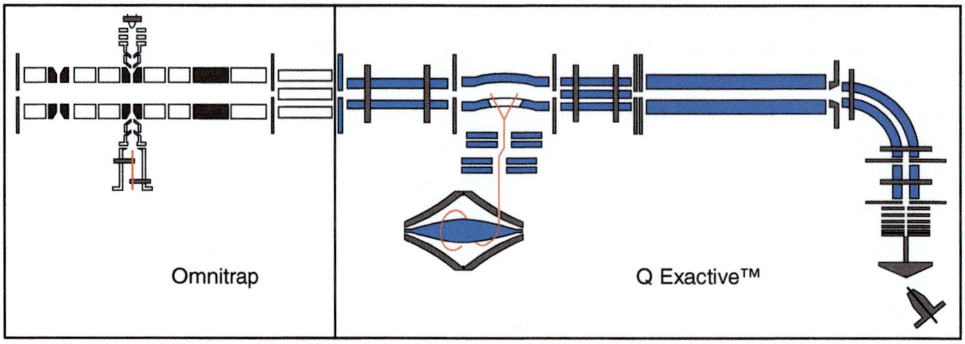

Figure 1.14 The Omnitrap platform implementation at the trailing end of a ThermoFisher Q Exactive™ instrument. Ions are received by the Q-Exactive, driven to the Omnitrap for processing and then sent back to the Orbitrap for analysis.

format with ThermoFisher's Q-Exactive™ or Exploris™ platforms, both employing an ESI ion source and an Orbitrap mass analyser (Figure 1.14). Ions are generated by the ThermoFisher Instrument's ESI source, and are typically transferred to the Omnitrap platformm for processing. A user-defined experimental sequence is then applied to the ion cloud, moving ions to the appropriate segments, taking the necessary actions, and inducing the desired fragmentation steps. At the end of the process the ions are delivered back to the ThermoFisher device for *m/z* measurement with the Orbitrap mass analyser. The extreme resolution of the Orbitrap mass analyser helps with the analysis of the information-rich spectra afforded by this instrument. The device can also be implemented with orthogonal TOF and FTICR mass analysers.

Combining the capabilities of the Omnitrap platform to handle high-mass ions and to apply the most appropriate fragmentation methods at each step makes this device a powerful tool for top-down proteomics. In the example in Figure 1.15, an intact protein (ubiquitin) [148] was introduced into a ThermoFisher device and analysed up to MS^2 mode in the Omnitrap platform using the electron beam at two different energy regimes to generate ECD and EID fragmentation. Both ECD and EID generate a near-complete sequence coverage. Figure 1.15(a) shows the annotated ECD mass spectrum and the corresponding sequence map. All types of primary fragment ions are being formed, including c/z• fragments generated throughout the backbone except next to proline residues, covering 96% of the amino acid sequence. A 91% sequence coverage is achieved by y-type fragments, while additional information throughout the backbone is obtained by the remaining a-, x-, and b-type fragments. Figure 1.15(b) shows the annotated EID mass spectrum of ubiquitin [M+8H]8+ ions. Complete sequence coverage is obtained for the 8+ charge state, while all other types of primary fragment ions are also observed.

The Omnitrap technology is an example of state-of-the art instrumentation developed through the collaboration of academic researchers with large and small commercial enterprises, with the support of public bodies (primarily the European Union). The technology has already provided exciting results in top-down protein structural characterization, intact antibody analysis and post-translational modification analysis, as well as the analysis of lipids, glycans, non-tryptic peptides and other important compound classes. One of the

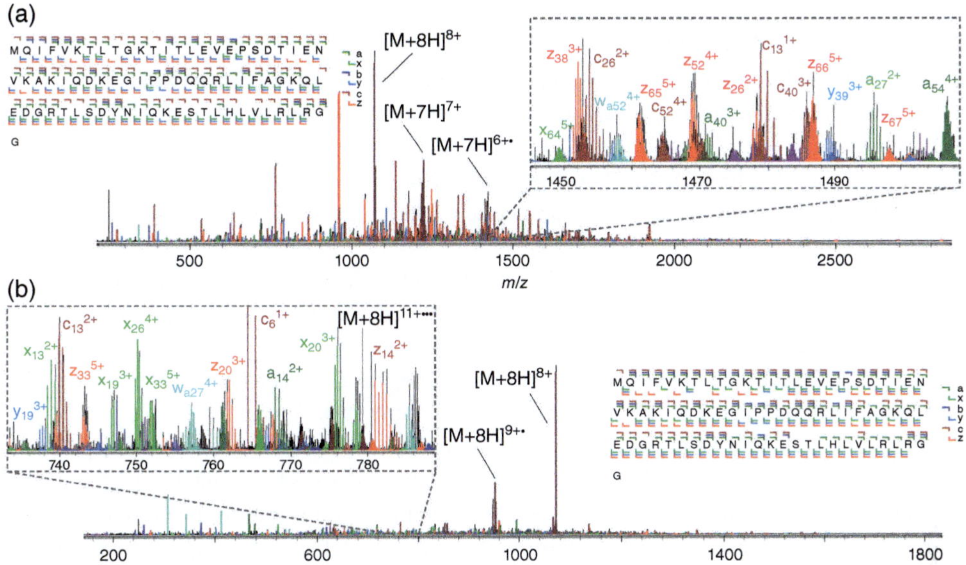

Figure 1.15 (a) Annotated MS^2 ECD mass spectrum of ubiquitin generated by exposing the $[M+8H]^{8+}$ ions to ~1 μA electron current for 150 ms. Complete sequence coverage is obtained with c/z• fragments covering 96% of the amino acid sequence. (b) Annotated MS^2 EID mass spectrum of ubiquitin generated by exposing the $[M+8H]^{8+}$ ions to ~2 μA electron current for 100 ms. Complete sequence coverage is obtained, with a-type fragments providing 92% sequence coverage. The mass spectra are generated by averaging the ion signal over ~500 scans.

most ambitious early-stage collaborative projects utilizing the power of the Omnitrap platform [150] is sponsored by the EU Future and Emerging Technologies (FET) programme and targets the comprehensive analysis of intact viruses (e.g. SARS-CoV-2) in breath samples. The success of such a project will deliver a powerful, direct, non-invasive, microbiology analysis method, adding to the long and distinguished sequence of scientific discoveries presented in this chapter.

References

1 Castellani, A. and Chalmers, J. (1919). *Manual of Tropical Medicine*. London: Baillière, Tindall and Cox.

2 Veillon, M.H. and Zuber, H. (1898). Recherches sur quelques microbes strictement anaerobies et leur role en pathologic. *Arch. Med. Exp. Anat. Pathol.* 10: 517–545.

3 Drasar, B.S. and Hill, M.J. (1974). *Human Intestinal Flora*. London: Academic Press.

4 Gram, C. (1884). The differential staining of Schizomycetes in tissue sections and in dried preparations. *Fortschitte der Medicin* 2: 185–189.

5 Bergey's Manual of Determinative Bacteriology (1923). *Society of American Bacteriologists*, 1923. Baltimore: Williams & Wilkins Co.

6 Cowan, S.T. and Steel, K.J. (1970). *Manual for the Identification of Medical Bacteria*. Cambridge: Cambridge University Press.

7 Owen, R.J., Hill, L.R., and Lapage, S.P. (1969). Determination of DNA base compositions from melting profiles in dilute buffers. *Biopolymers* 7: 503–516.

8 Shah, H.N. and Collins, M.D. (1988). Proposal for reclassification of *Bacteroides asaccharolyticus, Bacteroides gingivalis* and *Bacteroides endodontalis* in a new genus, Porphyromonas. *Int. J. Syst. Bacteriol.* 38: 128–131.

9 Shah, H.N. and Collins, M.D. (1989). Proposal to restrict the genus *Bacteroides* (Castellani and Chalmers) to *Bacteroides fragilis* and closely related species. *Int. J. Syst. Bacteriol.* 39: 85–97.

10 Shah, H.N. and Collins, M.D. (1990). *Prevotella,* a new genus to include *Bacteroides melaninogenicus* and related species formerly classified in the genus *Bacteroides*. *Int. J. Syst. Bacteriol.* 40: 205–208.

11 Shah, H.N. (1991a). The genus *Bacteroides* and related taxa. In: *The Prokaryotes*, 2e (ed. A. Balows, H.G. Trüper, M. Dworkin, et al.), 3593–3607. USA: Springer-Verlag.

12 Shah, H.N. (1991b). The genus *Porphyromonas* and related taxa. In: *The Prokaryotes*, 2e (ed. A. Balows, H.G. Trüper, M. Dworkin, et al.), 3608–3620. USA: Springer-Verlag.

13 Shah, H.N. and Collins, M.D. (1980). Fatty acid and isoprenoid quinone composition in the classification of *Bacteroides melaninogenicus* and related taxa. *J. Appl. Bacteriol.* 48: 75–84.

14 Shah, H.N. and Collins, M.D. (1983). Genus *Bacteroides* a chemotaxonomical perspective. *J. Appl. Bacteriol.* 55: 403–416.

15 Shah, H.N., van Steenbergen, T.J.M., Hardie, J.M., and de Graaff, J. (1982). DNA base compos-ition, DNA-DNA reassociation and isoelectric focusing of proteins of strains designated *Bacteroides oralis*. *FEMS Microbiol. Letts.* 13: 125–130.

16 Swindlehurst, C.A., Shah, H.N., Parr, C.W., and Williams, R.A.D. (1977). Sodium dodecyl sulphate-polyacrylamide gel electrophoresis of polypeptides from *Bacteroides melaninogenicus*. *J. Appl. Bacteriol.* 43: 319–324.

17 Shah, H.N., Gharbia, S.E., Al-Jalili, T.A.R. et al. (1988). Enzymes of diagnostic importance within the *Bacteroidaceae;* use as possible ecological markers. *Microb. Ecol. Health Dis.* 1: 115–122.

18 Shah, H.N., Chilton, C., Rajakaruna, L. et al. (2010). Changing concepts in the characterisation of microbes and the influence of mass spectrometry. In: *Mass Spectrometry for Microbial Proteomics* (ed. H.N. Shah and S.E. Gharbia), 3–34. Chichester: Wiley.

19 Shah, H.N. and Gharbia, S.E. (2017). A paradigm shift from research to front-line microbial diagnostics in MALDI-TOF and LC-MS/MS: a Laboratory's vision and relentless resolve to help develop and implement this new technology amidst formidable obstacles. In: *MALDI-TOF and Tandem MS for Clinical Microbiology* (ed. H.N. Shah and S.E. Gharbia), 110–276. Chichester: Wiley.

20 Collins, M.D. and Jones, D. (2008). Lipids in the classification and identification of Coryneform bacteria containing peptidoglycans based on 2, 4-diaminobutyric acid. *J. Appl. Microbiol.* 48: 459–470.

21 Goodfellow, M. and Minnikin, D.E. (1985). *Chemical Methods in Bacterial Systematics*, The Society for Applied Bacteriology Technical Series 20. New York and London: Academic Press.

22 Schleifer, K.H. and Kandler, O. (1972). Peptidoglycan types of bacterial cell walls and their taxonomic implications. *Bact. Revs.* 36: 407–477.

23 Dobson, G., Minnikin, D.E., Minnikin, S.M. et al. (1985). Systematics analysis of complex mycobacterial lipids. In: *Chemical Methods in Bacterial Systematics*, The Society for Applied Bacteriology Technical Series 20 (ed. M. Goodfellow and D.E. Minnikin), 237–266. New York and London: Academic Press.

24 Shah, H.N., Bonnet, R., Mateen, B., and Williams, R.A.D. (1979). The porphyrin pigmentation of subspecies of *Bacteroides melaninogenicus*. *Biochem. J* 180: 45–50.

25 Kamio, Y., Kanegasaki, S., and Takahashi, H. (1969). Occurrence of plasmalogens in anaerobic bacteria. *J. Gen. Appl. Microbiol.* 15: 439–451.

26 Kunsman, J.E. and Caldwell, D.R. (1974). Comparison of the sphingolipid content of rumen *Bacteroides* species. *Appl. Microbiol.* 28: 1088–1089.

27 White, D.C. and Tucker, A.N. (1970). Ceramide phosphorylglycerol phosphate a new sphingolipid found in bacteria. *Lipids* 5: 56–62.

28 Sasser, M. (1990). *Identification of Bacteria by Gas Chromatography of Cellular Fatty Acids*. Newark, USA: Database of Profiles. MIDI, Inc.

29 Collins, M.D., Shah, H.N., McKee, A.S., and Kroppenstedt, R.M. (1982). Chemotaxonomy of the genus *Capnocytophaga* (Leadbetter, Hot and Socransky). *J. Appl. Bacteriol.* 52: 409–415.

30 Schiller, J., Suss, R., Fuchs, B. et al. (2007). MALDI-TOF MS in lipidomics. *Front. Biosci.* 12: 2568–2579.

31 Solntceva, V., Kostrzewa, M., and Larrouy-Maumus1 G. (2021). Detection of species-specific lipids by routine MALDI TOF mass spectrometry to unlock the challenges of microbial identification and antimicrobial susceptibility testing. *Front. Cell. Infect. Microbiol.* 04 February 2021 https://doi.org/10.3389/fcimb.2020.621452.

32 Fuchs, B. (2012). Analysis of phospolipids and glycolipids by thin-layer chromatography-matrix-assisted laser desorption and ionization mass spectrometry. *J. Chromatogr.* 1259: 62–73.

33 Fuchs, B., Schiller, J., Süss, R. et al. (2007). A direct and simple method of coupling matrix-assisted laser desorption and ionization time-of-flight mass spectrometry (MALDI-TOF MS) to thin-layer chromatography (TLC) for the analysis of phospholipids from egg yolk. *Anal. Bioanal.Chem.* 389: 827–834.

34 Nakamura, K., Suzuki, Y., Goto-Inoue, N. et al. (2006). Structural characterization of neutral glycosphingolipids by thin-layer chromatography coupled to matrix-assisted laser desorption/ionization quadrupole ion trap time-of-flight MS/MS. *Anal. Chem.* 78: 5736–5743.

35 Calvano, C.D., Zambonin, C.G., and Palmisano, F. (2011). Lipid fingerprinting of gram-positive lactobacilli by intact–matrix-assisted laser desorption/ionization mass spectrometry using a proton sponge-based matrix. *Rapid Commun. Mass Spectrom.* 25: 1757–1764.

36 Dortet, L., Potron, A., Bonnin, R.A. et al. (2018). Rapid detection of colistin resistance in *Acinetobacter baumannii* using MALDI-TOF-based lipidomics on intact bacteria. *Sci. Rep.* 8: 16910.

37 Gonzalo, X., Broda, A., Drobniewski, F., and Larrouy-Maumus, G. (2021). Performance of lipid fingerprint-based MALDI-TOF for the diagnosis of mycobacterial infections. *Clin. Microbiol. Infect.* 27 (6): 912.e1–912.e5.

38 Larrouy-Maumus, G., Clements, A., Filloux, A. et al. (2016). Direct detection of lipid a on intact Gram-negative bacteria by MALDI-TOF mass spectrometry. *J. Microbiol. Meth.* 120: 68–71.

39 Larrouy-Maumus, G. and Puzo, G. (2015). Mycobacterial envelope lipids fingerprint from direct MALDI-TOF MS analysis of intact bacilli. *Tuberculosis* 95: 75–85.

40 Saromi, K., England, P., Tang, W. et al. (2020). Rapid Glycosyl-inositol-Phospho-ceramides fingerprint from filamentous fungi pathogens using the MALDI Biotyper Sirius system. *Rapid Commun. Mass Spectrom.* e8904. https://doi.org/10.1002/rcm.8904.

41 Sorensen, M., Chandler, C.E., Gardner, F.M. et al. (2020). Rapid microbial identification and colistin resistance detection via MALDI-TOF MS using a novel on-target extraction of membrane lipids. *Sci. Rep.* 10: 21536. https://doi.org/10.1038/s41598-020-78401-3.

42 Lellman, S. and Cramer, R. (2020). Bacterial identification by lipid profiling using liquid atmospheric pressure matrix-assisted laser desorption/ionization mass spectrometry. *Clin. Chem. Lab. Med.* 58: 930–938.

43 Leung, L.M., Fondrie, W.E., Doi, Y. et al. (2017). Identification of the ESKAPE pathogens by mass spectrometric analysis of microbial membrane glycolipids. *Sci. Rep.* 7: 6403–6409.

44 Sohlenkamp, C. and Geiger, O. (2016). Bacterial membrane lipids: diversity in structures and pathways. *FEMS Microbiol. Revs.* 40: 133–159.

45 Stübiger, G., Wuczkowski, M., Mancera, L. et al. (2016). Characterization of yeasts and filamentous fungi using MALDI lipid phenotyping. *J. Microbiol. Meth.* 130: 27–37.

46 Cain, T.C., Lubman, D.M., and Weber, W.J. (1994). Differentiation of bacteria using protein profiles from matrix-assisted laser desorption/ionization time-of-flight mass spectrometry. *Rapid Commun. Mass Spectrom.* 8: 1026–1030.

47 Holland, R.D., Wilkes, J.G., Rafii, F. et al. (1996). Rapid identification of intact whole bacteria based on spectral patterns using matrix-assisted laser desorption/ionization with time-of-flight mass spectrometry. *Rapid Commun. Mass Spectrom.* 10: 1227–1232.

48 Krishnamurthy, T. and Ross, P.L. (1996). Rapid identification of bacteria by direct matrix-assisted laser desorption/ionization mass spectrometric analysis of whole cells. *Rapid Commun. Mass Spectrom.* 10: 1992–1996.

49 Krishnamurthy, T., Ross, P.L., and Rajamani, U. (1996). Detection of pathogenic and non-pathogenic bacteria by matrix-assisted laser desorption/ionization time-of-flight mass spectrometry. *Rapid Commun. Mass Spectrom.* 10: 883–888.

50 Liang, X., Zheng, K., Qian, M.G., and Lubman, D.M. (1996). Determination of bacterial protein profiles by matrix-assisted laser desorption/ionization mass spectrometry with high-performance liquid chromatography. *Rapid Commun. Mass Spectrom.* 10: 1219–1226.

51 Welham, K.J., Domin, M.A., Scannell, D.E. et al. (1998). The characterization of micro-organisms by matrix-assisted laser desorption/ionization time-of-flight mass spectrometry. *Rapid Commun. Mass Spectrom.* 12: 176–180.

52 Claydon, M., Davey, S., Edwards-Jones, V., and Gordon, D.B. (1996). The rapid identification of intact microorganisms using mass spectrometry. *Nat. Biotechnol.* 14: 1584–1586.

53 Honisch, C., Chen, Y., and Hillenkamp, F. (2010). Comparative DNA sequence analysis and typing using mass spectrometry. In: *Mass Spectrometry for Microbial Proteomics* (ed. H.N. Shah and S.E. Gharbia), 443–462. Chichester: Wiley.

54 Bishop, C., Arnold, C., and Gharbia, S.E. (2010). Transfer of a traditional serotyping system (Kauffmann-White) onto a MALDI-TOF-MS platform for the rapid typing of salmonella isolates. In: *Mass Spectrometry for Microbial Proteomics* (ed. H.N. Shah and S.E. Gharbia), 463–496. Chichester: Wiley.

55 Shah, H.N., Keys, C., Gharbia, S.E. et al. (2000). The application of MALDI-TOF mass spectrometry to profile the surface of intact bacterial cells. *Microbial Ecol. Health Dis* 12: 241–246.

56 Shah, H.N., Keys, C.J., Schmid, O., and Gharbia, S.E. (2002). Matrix-assisted laser desorption/ionisation time of flight mass spectrometry and proteomics; a new era in anaerobic microbiology. *Clin. Infect. Dis.* 35: 58–64.

57 Keys, C.J., Dare, D.J., Sutton, H. et al. (2004). Compilation of a MALDI-TOF mass spectral database for the rapid screening and characterisation of bacteria implicated in human infectious diseases. *Infect. Genetics Evol.* 4: 221–242.

58 Rajakaruna, L., Hallas, G., Diane Dare, D. et al. (2009). High throughput identification of clinical isolates of *Staphylococcus aureus* using MALDI-TOF-MS of intact cells. *Infect. Genetics Evol.* 9: 507–513.

59 Kallow, W., Erhard, M., Shah, H.N. et al. (2010). MALDI-TOF MS for microbial identification: years of experimental development to an established protocol. In: *Mass Spectrometry for Microbial Proteomics* (ed. H.N. Shah and S.E. Gharbia), 255–276. Chichester: Wiley.

60 Shah, H.N. (2010). MALDI-TOF-MS of surface-associated and stable intracellular proteins for identification and resistance profiling of human pathogens. ECCMID, Session: 'MALDI-TOF in Clinical Microbiology'. Vienna, Austria (11 April 2010).

61 Kim, Y. and Kim, J.-S. (2017). ASTA's MicroID system and its MycoMp database for mycobacteria. In: *MALDI-TOF and Tandem MS for Clinical Microbiology* (ed. H.N. Shah and S.E. Gharbia), 110–276. Chichester: Wiley.

62 Fagerquist, C.K., Garbus, B.R., Williams, K.E. et al. (2009). Web-based software for rapid top-down proteomic identification of protein biomarkers, with implications for bacterial identification. *Appl. Environ. Microbiol.* 75: 4341–4353.

63 Dworzanski, J.P., Snyder, A.P., Chen, R. et al. (2004). Identification of bacteria using tandem mass spectrometry combined with a proteome database and statistical scoring. *Anal. Chem.* 76: 2355–2366.

64 Berendsen, E.M., Levin, E., Braakman, R. et al. (2017). Identification of microorganisms grown in blood culture flasks using liquid chromatography–tandem mass spectrometry. *Future Microbiol.* 12: 1135–1145.

65 Hale, O.J., Morris, M., Jones, B. et al. (2019). Liquid atmospheric pressure matrix-assisted laser desorption/ionization mass spectrometry adds enhanced functionalities to MALDI MS profiling for disease diagnostics. *ACS Omega* 4: 12759–12765.

66 Dai, Y., Shortreed, M.R., Scalf, M. et al. (2017). Elucidating *Escherichia coli* proteoform families using intact-mass proteomics and a global PTM discovery database. *J. Proteome Res.* 16: 4156–4165.

67 Dupré, M., Gilquin, B., Fenaille, F. et al. (2015). Multiplex quantification of protein toxins in human biofluids and food matrices using immunoextraction and high resolution targeted mass spectrometry. *Anal. Chem.* 87: 8473–8480.

68 Camara, J.E. and Hays, F.A. (2007). Discrimination between wild-type and ampicillin-resistant Escherichia coli by matrix-assisted laser desorption/ionization time-of-flight mass spectrometry. *Anal Bioanal. Chem.* 389 (5): 1633–1638.

69 Kaufmann, R., Kirsch, D., and Spengler, B. (1994). Sequencing of peptides in a time-of-flight mass spectrometer: evaluation of post-source decay following matrix-assisted laser desorption ionization (MALDI). *Int. J. Mass Spectrom. Ion Proc.* 131: 355–384.

70 Fenn, J.B., Mann, M., Meng, C.K. et al. (1989). Electrospray ionization for mass spectrometry of large biomolecules. *Science* 246: 64–71.

71 Anhalt, J.P. and Fenselau, C. (1975). Identification of bacteria using mass spectrometry. *Anal. Chem.* 47: 219–225.

72 Meuzelaar, H.L.C. and Kistemaker, P.G. (1973). Technique for fast and reproducible fingerprinting of bacteria by pyrolysis mass spectrometry. *Anal. Chem.* 45: 587–590.

73 Fenselau, C. (1993). *Mass Spectrometry for Characterization of Microorganisms*, ACS Symposium Series, vol. 541, 1–7. Washington, DC, Ch. 1: American Chemical Society.

74 Bland, C., Hartmann, E.M., Christie-Oleza, J.A. et al. (2014). N-terminal-oriented proteogenomics of the marine bacterium roseobacter denitrificans Och114 using (*N*-Succinimidyloxycarbonylmethyl)tris(2,4,6-trimethoxyphenyl) phosphonium bromide (TMPP) labeling and diagonal chromatography. *Mol. Cell. Proteomics* 13: 1369–1381.

75 Ma, W., Zhang, D., Li, G. et al. (2017). Antibacterialmechanismof daptomycin antibiotic against *Staphylococcus aureus* based on a quantitative bacterial proteome analysis. *J. Proteomics* 150: 242–251.

76 Ramos, S., Chafsey, I., Silva, N. et al. (2015). Effect of vancomycin on the proteome of the multiresistant *enterococcus faecium* SU18 strain. *J. Proteomics* 113: 378–387.

77 Cheng, K., Sloan, A., Li, X. et al. (2018). Mass spectrometry-based Shiga toxin identification: an optimized approach. *J. Proteomics* 180: 36–40.

78 Encheva, V., Wait, R., Gharbia, S.E. et al. (2006). Comparison of extraction procedures for proteome analysis of *Streptococcus pneumoniae* and a basic reference map. *Proteomics* 6: 3306–3317.

79 Encheva, V., Wait, R., Begum, S. et al. (2007). Protein expression diversity amongst serovars of *Salmonella enterica*. *Microbiology* 153: 4183–4193.

80 Karlsson, R., Chooneea, D., Carlsohn, E. et al. (2010). Characterization of bacterial membrane proteins using a novel combination of a lipid based protein immobilization technique with Mass Spectrometry. In: *Mass Spectrometry for Microbial Proteomics* (ed. H.N. Shah and S.E. Gharbia), 157–174. Chichester: Wiley.

81 Chooneea, D., Karlsson, R., Encheva, V. et al. (2010). Elucidation of the outer membrane proteome of *Salmonella enterica* serovar Typhimurium utilising a lipid-based protein immobilization technique. *BMC Microbiology* 10: 44.

82 Chilton, C.H., Gharbia, S.E., Fang, M. et al. (2014). Comparative Proteomic Analysis of *Clostridium difficile* Isolates of Varying Virulence. *J. Med. Microbiol.* 63: 489–503.

83 Misra, R., Gaulton, T., Rose, G. et al. (2016). Complete genome sequence of the hypervirulent bacterium, *Clostridium difficile* (027) str. *G46. Genome Announcements* 3 (2): e00073.

84 Al-Shahib, A., Misra, R., Ahmod, N. et al. (2010). Coherent pipeline for biomarker discovery using mass spectrometry and bioinformatics. *BMC Bioinformatics* 11: 437.

85 Misra, R.V., Ahmod, N.Z., Parker, R. et al. (2012). Developing an integrated proteo-genomic approach for the characterisation of biomarkers for the identification of *Bacillus anthracis*. *J. Microbiol. Methods* 88: 237–247.

86 Shah, H.N., Stephenson, J., Cardasis, H., Neil, J., Ping, Y., Ravela, S., Ritamo, I., Damsbo, M., Grist, R., Freeke, J., Stielow, B., Gvozdyak, O., Dukik, K., de Hoog, S., Damoc, E., Valmu, L., Cherkassky, A., Fang, M., Gaulton, T., Misra, R., Gharbia, S.E. (2015). A global diagnostic approach for microbial identification: accurate characterization of difficult to differentiate pathogens. Proceedings of the Congress of Clinical Microbiology and Infectious Diseases (ECCMID), Copenhagen, Denmark (25–28 April, 2015).

87 Karlsson, R., Thorsell, A., Gomila, M. et al. (2020). Discovery of species-unique peptide biomarkers of bacterial pathogens by tandem mass spectrometry-based proteotyping. *Mol. Cell. Proteomics* 19: 518–528.

88 Hayoun, K., Gaillard, J.-C., Pible, O. et al. (2020). High-throughput proteotyping of bacterial isolates by double barrel chromatography-tandem mass spectrometry based on microplate paramagnetic beads and phylopeptidomics. *J. Proteomics* 226: 103887.

89 Michalski, A., Cox, J., and Mann, M. (2011). More than 100,000 detectable peptide species elute in single shotgun proteomics runs but the majority is inaccessible to data-dependent LC–MS/MS. *J. Proteome Res.* 10: 1785–1793.

90 Ludwig, C., Gillet, L., Rosenberger, G. et al. (2018). Data-independent acquisition-based SWATH-MS for quantitative proteomics: a tutorial. *Mol. Syst. Biol.* 14: 1–23.

91 Purvine, S., Eppel, J.-T., Yi, E.C., and Goodlett, D.R. (2003). Shotgun collision-induced dissociation of peptides using a time of flight mass analyser. *Proteomics* 3: 847–850.

92 Venable, J.D., Dong, M.-Q., Wohlschlegel, J. et al. (2004). Automated approach for quantitative analysis of complex peptide mixtures from tandem mass spectra. *Nat. Methods* 1: 39–45.

93 Gillet, L.C., Navarro, P., Tate, S. et al. (2012). Targeted data extraction of the MS/MS spectra generated by data-independent acquisition: a new concept for consistent and accurate proteome analysis. *Mol. Cell. Proteomics* 11 (O111): 016717.

94 Curto, P., Santa, C., Allen, P. et al. (2019). A pathogen and a non-pathogen spotted fever group rickettsia trigger differential proteome signatures in macrophages. *Front. Cell. Infect. Microbiol.* 9: 43.

95 Coon, J.J. (2009). Collisions or electrons? Protein sequence analysis in the 21st century. *Anal. Chem.* 81: 3208–3215.

96 Mott, T.M., Everley, R.A., Wyatt, S.A. et al. (2010). Comparison of MALDI-TOF/MS and LC-QTOF/MS methods for the identification of enteric bacteria. *Int. J. Mass Spectrom.* 291: 24–32.

97 McFarland, M.A., Andrzejewski, D., Musser, S.M., and Callahan, J.H. (2014). Platform for identification of salmonella serovar differentiating bacterial proteins by top-down mass spectrometry: S. Typhimurium vs S. Heidelberg. *Anal. Chem.* 86: 6879–6886.

98 Karlsson, C., Malmström, L., Aebersold, R., and Malmström, J. (2012). Proteome-wide selected reaction monitoring assays for the human pathogen streptococcus pyogenes. *Nat. Commun.* 3 (1301): 1–9.

99 Charretier, Y., Dauwalder, O., Franceschi, C. et al. (2015). Rapid bacterial identification, resistance, virulence and type profiling using selected reaction monitoring mass spectrometry. *Sci. Rep.* 5: 13944.

100 Schmidt, A., Kochanowski, K., Vedelaar, S. et al. (2016). The quantitative and condition dependent *Escherichia coli* proteome. *Nat. Biotechnol.* 34: 104–110.

101 Vidova, V. and Spacil, Z. (2017). A review on mass spectrometry-based quantitative proteomics: targeted and data independent acquisition. *Anal. Chim. Acta* 964: 7–23.

102 Giebel, R., Worden, C., Rust, S.M. et al. (2010). Microbial fingerprinting using matrix-assisted laser desorption ionization time-of-flight mass spectrometry (MALDI-TOF MS) applications and challenges. *Adv. Appl. Microbiol.* 71: 149–184.

103 Jang, K.S. and Kim, Y.H. (2018). Rapid and robust MALDI-TOF MS techniques for microbial identification: a brief overview of their diverse applications. *J. Microbiol.* 56: 209–216.

104 Mellmann, A., Cloud, J., Maier, T. et al. (2008). Evaluation of matrix-assisted laser desorption ionization-time-of-flight mass spectrometry in comparison to 16S rRNA gene sequencing for species identification of nonfermenting bacteria. *J. Clin. Microbiol.* 46: 1946–1954.

105 Seng, P., Drancourt, M., Gouriet, F. et al. (2009). Ongoing revolution in bacteriology: routine identification of bacteria by matrix-assisted laser desorption ionization time-of-flight mass spectrometry. *Clin. Infect. Dis.* 49: 543–551.

106 Caprioli, R.M., Farmer, T.B., and Gile, J. (1997). Molecular imaging of biological samples: localization of peptides and proteins using MALDI-TOF MS. *Anal. Chem.* 69: 4751–4760.

107 Bradshaw, R. and Francese, S. (2021). MALDI-MSI in the forensic arena. In: *MALDI Mass Spectrometry Imaging: From Fundamentals to Spatial Omics* Chap. 15 (ed. T. Porta). Springer. ISBN: 978-1-83916-241-1.

108 Budowle, B., Schutzer, S.E., Einseln, A. et al. (2003). Building microbial forensics as a response to bioterrorism. *Science* 301: 1852–1853.

109 Robinson, J.M., Pasternak, Z., Mason, C.E., and Elhaik, E. (2021). Forensic applications of Microbiomics: a review. *Front Microbiol.* 11: 3455.

110 Tucker, T., Marra, M., and Friedman, J.M. (2009). Massively parallel sequencing: the next big thing in genetic medicine. *Am. J. Hum. Genet.* 85: 142–154.

111 Bouslimani, A., Porto, C., Rath, C.M. et al. (2015). Molecular cartography of the human skin surface in 3D. *Proc. Natl. Acad Sci.* 112 (17): E2120–E2129.

112 Kapono, C.A., Morton, J.T., Bouslimani, A. et al. (2018). Creating a 3D microbial and chemical snapshot of a human habitat. *Sci. Rep.* 8: 3669.

113 Alam, S.I., Kumar, B., and Kamboj, D.V. (2012). Multiplex detection of protein toxins using MALDI-TOF-TOF tandem mass spectrometry: application in unambiguous toxin detection from bioaerosol. *Anal. Chem.* 84: 10500–10507.

114 Kull, S., Pauly, D., Störmann, B. et al. (2010). Multiplex detection of microbial and plant toxins by immunoaffinity enrichment and matrix-assisted laser desorption/ionization mass spectrometry. *Anal. Chem.* 82: 2916–2924.

115 Boyer, A.E., Gallegos-Candela, M., Lins, R.C. et al. (2011). Quantitative mass spectrometry for bacterial protein toxins – a sensitive, specific, high-throughput tool for detection and diagnosis. *Molecules* 16: 2391–2413.

116 Kalb, S.R., Santana, W.I., Geren, I.N. et al. (2011). Extraction and inhibition of enzymatic activity of botulinum neurotoxins /B1, /B2, /B3, /B4, and /B5 by a panel of monoclonal anti-BoNT/B antibodies. *BMC Biochem.* 12: 58.

117 Kalb, R., Boyer, A.E., and Barr, J.R. (2015). Mass spectrometric detection of bacterial protein toxins and their enzymatic activity. *Toxins* 7: 3497–3511.

118 Pierce, C.Y., Barr, J.R., Woolfitt, A.R. et al. (2007). Strain and phase identification of the U.S. category B agent Coxiella burnetii by matrix assisted laser desorption/ionization time-of-flight mass spectrometry and multivariate pattern recognition. *Anal. Chim. Acta* 583: 23–31.

119 Pechal, J.L., Crippen, T.L., Benbow, M.E. et al. (2014). The potential use of bacterial community succession in forensics as described by high throughput metagenomic sequencing. *Int. J. Legal Med.* 128: 193–205.

120 Dell'Annunziata, F., Martora, F., Pepa, M.E.D. et al. (2022). Postmortem interval assessment by MALDI-TOF mass spectrometry analysis in murine cadavers. *J. Appl. Microbiol.* 132: 707–714. https://doi.org/10.1111/jam.15210. Epub 2021 Aug 8. PMID: 34251733.

121 Zhang, J., Wang, M., Qi, X. et al. (2021). Predicting the postmortem interval of burial cadavers based on microbial community succession. *Forensic. Sci. Int. Genet.* 52: 102488.

122 Lee, S.Y., Woo, S.K., Lee, S.M. et al. (2017). Microbiota composition and pulmonary surfactant protein expression as markers of death by drowning. *J. Forensic Sci.* 62: 1080–1088.

123 Foster, G., Whatmore, A.M., Dagleish, M.P. et al. (2019). Forensic microbiology reveals that *Neisseria animaloris* infections in harbour porpoises follow traumatic injuries by grey seals. *Sci. Rep.* 9 (1): 14338. https://doi.org/10.1038/s41598-019-50979-3.

124 Li, B., Comi, T.J., Si, T. et al. (2016). A one-step matrix application method for MALDI mass spectrometry imaging of bacterial colony biofilms. *J. Mass Spectrom.* 51: 1030–1035.

125 Si, T., Li, B., Zhang, K. et al. (2016). Characterization of *Bacillus subtilis* colony biofilms via mass spectrometry and fluorescence imaging. *J. Proteome Res.* 15: 1955–1962.

126 Wakeman, C.A., Moore, J.L., Noto, M.J. et al. (2016). The innate immune protein calprotectin promotes *Pseudomonas aeruginosa* and *Staphylococcus aureus* interaction. *Nat. Commun.* 7: 11951.

127 Phelan, V.V., Fang, J., and Dorrestein, P.C. (2015). Mass spectrometry analysis of *Pseudomonas aeruginosa* treated with Azithromycin. *J. Am. Soc. Mass. Spectrom.* 26: 873–877.

128 Boya, P.C.A., Fernández-Marín, H., Mejía, L.C. et al. (2019). Fungus-growing Ant's microbial interaction of Streptomyces sp. and Escovopsis sp. through molecular networking and MALDI imaging. *Nat. Prod. Commun.* 14: 63–66. https://doi.org/10.1177/1934578X1901400117.

129 Brockmann, E.U., Steil, D., Bauwens, A. et al. (2019). Advanced methods for MALDI-MS imaging of the chemical communication in microbial communities. *Anal. Chem.* 91: 15081–15089.

130 Geier, B., Sogin, E.M., Michellod, D. et al. (2020). Spatial metabolomics of in situ host-microbe interactions at the micrometre scale. *Nat. Microbiol.* 5: 498–510.

131 Spagnolo, E.V., Stassi, C., Mondello, C. et al. (2019). Forensic microbiology applications: a systematic review. *Leg. Med.* 36: 73–80.

132 Dunham, S.J.B., Ellis, J.F., Li, B., and Sweedler, J.V. (2017). Mass spectrometry imaging of complex microbial communities. *Acc. Chem. Res.* 50: 96–104.

133 Michelsen, F.C.H., Khademi, S.M., Krogh Johansen, H. et al. (2016). Evolution of metabolic divergence in Pseudomonas aeruginosa during long-term infection facilitates a proto-cooperative interspecies interaction. *ISME J.* 10: 1323–1336.

134 Englander, S.W., Mayne, L., Bai, Y., and Sosnick, T.R. (1997). Hydrogen exchange: the modern legacy of Linderstrøm-Lang. *Protein Sci.* 6: 1101–1109.

135 Hvidt, A. and Linderstrøm-Lang, K. (1954). Exchange of hydrogen atoms in insulin with deuterium atoms in aqueous solutions. *Biochim. Biophys. Acta* 14: 574–575.

136 Chait, B.T. (1994). Mass spectrometry – a useful tool for the protein *X*-ray crystallographer and NMR spectroscopist. *Structure* 2: 465–467.

137 Giladi, M. and Khananshvili, D. (2020). Hydrogen-deuterium exchange mass-spectrometry of secondary active transporters: from structural dynamics to molecular mechanisms. *Front. Pharmacol.* 11: https://doi.org/10.3389/fphar.2020.00070.

138 Oganesyan, I., Lento, C., and Wilson, D.J. (2018). Contemporary hydrogen deuterium exchange mass spectrometry. *Methods*, Mass Spectrometry-based Methods to Study Macromolecular Higher Order Structure and Interactions 144: 27–42.

139 Connelly, G.P., Bai, Y., Jeng, M.F., and Englander, S.W. (1993). Isotope effects in peptide group hydrogen exchange. *Proteins* 17: 87–92.

140 Weis, D.D. (2016). Hydrogen exchange: a sensitive analytical window into protein conformation and dynamics. In: *Hydrogen Exchange Mass Spectrometry of Proteins: Fundamentals, Methods, and Applications*, 1–17. Wiley https://doi.org/10.1002/9781118703748.

141 Huang, L., So, P.-K., Chen, Y.W. et al. (2021). Interdomain flexibility and interfacial integrity of β-lactamase inhibitory protein (BLIP) modulate its binding to class a β-lactamases. *J. Biol. Chem.* 297: 100980. https://doi.org/10.1016/j.jbc.2021.100980.

142 Reading, E., Ahdash, Z., Fais, C. et al. (2020). Perturbed structural dynamics underlie inhibition and altered efflux of the multidrug resistance pump AcrB. *Nat. Commun.* 11: 5565. https://doi.org/10.1038/s41467-020-19397-2.

143 Möller, I.R., Slivacka, M., Hausner, J. et al. (2019). Improving the sequence coverage of integral membrane proteins during hydrogen/deuterium exchange mass spectrometry experiments. *Anal. Chem.* 91: 10970–10978.

144 Paul, W. and Steinwedel, H. (1953). Ein neues Massenspektrometer ohne Magnetfeld. *RZeitschrift für Naturforschung A* 8 (7): 448–450.

145 Kosuke Hosoi, Masaji Furuta, Hideharu Shichi, Shosei Yamauchi, Kiyoshi Watanabe, Makoto Hazama, Kei Kodera, Shinichi Iwamoto, Koichi Tanaka (2019), *The Development of Miniature MALDI Digital Ion Trap Mass Spectrometer*, ASMS 2019.

146 Paul, W. (1990). Electromagnetic traps for charged and neutral particles. *Rev. Mod. Phys.* 62 (3): 531–540.

147 Schwartz, J.C., Senko, M.W., and Syka, J.E. (2002). A two-dimensional quadrupole ion trap mass spectrometer. *J. Am. Soc. Mass. Spectrom.* 13 (6): 659–669.

148 Papanastasiou, D., Kounadis, D., Lekkas, A. et al. (2022). The Omnitrap Platform: a versatile radio-frequency linear ion trap for multidimensional multiple-stage tandem mass spectrometry. *J. Am. Soc. Mass Spectrom.* 33: 1990–2007.

149 Brancia, F.L., McCullough, B., Entwistle, A. et al. (2010). Digital asymmetric waveform isolation (DAWI) in a digital linear ion trap. *J. Am. Soc. Mass. Spectrom.* 21 (9): 1530–1533.

150 Papanastasiou, Dimitris et al (2021) Redefining mass spectrometry – a breakthrough platform for real-time noninvasive breath analysis with single ion detection of intact viruses and bacteria and post-analysis molecular characterization. https://cordis.europa.eu/project/id/964553 (accessed 23 September 2022).

2

Machine Learning in Analysis of Complex Flora Using Mass Spectrometry

Luis Mancera[1], Manuel J. Arroyo[1], Gema Méndez[1], Omar Belgacem[2], Belén Rodríguez-Sánchez[3], and Marina Oviaño[4]

[1] *Clover Bioanalytical Software, Granada, Spain*
[2] *Ascend Diagnostics, Ltd, Manchester, UK*
[3] *Instituto de Investigación Sanitaria Gregorio Marañón, Madrid, Spain*
[4] *Complejo Hospitalario Universitario de A Coruña, A Coruña, Spain*

2.1 Introduction

Living organisms have a symbiotic relationship with a tremendous number of microorganisms, but these can become pathological if they colonize new areas of the body, if they displace the usual flora, causing opportunistic infections or if a pathogenic species enters the body by means of any kind of contamination. Bacteria are typically responsible for food poisoning, urinary tract and throat infections and diseases like tuberculosis. On the other hand, fungi are more complex eukaryote organisms which are responsible for causing conditions such as yeast infections, candidiasis, or meningitis. Like bacteria, fungi can play a beneficial role in our health, but they can also be a very serious threat, mainly in immunosuppressed patients.

Since Alexander Fleming discovered penicillin in 1928, antibiotics have been used to prevent and treat bacterial infections. The introduction of antimicrobials caused a severe drop in the number of deaths globally caused by microbial and fungal infections [1]. However, excessive use of antibiotics for treating these infections over recent decades has resulted in a substantial increase in antimicrobial resistance (AMR). Consequently, microorganisms have become increasingly resistant to multiple antimicrobials, making even simple infections and minor wounds potentially life-threatening. In 2019, AMR was the third leading global cause of death [2]. Consequently, AMR now stands in the WHO list of greatest threats to human health [3]. It is estimated that 30–50% of antibiotic treatments nowadays are started with the wrong antibiotics and without a proper diagnosis of the pathogen. This rises to 30–60% of cases in intensive care units [4]. Antibiotics are now failing to work in one out of every five patients suffering an infection even when proposed guidelines are followed [5]. If no action is taken, it has been estimated that drug-resistant

Microbiological Identification using MALDI-TOF and Tandem Mass Spectrometry: Industrial and Environmental Applications, First edition. Edited by Haroun N. Shah, Saheer E. Gharbia, Ajit J. Shah, Erika Y. Tranfield, and K. Clive Thompson.

infections will kill as many as 10 million people per year by 2050, although this number has been questioned and could be even higher [6].

Artificial intelligence (AI) was born soon after the development of the first computers, mainly by the work of pioneers like Allen Newell, Herbert Simon, John McCarthy, Marvin Minsky and others. The final goal of AI is to develop theories and software to explain and mimic human thinking using mathematics and computer science. Machine learning (ML) is a mathematical tool for leveraging technologies around AI, based on the development of software programs based on algorithms which can learn from examples and experience [7]. The goal of these programs, typically, is to take a decision, answer a question or respond to a situation based on a set of input features extracted from the data under analysis.

Machine-learning algorithms are largely divided in two groups: supervised and unsupervised learning. Supervised learning-based algorithms are fed with a large set of examples in the form of input features paired with expected answers. They use mathematical devices to explain the relationship between input and expected answers to be able to infer a new answer when an unprecedented input is presented to it. Examples of supervised ML tools are partial least-squares or support vector machines (SVM). On the other hand, unsupervised ML algorithms are provided only with a set of input data, without any expectations about the category they belong to or what decision to make. These algorithms typically try to separate the input examples into different groups based on their characteristics. They will then assign one of the groups to any new input. Examples of unsupervised learning algorithms are principal component analysis (PCA) or K-means clustering.

This chapter focuses on the use of matrix-assisted laser desorption/ionization time-of-flight (MALDI-TOF) mass spectrometry (MS) to identify AMR mechanisms in pathogenic or opportunistic microorganisms present in flora such as Enterobacterales using ML algorithms. We will describe several studies that have been performed to discriminate AMR mechanisms based on the detection of resistance-associated proteins in microorganism by analysing their entire protein profile. As examples, results related to two resistant bacteria will be presented, namely *Klebsiella pneumoniae* producing carbapenemase (KPC) and vancomycin-resistant *Enterococcus* (VRE); and two resistant fungi, azole-resistant *Aspergillus fumigatus* and the discrimination among *Cryptococcus neoformans* species complex and their interspecies hybrid. These studies have been carried out by two leading hospitals in Spain: Hospital General Universitario Gregorio Marañón (HGUGM) in Madrid and the Complejo Hospitalario Universitario in A Coruña.

The processing pipeline for these studies consists of three main phases. The first phase was the data collection. The aim was to obtain an appropriate collection of microorganism profiles, including well-balanced (where possible) sets of control and resistance showing samples. In the second phase, training and validation datasets were obtained by acquiring the MS raw data using the standard method for species-level identification that is used with MALDI-TOF MS. All the spectra in the dataset were then submitted to a pipeline of processing algorithms, resulting in a peak matrix suitable for statistical analysis. These peak matrices represent the input features for our ML algorithms, where the intensity of each relevant peak in the spectrum is considered an input to the algorithm, paired with an expected microbial category or subspecies if the algorithm is supervised. Finally, the third phase makes use of one or several ML algorithms to obtain the AMR predictor, with a special emphasis on evaluating and correctly interpreting the results. Our work indicates that ultra-fast, cost-effective tools for *in vitro* diagnosis of AMR mechanisms are feasible and can be integrated seamlessly into any microbiology laboratory equipped with a MALDI-TOF MS instrument.

2.2 An Improved MALDI-TOF MS Data Analysis Pipeline for the Identification of Carbapenemase-producing *Klebsiella pneumoniae*

2.2.1 Motivation

Various global organisations have recognized a health alarm caused by the increasing emergence of carbapenemase-producing *K. pneumoniae* (CPK) [6, 8–11]. Carbapenemases are β-lactamase enzymes with hydrolytic capacities. They can render many β-lactams antibiotics ineffective, such as penicillin, cephalosporins, monobactams and carbapenems. These are the antibiotics most used to treat infections caused by Enterobacterales [12].

Molecular techniques are currently used to detect this type of AMR [13]. However, this approach is time-consuming, labour-intensive and expensive. Methods that involve minimum sample preparation and provide rapid results are therefore urgently needed.

In this study [14], we evaluated the biological and technical variation associated with the detection of CPK isolates using MALDI-TOF MS. The aim was to design a standard operating procedure and a MALDI-TOF MS data analysis pipeline that could be implemented in routine screening in clinical microbiology laboratories at no additional cost and with no requirement for specific laboratory expertise. In clinical practice, it is of paramount importance to have consistent analysis. To achieve this, it is important to reduce the variability inherent in the sample analysis using MALDI-TOF MS, as with any other medical instrument or device. Biological variation (BV) is defined as the physiological fluctuation of the constituents of living organisms around a homeostatic point. Technical variation occurs due to the imprecision of the different steps and conditions throughout the entire analytical process.

2.2.2 Materials and Methods

We analysed a collection of 162 unique clinical isolates, including 100 CPK isolates, 93 of which were collected from 15 Spanish hospitals participating in a national survey promoted by the Spanish Society of Infectious Diseases and Clinical Microbiology (SEIMC) and by the Spanish Network for Research in Infectious Diseases (REIPI). The rest of the isolates came from a collection of the Complejo Hospitalario Universitario de A Coruña [15]. This study was performed in three different hospitals: Complejo Hospitalario Universitario de A Coruña (CHUAC), HGUGM and Hospital Universitario Puerta del Mar (HUPM). Isolates were screened for carbapenemase production according to recommendations by the EUCAST (meropenem or ertapenem MIC of >0.125 mg/l). The isolates coming from the Spanish survey were characterized using whole-genome sequencing and the ones from CHUAC using polymerase chain reaction.

The study was done in two stages (Figure 2.1): a reproducibility stage and a CPK validation stage. The reproducibility phase of the study was designed to assess the best extraction and peak integration methods envisioned to reduce the biological and technical variation associated with the entire analysis of CPK isolates. The CPK validation stage comprised the

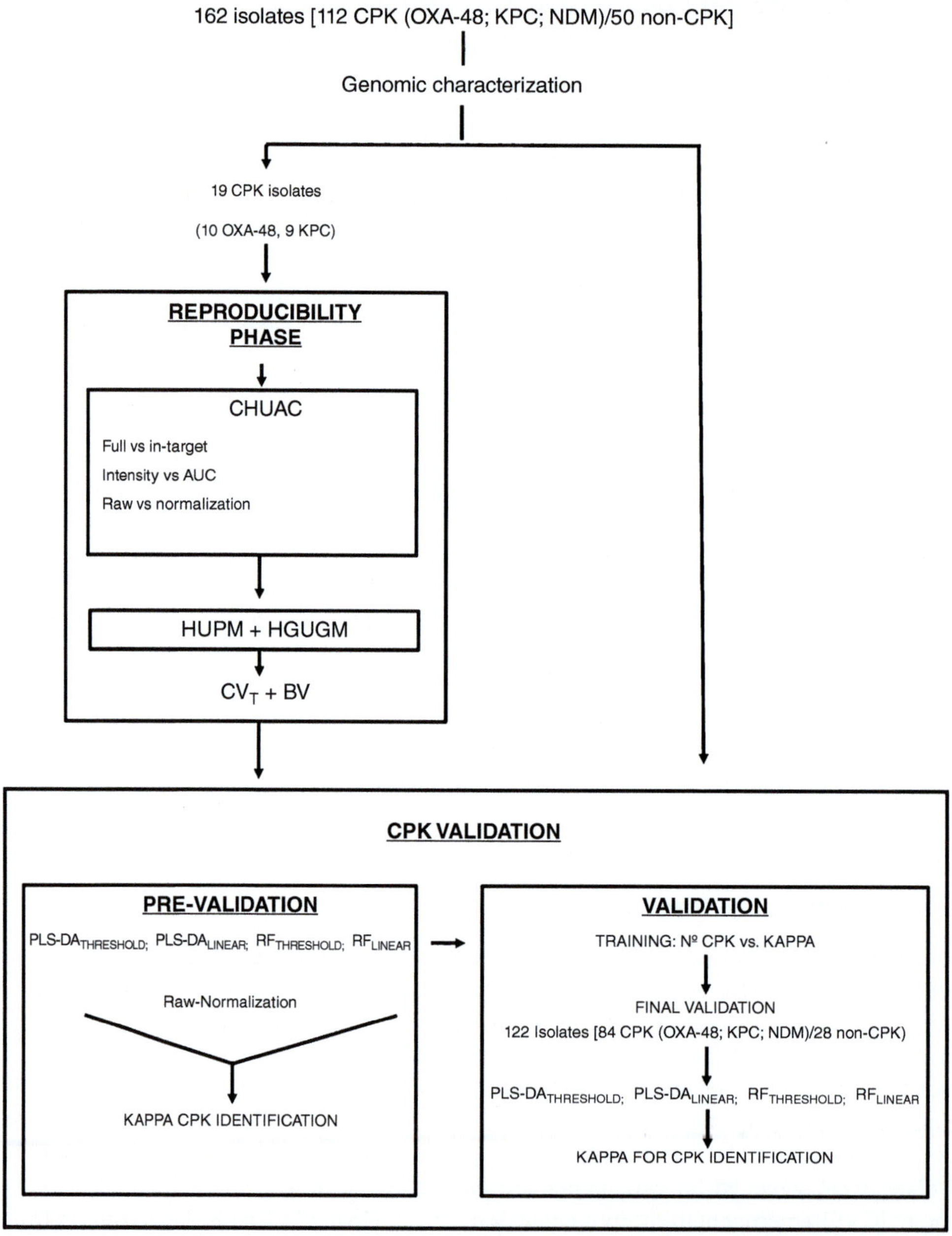

Figure 2.1 The different steps involved in formulation of the study, the reproducibility phase and the CPK identification, are in turn divided into pre-validation and validation steps. M_{LINEAR} and $M_{THRESHOLD}$, methods of generating mass spectra. *Source:* Reprinted from Gato et al. [14].

design of variable selection and validation in multivariate modelling by comparing two different statistical methods [partial least-squares discriminant analysis (PLS-DA) and random forest (RF)] in terms of accuracy of identification in relation to the reproducibility results obtained before.

The peak at *m/z* of 5380 was chosen as the reference peak for the reproducibility phase. This is because it represents 50S ribosomal protein L34, a highly conserved protein in *K. pneumoniae*. To express variability, we used the coefficient of variation (CV). We calculated three different CVs, namely variation related to the protein extraction method, sample spotting and generation of mass spectra using the MALDI-TOF MS. We used two different methods for protein extraction: full and on-target. All isolates were analysed in triplicate for each step (extraction, spotting and MALDI-TOF MS analysis) and over three different days to capture BV (Figure 2.2). CVs were calculated for both normalized and raw data. The testing of the BV was performed after 36 hours, and after previous thawing and re-subculturing.

The next step was to assess the accuracy of MALDI-TOF MS for the identification of CPK ($bla_{\text{OXA-48}}$, bla_{KPC} and bla_{NDM})-producing isolates. We used PLS-DA and RF as multivariate analytical tools that are implemented in the Clover MS Data Analysis software (Clover

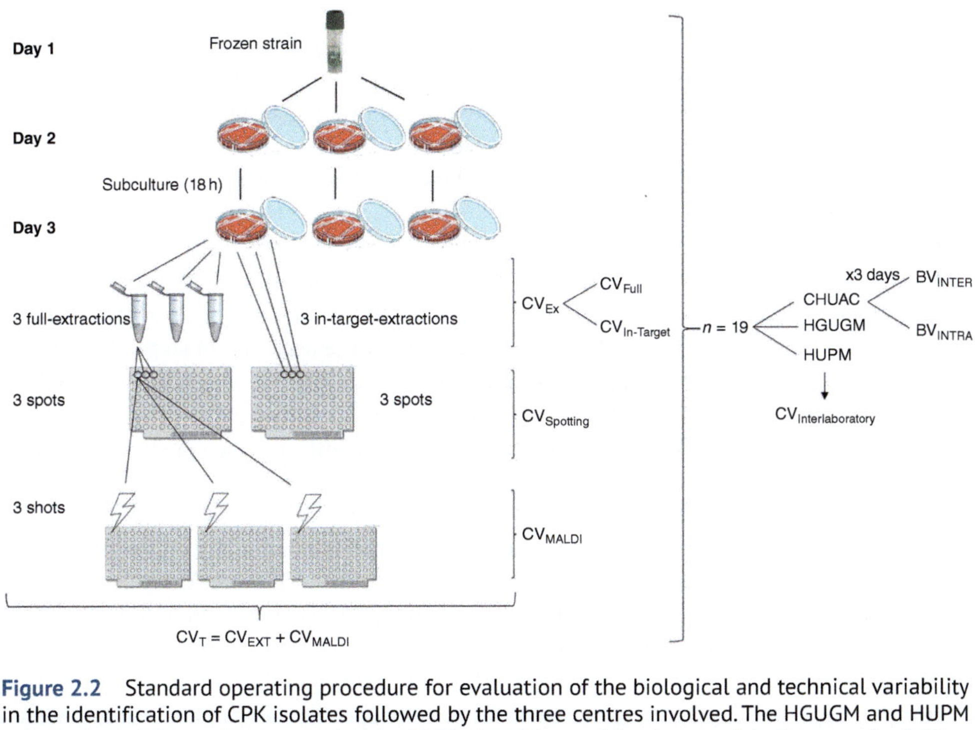

Figure 2.2 Standard operating procedure for evaluation of the biological and technical variability in the identification of CPK isolates followed by the three centres involved. The HGUGM and HUPM only evaluated the methods associated with the least imprecision to complete the reproducibility phase. The CVs associated with each step of the procedure are illustrated in the figure. CV refers to the level of imprecision associated with the extraction (CV_{EXT}) or the matrix-assisted laser desorption/ionization time-of-flight mass spectrometry (MALDI-TOF MS) process (CV_{MALDI}) and total or technical imprecision (CVT); BV_{INTER}, interindividual BV; BV_{INTRA}, intraindividual BV. *Source:* Gato et al. [14]/Reproduced from American Society for Microbiology.

Biosoft, Spain). This software analyses the similarities and differences between the mass peaks in the spectra and assigns a relative weight in the algorithm to each peak for final classification. Kappa (κ) values were obtained for the correct identification of CPK isolates. Non-carbapenemase-producing isolates were introduced as a category in the analysis, so κ values were also obtained for these isolates.

The pre-validation stage gave us an initial view of the state of the art of this technology for CPK identification, and if the results were very good ($\kappa > 80\%$) [8], a more extensive validation study would be performed. The possible correlation between the eight tools' combination of κ-values and their CVs calculated in the reproducibility stage was also studied.

At the validation stage, CPK isolates that were not used in the pre-validation stage were processed in CHUAC. Isolates were analysed by the same operator, using on-target extraction, the peak intensity method, and the same MALDI-TOF MS instrument. A training step was developed for the eight tool combinations to allow the software to learn to identify CPK correctly. We determined how many samples were required to obtain a κ peak for each of the eight groups. When the training step was completed, we performed the final validation step, in which the accuracy (κ) of the eight combinations in CKP identification was obtained.

2.2.3 Spectra Acquisition

All the isolates that were analysed were of the same age to control for senescence-associated variability. Isolates were submitted to a modified Hodge test to check for the presence of the carbapenemase enzyme [16]. We didn't find any discordance in relation to the phenotypic and genotypic annotation. To further reduce variability, the same operator performed the analysis of all samples in each participating laboratory by following a precise protocol and two different extraction methods.

All three laboratories were equipped with MALDI Biotyper (Bruker Microbiology & Diagnostics). MALDI-TOF MS spectra were acquired using a microflex LT/SH smart instrument with the FlexControl software version 3.4 in a linear positive ion mode within a mass range of 2–20 kDa. A total of 240 satisfactory laser shots were acquired in steps of 40 shots per spectrum using the spiral small movement. External calibration was performed using the bacterial test standard (BTS; Bruker Microbiology & Diagnostics) prior to each run. Species was confirmed by comparing the acquired spectra with the mass spectrum library using the MALDI Biotyper Compass software (version 4.1.100, Bruker Microbiology & Diagnostics).

After this initial acquisition and species-level identification, the data were prepared for subspecies analysis. All spectra were processed using the Clover MS Data Analysis Software The first step involved processing all the spectra by applying noise reduction with the Savitzky–Golay filter (smoothing filter with window length of 11 and polynomial order 3) and then subtracting the baseline by using the Top Hat filter (baseline removing filter with a factor of 0.02). Nine replicates (three spots with three spectra each) of each isolate were processed. For the data from the three laboratories, an average spectrum for each isolate was obtained by a two-step process of alignment. The first step consisted of aligning the spectra for all spots and then obtaining a unique average spectrum. The second step involved repeating this process but aligning the spectra obtained for each spot to produce the final average spectrum. Once this process was completed, only one average spectrum remained for each isolate. The main goal of performing an alignment-plus-average spectrum process was to minimize the variability between replicates of the same isolate. Once one spectrum per isolate was available in the software platform, all the spectra were again aligned to increase the accuracy of the next steps. All the alignment processes were performed by considering the most representative peaks of each sample included in the set to be aligned. These peaks were then used to form a reference peak list. Each spectrum peak was shifted within a linear tolerance of 2000 ppm to correspond to this list.

Two methods were explored to form peak matrices. The 'threshold' method retained all peaks with at least 1% of the maximum intensity of the spectrum considered. They were then merged into a common list within a linear tolerance of 2000 ppm The 'linear' method used the full spectra aligned into a common list of *m/z* values. Peak matrices were later normalized using the total ion current (TIC) method. Therefore, we obtained four different matrices: $M_{THRESHOLD\text{-}RAW}$, $M_{THRESHOLD\text{-}NORM}$, $M_{LINEAR\text{-}RAW}$ and $M_{LINEAR\text{-}NORM}$. Two supervised ML algorithms were then applied to these matrices: PLS-DA and RF. The main purpose of these analyses was to study how each of the eight different combinations of tools discriminated the different categories (OXA-48, KPC, NDM and non-CPK). The κ-statistic value was used to explain the efficacy of the combined results.

Statistical analysis was performed using GraphPad Prism 8.0 software (JMP Statistical Discovery, UK). The normalized distribution was tested with the D'Agostino–Pearson test. Continuous variables are presented as median values and interquartile ranges (IQRs) (25th–75th percentiles), and categorical data are displayed as counts and percentages. Analysis of differences between two groups was performed using the Wilcoxon matched-pairs signed-rank test. A two-sided P-value <0.05 was considered statistically significant for all calculations. The κ-value was the score of a k-fold ($k = 10$) cross-validation analysis, obtained as the relationship between the number of samples correctly classified and the total number of spectra under study [17].

2.2.4 Results

From the 112 CPK isolates, 11 different sequence types (STs) were detected: ST-11 ($n = 27$), ST-512 ($n = 27$), ST-15 ($n = 12$), ST-147 ($n = 13$), ST-307 ($n = 12$), ST-1961 ($n = 8$), ST-101 ($n = 6$), ST-273 ($n = 3$), ST-437 ($n = 2$), ST-395 ($n = 1$) and ST-405 ($n = 1$). A total of 47 of the 112 isolates carried the bla_{KPC} gene, 58 carried the $bla_{OXA\text{-}48}$ gene, and 7 carried the bla_{NDM} gene. Carbapenemase-non-producing *K. pneumoniae* isolates were not submitted to detailed genomic analysis.

Nineteen of the 112 representative and randomly selected CPK isolates were used in the reproducibility stage, including nine KPC-producing isolates (four belonging to ST11 and five to ST512) and 10 OXA-48-producing isolates (five belonging to ST11 and five to ST15). These isolates were examined in triplicate in the three laboratories (CHUAC, HGUGM and HUPM).

To study reproducibility, CHUAC calculated two statistical values for the peak at *m/z* of 5380: area under the curve (AUC) and intensity. The different values obtained for corresponding CVs are shown in Figure 2.3. These show that $CV_{INTENSITY}$ is slightly better than CV_{AUC}. Consequently, all peaks from the MALDI spectra for the rest of the study were integrated using the intensity mode. Regarding the extraction methods, a comparison between full and on-target extraction (Figure 2.3a) shows that the latter indicated better

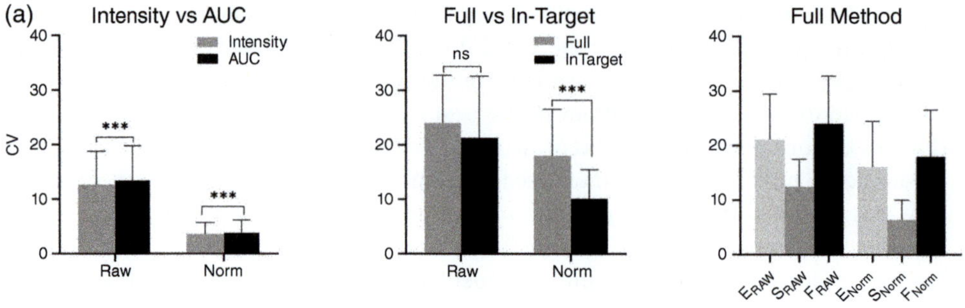

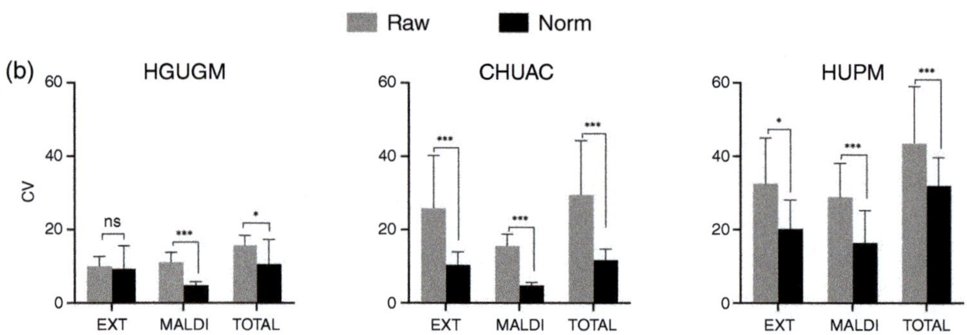

CV	HGUGM				CHUAC				HUPM			
	Raw	Norm	−Δ%	*P*	Raw	Norm	−Δ%	*P*	Raw	Norm	−Δ%	*P*
EXT	10.0 (6.7–12.7)	9.3 (7.3–15.6)	7	0.436	25.8 (18.4–40.2)	10.4 (7.2–13.9)	60	<0.001	32.5 (12.5–45.0)	20.2 (15.8–28.1)	38	0.049
MALDI	11.1 (7.5–13.8)	4.8 (4.2–5.8)	57	<0.001	15.6 (12.3–18.8)	4.8 (3.1–5.6)	69	<0.001	28.9 (20.2–38.1)	16.4 (5.6–25.2)	43	<0.001
Total	15.7 (12.4–18.4)	10.6 (8.1–17.3)	32	0.011	29.4 (24.6–44.3)	11.7 (9.2–14.7)	60	<0.001	43.5 (24.9–59.0)	32.0 (19.5–39.7)	26	<0.001

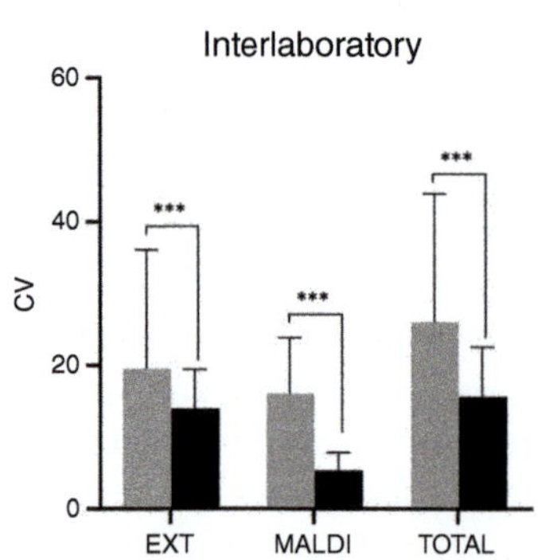

CV	INTERLABORATORY			
	Raw	Norm	−Δ%	*P*
EXT	19.5 (10.1–36.1)	13.9 (8.3–19.4)	29	<0.001
MALDI	16.0 (11.0–23.8)	5.4 (4.2–7.9)	66	<0.001
Total	26.0 (17.7–43.9)	15.6 (10.0–22.6)	40	<0.001

Figure 2.3 Reproducibility stage. (a) Initial study. Coefficients of variation (CVs) obtained in the CHUAC to compare intensity versus AUC methods of peak acquisition and full versus on-target methods of extraction. Norm, normalized data; E, extraction; S, spotting; F, full extraction. (b) Interlaboratory CVs obtained in three hospitals for *Klebsiella pneumoniae* isolates analyzed using MALDI-TOF MS (top) and considering the results from the three laboratories together (bottom). EXT, extraction step (on-target method); MALDI, MALDI-TOF MS process; TOTAL, the entire process (EXT+MALDI); interlaboratory represents overall CV from the three hospitals involved; −Δ%, % decrease in κ of normalized versus raw data. All data are presented as median values (bars) and interquartile ranges (error bars). All adjusted *P*-values were obtained by Wilcoxon matched-pairs signed-rank test. ns, $P \geq 0.05$; *, $P < 0.05$; **, $P < 0.005$; ***, $P < 0.001$. *Source:* Reprinted from Gato et al. [14].

performance for normalized data. Therefore, on-target extraction was used for the rest of the analyses. Finally, inter-laboratory variability was studied by analysing the same 19 isolates in both the HGUGM and the HUPM using on-target extraction and intensity integration. Three coefficients of variation were obtained, related to the extraction (CV_{EXT}), the spectra acquisition process in the instrument (CV_{MALDI}) and the technical imprecision (CV_T). Results for raw and normalized spectra are shown in Figure 2.3. This demonstrates that the lowest levels of imprecision were achieved in the HGUGM, and the highest in the HUPM. Also, normalization of the intensity was a significant factor in the reduction of imprecision across all laboratories. BV was calculated in the CHUAC for all isolates over three different days.

As for CPK identification, as mentioned earlier this was divided in two phases: pre-validation and validation. For the pre-validation stage, the results obtained in the PLS-DA-$M_{THRESHOLD}$ group showed a clear improvement for correct CPK identification after normalization of the mass spectra, except in the HGUGM, where the imprecision was the same for raw and normalized spectra (Figure 2.3a). In this case, technical imprecision CV appears to be inversely related to the accuracy of CPK identification (κ), but this was not perceived in other groups. The κ results for CPK identification were best for the RF-M_{LINEAR} case (Figure 2.3), and they were very similar among the three hospitals for all tool combinations. Reproducibility of the MALDI-TOF MS process across the three laboratories was confirmed. These promising results led us to continue with the next stages of the validation, using spectra obtained from all 110 isolates. The κ-values for the eight tool combinations were obtained using an increasing number of isolates (from 10 to 60 in steps of 10, selected randomly). Figure 2.3 shows that RF's performance peaked at 40 samples, using both the $M_{THRESHOLD}$ and the M_{LINEAR} input matrices. On the other hand, the PLS-DA algorithm had the best results with 50 isolates for M_{LINEAR} and with 50 for $M_{THRESHOLD}$. In all cases, CPK identification was better using normalized spectra. These results indicate that RF coupled with a linear peak selection strategy gave the best results for these analyses.

To unify criteria, a sampling size of 40 isolates was chosen for the final validation stage. The κ-value for the RF in normalized M_{LINEAR} yielded 97.5% accuracy. No specific mass peaks or biomarkers in the spectrum that were associated with a type of carbapenemase were identified, but the entire mass peak dataset was considered in the final algorithm in which each mass peak has its relative importance.

The final validation stage was carried out with the remaining 122 isolates, which consisted of 84 CPK and 38 non-CPK isolates. Among the CPK isolates, 45 of the 84 were OXA-48, 36 of the 84 were KPC, and 3 of the 84 were NDM. Thus, with a greater number of samples than in the previous stages, the relationship between tool combinations and imprecision depended similarly on both the type of spectrum acquisition (threshold vs linear) and the ML tool used for the analysis (PLSDA vs RF) (Figure 2.3). Therefore, we observed that for all the methods of analysis, there was a considerable increase in the κ-value for normalized versus raw spectra. By contrast, for RF-M_{LINEAR}, there was no such dependence, with this tool combination yielding similar κ-values regardless of the normalization of the spectra, especially in the identification of CPK.

2.2.5 Discussion

The main result of this study was the introduction of an improved MALDI-TOF MS data analysis pipeline for the identification of CPK. A method with adequate technical specifications should have an imprecision of about half the intraindividual biological variation (BV_{INTRA}) [17]. We achieved this when the spectra were normalized.

Many variables and decisions can affect the performance of ML algorithms on MALDI-TOF MS data. We chose standard settings for the instrument and sample preparation to minimize these variabilities and obtain a method agreeing to current standards. The fact that the direct spotting process proved to be more precise than full extraction methods is an indicator that CPK identification can be done in real time, as only one spectrum is used for the standard MALDI-TOF MS bacterial identification and for the detection of antibiotic resistance for each isolate. The variability of the process was shown to be significantly lower in more experienced laboratories. Normalization was shown in this study to be crucial to reduce this inter-laboratory variability.

In the pre-validation step for CPK identification, we concluded that when using ML, the imprecision was minimized as compared with using a fixed and definite set of peaks. In the final validation stage, we observed that the RF algorithm in an M_{LINEAR} matrix outperforms other options, being less dependent on the normalization procedure and increasing the κ-values. This indicates the convenience of using the whole spectrum for analysis.

Overall, our findings can be summarized as:

1) A large and well-characterized collection of non-epidemiologically related isolates has been used to build a training set.
2) Sum all intensities at corresponding mass peaks.
3) The on-target extraction method can be used not only to build the model but also for clinical validation and for routine use in clinical practice (need for preliminary study).
4) If the full extraction method is used for preparing the sample in the operating procedure, special attention should be paid at this step, as it can have a major impact on the variability.
5) Highly trained personnel operating MALDI-TOF MS generally provides the most accurate results.
6) When using specific biomarkers, ensure the extraction method is validated and users are trained in the operating procedure, as the results can be very imprecise otherwise.
7) For generating spectra, employ methods that use the maximum number of peaks that is possible, as these methods are less impacted by the variability in the processing steps.
8) The use of neural network analysis to detect AMR patterns is highly recommended, as the analysis is less influenced by the expression of certain peptides or proteins, variability is reduced and transferability is guaranteed.
9) Normalization helps to reduce the variability and is therefore strongly encouraged, especially when known biomarkers are used.

The use of ML algorithms in MALDI-TOF MS has proved to be valuable for assessing potential different groups of resistance using high-quality data focused on the presence or absence of definite spectral mass peaks to recognize antibiotic-resistant bacteria [18]. The whole procedure, presented herein, is designed to reduce variability, guarantee interlaboratory reproducibility and maximize the information obtained from the bacterial proteome. In contrast to approaches based on genomic assays, this approach is far less expensive, faster and simpler. This can potentially be used as an initial screening test for identification of CPK isolates in a clinical laboratory, as direct tracking can be performed at the same time as bacterial identification, using the same procedure and the same spectrum. With optimized software (Clover MS Data Analysis) this would only take minutes to apply, with a completely automated process fully integrated into the routine laboratory workflow and not requiring any specialized skill beyond the basic use of MALDI-TOF MS.

Identification of CPK contributes to the prescription of targeted treatments against specific classes of carbapenemases, such as the new β-lactam–β-lactamase inhibitors imipenem/relebactam and ceftazidime/avibactam, which are particularly valuable in treating KPC- and OXA-48-producing isolates, respectively. Replacement of molecular techniques by MALDI-TOF MS at this stage is still not feasible. However, the determination of the type of carbapenemase in a specific epidemiological setting, such as in areas with a high prevalence of OXA-48-producing *K. pneumoniae* isolates, could be very helpful until results from molecular techniques are available.

In the future, we will validate a larger collection of *K. pneumoniae* isolates before fully integrating the method into the clinical microbiology laboratory workflow. It would also be interesting to expand the application to the identification of carbapenemases to other Enterobacterales.

In conclusion, this study demonstrated that direct tracking of CPK isolates using MALDI-TOF MS is possible within minutes, reproducible with technical know-how available for all users, and significantly less expensive than genomic-based technologies. In a potential outbreak situation, MALDI-TOF MS-based identification of CPK isolates might initiate intensive healthcare and infection control measures, modifying the clinical outcome of these infections.

2.3 Detection of Vancomycin-Resistant *Enterococcus faecium*

2.3.1 Motivation

Enterococci are Gram-positive, facultative anaerobic cocci normally present in the human microbiota and often found in our environment. They are a common cause of infection in humans, and they are greatly adapted to find new ways of developing resistance mechanisms to avoid the effects of antibiotics. They have become a health threat due to their rapid spread, which causes outbreaks in hospital settings [20]. *Enterococcus faecium* was included in the ESKAPE group of highly resistant microorganisms because of its ability to avoid the action of conventional treatments [21]. Vancomycin is a glycopeptide antibiotic indicated for the treatment of serious, life-threatening infections by Gram-positive bacteria. Therefore, when enterococci develop resistance to vancomycin, there are very few alternatives left. More than 54 500 cases and 5400 estimated deaths due to this organism were reported in 2017 [9]. It is especially serious for patients who have been treated with vancomycin for a long time, who are recovering from surgery or who are immunosuppressed.

Detection of susceptibility to vancomycin is traditionally done either by a standard broth antimicrobial susceptibility testing microdilution method or by amplification of the *VanA*/*VanB* genes and subsequent analysis of the specific amplicons [22]. The first approach has a turnaround time of approximately two days. The latter has faster results in about one to three hours later, but its cost in laboratory reagents is high. The application of MALDI-TOF MS to detect AMR is not yet standardized [23], but in recent years it has been proposed as a fast and cost-efficient method [24].

2.3.2 Materials and Methods

In this study [25], we assessed the repeatability of MALDI-TOF MS for peak analysis and evaluated its ability to discriminate vancomycin-susceptible *Enterococcus* (VSE) from VRE isolates. We used a sample base of 178 unique *E. faecium* clinical isolates (92 VSE, 31 *VanA* VRE, 55 *VanB* VRE) collected during 2017–2019 in the HGUGM. Clonality was discarded by analysing the date, location and department where it was obtained. All strains were isolated from in-patient clinical samples, characterized and kept frozen at −80°C for further analysis. For this study, samples were thawed and cultured overnight at 37°C in Columbia blood agar.

Bacterial strains were analysed using MALDI-TOF MS in an MBT Smart MALDI Biotyper (Bruker Daltonics). A few bacterial colonies of each enterococcal isolate were spotted onto a MALDI target plate. On-plate protein extraction was performed by applying 1 μL of formic acid to each target spot and letting it dry at room temperature before adding 1 μL of HCCA matrix solution (Bruker Daltonics), following the manufacturer's instructions. Spectra were acquired in positive mode in the range of *m/z* 2000–20 000, applying default settings [26]. Two technical replicates were obtained per sample, and two technological replicates per spot, resulting in four protein spectra per isolate. These spectra were then aligned using the genus-specific peak at *m/z* 4428. Outlier spectra and zero lines were discarded.

After screening, all samples were processed using Clover MS Data Analysis software. Each spectrum was pre-processed to generate a peak matrix in the range of *m/z* 2000–20 000. Alignment of spectra to correctly form the matrix was done by creating an average spectrum per spot and then averaging spectra from each replicated spot. Finally, average spectra from different isolates were aligned together.

First, we performed a repeatability assay with 18 peaks common to both VSE and VRE with normalized intensities above 1% of the maximum peak. Twenty spectra were randomly selected (VSE, $n = 10$; VRE strains, $n = 10$, five *VanA* VRE and five *VanB* VRE). We found there was lower variability within the *m/z* 3000–9000 range.

Once the overall peak matrix was formed in that range and with the pre-processing described earlier, Clover MS Software was used to differentiate first between VSE and VRE strains and then from isolates hosting *VanA* and *VanB* resistance genes. To do this we evaluated three different methods: (i) a 'full-spectrum method', in which the matrix obtained included all TIC-normalized peak intensities from the spectrum separated by 0.5 Da regardless of their intensity; (ii) a 'threshold method' where peak matrices were formed only using peaks with intensities above 1.0% of the maximum peak intensity, and TIC normalization was applied before peak detection (TICp); and (iii) a similar method to (ii) but applying TIC normalization after peak detection (pTIC).

The resulting three peak matrices were then submitted to supervised ML algorithms: SVM, RF and PLS-DA. These algorithms were tested, first, to discriminate VSE and VRE isolates and, second, to differentiate VRE isolates hosting *VanA* and *VanB* genes. *k*-fold validation ($k = 10$) was used in all cases.

Support vector machines outperformed the other algorithms in terms of *k*-folding validation, reaching up to 80.9% differentiation between VSE and VRE and an F1 score (the harmonic mean of the sensitivity and the accuracy of the model) of 80.5% for the full-spectrum-TIC method (Table 2.1).

In a similar comparison between *VanA* and *VanB* VRE strain (Table 2.2) we found that the PLS-DA algorithm with TICp method provided 86.65% correct classification. It also achieved 89.09% of predictive value for identifying *VanB* VRE strains in a 10-fold cross-validation.

Further validation of the discrimination may be obtained if the peaks used to differentiate between VRE and VSE have biological significance. In this case, the analysis of the

Table 2.1 Discrimination of VSE from VRE isolates and, within the latter group, differentiation between the strains hosting the *VanA* and *VanB* resistance genes.

Actual classification	Predicted classification	
(a) SVM full spectrum	VRE	VSE
VRE	70	16
VSE	18	74
(b) PLS threshold TICp	*VanA*	*VanB*
VanA	25	6
VanB	6	49

The actual classification is shown in columns and the predictive classification in rows. (a) Results from the support vector machine (SVM) algorithm using the full-spectrum total ion current (TIC) method. Accuracy, 80.9%; F1 score, 80.46%; sensitivity, 81.4%; specificity, 80.43%; positive predictive value (PPV) or precision, 79.55%; negative predictive value (NPV), 82.22%. (b) Scores from the PLS-DA algorithm using the TIC normalization after peak finding (TICp) method. Accuracy, 86.05%; predictive value for *VanA* VRE, 80.65%; predictive value for *VanB* VRE, 89.09%.
Source: Candela et al. [25]/MDPI/CC BY 4.0.

Table 2.2 Accuracy of internal validation for the classification of VanA and VanB VRE strains as metric for algorithm scores.

	Full spectrum-TIC	Peaks-TIC	TIC-peaks
Partial least-squares (four components)	83.72%	83.72%	86.65%
Support vector machine	76.74%	80.23%	73.26%
Random forest	79.07%	80.23%	79.07%

Abbreviations: TIC-peaks, TIC normalization before peak detection; peaks-TIC, TIC normalization after peak detection.
Data obtained with *k*-fold cross-validation method where $k = 10$.
Source: Adapted from Candela et al. [25].

relevant peaks for differentiation revealed the presence of two potential resistance biomarkers at *m/z* 6891.33–5095.01. These two peaks showed AUC values > 0.8 in its receiver operating characteristic (ROC) curve (Figure 2.4). The peak at *m/z* 5095.01 was present in 82/86 of the VRE strains. The AUC for peak at *m/z* 6891.33 allowed the discrimination of VRE isolates hosting the *VanA* resistance mechanism from those carrying the *VanB* mechanism (Figure 2.4).

As a result, this study showed that MALDI-TOF MS, coupled with ML algorithms, can discriminate VSE from VRE in 80.9% of the cases, and *VanA* from *VanB* in 86.6% of the resistance cases. The meaning of the biomarkers found is still under debate. The peak at *m/z* of 5095.01 has been identified as hiracin, a secretory protein encoded by the hirJM79 gene, whose role in the vancomycin resistance mechanism remains unknown [27]. The signal at *m/z* of 6891.33 was found to be a biomarker for *VanA* VRE isolates. Both the AUC (0.831) and the CV values (22.99%) indicate the uniqueness and repeatability of this marker. Although this peak had been related before with different clonal complexes and STs, its correlation with *VanA* VRE isolates has not been reported so far [28].

The repeatability experiment performed in this study showed that this methodology seems to be reproducible and feasible for peak analysis, especially in the *m/z* range 3000–9000. In addition, 13 of the 18 most representative common peaks for all *E. faecium* are located within this range, as also are both biomarker peaks for VRE versus VSE and

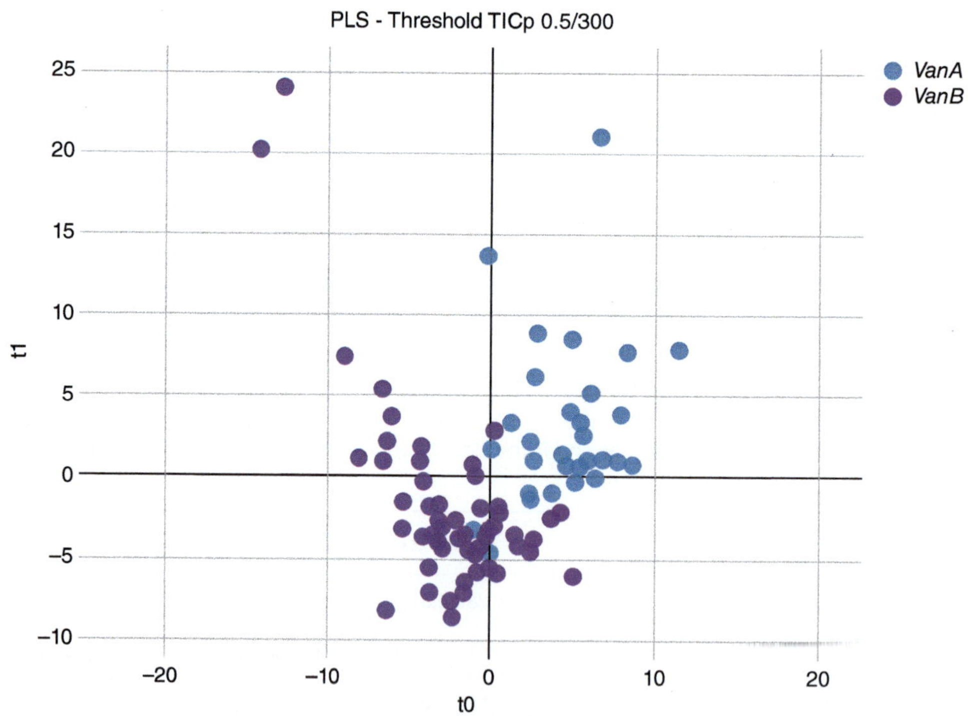

Figure 2.4 Distance plot of PLS-DA machine-learning algorithm for the discrimination of *VanA* from *VanB* VRE isolates using the total ion current normalization method applied before peak detection (TICp) method. *Source:* Reprinted from Candela et al. [25].

VanA versus *VanB*. The implementation of the methodology described in this study could provide standardization for data comparison with other studies analysing AMR with MALDI-TOF MS.

We found discrepancies in the peaks that we used for discrimination of the different *E. faecium* groups and those reported in two previous studies: none of the peaks included in the study by Nakano et al. [28] were found to be relevant in our models; besides, only the peak at *m/z* of 5095.01 (5094.7 *m/z* in [29]) was common to both studies for the discrimination of VRE from VSE. In our predictive models, the peak at *m/z* of 6603 proposed by Griffin et al. [29] for the discrimination of *VanA* from *VanB* VRE was not considered discriminative. Instead, the peak at *m/z* of 6891.78 served this purpose in the current work. Further studies are being carried out, but we propose the detection of peaks at *m/z* of 5095.01 and 6891.78 for the detection of suspected *VanA* VRE.

2.3.3 Results and Discussion

In this study we were limited by the lack of genomic background for the isolates analysed. The genes that were targeted for vancomycin resistance were confirmed using molecular methods. More comprehensive information about our VRE isolates could help to obtain higher discrimination power from the applied algorithms and explain the misclassifications from the current models.

The classification accuracy of the applied algorithms has shown to be $<90\%$ in all cases. Therefore, further studies with well-characterized isolates sourced from different geographic origins are needed to confirm the results obtained in the current work and to improve them if possible.

The current work indicates that MALDI-TOF MS has acceptable discrimination ability for *E. faecium* isolates beyond species assignment. Although further refining is required and isolates from different clones and origins must be included in predictive models to understand how they are discriminated by MALDI-TOF MS, protein profiling could become a suitable tool for the rapid detection of VRE in clinical microbiology laboratories. Its implementation could be key for the control of VRE isolates in hospital settings.

2.4 Detection of Azole Resistance in *Aspergillus fumigatus* Complex Isolates

2.4.1 Introduction

This study [30] was a collaboration between the Department of Clinical Microbiology and Infectious Diseases of the HGUGM and the team at Clover Bioanalytical Software, S.L., in Granada, Spain.

Aspergillus fumigatus complex is one of the most common airborne saprophytic fungi. It is the main agent that causes a wide spectrum of diseases collectively known as aspergillosis [31]. It used to be considered a weak pathogen, but with the increase in immunocompromised patients, it has become a more prevalent problem leading to severe and usually fatal infections [32]. The most relevant species within the complex is the *Ammophilus fumigatus* sensu stricto (s.s.), which has been identified as the main cause of invasive aspergillosis [33] but there are also other sibling species showing low azole susceptibility [11].

The only treatments that are currently available for invasive aspergillosis are amphotericin B (AmB) and azole antifungals. However, azole-resistance strains have been identified [10, 34]. This is one of the reasons why these drugs have a relatively low success rate. Rapid identification of this resistance and the subspecies causing the infection is therefore very important to be able to apply the appropriate treatment and containment measures as soon as possible.

Molecular identification is currently used to discriminate s.s. from cryptic species [35]. However, reliable phenotypic discrimination of complex species of *A. fumigatus* is still needed. Antifungal susceptibility testing is restricted to a limited number of filamentous fungi [36, 37]. The turnaround times for this is comparable to broth microdilution methods [38].

MALDI-TOF MS has proved to be promising for the identification of different species of *Aspergillus* [39]. Identification of fungal isolates using MALDI-TOF MS has been possible due to the availability of pre-processing sample methods and commercial or in-house reference libraries [40]. Updated fungal libraries have allowed up to 85.0% accurate species-level identification. However, closely related species may be misidentified. We have shown how MALDI-TOF MS coupled with ML techniques can be used to differentiate *A. fumigatus* s.s. from other species of the complex and to detect azole resistance in s.s. isolates without exposing them first to the drug.

2.4.2 Material and Methods

We used 868 samples from a national survey on azole resistance conducted in Spain ([41]). Most of these were isolates (812) from the lower respiratory tract. We assessed the susceptibility to various azoles by following the EUCAST procedure E.Def 9.3.2 [42]. All azole-susceptible isolates were s.s., while those showing phenotypic resistance to azoles all proved to be cryptic species. For these 20 s.s. isolates, the *cyp51a* gene was sequenced and several mutations were detected.

Species-level Identification of each sample was achieved by analysis of colonies grown on Sabouraud agar plates using a Microflex LT MALDI-TOF MS (Bruker Microbiology & Diagnostics). The details of sample preparation have been reported previously [19]. The databases used for identification were the Filamentous Fungi Library 2.0 (Bruker Microbiology & Diagnostics), the database of the National Institute of Health (NIH) [43] and an in-house library HGM filamentous fungi and yeasts (Gregorio Marañón Hospital, Spain). Strain identification was carried out in duplicate.

2.4.3 Results

To differentiate between spectra from *A. fumigatus* s.s. and cryptic species, we used Clover MS Data Analysis software. The first phase for any analysis with ML algorithms is to prepare the data for analysis, typically by reducing noise, aligning spectra, detecting relevant features and rescaling intensities from the raw spectra. The target is to calculate a peak matrix where each column corresponds to a m/z value where a relevant feature has been detected and each row corresponds to one of the samples under analysis.

In this case, we started by aligning technical replicates of each sample and then obtaining an average spectrum of these replicates. Relevant peaks were obtained from these average spectra by detecting local maxima and then discarding those peaks whose intensity is less than 1% of the maximum peak intensity found. Finally, these peaks were aligned across averaged spectra to form the peak matrix that was submitted for analysis.

To differentiate spectra (rows of the peak matrix) belonging to each of the groups (*A. fumigatus* s.s. from cryptic species), the peak matrix was analysed using PCA. Results are shown in Figure 2.5. PCA is a common computational technique that is used to maximize the inter-independence of each sample, thus classifying groups of similar data together. When coupled with a hierarchical clustering algorithm, it produced a dendrogram

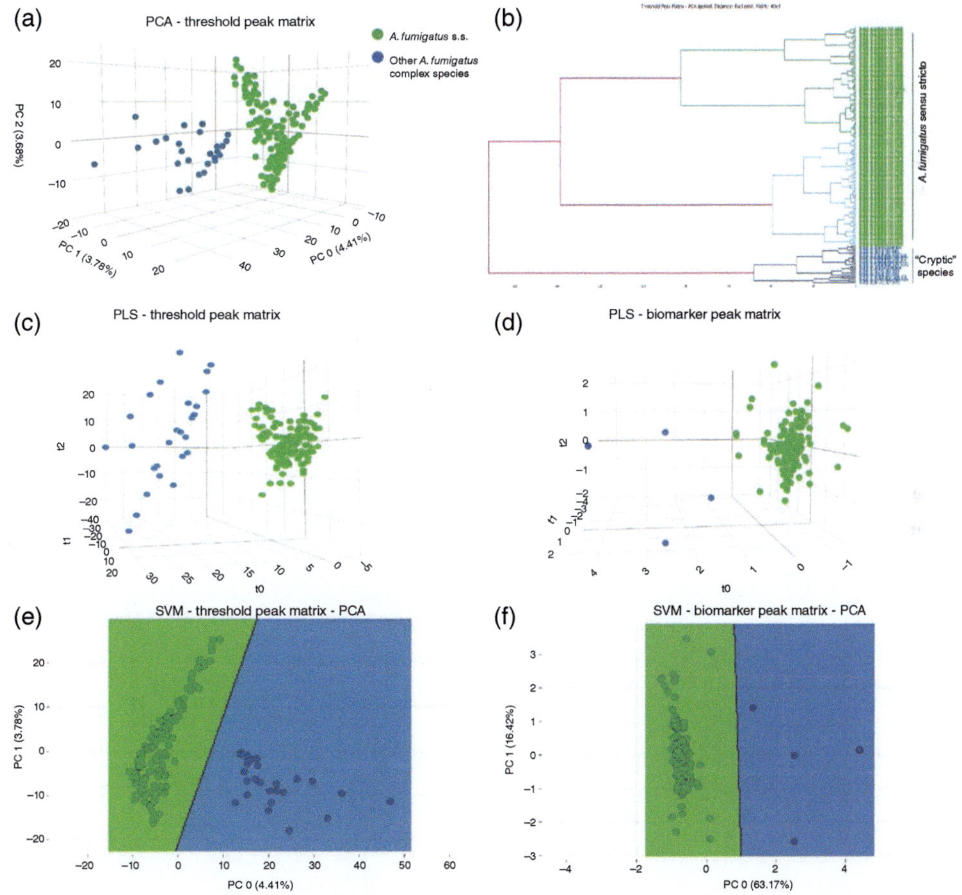

Figure 2.5 Discrimination of *Aspergillus fumigatus* sensu stricto from the cryptic species of the *A. fumigatus* complex. (a) Principal component analysis plot of the samples; (b) dendrogram showing the clustering of the two groups of *Aspergillus* isolates; (c) discrimination by PLS-DA applied to the peak matrix generated by thresholding; (d) discrimination by PLS-DA applied to the peak matrix built by selecting biomarker peaks; (e) discrimination by support vector machine applied to the peak matrix generated by thresholding; (f) discrimination by support vector machine applied to the peak matrix built by selecting biomarker peaks. *Source:* Reprinted from Zvezdanova et al. [19].

(Figure 2.6) where the grouping of each sample in its appropriate category can be evaluated. We achieved a 100% correct classification into both groups. This was also confirmed (Figure 2.7) by discrimination using SVM and PLS-DA [44]. Using SVM, 100% correct classification was obtained, and PLS-DA gave 99.43% accuracy. The discrimination was evaluated via *k*-fold repeated 20 times and calculating the mean squared error (MSE), which confirmed 100% accuracy for both models.

ROC curves with AUC were used to determine the sensitivity and specificity of candidate biomarker peaks. A search for category-specific biomarker peaks resulted in a potential candidate at *m/z* 4586.2, specifically present in all cryptic species isolates (AUC = 1.0). Four specific biomarkers for *A. fumigatus* s.s. were found at *m/z* 2999.9, 4578.1, 5023.7 and 7732.7 with AUC > 0.97 in all cases. A study to analyse the biological relevance of these biomarkers is pending.

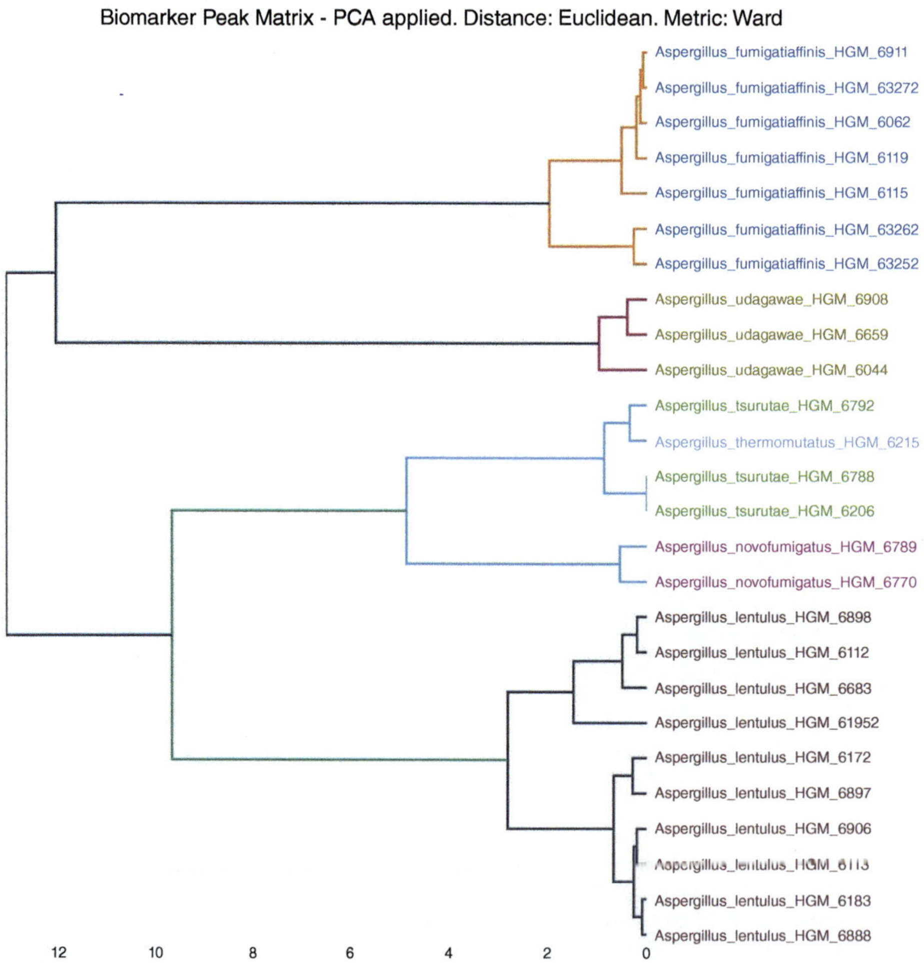

Figure 2.6 Clustering of cryptic species from the *Aspergillus fumigatus* complex by principal component analysis applied to a peak matrix built with biomarker peaks.

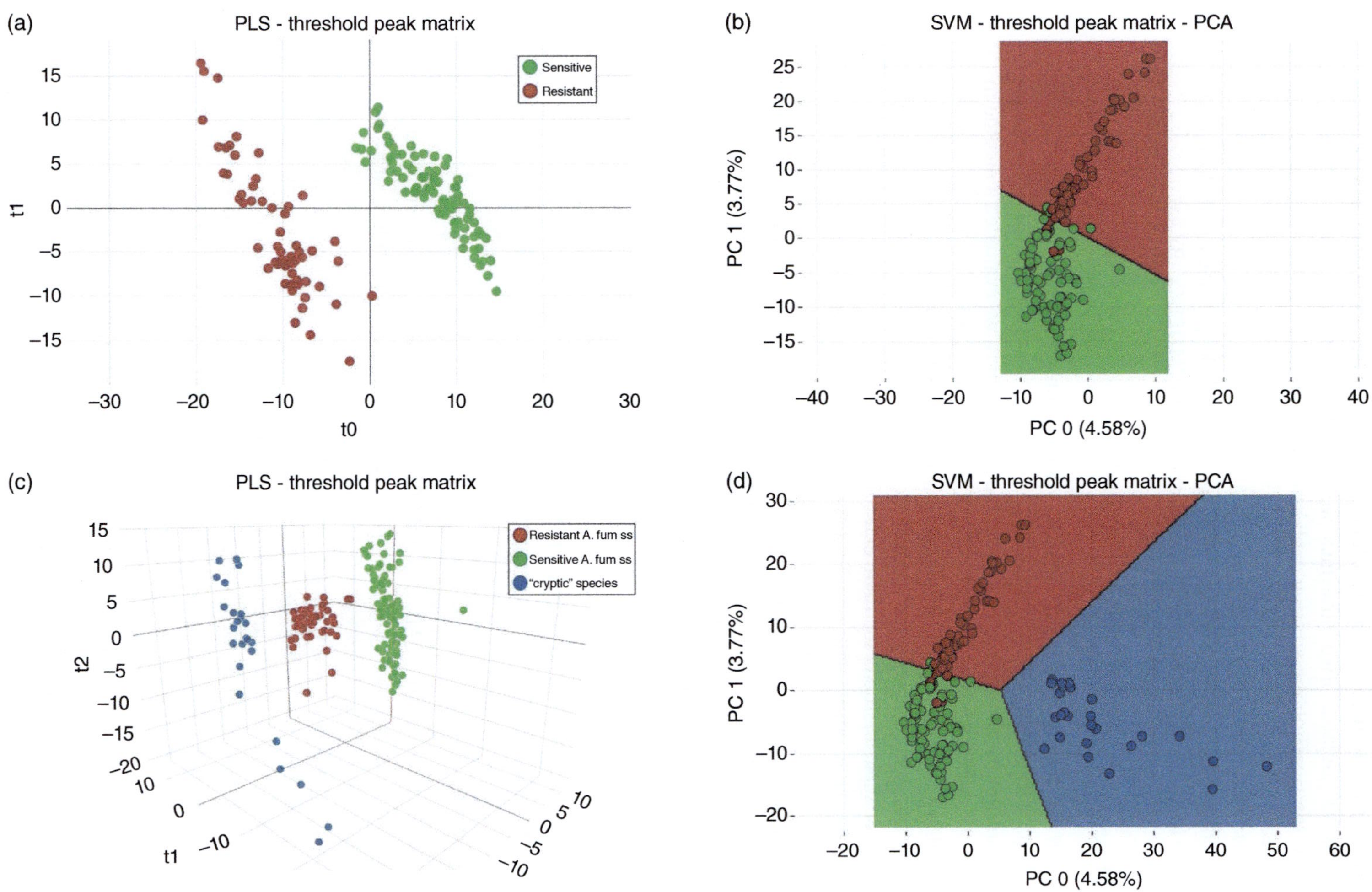

Figure 2.7 Differentiation of susceptible ($n = 91$) and resistant ($n = 40$) *Aspergillus fumigatus* sensu stricto isolates. (a) Discrimination using partial least-squares discriminant analysis. (b) Differentiation by support vector machine after principal component analysis. (c) Discrimination of the two previous categories from cryptic species using partial least-squares discriminant analysis. (d) Support vector machine after principal component analysis. Susceptible *A. fumigatus* sensu stricto are represented in green, resistant *A. fumigatus* s.s. isolates in red, and cryptic isolates from the *Aspergillus fumigatus* complex in blue. *Source:* Reprinted from Zvezdanova et al. [19].

New peak matrices were constructed to differentiate azole-susceptible from azole-resistant *A. fumigatus* s.s. isolates. Two strategies were evaluated to select relevant features (Figure 2.6): the thresholding strategy explained previously, and using a list of specific biomarker peaks from each analysed group [30]. We submitted the produced peak matrices to PLS-DA, which provided 100% accuracy in discrimination in the training set, with a validation score (k-fold with $k = 10$) of 98.5% (± 0.038). SVM provided a slightly lower performance in this case [19]. *Source:* Zvezdanova et al. [19]/MDPI/CC BY 4.0.

2.4.4 Discussion

De Carolis et al. [36] reported a MALDI-TOF MS procedure for antifungal susceptibility testing. The procedure requires an incubation period of at least three hours in the presence of different antifungal concentrations [37]. In the current study, we evaluated a method in which direct peak analysis from initial MALDI-TOF MS species-level identification was used. This allowed rapid screening of fungal isolates and to determine their susceptibility pattern. The increasing occurrence of azole-resistant *Aspergillus* isolates requires the development of rapid methods for their detection. The resulting software that has been developed is available online with an affordable price compared with the cost of conventional methods and their turnaround times. Peak analysis can be completed in less than 20 minutes. This methodology may provide clinically important information that would allow optimization of the management of infected patients. Further studies are needed to validate and improve the existing models and to determine whether these findings can be applied to a wider range of cryptic species and *A. fumigatus* s.s. with other resistance mechanisms.

2.5 Peak Analysis for Discrimination of *Cryptococcus neoformans* Species Complex and their Interspecies Hybrids

2.5.1 Motivation

To further explore the ability of ML and MALDI-TOF MS to discriminate subspecies of fungi, we carried out another study in collaboration with the Department of Clinical Microbiology and Infectious Diseases of the Gregorio Marañón Hospital in Madrid.

The genus *Cryptococcus* comprises two sibling species with great clinical importance: *C. neoformans* and *C. gattii*. They are the main agents causing cryptococcosis. The former is mainly associated with meningitis in immunosuppressed patients, while the latter can also affect immune-competent populations [30]. To differentiate between species, it is very important to establish the epidemiology, virulence and susceptibility pattern to antifungal drugs. Molecular techniques are the gold standard that are currently used for *Cryptococcus* spp. identification. However, the procedure is cumbersome and time-consuming, both of which delay the final identification. MALDI-TOF MS has been used to discriminate between *C. neoformans* and *C. gatti* complexes [45, 46], but the construction of an appropriate database is out of reach for most microbiology laboratories.

2.5.2 Material and Methods

Peak analysis and data modelling were applied to protein spectra of *Cryptococcus* isolates, with the objective of discriminating, at the species level, isolates belonging to the *C. neoformans* complex, characterized previously by amplified fragment length polymorphism (AFLP) and sequencing of the ITS1-5.8S-ITS2 region [30]. Our results showed a very high accuracy for differentiating *C. neoformans* from the interspecies hybrids. This optimized procedure can be automated and implemented into a laboratory routine.

We selected 70 *Cryptococcus* spp. isolates from clinical samples belonging to 67 patients admitted to Hospital Gregorio Marañón from 1994 to 2007. Isolates came from cerebrum spinal fluids (51%), blood (33%), respiratory samples (10%) and others (6%). Genotyping of the isolates detected 49% of *C. neoformans*, 41% of intersepecies hybrids and 10% *C. deneoformans*. Sample storage and preparation details can be found in Zvezdanova et al. [30].

Protein spectra were obtained for initial species-level identification on a LT Microflex MALDI-TOF MS (Bruker Microbiology & Diagnostics) using the default settings. The BDAL library contains 12 reference spectra from *C. neoformans* and 7 from *C. deneoformans*. We selected 26 isolates from the hospital collection to form an in-house library, and then challenged both databases with the remaining 44 isolates. The commercial database allowed the correct identification of 18/22 *C. neoformans* isolates (81.8%) and 1/3 *C. deneoformans* isolates (33.3%). The interspecies hybrids could not be identified in the commercial database due to the lack of reference spectra of this organism. However, when the database was enlarged with the HGM in-house library, all *C. neoformans* and *C. deneoformans* isolates were correctly identified, with 21/25 giving very high scores. But this only allowed for the correct identification of 12/19 interspecies hybrids, because of their high similarity with the other two species.

Therefore, to improve the identification of the interspecies hybrids and to discriminate *C. neoformans* and *C. deneoformans*, the spectra obtained from MALDI-TOF MS analysis were processed using Clover MS Data Analysis software. The peak matrices were restricted to m/z range 2400–12 000. In this range, replicates from the same isolate were averaged to form a single spectrum per isolate. Then, relevant peaks were selected from the spectrum for each isolate. Ten peaks were selected on the basis that the rate of presence of these for each species showed an AUC > 0.99. Finally, these 10 potential biomarker peaks were aligned between different spectra to form the peak matrices. These were submitted to statistical analysis via SVM to differentiate between the different species, and results were validated using k-fold confusion matrix. With a prior PCA step, it allowed 98.46% correct classification of the three *Cryptococcus* species (Figure 2.8). This separation was based on the presence of a signal at m/z 6688.24 found in *C. neoformans* isolates and peaks at m/z of 6576.08 and 7103.01 detected in the *C. deneoformans*.

2.5.3 Results and Discussion

The results of this work demonstrate that both the correct species-level identification of *C. neoformans* and *C. deneoformans* isolates using our in-house library and the discrimination of both groups plus hybrid interspecies agree with those reported in previous studies. The current study shows improved identification of *Cryptococcus* species based on using

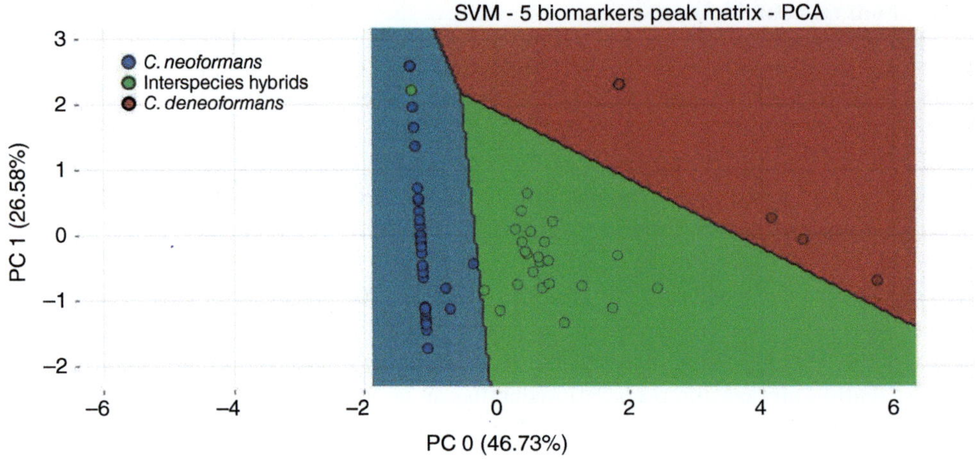

Figure 2.8 Classification of the three *Cryptococcus* species by SVM in the one-step approach, using five biomarker peaks.

MALDI-TOF MS extended with in-house libraries (see, e.g., [45, 46]). To the best of our knowledge, this is the first study where interspecies hybrids were fulfilled completely. The small number of *C. neoformans* included in this study is a limitation, but all isolates were reliably identified at the species level using the expanded database, and this indicates that even small improvements in the reference database can be useful to achieve correct identification of these species. These advancements, however, were not enough to differentiate the interspecies hybrids and this is where our approach of using ML and peak analysis plays a fundamental role. This analysis provided full discrimination of *C. neoformans*, *C. deneoformans* and interspecies hybrids.

2.6 Conclusions

Our results demonstrate the usefulness of ML to pinpoint extremely weak or evasive signals in the MALDI-TOF MS spectra which allows the rapid and reliable identification of complex bacterial and fungal subspecies in the microbiology laboratory. Our studies add value to the initial MALDI-TOF MS-based species identification performed using commercial instruments by complementing those with streamlined AMR information in just a few minutes. The process is fast, requires no special skills and can be very easily integrated into a microbiology laboratory workflow. We have shown application of our strategy to four cases, namely CPK, VRE, azole-resistant *A. fumigatus* and separation of *C. neoformans*, *C. deneoformans* and interspecies hybrids. We have also shown how to reduce the variability in the acquisition of spectra from different samples and different laboratories, which is one of the challenges of MALDI-TOF MS-based tools. These techniques can be extended to many other AMR and subtyping cases. For interested readers, Weis et al. [44] shows a systematic review of studies employing ML to refine the identification of microbial species, and many other studies have been published since and continue to be published. As stated in the review by Weiss et al., there are still a lot of advances in ML that can be leveraged to advance our microbiology tools. With the coming of new software and the in vitro diagnostic approval of these tools, we foresee a high impact on how AMR diagnosis and information are managed in microbiology laboratories. The overarching goal of these advances is to provide the clinician with a rapid diagnostic tool and a data-driven approach to AMR identification. This will lead to a paradigm shift to statistically backed-up clinical decisions.

For the next phase of the work, we need to extend our studies to larger sample sizes coming from wider geographical areas and include external validation. In the coming months and years, we will continue to do that by expanding our network of collaborating hospitals and combining their know-how and resources with the computational skills of our R&D team.

We are aware of the serious societal challenges that are posed by AMR, especially in nosocomial infections acquired in the hospital. We will push forwards to clinical practice the tools described in this work and develop new technologies for the diagnosis of AMR to bring forward a new generation of tools to fight AMR more efficiently, while simultaneously saving costs and lives.

References

1. Aminov, R.I. (2010). A brief history of the antibiotic era: lessons learned and challenges for the future. *Front. Microbiol.* 1: 134.
2. Murray, C.J.L. (2022). Global burden of bacterial antimicrobial resistance in 2019: a systematic analysis. *Lancet* 399 (10325): 629–655.
3. World Health Organization. *Ten Global Health Issues to Track in 2021*. https://www.who.int/ne-room/spotlight/10-global-health-issues-to-track-in-2021 (accessed 9 March 2022) 2021.
4. Ventola, C.L. (2015). The antibiotic resistance crisis. *Pharm. Ther.* 40 (4): 277–283.
5. Tillotson, G., Lodise, T., Classi, P. et al. (2020). Antibiotic treatment failure and associated outcomes among adult patients with community-acquired pneumonia in the outpatient setting: a real-world US insurance claims database study. *Open Forum Infect. Dis.* 3: 1–7.
6. de Kraker, M.E.A., Stewardson, A.J., and Harbarth, S. (2016). Will 10 million people die a year due to antimicrobial resistance by 2050? *PLoS Med.* 13 (11): e1002184. https://doi.org/10.1371/journal.pmed.1002184.
7. Norvig, P. and Russell, S.J. (2020). *Artificial Intelligence: A Modern Approach*, 4e. Prentice Hall.
8. Altman, D.G. (1991). *Practical Statistics for Medical Research*. New York: Chapman and Hall.
9. Centers for Disease Control and Prevention (CDC) (2019). *Antibiotic Resistance Threats in the United States, 2019*. https://www.cdc.gov/drugresistance/biggest-threats.html#van (accessed 23 September 2022).
10. Denning, D.W., Venkateswarlu, K., Oakley, K.L. et al. (1997). Itraconazole resistance in *Aspergillus fumigatus*. *Antimicrob. Agents Chemother.* 41 (6): 1364–1368.
11. Dos Santos, R.A.C., Steenwyk, J., Rivero-Menendez, O. et al. (2020). Genomic and phenotypic heterogeneity of clinical isolates of the human pathogens *Aspergillus fumigatus*, *Aspergillus lentulus*, and *Aspergillus fumigatiaffinis*. *Front. Genet.* 11: 459.

12 Rodríguez-Baño, J., Gutiérrez-Gutiérrez, B., Machuca, I., and Pascual, A. (2018). Treatment of infections caused by extended-spectrum-beta-lactamase-, AmpC-, and carbapenemase-producing Enterobacteriaceae. *Clin. Microbiol. Rev.* 31: e00079.

13 Magiorakos, A.P., Burns, K., Rodríguez Baño, J. et al. (2017). Infection prevention and control measures and tools for the prevention of entry of carbapenem-resistant Enterobacteriaceae into healthcare settings: guidance from the European Centre for Disease Prevention and Control. *Antimicrob. Resist. Infect. Control* 6: 113.

14 Gato, E., Constanso, I.P., Candela, A. et al. (2021). An improved matrix-assisted laser desorption ionization-time of flight mass spectrometry data analysis pipeline for the identification of Carbapenemase-producing *Klebsiella pneumoniae*. *J. Clin. Microbiol.* 59 (7): e0080021.

15 Vázquez-Ucha et al. (2021). Activity of imipenem/relebactam against a Spanish nationwide collection of carbapenemase-producing Enterobacterales. *J. Antimicrob. Chemother.* 76 (6): 1498–1510.

16 Vasoo, S., Cunningham, S.A., Kohner, P.C. et al. (2013). Comparison of a novel, rapid chromogenic biochemical assay, the Carba NP test, with the modified Hodge test for detection of carbapenemase-producing Gram-negative bacilli. *J. Clin. Microbiol.* 51 (9): 3097–3101.

17 Kuhn, M. and Johnson, K. (2018). *Applied Predictive Modelling*. New York: Springer-Verlag.

18 Lau, A.F., Wang, H., Weingarten, R.A. et al. (2014). A rapid matrix-assisted laser desorption ionization–time of flight mass spectrometry-based method for single-plasmid tracking in an outbreak of carbapenem-resistant Enterobacteriaceae. *J. Clin. Microbiol.* 52: 2804–2812.

19 Zvezdanova, M.E., Arroyo, M.J., Méndez, G. et al. ASPEIN group(2022). Detection of azole resistance in *Aspergillus fumigatus* complex isolates using MALDI-TOF mass spectrometry. *Clin. Microbiol. Infect.* 28: 260–266.

20 Arias, C.A. and Murray, B.E. (2012). The rise of the *Enterococcus*: beyond vancomycin resistance. *Nat. Rev. Microbiol.* 10 (4): 266–278.

21 Rice, L.B. (2008). Federal Funding for the study of antimicrobial resistance in nosocomial pathogens: no ESKAPE. *J. Infect. Dis.* 197 (8): 1079–1081.

22 Mirzaei, B., Farivar, T.N., Juhari, P. et al. (2013). Investigation of the prevalence of vanA and vanB genes in vancomycin resistant enterococcus (VRE) by Taq man real time PCR assay. *J. Microbiol. Infect. Dis.* 3 (04): 192–198.

23 Seng, P., Drancourt, M., Gouriet, F. et al. (2009). Ongoing revolution in bacteriology: routine identification of bacteria by matrix-assisted laser desorption ionization time-of-flight mass spectrometry. *Clin. Infect. Dis.* 49 (4): 543–551.

24 Oviaño, M., Gómara, M., Barba, M.J. et al. (2017). Towards the early detection of β-lactamase-producing Enterobacteriaceae by MALDI-TOF MS analysis. *J. Antimicrob. Chemother.* 72 (8): 2259–2262.

25 Candela, A., Arroyo, M.J., Sánchez-Molleda, A. et al. (2022). Rapid and reproducible MALDI-TOF-based method for the detection of vancomycin-resistant *Enterococcus faecium* using classifying algorithms. *Diagnostics* 12 (2): 328.

26 Rodríguez-Sánchez, B., Marín, M., Sánchez-Carrillo, C. et al. (2014). Improvement of matrix-assisted laser desorption/ionization time-of-flight mass spectrometry identification of difficult-to-identify bacteria and its impact in the workflow of a clinical microbiology laboratory. *Diag. Microbiol. Infect. Dis.* 79 (1): 1–6.

27 Brackmann, M., Leib, S.L., Tonolla, M. et al. (2020). Antimicrobial resistance classification using MALDI-TOF-MS is not that easy: lessons from vancomycin-resistant *Enterococcus faecium*. *Clin. Microbiol. Infect.* 26: 391–393.

28 Nakano, S., Matsumura, Y., Kato, K. et al. (2014). Differentiation of vanA-positive *Enterococcus faecium* from vanA-negative *E. faecium* by matrix-assisted laser desorption/ ionisation time-of-flight mass spectrometry. *Int. J. Antimicrob. Agents* 44 (3): 256–259.

29 Griffin, P.M., Price, G.R., Schooneveldt, J.M. et al. (2012). Use of matrix-assisted laser desorption ionization-time of flight mass spectrometry to identify vancomycin-resistant enterococci and investigate the epidemiology of an outbreak. *J. Clin. Microbiol.* 50: 2918–2931.

30 Zvezdanova, M.E., Arroyo, M.J., Méndez, G. et al. (2020). Implementation of MALDI-TOF mass spectrometry and peak analysis: application to the discrimination of *Cryptococcus neoformans* species complex and their interspecies hybrids. *J. Fungi* 6 (4): 330.

31 Latgé, J.P. (1999). *Aspergillus fumigatus* and *Aspergillosis*. *Clin. Microb. Rev.* 12 (2): 310–350.

32 Lamoth, F. (2016). *Aspergillus fumigatus*-related species in clinical practice. *Fron. Microbiol.* 7: 683.

33 Falces-Romero, I., Quiles-Melero, I., and García-Rodríguez, J. (2017). Sensibilidad in vitro de especies crípticas de *Aspergillus fumigatus* a isavuconazol, itraconazol y voriconazol mediante E-test. *Rev. Iberoam. Micol.* 35 (2): 113–114.

34 Chryssanthou, F. (1997). in vitro susceptibility of respiratory isolates of *Aspergillus species* to itraconazole and Amphotericin B. acquired resistance to itraconazole. *Scand. J. Infec. Dis.* 29 (5): 509–512.

35 Alastruey-Izquierdo, A., Alcazar-Fuoli, L., and Cuenca-Estrella, M. (2014). Antifungal susceptibility profile of cryptic species of *Aspergillus*. *Mycopathologia* 178: 427–433.

36 De Carolis, E., Vella, A., Florio, A.R. et al. (2021). Use of matrix-assisted laser desorption ionization-time of flight mass spectrometry for caspofungin susceptibility testing of Candida and *Aspergillus species*. *J. Clin. Microbiol.* 50: 2479–2483.

37 Gitman, M.R. et al. (2013). Antifungal susceptibility testing by matrix-assisted laser desorption ionization-time of flight mass spectrometry analysis. *J. Clin. Microbiol* 51: 2964–2969.

38 Gitman, M.R., McTaggart, L., Spinato, J. et al. (2017). Antifungal susceptibility testing of *Aspergillus* spp. by using a CCI-based MALDI-ToF MS method appears to not offer benefit over the traditional broth microdilution testing. *J. Clin. Microbiol.* 55 (7): 2030–2034.

39 Imbert, S., Normand, A.C., Gabriel, F. et al. (2019). Multi-centric evaluation of the online MSI platform for the identification of cryptic and rare species of *Aspergillus* by MALDI-TOF. *Med. Mycol.* 57: 962–968.

40 Patel, R.A. (2019). A moldy application of MALDI: MALDI-ToF mass spectrometry fungal identification. *J. Fungi* 5: 4.

41 Escribano, P., Rodríguez-Sánchez, B., Díaz-García, J. et al. ASPEIN study group(2021). Azole resistance survey on clinical *Aspergillus fumigatus* isolates in Spain. *Clin. Microbiol. Infect.* 8: 1170.e1–1170.e7.

42 Subcommittee on Antifungal Susceptibility Testing of the ESCMID European Committee for Antimicrobial Susceptibility Testing (2008). EUCAST Technical Note on the method for the determination of broth dilution minimum inhibitory concentrations of antifungal agents for conidia-forming moulds. *Clin. Microbiol. Infect.* 14 (10): 982–984.

43 Lau, A.F., Drake, S.K., Calhoun, L.B. et al. (2019). Development of a clinically comprehensive database and a simple procedure for identification of molds from solid media by matrix-assisted laser desorption ionization-time of flight mass spectrometry. *Med. Micol.* 57: 63–70.

44 Weis, C.V., Jutzeler, C.R., and Borgwardt, K. (2020). Machine learning for microbial identification and antimicrobial susceptibility testing on MALDI-TOF mass spectra: a systematic review. *Clin. Microbiol. Infect.* 26: 1310–1327.

45 Firacative, C., Trilles, L., and Meyer, W. (2012). MALDI-TOF MS enables the rapid identification of the major molecular types within the *Cryptococcus neoformans/C. gattii* species complex. *PLoS One* 7 (5): e37566.

46 Posteraro, B., Vella, A., Cogliati, M. et al. (2012). Matrix-assisted laser desorption ionization-time of flight mass spectrometry-based method for discrimination between molecular types of *Cryptococcus neoformans* and *Cryptococcus gattii*. *J. Clin. Microbiol.* 50: 2472–2476.

3

Top-down Identification of Shiga Toxin (and Other Virulence Factors and Biomarkers) from Pathogenic *E. coli* using MALDI-TOF/TOF Tandem Mass Spectrometry

Clifton K. Fagerquist

Produce Safety & Microbiology, Western Regional Research Center, Agricultural Research Service, U.S. Department of Agriculture, CA, USA

3.1 Introduction

Matrix-assisted laser desorption/ionization time-of-flight mass spectrometry (MALDI-TOF-MS) is now routinely used to identify microorganisms worldwide [1, 2]. The popularity of this technique is its speed, ease of use, and the minimal training required in MS to operate such platforms.

MALDI-TOF technology continues to advance and has found applications in synthetic polymer analysis [3] and imaging mass spectrometry (IMS) [4]. With the development of tandem TOF platforms (TOF/TOF) [5], it became possible to perform tandem mass spectrometry (MS/MS) of specific precursor ions detected in MS mode. Although MALDI-TOF and TOF/TOF excel at ionization and fragmentation of peptides, especially peptides from tryptic digestion, it was discovered that intact proteins can also be fragmented and identified [6–8]. In fact, protein ion fragmentation, by the aspartic acid effect (AAE) mechanism [6, 9–11], was first discovered on MALDI-TOF instruments using special scanning techniques and post-source decay (PSD) [6]. Further advances, utilized in-source decay (ISD) which cleaves residues primarily from the N- and C-termini of a protein, analogous to Edman sequencing, with subsequent MS/MS of protein ion fragments. This technique is referred to as T3 sequencing [12, 13]. It is hypothesized that electron/proton recombination reactions occur in the source, resulting in cleavages of the polypeptide backbone by a non-ergodic mechanism similar to that of electron capture dissociation (ECD) [14–16] or electron transfer dissociation (ETD) [17] of multiply charged protein ions generated by electrospray ionization (ESI). A serious disadvantage of this technique is that (like Edman sequencing) only a pure protein can be analyzed. The sample cannot be a mixture of proteins because precursor ion isolation is not possible by ISD, and thus a specific fragment ion cannot be definitively attributed to a specific precursor ion. This is referred to as precursor ion-fragment ion attribution. By contrast, MALDI-TOF/TOF-MS/MS allows

Microbiological Identification using MALDI-TOF and Tandem Mass Spectrometry: Industrial and Environmental Applications, First edition. Edited by Haroun N. Shah, Saheer E. Gharbia, Ajit J. Shah, Erika Y. Tranfield, and K. Clive Thompson.

precursor-fragment ion attribution because the precursor ion is isolated prior to fragmentation. Protein ion fragmentation by MALDI-TOF/TOF-MS/MS can occur by high-energy collision-induced dissociation (HE-CID) using a target gas [18] or by PSD with no target gas [6].

3.2 Decay of Metastable Peptide and Protein Ions by the Aspartic Acid Effect

The AAE mechanism of peptide and protein ion fragmentation has been studied both experimentally and theoretically for more than 25 years [6, 9–11]. It has been observed in the fragmentation of peptide and protein ions generated by ESI and MALDI [6, 10]. The AAE mechanism is a molecular ion rearrangement involving the side-chain of acidic residues, i.e. aspartic (D) and glutamic acid (E) residues (and more recently asparagine (N) residues [19]), whereby the proton of the acidic side-chain is transferred to the polypeptide backbone on its C-terminal side, resulting in polypeptide backbone cleavage (PBC) with simultaneous formation a b-type fragment with a five- or six-membered ring end-group at its C-terminus and a y-type fragment with the standard H- end-group at its N-terminus.

Polypeptide backbone cleavage by PSD and AAE is relatively slow (~10 μs), ergodic, low-energy, statistical dissociation. Energy is deposited into a peptide or protein ion where it is randomized among the vibrational and rotational degrees-of-freedom of the molecule before a low-energy dissociation channel is achieved by molecular ion rearrangement. PSD is only one example of an ergodic dissociation technique that can lead to AAE. Others include sustained off-resonance ionization collision-induced dissociation (SORI-CID) [20], infrared multiphoton dissociation (IRMPD) [20–23], blackbody infrared dissociation (BIRD) [20] and surface-induced dissociation (SID) [20].

Ergodic dissociation (and ergodic dissociation techniques) has advantages and disadvantages for analysis of peptide and protein ions. As ergodic dissociations involve increasing the internal energy of the molecule such that it fragments at some later time (μs to s), the fragmentation, when it occurs, is often highly specific as it is the result of a low-energy dissociation channel which is the most energetically accessible. Although higher-energy dissociation channels are possible, they are statistically less likely due to energy redistribution and randomization. A disadvantage of ergodic dissociations is that certain post-translational modifications (PTMs), often those attached to residue side-chains, may undergo dissociative loss before their attachment site is determined. Thus, the utility of non-ergodic dissociation techniques (ECD [14–16], ETD [17], ultraviolet photodissociation [24], etc.) is that they cleave the polypeptide backbone by a mechanism that does not involve "heating" of the molecule. In addition, it has been shown that these non-ergodic techniques also have the advantage of providing increased sequence coverage, identifications of PTMs and their attachment sites.

Given the advantages of non-ergodic dissociation techniques (typically present on high-resolution mass spectrometers), it might seem as if dissociation techniques available on MALDI-TOF and TOF/TOF instruments (i.e. ISD/T3-sequencing, PSD and HE-CID), would be severely limiting. However, MALDI-TOF and TOF/TOF platforms (and their associated dissociation techniques) have continued to show their utility in applications as varied as such peptide mass fingerprinting (PMF), taxonomic identification of microorganisms, analysis of synthetic polymers and IMS. The reason for the continued popularity of MALDI-TOF (TOF/TOF) platforms is that they are extremely rapid (compared with other techniques), easy to operate with real-time analysis of data and a simple sample preparation. Technical replicates are also easy to accomplish compared with replicates involving chromatography, e.g. liquid chromatography-MS (LC-MS). This is not to say that all applications in biological MS will work on MALDI-TOF (or TOF/TOF) platforms but, given the ease and speed of analysis, there is an advantage in using techniques that provide a quick result even if the result is not necessarily definitive.

When confronting a new research problem in biological MS, it is not unusual to assume that only the most powerful mass spectrometer, i.e. the highest resolution, sensitivity and mass accuracy, will be adequate, and, by inference, any "failure" of an experiment, when performed on a lower-end instrument, must be due to the inadequacy of the instrument and not to other factors, e.g. sample preparation. However, experience has shown that if other factors have been adequately addressed, success can be achieved with a platform having lower resolution and mass accuracy.

With the increasing accessibility of whole-genome sequencing (WGS) data in public databases, it is possible to rapidly scan the trove of information provided. For proteomic scientists, the amino acid sequences of all identified open reading frames provide the foundational bedrock and starting point for proteomic investigations. Although there are many commercial proteomic search engines (e.g. MASCOT) that provide the proteomic identifications in a user-friendly format, most of these highly useful tools are geared to analysis of digested proteins often with trypsin or other commonly available enzymes used for bottom-up proteomic investigations [25]. At the other pole of proteomic research, there are commercial software tools that facilitate top-down proteomic identifications, e.g. ProSight [26]. Most of the top-down proteomic research is performed with ESI (or nano-ESI) on high-resolution mass analyzers (Fourier transform ion cyclotron resonance, Orbitrap, etc.) that often have numerous ergodic and non-ergodic dissociation techniques, mentioned previously.

Top-down proteomic investigations using MALDI-TOF/TOF are far less prevalent given the limitations of the platform in terms of sensitivity, resolution, and mass accuracy. This lack of popularity of MALDI-TOF (TOF/TOF) for top-down proteomic research has resulted in fewer software options. However, as nature abhors a vacuum, a few researchers have ventured into this relatively unexplored area of research using MALDI-TOF/TOF platforms and have developed in-house software programs [27–30].

The MALDI-TOF/TOF instruments first became commercially available early in the twenty-first century and several reports quickly appeared demonstrating the ability of these instruments to fragment intact, undigested protein ions by HE-CID or PSD [7, 8]. The results obtained on the new MALDI-TOF/TOF platforms were also consistent with earlier experiments performed on MALDI-TOF instruments that performed a time-consuming quasi-MS/MS experiment involving stepwise adjustment of the acceleration voltage and stitching together these sequential spectra to reveal dissociation of metastable protein ions. It was in one of these early studies that the AAE was first detected on protein ions on a MALDI-TOF instrument [6].

One of the first studies to use MALDI-TOF/TOF and top-down proteomic for identification of protein biomarkers of microorganisms was by Demirev et al. [27]. In this pioneering work, they demonstrated identification of intact (non-digested) proteins and their source microorganisms using MALDI-TOF/TOF and top-down proteomic identification using software developed in-house. The *m/z* of MS/MS-PSD fragment ions were compared with the *m/z* of *in silico* fragment ions generated from PBC at *every* residue of the bacterial protein sequences derived from WGS. Bacterial protein sequences were downloaded from public databases, e.g. ExPASy. *In silico* fragment ions included a-, b-, y-type fragment ions including small neutral losses of water (−18 Da) and/or ammonia (−17 Da). A sophisticated algorithm (*P*-value) was used to score identifications with respect to their likelihood of being incorrect. Their approach was very successful as it allowed other researchers to exploit bacterial protein sequences in public or private databases for identification using the relatively new MALDI-TOF/TOF platforms [27]. The AAE played a dominant role in protein ion fragmentation by MALDI-MS/MS-PSD and the most abundant fragment ions were the result of PBC on the C-terminal side of D-, E-residues and on the N-terminal side of P-residues. However, the in-house software compared fragment ions from PBC at all residues not just at D-, E-, or P-residues.

Following in the footsteps of this early work, our laboratory developed a user-friendly web-based software (written in Java) that could be used to compare MALDI-TOF/TOF-MS/MS-PSD data of bacterial proteins with a database of *in silico* fragment ions derived from bacterial protein sequences downloaded from ExPASy [29]. Our laboratory incorporated Demirev's *P*-value algorithm into our software. We also developed a much more simplistic scoring method, i.e. percentage of matched MS/MS fragment ions (or USDA score). Finally, the software also included a residue-specific option to limit the *in silico* fragment ions to be matched to specific residues instead of all residues. This approach was based on the observation that since the most abundant fragment ions of MS/MS-PSD of metastable protein ions were the result of the AAE, it is logical to limit the MS/MS-to-*in silico* comparison to *in silico* fragment ions at specific residues, e.g. D, E, and/or P. [29]. Using this approach, our group identified several highly conserved cytoplasmic and periplasmic proteins of food-borne pathogens [28]. We later used this same approach to identify types and subtypes of Shiga toxin produced by STEC using antibiotic induction and MALDI-MS/MS-PSD [31–37]. The STEC work will be discussed in detail in a later section.

In examining thousands of MS/MS-PSD spectra of the fragmentation of metastable protein ions on our MALDI-TOF/TOF instrument, it became increasingly apparent that the AAE fragmentation mechanism was a very reliable predictor of PBC sites (and their fragment ions) of singly charged (protonated) metastable protein ions. Regardless of the fragmentation efficiency of a protein ion, whether the MS/MS data are of excellent or marginal quality, the most abundant fragment ions were always the result of PBC on the C-terminal side of D- and/or E-residues, and especially D-residues. As such it was possible to focus our attention on only a handful of relatively abundant MS/MS-PSD fragment ions for a protein identification, or, at least, a narrowing of possible sequences. Another factor was the increasing availability of genomic data of the specific bacterial strain being studied. As WGS data become more easily obtained by rapid sequencing technologies and readily available in public databases, protein identification can be narrowed to the specific bacterial strain that produced it without having to scan the entire bacterial proteome landscape.

Following this reasoning, our laboratory developed a new software program (Protein Biomarker Seeker) for top-down identification of intact protein ions produced from organisms that have been genomically sequenced [38, 39]. Once again, MALDI-TOF/TOF-MS/

MS-PSD is used for identification, however the software was designed deliberately to restrict the MS/MS-to-*in silico* comparison to only PBC on the C-terminal side of D- and/or E- and/or N-residues and/or the N-terminal side of P-residues. After collection of MS and MS/MS data, the operator manually enters the *m/z* of fragment ions they deem are significantly abundant as to be the result of PBC on the C-terminal side of D- and/or E- and/or N-residues and/or the N-terminal side of P-residues. From MS analysis, the protein mass and the associated mass tolerance are also entered into the software. Finally, the software performs an iterative sequence truncation of every protein sequence on the possibility that the mature protein may be truncated, and the truncation can be anywhere in the sequence at either N- or C-termini (or both). The only requirement is that the mass of a truncated *in silico* protein sequence matches the measured mass of the protein (or falls within its associated mass tolerance). Once these search parameters are entered, a text file containing all the proteins of a microorganism is selected by the operator and scanned by the software. Protein (or protein fragment) sequences that match the search criteria are then returned by the software. The operator then manually inspects each sequence to determine whether the major and minor MS/MS-PSD fragment ions match the *in silico* fragment ions of the candidate sequence(s).

The software does not include a scoring algorithm for identifications [39], and therefore the correctness of any candidate sequence must be confirmed by manual inspection. In consequence, a manageable number of sequences must be returned for confirmation, e.g. 10 or fewer. Interestingly, it was discovered early that *only* two or three non-complementary fragment ions (corresponding to two or three PBC) are necessary to identify the correct protein (or protein fragment) sequence out of a bacterial proteome of approximately 5000 proteins that result in ~200 000 protein fragments that match the protein mass criteria. This was a striking result and demonstrated the high specificity provided by AAE for analysis of bacterial proteins. Presumably, the number and location of D- and E-residues in bacterial proteins are sufficiently unique across its proteome to provide such high chemical specificity [39].

3.3 Energy Deposition during Desorption/Ionization by MALDI

In an early paper, Demirev et al., utilizing MALDI-TOF/TOF (Ultraflex; Bruker Microbiology & Diagnostics) for top-down proteomic identification of bacterial proteins, noted the importance of using high laser fluence (significantly above threshold fluence) as well as matrix selection to increase fragmentation efficiency of intact protein ions. The reasons why high laser influence was critical to protein ion fragmentation was discussed, but a definitive explanation was not firmly established [27]. Following this early work, we similarly observed a link between high laser fluence and the efficiency of protein ion fragmentation using a different MALDI-TOF/TOF (4800; Sciex) [28].

As with all commercial MALDI-TOF and TOF/TOF platforms, the firing of the laser pulse (pulse width ~5 ns) event is decoupled from ion acceleration event which was found to improve resolution and mass accuracy. This is referred to as delayed-ion extraction or pulsed-ion extraction. There is a delay time (~100–600 ns) between firing of the laser pulse and the acceleration of ions from the source. Although it is possible that direct photon absorption by protein analytes embedded in the MALDI matrix may subsequently contribute to PSD (referred to as laser-induced dissociation, or LID [40]), it is more likely that collisions between protein ions and matrix molecules before and/or during ion acceleration are responsible for depositing energy into the protein ion. It is also logical to assume a much greater density of matrix molecules in the expanding MALDI plume at high laser fluence than at lower or threshold laser fluence. On the 4800 MALDI-TOF/TOF, the delay time between the laser pulse and the ion extraction pulse allows the MALDI plume to expand adiabatically in the narrow gap (~2 mm) between the flat stainless-steel surface of the MALDI target plate (referred to as "Source 1") and the first ion optic (referred to as "Grid 1"), a cone-shaped ion optic with a spherical orifice at the apex through which the laser beam and ions traverse (but not at the same time). Upon pulsed ion extraction, a potential difference (~800 V) is applied between Source 1 and Grid 1, accelerating all positively charged ions toward the orifice of Grid 1. Assuming a matrix desorption velocity of 700 m/s and a delay time of 500 ns, the MALDI plume travels 0.35 mm above the surface of Source 1 before the ion extraction pulse is triggered. With a potential difference of 800 V between the Source 1 and Grid 1, a singly charged 10-kDa protein will attain a velocity of ~4000 m/s before passing through the orifice of Grid 1. In consequence, matrix neutrals are relatively stationary compared with the rapidly accelerating protein ions. Any matrix neutrals blocking the path of an accelerating protein ion will result in inelastic collisions and energy deposition into the protein ion.

To test this hypothesis, a simple experiment was conducted whereby the grid-to-source voltage ratio (grid/source) was varied while observing its effect on fragment ion intensities of the B-subunit of Shiga toxin by PSD as shown in Figure 3.1 [34]. As can be seen, as the grid-to-source ratio decreases from 0.9222 to 0.9110 to 0.9000 (increasing the potential difference between grid and source), fragment ion intensity increases. This suggests that as protein ion velocity is increased in this initial stage of ion acceleration, while the protein ion is still within the desorption plume of matrix molecules, more fragmentation is obtained as a result of more energetic protein ion/matrix molecule collisions.

3.4 Protein Denaturation and Fragmentation Efficiency of PSD

Protein denaturation, i.e. protein unfolding, plays a significant role in the fragmentation of singly and multiply charged protein ions in the gas phase. Early in the application of ECD to top-down proteomic analysis of multiply protonated protein ions, the number of fragment ions generated by ECD (a non-ergodic dissociation technique) was at times found to be unexpectedly low, with only charge reduced precursor ions predominating, which at least confirmed that proton/electron recombination had occurred [41]. It was hypothesized that intramolecular noncovalent bonds, e.g. hydrogen bonds and/or hydrophobic interactions of the secondary and tertiary structure of the protein, may thwart detachment of fragment ions after PBC [21, 22, 41]. Further experiments confirmed that if a protein was subjected to IRMPD (an ergodic technique that heats up and unfolds the protein), before or after treatment by ECD, a greater number of fragment ions were detected [21–23, 41]. These results demonstrated that intramolecular noncovalent bonds, although relatively weak, can be cumulative in their effects, resulting in retention of secondary and tertiary structures of the protein and thwarting fragment ion detachment.

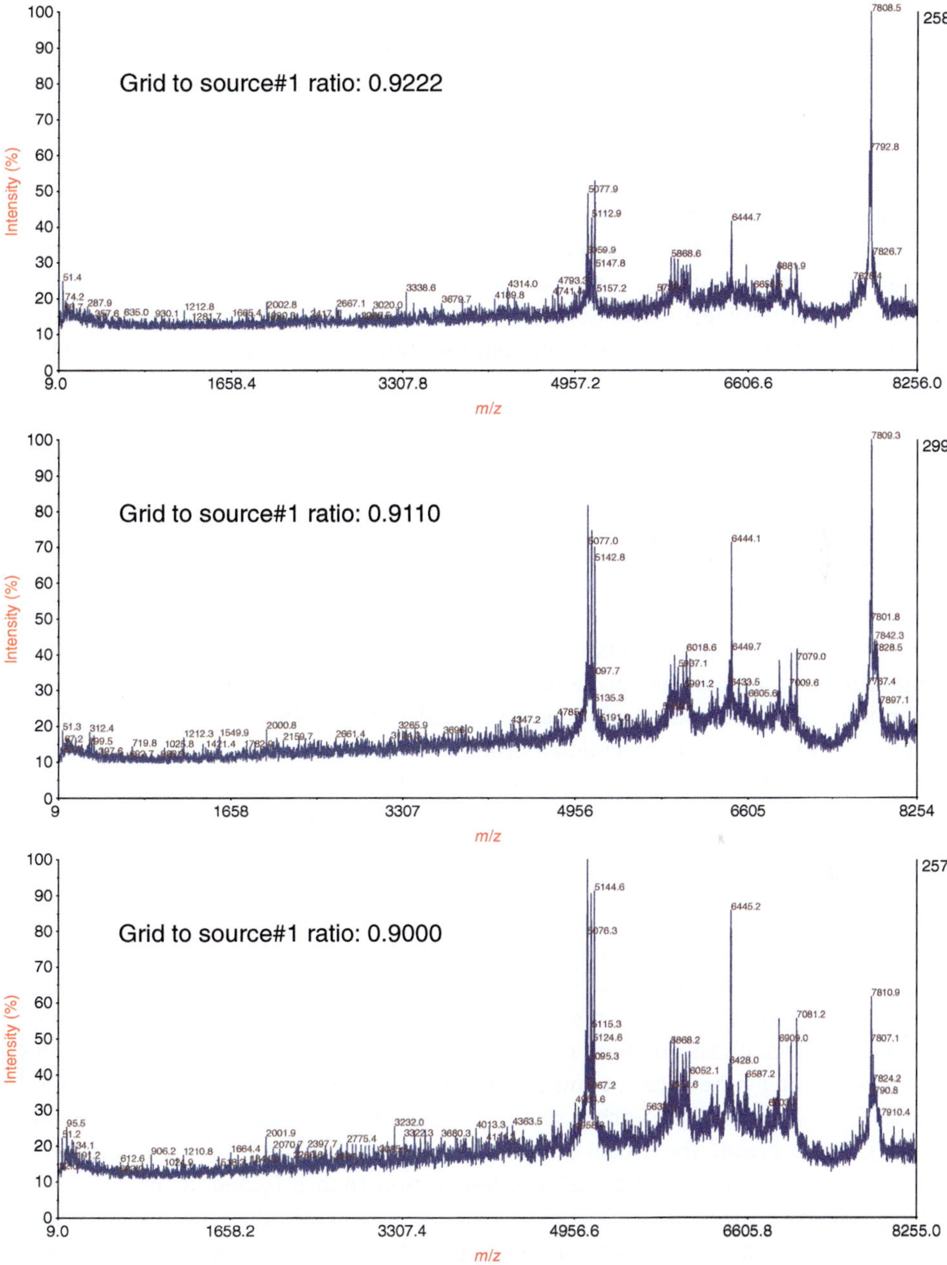

Figure 3.1 MS/MS-PSD of the singly charged $[M+H]^+$ of B-subunit of Shiga toxin 2 at different grid-to-source 1 voltage ratios. A decreasing ratio corresponds to an increasing velocity and kinetic energy of the precursor ion between the MALDI plate surface (i.e. source #1) and the first ion optic (i.e. grid). *Source:* Fagerquist and Zaragoza [34]/Elsevier/CC BY-4.0.

This hypothesis, postulating the effect of protein secondary and tertiary structures on the efficiency of detecting fragment ions, was based on the dissociation of multiply protonated protein ions generated by ESI (or micro-ESI or nano-ESI) analyzed using high-resolution MS instruments, e.g. Fourier transform ion cyclotron resonance (FTICR) or Orbitrap. Little attention has been paid to the efficiency of protein ion fragmentation ionized by MALDI analyzed by TOF. This is not surprising, as the fragmentation efficiency of protein ions using MALDI-TOF instruments has been found to be quite limited [6, 13]. Because the protein analyte (after solubilization in aqueous/organic/acid solution) is embedded in a dried solid matrix of a weak organic acid, little is known about its secondary and tertiary structures. As conventional MALDI generates primarily lower charge state ions (+1 and +2), the charge state envelope generated by MALDI gives no information as to whether the protein is folded, partially folded or unfolded, unlike the higher charge state envelopes generated by ESI [42]. However, it can be said with certainty that the low charge states generated by MALDI result in little or no Coulomb repulsion that might facilitate protein unfolding or fragmentation. In consequence, protein unfolding must be brought about prior to desorption/ionization of the protein into the gas phase or by depositing energy into the protein during desorption/ionization in the source, e.g. protein/matrix collisions.

It is assumed that only a small fraction (perhaps as little as ~1%) of protein ions fragment on the timescale of TOF instruments (~10 μs). One can also assume that these decompositions of metastable protein ions [20] are either completely unfolded or partially unfolded as they have sufficient internal energy to fragment on these timescales by PSD [6]. The degree to which a protein is unfolded can also affect subsequent events after the firing of the laser pulse. All state-of-the-art MALDI-TOF and TOF/TOF instruments incorporate a time delay between the laser pulse and the pulsed acceleration of ions from the source. This delay allows expansion of the MALDI plume before ion acceleration which improves resolution and mass accuracy as ion acceleration is decoupled from the laser ionization/desorption event. Upon triggering of the pulsed acceleration voltage at 8–20 kV, ions in the MALDI plume begin their rapid transit across the region between the stainless steel MALDI plate and the first ion optic. It is commonly known, but not often mentioned, that most of the desorbed material in a MALDI plume are neutrals not ions. This is also true for ESI. In consequence, as ions are rapidly accelerated by the acceleration pulse, they will undergo collisions with neutral molecules (as discussed in the earlier section), i.e. matrix neutrals (as they are the most abundant). Once again, the issue of protein denaturation becomes relevant as an unfolded protein has a greater collision cross-section than a folded-up protein. As a result, an unfolded protein ion will experience a greater number of collisions with matrix molecules than the same protein ion in its folded-up state. Such collisions deposit energy into the protein ion, and, after energy redistribution, result in statistical dissociation at the most accessible, i.e. lowest energy dissociation channels, which is also often preceded by rearrangement of the ion. Excess internal energy in an unfolded protein ion is more likely to result in fragmentation, whereas excess internal energy in a folded-up protein is only likely to result in more unfolding.

3.5 Arginine and its Effect on Fragment Ion Detection and MS/MS Spectral Complexity

Arginine (R) is the most basic of the 20 natural amino acids due to its side-chain having a guanadino functional group. The gas phase basicity (GB) of R is 237.0 kcal/mol (10.30 eV) [43]. One of the next highest is lysine (K) with a GB of 221.8 kcal/mol (9.64 eV) [43]. Typically, K-residues are far more numerous than R-residues. Conventional MALDI is a low charge state ionization technique. In positive ion mode, the number of basic residues in a protein ion far exceeds the number of ionizing protons. In consequence, the ionizing proton(s) is (are) sequestered at the side-chain of basic residues (K or R), as these locations have the highest GB in the protein. It had been previously noted that tryptic peptides with an R-residue have a higher ionization efficiency than tryptic peptides containing K-residue [44]. It is logical to assume that the GB plays a significant role in this phenomenon. It is also logical to assume that this phenomenon could also play a role in the ionization of small proteins. The only difference is that small proteins have many more basic residues (and potential ionization sites) than tryptic peptides which only have one basic residue at the C-terminus (if one assumes trypsin digestion results in no missed cleavages).

Given the preferential ionization of R-residues over K-residues for tryptic peptides, one might expect to observe an analogous effect on the fragment ions detected in the dissociation of metastable protein ions having zero R-residues versus metastable protein ions having one R-residue. This clearly appears to be the case, as shown in Figure 3.2 [45], for MS/MS-PSD of a singly charged metastable YahO protein ion which has no R-residues and 11 K-residues. PBC occurs throughout the YahO protein chain, but the fragment ions detected suggest that the single ionizing proton is not at a single location in the metastable YahO precursor ions but at multiple locations, i.e. K-residues. In addition, eight complementary fragment ion pairs (CFIP or b/y) are also detected, which supports the conclusion that the ionizing proton is located at multiple locations in the packet of YahO precursor ions being analyzed by MS/MS.

By contrast, MS/MS-PSD of cold-shock protein E (CspE), shown in Figure 3.3 [45] highlights that all fragment ions (except a minor fragment ion b_{45}) include the only R-residue in their sequence even though there are seven K-residues also present. Note the abrupt termination in the series of y-type fragment ions at the R-residue. Except for b_{45}, all b-type fragment ions include the R-residue.

Similarly, the B-subunit of Shiga toxin 2 (B-Stx2), has one R-residue and six K-residues, and all (or nearly all) fragment ions detected by MS/MS-PSD include the R-residue in their sequence. MS/MS of B-Stx2 will be discussed in detail in the next section.

MS/MS of CspE suggests that the single ionizing proton is located exclusively (or nearly exclusively) at the side-chain of the only R-residue. A metastable protein ion with only a single ionization site will restrict the number of detectable fragment ions, as every fragment must contain the R-residue to be detected. In addition, CFIP will not be detected as complementary b/y fragment ions cannot both have the arginine residue in their sequence. The limited number of detectable fragment ions should not be construed to suggest that the

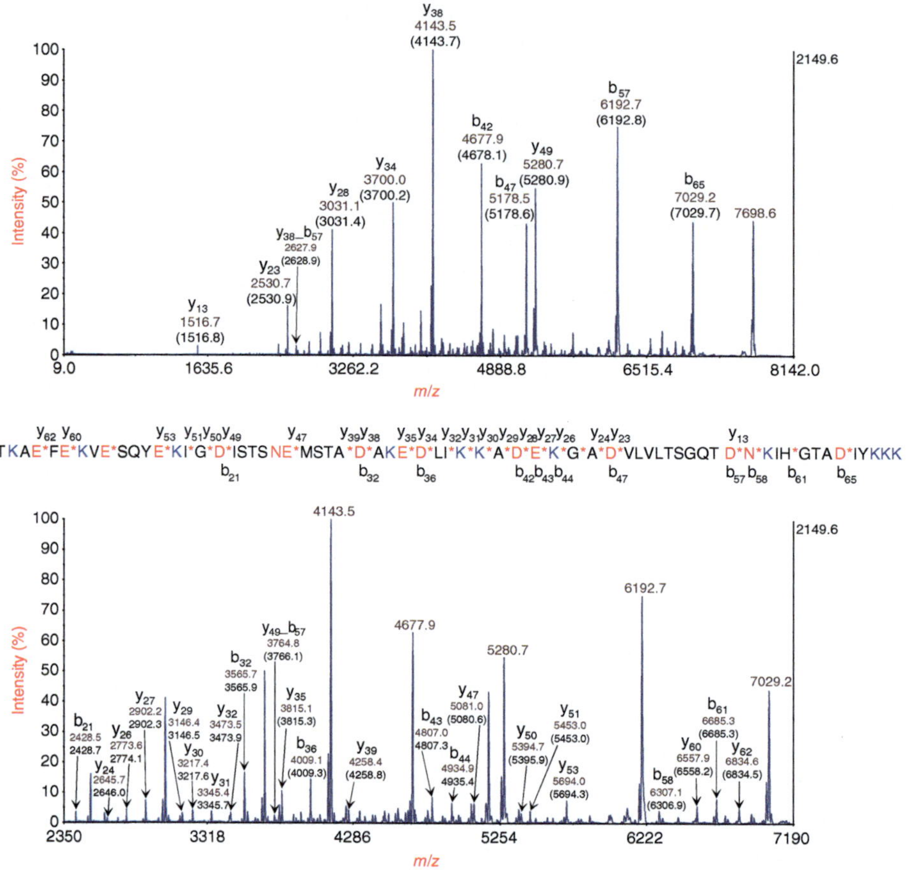

Figure 3.2 MS/MS-PSD of the singly charged $[M+H]^+$ of YahO protein ion. Average theoretical *m/z* values of fragment ions are given in parentheses. The bottom panel is an expanded *m/z* region of the top panel. Amino acid sequence shown above with backbone cleavage sites (*) and their corresponding fragment ions. Double cleavages are represented with y_b. *Source:* Fagerquist and Dodd [45]/With permission of Elsevier.

fragmentation efficiency of the protein is diminished. The single ionizing proton does not appear to play a significant role in protein ion fragmentation regardless of its location: single or multiple sites. Only the internal energy of the metastable protein ion and the number and location of D-, E-, N- and P-residues affect the fragmentation efficiency via AAE.

For protein ions having two (or more) R-residues, the number of possible ionization sites is increased (and perhaps its ionization efficiency as well) proportional to the increased number of R-residues. A greater number of detectable fragment ions will be possible so long as they have at least one of the R-residues in their sequence. CFIP will also be detected once again.

Another consequence of the presence of multiple R-residues in metastable protein ions is dissociative loss of ammonia (NH_3) from the side-chain of R-residues. Although it has been well established that small neutral dissociative losses (H_2O and/or NH_3) occur for

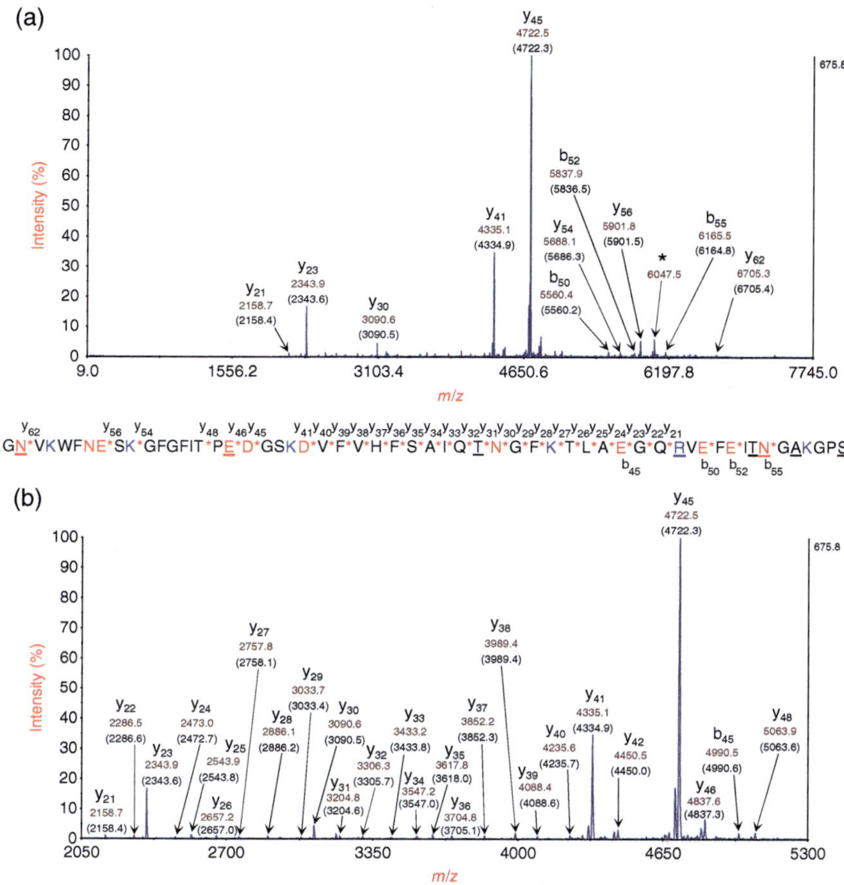

Figure 3.3 MS/MS-PSD of the singly charged [M+H]$^+$ of CspE protein ion. Average theoretical *m/z* values of fragment ions are given in parentheses. Panel (b) is an expanded *m/z* region of (a). Amino acid sequence shown above with backbone cleavage sites (*) and their corresponding fragment ions. *Source:* Fagerquist and Dodd [45]/With permission of Elsevier.

metastable protein ions ionized by MALDI and analyzed by MS/MS-PSD [6, 27], our laboratory has more firmly established the link between three (or more) R-residues and multiple dissociative losses of NH_3, resulting in a significant increase in MS/MS spectral complexity. This is highlighted in MS/MS-PSD of the cold-shock protein CsbD shown in Figure 3.4 [38]. As can be seen, in addition to fragment ions from PBC there are many fragment ions that include multiple losses separated by −17 Da. PBC with small neutral dissociative losses are so dominant that the fragment ion from PBC (and no losses) is not the most abundant fragment ion of a cluster of fragment ions. Contrast this with MS/MS-PSD of metastable YahO (which has no R-residues) (Figure 3.2) where the most abundant fragment ions are the result of PBC only. Fragment ions from a neutral loss do occur but are less abundant than fragment ions from PBC only. In conclusion, three (or more) R-residues in a singly charged (protonated) metastable protein ion will result in increased MS/MS spectral complexity due to facile dissociative loss of NH_3 from the side-chain of R-residues.

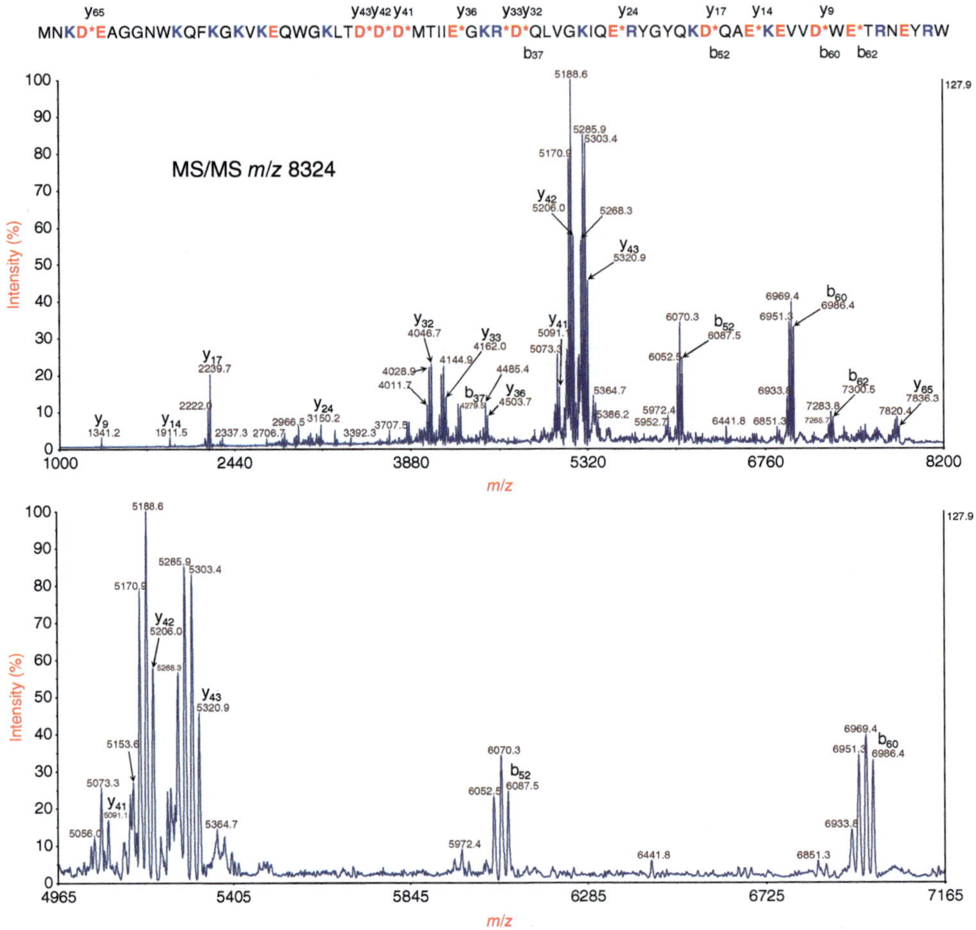

Figure 3.4 MS/MS-PSD of the singly charged $[M+H]^+$ of CsbD protein ion. The bottom panel is an expanded *m/z* region of the top panel. Amino acid sequence shown above with backbone cleavage sites (*) and their corresponding fragment ions. *Source:* Fagerquist et al. [38]/With permission of Elsevier.

3.6 Inducing Gene Expression in Wild-type Bacteria for Identification by Top-Down Proteomic Analysis

Although their lifetime is brief, each generation of bacteria may confront survival challenges. Bacteria can be subjected to stress by several environmental conditions, e.g. DNA damage by antibiotics or UV exposure, nutrient deprivation, iron-deprivation, oxidative or osmotic stress, and competition from other bacteria that occupy the same environmental niche. These stressors may trigger a bacterial response, resulting in changes in gene expression. One of the most well-known and intensively studied bacterial responses is the SOS response which is triggered by DNA damage by exposure to antibiotics, e.g. mitomycin-C or ciprofloxacin, or UV radiation [46]. For bacteria harboring prophage or plasmids, DNA damage triggers the LexA/RecA response system which results in the de-repression of promoters of phage and plasmid genes, allowing their expression and resulting in production of phage, toxins as well as bactericidal proteins. Such bacterial responses can result in a dramatic shift in expression priorities allowing detection and identification of proteins not normally produced or not produced in high abundance.

A well-known example of the SOS response involves STEC, which, when exposed to DNA-damaging antibiotics, can trigger expression of prophage genes including *stx* when present in the prophage genome [46, 47]. Shiga toxin and STEC strains have been studied for several decades because of its impact on food safety and public health [48–50]. Although it is possible to detect Shiga toxin produced from STEC strains by other induction techniques (e.g. minimal media), our laboratory has found that antibiotic induction (with mitomycin-C or ciprofloxacin) is the most reliable method for producing detectable levels of Stx for MS detection without sample fractionation.

3.7 Top-down Proteomic Identification of B-Subunit of Shiga Toxin from STEC Strains

The ease-of-use and versatility of MALDI-TOF/TOF platforms allowed rapid MS/MS analysis of peptides and small proteins. The ability to fragment small undigested proteins (under 20 kDa) was not unexpected, as proteins had been fragmented on MALDI-TOF instruments previously. However, that process was laborious and not easy to implement on a regular basis. The arrival of commercial TOF/TOF instruments by multiple vendors allowed MS/MS to be accomplished in a straightforward manner.

Because of recurrent STEC outbreaks in the USA (and worldwide), our laboratory began to investigate whether MALDI-TOF/TOF and top-down proteomic analysis could be used to detect, identify, and distinguish types and subtypes of Shiga toxin produced from clinical and environmental STEC strains. We quickly found that it was feasible to detect and identify the B-subunit of Shiga toxin; however, it was necessary to employ antibiotic induction to detect the toxin from an unfractionated sample. Broth cultures contain salts and small molecules that interfere with ionization by MALDI and, in consequence, STEC strains were cultured on solid agar supplemented with antibiotic. Another benefit of agar media is that it allowed visual inspection of colony morphology as a function of antibiotic concentration. Antibiotic inhibition is often easily detected by inspection of agar plates. For example, Figure 3.5 shows *E. coli* O157:H7 strain EDL933 grown overnight at 37°C on LBA plates supplemented at different concentrations (ng/ml) of ciprofloxacin [32]. The corresponding MALDI-TOF-MS of the lysate of cells harvested from each plate is shown. The singly charged B-subunit of Stx2a appears at $m/z \sim 7819$. Its average theoretical value is 7815.4 Da (without its signal peptide and with the intramolecular disulfide bond intact). The highest ion intensity of B-Stx2a occurs at the antibiotic concentration of 20–30 ng/ml for this strain.

Figure 3.6 shows MS/MS-PSD of the B-subunit of Stx2a detected at m/z 7819 in Figure 3.5 [32]. PBC is observed on the C-terminal side of D- and E-residues (as well as at other locations along the polypeptide backbone). The sample was not reduced. In consequence, fragment ion triplets are observed, resulting from PBC between the two cysteine

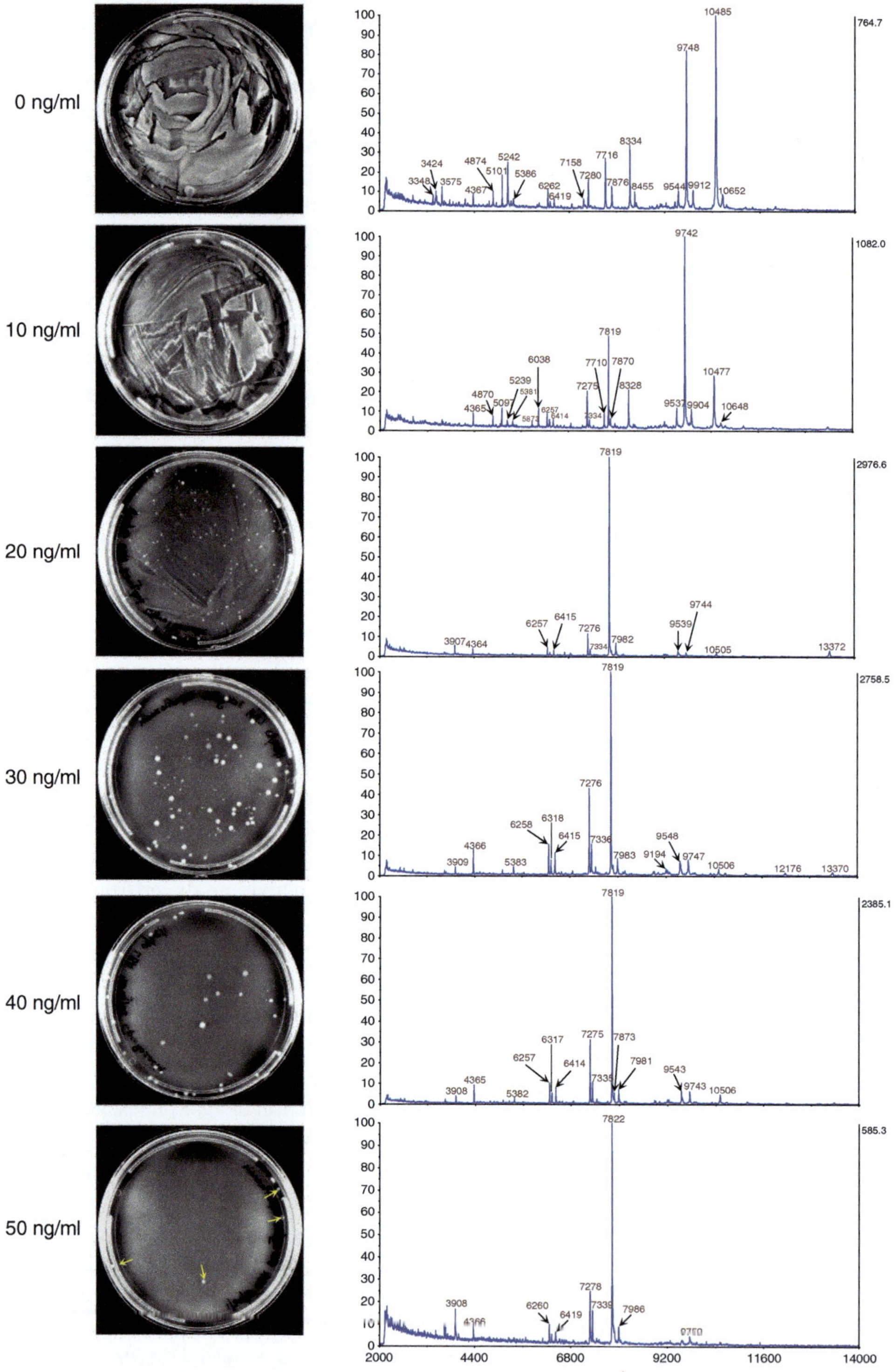

Figure 3.5 Agar plates of *Escherichia coli* O157:H7 strain EDL933 cultured overnight on growth media supplemented at varying concentrations of ciprofloxacin. MALDI-TOF-MS of cell lysates harvested from each plate. Maximum Shiga toxin detection occurred with 20 ng/ml of ciprofloxacin in agar. *Source:* Fagerquist and Sultan [32]/Reproduced with permission from Royal Society of Chemistry.

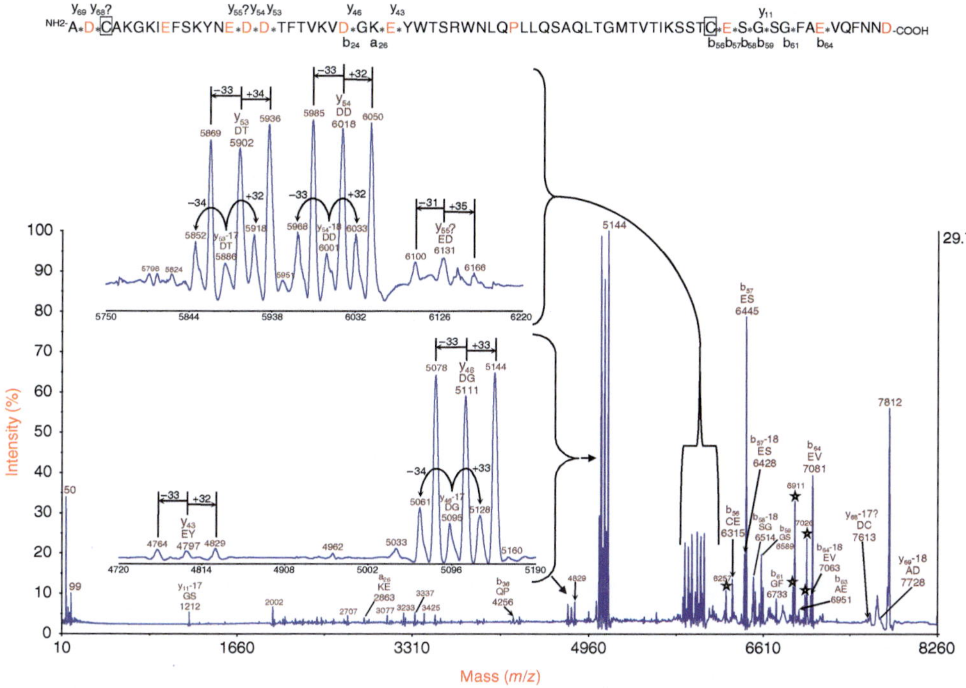

Figure 3.6 MS/MS-PSD of the singly charged $[M+H]^+$ of the disulfide bond-intact B-subunit of Shiga toxin 2a at *m/z* 7819 in Figure 3.5. Amino acid sequence shown above. Insets highlight fragment ion triplets resulting from backbone cleavage between the two cysteine residues of the intramolecular disulfide bond and symmetric and asymmetric cleavage of the disulfide bond. *Source:* Fagerquist and Sultan [32]/with permission from Royal Society of Chemistry.

residues that form an intramolecular disulfide bond, and which undergo symmetric and asymmetric cleavage, resulting in three adjacent fragment ions separated by $\Delta m/z \sim 33$. PBC outside the two cysteine residues do not show fragment ion triplets as expected.

Figure 3.7 shows MS/MS of the disulfide-reduced B-subunit of Stx2a [32]. As expected, the fragment ion triplets no longer appear, and the most abundant fragment ions are the result of PBC on the C-terminal side of D-residues and, to a lesser extent, on the C-terminal side of E-residues. The reduced B-Stx2a has a linear chain structure in contrast to the large looped-structure of disulfide-intact B-Stx2a. In consequence, the MS/MS spectra have similarities but also significant differences.

3.8 Furin-digested Shiga Toxin and Middle-down Proteomics

Unlike top-down proteomics, where there is no proteolytic digestion of the target protein(s) prior to MS analysis, middle-down proteomics involves use of a proteolytic enzyme that cleaves the target protein, resulting in large protein fragments in contrast to the relatively short peptides produced by trypsin digestion commonly used in bottom-up proteomics.

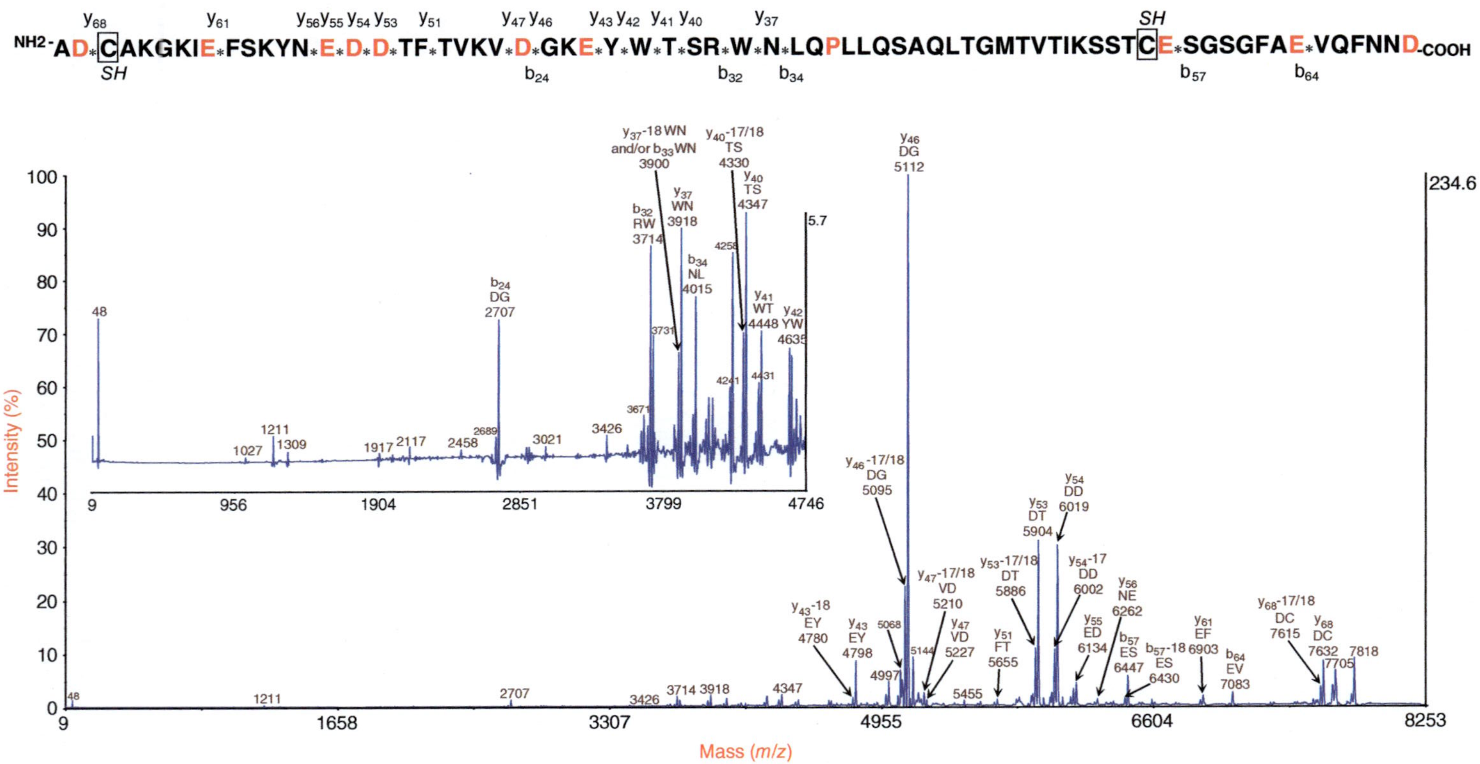

Figure 3.7 MS/MS-PSD of the singly charged $[M+H]^+$ of the disulfide bond-reduced B-subunit of Shiga toxin 2a. Amino acid sequence shown above. Fragment ion triplets are absent as expected and the linear chain structure of this protein results in a more simplified MS/MS spectrum. *Source:* Fagerquist and Sultan [32]/with permission from Royal Society of Chemistry.

Thus, middle-down proteomics represents an analytical compromise between the no-digestion approach of top-down versus the bottom-up approach of digestion with trypsin resulting in short peptides having a basic residue at the C-terminus. The utility of middle-down is determined by the goal of the experiment. A large protein fragment may allow identification of more PTMs compared with bottom-up analysis, which can result in gaps in sequence coverage. Sequence coverage of 50% is considered more than adequate for bottom-up identification, and the non-detection of certain peptides is either due to poor ionization efficiency caused by co-eluting peptides during liquid chromatography (LC) separation or non-binding of the peptide to the trap used for sample clean-up and pre-concentration. Ultimately the decision to digest (and to what extent) is often determined by the resolution, mass accuracy and sensitivity of the mass spectrometer available for the research. Another important consideration is the fact that peptides are easier to separate chromatographically and easier to ionize as well as fragment. From a theoretical standpoint, simulations of peptide fragmentation are computationally more tractable than proteins and thus easier to rationalize and predict. Secondary and tertiary structures from protein folding can be retained even after transfer to the gas phase which can hinder efficient formation and detection of fragment ions regardless of the ionization and/or dissociation technique used.

Furin is an endopeptidase proteolytic enzyme common in eukaryotic cells (including human cells) and plays a critical role in the processing of the Stx holotoxin. After attachment at the cell surface by a Gb3 receptor, Stx holotoxin (Figure 3.8) is enveloped by the cell after which it follows a retrograde pathway to the Golgi apparatus and the endoplasmic reticulum (ER) [47]. At the ER, a highly exposed 20-residue loop of the A-subunit is cleaved by furin at a recognition site (RXXR) located in the middle of the loop which is the boundary between A1 and the A2 fragments. A2 provides stability to the quaternary B-subunit pentamer by occupying the pentamer "donut hole". At the base of this 20-residue loop is an intramolecular disulfide bond (Figure 3.8). After furin cleavage and disulfide reduction (by a disulfide reductase), A1 is released where it travels to the cytoplasm, disabling the ribosomal machinery and resulting in cell death [47].

The mass of the A-subunit of Shiga toxin 2 is 35.7 kDa. After removal of its N-terminal signal peptide, its mass is 33.2 kDa. Furin cleavage and disulfide reduction of the A-subunit result in A1 and A2 fragments with masses of 27.9 and 5.3 kDa, respectively. Figure 3.9 schematically details this process [31]. Figure 3.10 shows the detection of A2 from a furin-digested cell lysate of *E. coli* O157:H7 strain EDL933 for different digestion times [31].

Figure 3.11 shows MS/MS of the putative A2 fragment of the A-subunit at m/z 5286 $[M+H]^+$ shown in the bottom panel of Figure 3.10 [31]. Backbone cleavage of this protein fragment occurs at numerous sites although the most abundant fragment ions are the result of backbone cleavage on the C-terminal side D- and E-residues.

3.9 Top-down Identification of an Immunity Cognate of a Bactericidal Protein Produced from a STEC Strain

Antibiotic induction has also led to the expression of plasmid-encoded genes in some STEC strains. For example, *E. coli* O113:H21 strain RM7788 is an environmental isolate that carries both prophage and plasmids. Culturing this strain overnight on LBA supplemented

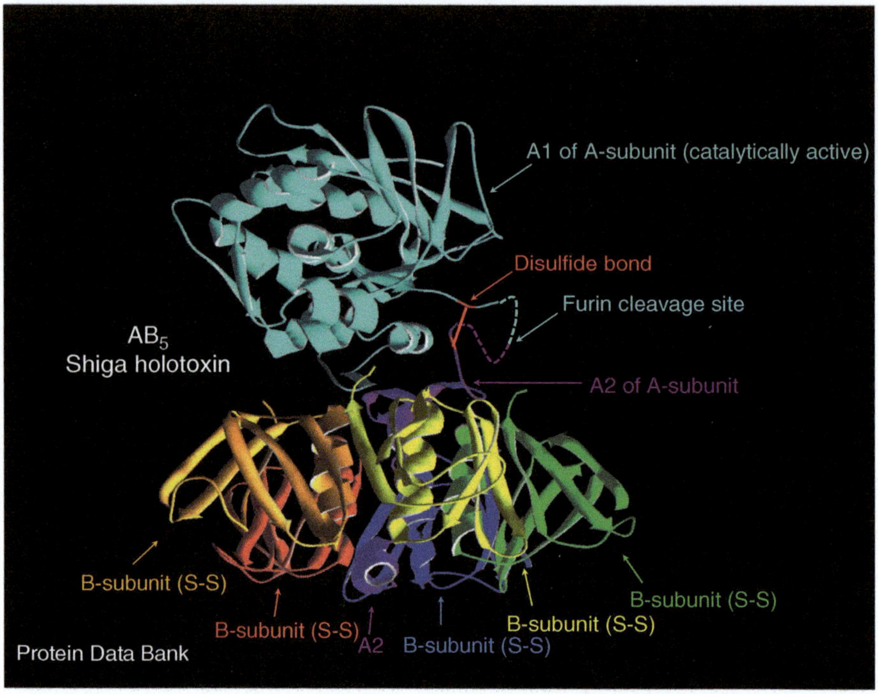

Figure 3.8 Three-dimensional image of the AB_5 Shiga holotoxin. The five identical B-subunits (yellow, gold, red, blue, and green) form a quaternary donut-shaped structure. The A-subunit (cyan) sits above and part of the A-subunit (A2, violet) fits within the donut hole anchoring the A-subunit as well as providing stability to the pentamer structure. A highly exposed 20-residue loop (denoted by dashes) of the A-subunit has a furin recognition/cleavage site. At the base of this loop is an intramolecular disulfide bond (red). The catalytically active A-subunit (A1) is released from A2 upon furin digestion and reduction of the disulfide bond. *Source:* Adapted from Protein Data Bank ID:1R4P from Fraser et al. [51].

with 400 ng/ml of mitomycin-C resulted in the MS spectrum shown in Figure 3.12 (top panel) [39]. Several proteins were previously identified (e.g. CspC, CspE, and a plasmid protein of unknown function). The protein ion at m/z 9780 was analyzed using MS/MS (bottom panel). The protein was identified as the cognate immunity protein (Im3) of the bactericidal protein: colicin E3.

An examination of the plasmid genome shown in Figure 3.13 [39] revealed the inverted repeat of the SOS box upstream of the colicin E3 and Im3 genes. A repressor (LexA) binds to the SOS box blocking expression of genes downstream. However, upon DNA damage, LexA undergoes self-cleavage, allowing expression of genes downstream.

3.10 LC-MALDI-TOF/TOF

Off-line LC has been coupled with MALDI-TOF/TOF for top-down proteomic identification of bacterial proteins [30]. The main advantage of this approach is that, by reducing sample complexity, it allows more proteins to be detected and identified than is possible

MKCILFKWVLCLLLGFSSVSYSREFTIDFSTQQSYVSSLNSIRTEISTPLEHISQGTTSVSVINHT
PPGSYFAVDIRGLDVYQARFDHLRLIIEQNNLYVAGFVNTATNTFYRFSDFTHISVPGVTTVSMT
TDSSYTTLQRVAALERSGMQISRHSLVSSYLALMEFSGNTMTRDASRAVLRFVTVTAEALRFR
QIQREFRQALSETAPVYTMTPGDVDLTLNWGRISNVLPEYRGEDGVRVGRISFNNISAILGTVA
VILNCHHQGARSVR*AVNEESQPECQITGDRPVIKINNTLWESNTAAAFLNRKSQFLYTTGK
S...S

↓ **Furin digestion & disulfide reduction**

REFTIDFSTQQSYVSSLNSIRTEISTPLEHISQGTTSVSVINHT
PPGSYFAVDIRGLDVYQARFDHLRLIIEQNNLYVAGFVNTATNTFYRFSDFTHISVPGVTTVSMT
TDSSYTTLQRVAALERSGMQISRHSLVSSYLALMEFSGNTMTRDASRAVLRFVTVTAEALRFR
QIQREFRQALSETAPVYTMTPGDVDLTLNWGRISNVLPEYRGEDGVRVGRISFNNISAILGTVA
VILNCHHQGARSVR AVNEESQPECQITGDRPVIKINNTLWESNTAAAFLNRKSQFLYTTGK
-SH -SH

Figure 3.9 Full amino acid sequence of the A-subunit of Shiga toxin 2 outlining the process of furin digestion followed by disulfide reduction resulting in the separation of the A-subunit into protein fragments: catalytically active A1 and the smaller A2 which is detected by MS and identified by MS/MS-PSD. The furin recognition site is boxed, as are the cysteine residues involved in the disulfide bond. An N-terminal signal peptide (shown in outline) is removed from the mature A-subunit. *Source:* Adapted from Fagerquist and Sultan [31].

from an unfractionated sample. Another advantage of this approach is that it can result in protein denaturation which can facilitate protein ion fragmentation during MS/MS. Of course, LC separation would add an extra analytical step. With this technique, a protein biomarker (DNA-binding HU protein) was identified that correlated with β-lactamase antibiotic resistance in non-genomically sequenced pathogenic strains of *E. coli*. It was concluded that HU and β-lactamase genes were probably located on a plasmid carried by pathogenic host strains [30].

3.11 Conclusions

Although top-down proteomic analysis has been achieved primarily using ESI-HR-MS, MALDI-TOF/TOF platforms can also provide top-down identification if the protein mass is below ~20 kDa. An important advantage of MALDI-TOF and TOF/TOF is that unfractionated samples can be analyzed directly because the low-charge states produced by MALDI reduce (although do not eliminate completely) the likelihood of *m/z* overlap of multiple protein analytes. It is this phenomenon that allows MALDI-TOF to be utilized for taxonomic identification of bacteria at the genus and species level analyzed from unfractionated samples. Despite limitations in resolution and mass accuracy, the continued popularity of MALDI-TOF and TOF/TOF platforms is due, in part, to the fact that these instruments are much easier to operate and maintain than most ESI-HR-MS instruments, which is an important consideration if users have no formal training in MS. In addition, MALDI sample preparation and data analysis are often less complicated and labor-intensive than for ESI-HR-MS platforms.

Regardless of the MS platform used, the success of top-down proteomic analysis is significantly increased if the protein of interest can be overexpressed (e.g. induction) because

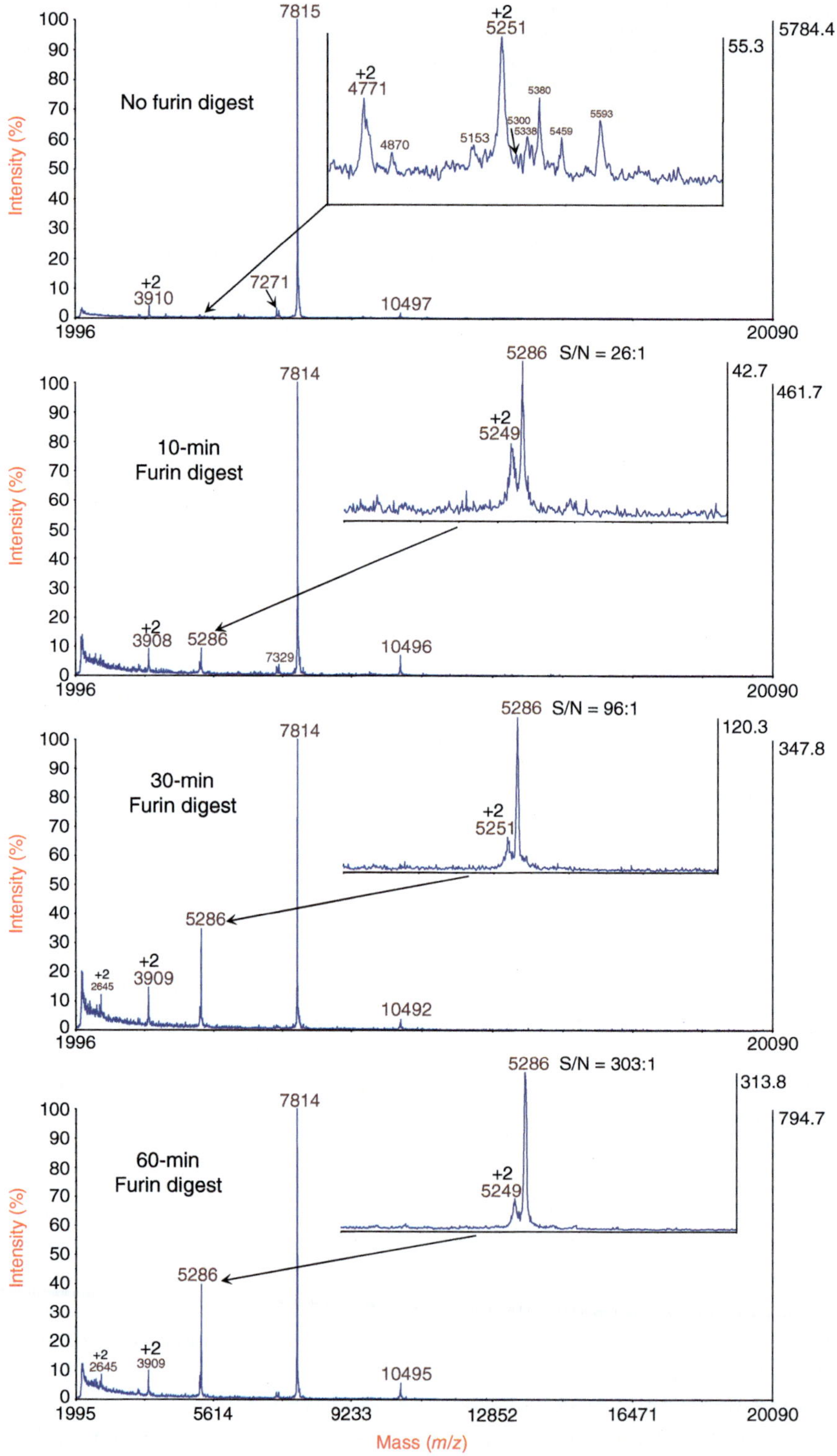

Figure 3.10 Detection of the singly charged $[M+H]^+$ A2 protein fragment of the A-subunit produced at increasing furin digestion times. *Source:* Fagerquist and Sultan [31]/Hindawi/CC BY-3.0.

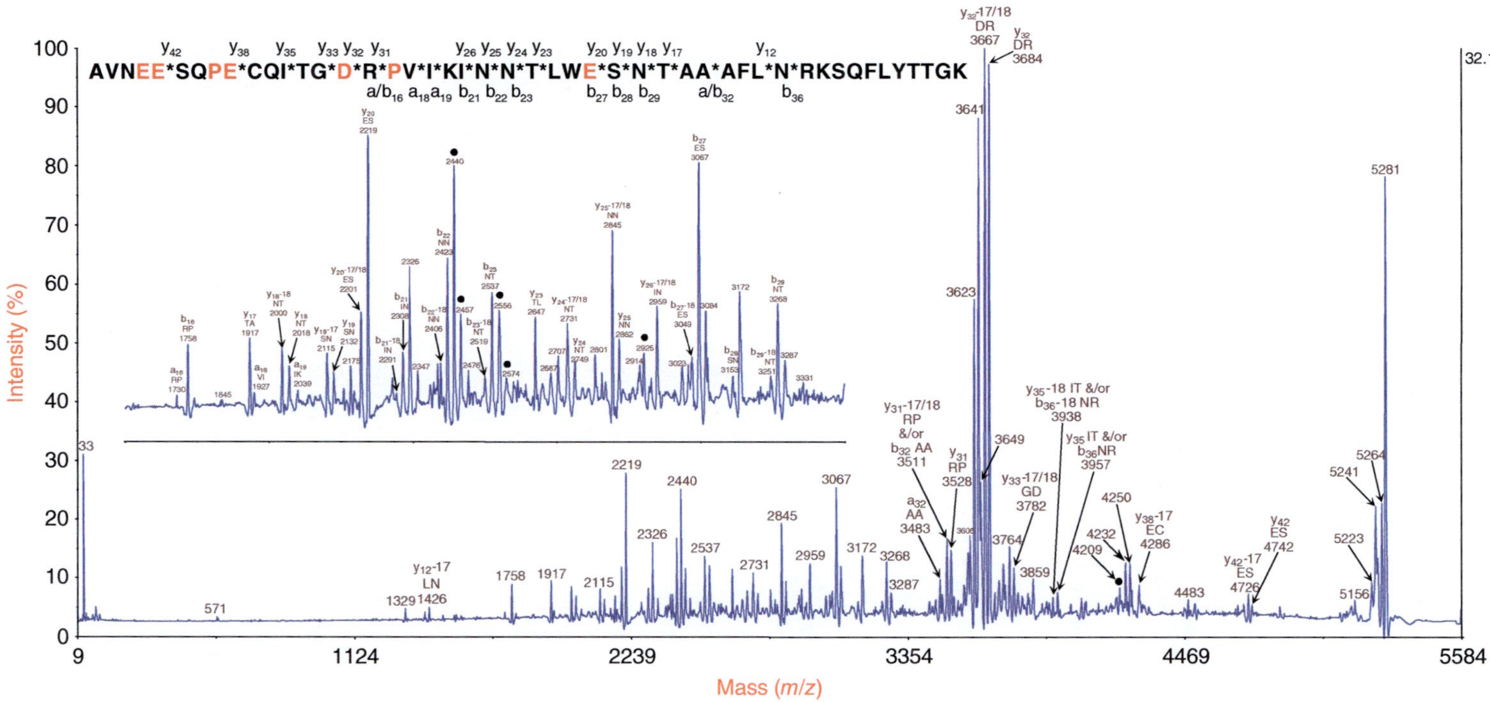

Figure 3.11 MS/MS-PSD of the A2 protein ion. A2 sequence shown above with backbone cleavage sites (*) and resulting fragment ions. Inset shows an expanded *m/z* region. *Source:* Fagerquist and Sultan [31]/Hindawi/CC BY-3.0.

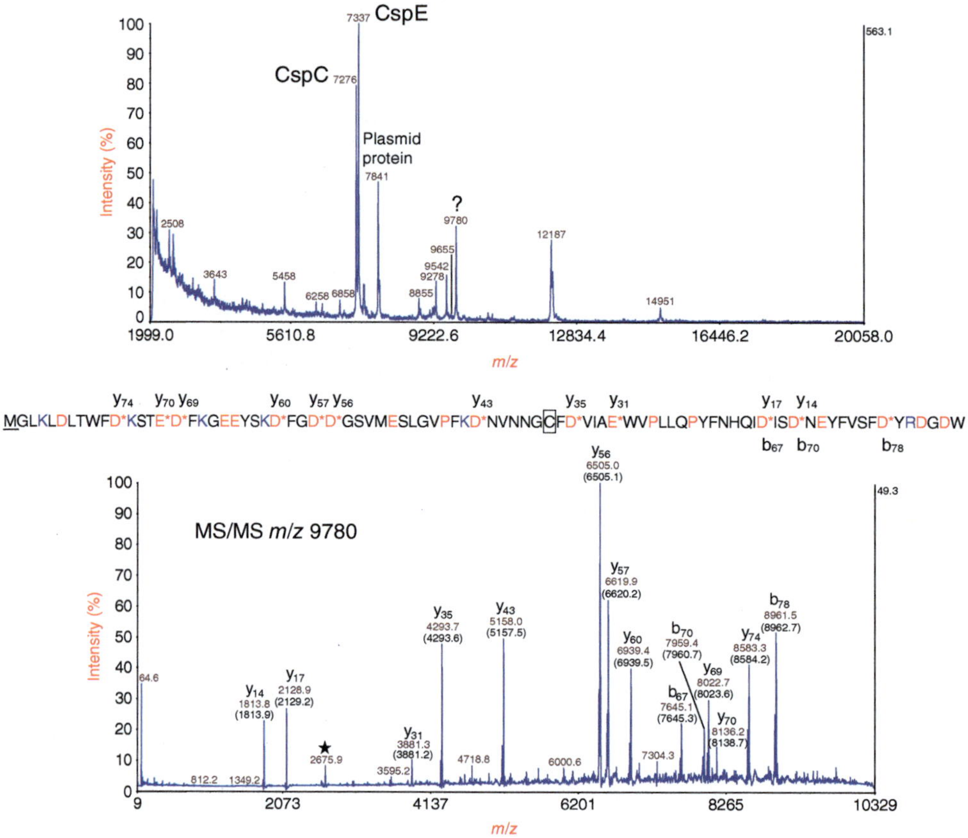

Figure 3.12 Top panel: MS analysis of *E. coli* O113:H21 strain RM7788 cultured overnight on agar supplemented with mitomycin-C. Previously identified proteins are labeled. Bottom panel: MS/MS-PSD of the protein ion at *m/z* 9780 shown in the top panel. The sequence of the protein ion is shown above, i.e. immunity protein (Im3) of colicin E3. N-terminal methionine (underlined) is removed in the mature Im3 protein. An unidentified fragment ion is indicated with a star. *Source:* Adapted from Fagerquist and Rojas [39].

proteins of high abundance are easier to characterize than low-abundance proteins. We have demonstrated the usefulness of antibiotic induction in the overexpression of certain bacterial genes, e.g. *stx*. However, induction of specific genes depends on the nature of promoters upstream of the target genes. The top-down proteomic identification of cognate immunity proteins of bactericidal proteins by antibiotic induction resulted in the subsequent discovery of a repressed promoter (SOS box) upstream of their plasmid-encoded genes, and de-repression of the promoter occurs because of the induction process i.e. damage of the bacterial DNA. It is clear that combining gene induction, top down proteomic analysis and WGS can be used to characterize bacterial proteins and advance biological understanding of pathogenic microorganisms [39].

```
ORIGIN
      1 ccgccaacac atcacgggcc acaaaatttt ttgtggcccg ctctgcgttt tctaagtgtt
                                                 -35
     61 atccctcctg atttctaaaa aattttccac ctgagcttga caagaaaaac gatgacgggt
          PB              SOS Box →        ← SOS Box
    121 acttttgat ctgtacataa aaccagtggt tttatgtaca gtattaatca tgtaattaat
                       SD-RBS           Col E3
    181 tgttttaacg cttaaaagag ggaattttta tgagcggtgg tgatggacgc ggccataaca
    241 cggcgcgca tagcacaagt ggtaacatta atggtggccc gaccgggctt ggtgtaggtg
    301 gtggtgcttc tgatggttcc gggtggagtt cggaaaataa cccgtggggt ggtggttccg
    361 gtagcggcat tcactgggga ggtggttccg gtcatggtaa tggcgggggg aatggtaatt
    421 ccggtggtgg ctcgggaaca ggcggtaatc tgtcagcagt agctgcgcca gtggcatttg
    481 gttttccggc actttccact ccaggagctg gcggtctggc ggtcagtatt tcagcgggag
    541 cattatcggc agctattgct gatattatgg ctgccctgaa aggaccgttt aaatttggtc
    601 tttgggggt ggctttatat ggtgtattgc catcacaaat agcgaaagat gaccccaata
    661 tgatgtcaaa gattgtgacg tcattacccg cagatgatat tactgaatca cctgtcagtt
    721 cattacctct cgataaggca acagtaaacg taaatgttcg tgttgttgat gatgtaaaag
    781 acgaacgaca gaatatttcg gttgtttcag gtgttccgat gagtgttccg gtggttgatg
    841 caaaacctac cgaacgtccg ggtgttttta cggcatcaat tccaggtgca cctgttctga
    901 atatttcagt taataacagt acgccagcag tacagacatt aagcccaggt gttacaaata
    961 atactgataa ggatgttcgc ccggcaggat ttactcaggg tggtaatacc agggatgcag
   1021 ttattcgatt cccgaaggac agcggtcata atgccgtata tgtttcagtg agtgatgttc
   1081 ttagccctga ccaggtaaaa caacgtcagg atgaagaaaa tcgccgtcag caggaatggg
   1141 atgctacgca tccggttgaa gcggctgagc gaaattatga acgcgcgcgt gcagagctga
   1201 atcaggcaaa tgaagatgtt gccagaaatc aggagcgaca ggctaaagct gttcaggttt
   1261 ataattcgcg taaaagcgaa cttgatgcag cgaataaaac tcttgctgat gcaatagctg
   1321 aaataaaaca atttaatcga tttgcccatg acccaatggc tggcggtcac agaatgtggc
   1381 aaatggccgg gcttaaagcc cagcgggcgc agacggatgt aaataataag caggctgcat
   1441 ttgatgctgc tgcaaaagag aagtcagatg ctgatgctgc attgagttct gctatggaaa
   1501 gcaggaagaa gaaagaagat aagaaaagga gtgctgaaaa taatttaaac gatgaaaaga
   1561 ataagcccag aaaaggtttt aaagattacg ggcatgatta tcatccagct ccgaaaactg
   1621 agaatattaa agggcttggt gatcttaagc ctgggatacc aaaaacacca aagcagaatg
   1681 gtggtggaaa acgcaagcgc tggactggag ataaagggcg taagatttat gagtgggatt
   1741 ctcagcatgg tgagcttgag ggtatcgtg ccagtgatgg tcagcatctt ggctcatttg
   1801 accctaaaac aggcaatcag ttgaaaggtc cagatccgaa acgaaatatc aagaaatatc
          SD-RBS    Im3
   1861 tttgagagga agttatggga cttaaattgg atttaacttg gtttgataaa agtacagaag
   1921 atttaaggg tgaggagtat tcaaaagatt ttggagatga cggttcagtt atggaaagtc
   1981 taggtgtgcc ttttaaggat aatgttaata acggttgctt tgatgttata gctgaatggg
   2041 tacctttgct acaaccatac tttaatcatc aaattgatat ttccgataat gagtattttg
   2101 tttcgtttga ttatcgtgat ggtgattggt gatcaaatat tatcagggat gagttgatat
   2161 acgggcttct agtgttcatg gatgaacgct ggagcctcca aatgtagaaa tgttatattt
   2221 ttttattggg ttcttggtta taattgctcc gcaatgattt aaataagcat tatttaaaac
             SD-RBS     Im-Bac
   2281 attctcagga gaggtgaagg tggagctaaa aaaaagtatt ggtgattaca ctgaaaccga
   2341 attcaaaaaa attattgaag acatcatcaa ttgtgaaggt gatgaaaaaa aacaggatga
   2401 taacctcgag tattttataa atgttactga gcatcctagt ggttctgatc tgatttatta
   2461 cccagaaggt aataatgatg gtagccctga aggtgttatt aaagagatta aagaatggcg
   2521 agccgctaac ggtaagtcag gatttaaaca gggctgaaat atgaatgccg gttgtttatg
                                                     Lysis
   2581 gatgaatggc tggcattctt tcacaacaag gagtcgttat gaaaaaaata acagggatta
   2641 ttttattgct tcttgcagtc attattctgt ctgcatgtca ggcaaactat atccgggatg
   2701 ttcagggcgg gaccgtatct ccgtcatcaa cagctgaagt gaccggatta gcaacgcagt
   2761 aacccgaaat cctctttgac aaaaacaaag cgtgtcaggc tgattctgat gcgctttttt
```

Figure 3.13 A portion of the genome of a plasmid carried by *E. coli* O113:H21 strain RM7788. Genes for bactericidal colicin E3 and its immunity protein Im3 are highlighted in yellow. Upstream of these genes is the inverted repeat of an SOS box. *Source:* Adapted from Fagerquist and Rojas [39].

References

1 Angeletti, S. and Ciccozzi, M. (2019). Matrix-assisted laser desorption ionization time-of-flight mass spectrometry in clinical microbiology: an updating review. *Infect. Genet. Evol.* 76: 104063.

2 Sandrin, T.R., Goldstein, J.E., and Schumaker, S. (2013). MALDI TOF MS profiling of bacteria at the strain level: a review. *Mass Spectrom. Rev.* 32: 188–217.

3 Yoo, H.J., Kim, D.H., Shin, D. et al. (2020). Recent developments in pre-treatment and analytical techniques for synthetic polymers by MALDI-TOF mass spectrometry. *Anal. Methods* 12: 5767–5800.

4 Caprioli, R.M. (2019). Imaging mass spectrometry: a perspective. *J. Biomol. Tech.* 30: 7–11.

5 Vestal, M.L. and Campbell, J.M. (2005). Tandem time-of-flight mass spectrometry. *Methods Enzymol.* 402: 79–108.

6 Yu, W., Vath, J.E., Huberty, M.C., and Martin, S.A. (1993). Identification of the facile gas-phase cleavage of the Asp-Pro and Asp-Xxx peptide bonds in matrix-assisted laser desorption time-of-flight mass spectrometry. *Anal. Chem.* 65: 3015–3023.

7 Lin, M., Campbell, J.M., Mueller, D.R., and Wirth, U. (2003). Intact protein analysis by matrix-assisted laser desorption/ionization tandem time-of-flight mass spectrometry. *Rapid Commun. Mass Spectrom.* 17: 1809–1814.

8 Liu, Z. and Schey, K.L. (2005). Optimization of a MALDI TOF-TOF mass spectrometer for intact protein analysis. *J. Am. Soc. Mass. Spectrom.* 16: 482–490.

9 Gu, C., Tsaprailis, G., Breci, L., and Wysocki, V.H. (2000). Selective gas-phase cleavage at the peptide bond C-terminal to aspartic acid in fixed-charge derivatives of Asp-containing peptides. *Anal. Chem.* 72: 5804–5813.

10 Herrmann, K.A., Wysocki, V.H., and Vorpagel, E.R. (2005). Computational investigation and hydrogen/deuterium exchange of the fixed charge derivative tris(2,4,6-trimethoxyphenyl) phosphonium: implications for the aspartic acid cleavage mechanism. *J. Am. Soc. Mass. Spectrom.* 16: 1067–1080.

11 Rozman, M. (2007). Aspartic acid side chain effect-experimental and theoretical insight. *J. Am. Soc. Mass. Spectrom.* 18: 121–127.

12 Resemann, A., Wunderlich, D., Rothbauer, U. et al. (2010). Top-down de Novo protein sequencing of a 13.6 kDa camelid single heavy chain antibody by matrix-assisted laser desorption ionization-time-of-flight/time-of-flight mass spectrometry. *Anal. Chem.* 82: 3283–3292.

13 Suckau, D. and Resemann, A. (2003). T3-sequencing: targeted characterization of the N- and C-termini of undigested proteins by mass spectrometry. *Anal. Chem.* 75: 5817–5824.

14 Kelleher, N.L., Zubarev, R.A., Bush, K. et al. (1999). Localization of labile posttranslational modifications by electron capture dissociation: the case of gamma-carboxyglutamic acid. *Anal. Chem.* 71: 4250–4253.

15 Mirgorodskaya, E., Roepstorff, P., and Zubarev, R.A. (1999). Localization of O-glycosylation sites in peptides by electron capture dissociation in a Fourier transform mass spectrometer. *Anal. Chem.* 71: 4431–4436.

16 Zubarev, R.A., Horn, D.M., Fridriksson, E.K. et al. (2000). Electron capture dissociation for structural characterization of multiply charged protein cations. *Anal. Chem.* 72: 563–573.

17 Syka, J.E., Coon, J.J., Schroeder, M.J. et al. (2004). Peptide and protein sequence analysis by electron transfer dissociation mass spectrometry. *Proc. Natl. Acad. Sci. U S A.* 101: 9528–9533.

18 Medzihradszky, K.F., Campbell, J.M., Baldwin, M.A. et al. (2000). The characteristics of peptide collision-induced dissociation using a high-performance MALDI-TOF/TOF tandem mass spectrometer. *Anal. Chem.* 72: 552–558.

19 Fagerquist, C.K. (2020). Polypeptide backbone cleavage on the C-terminal side of asparagine residues of metastable protein ions analyzed by MALDI-TOF-TOF-MS/MS and post-source decay. *Int. J. Mass spectrom.* 457: 116433.

20 Sleno, L. and Volmer, D.A. (2004). Ion activation methods for tandem mass spectrometry. *J. Mass Spectrom.* 39: 1091–1112.

21 Chalmers, M.J., Hakansson, K., Johnson, R. et al. (2004). Protein kinase A phosphorylation characterized by tandem Fourier transform ion cyclotron resonance mass spectrometry. *Proteomics* 4: 970–981.

22 Patrie, S.M., Charlebois, J.P., Whipple, D. et al. (2004). Construction of a hybrid quadrupole/Fourier transform ion cyclotron resonance mass spectrometer for versatile MS/MS above 10 kDa. *J. Am. Soc. Mass. Spectrom.* 15: 1099–1108.

23 Song, H. and Hakansson, K. (2012). Electron detachment dissociation and negative ion infrared multiphoton dissociation of electrosprayed intact proteins. *Anal. Chem.* 84: 871–876.

24 Brodbelt, J.S., Morrison, L.J., and Santos, I. (2020). Ultraviolet photodissociation mass spectrometry for analysis of biological molecules. *Chem. Rev.* 120: 3328–3380.

25 Matthiesen, R. (2013). Algorithms for database-dependent search of MS/MS data. *Methods Mol. Biol.* 1007: 119–138.

26 DeHart, C.J., Fellers, R.T., Fornelli, L. et al. (2017). Bioinformatics analysis of top-down mass spectrometry data with ProSight Lite. *Methods Mol. Biol.* 1558: 381–394.

27 Demirev, P.A., Feldman, A.B., Kowalski, P., and Lin, J.S. (2005). Top-down proteomics for rapid identification of intact microorganisms. *Anal. Chem.* 77: 7455–7461.

28 Fagerquist, C.K., Garbus, B.R., Miller, W.G. et al. (2010). Rapid identification of protein biomarkers of *Escherichia coli* O157:H7 by matrix-assisted laser desorption ionization-time-of-flight-time-of-flight mass spectrometry and top-down proteomics. *Anal. Chem.* 82: 2717–2725.

29 Fagerquist, C.K., Garbus, B.R., Williams, K.E. et al. (2009). Web-based software for rapid top-down proteomic identification of protein biomarkers, with implications for bacterial identification. *Appl. Environ. Microbiol.* 75: 4341–4353.

30 Maus, A., Bisha, B., Fagerquist, C., and Basile, F. (2020). Detection and identification of a protein biomarker in antibiotic-resistant *Escherichia coli* using intact protein LC offline MALDI-MS and MS/MS. *J. Appl. Microbiol.* 128: 697–709.

31 Fagerquist, C.K. and Sultan, O. (2010). Top-down proteomic identification of furin-cleaved alpha-subunit of Shiga toxin 2 from *Escherichia coli* O157:H7 using MALDI-TOF-TOF-MS/MS. *J. Biomed. Biotechnol.* 2010: 123460.

32 Fagerquist, C.K. and Sultan, O. (2011). Induction and identification of disulfide-intact and disulfide-reduced beta-subunit of Shiga toxin 2 from *Escherichia coli* O157:H7 using MALDI-TOF-TOF-MS/MS and top-down proteomics. *Analyst* 136: 1739–1746.

33 Fagerquist, C.K. and Zaragoza, W.J. (2016). Bacteriophage cell lysis of Shiga toxin-producing *Escherichia coli* for top-down proteomic identification of Shiga toxins 1 & 2 using matrix-assisted laser desorption/ionization tandem time-of-flight mass spectrometry. *Rapid Commun. Mass Spectrom.* 30: 671–680.

34 Fagerquist, C.K. and Zaragoza, W.J. (2019). Top-down and middle-down proteomic analysis of Shiga toxin using MALDI-TOF-TOF mass spectrometry. *MethodsX* 6: 815–826.

35 Fagerquist, C.K., Zaragoza, W.J., and Carter, M.Q. (2019). Top-down proteomic identification of Shiga toxin 1 and 2 from pathogenic *Escherichia coli* using MALDI-TOF-TOF tandem mass spectrometry. *Microorganisms* 7: 488.

36 Fagerquist, C.K., Zaragoza, W.J., Lee, B.G. et al. (2019). Clinically-relevant Shiga toxin 2 subtypes from environmental Shiga toxin-producing *Escherichia coli* identified by top-down/middle-down proteomics and DNA sequencing. *Clin. Mass Spectrom.* 11: 27–36.

37 Fagerquist, C.K., Zaragoza, W.J., Sultan, O. et al. (2014). Top-down proteomic identification of Shiga toxin 2 subtypes from Shiga toxin-producing *Escherichia coli* by matrix-assisted laser desorption ionization-tandem time of flight mass spectrometry. *Appl. Environ. Microbiol.* 80: 2928–2940.

38 Fagerquist, C.K., Lee, B.G., Zaragoza, W.J. et al. (2019). Software for top-down proteomic identification of a plasmid-borne factor (and other proteins) from genomically sequenced pathogenic bacteria using MALDI-TOF-TOF-MS/MS and post-source decay. *Int. J. Mass spectrom.* 438: 1–12.

39 Fagerquist, C.K. and Rojas, E. (2021). Identification of antibacterial immunity proteins in *Escherichia coli* using MALDI-TOF-TOF-MS/MS and top-down proteomic analysis. *J. Vis. Exp.* 1–22.

40 Suckau, D., Resemann, A., Schuerenberg, M. et al. (2003). A novel MALDI LIFT-TOF/TOF mass spectrometer for proteomics. *Anal. Bioanal.Chem.* 376: 952–965.

41 Horn, D.M., Ge, Y., and McLafferty, F.W. (2000). Activated ion electron capture dissociation for mass spectral sequencing of larger (42 kDa) proteins. *Anal. Chem.* 72: 4778–4784.

42 Gross, D.S., Schnier, P.D., Rodriguez-Cruz, S.E. et al. (1996). Conformations and folding of lysozyme ions in vacuo. *Proc. Natl. Acad. Sci. U S A.* 93: 3143–3148.

43 Harrison, A.G. (1997). The gas-phase basicities and proton affinities of amino acids and peptides. *Mass Spectrom. Rev.* 16: 201–217.

44 Krause, E., Wenschuh, H., and Jungblut, P.R. (1999). The dominance of arginine-containing peptides in MALDI-derived tryptic mass fingerprints of proteins. *Anal. Chem.* 71: 4160–4165.

45 Fagerquist, C.K. and Dodd, C.E. (2022). Sequestration of the ionizing proton in singly charged metastable protein ions generated by MALDI. *Int. J. Mass Spectrom.* 471: 116736.

46 Michel, B. (2005). After 30 years of study, the bacterial SOS response still surprises us. *PLoS Biol.* 3: e255.

47 Johannes, L. and Romer, W. (2010). Shiga toxins–from cell biology to biomedical applications. *Nat. Rev. Microbiol.* 8: 105–116.

48 Boisen, N., Melton-Celsa, A.R., Scheutz, F. et al. (2015). Shiga toxin 2a and Enteroaggregative *Escherichia coli*–a deadly combination. *Gut Microbes.* 6: 272–278.

49 Melton-Celsa, A.R. (2014). Shiga toxin (Stx) classification, structure, and function. *Microbiol. Spectr.* 2, EHEC-0024-2013: 1–13.

50 Melton-Celsa, A.R. and O'Brien, A.D. (2014). New therapeutic developments against Shiga toxin-producing *Escherichia coli*. *Microbiol. Spectr.* 2: 1–14.

51 Fraser, M.E., Fujinaga, M., Cherney, M.M. et al. (2004). Structure of Shiga toxin type 2 (Stx2) from *Escherichia coli* O157:H7. *J. Biolumin. Chemilumin.* 279: 27511–27517.

4

Liquid Atmospheric Pressure (LAP) – MALDI MS(/MS) Biomolecular Profiling for Large-scale Detection of Animal Disease and Food Adulteration and Bacterial Identification

Cristian Piras[1] *and Rainer Cramer*[2]

[1] *Department of Health Sciences, Magna Graecia University, Catanzaro, Italy*
[2] *Department of Chemistry, University of Reading, Reading, UK*

4.1 Introduction

Ambient ionization mass spectrometry (AIMS) permits rapid, high-throughput and in situ analysis in biomedicine, pharmaceutics, and analytical sciences [1]. Analytes from solid, liquid, and gaseous biomatrices can be analyzed with very rapid or no sample preparation. Recent advances in AIMS include desorption electrospray ionization (DESI) [2] and direct analysis in real time (DART) [3], also evolving into many related branches that are now further explored for high-throughput analysis.

However, ambient ionization can also be achieved with the use of sample chromophores, absorbing laser irradiation. Atmospheric pressure (AP) MALDI produces results that are very similar to those obtained using vacuum MALDI, but with the advantage of interchangeability with ESI allowing the use of hybrid instruments for MS/MS experiments and the possibility to use liquid matrices. Liquid atmospheric pressure (LAP) MALDI is capable of generating multiply charged peptides and proteins [4] whereas conventional vacuum MALDI generates mainly singly charged ions.

Avoiding sample preparation and sample pre-fractionation is a possible further advantage of LAP-MALDI, which is useful for gaining speed and making this technique applicable to high- throughput analyses. However, the generated datasets, especially due to the creation of multiple charge states, become extremely complex for data processing and mining. For this reason, such a dataset requires advanced data-mining tools like machine learning algorithms [5]. These computational tools simplify the dataset providing the possibility to rapidly classify specimens of different quality or nature according to their mass spectral features that are linked to their molecular fingerprint.

The combination of LAP-MALDI MS and machine learning has lately been employed for bacterial identification [6], the detection of food adulteration [5], and detection of animal diseases [7].

Microbiological Identification using MALDI-TOF and Tandem Mass Spectrometry: Industrial and Environmental Applications, First edition. Edited by Haroun N. Shah, Saheer E. Gharbia, Ajit J. Shah, Erika Y. Tranfield, and K. Clive Thompson.

4.2 Background to LAP-MALDI MS

Most MS-based analytical methods involve time-consuming, laborious, and complex sample preparation and separation methods, e.g. liquid chromatography (LC), followed by a minimum of several minutes of mass spectrometric analysis time per sample in the case of ESI MS(/MS).

Sample preparation for LC-ESI MS/MS of proteinaceous samples typically requires enzymatic digestion (bottom-up protein analysis) which results in the partial loss of information about post-translational modifications. Moreover, subsequent analytical steps are also laborious and demanding from a computational point of view, particularly if the information about the relative expression/abundance of each protein species needs to be obtained [8]. Technological advances are, on the other hand, pushing the way toward the analysis of specific proteoforms via LC-ESI MS/MS, often through top-down approaches [9, 10]. However, these approaches still require an extremely clean sample and still lack efficiency (sensitivity), particularly in the case of complex protein mixtures.

MALDI MS profiling using axial linear TOF instrumentation, as in microbiological testing/biotyping, is substantially quicker using simpler sample preparation and it still provides an informative fingerprint of the complex mixtures of proteoforms in the analyzed specimens. For MALDI MS profiling of bacterial cultures, these mixtures typically consist of ribosomal proteins, displaying differences between the bacterial species. MALDI-TOF MS profiling started a new era for microbiology laboratories once the clinical protocols for quick sample preparation and adequate spectral libraries had been established. These advances and subsequent implementations in clinical diagnostics now allow the rapid and direct identification of microorganisms from clinical samples with a positive impact on public healthcare. Nonetheless, currently employed MALDI-TOF MS biotyping platforms suffer from some disadvantages, such as high shot-to-shot ion signal variability, extremely high non-analyte ion signals (particularly in the low m/z range), and the generally lower performance of axial linear TOF instruments with their relatively low mass accuracy and resolution. The use of the latter is closely connected to the low-charge state of the ions produced by conventional solid MALDI (mainly single charged), which generally prohibits the use of high-performing instruments with quadrupoles (e.g. Q-TOF) or ion traps (e.g. Orbitrap) for biomolecules of greater mass.

On the other hand, ESI used modern hybrid instruments can produce ions with higher charge states, which are typically detectable in m/z ranges < 2000, allowing the employment of high-performing instruments such Q-TOF and Orbitrap mass analysers. Moreover, higher-charge state ions are also more suitable for MS/MS analysis because of a superior fragmentation pattern when compared with singly charged ions [11].

Atmospheric pressure-MALDI [12] and other AP-LDI methods have recently been employed for generating (often multiply charged) analyte ions through laser irradiation [13, 14].

Among those, LAP-MALDI, with the synergistic effect of a chromophore, allows the formation of multiply charged protein and peptide ions. It is based on the use of truly liquid samples, thus providing superior sample homogeneity while keeping laser energies at very low, conventional MALDI levels. These are important advantages and are in contrast to other AP-LDI techniques such as laser spray ionization with its laser fluence range of 40–150 kJ/m^2 [15, 16], which typically causes extremely rapid depletion of the sample.

LAP-MALDI using UV lasers (LAP-UV-MALDI) cannot only be used with lower fluences but also permits substantially less sample consumption with a more durable and stable (shot-to-shot) ion yield [4] [17].

In LAP-MALDI MS, the application of a heated (up to 400°C) ion transfer tube enables high yields of multiply charged ion species. As with conventional MALDI MS, using chromophores such as 2,5-dihydroxybenzoic acid (DHB) or α-cyano-4-hydroxycinnamic acid (CHCA), pulsed (ns) UV laser irradiation with a maximum fluence of around 2 kJ m^{-2} facilitates the detection of intact peptide and protein ions. Figure 4.1 shows a LAP-UV-MALDI mass spectrum of MK-bradykinin (sequence: MKRPPGFSPFR), revealing singly, doubly, and triply charged analyte ions.

High-charge states greatly facilitate peptide fragmentation and thus peptide/protein sequencing. In Figure 4.2, the collision-induced dissociation (CID) MS/MS fragment spectra of MK-bradykinin is shown for the doubly and triply charged ion species.

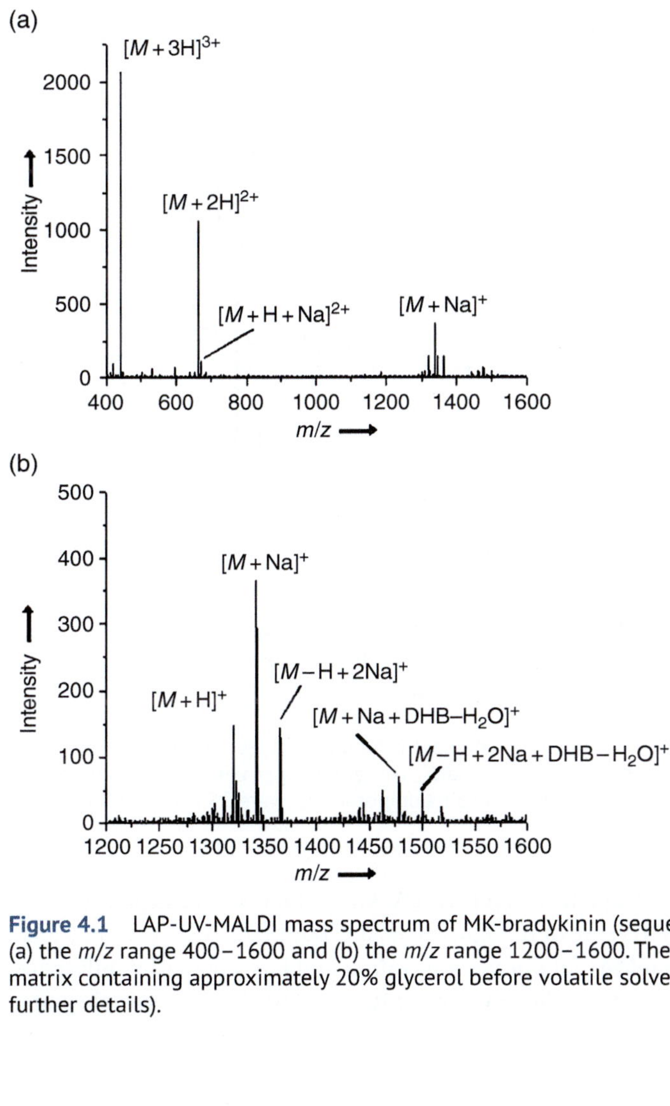

Figure 4.1 LAP-UV-MALDI mass spectrum of MK-bradykinin (sequence: MKRPPGFSPFR), showing (a) the *m/z* range 400–1600 and (b) the *m/z* range 1200–1600. The matrix was a DHB-based liquid matrix containing approximately 20% glycerol before volatile solvent evaporation (see [4] for further details).

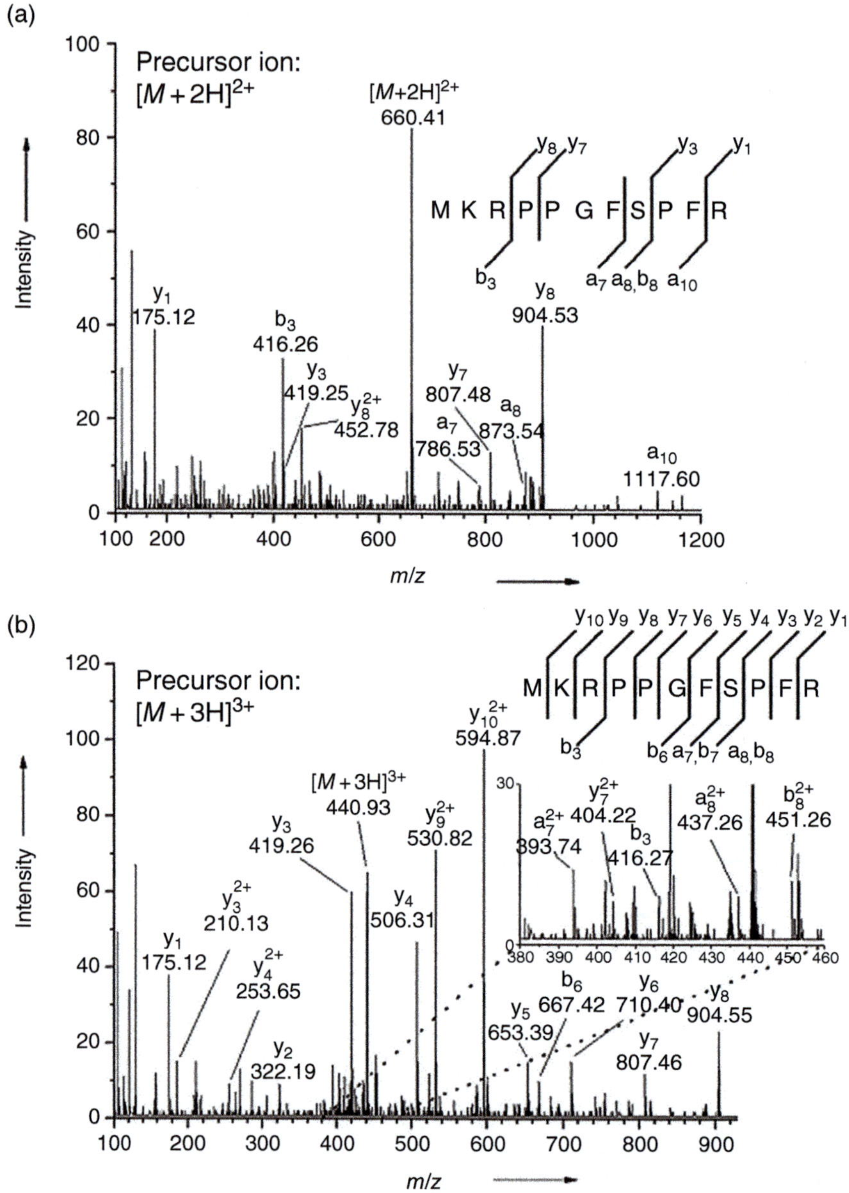

Figure 4.2 LAP-UV-MALDI CID MS/MS spectra of the doubly (a) and triply (b) protonated MK-bradykinin ions. The matrix was a DHB-based liquid matrix containing approximately 20% glycerol before the evaporation of volatile solvents. The CID collision potentials were 35 V (a) and 20 V (b) (see [4] for further details).

The general configuration and a picture of a LAP-UV-MALDI ion source is presented in Figure 4.3a and b, respectively. As shown in Figure 4.3a, it is built with a heated ion transfer tube (> 200°C) for improved MALDI plume desolvation and has a potential difference of approximately 3 kV applied between the ion transfer tube and the MALDI sample plate.

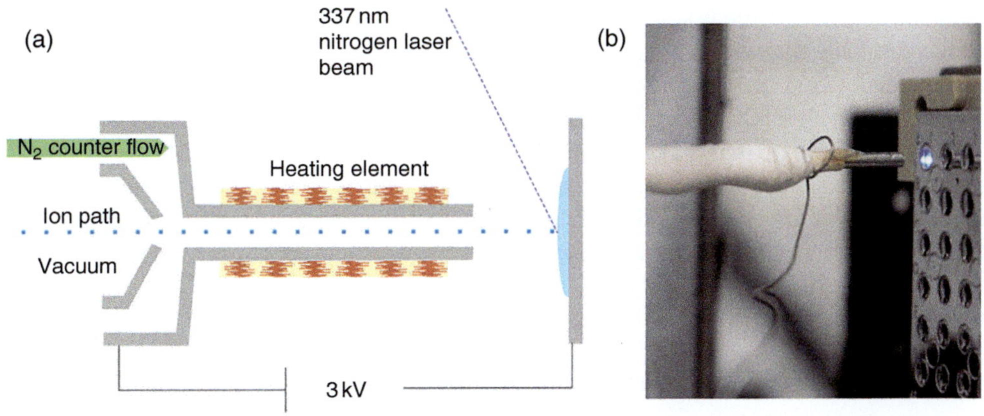

Figure 4.3 (a) Schematic of a LAP-UV-MALDI source. (b) Close-up image of the LAP-UV-MALDI sample irradiation, showing the ion transfer tube heated by a resistance wire opposite the sample plate with several liquid MALDI sample droplets. The irradiated sample droplet shows strong fluorescence.

The sample is irradiated with a UV laser such as a pulsed nitrogen laser (337 nm, MNL 103 LD; LTB Lasertechnik GmbH, Berlin, Germany), whose beam is typically focused down to a diameter of approximately 100–150 μm onto the surface of the sample droplet and centered on the axis of the ion transfer tube. The laser energy/fluence and pulse repetition rate can vary according to the experiment to be performed but is generally close to conventional solid MALDI levels.

Typically, the configuration in Figure 4.3 can produce an extremely heterogenous spectrum profile that includes metabolites (or other small molecules), lipids, peptides, and entire proteins. Figure 4.4 presents an example of cow's milk analyzed with the previously described ion source coupled with a high-performing Q-TOF hybrid instrument [8].

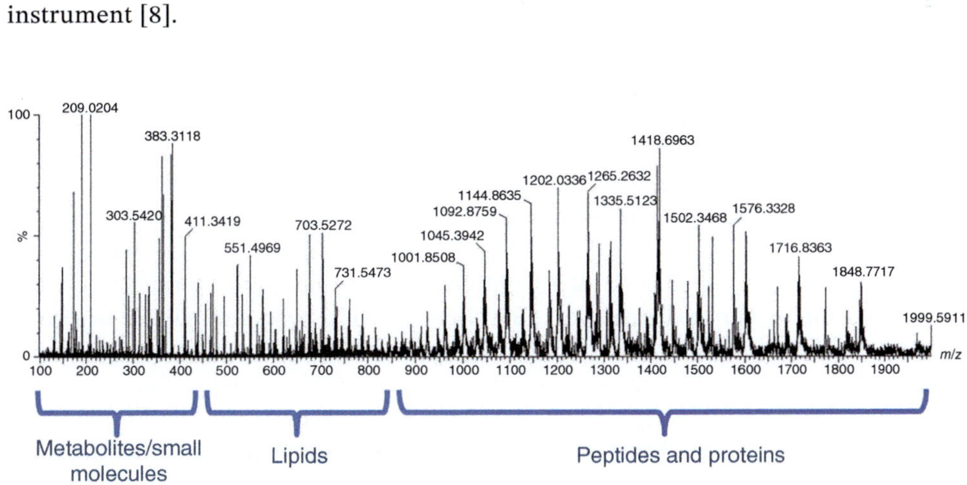

Figure 4.4 LAP-UV-MALDI MS spectrum (*m/z* 100–2000) obtained by acquiring data over one minute.

Such a spectral profile is suitable for multivariate analysis that takes into consideration all the variables detected in complex biological samples. Algorithms such as linear discriminant analysis (LDA) or partial least-squares discriminant analysis (PLS-DA) can be applied to provide a rapid or on-time classification of the analyzed samples [5, 7, 18].

This extremely fast analysis can be achieved through algorithms designed to bin together the intensities of the different *m/z* intervals which can be subsequently analyzed with unsupervised or supervised multivariate analysis. Supervised analytical methods can be applied through the analysis of a representative sample set and validated by the analysis of another independent sample set [5, 7].

4.3 Bacterial Identification by LAP-MALDI MS

Traditional bacteriological methods like culture growth, colony observation, staining methods, and biochemical assays have been widely replaced by MS biotyping-based approaches. The latter are more advantageous as they do not require any major additional time after the primary culture [19]. Among these, the MALDI MS biotyping approach is probably the most employed [20]. This technique is now invaluable in the clinical practice and capable of providing bacterial identification in one hour after primary culture growth [20]. The mass spectral profile is obtained by recording mostly singly charged protein ions in the *m/z* range of 2000–12 000. The classification of these profiles can recognize bacterial species after querying the available mass spectral profile databases and has been proven successful at detecting even antimicrobial resistance (AMR) profiles [21]. However, this technique does not take into account the lower *m/z* ranges (≤2000) where lipids and other small molecules are present and potentially useful for classification. LAP-MALDI MS, on the other hand, is capable of efficiently detecting small biomolecules such as lipids in the low *m/z* range with the advantage of including peptides and proteins/proteoforms as multiply charged ion species, thus recording a highly informative spectrum as mentioned in the previous section.

In Figure 4.5a, a solid AP-UV-MALDI mass spectrum of an *Escherichia coli* sample is presented, using the same simple analyte extraction as for LAP-UV-MALDI MS. In this spectrum, it is possible to see MALDI matrix-related ions but it is not possible to detect peptides or proteins. By contrast, the employment of LAP-UV-MALDI for the analysis of the same sample generates a panel of molecules, including lipids (*m/z* 500–850) and larger peptides detected as multiply charged ions with a deconvoluted mass of ~7180 Da (see Figure 4.5b).

This type of mass spectral fingerprint is highly informative and specific for each bacterial species. With the proper analytical workflow, it can provide rapid and reliable bacterial identification at the species level.

Escherichia coli, *Klebsiella pneumoniae*, *Campylobacter jejuni*, *Staphylococcus aureus*, *Staphylococcus epidermidis*, *Streptococcus pyogenes*, *Lactobacillus brevis*, *Enterococcus faecalis*, *Enterococcus hirae*, and *Pseudomonas aeruginosa* all have lipid profiles which are very different and allow the rapid discrimination between these bacterial species. In Figure 4.6, the representative spectral profiles of each bacterial species and the differences between their profiles can be seen.

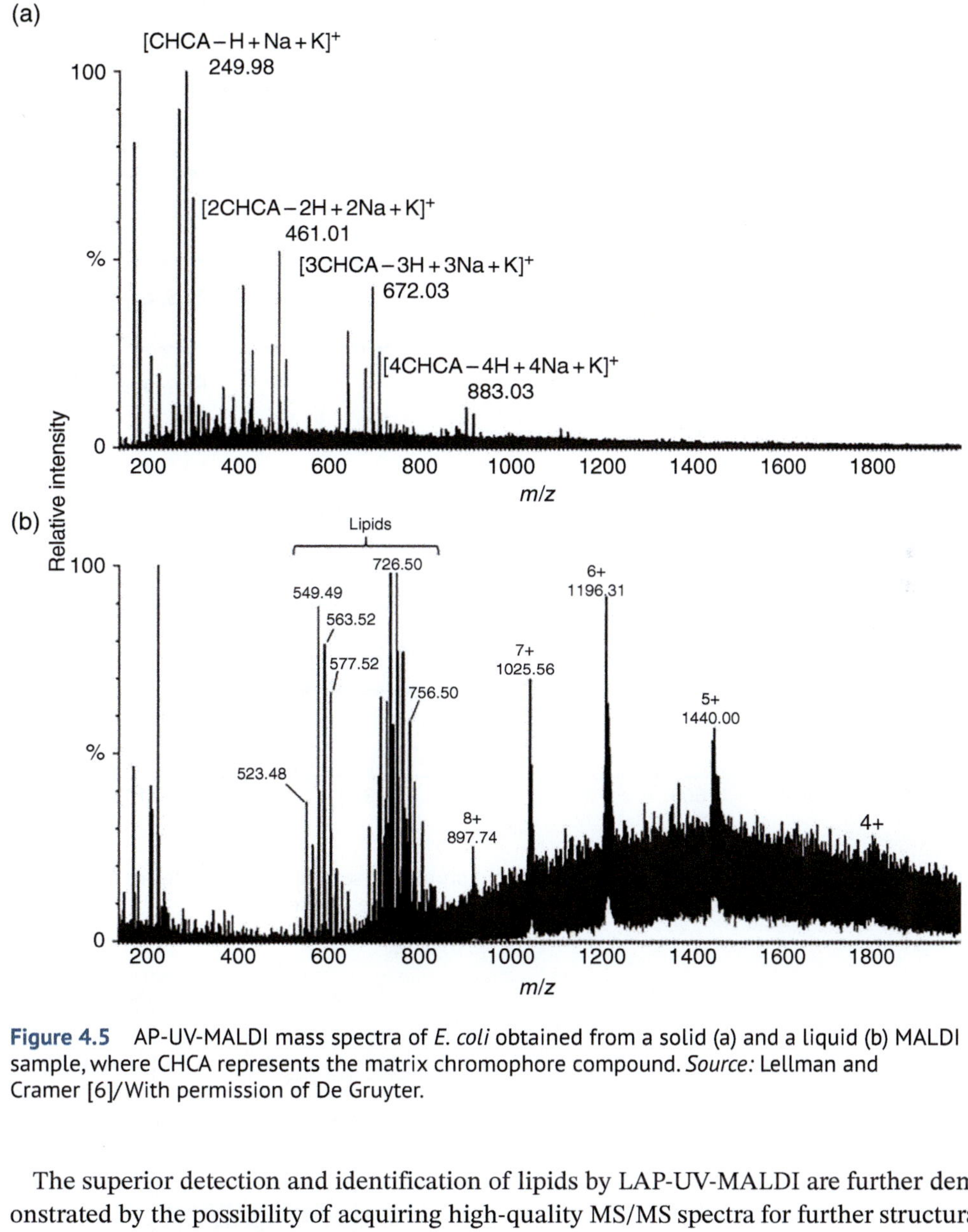

Figure 4.5 AP-UV-MALDI mass spectra of *E. coli* obtained from a solid (a) and a liquid (b) MALDI sample, where CHCA represents the matrix chromophore compound. *Source:* Lellman and Cramer [6]/With permission of De Gruyter.

The superior detection and identification of lipids by LAP-UV-MALDI are further demonstrated by the possibility of acquiring high-quality MS/MS spectra for further structural elucidation, as shown in Figure 4.7.

The previously mentioned 10 different bacterial species can easily be classified by the application of principal component analysis (PCA) as shown in Figure 4.8. A multivariate supervised statistics model using all LAP-UV-MALDI MS profile features over the *m*/*z* range 400–1100 yielded a correct classification rate of 97.78% for the above study [6].

In summary, the above LAP-UV-MALDI MS profiling approach of bacterial cultures provides the possibility of rapidly biotyping bacterial species after the primary culture step and a sample preparation of around five minutes [6].

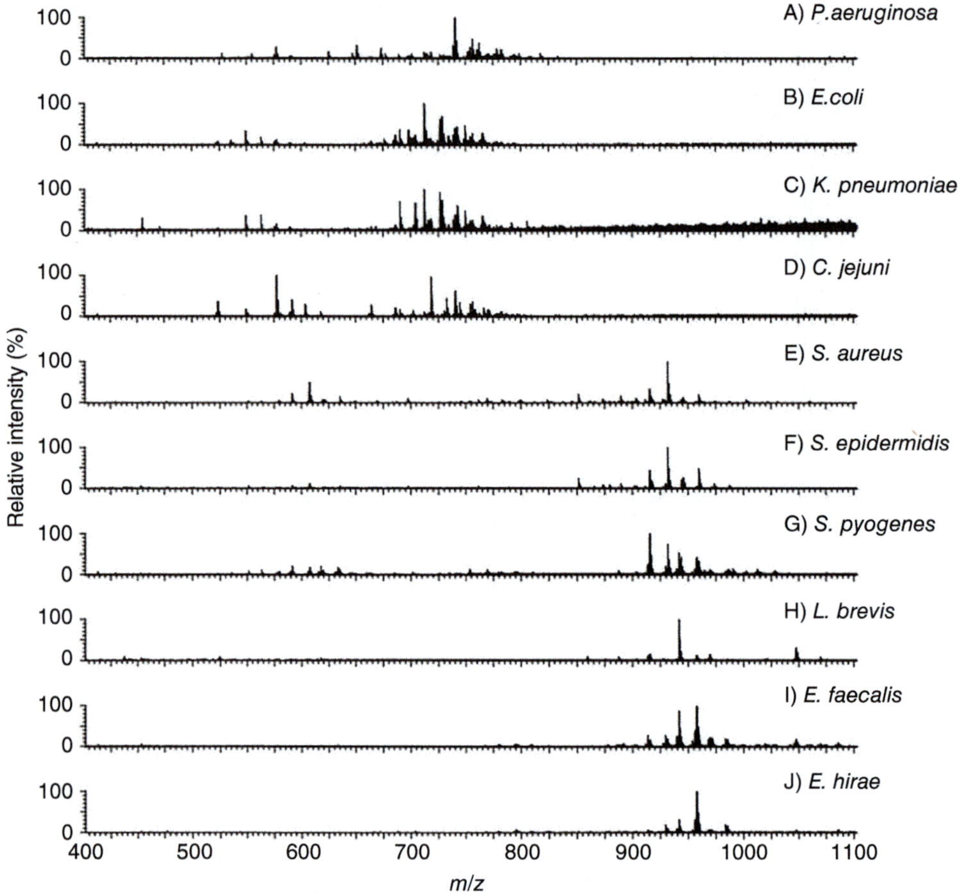

Figure 4.6 LAP-UV-MALDI mass spectra of the lipid profiles of several bacterial species in positive ion mode. *Source:* Lellman and Cramer [6]/With permission of De Gruyter.

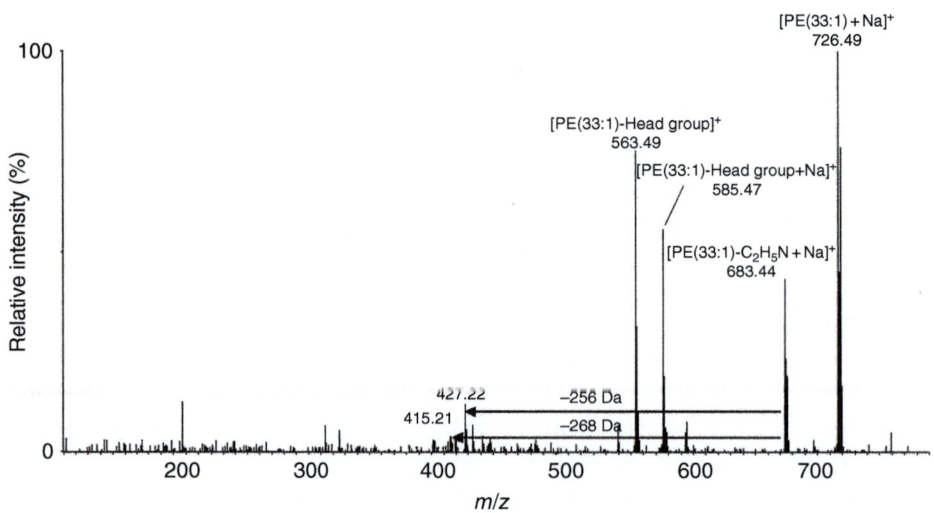

Figure 4.7 LAP-UV-MALDI MS/MS spectrum of the precursor ion at *m/z* 726.5, detected in the lipid MS profile of *E. coli*. *Source:* Lellman and Cramer [6]/With permission of De Gruyter.

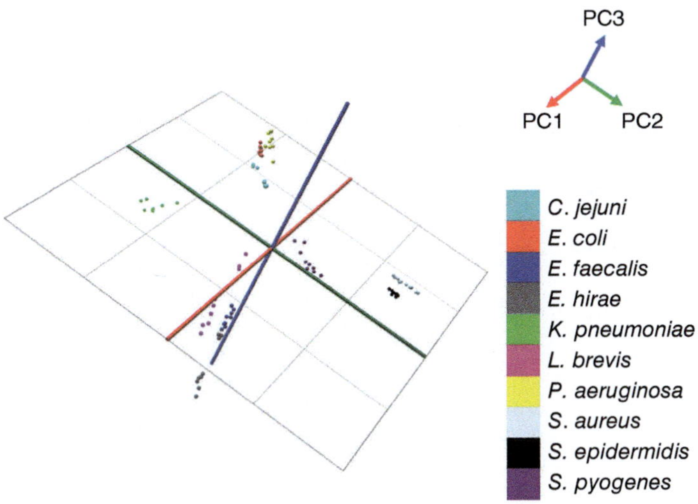

Figure 4.8 Plot of the PCA data points for the first three principal components (PC1, PC2, and PC3) using LAP-MALDI MS profile data from 10 bacterial species. *Source:* Lellman and Cramer [6]/With permission of De Gruyter.

4.4 Food Adulteration and Milk Quality Analysis by LAP-MALDI MS

Food production and retail need rapid, reliable, and cost-effective methods for the quality and safety assessment of food. Among all the frauds in food supply, milk adulteration probably represents the most common, with the addition of less expensive and more available cow's milk to milk of other species [22]. This adulteration can be life-threatening in the case of individuals being affected by cow's milk allergies. Rapid analysis of milk quality is also necessary for quality assessment before milk is further processed to other dairy products and before milk's use in animal production. For example, good quality colostrum administration during the early stages of the animal's life is necessary to rapidly transfer passive immunity to the offspring to avoid pathogen infection. The rapid evaluation of colostrum quality is therefore important before its distribution, sale, or administration [23].

In the above context, conventional solid MALDI-TOF MS analysis has already been successfully used for the detection of cow's milk as contaminant or adulterant (see [24] for a review). As recently shown, LAP-MALDI MS can also be applied to the detection of food adulterations [5].

Based on lipid profiles only, LAP-MALDI MS can classify milk from cows, camels, goats and sheep with 100% accuracy, using data acquisitions of 10 seconds per sample (Figure 4.9a) and can provide a list of ions that are species-specific for this classification, and thus possibly to be used as specific biotype markers (Figure 4.9b).

Results like these are achievable with only a single 1 : 5 dilution of the milk sample in hexane/isopropanol (3 : 2, v/v) and subsequent vortexing for five seconds. After this simple sample preparation step, the sample only needs to be mixed with the liquid support matrix for LAP-MALDI MS and analyzed for 10 seconds [5].

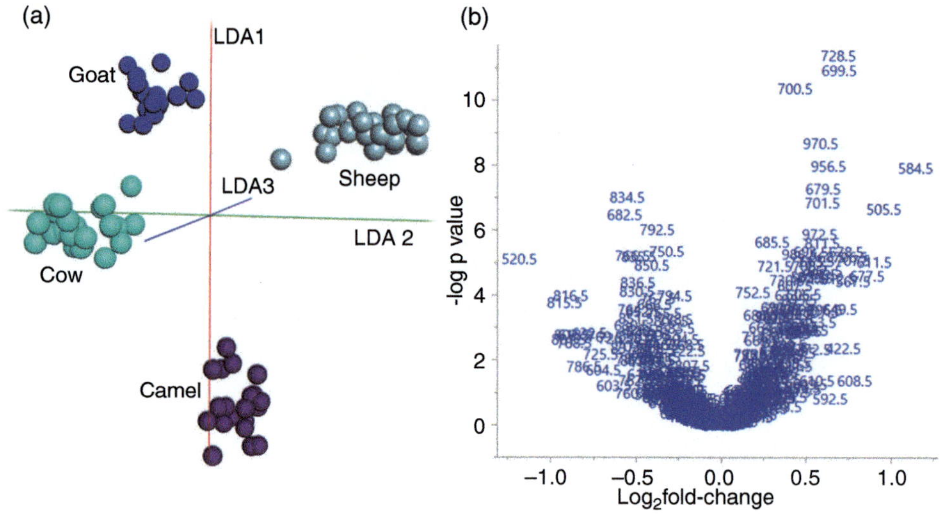

Figure 4.9 (a) Linear discriminant analysis (LDA) plot obtained with Abstract Model Builder (AMX) from the LAP-MALDI MS profile data of sample sets of four different mammals, using the *m/z* range 400–1000. (b) Volcano plot displaying the *m/z* values of ions linked to cow's milk classification.

With the addition of a precipitation step to the sample preparation, and with an MS scan time of 30 seconds, it is possible to detect up to 5% cow's milk in goat's milk with 92.5% sensitivity and 94.5% specificity (Table 4.1) [5]. A higher percentage of 10% cow's milk in goat's milk is detectable with a sensitivity of 99.2% and a specificity of 99.1% (Table 4.2).

The different sample preparation allows the detection of multiply charged proteins of milk components, improving the correct classification rate. In Figure 4.10, two representative spectra of cow's milk and goat's milk, using a data acquisition time of five minutes, are shown to highlight the differences between the two species. These differences, also within the mixtures of milk from the two species, can be easily detected in 30 seconds of acquisition time as used for the data of Tables 4.1 and 4.2.

Table 4.1 Confusion matrix for the detection of 5% cow's milk in goat's milk using an LDA prediction model based on LAP-MALDI MS profiles. The "condition" columns indicate the real nature of the samples analyzed. The rows "test results" indicate the results obtained by the classification model. Positives are defined as milk samples with 5% cow's milk. Negatives are defined as milk samples without cow's milk.

		Condition		
		Positive (5% cow's milk in goat milk)	**Negative (100% goat's milk)**	
Test results	Positive (5% cow's milk in goat's milk)	TP = 98	FP = 5	Positive predictive value ~ 95.1%
	Negative (100% goat's milk)	FN = 8	TN = 107	Negative predictive value ~ 93.0%
		Sensitivity ~ 92.5%	**Specificity ~ 94.5%**	

Abbreviations: TP, true positives; FP, false positives; FN, false negatives; TN, true negatives.

Table 4.2 Confusion matrix for the detection of 10% cow's milk in goat's milk using an LDA prediction model based on LAP-MALDI MS profiles. The "condition" columns indicate the real nature of the samples analyzed. The rows "test results" indicate the results obtained by the classification model. Positives are defined as milk samples with 10% cow's milk. Negatives are defined as milk samples without cow's milk.

		Condition		
		Positive (10% cow's milk in goat milk)	**Negative (100% goat's milk)**	
Test results	Positive (10% cow's milk in goat's milk)	TP = 111	FP = 1	Positive predictive value ~ 99.2%
	Negative (100% goat's milk)	FN = 1	TN = 117	Negative predictive value ~ 99.1%
		Sensitivity ~ 99.2%	**Specificity ~ 99.1%**	

Abbreviations: TP, true positives; FP, false positives; FN, false negatives; TN, true negatives.

For instance, the ion signal of protein ions (e.g. at *m/z* 1265 and 1335) increases with the concentration of cow's milk, demonstrating that these peaks are clearly correlated with cow's milk addition. The inset in Figure 4.10a magnifies the *m/z* region of 1190–1220 with strong signals of highly charged analyte ions. These ions represent proteoforms of bovine β-casein with 20 positive charges [25], which are not present in the mass spectrum of goat's milk (insert of Figure 4.10b).

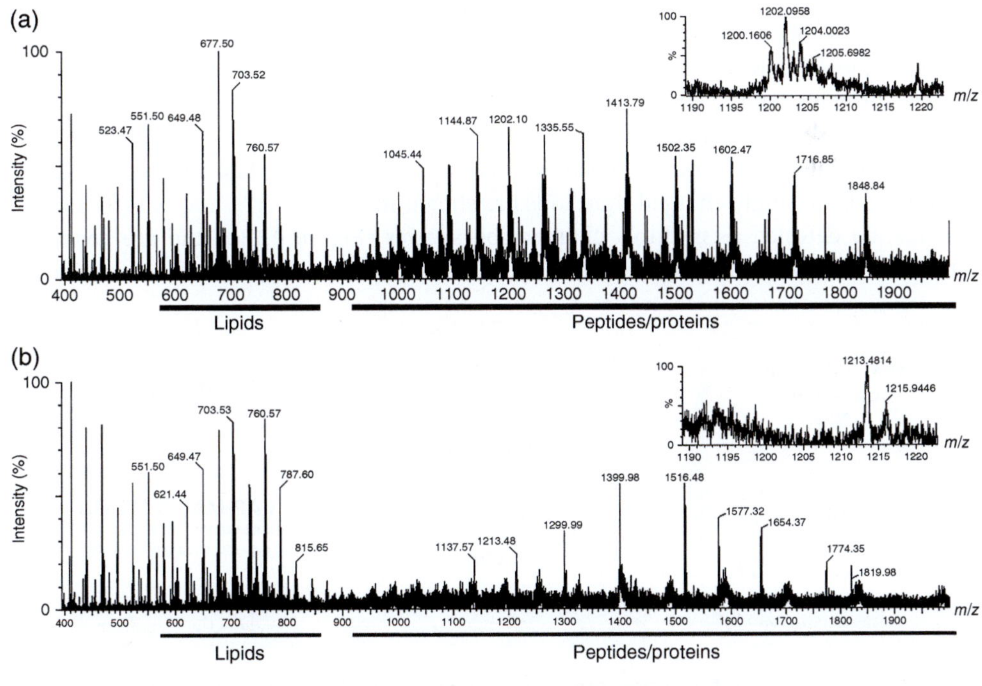

Figure 4.10 LAP-MALDI mass spectra of cow's milk (a) and goat's milk (b), each acquired over five minutes.

The application of machine learning tools greatly facilitates and accelerates the correct recognition of the fingerprints typical of each sample or sample mixture, allowing on-time scanning and class assignment.

4.5 Animal Disease Detection by LAP-MALDI MS

Large-scale screening for early detection of animal diseases is a key feature for improving animal (and human) welfare under the "One Health" approach. Diagnostic techniques should be cost-effective, user-friendly, and versatile. In this context, MALDI MS profiling has been successfully employed in clinical microbiology, but less so in early detection of disease [26].

Liquid atmospheric pressure-MALDI MS profiling of raw cow's milk collected during routine milking has recently been shown to detect bovine mastitis two days before its clinical manifestation with a sensitivity of up to 70% and a specificity of up to 100%. This highly accurate, pre-symptomatic detection was demonstrated by using a large set of milk samples collected weekly over six months from approximately 500 dairy cows [7].

The sample preparation for these analyses is very rapid and can be performed in 96-well microtiter plates. The required starting amount of milk is only 20 μl, which is prepared using a simple protein precipitation step with 100 μl of 5% trichloroacetic acid (TCA). Afterwards, the pellet can be solubilized with a mixture of water, acetonitrile and isopropanol. The obtained sample is then ready to be analyzed with LAP-MALDI MS (Figure 4.11).

With this sample preparation protocol, multiple samples can be analyzed on the same MALDI sample plate, and the signal drop in the total ion chromatogram separates the different samples (Figure 4.12). As before, the spectral profiles include metabolites, lipids, and peptides/proteins, and LAP-MALDI MS allows the employment of hybrid instruments with high mass accuracy and resolution that can facilitate the assignment and identification of potential biomarkers. For the above case, the mass spectral data and possible identifications of the singly charged lipid ions are shown in Figure 4.13a and b, respectively. Proteins are detected as multiply charged ion species (Figure 4.13c) and are generally abundant milk proteins (Figure 4.13d) but can be truncated or modified to specific

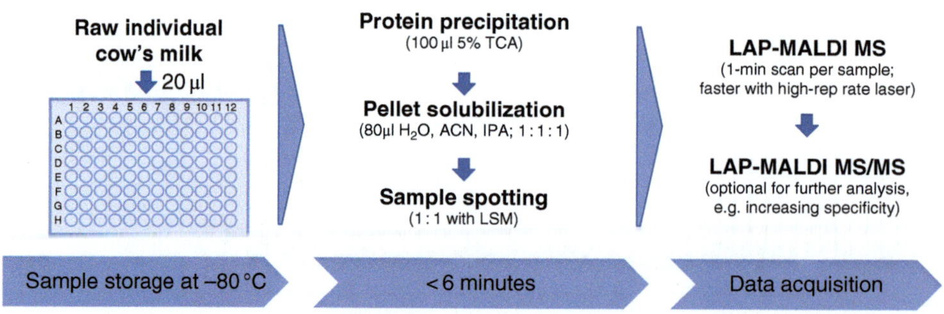

Figure 4.11 Workflow for the sample preparation and LAP-MALDI MS analysis of raw cow's milk (TCA, trichloroacetic acid; ACN, acetonitrile; IPA, isopropanol; LSM, liquid support matrix).

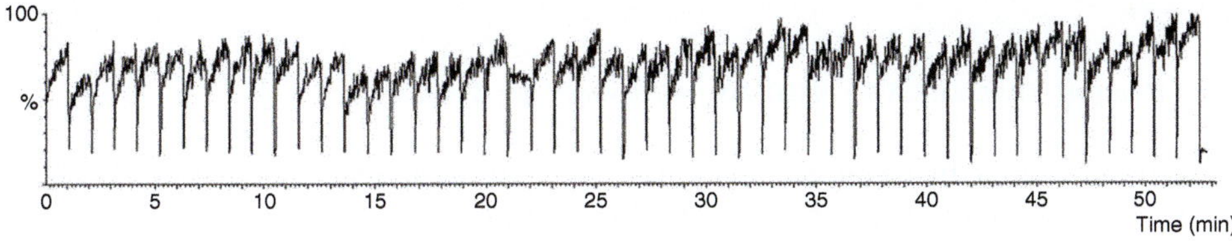

Figure 4.12 Total ion chromatogram recorded for 50 separate milk samples analyzed by LAP-MALDI MS.

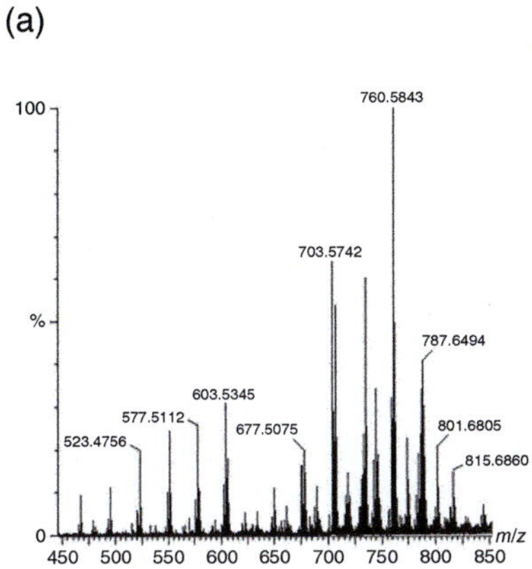

(b)

Ion	Matched mass	Delta	Name	Formula	Ion
523.4726	523.4721	0.0005	DG 30:0	$C_{33}H_{64}O_5$	$[M+H-H_2O]^+$
551.5032	551.5034	0.0002	DG O-32:2	$C_{35}H_{66}O_4$	$[M+H]^+$
577.5186	577.5190	0.0004	DG 34:1	$C_{37}H_{70}O_5$	$[M+H-H_2O]^+$
603.5339	603.5352	0.0013	DG 36:4	$C_{39}H_{70}O_4$	$[M+H]^+$
703.5654	703.5748	0.0094	SM 34:1;O2	$C_{39}H_{79}N_2O_6P$	$[M+H]^+$
706.5406	706.5381	0.0025	PC 30:0	$C_{38}H_{76}NO_8P$	$[M+H]^+$
734.5678	734.5694	0.0016	PC 32:0	$C_{40}H_{80}NO_8P$	$[M+H]^+$
760.5860	760.5851	0.0009	PC 34:1	$C_{42}H_{82}NO_8P$	$[M+H]^+$
787.6425	787.6615	0.0190	SM 40:1	$C_{45}H_{91}N_2O_6P$	$[M+H]^+$
801.6801	801.6844	0.0043	SM 41:1;O2	$C_{46}H_{93}N_2O_6P$	$[M+H]^+$
815.6811	815.7000	0.0189	SM 42:1;O2	$C_{47}H_{95}N_2O_6P$	$[M+H]^+$

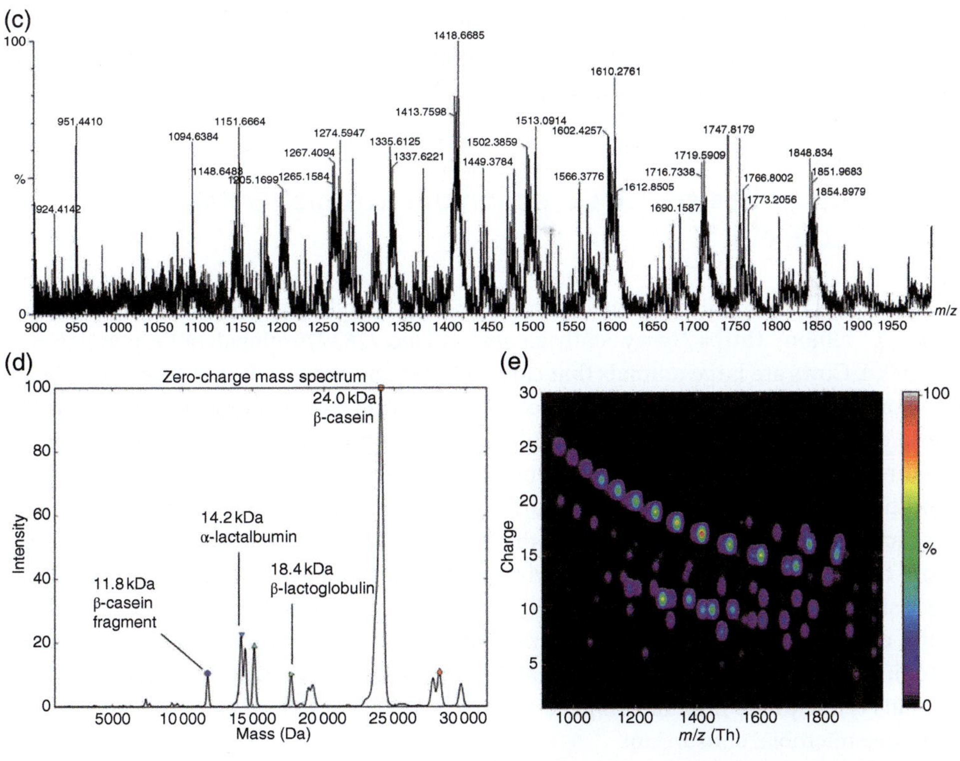

Figure 4.13 (a) LAP-MALDI mass spectrum of dairy milk, displaying the *m/z* range 450–850. (b) Putative identification details for the most abundant lipid species detected in the mass spectrum shown in (a) (based on mass accuracy, literature, and occurrence in the MCDB database at mcdb.ca). (c) LAP-MALDI mass spectrum of dairy milk, displaying the *m/z* range 900–2000, where most proteins and peptides are detected as multiply charged ions. (d) Zero-charge mass spectrum obtained from the deconvolution of the mass spectrum shown in (c). (e) *m/z* versus charge ion intensity signal plot showing the charge train distributions for the ion signals of the mass spectrum shown in (c).

proteoforms, resulting in a valuable biomarker. Their representative charge state distributions are like those typically obtained by ESI (Figure 4.13e).

The spectral profiles can be subsequently analyzed by the application of multivariate statistics for sample stratification and classification. Machine learning tools allow this analysis with good efficiency with the possibility to train class prediction models, which can then be validated by the analysis of a set of unknown samples. For the above study, the results obtained from a set of unknown samples yielded a sensitivity of > 56% and a specificity of 100% for the detection of pre-clinical mastitis 48 hours before the clinical event [7]. Figure 4.14a and b show the biplots of the multivariate models applied for the recognition of unknown samples in clinical and pre-clinical mastitis, respectively.

One of the most important molecules detected for the classification of mastitic versus non-mastitic samples is an α-s_1-casein fragment that contains the isracidin sequence. This peptide contributes with all its charge states to the correct classification of clinical and pre-clinical mastitic samples. Figure 4.14c and d show the relative abundance of the isracidin-containing peptide ion among the experimental groups, which also include samples from cows with high somatic cell counts (hSCC) in addition to low somatic cell counts (lSCC).

As discussed before, LAP-MALDI improves MS/MS analysis and thus the sequencing of possible protein (or other molecular) biomarkers. The isracidin-containing peptide sequence of the detected biomarker, together with its MS/MS spectrum obtained from milk as prepared for this study, is shown in Figure 4.15.

4.6 Antibiotic Resistance Detection of Microbial Consortia by LAP-MALDI MS

Throughout 2012–2020, the estimated number of cows worldwide was constantly stable at around 1 billion (https://www.statista.com/statistics/263979/global-cattle-population-since-1990). Cows are large animals that carry different biotopes rich in diverse microbes/microbiomes. Therefore, they can be carriers of high numbers of bacteria with AMR. AMR bacteria commonly do not present a problem for dairy products because of the hygiene procedures adopted during food processing before sale (pasteurization, heat treatments, microfiltration, and fermentation, e.g. during cheese-making) [27]. However, inter- and intraspecific recombination may lead to the creation of single- and multi-drug-resistant bacteria that might be harmful for the environment and ultimately human health [28–30]. Once resistance genes are introduced inside an organism, it is difficult to track their flow because of their high rate of genetic recombination [31]. Therefore, it also becomes difficult to link this genetic flow to an eventual AMR infection that occurs in humans or animals. This is due to the high rate of gene transfer and exchange between bacterial species of the same microbial consortium.

The detection of the bacterial AMR profile of a bacterial consortium can be done only through time-consuming and very expensive work that requires the employment of extremely well-trained personnel at all stages of the experimental procedures. This can be achieved through metagenomics and metaproteomics studies by searching for gene or

(a)

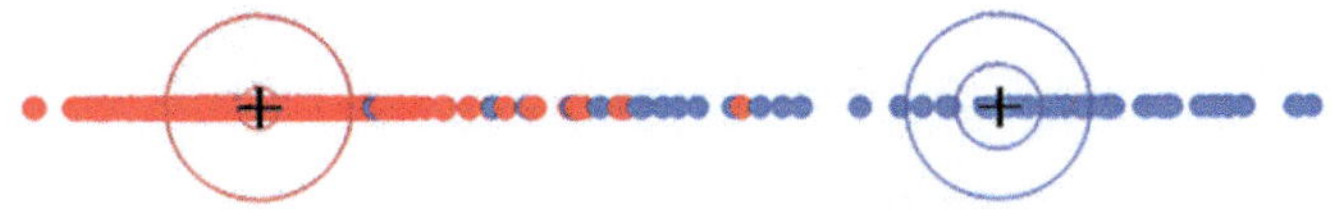

	Training dataset		Test dataset	
Class	Control	Mastitis	Control	Mastitis
Model prediction: Control	164	4	163	1
Model prediction: Mastitis	1	36	1	19

(b)

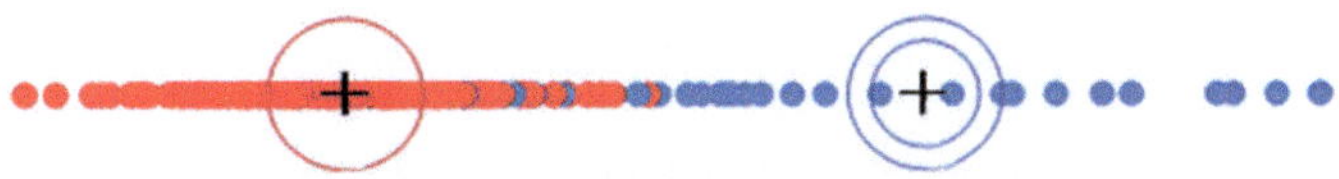

	Training dataset		Test dataset	
Class	Control	Pre-clinical mastitis	Control	Pre-clinical mastitis
Model prediction: Control	312	2	102	7
Model prediction: Mastitis	1	15	0	9

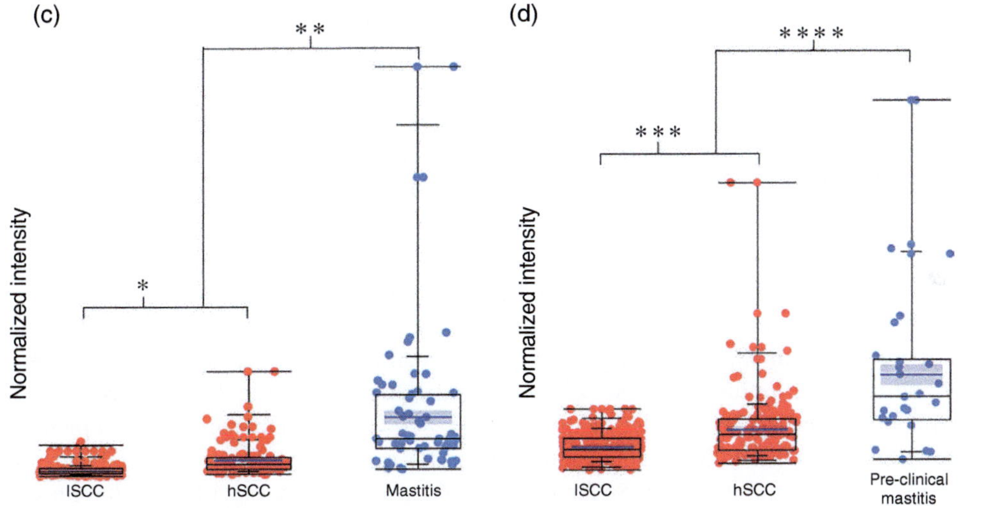

Figure 4.14 (a, b) Graphic representation (biplot) of the LDA model and the confusion matrix for its prediction of clinical mastitis (a) and pre-clinical mastitis (b). The observations and the multivariate means of each group are represented as points on the biplot. They are expressed in terms of the first two canonical variables. The point corresponding to each multivariate mean is denoted by a plus (+) marker. A 95% confidence level circle is plotted for each mean. If two groups differ significantly, the confidence circles tend not to intersect. A circle denoting a 50% contour is also plotted for each group. This depicts a region in the space of the first two canonical variables that contains ~50% of the observations, assuming normality. (c, d). Normalized signal intensities of the AMX mass bin at *m/z* 724.5, which contains the $[M + 6H]^{6+}$ ion species of isracidin-like peptide, using the data from the classification of clinical (c) and pre-clinical (d) mastitis. $^{*}P = 8.04 \times 10^{18}$; $^{**}P = 4.84 \times 10^{34}$, fold-change = 5.81; $^{***}P = 3.43 \times 10^{17}$; $^{****}P = 4.09 \times 10^{26}$, fold change = 2.12. See text for more details.

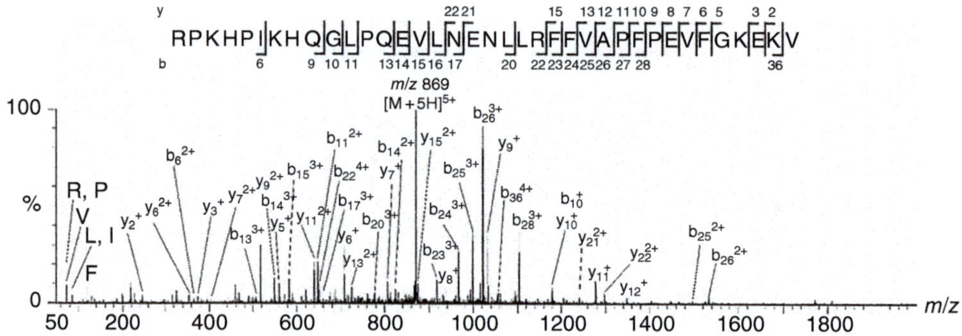

Figure 4.15 LAP-MALDI-CID MS/MS spectra of MS profile ion peaks. MS/MS spectrum of the diagnostically important precursor ion [M + 5H]5+ (~ *m/z* 869). The sequence is displayed above the spectrum and indicate N-terminal fragments (R1- V37) derived from α-s1-casein (P02662).

gene products involved in AMR. Metaproteomics rather than metagenomics may be more accurate in detecting the biological functions, pathways, and metabolisms accomplished by a bacterial consortium because of the measurement of the final gene products, the biological effectors.

However, such a technique is practically not applicable because it is too time-consuming, negating any timely pharmacological treatment or isolation of the animal. On the other hand, MALDI MS profiling coupled with machine learning can rapidly detect a molecular fingerprint representative of a bacterial consortium, potentially with molecular features of AMR. As shown in the earlier sections, LAP-MALDI MS analysis can easily classify unknown samples according to the measured mass spectral profile of metabolites, lipids, and proteins/peptides [5]. However, it is also possible for LAP-MALDI MS to detect, in a maximum of four hours, the cleavage of ampicillin (and other antibiotics) induced by cow's milk bacterial consortia. As shown in Figure 4.15, LAP-MALDI MS rapidly detects both the activity of spiked β-lactamase at different concentrations in bovine milk and the endogenous β-lactamase activity of cow's milk bacterial consortia (Figure 4.16). The simplicity of such an approach can be exploited using machine learning-aided auto-recognition or, as in the presented case, a manual computational tool that measures the ratio between the decarboxylated hydrolyzed (*m/z* 324) and intact (*m/z* 350) ampicillin (AMP). Such a methodology can be easily applied for rapid screening of bacterial consortia for their β-lactamase activity. So far, only LAP-MALDI MS assays with AMP have been reported, but theoretically all antibiotics that undergo enzymatic inactivation with changes in their mass spectral detection are amenable to this approach (fluoroquinolones [32], macrolides [33, 34], aminoglycosides [35, 36], cephalosporins [37], and sulfomamides [38]).

The possibility of applying these rapid screening tools that might reflect the AMR potential of bacterial communities may represent an important tool for the future. Being able to measure the microbiome's antibiotics cleavage potential in an easier and more user-friendly way will certainly help in limiting or avoiding the use of antibiotics.

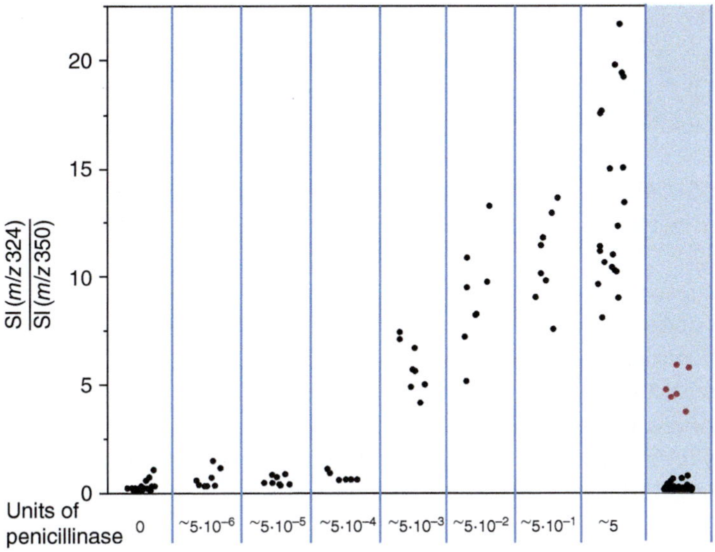

Figure 4.16 Detection of β-lactamase activity in bovine milk with spiked (left columns) and endogenous (right column) β-lactamases. SI, signal intensity.

4.7 Future Directions for LAP-MALDI MS Applications

Liquid atmospheric pressure-MALDI MS offers the possibility to rapidly obtain a molecular fingerprint of biological specimens. The molecular fingerprint is based on ionizable molecules such as metabolites, lipids, and proteins and can be obtained after simple (one-pot/two-steps) and fast (≤ 5 minutes) sample preparation and MS analysis. Modern machine learning tools can perform time-effective (on-the-fly) classification of the obtained data using minimal computational and human resources.

This biomolecular profiling has already been demonstrated for large-scale detection of animal diseases, for the detection of food adulteration, and for bacterial identification. Future applications of this methodology could be applied in all fields, where molecular profiling can distinguish certain conditions, including human medicine.

High-throughput (prote)omics analysis has been pursued by the scientific community for years but the pre-analytical purification and prefractionation steps are major limiting factors for speed and data analysis. In this context, laser-based ionization methods have the advantage of being more tolerant to sample contaminants compared with ESI MS and being more rapid in moving from one sample to the next. LAP-MALDI MS, in particular, can achieve this speed while providing many of the advantages of ESI MS such as the formation of multiply charged analyte ions and the use of high-performing instruments with superior MS/MS capabilities. These technical developments also provide a decisive advantage in sample analysis speed that is applicable to the omics sciences at large [39].

References

1 Kuo, T.-H., Dutkiewicz, E.P., Pei, J., and Hsu, C.-C. (2019). Ambient ionization mass spectrometry today and tomorrow: embracing challenges and opportunities. *Anal. Chem.* 92: 2353–2363.

2 Takats, Z., Wiseman, J.M., Gologan, B., and Cooks, R.G. (2004). Mass spectrometry sampling under ambient conditions with desorption electrospray ionization. *Science* 306: 471–473.

3 Cody, R.B., Laramée, J.A., and Durst, H.D. (2005). Versatile new ion source for the analysis of materials in open air under ambient conditions. *Anal. Chem.* 77: 2297–2302.

4 Cramer, R., Pirkl, A., Hillenkamp, F., and Dreisewerd, K. (2013). Liquid AP-UV-MALDI enables stable ion yields of multiply charged peptide and protein ions for sensitive analysis by mass spectrometry. *Angew. Chemie Int. Ed.* 52: 2364–2367.

5 Piras, C., Hale, O.J., Reynolds, C.K. et al. (2021). Speciation and milk adulteration analysis by rapid ambient liquid MALDI mass spectrometry profiling using machine learning. *Sci. Rep.* 11: 1–9.

6 Lellman, S.E. and Cramer, R. (2020). Bacterial identification by lipid profiling using liquid atmospheric pressure matrix-assisted laser desorption/ionization mass spectrometry. *Clin. Chem. Lab. Med.* 58: 930–938.

7 Piras, C., Hale, O.J., Reynolds, C.K. et al. (2022). LAP-MALDI MS coupled with machine learning: an ambient mass spectrometry approach for high-throughput diagnostics. *Chem. Sci.* 13 (6): 1746–1758.

8 Schaffer, L.V., Millikin, R.J., Miller, R.M. et al. (2019). Identification and quantification of proteoforms by mass spectrometry. *Proteomics* 19: 1800361.

9 Donnelly, D.P., Rawlins, C.M., DeHart, C.J. et al. (2019). Best practices and benchmarks for intact protein analysis for top-down mass spectrometry. *Nat. Methods* 16: 587–594.

10 Zhang, L., Vasicek, L.A., Hsieh, S. et al. (2018). Top-down LC–MS quantitation of intact denatured and native monoclonal antibodies in biological samples. *Bioanalysis* 10: 1039–1054.

11 Cramer, R. and Corless, S. (2001). The nature of collision-induced dissociation processes of doubly protonated peptides: comparative study for the future use of matrix-assisted laser desorption/ionization on a hybrid quadrupole time-of-flight mass spectrometer in proteomics. *Rapid Commun. Mass Spectrom.* 15: 2058–2066.

12 König, S., Kollas, O., and Dreisewerd, K. (2007). Generation of highly charged peptide and protein ions by atmospheric pressure matrix-assisted infrared laser desorption/ionization ion trap mass spectrometry. *Anal. Chem.* 79: 5484–5488.

13 Hiraoka, K., Konosuke, M., Keiichi, A. et al. (1997). Explosive vaporization of a liquid water beam by irradiation with a 10.6 μm infrared laser. *Rapid Commun. Mass Spectrom.* 11: 474–478.

14 Trimpin, S.A. (2010). A perspective on MALDI alternatives – total solvent-free analysis and electron transfer dissociation of highly charged ions by laserspray ionization. *J. Mass Spectrom.* 45: 471–485.

15 Inutan, E.D., Wager-Miller, J., Mackie, K., and Trimpin, S. (2012). Laserspray ionization imaging of multiply charged ions using a commercial vacuum MALDI ion source. *Anal. Chem.* 84: 9079–9084.

16 Trimpin, S., Wang, B., Inutan, E.D. et al. (2012). A mechanism for ionization of nonvolatile compounds in mass spectrometry: considerations from MALDI and inlet ionization. *J. Am. Soc. Mass Spectrom.* 23: 1644–1660.

17 Rohlfing, A., Leisner, A., Hillenkamp, F., and Dreisewerd, K. (2010). Investigation of the desorption process in UV matrix-assisted laser desorption/ionization with a liquid 3-nitrobenzyl alcohol matrix by photoacoustic analysis, fast-flash imaging, and UV-laser postionization. *J. Phys. Chem. C* 114: 5367–5381.

18 Black, C., Chevallier, O.P., Cooper, K.M. et al. (2019). Rapid detection and specific identification of offals within minced beef samples utilising ambient mass spectrometry. *Sci. Rep.* 9: 1–9.

19 Carbonnelle, E., Mesquita, C., Bille, E. et al. (2011). MALDI-TOF mass spectrometry tools for bacterial identification in clinical microbiology laboratory. *Clin. Biochem.* 44: 104–109.

20 Jang, K.-S. and Kim, Y.H. (2018). Rapid and robust MALDI-TOF MS techniques for microbial identification: a brief overview of their diverse applications. *J. Microbiol.* 56: 209–216.

21 Florio, W., Baldeschi, L., Rizzato, C. et al. (2020). Detection of antibiotic-resistance by MALDI-TOF mass spectrometry: an expanding area. *Front. Cell. Infect. Microbiol.* 10: 616.

22 Nicolaou, N., Xu, Y., and Goodacre, R. (2010). Fourier transform infrared spectroscopy and multivariate analysis for the detection and quantification of different milk species. *J. Dairy Sci.* 93: 5651–5660.

23 Piras, C., Ceniti, C., Hartmane, E. et al. (2020). Rapid liquid AP-MALDI MS profiling of lipids and proteins from goat and sheep milk for speciation and colostrum analysis. *Proteomes* 8: 20.

24 Poonia, A., Jha, A., Sharma, R. et al. (2017). Detection of adulteration in milk: a review. *Int. J. Dairy Technol.* 70: 23–42.

25 Hale, O.J., Morris, M., Jones, B. et al. (2019). Liquid atmospheric pressure matrix-assisted laser desorption/ionization mass spectrometry adds enhanced functionalities to MALDI MS profiling for disease diagnostics. *ACS Omega* 4: 12759–12765.

26 Greco, V., Piras, C., Pieroni, L. et al. (2018). Applications of MALDI-TOF mass spectrometry in clinical proteomics. *Expert Rev. Proteomics* 15: 683–696.

27 Noyes, N.R., Yang, X., Linke, L.M. et al. (2016). Resistome diversity in cattle and the environment decreases during beef production. *Elife* 5: e13195.

28 Cuny, C., Arnold, P., Hermes, J. et al. (2017). Occurrence of cfr-mediated multiresistance in staphylococci from veal calves and pigs, from humans at the corresponding farms, and from veterinarians and their family members. *Vet. Microbiol.* 200: 88–94.

29 Michael, G.B., Freitag, C., Wendlandt, S. et al. (2015). Emerging issues in antimicrobial resistance of bacteria from food-producing animals. *Future Microbiol.* 10: 427–443.

30 Rovira, P., McAllister, T., Lakin, S.M. et al. (2019). Characterization of the microbial resistome in conventional and "raised without antibiotics" beef and dairy production systems. *Front. Microbiol.* 10: 1980.

31 Liu, J., Taft, D.H., Maldonado-Gomez, M.-X. et al. (2019). The fecal resistome of dairy cattle is associated with diet during nursing. *Nat. Commun.* 10: 1–15.

32 Maia, A.S., Ribeiro, A.R., Amorim, C.L. et al. (2014). Degradation of fluoroquinolone antibiotics and identification of metabolites/transformation products by liquid chromatography–tandem mass spectrometry. *J. Chromatogr. A* 1333: 87–98.

33 Nakajima, Y. (1999). Mechanisms of bacterial resistance to macrolide antibiotics. *J. Infect. Chemother.* 5: 61–74.

34 Zieliński, M., Park, J., Sleno, B., and Berghuis, A.M. (2021). Structural and functional insights into esterase-mediated macrolide resistance. *Nat. Commun.* 12: 1–9.

35 Liu, Y., Chang, H., Li, Z. et al. (2017). Biodegradation of gentamicin by bacterial consortia AMQD4 in synthetic medium and raw gentamicin sewage. *Sci. Rep.* 7: 1–11.

36 Ramirez, M.S. and Tolmasky, M.E. (2010). Aminoglycoside modifying enzymes. *Drug Resist. Updat.* 13: 151–171.

37 Demain, A.L., Walton, R.B., Newkirk, J.F., and Miller, I.M. (1963). Microbial degradation of cephalosporin C. *Nature* 199: 909–910.

38 Chen, J. and Xie, S. (2018). Overview of sulfonamide biodegradation and the relevant pathways and microorganisms. *Sci. Total Environ.* 640: 1465–1477.

39 Cramer, R. (2020). High-speed analysis of large sample sets–how can this key aspect of the omics be achieved? *Mol. Cell. Proteomics* 19: 1760–1766.

5

Development of a MALDI-TOF Mass Spectrometry Test for Viruses

Ray K. Iles, Jason K. Iles, and Raminta Zmuidinaite

MAP Sciences Limited, iLab, Priory Park, Bedfordshire, UK

5.1 Introduction

Since the beginning of this century there have been eight major viral disease outbreaks, but SARS-CoV-2, via it associated syndrome COVID-19, has been the most widespread (Table 5.1).

As we move into a third year since COVID-19 was detected, four waves of variants have traveled around the world. In the UK, alpha, delta, and omicron have been the most dominant strains to have affected the UK population [1]. The combined effect of these will probably match the UK death toll of influenza at the beginning of the twentieth century in 1918 (Figure 5.1).

The year 2020 has marked the beginning of a new era in human viral infections in which the coronavirus SARS-CoV-2 has emerged as the defining contagious disease of the twenty-first century. It is going to be essential that we can detect viral-infected individuals and trace contacts if future outbreaks are to be successfully managed. As nations move from lockdown to staged "back to work" it will be imperative that we are able to rapidly contain subsequent outbreaks of infection.

It took four months for the UK to achieve sufficient national testing systems, based on polymerase chain reaction (PCR) technology (first notification of COVID-19 epidemic was 31 December 2019), to effectively diagnose and monitor the disease. It took a full year before immunoassay-based rapid lateral flow tests (LFTs), recognizing viral protein antigens, were sufficiently robust enough to be adopted by the UK government [2].

For early detection of asymptomatic and pre-symptomatic infection, reliance is currently dependent on PCR technology. Immunoassay-based LFTs are widely available. However, their ability to detect the disease early, and therefore to effect a public health containment of outbreaks effectively is a highly contentious debate [3–5].

The reverse transcription PCR (RT-PCR)-based tests are specific for SARS-CoV-2 strains and highly sensitive. But the test cannot be used for a new virus, or it can give false

Microbiological Identification using MALDI-TOF and Tandem Mass Spectrometry: Industrial and Environmental Applications, First edition. Edited by Haroun N. Shah, Saheer E. Gharbia, Ajit J. Shah, Erika Y. Tranfield, and K. Clive Thompson.

Table 5.1 Global viral disease outbreaks since the beginning of the twentieth century.

Year	Virus	Description of outbreak
1999–2002	West Nile Virus	The first account of the West Nile Virus in the US was in New York City in the summer of 1999. The virus was transmitted by mosquito bites, concentrating in areas with warmer temperatures; as the number of human cases grew, so did the number of bird deaths. Typical symptoms include breaking out in a high fever, a stiff neck and headaches, although most people who are infected do not show any symptoms unless the infection is serious. During a three-year period, WNV spread to 44 states, amounting to 4156 cases and 284 deaths.
2003	SARS-CoV	Severe acute respiratory syndrome, a viral respiratory illness caused by a coronavirus, was first diagnosed in Asia in February 2003, later reaching the US in that same year. Common symptoms included high fevers, coughing, and shortness of breath. A total of 8098 people worldwide became sick, with only eight of those cases being in the US, and 774 people died.
2006	Mumps	Mumps spreads through close-contact activities such as sports, dancing, kissing, etc. From 2009 to 2010, the largest outbreak reported 3000 cases in a close-knit religious community in New York City after an infected student returned from the UK where there was another outbreak. Notable symptoms include fever, puffy cheeks and a swollen, tender jaw. The US saw multi-state outbreaks of mumps reporting more than 6500 cases in 2006, concentrated mainly in Midwestern college campuses.
2009	H1N1 virus (swine flu)	In April 2009, 20 cases of swine flu were reported in the US. The swine flu was transmitted by pigs as a respiratory disease, usually through contact with farm raisers or participation in fairs. It eventually became a pandemic flu, reaching 208 countries by December and resulting in at least 12 220 deaths.
2012	Pertussis	Also known as whooping cough, this respiratory disease surmounted to 48 277 cases in the US in 2012. Coughing fits last up to 10 weeks or more and can be life-threatening for infants. Worldwide, the estimation of whooping cough cases is 24.1 million with about 160 700 deaths per year.
2012	MERS-CoV	Middle East Respiratory Syndrome was first found in the Arabian Peninsula in 2012 and was also associated with the coronavirus once the illness became viral. Only two cases were found in the US in 2014, one in Indiana and another in Florida, linked to healthcare providers who previously worked in Saudi Arabia. Up until 2019, 2494 cases had been reported, including 858 deaths.
2014	Ebola	Ebola was found in Africa near the Ebola River in 1976, but the outbreak that recently caught the world's attention started in Guinea before spreading to neighboring countries. Ebola reached the US but only four cases were recorded and one fatality, and in 2015 the CDC announced a vaccine against the disease. The most recent outbreaks of Ebola occurred in the Democratic Republic of Congo in 2018. The highest number of cases in a single site outbreak occurred in Sierra Leone, with 14,214 infected, of whom 3,956 (28%) died.

Table 5.1 (Continued)

Year	Virus	Description of outbreak
2016	Zika	The most at-threat potential victims for this mosquito-transmitted disease are pregnant women, whose pregnancy, if bitten, could result in stillbirth, preterm birth, or fetal loss. In November 2016, 84 countries were reported to have Zika virus cases in the Americas, with the first cases reported in Brazil, then spreading in high numbers to New York, Florida and Texas.
2020	SARS-CoV-2 (COVID-19)	On 30 January2020, the World Health Organization declared a "public health emergency of international concern" for the new coronavirus, COVID-19, originating from Wuhan, Hubei Province, in China. Like SARS and MERS, this virus originates from certain animals before infecting humans, and spread quickly around the world. By February 2022, it was estimated that there had been 358 516 581 confirmed cases and 5 661 144 deaths.

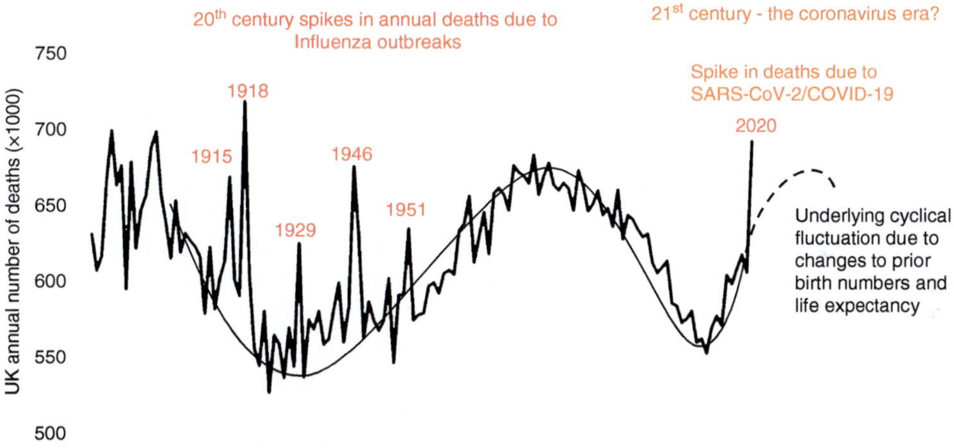

Figure 5.1 Periods of excess deaths corresponding to viral epidemics/pandemics in the UK, above the corresponding cyclic increases and decreases of the birth rate (and life expectancy). *Source:* D. Clark, Number of deaths in the United Kingdom from 1887 to 2020. Available at https://www.statista.com/statistics/281488/number-of-deaths-in-the-united-kingdom-uk/.

negatives when variants arise: i.e. if a base-pair mutation destroys a primer recognition sequence, as has happened for SARS-CoV-2 spike protein gene sequence primers for delta and the omicron variants. In addition, it is not easily adapted to measure viral load as an affordable option.

The economics of testing a population on such huge scales are staggering; for PCR-based tests the cost is high (up to £150 to individuals) and, worryingly, dependent on other countries for multiple component reagents.

Lateral flow devices/tests are considerably cheaper (although over £4 billion of the UK tax budget has been spent on them) and easier to use, even allowing point-of-care testing

(POCT) by the general public. However, sensitivity is an issue. This has given rise to questions about inadequate sampling, and when combined with the reduced sensitivity compared with PCR, whether its use should be confined to being only a definitive or confirmatory test of infection in those with clinical symptoms of an upper respiratory tract infection (pyrexia, fatigue, cough, loss of sense of smell, and cold/flu -like symptoms). LFTs' limited ability to reliably detect infection in asymptomatic carriers or most pre-symptomatic infections is its greatest weakness if used as a population test to prevent contact spread [6–9].

Thus, the clinical need is for a home-sampled test, where the efficiency of sampling can be confirmed: that is, a test that is as sensitive as PCR but not so specific that mutation variants or indeed other emerging viral infections could not be simultaneously detected; which is cheaper, requires less reagent, is not dependent on a skilled operator, and is fast and identifies asymptomatic carrier.

In our development of a MALDI-TOF mass spectrometry (MS) test for enveloped viruses, for SARS-CoV-2, we have attempted to address all these challenges.

5.2 Understanding the Systems Biology of the Virus and Viral Infections

From a clinical analytical point of view, the first consideration is: what molecules are you going to look for? Ideally these are going to be specific and unique to the pathogens. With MALDI-TOF MS analysis of bacteria, this is the detection of abundant and species-specific ribosomal proteins and other associated housekeeping proteins. These are relatively small simple polypeptides (e.g. not glycosylated) which can be resolved easily and with high definition using MALDI-TOF MS. Furthermore, bacteria can be grown on artificial media and sampled as near-pure colonies without contamination from other proteins arising from the host cells. The first hurdle in a viral MS test is that no equivalent housekeeping metabolic proteins exist for the viruses. Indeed, the main feature is that they cannot be propagated upon artificial media *in vitro* and therefore enriched in quite the same way. Propagation must be on a host cell culture. Although several unique proteins are expressed, by hijacking the host cells' housekeeping and protein synthesis systems, the only stage at which a host cell contamination-free sample can be found is if the virion particle can be simply isolated from fragments of host cells.

Virion particles contain a limited but distinctive number of proteins, as functionally it does not require any of the multitude of proteins that sustain the characteristics of living organisms other than the ability to replicate. In this respect, the only proteins required are those that protect its genome (nucleocapsid) and those that enable it to infect other cells (Figure 5.2).

Thus, in a MS detection assay in a tissue sampling-based system, the viral proteins will be among a forest of host cell proteins, but in a virion-targeted or -enriched fraction the viral-specific proteins, although more limited, have a much higher chance of being resolved and detected (Figure 5.3). The resolution power of liquid chromatographic (LC) coupled to electrospray ionization (ESI) MS is extremely good and resolving characteristic molecules, even within a crowded forest situation, is better than a MALDI-based system. However, this is limited to proteins of <2–5 kDa [11–13].

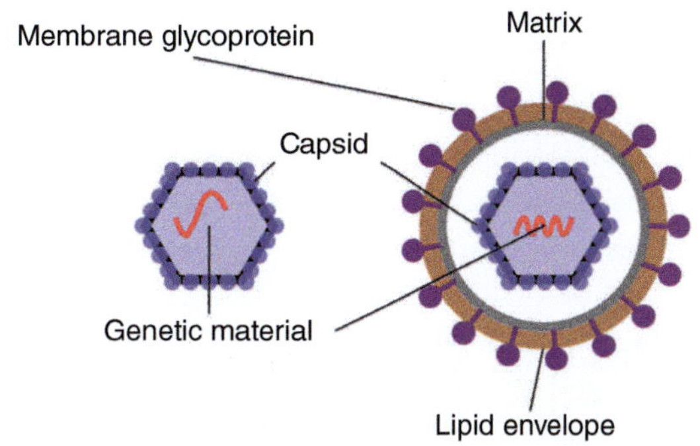

Figure 5.2 Illustration of the limited composition of unique proteins found in the isolated non-enveloped and enveloped virion particles.

5.3 Understanding the Nature of Viral Proteins and Molecular Biology

A distinguishing feature of the unique proteins of a virus, whether expressed as part of a virion particle or the infective process within the host cell, is their large size. The advantage of MALDI-TOF MS is the wide dynamic range of masses it can resolve. ESI-based MS cannot very precisely resolve molecules with high mass, > 2000 Da. To address this issue, LC-ESI-based analytical approaches are based on using enzyme(s) and or chemicals to digest the sample before LC-MS/MS analysis. However, this necessary step adds a considerable amount of time, costs, and the requirement of technical ability to any developed test [11].

Having said this, even with MALDI-TOF MS you must consider the nature and relative abundance of the target characteristic proteins and the matrix used for both facilitating desorption and ionization of the analyte. In microbiology, alpha-cyano-4-hydroxycinnamic acid (CHCA) is the standard matrix. It is also a universal matrix that is used for the analysis of small molecules and largely unmodified peptides/small proteins up to ~10 000 Da in mass [14, 15].

The nucleocapsid proteins are much larger proteins and form a complex tessellation of polymeric shells to protectively encase, while also holding in a stable formation, the nucleic acid viral genome. For the non-enveloped viruses, within this capsid tessellation are also specific protein elements that effect infection of the target host cells (Figure 5.3). The protein components of the nucleocapsid tend to carry phosphorylation and other post-translational modification such as covalent linkages with small ubiquitin-like modifier proteins, (SUMOylation) [16, 17]. This increases their final mass as determined by a mass spectrometer and adds variability/peak spreading around the mass peak maxima.

The variability in mass of characteristic proteins found in the enveloped viruses resolved by MS can be very significant. Lipid membrane embedded proteins of the enveloped viruses undergo post-translation modification by glycosylation and lipid moiety attachments (e.g. palmitoylation). Hydrophobic domains and lipid attachments embed and orientate the external protruding glycosylated regions of these unique viral genome envelope proteins which mediate attachment and invasion of the target host cells.

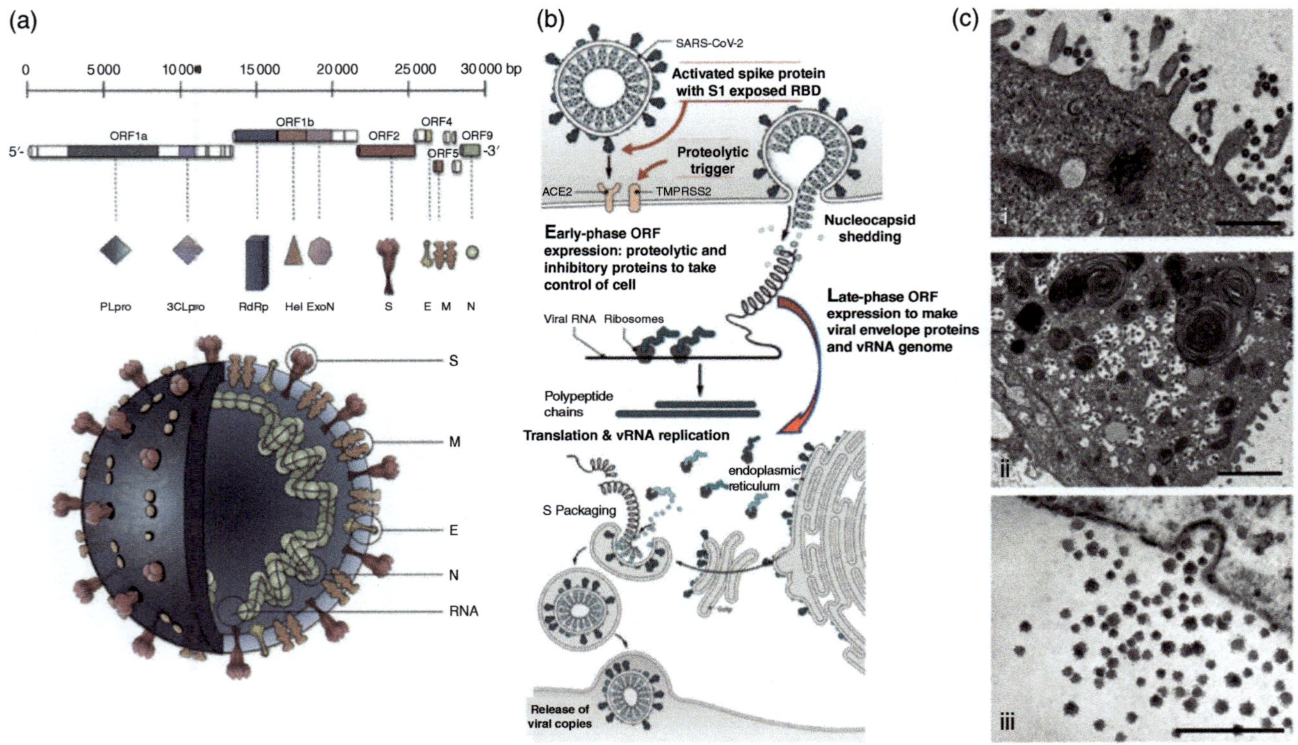

Figure 5.3 Molecular cellular pathology of SARS-CoV-2 infection. (a) Schematic illustration of SARS-CoV-2 viral RNA genome in relation to order of expression and functional incorporation with the virion particle. The genome comprises a 5′-untranslated region (5′-UTR), open reading frames (ORFs) 1a and 1b encoding nonstructural proteins 3-chymotrypsin-like protease (3CLpro), papain-like protease (PLpro), helicase (Hel), and RNA-dependent RNA polymerase (RdRp) as well as accessory proteins, and the structural S protein (S), E protein (E), M protein (M), and nucleocapsid protein (N) in ORFs 2, 4, 5 and 9. (b) Schematic of virion fusion mediated by spike protein attachment to the ACE2 receptor and accessory cleavage of the spike protein quaternary complex, resulting in fusion peptide exposure and binding to target cell plasma membrane. Focal, host cell and viral envelope lipid bilayer membrane fusion and viral RNA ingress of the target host cell. Shedding of nucleocapsid and early ORF expression to take control of the cell and inactivate internal antiviral proteins (e.g. ubiquitin and RNase). Late ORF expression: replication of viral RNA genome and expression of, via post-translational processing mechanisms of the endoplasmic reticulum and Golgi apparatus (e.g. glycosylation), membrane-embedded proteins that are responsible for infectivity including the spike protein complex. Copies of SARS-CoV-2 vRNA are packaged into lipid bilayer membrane envelopes containing multiple copies of S, E, M and N, glyco- and phosphoproteins. Inclusion transport vesicle, packed with enveloped virions, fuses with the host cell membrane, releasing infective SARS-CoV-2 virion. *Source:* Iles et al. [10]/MDPI/CC BY-4.0. (c) Electron micrographs of multiple virions attacking the cell membrane (i), viral genome expression subsuming all functions of the cell to form thousands of virion copies within inclusion vesicles (ii) and, finally, release of multiple virions by fusion of inclusion packaging vesicles with the cell surface plasma membrane of the infected host cell (iii). *Source:* Iles et al. [10]/MDPI/CC BY 4.0.

The large and polymeric nature of nucleocapsid proteins may produce very large polymer fragments that are beyond even the mass range of the MALDI-TOF MS. This can be overcome relatively simply by denaturing and reducing conditions that liberate the monomers and hence resolve them as single peaks in a mass range that are more easily detected. Although still large, these require matrices capable of ionization of large, predominately water-soluble proteins. Many of these are within the operational mass range of MALDI-TOF MS that is currently certified and used for clinical diagnostics (i.e. bacterial identification). Theoretically proteins of up to 1 million Da can be detected, but in our experience these instruments are usually not sufficiently sensitive at > 250 000 Da (i.e. can detect picomoles of the chosen marker protein > 250 000 Da on the target plate well). Also, in our experience, sinapinic acid (SA) is the most versatile matrix for ionization of large post-translationally modified proteins, but this does have its own unique problems compared with CHCA (see the operational parameters of the chosen mass spectrometer).

The enveloped virus – the equivalent of pirate exosomes – presents an added complication of liberation and solubilization from the lipid membrane. The combination of size, solubilization, and ionization is a significant hurdle in sample processing prior to MALDI-TOF MS analysis.

5.4 Virion Protein Solubilization and Extraction

Protein denaturation in water or in combination with volatile organic chemicals that evaporate leaving no residues (such as acetonitrile) can effectively denature proteins, while high-concentration salt-based denaturation will suppress ionization and/or cause significant adducts with the target proteins. Most reducing agents, such as dithiothreitol, beta-mercaptoethanol, and tris(2-carboxyethyl)phosphine, are water-soluble and thus complete denaturation and fragmentation of a complex protein structure (where subunits or proteolytic activated complexes are held together by disulfide bridges) can be achieved without compromising ionization and hence sensitivity.

Water-based solubilization of hydrophobic proteins and extraction from lipid membranes is usually achieved by addition of a high concentration of detergents and surfactants such as sodium dodecyl sulfate, Triton, and 3-[(3-cholamidopropyl)dimethylammonio]-1-propanesulfonate (CHAPS). However, these significantly supress ionization. This has been overcome by the use of a series of non-ionization-suppressing surfactants that have been developed by MAPSciences to be used at low concentrations. In our work on enveloped viruses, one termed LBSD -X has proved to be extremely versatile and efficient (Figure 5.4).

5.5 Sampling and Virion Enrichment

The prevailing sampling method for upper respiratory viruses is nasal-pharyngal swabs. Direct application of swabs onto MALDI-TOF MS plates followed by addition of matrix has been claimed to result in distinctive mass patterns correlating with SARS-CoV-2 infection at masses < 20 000 m/z [18, 19]. As detailed earlier, it is difficult to imagine that these are arising specifically and therefore characteristically from the SARS-CoV-2 virus itself.

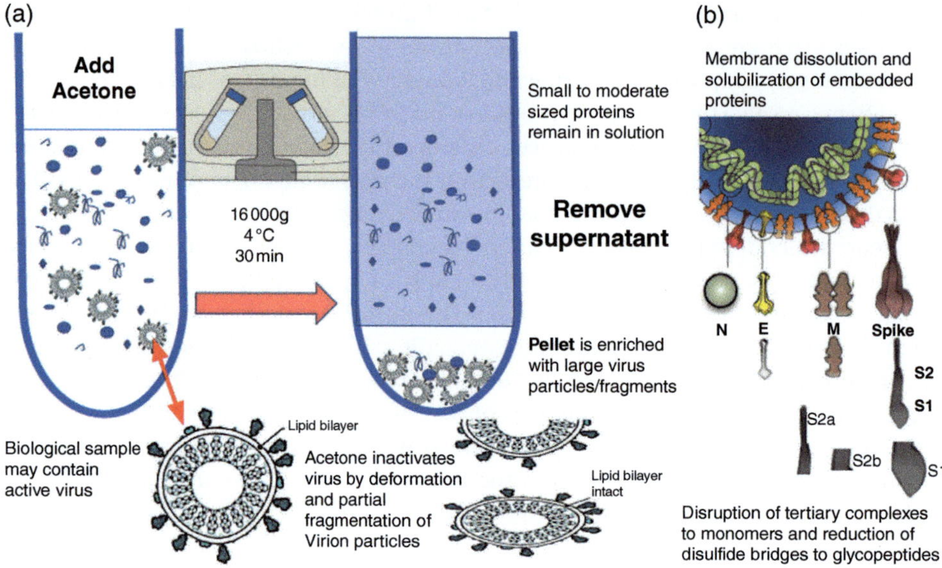

Figure 5.4 Process of virion enrichment within a biological sample and extraction/solubilization of spike (S) and other viral envelope protein for MALDI-TOF MS analysis. (a) Biological samples containing virus are mixed 1 : 1 v/v with ice-cold acetone and centrifuged at 16 000 × *g* for 30 min at 4°C. The supernatant containing smaller non-precipitated solutes is discarded. The pellet is enriched with viral particles and kept for analysis. Acetone treatment inactivates envelope virus by deformation and partial fragmentation of the viral envelope and embedded protein Structures A. (b) The pellet is resuspended in 10–100 μL of MALDI -TOF MS-compatible dissolution and solubilization buffer. Termed LBSD-X, this buffer does not suppress ionization and contains a detergent at a concentration optimized to release viral envelope-embedded proteins together with non-embedded viral proteins. It also contains dithiothreitol in order to further reduce disulfide bonds so that quaternary and tertial structures are fully disrupted, and monomers, polypeptide chains and glycopolypeptides are liberated for detailed mass analysis. *Source:* Iles et al. [10]/MDPI/CC BY-4.0.

Indeed, the swab sample will contain many resident bacterial and human proteins within this mass range, particularly if not chemically modified, including inflammatory cytokines. Thus, it is very doubtful that any such associating marker MS peaks would be specific for SARS-CoV-2, from direct application of a swab sample onto a MALDI-TOF plate and addition of CHCA alone, as has been claimed in several studies [18, 19]. In these studies, the comparison has been with PCR-positive versus PCR-negative samples only, with complex bioinformatics establishing a peak profile correlating with infection. However, these peaks probably arise due to any upper respiratory infection and correspond to immune response molecules/small proteins such as cytokines. Thus, specificity is a major issue in this approach.

In order to achieve specificity, some form of selection or enrichment of the virus is required prior to MS. In addition, the MS technique adopted should be directed against the unique proteins of the virion causing the infectious disease. Clearly having an *in vitro* culture step within the clinical test protocol is neither cost-effective nor feasible for mass testing purposes. There are various approaches to direct enrichment of virions from a sample. These include highly selective solid phase (i.e. magnetic bead conjugated) antibodies and aptamers for semi-selective precipitation of large particles [20, 21]. Being over-selective has disadvantages in that variants or other viruses are not captured. In semi-selective enrichment, all viruses present can be captured and not only detected but also identified from the mass spectral pattern resolved. Thus, not only is the test detecting a target virus infection, but it can also differentiate, in one sample analysis, the precise virus cause of an infection from the mass spectral pattern obtained. Furthermore, this allows for variation arising from mutation, giving rise to antigenic drift, and new human pathogenic viruses, arising from zoonosis, still to be detected by the system.

In our work on SARS-CoV-2 we found that simple acetone precipitation of the virions from the sample not only enriched them but also rendered them inert [10].

5.6 Peak Identification: Quantification and Bioinformatics

Automated and machine learning are key features of MS diagnostic of infectious disease and other human disorders. Peak mass and peak height/intensity are key components fed into bioinformatic routines direct from a mass spectrometer's output data. However, unlike the mass spectral peaks that arise from distinct biomolecules and even peptides, such as those characterized in current MALDI-TOF MS of bacterial samples, the shape and resolution of viral-specific proteins are not as precise (Figure 5.5). The broadening of the peak is partly because of the sheer number of isotopic combinations of the many hundreds of carbon, hydrogen, and nitrogen ions which are joined by the even more frequent isotopic variants of sulfur. Of far more significance to our studies of viruses is the variability in mass due to post-translational modifications [22]. For example, N- and O-linked sugar moieties can be mono-, di-, and tri-antennary, and also may or may not be partially degraded and missing terminal residues, depending on enzymes and the chemical environment: terminal residue sialic acid glycans alone are easily lost by acidic extractions [23, 24]. At c. 200 m/z each, loss of terminal sialic acids from a triantennary N-linked glycan can vary the mass of a single peptoglycan chain by 600 Da.

This alters the mass spectral characteristics from a relatively sharp peak of defined precise mass and peak intensity to a broad asymmetric peak, where a maximum is within a range and peak area. As a result, rather than peak height, peak area has to be calculated to characterize relative intensities (Figure 5.5). Thus, calculation of mass peak characteristic of viral proteins has to allow for post-translational modification and isotopic variation.

Viral nucleocapsid proteins tend not to be glycosylated, but present with broadened mass spectra by the sheer number of isotopic combinations, ADP-ribosylation and SUMOlyation (not seen in recombinant-bacterial expression). Enveloped viruses also contain variably post-translationally modified and heavily glycosylated proteins embedded with a lipid membrane [17, 25–27].

Much of the proteomic bioinformatic tools available have been written to recognize and process sharp, intense spectral mass peaks. Thus, given that clean, well-defined single peaks are no longer resolved, modified and new bioinformatics tools are required for large proteins, and particularly large post-translationally modified proteins, i.e. mass spectra, where a broad mass area encompassing the mass variability (spread) of the

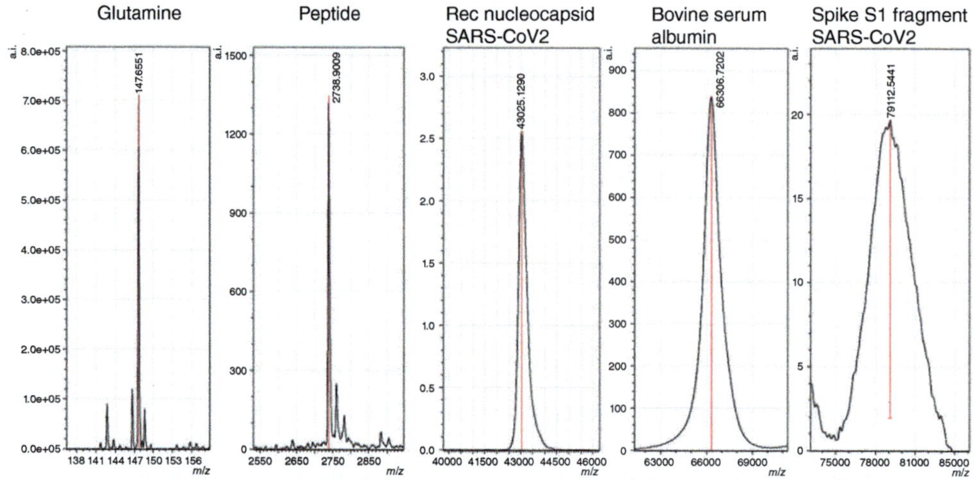

	m/z	Intensity (peak height)	FWHM
Amino acid (glutamine)	147.7	711 291.6	0.2 (0.14%)
Peptide	2738.9	1347.8	5.2 (0.19%)
Recombinant protein SARS-CoV2 nucleocapsid	43 025.1	2.6	379.2 (0.88%)
Large serum protein (bovine serum albumin)	66 305.7	839.6	1166.3 (1.75%)
Large glycosylated protein (S1 fragment SARS-CoV2 Spike)	79 112.5	17.8	3382.0 (4.27%)

Figure 5.5 Illustration of mass spectral peak spreading due to increased complexity of combined isotopic composition and post-translational modification of large proteins.

marker protein peak has to be identified and a defined area under the peak region recorded, rather than a specific mass peak height. These tools need to report a peak maximum within a comparatively widely defined range and then calculate the peak area, before applying machine learning for pattern recognition [28, 29]. In addition, when applying machine learning to a clinical diagnostic as opposed to a laboratory-based problem in MS pattern recognition, a high degree of uncertainty has to be allowed for in the training sets [30].

5.7 Promise and Pitfalls of Machine Learning Bioinformatics

The two-dimensional simultaneous measurement of multiple marker, such as that afforded by MALDI-TOF MS, needs bioinformatics not only to interpret but also to sort the significant spectral features [10]. The ability to match spectral patterns of clinical patient samples to a disease or medical condition, with a high degree of accuracy, is the goal of clinical MS, and machine learning systems (MLS) are the way forward. However, although overfitting is a recognized problem, mathematicians need to take into account the fact that nothing is 100% certain and this is particularly true in the diagnosis of a viral infection. If your machine learning algorithm matches (96–100%) the result of a given pre-existing diagnostic test for an infection, but the defining test is only 85–90% accurate in diagnosing that infection, what does this say about the clinical utility of your MS and MLS?

The COVID-19 pandemic has acutely highlighted the problems of validation of a test and the issues surrounding the defining a clinical diagnosis on a test result alone with no other confirmation, including clinical symptoms: i.e. identifying who is or is not infected with the SARS-CoV-2 virus. In the case of RT-PCR testing for SARS-CoV-2, this centers around the now widely used definition of "asymptomatic carriers" as not ever developing symptoms but being just an RT-PCR-positive qualitative result, and at whatever threshold of detection a test has been specified in their Emergency Use Authorization (EUA) submission. This has been taken to be the gold standard of defining a COVID-19 infection and against which all other tests are compared and validated.

However, comparative studies of RT-PCR tests based on saliva versus nasopharyngeal swab sampling had to address this problem and, even with this self-defining of "asymptomatic COVID-19 infection," they concluded that RT-PCR had an overall clinical efficacy of 84.8% sensitivity and 98.9% specificity [31]. The realization is that, even among those patients who develop symptoms, the RT-PCR test is not 100% sensitive. Indeed, one early study of the population prevalence of anti-SARS-CoV-2 immunoglobulin G (IgG) antibodies found that in participants with a previous negative RT-PCR test, 8% had developed IgG antibodies [32]. But the issue of false-positives and true asymptomatic individuals is a more significant one. Why all SARS-CoV-2 RT-PCR test-positives, who do not develop symptoms, are immediately classified as asymptomatic carriers, and at least some are not defined as clinical false-positives, is a poorly debated point. The estimates of pre-symptomatic and asymptomatic SARS-CoV-2 infections from a wide variety of general population screening studies range from as low as 4% to more than 80%. However, studies that specifically differentiated pre-symptomatic from asymptomatic detection rarely estimated an asymptomatic fraction greater than 50% [33–35].

It is very unsafe to assume that there is not a false-positive rate to RT-PCR testing of clinical samples for SARS-CoV-2 infection. Yet, Tran et al. [19] describe the performance characteristics of an MLS termed MILO applied to MALDI-TOF mass spectra of 226 nasal swab extracts from individuals also tested for SARS-CoV-2 infection by RT-PCR only. Of these, 107 were RT-PCR-positive and 159 were RT-PCR-negative. The results show a 100% agreement with RT-PCR-positive samples and 96% agreement on RT-PCR-negatives [19]. There are several fundamental false assumptions in this study that need to be addressed, as has already been highlighted, but the machine learning issue is an apt example.

Assuming an 85% sensitivity and 5% false-positive rate for the RT-PCR tests on the clinical samples used, a reasonable estimate is that 10% of the samples in this study were misclassified by the RT-PCR test. The biological markers measured using MALDI-TOF MS are proteins, whereas RT-PCR is priming from viral RNA; these are different but correlating markers of the infection. Given they are measuring different markers and given each method's inherent experimental errors, the probability of exact concordance in all samples is virtually zero. Thus, to report an accuracy as agreement with the RT-PCR classification of > 80% indicates over-fitting.

There is an often-repeated and very harsh saying in bioinformatics of "garbage in garbage out." In clinical MALDI-TOF MS initially a lot of this "garbage" contamination of data is due to inconsistent protocols, poor calibration, detector and sample ionization drift. Analytical problems need to be addressed methodically, and many of these analytical issues can be solved by good laboratory science and practice. However, the final, poorly recognized problem for machine learning and contamination of data is the correct classification of cases as categorically infected or not infected. Machine learning algorithms pivot on this one definition and the system will be geared to correctly identifying all classified positives and negatives with 100% accuracy.

As clearly demonstrated during the COVID-19 pandemic, we do not, in reality, have the ability to categorically say yes or no that a sample is from a person who is infected (or infective) – only the probability that they are. Currently, the best probability tests are RT-PCR systems, but the probability of being correctly classified varies and should not be absolutely defined by RT-PCR "positivity" alone. Thus any given MLS reporting near-perfect accuracy for identifying COVID-19, when the confidence that the sensitivity of the defining test to diagnose a SARS-Cov2 infection correctly (i.e. RT-PCR) is, at its very best, 85%, is open to question. Indeed, if the MALDI-TOF MS–MLS system reported was to improve on RT-PCR diagnosis of SARS-CoV-2 infection, you would not expect it to match the RT-PCR result with such conformity. Thus, greater confidence should be given to a trimmed machine learning algorithm matching at a level of 50–80% with an existing test not near 100%.

When attempting to achieve a machine learning algorithm with 98–100% accuracy, the confidence of yes/no that the sample is from someone with and without the disease also has to be 98–100%. Not many single diagnostic tests, if any, matches this expectation. Thus, supervised machine learning needs to account for the level of uncertainty and probability of correct classification of the real-world sample datasets [36, 37].

5.8 Accelerating MALDI-TOF Assay Protocol Development Using Pseudotypes/pseudoviruses

The ability to use and distribute a non-replicating form of virion particle constructed to express target viral envelope proteins (VEPs) post-translationally modified, in a similar manner to those of the wild-type virus *in vivo*, allows correct extraction and mass peak determination (Figure 5.6). Thus, the use of "pseudovirus" constructs is invaluable in both the initial development and subsequent quality control of MALDI-TOF virus tests [39]. However, live virus testing is an essential final stage of test development, and working with culture fluid containing virion particles will require category 3 facilities and/or robust processing to render the virus inactive (Figure 5.7). Fortunately, UV irradiation at 280 nm and

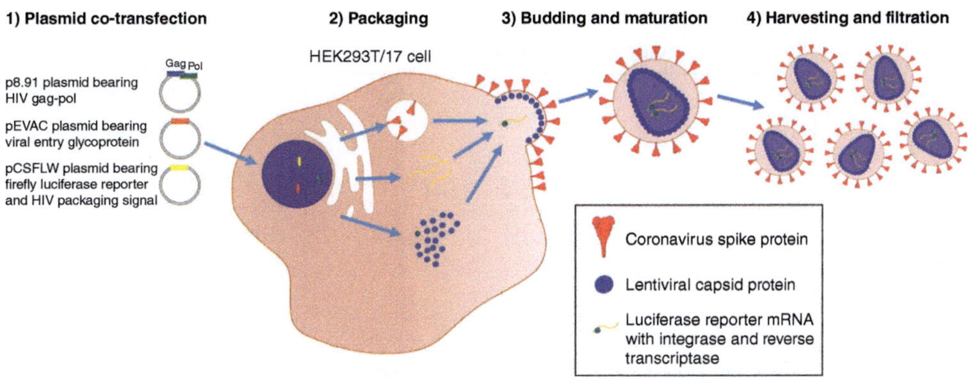

Figure 5.6 Graphic illustration of production of "pseudotype virus" expressing envelope proteins such as the coronavirus spike protein which are correctly post-translationally modified. *Source:* Adapted from [38].

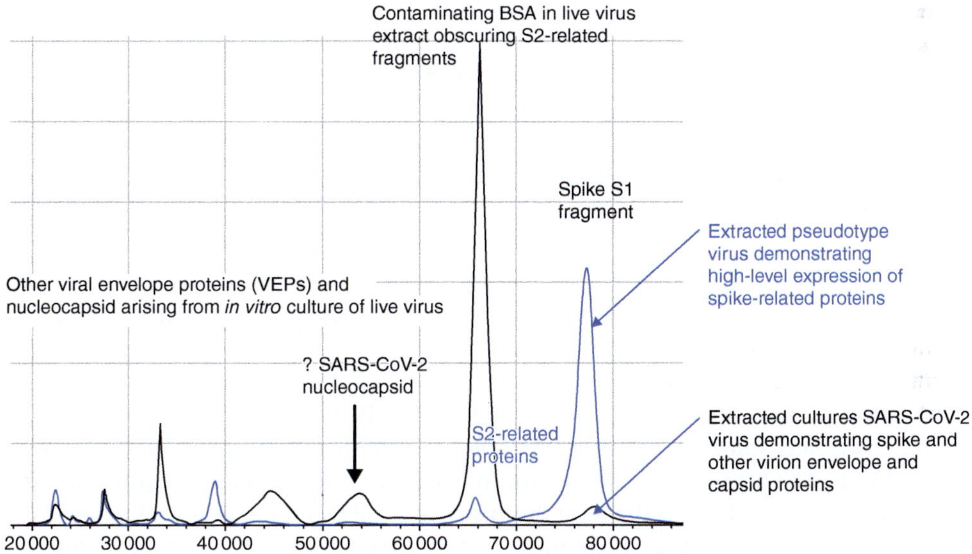

Figure 5.7 Illustration of full profile of virion proteins detected by MALDI-TOF not seen in pseudotype virus.

acetone precipitation are both highly effective at deactivating enveloped viruses, including SARS-CoV-2 [40, 41].

In addition, analysis of the biological sample is needed, as residual or co-enriched proteins from the biological sample, such as antibodies, can complicate the spectra and limit what spectral features can be resolved without interference (Figure 5.8). Nevertheless, increased antibody protein response can contribute as a biological marker of an infection [10].

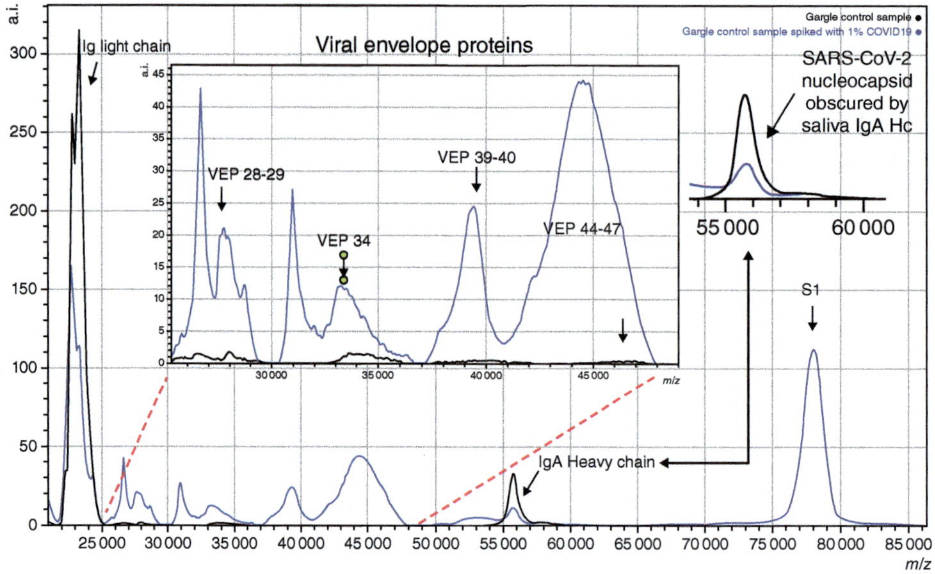

Figure 5.8 Spectra illustrating how endogenous sample proteins from gargle-saliva can hide key virion proteins such as nucleocapsid peaks by co-enrichment of endogenous IgA.

5.9 Understanding the Operational Parameters of your MALDI-TOF MS

Within the clinical field of analysis, unlike for research instruments, MALDI-TOF MS instruments are used continuously. The two critical components that account for deterioration in analytical performance are the accumulation of desorption matter/ionization spray fouling of the source/primary lens and the deterioration of the detector.

The rate of primary lens fouling is a function of matrix, the sample (i.e. urine, serum/plasma or extract) and how much energy you need to apply from the laser to achieve ionization of your target molecules within the sample. CHCA, which requires a low-energy burst from the laser is the least fouling matrix formulation, while a high-energy burst to the SA matrix is one of the most fouling. Thus, the analysis of glycoproteins embedded in lipid-enveloped virion particles is going to have much higher primary lens fouling than analysis of bacterial housekeeping proteins such as those of ribosomes [10].

The additional problem is detector response to slower-moving large biomolecules, many of which may have lost their gained positive charge, after initial acceleration, by the time they hit the detector at the end of a flight tube. In order to achieve good responses, the voltage gain across the detector often has to be raised. This shortens the operation life of the detector [42].

Thus, in the operationalization of a MALDI-TOF system for viral detection, the degree to which the primary/source lens can be regularly cleaned and accessed for manual maintenance is an important feature. Furthermore, detector gain will need to be regularly monitored for loss of signal intensity. The primary acceleration voltage will play a part in the longevity of the detector as current instruments' initial voltage acceleration is set at either

15 or 20 kV. This dramatically affects if, and with what increased voltage gain, a detector on an instrument needs to be set to record large and partially hydrophobic glycoproteins in the MALDI-TOF MS [10, 42, 43].

5.10 Understanding the Operational Requirements of the Clinical Testing Laboratory: Validation and International Accreditation

As alluded to earlier, validation data are required for authorization of a test to be used in clinical diagnostics [44]. This also has to consider the operational requirement of testing (i.e. within category 2 or 3 laboratory facilities) after making the sample inert. Accreditation then has to consider the efficacy proof required for EUA, versus post-pandemic adoption into clinical practice via laboratory-developed test (LDT) legislation, and then the much more rigorous requirements of the U.S. Food and Drug Administration (FDA) and *in vitro* diagnostic medical devices legislation.

Although most tests for SARS-CoV-2 infection (COVID-19) have received EUA approval, this is not a good strategy for a viral screen test that has greater utility than just SARS-CoV-2 detection, and where there is a view to post-pandemic utility in seasonal viral infection screening and biosecurity such as rapid boarder/aviation testing of passengers.

The EUA allows the FDA and other nations' regulatory bodies to help strengthen public health protections against CBRN threats ("CBRN" is the abbreviation commonly used to describe the malicious use of chemical, biological, radiological, and nuclear materials or weapons with the intention to cause significant harm or disruption). By international agreement, EUAs, by facilitating the availability and use of medical counter-measures (i.e. diagnosis and screening), may also be required and issued during public health emergencies arising from naturally occurring and emerging diseases, such as with COVID-19. However, approval for use of such tests could be withdrawn if such a state of emergency is suspended or terminated [45].

The FDA recently clarified that, when it grants an EUA for a point-of-care test, that test is deemed to be Clinical Laboratory Improvement Amendments (CLIA)-waived. For the duration of the national emergency declaration for COVID-19, such tests can be performed in any patient care setting that operates under a CLIA Certificate of Waiver or Certificate of Compliance/Certificate of Accreditation. In addition, the FDA clarified that tests for SARS-CoV2 that are offered prior to or without an EUA have not been reviewed by the FDA, are not FDA-authorized and have not received a CLIA categorization external icon. Thus, those tests are considered of high complexity by default until they receive an EUA or other FDA approval that indicates they may be performed as moderate complexity or waived tests. So, such allowance to use could be terminated, as they are not FDA-approved, and is conditional on the level of the health threat to the public. The vast majority of COVID-19 tests are used under EUAs, which is not the same as being FDA-approved, and these will be removed at some point.

5.10.1 Limitation and Advantages of CLIA LDTs

The CLIA regulate laboratory testing and require clinical laboratories to be certified by the Centers for Medicare & Medicaid Services (CMS) before they can accept human samples for diagnostic testing [46]. Laboratories can obtain multiple types of CLIA certificates, based on the kinds of diagnostic tests they conduct. Three federal agencies are responsible for CLIA: the FDA, the CMS, and the Centers for Disease Control and Prevention (CDC). Each agency has a unique role in assuring quality laboratory testing.

A laboratory developed test (LDT) is a type of a diagnostic test that is designed, manufactured, and used within a single CLIA-certified laboratory. LDTs can be used to measure or detect a wide variety of analytes (substances such as proteins) in a sample taken from a human body. Some LDTs are relatively simple tests that measure single analytes, whereas others are complex and may measure or detect many analytes simultaneously such as by using MALDI-TOF MS [46]. The regulations and guidelines state that the use of LDTs is the same as using an FDA-cleared or -approved *in vitro* diagnostic tests.

We applied to the FDA for our MALDI-TOF test for virus infection by SARS-CoV-2 to be recognized as an LDT-validated, high-complexity test governed by FDA CLIA guidelines in a CLIA-certified laboratory. It was validated and compared against an FDA EUA-approved SARS-CoV-2 RT-PCR test. Such a validation will not have the same risk of possible suspension in the future as does an EUA, but it constrains use to a CLIA laboratory and requires internal re-validation at each CLIA laboratory adopting the MALDI-TOF MS viral infection detection technique.

5.11 MALDI-TOF MS Screening Test for SARS-CoV-2s

The procedural protocol developed in the UK at the laboratories of Viral Zoonosis, University of Cambridge, and MAPSciences, The iLab, Bedford UK, is fully described by Iles et al. [10]. This was reproduced and utilized at the University of Northern Illinois [47]. The basic protocol is detailed in the following.

5.11.1 Prepare Positive Control

Take 300 μl of ultra-pure water and add to the freeze-dried positive control SARS-Cov-2 pseudovirus tube vortex until there are no visible clumps in the tube. Spike 150 μl into 4.85 ml of saline. Process as for samples.

5.11.2 Prepare Gargle-saliva Samples

Irradiate collected gargle-saliva samples with UV light at 280 nm for 15 minutes. Collect the liquid into a lure lock syringe attached with a 21G, 1.5″ needle. Once the liquid has been collected into the syringe, ensure to create an air bubble to keep the liquid in place while removing the needle from the syringe. Add the plastic cover and dispose of it into a sharps box. Apply a 0.45-μm filter onto the syringe and push 5 ml of the sample/spiked saline through the filter into a clean 10 ml centrifuge tube.

5.11.3 Viral Particle Enrichment

Add ice-cold acetone (4°C) to match the amount of sample in the tube (1 : 1, v/v); it should be 5 ml giving a total volume of 10 ml. Centrifuge the samples at 16 000 RCF at 4°C, for 30 minutes. Discard the supernatant completely into a suitable container. Note that no visible pellet can be observed.

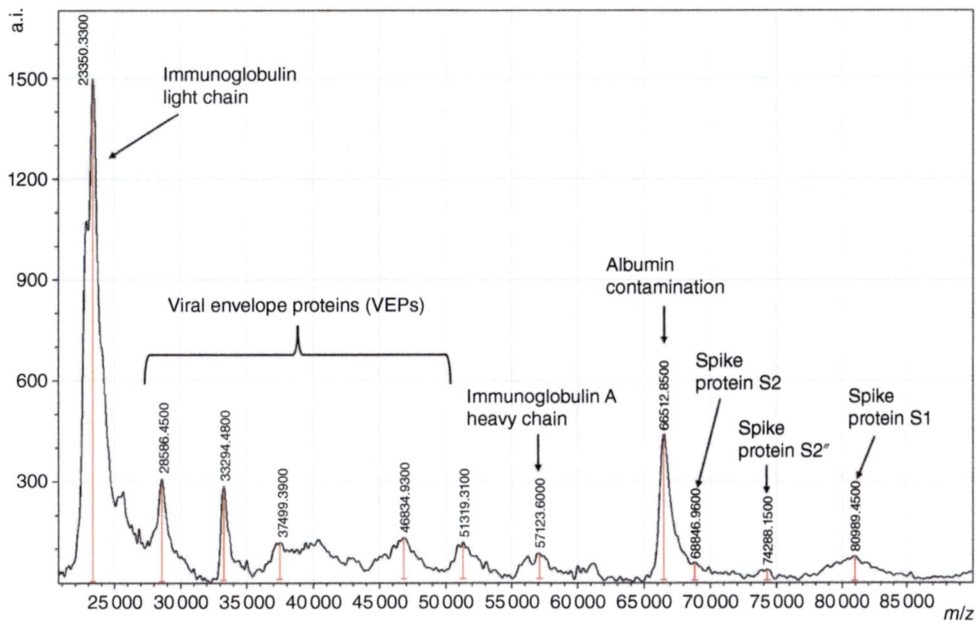

Figure 5.9 Example of an asymptomatic SARS-CoV-2-positive gargle-saliva sample mass spectra.

5.11.4 Dissolution of Virions and Solubilization of Viral Proteins

Add 50 μl of the buffer LBSD-X working solution made up with 20 mM Tris Carboxy Ethyl Phosphene (TCEP). Manually, with the pipette tip, scrape the side of the container and pump the mix with a pipette. Vortex the sample well and incubate for 15 minutes at a room temperature.

5.11.5 MALDI-TOF MS

After 15 minutes of sample incubation with the buffer LBSD-X and TCEP, vortex the sample again to ensure homogeneity. Transfer 1 μl to the MALDI target plate and overlay with SA (15 mg/ml).

A typical spectra generated will detect viral spike proteins and other VEPs along with salivary-mucosal immunoglobulin A (IgA) antibody light and heavy chains in a patient sample (Figure 5.9).

5.12 CLIA LDT Validation of a MALDI-TOF MS Test for SARS-CoV-2

Gargle-saliva samples are analyzed on MALDI-ToF mass spectrometers in a CLIA laboratory. The criteria for validation of LDT are as follows:

1) Limit of Detection (LoD)
2) Interfering substances and specificity
3) Clinical performance evaluation
4) Reproducibility
5) Stability.

5.12.1 Limit of Detection

Limit of detection is determined as detection of S1, S2, S2′ peaks above combined threshold; the lowest amount of analyte in a sample that can be detected needs to be established. The LoD is considered the lowest concentration that triggers a positive result 95% of the time. LoD is determined by diluting a sample with known concentration of virus and testing until the assay no longer detects > 95% positive rate (Table 5.2).

In a biological assay of infectious SARS-CoV-2 (Wuhan isolate), MALDI-TOF spectral signal was no longer detectable at 2 plaque-forming units (PFU)/ml. A signal would be seen 95% of the time at about 700 PFU/ml.

In an RT-PCR assay comparison, a MALDI-TOF spectral signal was no longer detectable for a sample corresponding to 77 copies/ml as quantitated by RT-PCR. A signal was seen 95% of the time at about 400 copies/ml.

Note that discrepancies exist between RT-PCR tests in their estimation of copy numbers, and serial dilution of culture media of the Wuhan isolate had indicated that 1 PFU equated to 88 copies of SARS-CoV-2 genome in an in-house RT-PCR. Non-functional fragments are often a cause of these discrepancies.

5.12.2 Interfering Substances and Specificity

Interference of endogenous compounds that can be found in saliva was tested to evaluate interference with results of the test. The following substances were tested for interference with the assay: three pools of blank saliva samples and three pools of positive saliva samples

Table 5.2 (a) 1. Limit of detection based on biological assay (PFU) of SARS-2 viral levels from SARS-CoV-2 (Wuhan) cultures spiked into negative gargle-saliva samples. (b) LoD based on quantitative RT-PCR-determined viral copy number in an (RT-PCR) saliva sample positive for SARS-CoV-2.

(a)		
Concentration – biological activity	**Positive/negative**	**% positive**
1000 PFU (88 000 copies/ml)	10/10	100%
750 PFU (66 000 copies/ml)	10/10	100%
500 PFU (440 000 copies/ml)	5/10	50%
10 PFU (888 copies/ml)	0/10	0%
(b)		
Concentration – PCR-determined copy number (copies/ml)	**Positive/negative**	**% positive**
2304	10/10	100%
1502	10/10	100%
576	10/10	100%
288	9/10	90%
144	5/10	50%
77	0/10	0%

were tested in replicates of six. No interference was defined as spiked blank samples testing negative and spiked positive samples testing positive. Interference was defined as blank samples testing positive and positive samples testing negative. In these experiments a positive was determined as detection of spike S1, S2, and S2′ mass spectral peaks that were characteristic of the SARS-CoV-2 virus (Table 5.3).

Table 5.3 Determination of interfering substance and cross-reaction with other virus in the MALDI-TOF MS test for SARS-CoV-2.

Potential interfering substance	Concentration of substance	Positive sample pools			Negative sample pools		
(SARS-COV2)		1	2	3	1	2	3
Toothpaste				Not evaluated			
Mouthwash[a]	0.1% v/v benzalkonium chloride and/or 0.1% chlorhexidine	Neg	Neg	Neg	Neg	Neg	Neg
Nicotine				Not evaluated			
Benzocaine/menthol							
Caffeine							
No interfering substance (controls)	(Saline)	Pos	Pos	Pos	Neg	Neg	Neg
SARS-1 virus	Pseudotype at 10^3 FU/μl	Pos	Pos	Pos	Neg	Neg	Neg
Rhino virus	Infected individuals gargle (concentration unknown)	Pos	Pos	Pos	Neg	Neg	Neg
Human coronavirus MERS	Pseudotype at 10^3 FU/μl	Pos	Pos	Pos	Neg	Neg	Neg
Human coronavirus OC43[b]	Pseudotype at 10^3 FU/μl	Pos	Pos	Pos	Pos	Pos	Neg
Human coronavirus 229E[b]	Pseudotype at 10^3 FU/μl	Pos	Pos	Pos	Neg	Neg	Pos
Human coronavirus NL63	Pseudotype at 10^3 FU/μl	Pos	Pos	Pos	Neg	Neg	Neg
Human coronavirus HKU1	Pseudotype at 10^3 FU/μl	Pos	Pos	Pos	Neg	Neg	Neg
Influenza A virus (H1N1 swine flu)	1000 PFU	Pos	Pos	Pos	Neg	Neg	Neg
Influenza A virus (H1N21 Puerto Rico)	1000 PFU	Pos	Pos	Pos	Neg	Neg	Neg

Functional infectivity assay determination of concentration:
FU, fluorescence unit: This is a functional transfection measure of light detection from reporter luciferase gene being expressed as a result of pseudotype virus successfully infecting target HEK293 cells *in vitro*. FU is determined from serial dilution of the pseudovirus samples' infection of HEK293 cells *in vitro*.
PFU, plaque-forming units. A biological measure of functional pathogenic virus-infected Vero cells from co-culture with live virus containing samples. PFU is determined from serial dilution of the virus samples' plaque formation on a Vero cell *in vitro* culture monolayer.
[a] Persons who are providing samples to be analyzed by the test must abstain from mouthwash for one hour prior to sample collection.
[b] Phylogeny-related beta-coronaviruses OC43 and alpha-coronavirus 229E have similar characteristic S1, S2, and S2′ mass peaks which can score positive on this evaluation.

5.12.3 Clinical Performance Evaluation

The intended purpose is mass population screening to identify pre-symptomatic and asymptomatic carriers of SARS-CoV-2. Clinical evaluation was specific to this purpose.

- No other test for SARS-CoV-2 testing with EUA has been evaluated for this purpose in the clinical performance evaluations.
- All previous EUA *in vitro* diagnostic tests have been evaluated on hospitalized, clinically confirmed, patient samples only.

Asymptomatic people are those who have caught the infectious agent, who do not develop symptoms, and who infect others who will develop the disease. Identifying these is a difficult process. Although the purpose of population screening is to identify pre-symptomatic and asymptomatic people via a simple affordable test, the problem of false-positives is often ignored by the need for near 100% sensitivity. Clinically those who score positive on an RT-PCR and do not develop symptoms are assumed to be asymptomatic with no other biomarker or clinical verification. Even with self-defining of "asymptomatic infection," a recent study concluded that RT-PCR had an overall clinical efficacy of 84.8% sensitivity [31]. But the issue of clinical false-positives is ignored. It is recognized that, on average, 30–50% of positive SARS-CoV-2 RT-PCR test results identify individuals with no symptoms and who never developed COVID-19 symptoms, however mild. It is not demonstrated whom of these pass on the infection. The estimates of asymptomatic SARS-CoV-2 infections from a wide variety of studies ranges from as low as 4% to more than 80%. However, studies that specifically differentiated pre-symptomatic from asymptomatic detection rarely estimated an asymptomatic fraction > 50% [33–35]. Nevertheless, despite evidence that lateral flow devices can detect viral antigens with an acceptable degree of reproducibility on mock samples, real-world testing indicates that the Innova lateral flow SARS-CoV-2 antigen device has only a 3–4% sensitivity for RT-PCR-positive pre-symptomatic/asymptomatic samples when screening the general population [48]. Given that the RT-PCR test is not a perfect gold standard upon which to base an evaluation, it is the best we have. Thus, although a strong correlation is going to be extremely unlikely for any alternate testing modality, the magnitude of discrepancy between the Innova lateral flow and nasopharyngeal swab RT-PCR testing of a symptomless population for SARS-CoV-2 infection is alarming. Accepting that only broad comparison can be made, in the case of the Innova test, the specificity is probably close to 99.9% but sensitivity is very low for the purpose of identifying pre-symptomatic and asymptomatic carriers. Conversely, COVID-19 RT-PCR tests may have high sensitivity for asymptomatic cases, possibly an 80–90% detection rate, but specificity at the minimum LoD thresholds [high cycle threshold (Ct)] is likely to be much lower but impossible to determine. What published data there is indicate that the clinical specificity of high-Ct single-primer RT-PCR SARS-CoV-2 could potentially be as low as 70% when applied to a symptomless population screening for the pre-symptomatic/asymptomatic individuals spreading the virus.

In light of the above issues the gargle-saliva MALDI-TOF MS test incorporated both signature mass spectral peak detection characteristic of the SARS-CoV-2 virus and a marker of elevated immunological activity with the oral mucosa of the upper respiratory tract (i.e. IgA heavy chain mass spectra peak measurement). This was then compared against general population screening in order to give an indication as to its ability to achieve non-invasive, easy-sampling mass population screening for pre-symptomatic and clinically significant asymptomatic individuals.

5.12.3.1 Establishing Operational Cut-off Values

A negative population was evaluated, consisting of 150 samples three months after recovery from SARS-CoV-2 infection (123 medical staff with mild COVID-19 symptoms) and convalescent patients (27 intensive therapy unit patients after discharge), all with confirmed serological-positive SARS-Cov-2 antibodies. None of the 150 exceeded the threshold of positive testing in the combined S1, S2, S2′ peak measure and IgA heavy chain on the Shimadzu 8020 mass spectrometer.

In a second control population analysis, 300 members of the UK general population (volunteers from London and Bedford, UK) gave gargle-saliva samples. Eleven scored positive on the gargle-saliva MALDI-TOF MS test at a conservative threshold. When the positive thresholds were increased by a factor of 2, seven of the previous 11 scored positive. One individual (marked with f) reported having experienced COVID-19 symptoms requiring bedrest two weeks prior to giving a sample Figure 5.10. Thus, the expected RT-PCR-positive rate for pre-symptomatic and asymptomatic carriers was 8–11. (Note that the UK government RT-PCR SARS-CoV-2 testing conforms to a minimum threshold of two out of three primer pair positives for a Ct < 35 to be recorded as RT-PCR-positive for COVID-19 infection.)

The extrapolated estimated detection of "pre-symptomatic and asymptomatic infection" therefore varied from 64–88% to 100–138% depending on the threshold applied of the expected RT-PCR detection rate.

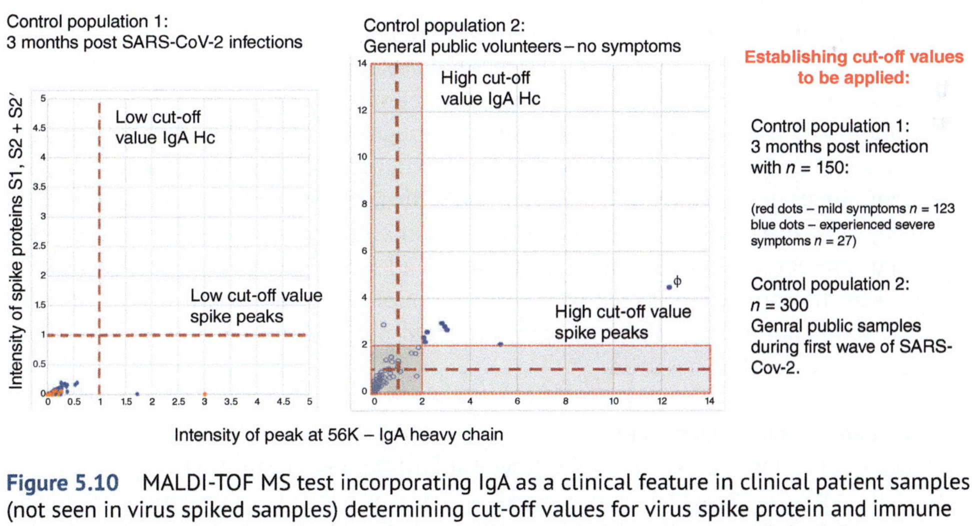

Figure 5.10 MALDI-TOF MS test incorporating IgA as a clinical feature in clinical patient samples (not seen in virus spiked samples) determining cut-off values for virus spike protein and immune response IgA (heavy chain) combined. The prevailing population estimates of infection among those not reporting symptoms for Bedfordshire and north London (UK) varied from 2.5% to 3.5% (for London, UK) at the time of sample collection (November 2020 to January 2021; Office of National Statistics, UK government).

5.12.3.2 Direct comparison with an RT-PCR SARS-CoV-2 test

As RT-PCR is the mainstay of the global diagnosis and population screening of the SARS-CoV-2 pandemic, all other clinical testing technology has to be compared against "PCR"-based assessments. The EUA-approved Abbott SARS2 RT-PCR test was used as comparator in this evaluation.

The study group came from 550 samples from student athletes at Northern Illinois University, taken between November and December 2020. All were nasopharyngeal swab sample-tested using the Abbott RT-PCR SARS-2 test. Of these, 77 were RT-PCR-positive, 10 of whom developed symptoms. That is a positivity rate of 14%, of whom 66 (85.7%) are defined as RT-PCR "asymptomatic." Thus, overall among this population, 12% were RT-PCR asymptomatic carriers [49].

One hundred and fifty-two gargle-saliva samples were collected simultaneously with nasopharyngeal swabs. A comparative evaluation study was performed to evaluate the performance of the MALDI-TOF MS COVID-19/SARS-CoV-2 test detection of pre-symptomatic and asymptomatic people using the Shimadzu Axima MALDI-TOF mass spectrometer. Of the 152 gargle-saliva samples analyzed, three were pre-symptomatic RT-PCR-positive and 57 were RT-PCR-positive, determined asymptomatic; 92 were RT-PCR-negative samples. In this evaluation, both the virus S1, S2, S2′ spectral mass peaks combined threshold and that of the IgA heavy chains must be exceeded to score positive (Table 5.4).

Overall agreement with the Abbott SARS2 RT-PCR classification of positive or negative was 70%. Of the Abbott SARS2 RT-PCR positive classification of samples, this varied from 53 to 77% depending on thresholds applied.

5.12.3.3 Internal Sampling Quality Control

In order to determine if a negative is a true negative and not due to poor sample collection, non-digestive (ND) non-mucin mucus gel structural proteins found in oral and respiratory mucus and saliva were monitored in all clinical samples. The peaks at m/z 10 900 –11 900 correspond to ND non mucin proteins co-precipitating with MUC-glycoproteins during acetone extraction. Gastric and intestinal mucus contains only smaller, 6 m/z or fewer, non-mucin mucus gel proteins [50]. They are only liberated from the viscous mucus gel complex by extensive disulfide reduction. The peaks at m/z 10 900 –11 900 are attributable endogenous proteins of oral respiratory tract mucus and saliva. These peaks should always be present in the correctly collected gargle-saliva sample mass spectra. If a peak at m/z 10 900 –11 900 is not present, then the sample should not be considered to be in a good condition and the result is not valid.

5.12.3.4 Daily System Quality Control

The positive control is SARS-CoV-2 pseudotype at [10^3 fluorescent units (FU)/μl] or heat-inactivated SARS-CoV-2, which is processed as if it was a gargle-saliva sample. Pooled human pre-2019 saliva is used as the negative control. A confirmed positive (positive control) and confirmed negative (negative control) sample must be analyzed with each plate run on the MALDI-TOF MS test. The positive sample must be positive and the negative sample must be negative for results of any given sample plate to be accepted. If either control is out, samples must be re-analyzed.

Table 5.4 Tabulated results of a comparison of the MALDI-TOF MS test for SARS-CoV-2 and an EUA RT-PCR test with a low threshold and a high threshold.

		Abbott SARS2 RT-PCR		
Low cut-off threshold		**Pre-symptomatic positive**	**RT-PCR asymptomatic positive**	**RT-PCR-negative**
MALDI-TOF method	Pre-symptomatic positive	3/3	57	92
	Asymptomatic positive	47	43/57	18
	Negative	102	5	60/92
Positive % agreement	Pre-symptomatic cases – 100%, asymptomatic cases – 75%			
Negative % agreement	65% agreement on negatives			
		Abbott SARS2 RT-PCR		
High cut-off threshold		**Pre-symptomatic positive**	**RT-PCR asymptomatic positive**	**RT-PCR-negative**
MALDI-TOF method	Pre-symptomatic positive	3/3	57	92
	Asymptomatic positive	47	29/57	18
	Negative	102	5	74/92
Positive % agreement	Pre symptomatic cases – 100%, asymptomatic cases – 51%			
Negative % agreement	80% agreement on negatives			

Overall agreement with the Abbott SARS2 RT-PCR classification of positive or negative is 70%.
Of the Abbott SARS2 RT-PCR-positive classification of samples, this varied from 53 to 77% depending on thresholds applied.
Note that this RT-PCR test requires only one of two SARS-CoV-2 genome (RdRp and nucleocapsid) sequence primers to be positive at <40 Ct to be recorded as qualitatively positive.

5.12.4 Reproducibility

Reproducibility shows the ability of the assay to generate the same result from the same sample. Reproducibility of the assay is demonstrated by injecting a minimum of four positive samples and four negative samples for four days. Intra-day reproducibility will be determined as result of a single day. Inter-day will be the total over the four days. Reproducibility is calculated as positive measured/positive expected ×100 and negative measured/negative expected ×100 (Table 5.5).

5.12.5 Stability

Storage sample stability was determined by storing spiked samples (positive control CDC inactivated virus and pseudotype virus), frozen at −18 to −20°C, in the same conditions in which patient samples were stored and for periods of zero, one, two, three, and four weeks.

Table 5.5 Tabulated determination of repeat positive samples and repeat negative samples to be detected over 4 days.

	Day 1	Day 2	Day 3	Day 4	Total
	Positive/ negative	Positive/ negative	Positive/ negative	Positive/ negative	Positive/ negative
Positive (PCR)	4/0	4/0	4/0	4/0	16/0
Negative (PCR)	0/4	0/4	0/4	0/4	0/16

Table 5.6 Composite tabulation of storage stability of gargle-saliva spiked samples containing inactivated SARS-CoV-2 virus (CDC inactivated virus) and pseudotype expressing SARS-2 spike protein, as complete frozen mock "sample" (a) and as processed acetone precipitate "extract" (b).

(a) Storage sample stability					
CDC inactivated virus					
	Week 1	**Week 2**	**Week 3**	**Week 4**	**Total**
	Positive/ negative	**Positive/ negative**	**Positive/ negative**	**Positive/ negative**	**Positive/ negative**
Positive (PCR)	5/0	5/0	5/0	5/0	20/0
Negative (PCR)	0/5	0/5	0/5	0/5	0/20
Pseudotype virus					
	Week 1	**Week 2**	**Week 3**	**Week 4**	**Total**
	Positive/ negative	**Positive/ negative**	**Positive/ negative**	**Positive/ negative**	**Positive/ negative**
Positive (PCR)	5/0	5/0	5/0	5/0	20/0
Negative (PCR)	0/5	0/5	0/5	0/5	0/20
(b) Extract stability					
CDC inactivated virus					
	0 h	**12 h**	**24 h**	**48 h**	**Total**
	Positive/ negative	**Positive/ negative**	**Positive/ negative**	**Positive/ negative**	**Positive/ negative**
Positive (PCR)	5/0	5/0	5/0	—	15/0
Negative (PCR)	0/5	0/5	0/5	—	0/15
Pseudotype virus					
	0 h	**12 h**	**24 h**	**48 h**	**Total**
	Positive/ negative	**Positive/ negative**	**Positive/ negative**	**Positive/ negative**	**Positive/ negative**
Positive (PCR)	10/0	10/0	10/0	10/0	50/0
Negative (PCR)	0/10	0/10	0/10	0/10	0/50

Extract stability was determined by leaving samples on the MALDI-TOF sample plate after time 0 analysis and rerunning at 12, 24, and 48 hours. Acceptance for stability of all stability samples was no change in result (Table 5.6).

5.12.6 Validation Disposition

In this CLIA LDT validation, we recommend that the test is used as a general screen of the population at large in order to identify pre-symptomatic and asymptomatic carriers of SARS-CoV-2 infection. Set at a low threshold confirmation using RT-PCR for those scoring positive is recommended for increased probability of maximal asymptomatic carrier diagnosis.

Repeat regular testing of gargle-saliva samples is recommended for continued surveillance and follow-up.

The sample is a 30-second gargle-saliva sample. Mouthwash prior to sampling is an interfering substance to the process (as it disrupts the viral particle structure prior to size partition and concentration).

Cross-identification of coronavirus OC43 and 229E can occur. Confirmation of positives with RT-PCR is recommended.

Rapid first screen using MALDI-TOF MS is cheaper and home sampling will reduce cost and resource pressures on RT-PCR services, which are 10 times higher.

5.12.6.1 Global Biosecurity

The UK and many other nations cannot afford another viral pandemic within a short time span. Given increasing global connectivity and incidence of viral species jumping (zoonotic adaptation), the probability of new viral outbreaks is higher and they are likely to occur more frequently.

There have been three coronavirus zoonotic transitions in the last 17 years: MERS, SARS and SARS-CoV-2. Ebola kills 30% of those infected, but transmission is via "bush meat" and physical contact with body fluids [50]. SARS-1 kills 10% of those infected, but has a relatively low airborne human transmission rate. SARS -CoV-2 kills 2% of those infected but infects many via aerosol transmission. This coronavirus species transition pandemic was predicted in 2007 [51].

What is needed for global biosecurity is a test that:

- can readily identify the emergence of novel viral infections;
- produces results within an hour;
- detects asymptomatic and pre-symptomatic infection;
- can be deployed at ports, airports, etc. as well as hospitals;
- can be rapidly applied to mass screening;
- is not expensive;
- is not critically dependent on limited reagents;
- requires non-invasive or minimal invasive sampling of the population.

MALDI-TOF MS-based tests are going to play a significant role in future diagnostic surveillance and screening for viral outbreaks. In order to do so, multidisciplinary research groups are required to achieve further advances in virion particle enrichment, mass spectral bioinformatics, and MALDI-TOF instrument design to enable robust, high-throughput systems to be deployed [52, 53].

References

1 Wassenaar, T.M., Wanchai, V., Buzard, G., and Ussery, D.W. (2022). The first three waves of the Covid-19 pandemic hint at a limited genetic repertoire for SARS-CoV-2. *FEMS Microbiol. Rev.* 46 (3): fuac003.

2 Vandenberg, O., Martiny, D., Rochas, O. et al. (2021). Considerations for diagnostic COVID-19 tests. *Nat. Rev. Microbiol.* 19: 171–183.

3 Deeks, J.J., Dinnes, J., Davenport, C. et al. (2021). Letter to the editor regarding Peto T; UK COVID-19 lateral flow oversight team: COVID-19: rapid antigen detection for SARS-CoV-2 by lateral flow assay. *EClinicalMedicine* 38: 101037.

4 Lamb, G., Heskin, J., Randell, P. et al. (2021). Real-world evaluation of COVID-19 lateral flow device (LFD) mass-testing in healthcare workers at a London hospital; a prospective cohort analysis. *J. Infect.* 83 (4): 452–457.

5 Leber, W., Lammel, O., Siebenhofer, A. et al. (2021). Comparing the diagnostic accuracy of point-of-care lateral flow antigen testing for SARS-CoV-2 with RT-PCR in primary care (REAP-2). *EClinicalMedicine* 38: 101011.

6 De la Matta, M., Delgado-Sánchez, J.M., Gutiérrez, G.M. et al. (2021). Utility of preoperative polymerase chain reaction testing during SARS-CoV-2 pandemic: the challenge of evolving incidence. *Rev. Esp. Anestesiol Reanim. (Engl. Ed.)* 68 (6): 346–352.

7 Ferguson, J., Dunn, S., Best, A. et al. (2021). Validation testing to determine the sensitivity of lateral flow testing for asymptomatic SARS-CoV-2 detection in low prevalence settings: testing frequency and public health messaging is key. *PLoS Biol.* 19 (4): e3001216.

8 Subudhi, P. D., Rooge, S., Thangriyal, S., Aggarwal, R., Gupta, E., Baweja, S. (2021). Prolonged detection of SARS CoV2 RNA in extracellular vesicles in nasal swab RT-PCR Negative Patients. https://www.medrxiv.org/content/10.1101/2021.12.16.21267912v2 (accessed 22 September 2022).

9 Tanriover, M.D., Fathi, A., Raspe, M. et al. (2022). Characteristics and management of asymptomatic SARS-CoV-2 infections. *J. Basic Clin. Physiol. Pharmacol.* 33 (1): 1–7.

10 Iles, R.K., Zmuidinaite, R., Iles, J.K. et al. (2020). Development of a clinical MALDI-ToF mass spectrometry assay for SARS-CoV-2 rational design and multi-disciplinary team work. *Diagnostics* 10 (10): 746.

11 Donnelly, D.P., Rawlins, C.M., DeHart, C.J. et al. (2019). Best practices and benchmarks for intact protein analysis for top-down mass spectrometry. *Nat. Methods* 16: 587–594.

12 Kang, L., Weng, N., and Jian, W. (2020). LC–MS bioanalysis of intact proteins and peptides. *Biomed. Chromatogr.* 34 (1): e4633.

13 Lefebvre, D., Blanco-Valle, K., Feraudet-Tarisse, C. et al. (2021). Quantitative determination of *Staphylococcus aureus* enterotoxins types a to I and variants in dairy food products by multiplex Immuno-LC-MS/MS. *J. Agri. Food Chem.* 69 (8): 2603–2610.

14 Monopoli, A., Nacci, A., Cataldi, T.R., and Calvano, C.D. (2020). Synthesis and matrix properties of α-Cyano-5-phenyl-2, 4-pentadienic acid (CPPA) for intact proteins analysis by matrix-assisted laser desorption/ionization mass spectrometry. *Molecules* 25 (24): 6054.

15 Sogawa, K., Watanabe, M., Sato, K. et al. (2011). Use of the MALDI BioTyper system with MALDI–TOF mass spectrometry for rapid identification of microorganisms. *Anal. Bioanal. Chem.* 400 (7): 1905–1911.

16 Cubuk, J., Alston, J.J., Incicco, J.J. et al. (2021). The SARS-CoV-2 nucleocapsid protein is dynamic, disordered, and phase separates with RNA. *Nat. Commun.* 12 (1): 1936.

17 Fung, T.S. and Liu, T.X. (2018). Post-translational modifications of coronavirus proteins: roles and function. *Future Virol.* 13 (6): 405–430.

18 Nachtigall, F.M., Pereira, A., Trofymchuk, O.S. et al. (2020). Detection of SARS-CoV-2 in nasal swabs using MALDI-MS. *Nat. Biotechnol.* 38 (10): 1168–1173.

19 Tran, N.K., Howard, T., Walsh, R. et al. (2021). Novel application of automated machine learning with MALDI-TOF-MS for rapid high-throughput screening of COVID19: a proof of concept. *Sci. Rep.* 11: 8219.

20 Armengaud, J. (2021). Unleashing immuno-mass spectrometry superpowers to detect SARS-CoV-2. *EBioMedicine* 69: 103480.

21 Chavan, S., Mangalaparthi, K.K., Singh, S. et al. (2021). Mass spectrometric analysis of urine from COVID-19 patients for detection of SARS-CoV-2 viral antigen and to study host response. *J. Proteome Res.* 20 (7): 3404–3413.

22 Witze, S.E., Old, W.M., Resing, A.K., and Ahn, N.G. (2007). Mapping protein post-translational modifications with mass spectrometry. *Nat. Methods* 4: 798–806.

23 Nishikaze, T. (2019). Sialic acid derivatization for glycan analysis by mass spectrometry. *Proc. Jpn. Acad. Ser. B Phys. Biol. Sci.* 95 (9): 523–537.

24 Peng, Y., Gu, B., Sun, Z. et al. (2021). Linkage-selective derivatization for glycosylation site-and glycoform-specific characterization of sialic acid isomers using mass spectrometry. *Chem. Comm.* 57 (75): 9590–9593.

25 Bouwman, K.M., Tomris, I., Turner, H.L. et al. (2021). Multimerization- and glycosylation-dependent receptor binding of SARS-CoV-2 spike proteins. *PLoS Pathog.* 17 (2): e1009282.

26 Pikora, C.A. (2004). Glycosylation of the ENV spike of primate immunodeficiency viruses and antibody neutralization. *Curr. HIV Res.* 2 (3): 243–254.

27 Zhang, M., Gaschen, B., Blay, W. et al. (2004). Tracking global patterns of N-linked glycosylation site variation in highly variable viral glycoproteins: HIV, SIV, and HCV envelopes and influenza hemagglutinin. *Glycobiology* 14 (12): 1229–1246.

28 Pais, R.J., Iles, R.K., and Zmuidinaite, R. (2021). MALDI-ToF mass spectra Phenomic analysis for human disease diagnosis enabled by cutting-edge data processing pipelines and bioinformatics tools. *Curr. Med. Chem.* 28 (32): 6532–6547.

29 Seethi, V. D. R., LaCasse, Z., Chivte, P., Gaillard, E. R., Bharti, P. (2021). An explainable-AI approach for diagnosis of COVID-19 using MALDI-ToF mass spectrometry. https://arxiv.org/abs/2109.14099 (accessed 26 September 2022).

30 Kononenko, I. (1993). Inductive and Bayesian learning in medical diagnosis. *Appl. Artif. Intell. Int. J.* 7 (4): 317–337.

31 Butler-Laporte, G., Lawandi, A., Schiller, I. et al. (2021). Comparison of saliva and nasopharyngeal swab nucleic acid amplification testing for detection of SARS-CoV2: a systematic review and meta-analysis. *JAMA Intern Med.* 181 (3): 353–360.

32 Xu, X., Sun, J., Nie, S. et al. (2020). Seroprevalence of immunoglobulin M and G antibodies against SARS-CoV-2 in China. *Nat. Med.* 26: 1193–1195.

33 Buitrago-Garcia, D., Egli-Gany, D., Counotte, M.J. et al. (2020). Occurrence and transmission potential of asymptomatic and presymptomatic SARS-CoV-2 infections: a living systematic review and meta-analysis. *PLoS Med.* 17: e1003346.

34 Byambasuren, O., Cardona, M., Bell, K. et al. (2020). Estimating the extent of asymptomatic COVID-19 and its potential for community transmission: systematic review and meta-analysis. *JAMMI* 5 (4): 223–234.

35 Meyerowitz, E.A., Richterman, A., Isaac, I. et al. (2020). Towards an accurate and systematic characterisation of persistently asymptomatic infection with SARS-CoV-2. *Lancet Infect. Dis.* 21 (6): e163–e169.

36 Dietterich, T. (1995). Overfitting and under computing in machine learning. *ACM Comput. Surv.(CSUR)* 27 (3): 326–327.

37 Peng, Y. and Nagata, M.H. (2020). An empirical overview of nonlinearity and overfitting in machine learning using COVID-19 data. *Chaos, Solitons & Fractals* 139: 110055.

38 Carnell, G., Grehan, K., Ferrara, F. et al. (2017). An optimized method for the production using PEI, titration and neutralization of SARS-CoV spike luciferase pseudotypes. *Bio Protoc.* 7 (16): e2514.

39 Bentley, E.M., Mather, S.T., and Temperton, N.J. (2015). The use of pseudotypes to study viruses, virus sero-epidemiology and vaccination. *Vaccine* 33 (26): 2955–2962.

40 Anderson, M. (2020). The ultraviolet offense: germicidal UV lamps destroys vicious viruses. New tech might put them many more places without harming humans. *IEEE Spectrum* 57 (10): 50–55.

41 Ruetalo, N., Businger, R., and Schindler, M. (2021). Rapid, dose-dependent and efficient inactivation of surface dried SARS-CoV-2 by 254 nm UV-C irradiation. *Euro. Surveill.* 26 (42): 2001718.

42 Chen, C.H., Lin, J.L., Chu, M.L., and Chen, C.H. (2010). MALDI ion trap mass spectrometer with charge detector for large biomolecule detection. *Anal. Chem.* 82 (24): 10125–10128.

43 Sage, E., Brenac, A., Alava, T. et al. (2015). Neutral particle mass spectrometry with nanomechanical systems. *Nat. Commun.* 6 (1): 1–5.

44 van Belkum, A., Chatellier, S., Girard, V. et al. (2015). Progress in proteomics for clinical microbiology: MALDI-TOF MS for microbial species identification and more. *Expert Rev. Proteomics* 12 (6): 595–605.

45 Lefferts, J.A., Gutmann, E.J., Martin, I.W. et al. (2020). Implementation of an emergency use authorization test during an impending national crisis. *J. Mol. Diagn.* 22 (7): 844–846.

46 Chan, D.W., Semmes, O.J., Petricoin, E.F. et al. (2006). *National Academy of Clinical Biochemistry Guidelines: The Use of MALDI-TOF Mass Spectrometry Profiling to Diagnose Cancer.* American Association for Clinical Chemistry.

47 Chivte, P., LaCasse, Z., Seethi, V.D.R. et al. (2021). MALDI-ToF protein profiling as a potential rapid diagnostic platform for COVID-19. *J. Mass Spectrom. Adv. Clin Lab.* 21: 31–41.

48 Meldrum, O.W., Yakubov, G.E., Bonilla, M.R. et al. (2018). Mucin gel assembly is controlled by a collective action of non-mucin proteins, disulfide bridges, Ca^{2+} mediated links, and hydrogen bonding. *Sci. Rep.* 8: 5802.

49 Ferguson, J., Dunn, S., Best, A., Mirza, J., Percival, B., Mayhew, M., Megram, O., Ashford, F., White, T., Moles-Garcia, E., Crawford, L., Plant, T., Bosworth, A., Kidd, M., Richter, A., Deeks, J., McNally, A. (2020). Validation testing to determine the effectiveness of lateral flow testing for asymptomatic SARS-CoV-2 detection in low prevalence settings. https://www.medrxiv.org/content/10.1101/2020.12.01.20237784v2 (accessed 26 September 2022).

50 Bhattacharyya, D., Kumari, N.M.J., Joshua, E.S.N., and Rao, N.T. (2020). Advanced empirical studies on group governance of the novel Corona virus, MERS, SARS and EBOLA: a systematic study. *Int. J. Curr. Res. Rev* 12: 35–41.

51 Cheng, V.C., Lau, S.K., Woo, P.C., and Yuen, K.Y. (2007). Severe acute respiratory syndrome coronavirus as an agent of emerging and reemerging infection. *Clin. Microbiol. Rev.* 20 (4): 660–694.

52 Griffin, J.H. and Downard, K.M. (2021). Mass spectrometry analytical responses to the SARS-CoV2 coronavirus in review. *Trends Analyt. Chem.* 142: 116328.

53 Sivanesan, I., Gopal, J., Vinay, R.S. et al. (2022). Consolidating the potency of matrix-assisted laser desorption/ionization-time of flight mass spectrometry (MALDI-TOF MS) in viral diagnosis: extrapolating its applicability for COVID diagnosis? *Trends Analyt. Chem.* 150: 116569.

6

A MALDI-TOF MS Proteotyping Approach for Environmental, Agricultural and Food Microbiology

Hiroto Tamura

Department of Environmental Bioscience, Laboratory of Environmental Microbiology, Meijo University, Nagoya, Japan

6.1 Introduction

The environment is a source of foodborne pathogens which are ubiquitously detected in agricultural environments, food animals, food, and food-processing environments. As foodborne pathogens cause severe life-threatening diseases, they are a great and global concern for public health, food industry, and regulatory authorities as one of the most important public health hazards. Those pathogens also become an economic hazard because they cause significant economic loss through product recalls and withdrawals of suspected food in the distribution chain. In those routine microbiology laboratories, therefore, a method for the rapid and cost-effective identification of pathogenic and/or opportunistic bacteria has been getting more important than ever. In terms of sustainable development goals (SDGs), its establishment brings the following benefits: decreasing food waste in association with maintaining resources of food supply and unnecessary recalls, decreasing risk of starvation.

Despite substantial efforts by the food industry and public health and regulatory authorities to prevent (control) their infections and/or outbreak, it is hard to eradicate foodborne pathogens because of their ubiquitous presence in the environment and the many potential avenues for escape to invade the human body, particularly via ready-to eat (RTE) foods. In particular, the US Food and Drug Administration (FDA) upholds its policy of zero-tolerance for *Listeria monocytogenes* in RTE foods [1]. Moreover, foodborne pathogens are specific serovar variants or highly related species (lineages) carrying toxin genes in the environmentally ubiquitous reservoir. However, physiological and pathobiochemical tests, and the 16S rRNA gene sequencing technique are still used widely for routine bacterial identification, even though the use of conventional microbiological methods is limited from family to species and is not applicable to discriminating pathogenic and/or opportunistic bacteria at the strain or serovar level. Moreover, although a variety of the molecular typing methods

Microbiological Identification using MALDI-TOF and Tandem Mass Spectrometry: Industrial and Environmental Applications, First edition. Edited by Haroun N. Shah, Saheer E. Gharbia, Ajit J. Shah, Erika Y. Tranfield, and K. Clive Thompson.

have been developed, including pulsed-field gel electrophoresis (PFGE), multilocus sequence-based typing (MLST), ribotyping, and phage-typing as well as the use of the antisera and/or antibody (summarized in reviews in [2–4]). These classical methods take up to about one week to confirm results for bacteria responsible for an infection. Therefore, in terms of the human health protection and food safety, a crucial key for diagnosis and prevention of foodborne illness is the establishment of a rapid and accurate discrimination method for foodborne pathogens at the strain or serovar level. The development of a fully satisfactory discrimination method for foodborne bacteria has become one of the most challenging and state-of-the art goals in those fields.

With the arrival of mass spectrometry (MS) in bacterial identification, using unique signatures for bacterial characterization obtained from bacterial extracts and low molecular biomarkers [5, 6], this robust MS approach has opened a new window for bacterial identification and has received much attention as a promising detection method for microorganisms.

Today, as one of the most reliable MS techniques, matrix-assisted laser desorption/ionization time-of-flight mass spectrometry (MALDI-TOF MS) analysis of whole cells by soft ionization has been developed and successfully applied to identifying not only clinically important microorganisms in diagnostic laboratories but also foodborne bacteria in food industries and institutes for health [7–9]. In the MALDI-TOF MS analysis, a microorganism sample derived from the single colony is mixed with just a few microliters of matrix solution and the resultant mixture is put on plate for analysis, which takes less than one minute per sample. The feature of the MALDI-TOF MS analysis is its ease of use and the possibility of treating many specimens at once, resulting in reductions in the time and cost of bacterial identification. The technique most utilized in bacterial identification by MALDI-TOF MS is the fingerprint method, which can identify a great variety of bacteria at the species level more easily by comparing the mass spectra of the unknown target isolates with reference spectra of well-characterized known species included in the commercially available database. The fingerprint method is easier, more rapid, higher-throughput and of lower cost than conventional bacterial identification techniques [10–15], leading to a significant rekindling of interest in bacterial identification. Since bacterial identification by MALDI-TOF MS was authorized by the US FDA in 2013, the fingerprint method has been burgeoning, especially in routine clinical microbiology laboratories. However, one important demerit of the MALDI-TOF MS fingerprint method is the limitation in its discriminability at the strain or serovar level, because the closely related pathogenic and/or opportunistic microorganisms give very similar mass spectra. Accordingly, the discriminatory power of the fingerprint method is of growing concern, owing to the inadequacy of the conventional fingerprint approach for the typing of foodborne bacteria such as *Salmonella enterica* or *Escherichia coli* at the serovar level in outbreak investigations. The use of selected masses by statistical algorithmic methods [16–19] is one possibility to address the optimization of reliable routine bacterial discrimination at the strain or serovar level.

Fortunately, the most informative masses with molecular weight range (m/z) 4000–15 000 observed by MALDI-TOF MS analysis are derived mostly from ribosomal and other housekeeping proteins [20–22].

Given that their observed masses can be theoretically deduced from their corresponding amino acid sequences associated with the target genes, the masses as biomarkers give us a useful clue to develop a bioinformatics-based approach for reliable discrimination of closely related foodborne bacteria at the strain or serovar level [8, 21–25]. Although the technology of next-generation pyrosequencing has provided a critical boost to whole-genome sequencing, it is still essential for user-friendliness to establish and standardize simple and reliable construction procedures for the database of ribosomal proteins as a biomarker mines database by the conventional DNA sequencer.

To address the solution to this challenge, the *S10-spc-alpha* operon was selected for biomarker mines for the following reasons: (i) this operon encodes more than half of the bacterial ribosomal proteins; (ii) it is highly conserved among bacterial genomes; and (iii) its genome size is approximately 15–18 kb [26, 27].

Accordingly, the use of *in silico*-calculated masses as theoretical mass-to-ion ratios (*m/z*) on the basis of the DNA sequence information in this operon holds the promise of specific candidate biomarkers for the discrimination of bacteria at the strain or serovar level.

In the absence of sequenced strains, the DNA sequence analysis of the operon can be performed by designed primers using the consensus DNA sequences among the target genus and/or species of the bacteria. The important features are the post-translational modifications such as *N*-terminal methionine loss, methylation, and acetylation [22, 28–30], and incorrect annotations in the sequenced genomes in the databases, giving the differences between the calculated and observed *m/z* of the ribosomal proteins as biomarkers [29, 31–33].

To establish and standardize a simple and reliable discrimination method for bacteria at the strain or serovar level, the *S10*-GERMS (*S10-spc-alpha* operon Gene-Encoded Ribosomal protein Mass Spectrum) method was developed as the standardized proteotyping method that combines genomics and proteomics (Figure 6.1). The *S10*-GERMS offers an accurate method to construct a database by comparing the experimentally observed *m/z* values of the selected biomarkers with their theoretically calculated *m/z* values, and has been employed as a typing method for various taxa. The *S10*-GERMS method can discriminate the target bacteria at strain and serovar level in minutes.

The theory of the standardized *S10*-GERMS method and its practical use of successful discrimination of environmental and agricultural bacteria were well documented in our previous book [35], as follows: *Pseudomonas putida* at the strain level [36], *Pseudomonas syringae* at the pathovar level [34], the genera *Bacillus* [37] and *Sphingopyxis* which had only two and one base differences in the 16S rRNA gene sequence [38], *Lactobacillus casei* at the subspecies level [39], and the enterohemorrhagic *Escherichia coli* (EHEC) O157, O26, and O111 serovars with a high level of confidence [40, 41].

In this chapter, the *S10*-GERMS method focused on the characterization of important foodborne bacteria at the strain or serovar level in the food environment to improve control and risk management of foodborne infection, as well as on the identification of the isolates with octylphenol polyethoxylate ($OPEO_n$)-degrading capability in the soil environment. This proteotyping method by a reliable MS technique may contribute to the development of a fully satisfactory discrimination method for bacteria, one of the most challenging and state-of-the art goals in those fields.

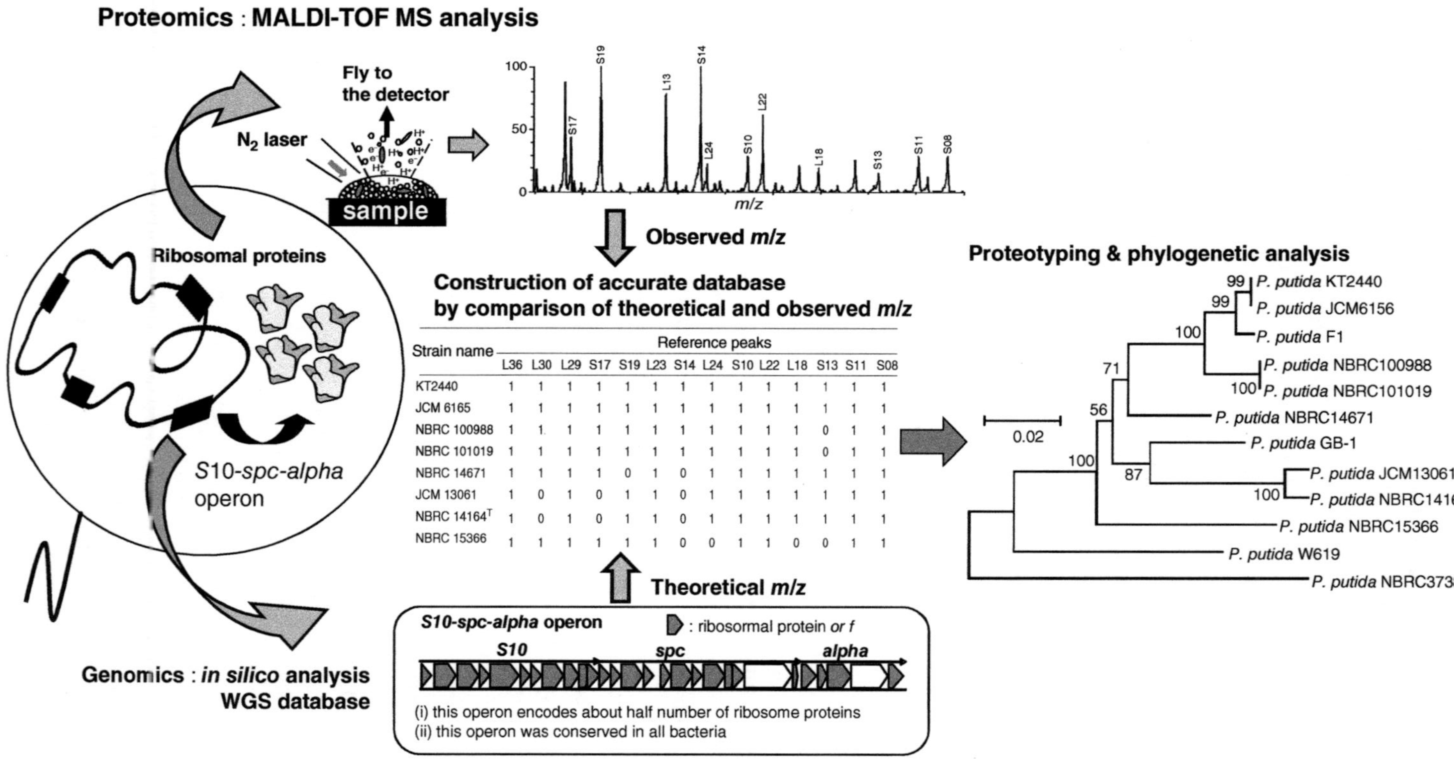

Figure 6.1 Concept of the *S10*-GERMS method for bacterial proteotyping. The theoretical *m/z* can be more conveniently obtained from WGS by enhancing the WGS database. *Source:* Adapted from Tamura et al. [34].

6.2 Serotyping of *Salmonella enterica* Subspecies *enterica*

The genus *Salmonella* has the ability to persist in various environments, including hosts such as plants, animals, and humans. Most notably, the foodborne bacterium *S. enterica* subsp. *enterica* causes severe outbreaks and life-threatening diseases such as salmonellosis in humans as well as a significant economic losses in food industries worldwide. However, in *S. enterica* there are more than 2600 serovars, and fewer than 100 serovars have been reported as causative bacteria in human health [42]. Among them, serotypes Enteritidis, Typhimurium, and Newport had been identified for nearly half of human isolates in the USA [43], and Enteritidis is the one that is most often isolated as a prevalent serotype, followed by Typhimurium, while Infantis and Saintpaul are uncommon in Japan (https://www.niid.go.jp/niid/ja).

Therefore, serotyping of *Salmonella* has been the crucial key in human health protection and the quality control of food products in industry.

Given the ever-growing requirement in routine microbiology laboratories, the rapid and cost-effective discrimination method of *S. enterica* subsp. *enterica* at the serovar level is increasingly important.

Traditionally, the classification of the genus *Salmonella* into serovars known as the serotype is characterized based on three surface antigens, including the flagellar (H), the oligosaccharide (O), and the polysaccharide Vi antigens, based on agglutination reactions with specific antibodies [44]. To operate conventional serotyping, however, more than 250 different antisera and 350 antigens are required in routine microbiology laboratories and it takes several days to obtain the result, which is sometimes false-positive because of its unsatisfactory discrimination ability. A number of research studies on the discrimination of foodborne bacteria have used traditional methods such as enzyme-linked immunosorbent assay (ELISA) and nucleic acid-based assays, as follows: PFGE, MLST, multilocus variable-number tandem repeat analysis (MLVA) (see, e.g., review in [4]. However, their major limitations lie in their poor discrimination ability and, in particular, in the fact that they are burdensome tasks.

To overcome the limit of the conventional serotyping of *Salmonella*, extensive research into serotyping of *S. enterica* subsp. *enterica* by MALDI-TOF MS has been performed and several specific biomarker peaks for Enteritidis, Typhimurium, Virchow, Infantis, Hadar, and so on have been found as an excellent solution to serotyping *S. enterica* subsp. *enterica* [45]. The major *Salmonella* serotypes were identified by the proposed hierarchical decision tree network-like approach.

As the *S10*-GERMS method is theoretically supported by genomics and proteomics approaches [34, 36] and provides an accurate mass (m/z) database for serotype-specific biomarkers, the proteotyping of isolate can be performed using the results of mass-matching profiles of the registered biomarkers in the Strain Solution™ v.2 software (Shimadzu Corporation, Kyoto, Japan), which is an advanced format program that is automatic and highly reproducible and can be used for detailed classification of microorganisms at the strain or serotype level, beyond the conventional MALDI-TOF MS fingerprinting method. Therefore, *S10*-GERMS was promising for the analysis of closely related species because bacteria with only one base difference in the 16S rRNA gene sequence were able to be discriminated using the m/z difference of the selected biomarker [38].

To achieve automatic and systematic proteotyping of *S. enterica* at the serotype level by MALDI-TOF MS analysis, first, accurate mass databases for *S. enterica* based upon the principle of the *S10*-GERMS method (Figure 6.1) are constructed, as follows. (i) The genome-sequenced strains are analyzed by MALDI-TOF MS to obtain "the observed *m/z* values." (ii) "The theoretical *m/z* values" of ribosomal proteins encoded in *S10-spc-alpha* operon are calculated by using genome-sequenced strains from the NCBI databank and, as an *in silico* study, the candidate biomarkers of proteotyping at the serotype level are selected by comparison with "the theoretical *m/z* values" of each ribosomal protein. When there is no available genome sequence information of the candidate biomarkers in the NCBI databank, their *m/z* values are determined by DNA sequence using designed primers against the consensus sequences. The DNA primers for sequencing experiments are listed in Table 6.1. (iii) The reliable *m/z* values of serotype-specific biomarkers are corrected and confirmed by comparing "the observed *m/z* values" of the candidate biomarkers with their "theoretical *m/z* values." (iv) Finally, the reliable and accurate *m/z* values of the biomarkers specific for serotyping of *S. enterica* subsp. *enterica* (Table 6.2) were registered on the Strain Solution ver. 2 software [46]. The constructed database, using a total of 31 strains with different serotypes in Strain Solution ver. 2 software, was further evaluated using 116 isolates of *S. enterica* strains. Analytical conditions of the MALDI-TOF MS AXIMA Microorganism Identification System (Shimadzu Corporation) were almost the same as described in a previous report [41], except for the modified matrix solution, which was optimized for *S. enterica* analysis by Dieckmann and Malorny [45]. Briefly, a 1.2 μl droplet of a mixture of 10 μl of matrix solution [25 mg/ml sinapic acid (SA), 0.6% (v/v) trifluoroacetic acid (TFA), 50% (v/v) acetonitrile] and bacteria scratched from colony was spotted onto an analytical metal plate pre-layered with 0.5 μl of saturated SA solution dissolved in ethanol. According to the software instructions, the obtained analytical data were incorporated into the Strain Solution ver. 2 software and scanned with default settings, as follows: mass tolerance at 500 ppm, auto peak selection mode. Before scanning, with the addition of the usual daily calibration by *E. coli*, the calibration of mass spectra was performed by the common peaks with identical *m/z* values observed in all *S. enterica* strains, which are eight ribosomal proteins and DNA-binding protein, H-NS: *m/z* $[M+H]^+$ 4265.34 (L36), 6383.60 (L30), 8369.71 (S21), 10 286.07 (S19), 12 770.64 (L18), 15 412.50 (H-NS), 16 019.52 (L13), 17 473.18 (S5), and 23 354.87 (S4). For fingerprinting analysis, α-cyano-4-hydroxycinnamic acid (CHCA) matrix at 20 mg/ml containing 1% (v/v) TFA and the SARAMIS database v.3.5 (VITEK MS; bioMérieux) were used.

Notwithstanding no DNA sequence diversity in most of the genes encoded in the operons, some ribosomal proteins such as L16 in *S10* operon, L24 in *spc* operon, S8 in *spc* operon, S5 in *spc* operon, L15 in *spc* operon, and L17 in *alpha* operon, had specific mutations leading to their mass differences as biomarker candidates. In particular, ribosomal proteins S8 in *spc* operon (*m/z* 13 996.36, or 14 008.414 specific to serotype Rissen), L15 in *spc* operon (*m/z* 14 967.38 or 14 948.33 specific to serotype Typhimurium), and L17 in *alpha* operon (*m/z* 14 395.61 or 14 361.59 specific to serotype Pullorum-Gallinarum) were selected as reproducibly detectable biomarkers by MALDI-TOF MS. However, those three biomarkers were insufficient to proteotype *S. enterica* subsp. *enterica* at the serovar level by MALDI-TOF MS analysis. As a recent development of whole-genome sequencing

Table 6.1 Primers used for sequencing ribosomal proteins encoded in the *S10-spc-alpha* operon [46].

Sequence	Description
AAGAACGGTTACACTCTCCC	Amplification of the *S10* region and sequencing
ATGTCGCCTACGCCTGCGTAGCG	Amplification of the *S10* region and sequencing
GCTGGCATGATTCGTGAAGAACG	Sequencing of the *S10* region
TGCTGAAGTAACTGGTTCCGG	Sequencing of the *S10* region
CATAACGTAGAAATGAAACCAGG	Sequencing of the *S10* region
AGCGTCGCTGATGTTACAAC	Sequencing of the *S10* region
ATCAATCGTAATGGGTGTGAG	Sequencing of the *S10* region
AAGCCGGAAGCGGGTCACTTTG	Sequencing of the *S10* region
CTTTTTGACTGTGGTCCTGC	Sequencing of the *S10* region
CGTTCTCTCAAGAAAGGTCC	Sequencing of the *S10* region
GCCGATTACACCATAAGTGG	Sequencing of the *S10* region
TAGCACGACGTGCTGCTTCGAT	Sequencing of the *S10* region
TCCGTACTCTGCAAGGTCGCGTG	Sequencing of the *S10* region
ACTCCGGTGCACGTCGCGTAATG	Amplification of the *spc* resion and sequencing
AGCAGTCTGCGTTTCAGCTC	Amplification of the *spc* resion and sequencing
ATTGTTGAAGGTATCAACCTG	Sequencing of the *spc* region
TTACCGGTTAACACGATAAC	Sequencing of the *spc* region
TCGTGGTAACTACAGCATG	Sequencing of the *spc* region
ACCATGCCTTCCTCCAAG CT	Sequencing of the *spc* region
ATGCTGCCCGTGAAGCTGGC	Sequencing of the *spc* region
ATCGGTCGTCTGCCGAAACAC	Sequencing of the spc region and amplification of the *alpha* region
GTCACCATGCCTTCCTCCAAG	Sequencing of the *spc* region
AATGTGTATCAAGGTTCTGGG	Sequencing of the *spc* region
CTTCACTAAGAAGCTGCAGCTG	Sequencing of the *spc* region
ATATCCAAACCACGAACGCG	Sequencing of the *spc* region
GATGTGAGCCATCTTACACCTCT	Sequencing of the *spc* region
TGCGGACATTAACGAACACCTG	Sequencing of the *alpha* region
TGCCTACAATGTTGAAGCAGCG	Sequencing of the *alpha* region
AATCATCAAGACGACCCTGCC	Sequencing of the *alpha* region
CGCGATAGCAACCAAGATCCA	Sequencing of the *alpha* region
CGCGAACGCCAGACTTAAGGA	Sequencing of the *alpha* region
ACTCTGTCACAGAACCCTGCA	Sequencing of the *alpha* region

(*Continued*)

Table 6.1 (Continued)

Sequence	Description
TGTACTGGCTGATCTTACGACG	Sequencing of the *alpha* region
GGTTCATCCAACGCTGGCAG	Sequencing of the *alpha* region
ATATGACGCTCTGCCCACTG	Amplification of the *alpha* region and sequencing
GCCGTAACGTTTATAACCCTGG	Amplification of the SodA gene
ACACGTCAACCGGATAATGCA	Amplification of the SodA gene
ATGAACCAACTGCTTACGC	Sequencing of the SodA gene
CGTTACCGTATGCGTTGTGT	Amplification of the L21 gene
GCACAGCCACACGGGTATAT	Amplification of the L21 gene
CACTTGTCAAAGCCGTGCAA	Sequencing of the L21 gene
GTAAAAAGCCCCGCAACAGG	Sequencing of the L21 gene
GCGATTGAAGGTAACCGCTT	Amplification and sequencing of the S7 gene
ATGTGCGCACTGATACCGAT	Amplification and sequencing of the S7 gene
GTCGTCAGGGTTGACTATACT	Amplification and sequencing of the S7 gene
CATGACGACACTGTCTTATTGC	Amplification and sequencing of the gns gene
TCGGTAAACCAGTCACCACT	Amplification and sequencing of the gns gene
TAAACTCAAATAAGCGRCCGCG	Amplification and sequencing of the YibT gene
TACGCCATGCAAATTCAGCGC	Amplification and sequencing of the YibT gene
AAAATCAAGCAGACGATGTAGGC	Amplification and sequencing of the PPIase gene
GTCTGACGGCCTAAGGAGTGAT	Amplification and sequencing of the PPIase gene
TCGACTATGAAGGCCGCTTC	Amplification and sequencing of the L25 gene
GGGCGGGGTTTTTCTTCAAC	Amplification of the L25 gene
ATTATTGACGCGACGAAAGC	Amplification and sequencing of the YaiA gene
AATGTTTCAAGAATGTGATGGAGA	Amplification and sequencing of the YaiA gene
ATAGTCACGGCGGACGTAAA	Amplification and sequencing of the YciF gene
CCTCAATGCGTGATAATCTGC	Amplification and sequencing of the YciF gene

Source: Ojima-Kato et al. [46]/With permission of Springer Nature.

(WGS), the molecular-based methods open the way for serotyping of foodborne pathogens like *S. enterica* serovars [4]. By using WGS data, our method can also extend the biomarker pipeline from the *S10-spc-alpha* operon to the bacterial whole genome. By matching mass spectra observed by MALDI-TOF MS analysis against genetic information from the web-accessible WGS database of target bacteria, nine further

Table 6.2 Theoretical *m/z* ([M+H]$^+$) values of selected biomarkers for serotyping of *S. enterica* [46].

			Biomarker											
Group	**Serotype**	**Strain**	**Gns**	**YaiA**	**YibT**	**PPIase**	**L25**	**L21**	**S8**	**L17**	**L15**	**S7**	**YciF**	**SodA**
1	Enteritidis	GTC08914	6483.51 (I)	7110.89 (I)	7993.05 (I)	10 198.07 (I)	10 542.19 (I)	11 579.36 (I)	13 996.36 (I)	14 395.61 (I)	14 967.38 (I)	17 474.18 (I)	18 643.13 (I)	23 010.84 (I)
2	Typhimurium	NBRC13245	6483.51 (I)	7096.82 (III)	7993.05 (I)	10 198.07 (I)	10 542.19 (I)	11 579.36 (I)	13 996.36 (I)	14 395.61 (I)	14 967.38 (I)	17 460.15 (II)	18 653.16 (II)	22 948.82 (VI)
	UN	GTC09549	6483.51 (I)	7096.82 (III)	7993.05 (I)	10 198.07 (I)	10 542.19 (I)	11 579.36 (I)	13 996.36 (I)	14 395.61 (I)	14 967.38 (I)	17 460.15 (II)	18 653.16 (II)	22 948.82 (VI)
3	Minnesota	NBRC15182	6483.51 (I)	7110.89 (I)	7993.05 (I)	10 198.07 (I)	10 542.19 (I)	11 565.33 (II)	13 996.36 (I)	14 395.61 (I)	14 967.38 (I)	17 460.15 (II)	18 643.13 (I)	22 976.83 (IV)
	Infantis	ATCC BAA-1675	6483.51 (I)	7110.89 (I)	7993.05 (I)	10 198.07 (I)	10 542.19 (I)	11 565.33 (II)	13 996.36 (I)	14 395.61 (I)	14 967.38 (I)	17 460.15 (II)	18 643.13 (I)	22 976.83 (IV)
	Brandenburg	jfrlSe1402–3	6483.51 (I)	7110.89 (I)	7993.05 (I)	10 198.07 (I)	10 542.19 (I)	11 565.33 (II)	13 996.36 (I)	14 395.61 (I)	14 967.38 (I)	17 460.15 (II)	18 643.13 (I)	22 976.83 (IV)
4	Rissen	jfrlSe1402–6	6483.51 (I)	7110.89 (I)	7993.05 (I)	10 198.07 (I)	10 542.19 (I)	11 579.36 (I)	14 008.41 (II)	14 395.61 (I)	14 967.38 (I)	17 474.18 (I)	18 643.13 (I)	23 004.88 (II)
5	Infantis	jfrlSe1402–1	6511.56 (II)	7110.89 (I)	7993.05 (I)	10 198.07 (I)	10 542.19 (I)	11 579.36 (I)	13 996.36 (I)	14 395.61 (I)	14 967.38 (I)	17 460.15 (II)	18 643.13 (I)	22 976.83 (IV)
6	Schwarzengrund	HyogoSO12004	6483.51 (I)	7110.89 (I)	7993.05 (I)	10 198.07 (I)	10 542.19 (I)	11 565.33 (II)	13 996.36 (I)	14 395.61 (I)	14 967.38 (I)	17 474.18 (I)	18 643.13 (I)	22 948.77 (VII)
	Montevideo	jfrlSe1409–6	6483.51 (I)	7110.89 (I)	7993.05 (I)	10 198.07 (I)	10 542.19 (I)	11 565.33 (II)	13 996.36 (I)	14 395.61 (I)	14 967.38 (I)	17 474.18 (I)	18 643.13 (I)	22 948.77 (VII)
7	Mbandaka	jfrlSe1402–13	6511.56 (II)	7110.89 (I)	7993.05 (I)	10 198.07 (I)	10 542.19 (I)	11 579.36 (I)	13 996.36 (I)	14 395.61 (I)	14 967.38 (I)	17 474.18 (I)	18 653.16 (II)	22 976.83 (IV)
	Mbandaka	jfrlSe1402–14	6511.56 (II)	7110.89 (I)	7993.05 (I)	10 198.07 (I)	10 542.19 (I)	11 579.36 (I)	13 996.36 (I)	14 395.61(I)	14 967.38(I)	17 474.18(I)	18 653.16(II)	22 976.83(IV)
8	Orion	jfrlSe1402–7	6483.51(I)	7110.89(I)	7993.05(I)	10 216.11(II)	10 542.19(I)	11 579.36(I)	13 996.36(I)	14 395.61(I)	14 967.38(I)	17 460.15(II)	18 643.13(I)	22 948.82(VI)
9	Pullorum_ Gallinarum	NBRC3163	6483.51(I)	7109.91(II)	7993.05(I)	10 198.07(I)	10 542.19(I)	11 579.36(I)	13 996.36(I)	14 381.59(II)	14 967.38(I)	17 474.18(I)	18 643.13(I)	23 010.84(I)
10	Abony	NBRC100797	6483.51(I)	7110.89(I)	7993.05(I)	10 198.07(I)	10 542.19(I)	11 579.36(I)	13 996.36(I)	14 395.61(I)	14 967.38(I)	17 460.15(II)	18 653.16(II)	23 010.84(I)
11	Chloeraesuis	NBRC105684	6483.51(I)	7110.89(I)	7993.05(I)	10 198.07(I)	10 542.19(I)	11 579.36(I)	13 996.36(I)	14 395.61(I)	14 967.38(I)	17 432.10(III)	18 653.16(II)	22 918.79(VIII)
12	UN	JCM3919	6483.51(I)	7110.89(I)	7993.05(I)	10 198.07(I)	10 542.19(I)	11 579.36(I)	13 996.36(I)	14 395.61(I)	14 967.38(I)	17 460.15(II)	18 653.16(II)	22 976.83(IV)
13	Saintpaul	ATCC9712	6483.51(I)	7110.89(I)	7993.05(I)	10 198.07(I)	10 542.19(I)	11 579.36(I)	13 996.36(I)	14 395.61(I)	14 967.38(I)	17 460.15(II)	18 653.16(II)	22 948.82(VI)
	UN	jfrlSe1409–8	6483.51(I)	7110.89(I)	7993.05(I)	10 198.07(I)	10 542.19(I)	11 579.36(I)	13 996.36(I)	14 395.61(I)	14 967.38(I)	17 460.15(II)	18 653.16(II)	22 948.82(VI)
14	Braenderup	GTC09492	6483.51(I)	7110.89(I)	8023.08(II)	10 198.07(I)	10 528.17(II)	11 579.36(I)	13 996.36(I)	14 395.61(I)	14 967.38(I)	17 474.18(I)	18 653.16(II)	22 962.80(V)
15	Pakistan	GTC09493	6483.51(I)	7110.89(I)	7993.05(I)	10 198.07(I)	10 542.19(I)	11 579.36(I)	13 996.36(I)	14 395.61(I)	14 967.38(I)	17 460.15(II)	18 659.13(III)	22 996.82(III)
16	Thompson	ATCC BAA-1738	6483.51(I)	7110.89(I)	8023.08(II)	10 198.07(I)	10 542.19(I)	11 579.36(I)	13 996.36(I)	14 395.61(I)	14 967.38(I)	17 474.18(I)	18 653.16(II)	22 976.83(IV)
17	Altona	jfrlSe1409–1	6483.51(I)	7110.89(I)	7993.05(I)	10 198.07(I)	10 542.19(I)	11 579.36(I)	13 996.36(I)	14 395.61(I)	14 948.33(II)	17 474.18(I)	18 643.13(I)	22 976.83(IV)
18	Istanbul	jfrlSe1409–2	6483.51(I)	7110.89(I)	7863.94(IV)	10 198.07(I)	10 542.19(I)	11 579.36(I)	13 996.36(I)	14 395.61(I)	14 967.38(I)	17 474.18(I)	18 653.16(II)	22 976.83(IV)
19	Manhattan	HyogoSO11001	6511.56(II)	7110.89(I)	7993.05(I)	10 198.07(I)	10 528.17(II)	11 579.36(I)	13 996.36(I)	14 395.61(I)	14 967.38(I)	17 474.18(I)	18 711.20(V)	22 976.83(IV)

(Continued)

Table 6.2 (Continued)

Group	Serotype	Strain	Biomarker											
			Gns	YaiA	YibT	PPIase	L25	L21	S8	L17	L15	S7	YciF	SodA
20	UN	jfrlSe1409-4	6483.51(I)	7110.89(I)	7993.05(I)	10198.07(I)	10542.19(I)	11579.36(I)	13996.36(I)	14395.61(I)	14967.38(I)	17474.18(I)	18653.16(II)	22976.83(IV)
21	Senftenberg	jfrlSe1409-3	6511.56 (II)	7110.89 (I)	7935.02 (III)	10198.07 (I)	10542.19 (I)	11579.36 (I)	13996.36 (I)	14395.61 (I)	14967.38 (I)	17474.18 (I)	18605.1 (IV)	22976.83 (IV)
	UN	jfrlSe1409-7	6511.56 (II)	7110.89 (I)	7935.02 (III)	10198.07 (I)	10542.19 (I)	11579.36 (I)	13996.36 (I)	14395.61 (I)	14967.38 (I)	17474.18 (I)	18605.1 (IV)	22976.83 (IV)
	UN	jfrlSe1409-11	6511.56 (II)	7110.89 (I)	7935.02 (III)	10198.07 (I)	10542.19 (I)	11579.36 (I)	13996.36 (I)	14395.61 (I)	14967.38 (I)	17474.18 (I)	18605.1 (IV)	22976.83 (IV)
	UN	jfrlSe1409-5	6511.56 (II)	7110.89 (I)	7935.02 (III)	10198.07 (I)	10542.19 (I)	11579.36 (I)	13996.36 (I)	14395.61 (I)	14967.38 (I)	17474.18 (I)	18605.1 (IV)	22976.83 (IV)
22	Amsterdam	jfrlSe1409-21	6511.56 (II)	7110.89 (I)	7993.05 (I)	10198.07 (I)	10528.17 (II)	11579.36 (I)	13996.36 (I)	14395.61 (I)	14967.38 (I)	17460.15 (II)	18653.16 (II)	22976.83 (IV)

In parentheses, different mass patterns of each biomarker were indicated as Roman numerals. The theoretical *m/z* value of Enteritidis was expressed as the Roman numeral I as the serotype standard for each biomarker.
Source: Ojima-Kato et al. [46]/With permission of Springer Nature.

promising biomarkers were selected as follows: Gns protein (*m*/*z* 6483.51 or 6511.56), uncharacterized protein YaiA (*m*/*z* 7110.89, 7109.91, or 7096.82), uncharacterized protein YibT (*m*/*z* 7993.05, 8023.08, 7935.02, or 7863.94), peptidyl-prolyl *cis-trans* isomerase C (PPIase C; *m*/*z* 10 198.07 or 10 216.11), ribosomal protein L25 (*m*/*z* 10 542.19 or 10 528.17), ribosomal proteins L21 (*m*/*z* 11 579.36 or 11 565.33), S7 (*m*/*z* 17 460.15, 17 474.18, or 17 432.10), uncharacterized protein YciF (*m*/*z* 18 643.13, 18 653.16, 18 659.13, 18 605.10, or 18 711.20), and Mn-cofactor-containing superoxide dismutase SodA (*m*/*z* 23010.84, 23 004.88, 22 996.82, 22 976.83, 22 962.80, 22 948.82, 22 948.77, or 22 918.79). While Dieckmann and Malorny described that the *m*/*z* 6036 peak derived from a putative uncharacterized protein is important for differentiating Enteritidis from other serotypes, this peak was unidentified in 8/35 (23%) Enteritidis isolates in our study. Therefore, it was excluded from our database for the proteotyping of *S. enterica* subsp. *enterica* strains at the serotype level.

According to the mass patterns of 12 biomarkers selected for *Salmonella* serotyping, 31 strains with different serotypes were classified into 22 distinct groups (Table 6.2). Because of mass differences of biomarkers Gns and L21, the serotype Infantis had two diverse groups: groups 3 and 5. However, due to their identical proteotypes, the following serotypes comprised the same groups, as follows: group 3 includes Minnesota, Infantis, and Brandenburg; group 6 includes Schwarzengrund and Montevideo; group 13 includes Saintpaul and UN-O18; and group 21includes Senftenberg and UN-O1,3,19. Those biomarkers were registered on the database of Strain Solution ver. 2 software.

To evaluate the availability of biomarkers and the performance of Strain Solution ver. 2 software for *Salmonella* serotyping, the serovar known 116 isolates of *S. enterica* subsp. *enterica* obtained from various culture collections and natural isolates from patients and foods were analyzed.

In the comparison of Strain Solution ver. 2 software and SARAMIS fingerprinting, for SARAMIS fingerprinting 115 isolates were identified as "*S. enterica* subsp. *enterica*" and one isolate, *S. enterica* subsp. *enterica* Typhimurium NBRC13245, was simply identified as "*Salmonella* sp." while for the Strain Solution ver. 2 software, 109 isolates (94% of all) were successfully proteotyped as their correct serogroups with the highest hit score for the registered biomarkers although 16/109 isolates had the same scores as other groups Figure 6.2.

By contrast, the remaining seven isolates – NBRC14194 (Typhimurium), NBRC14212 (Typhimurium), jfrlSe1409-17 (UN), HyogoSO13004 (Saintpaul), HyogoSO12001 (Saintpaul), jfrlSe1507-4 (UN), and jfrlSe1507-6 (UN) – were discriminated as groups 13 (Saintpaul or UN), 13, 4 (Rissen), 12 (UN), 12, 5 (Infantis), and 1 (Typhimurium), respectively. Taken together, according to the resultant output of biomarker hit profiles and classification with the highest hit scores by Strain Solution scanning (Figure 6.3), the correct identification rates were as follows: Enteritidis (35/35, 100%), Typhimurium (11/13, 85%), Rissen (9/9, 100%), Minnesota (7/7, 100%), Infantis (6/6, 100%), Thompson (5/5, 100%), and Mbandaka (5/5, 100%). Although two Typhimurium isolates, NBRC14194 and NBRC14212, were classified into UN or Saintpaul (group 13 in Table 6.2) due to a mass difference of the YaiA peak which is crucial to differentiating between groups 2 and 13, the cluster analysis by the neighbor-joining method indicates that most serotypes make up their own clusters (Figure 6.2).

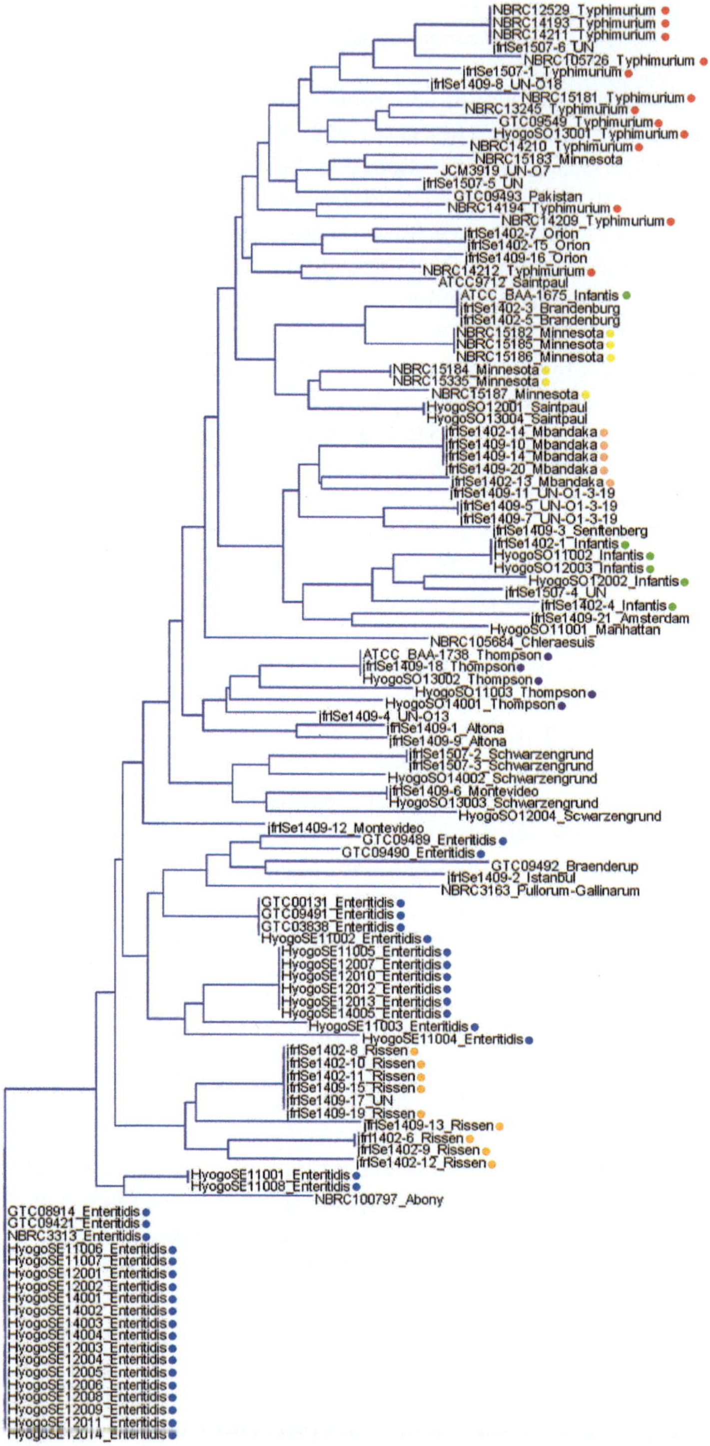

Figure 6.2 Cluster analysis of *S. enterica* subsp. *enterica* by Strain Solution ver. 2 using 12 serotyping specific biomarkers [46]. The biomarker hit profile of 116 *S. enterica* strains scanned by Strain Solution ver. 2 was converted to a dendrogram using the neighbor-joining method. The main serotypes analyzed in this study are marked with colored circles: Typhimurium (red), Enteritidis (blue), Rissen (orange), Infantis (green), Thompson (purple), Mbandaka (pink), Minnesota (yellow). Biomarker hit profiles are from Figure 6.3. *Source:* Ojima-Kato et al. [46]/With permission of Springer Nature.

	Biomarker *m/z* [M + H]⁺																																					Identification	
	Gns		YaiA			YibT				PPIase		L25		L21		S8		L17		L15		S7			YciF						SodA								
	6483.51	6511.56	7110.89	7109.91	7096.82	7993.05	8023.08	7935.02	7863.94	10198.07	10216.11	10542.19	10528.17	11579.36	11565.33	13996.36	14008.41	14395.61	14381.59	14967.38	14948.33	17474.18	17460.15	17432.10	18643.13	18653.16	18659.13	18605.10	18711.20	18634.12	23010.84	23004.88	22996.82	22976.83	22962.80	22948.82	22948.77	22918.79		
Strain Name	I	II	I	II	III	I	II	III	IV	I	II	I	II	I	II	I	II	I	II	I	II	I	II	III	I	II	III	IV	V	VI	I	II	III	IV	V	VI	VII	VIII	Group	Score
NBRC13245_Typhimurium	1	0	0	0	1	1	0	0	0	1	0	1	0	1	0	1	0	1	0	1	0	0	1	0	0	0	0	0	0	0	0	0	0	0	0	1	0	0	2	11
GTC00131_Enteritidis	1	0	1	0	0	0	0	0	0	1	0	1	0	1	0	1	0	1	0	1	0	1	0	0	0	0	0	0	0	0	1	0	0	0	0	0	0	0	1	10
GTC09491_Enteritidis	1	0	1	0	0	0	0	0	0	1	0	1	0	1	0	1	0	1	0	1	0	1	0	0	0	0	0	0	0	0	1	0	0	0	0	0	0	0	1	10
GTC03838_Enteritidis	1	0	1	0	0	0	0	0	0	1	0	1	0	1	0	1	0	1	0	1	0	1	0	0	0	0	0	0	0	0	1	0	0	0	0	0	0	0	1	10
GTC08914_Enteritidis	1	0	1	0	0	1	0	0	0	1	0	1	0	1	0	1	0	1	0	1	0	1	0	0	1	0	0	0	0	0	1	0	0	0	0	0	0	0	1	12
GTC09421_Enteritidis	1	0	1	0	0	1	0	0	0	1	0	1	0	1	0	1	0	1	0	1	0	1	0	0	1	0	0	0	0	0	1	0	0	0	0	0	0	0	1	12
GTC09489_Enteritidis	1	0	0	1	0	0	0	0	0	1	0	1	0	1	0	1	0	1	0	1	0	1	0	0	0	0	0	0	0	0	1	0	0	0	0	0	0	0	1	10
NBRC3163_Pullorum-Gallinarum	1	0	0	0	0	0	1	0	0	1	0	1	0	1	0	1	0	0	1	1	0	1	0	0	0	0	0	0	0	0	1	0	0	0	0	0	0	0	9	10
NBRC3313_Enteritidis	1	0	1	0	0	1	0	0	0	1	0	1	0	1	0	1	0	1	0	1	0	1	0	0	1	0	0	0	0	0	1	0	0	0	0	0	0	0	1	12
NBRC12529_Typhimurium	1	0	0	0	1	1	0	0	0	1	0	1	0	1	0	1	0	1	0	1	0	0	1	0	0	1	0	0	0	0	0	0	0	0	0	0	1	0	2	12
NBRC14193_Typhimurium	1	0	0	0	1	1	0	0	0	1	0	1	0	1	0	1	0	1	0	1	0	0	1	0	0	1	0	0	0	0	0	0	0	0	0	0	1	0	2	12
NBRC14194_Typhimurium	1	0	1	0	0	0	0	1	0	1	0	1	0	1	0	1	0	1	0	1	0	0	1	0	1	0	0	0	0	0	0	0	0	0	0	0	1	0	13	12
NBRC14209_Typhimurium	1	0	0	0	0	0	0	1	0	1	0	1	0	1	0	1	0	1	0	1	0	0	1	0	0	0	0	0	1	0	0	0	0	0	0	0	1	0	2 or 13	9
NBRC14210_Typhimurium	1	0	0	0	1	1	0	0	0	1	0	1	0	1	0	1	0	1	0	1	0	0	1	0	0	0	0	0	0	0	0	0	0	0	1	0	0	0	2	11
NBRC14211_Typhimurium	1	0	0	0	1	1	0	0	0	1	0	1	0	1	0	1	0	1	0	1	0	0	1	0	0	1	0	0	0	0	0	0	0	0	0	0	1	0	2	12
NBRC14212_Typhimurium	1	0	1	0	0	0	0	0	0	1	0	1	0	1	0	1	0	1	0	1	0	0	1	0	0	0	0	0	0	0	0	0	0	0	0	1	0	0	13	10
NBRC15181_Typhimurium	0	1	0	0	1	1	0	0	0	1	0	1	0	1	0	1	0	1	0	1	0	0	1	0	1	0	0	0	0	0	0	0	0	0	0	0	1	0	2	11
NBRC15182_Minnesota	1	0	1	0	0	1	0	0	0	1	0	1	0	0	1	1	0	1	0	1	0	0	1	0	0	0	0	0	0	0	0	0	0	1	0	0	0	0	3	11
NBRC15183_Minnesota	1	0	0	0	0	0	0	0	0	1	0	1	0	1	0	1	0	1	0	1	0	0	1	0	0	0	0	1	0	0	0	0	0	1	0	0	0	0	3 or 12	10
NBRC15184_Minnesota	1	0	1	0	0	1	0	0	0	1	0	1	0	1	0	1	0	1	0	1	0	0	1	0	0	0	0	0	0	0	0	0	0	1	0	0	0	0	3 or 12	11
NBRC15185_Minnesota	1	0	1	0	0	1	0	0	0	1	0	1	0	0	1	1	0	1	0	1	0	0	1	0	0	0	0	0	0	0	0	0	0	1	0	0	0	0	3	11
NBRC15186_Minnesota	1	0	1	0	0	1	0	0	0	1	0	1	0	0	1	1	0	1	0	1	0	0	1	0	0	0	0	0	0	0	0	0	0	1	0	0	0	0	3	11
NBRC15187_Minnesota	1	0	1	0	0	1	0	0	0	1	0	1	0	1	0	1	0	1	0	1	0	0	1	0	0	0	0	0	0	1	0	0	0	1	0	0	0	0	3 or 5	12
NBRC100797_Abony	1	0	1	0	0	1	0	0	0	1	0	1	0	1	0	1	0	1	0	1	0	0	1	0	0	0	0	0	0	0	1	0	0	0	0	0	0	0	10	11
NBRC105684_Chleraesuis	1	0	1	0	0	0	0	0	0	1	0	1	0	1	0	1	0	1	0	1	0	0	0	1	0	0	0	0	0	0	0	0	0	0	0	0	0	1	11	10
NBRC105726_Typhimurium	1	0	0	0	1	1	0	0	0	1	0	1	0	1	0	1	0	1	0	0	0	0	1	0	0	1	0	0	0	0	0	0	0	0	0	0	1	0	2	11
JCM3919_UN-O7	1	0	0	0	0	0	0	0	0	1	0	1	0	1	0	1	0	1	0	1	0	0	1	0	0	0	0	0	0	0	0	0	0	1	0	0	0	0	12	9
NBRC15335_Minnesota	1	0	1	0	0	1	0	0	0	1	0	1	0	1	0	1	0	1	0	1	0	0	1	0	0	0	0	0	0	0	0	0	0	1	0	0	0	0	3 or 12	11
GTC09490_Enteritidis	1	0	0	1	0	1	0	0	0	1	0	1	0	1	0	1	0	1	0	1	0	1	0	0	0	0	0	0	0	0	1	0	0	0	0	0	0	0	1	11
GTC09492_Braenderup	1	0	0	1	0	0	0	0	0	1	0	0	1	1	0	1	0	1	0	1	0	1	0	0	0	0	0	0	0	0	0	0	0	0	1	0	0	0	3	10
GTC09493_Pakistan	1	0	0	0	0	0	0	0	0	1	0	1	0	1	0	1	0	1	0	1	0	0	1	0	0	0	0	0	0	0	0	0	1	0	0	0	0	0	10 or 15	9
GTC09549_Typhimurium	1	0	0	0	1	1	0	0	0	1	0	1	0	1	0	1	0	1	0	1	0	0	1	0	0	0	1	0	0	0	0	0	0	0	0	1	0	0	2	12
ATCC BAA-1675_Infantis	1	0	1	0	0	1	0	0	0	1	0	1	0	0	1	1	0	1	0	1	0	0	1	0	1	0	0	0	0	0	0	0	0	1	0	0	0	0	3	12
ATCC BAA-1738_Thompson	1	0	1	0	0	1	0	0	0	1	0	1	0	1	0	1	0	1	0	1	0	1	0	0	0	1	0	0	0	0	0	0	0	1	0	0	0	0	16 or 20	12
ATCC9712_Saintpaul	1	0	1	0	0	1	0	0	0	1	0	1	0	1	0	1	0	1	0	1	0	0	1	0	0	1	0	0	0	0	0	0	0	0	0	1	0	0	13	12
jfrlSe1402-1_Infantis	0	1	1	0	0	1	0	0	0	1	0	1	0	1	0	1	0	1	0	1	0	0	1	0	0	0	0	0	0	1	0	0	0	1	0	0	0	0	5	12
jfrlSe1402-3_Brandenburg	1	0	1	0	0	1	0	0	0	1	0	1	0	0	1	1	0	1	0	1	0	0	1	0	1	0	0	0	0	0	0	0	0	1	0	0	0	0	3	12
jfrlSe1402-4_Infantis	0	1	0	1	0	1	0	0	0	1	0	1	0	1	0	1	0	1	0	1	0	0	1	0	0	0	0	0	0	1	0	0	0	1	0	0	0	0	5	12
jfrlSe1402-5_Brandenburg	1	0	1	0	0	1	0	0	0	1	0	1	0	0	1	1	0	1	0	1	0	0	1	0	1	0	0	0	0	0	0	0	0	1	0	0	0	0	3	12
jfrl1402-6_Rissen	1	0	1	0	0	1	0	0	0	1	0	1	0	1	0	0	1	1	0	1	0	1	0	0	1	0	0	0	0	0	0	0	1	0	0	0	0	0	4	12
jfrlSe1402-7_Orion	1	0	1	0	0	1	0	0	0	0	1	1	0	1	0	1	0	1	0	1	0	0	1	0	0	0	0	0	0	0	0	0	0	0	0	1	0	0	8	11
jfrlSe1402-8_Rissen	1	0	1	0	0	1	0	0	0	1	0	1	0	1	0	0	1	1	0	1	0	1	0	0	1	0	0	0	0	0	1	0	0	0	0	0	0	0	4	12
jfrlSe1402-9_Rissen	1	0	1	0	0	1	0	0	0	1	0	1	0	1	0	0	1	1	0	1	0	1	0	0	1	0	0	0	0	0	0	0	1	0	0	0	0	0	4	12
jfrlSe1402-10_Rissen	1	0	1	0	0	1	0	0	0	1	0	1	0	1	0	0	1	1	0	1	0	1	0	0	1	0	0	0	0	0	1	0	0	0	0	0	0	0	4	12
jfrlSe1402-11_Rissen	1	0	1	0	0	1	0	0	0	1	0	1	0	1	0	0	1	1	0	1	0	1	0	0	1	0	0	0	0	0	1	0	0	0	0	0	0	0	4	12
jfrlSe1402-12_Rissen	1	0	1	0	0	1	0	0	0	1	0	1	0	1	0	0	1	1	0	1	0	1	0	0	1	0	0	0	0	0	0	1	0	0	0	0	0	0	4	12
jfrlSe1402-13_Mbandaka	0	1	1	0	0	1	0	0	0	1	0	1	0	1	0	1	0	1	0	1	0	1	0	0	0	0	1	0	0	0	0	0	0	1	0	0	0	0	7	12
jfrlSe1402-14_Mbandaka	0	1	1	0	0	1	0	0	0	1	0	1	0	1	0	1	0	1	0	1	0	1	0	0	0	1	0	0	0	0	0	0	0	1	0	0	0	0	7	12
jfrlSe1402-15_Orion	1	0	1	0	0	1	0	0	0	0	1	1	0	1	0	1	0	1	0	1	0	0	1	0	1	0	0	0	0	0	0	0	0	0	0	1	0	0	8	12
jfrlSe1409-1_Altona	1	0	1	0	0	1	0	0	0	1	0	1	0	1	0	1	0	1	0	0	1	1	0	0	0	0	0	0	0	0	0	0	0	1	0	0	0	0	17	11
jfrlSe1409-2_Istanbul	1	0	0	1	0	0	0	0	1	1	0	1	0	1	0	1	0	1	0	1	0	1	0	0	0	0	0	0	0	0	0	0	0	1	0	0	0	0	18	11
jfrlSe1409-3_Senftenberg	0	1	1	0	0	0	0	1	0	1	0	1	0	1	0	1	0	1	0	1	0	1	0	0	0	0	0	1	0	0	0	0	0	1	0	0	0	0	21	12
jfrlSe1409-4_UN-O13	1	0	1	0	0	1	0	0	0	1	0	1	0	1	0	1	0	1	0	1	0	1	0	0	0	0	0	0	0	0	0	0	0	1	0	0	0	0	20	11
jfrlSe1409-5_UN-O1-3-19	0	1	1	0	0	0	0	1	0	1	0	1	0	1	0	1	0	1	0	1	0	1	0	0	0	0	0	0	0	0	0	0	0	1	0	0	0	0	7 or 21	11
jfrlSe1409-6_Montevideo	1	0	1	0	0	1	0	0	0	1	0	1	0	0	1	1	0	1	0	1	0	1	0	0	1	0	0	0	0	0	0	0	0	0	0	1	0	0	6	12
jfrlSe1409-7_UN-O1-3-19	0	1	1	0	0	0	0	1	0	1	0	1	0	1	0	1	0	1	0	1	0	1	0	0	0	0	0	0	0	0	0	0	0	1	0	0	0	0	7 or 21	11
jfrlSe1409-8_UN-O18	1	0	0	0	0	1	0	0	0	1	0	1	0	1	0	1	0	1	0	1	0	0	1	0	0	0	0	0	0	0	0	0	0	0	0	0	1	0	2 or 13	10
jfrlSe1409-9_Altona	1	0	1	0	0	1	0	0	0	1	0	1	0	1	0	1	0	1	0	0	1	1	0	0	1	0	0	0	0	0	0	0	0	1	0	0	0	0	17	12
jfrlSe1409-10_Mbandaka	0	1	1	0	0	1	0	0	0	1	0	1	0	1	0	1	0	1	0	1	0	1	0	0	0	1	0	0	0	0	0	0	0	1	0	0	0	0	7	12
jfrlSe1409-11_UN-O1-3-19	0	1	1	0	0	1	0	0	0	1	0	1	0	1	0	1	0	1	0	1	0	1	0	0	0	0	0	1	0	0	0	0	0	1	0	0	0	0	7 or 21	11
jfrlSe1409-12_Montevideo	1	0	1	0	0	1	0	0	0	1	0	1	0	1	0	1	0	1	0	1	0	1	0	0	1	0	0	0	0	0	0	0	0	0	0	0	1	0	6	12
jfrlSe1409-13_Rissen	1	0	0	1	0	1	0	0	0	1	0	1	0	1	0	0	1	1	0	1	0	1	0	0	1	0	0	0	0	0	1	0	0	0	0	0	0	0	4	12
jfrlSe1409-14_Mbandaka	0	1	1	0	0	1	0	0	0	1	0	1	0	1	0	1	0	1	0	1	0	1	0	0	0	1	0	0	0	0	0	0	0	1	0	0	0	0	7	12
jfrlSe1409-15_Rissen	1	0	1	0	0	1	0	0	0	1	0	1	0	1	0	0	1	1	0	1	0	1	0	0	1	0	0	0	0	0	1	0	0	0	0	0	0	0	4	12
jfrlSe1409-16_Orion	1	0	1	0	0	1	0	0	0	0	1	1	0	1	0	1	0	1	0	1	0	0	1	0	1	0	0	0	0	0	0	0	0	0	0	0	1	0	8	12
jfrlSe1409-17_UN	1	0	1	0	0	1	0	0	0	1	0	1	0	1	0	0	1	1	0	1	0	1	0	0	1	0	0	0	0	0	1	0	0	0	0	0	0	0	4	12
jfrlSe1409-18_Thompson	1	0	1	0	0	1	0	0	0	1	0	1	0	1	0	1	0	1	0	1	0	1	0	0	0	1	0	0	0	0	0	0	0	1	0	0	0	0	16 or 18	12
jfrlSe1409-19_Rissen	1	0	1	0	0	1	0	0	0	1	0	1	0	1	0	0	1	1	0	1	0	1	0	0	1	0	0	0	0	0	1	0	0	0	0	0	0	0	4	12
jfrlSe1409-20_Mbandaka	0	1	1	0	0	1	0	0	0	1	0	1	0	1	0	1	0	1	0	1	0	1	0	0	0	1	0	0	0	0	0	0	0	1	0	0	0	0	7	12
jfrlSe1409-21_Amsterdam	0	1	1	0	0	0	0	0	0	1	0	0	1	1	0	1	0	1	0	1	0	0	1	0	0	0	0	0	0	0	0	0	0	1	0	0	0	0	22	10
HyogoSE11001_Enteritidis	1	0	1	0	0	1	0	0	0	1	0	1	0	1	0	1	0	1	0	1	0	1	0	0	0	0	0	0	0	0	1	0	0	0	0	0	0	0	1	11

Figure 6.3 Biomarker hit profiles and classification results of 116 *S. enterica* strains by Strain Solution ver. 2 using 12 biomarkers [46]. The output from the software was converted as a binary table with presence (1, colored) and absence (0) of the peaks. Roman numerals of biomarkers are from Table 6.2. The two columns on the right show the identification results, and the group numbers correspond to those in Table 6.2. Score refers to the number of hit biomarkers. *Source:* Ojima-Kato et al. [46]/With permission of Springer Nature.

Although all tested isolates had YciF genes and their mass values were varied (Table 6.2) and YibT was characteristic of some serotypes (Braenderup, Thompson, Istanbul, Senftenberg, and UN), the fact that their peaks were not stably detected in some samples resulted in the overall accuracy of serovar prediction with high reproducibility. Therefore, a set of biomarkers with 11 or 10 biomarkers excluding YciF or both YciF and YibT would be sufficient for preliminary serotyping of *S. enterica,* especially to discriminate the major serotypes, because they did not affect the proteotyping results of major serotypes such as Enteritidis, Typhimurium, and Infantis (Table 6.3).

SodA was also an important biomarker for proteotyping to discriminate each serogroup (Table 6.2) because of a variety of mass values (m/z 23 010.84, 23 004.88, 22 996.82, 22 976.83, 22 962.80, 22 948.82, 22 948.77, or 22 918.79) in conjunction with the serotypes. However, in using CHCA as matrix reagent, SodA became undetectable because of its high molecular weight. By contrast, SA as the matrix reagent contributed to success in the consistent detection of the SodA peak. In general, high-accuracy and high-precision mass spectra are required for a database to decrease misidentification because SodA plays a key role in highly reproducible results of proteotyping of the *Salmonella* serotype. As the amount of bacteria spotted on the target and the homogeneity of the smear across the surface also affected the reproducibility of the spectrum, an optimized procedure of MALDI-TOF MS analysis for detecting the biomarker with high sensitivity and reproducibility was examined (Figure 6.4) [47]. It was shown that, to achieve a stable detection of the biomarker spectrum, the following steps were particularly effective: (i) controlling bacterial concentration at 1×10 to 1×10^2 MFU (6×10^6 to 3×10^7 CFU/μl calculated from the MFU according to McFarland standards); (ii) using the supernatant of the colony suspension by centrifugation (Figure 6.5); and (iii) using matrix additives such as methylenediphosphonic acid and *N*-decyl-*beta*-D-maltopyranoside. We propose that the method including those steps is one of the best for detecting the biomarkers with high sensitivity and reproducibility. Those procedures are expected to contribute to the detection of not only SodA biomarker but also 11 other biomarkers in MALDI-TOF MS analysis.

In comparison to the traditional *Salmonella* serotyping with an accuracy of 73% [4, 48], this method based on *S10*-GERMS had an accuracy of 94% when 116 isolates were successfully proteotyped as their correct serogroups. This value is close to the value (96%) of the

Table 6.3 Summary of serotyping results of 116 *S. enterica* strains by Strain Solution ver. 2 [46].

	Number of biomarkers		
	12	**11[a]**	**10[b]**
Correctly identified	109 (16)	108 (15)	111 (23)
Misidentified or uncharacterized	7	8	5

Numbers of strains identified as more than two groups due to the same biomarker hit score are indicated in parentheses.
[a] YciF was excluded.
[b] YciF and YibT were excluded.
Source: Ojima-Kato et al. [46]/With permission of Springer Nature.

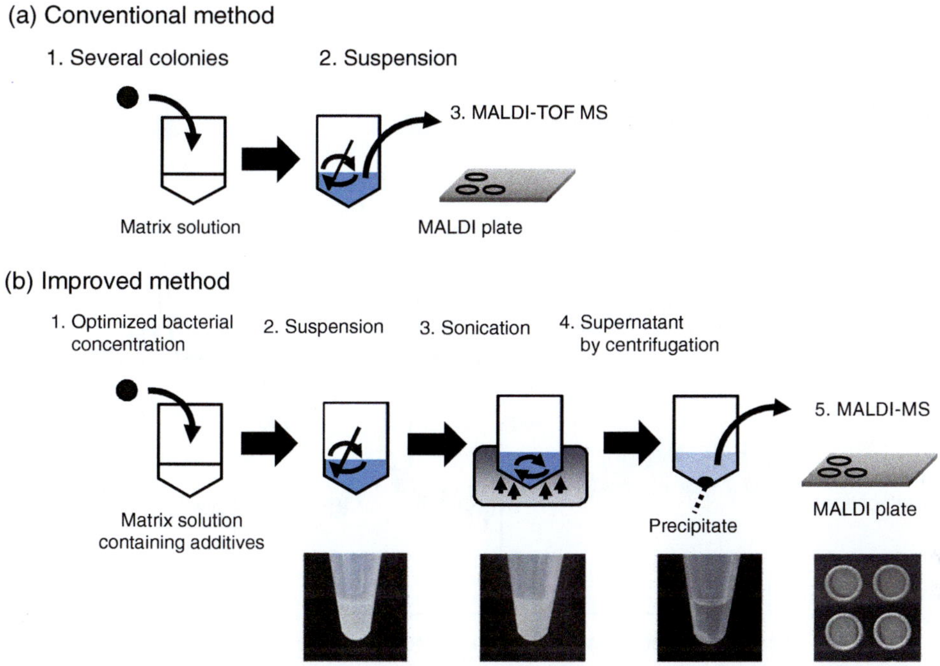

Figure 6.4 Sample preparation process for detecting the biomarker with high sensitivity and reproducibility: (a) a conventional method; and (b) an improved method in this study [47]. *Source:* Fukuyama et al. [47]/With permission from John Wiley & Sons.

molecular method based on WGS [4, 49], suggesting that the proteotyping by MALDI-TOF MS may be more rapid, reliable, and easy to use than traditional *Salmonella* serotyping. Moreover, the entire cost of WGS-based routine use limits their serotyping in the food industry whereas WGS has proven effective for discrimination of foodborne pathogens at the serovar level.

In conclusion, despite the fact that further studies are needed to improve the ability of the database to proteotype foodborne bacteria (*S. enterica* subsp. *enterica*) at the species or serovar level, Strain Solution ver. 2 software integrated with the accurate mass database will be useful for bacterial proteotyping by MALDI-TOF MS-based microbial classification in clinical and food safety laboratories.

6.3 Discrimination of the Lineages of *Listeria monocytogenes* and Species of *Listeria*

The genus *Listeria* is Gram-positive, non-spore-forming, facultative anaerobic bacterium that can grow even in saline and cold environments [50]. At present, 19 species within the genus *Listeria* belong to two groups, which primarily consist of *L. monocytogenes*, *L. innocua*, *L. welshimeri*, *L. seeligeri*, *L. ivanovii*, and *L. marthii* [51]. Among these, *L. monocytogenes* and *L. ivanovii* are known as pathogens of warm-blooded hosts and, recently, *L. innocua* was

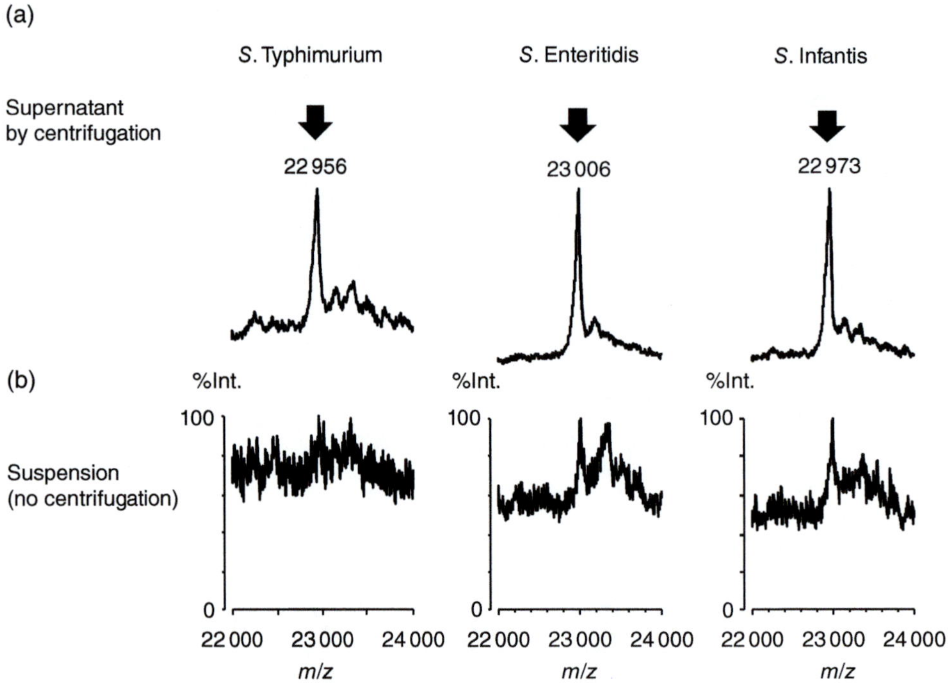

Figure 6.5 Mass spectral comparison of SodA for *S.* Typhimurium (NBRC13245), *S.* Enteritidis (NBRC 3313), and *S.* Infantis (jfrlSe 1402–4) by the improved method using: (a) the supernatant by centrifugation; and (b) the suspension without centrifugation step [47]. *Source:* Fukuyama et al. [47]/With permission from John Wiley & Sons.

also recognized as a potentially pathogenic bacteria [52]. Most notably, *L. monocytogenes* causes human listeriosis through infections from RTE foods [53] and is prevalent in the retail produce environment because its psychotropic nature allows *L. monocytogenes* to grow in refrigerated storage of RTE foods [54, 55]. Therefore, many countries around the world have developed their policies for *L. monocytogenes* in food, ranging from 100 cfu/g for low-risk food to zero tolerance for all RTE food, which means no detectable level of viable pathogens in 25 g of food in the US [1, 56]. Product recalls and market withdrawals cause significant economic losses as well as having an impact on public health. Therefore, a crucial concern for clinical and food safety has been the detection of *Listeria* species and serotyping of *L. monocytogenes*. However, traditional culture-based methods using chromogenic media utilized in food laboratories differentiate *L. monocytogenes* from other *Listeria* species [57]. Using immunological methods with antisera [58], *L. monocytogenes* is divided into 13 serotypes, and then classified into four evolutionary lineages using DNA sequencing and PFGE [59], as follows: lineage I (serotypes 1/2b, 3b, 4b, 4d, and 4e), lineage II (serotypes 1/2a, 1/2c, 3a, and 3c), and lineages III (serotypes 4a and 4c) and IV (serotypes 4a, 4c, and atypical 4b) [60, 61]. Of these, *L. monocytogenes* serotypes 1/2a, 1/2b, and 4b, which belong to lineages I and II, are involved in most cases of human disease [62] and are isolated from environmental and food sources as well as human patients, while lineages III and IV are isolated from ruminants [60, 63]. Therefore, serotyping of *L. monocytogenes* plays a critical role in epidemiological experiments for outbreak investigation [64]. Although the alternative advanced serotyping methods such as multilocus sequence typing [65, 66] and DNA microarray [67] have been developed and provide reliable serotyping of *L. monocytogenes* [2], they are laborious and time-consuming in routine microbiology laboratories which push for a simple, rapid, and reliable analytical system.

With the arrival of MALDI-TOF MS in routine microbiology laboratories, it has become a prevailing tool that can provide much easier, rapid, and reliable identification of a great variety of bacteria by comparing the mass spectra of the unknown isolates with reference spectra of known species included in the commercially available database. However, as the performance of MALDI-TOF MS is mostly limited to the bacterial identification at species level, the precise discrimination of *Listeria* species and its lineages has remained a crucial issue in clinical and food safety microbiology laboratories.

According to previous research [68–71], *L. monocytogenes*, *L. innocua*, *L. welshimeri*, *L. ivanovii*, and *L. seeligeri* could be identified by a MALDI TOF-MS fingerprinting method, and MS peaks at m/z 5590 and 11 179 could differentiate between serotype 4a and 4c of *L. monocytogenes* [68]. Farfour et al. revealed that identification of *Listeria* spp. could only be achieved at the genus level, except for *L. grayi* using the MALDI-TOF MS Andramas system [69]. By contrast, Hsueh et al. found that five MS peaks observed at m/z 5594.85, 5601.21, 6184.39, 11 199.33, and 11 871.31 played an important role in the discrimination of serotypes 1/2a (lineage II), and 1/2b and 4b (lineage I) of *L. monocytogenes* [70], suggesting that MALDI-TOF MS analysis has strong potential to discriminate the serotype of *L. monocytogenes*, although the observed proteins were not fully identified, potentially leading to data inconsistency.

Because the observed mass spectra by MALDI-TOF MS analysis consist mainly of the conserved ribosomal proteins [20, 22], the *S10*-GERMS method can provide an accurate and reliable m/z database of biomarker proteins, the m/z values of which are confirmed by genetic information (i.e. genomics) and observed MS peaks by MALDI TOF-MS analysis (i.e. proteomics) [36]. Therefore, the *S10*-GERMS method can utilize the subtle m/z differences that occur with even a single amino acid substitution within the selected biomarker proteins of closely related species, implying that the *S10*-GERMS method may discriminate the serotype of *L. monocytogenes* at a higher accuracy than the conventional fingerprinting method. Accordingly, our *S10*-GERMS proteotyping method was applied to discriminate the lineages of *L. monocytogenes* and the species of *Listeria* (*L. monocytogenes*, *L. innocua*, *L. welshimeri*, *L. seeligeri*, *L. ivanovii*, *L. grayi*, and *L. rocouritiae*) using the external software Strain Solution ver. 1.0.0 for an automatic analysis [72].

Among the theoretically calculated ribosomal proteins (Figure 6.6), 15 consisted of L3, L4, L23, L2, L24, L6, L18, S5, L15, S13, and S11 in the *S10-spc-alpha* operon and the three additional ribosomal proteins L10, L13, and S9 were selected as candidate biomarkers for serotyping of *L. monocytogenes*. When the theoretically calculated m/z of S9 was compared with its observed m/z value, the observed m/z value of S9 gained +43, indicating an acetylated S9. Candidate biomarkers L3, L4, L23, L2, L10, and L21 were not detected precisely in the MALDI-TOF MS analysis. As the peaks of S5 with m/z 17 325.94 or 17 324.95, and L13 with m/z 16 200.61, 16 184.61, or 16 187.61 were not distinguishable in the MALDI-TOF MS system under the current MS detection tolerance, i.e. >500 ppm, these two

		Lineage	II						I							III
		Serotype	1/2a	1/2a	1/2a	1/2c	3a	3c	1/2b	4b	3b	3b	4b	4d	4d	4a
		Strain	ATCC15313^T	JCM7671	ATCC51772	JCM7672	JCM7673	JCM7678	JCM7676	JCM7675	JCM7677	JCM7683	ATCC19115	JCM7680	JCM2873	JCM7674
Coded operon	Detection	Protein														
S10	○	S10	11551.43	11551.43	11551.43	11551.43	11551.43	11551.43	11551.43	11551.43	11551.43	11551.43	11551.43	11551.43	11551.43	11551.43
	×	L3	22695.18	22695.18	22695.18	22695.18	22695.18	22695.18	22695.18	22695.18	22695.18	22695.18	22695.18	22695.18	22695.18	22695.18
	×	L4	22472.97	22472.97	22472.97	22472.97	22472.97	22472.97	22472.97	22472.97	22472.97	22472.97	22472.97	22472.97	22472.97	22472.97
	×	L23	10916.61	10916.61	10916.61	10916.61	10916.61	10916.61	10916.61	10916.61	10916.61	10916.61	10916.61	10916.61	10916.61	10930.64
	×	L2	30374.96	30374.96	30374.96	30374.96	30374.96	30303.88	30374.96	30374.96	30374.96	30374.96	30374.96	30374.96	30374.96	30374.96
	○	S19	10344.86	10344.86	10344.86	10344.86	10344.86	10344.86	10344.86	10344.86	10344.86	10344.86	10344.86	10344.86	10344.86	10344.86
	○	L22	12743.72	12743.72	12743.72	12743.72	12743.72	12743.72	12743.72	12743.72	12743.72	12743.72	12743.72	12743.72	12743.72	12743.72
	×	S3	24411.84	24411.84	24411.84	24411.84	24411.84	24411.84	24411.84	24411.84	24411.84	24411.84	24411.84	24411.84	24411.84	24411.84
	×	L16	16138.96	16138.96	16138.96	16138.96	16138.96	16138.96	16138.96	16138.96	16138.96	16138.96	16138.96	16138.96	16138.96	16138.96
	○	L29	7402.53	7402.53	7402.53	7402.53	7402.53	7402.53	7402.53	7402.53	7402.53	7402.53	7402.53	7402.53	7402.53	7402.53
	×	S17	9905.45	9905.45	9905.45	9905.45	9905.45	9905.45	9905.45	9905.45	9905.45	9905.45	9905.45	9905.45	9905.45	9905.45
spc	×	L14	13227.39	13227.39	13227.39	13227.39	13227.39	13227.39	13227.39	13227.39	13227.39	13227.39	13227.39	13227.39	13227.39	13227.39
	○	L24	11180.22	11180.22	11180.22	11180.22	11180.22	11180.22	11194.25	11194.25	11194.25	11194.25	11194.25	11194.25	11254.35	11180.22
	×	L5	19996.21	19996.21	19996.21	19996.21	19996.21	19996.21	19996.21	19996.21	19996.21	19996.21	19996.21	19996.21	19996.21	19996.21
	○	S14	7016.49	7016.49	7016.49	7016.49	7016.49	7016.49	7016.49	7016.49	7016.49	7016.49	7016.49	7016.49	7016.49	7016.49
	○	S8	14513.89	14513.89	14513.89	14513.89	14513.89	14513.89	14513.89	14513.89	14513.89	14513.89	14513.89	14513.89	14513.89	14513.89
	○	L6	19270.08	19270.08	19270.08	19270.08	19270.08	19270.08	19256.01	19256.01	19256.01	19256.01	19256.01	19256.01	19256.01	19270.08
	○	L18	13096.86	13096.86	13096.86	13096.86	13096.86	13096.86	13096.86	13096.86	13096.86	13096.86	13096.86	13096.86	13096.86	13096.86
	△	S5	17325.94	17325.94	17325.94	17325.94	17325.94	17325.94	17325.94	17325.94	17325.94	17325.94	17325.94	17325.94	17325.94	17325.94
	○	L30	6362.55	6362.55	6362.55	6362.55	6362.55	6362.55	6362.55	6362.55	6362.55	6362.55	6362.55	6362.55	6362.55	6362.55
	○	L15	15782.02	15782.02	15782.02	15782.02	15782.02	15782.02	15782.02	15782.02	15782.02	15782.02	15782.02	15782.02	15782.02	15782.02
alpha	○	L36	4324.42	4324.42	4324.42	4324.42	4324.42	4324.42	4324.42	4324.42	4324.42	4324.42	4324.42	4324.42	4324.42	4324.42
	△	S13	13578.69	13578.69	13578.69	13578.69	13578.69	13578.69	13578.69	13578.69	13578.69	13578.69	13578.69	13578.69	13578.69	13578.69
	○	S11	13655.65	13655.65	13655.65	13655.65	13655.65	13655.65	13655.65	13655.65	13655.65	13655.65	13655.65	13655.65	13655.65	13655.65
	○	L17	15084.51	15084.51	15084.51	15084.51	15084.51	15084.51	15084.51	15084.51	15084.51	15084.51	15084.51	15084.51	15084.51	15084.51
Others	×	L10	17433.93	17563.04	17563.04	17563.04	17563.04	17563.04	17577.07	17577.07	17577.07	17577.07	17577.07	17577.07	17577.07	17577.07
	△	L13	16200.61	16200.61	16200.61	16200.61	16200.61	16200.61	16200.61	16184.61	16200.61	16200.61	16184.61	16184.61	16200.61	16200.61
	○	S9+Ac	14283.40	14283.40	14283.40	14283.40	14283.40	14283.40	14359.50	14359.50	14359.50	14359.50	14359.50	14359.50	14359.50	14359.50

Figure 6.6 The theoretical *m/z* values of the candidate ribosomal proteins in the *S10-spc-alpha* operon and an additional three biomarker candidates [72]. The symbols in the detection column are as follows: open circle, corresponding peaks detectable in all strains with default analytical condition (threshold offset: 0.015 mV; threshold response: 1.200) in the AXIMA system; triangle, differences of *m/z* value in each strain or the other protein peaks could not be distinguished from one another within 500 ppm tolerance, although putative peaks were detected; multiplication, undetected or weak peaks with low reproducibility. Characteristic *m/z* values are highlighted in color. *Source:* Ojima-Kato et al. [72]/PLOS/CC BY-4.0.

proteins were not suitable as biomarkers in this study. S13 was also deleted from the candidate biomarker list because one of the peaks of S13 (*m/z* 13 578.69 or 13 552.65) overlapped with the peaks of L20 (*m/z* 13 552.08). Then, as the performance of MALDI-TOF MS analysis depends on the quality of the spectrum obtained for a given isolate, out of 15 candidate biomarkers, six potential biomarkers (L24, L6, L18, L15, S11, and S9) with high reproducibility under the current MS detection tolerance were determined as specific biomarkers for proteotyping *L. monocytogenes* at the serovar level. Therefore, *m/z* values of the six promising biomarkers [L24 (*m/z* 11 180.22, 11 194.25, or 11 254.35), L6 (*m/z* 19 270.08 or 19 256.01), L18 (*m/z* 13 096.86 or 13 110.89), L15 (*m/z* 15 782.02, 15 797.08, or 15 668.86), S13 (*m/z* 13 578.69 or 13 552.65), S11 (*m/z* 13 655.65 or 13 674.66), and acetylated S9 (*m/z* 14 283.40, 14 359.50, or 14 302.45)] were registered on the Strain Solution software, and the mass spectrum data obtained with SA as matrix reagent were imported into the software. As shown in Figure 6.7, the lineages of *L. monocytogenes* were classified into five groups (A to E) as follows: lineage I [registered group B (serotypes 1/2b, 3b, 4b, 4d, and 4e) and group C (serotype 4d)], lineage II [group A (serotypes 1/2a, 1/2c, 3a, and 3c)] and lineage III [group D (serotype 4a) and group E (serotype 4c)]. Thus, these six promising biomarkers are useful in discriminating between lineages or serotypes of *L. monocytogenes* in MALDI-TOF MS analysis using the software.

In addition, L31 type B (*m/z* 9259.36, 9290.34, 9327.44, or 9271.3) and S16 (*m/z* 10 234.94, 10 252.97, 10 230.88, or 10 003.54) could be used to improve the discrimination of *Listeria* species, including *L. monocytogenes* (Figure 6.7), whereas the conventional fingerprinting method could not distinguish between *L. innocua* and *L. ivanovii* correctly due to their very similar peak profiles (Table 6.4). It is notable that the novel six biomarkers with high *m/z* values were successfully detected in this study because SA was used as a matrix reagent, whereas conventional fingerprint methods could not stably detect those high-molecular-weight biomarkers because CHCA is a recommended matrix reagent.

Our research revealed that previously reported peaks of *m/z* 11 179, 11 871.31, or 11 199.33 by MALDI-TOF MS fingerprint analysis [68, 70] corresponded to the ribosomal protein L24 with *m/z* 11 180.22, 11 194.25, or 11 254.35, respectively. The differences in *m/z* values between previous reports and this study may arise from the accuracy of the experimental procedure.

In addition, the peaks with *m/z* 5590, 5594.85, 5601.21, and 6184.39 observed by previous reports for serotyping *L. monocytogenes* [68, 70] were also detected in our study as *m/z* 5590 (lineages II and III) and *m/z* 5595 (lineage I). However, serotype 4d strains had insufficient peak intensities of the corresponding peaks in our study. Therefore, this unknown protein was not used as a biomarker.

Listeria grayi is known to be evolutionarily distant from other *Listeria* spp. [73], and *L. rocouritiae* is a recent isolate from lettuce in Australia [74]. The *m/z* values of the eight biomarkers in these two species are significantly different from those of the other *Listeria* spp. (Figure 6.7). This suggests that evolution of ribosomal protein is well associated with bacterial species [75]. Therefore, the accuracy of *m/z* values of detected peaks is important to proteotype the bacteria by MALDI-TOF MS. Our method plays a significant role in the success of discrimination of foodborne bacteria even at the serovar level because our proteotyping data are backed up by genomics and proteomics.

	Group												
	A	B	C	D	E	F	G	H	I	J	K	L	M
Species	*L. monocytogenes*				*L. innocua*		*L. ivanovii*		*L. seeligeri*		*L. welshimeri*	*L. rocourtiae*	*L. grayi*
Subspecies							*ivanovii*	*londiniensis*					
Strain					ATCC33090^T	GTC02960	JCM7681	ATCC49954	JCM7679 JCM7682	ATCC35967^T	GTC02963	GTC16429^T	ATCC19120^T
Serotype	1/2a, 1/2c, 3a, 3c	1/2b, 3b, 4b, 4d, 4e	4d	4a	6a								
Lineage	II	I		III									
Biomarker: L24	11180.22	11194.25	11254.35	11180.22	11194.25	11194.25	11194.25	11194.25	11194.25	11194.25	11194.25	11558.65	11180.14
L6	19270.08	19256.01	19256.01	19270.08	19270.08	19270.08	19270.08	19270.08	19270.04	19270.04	19270.08	19097.81	19371.01
L18	13096.86	13096.86	13096.86	13096.86	13082.84	13096.86	13096.86	13082.84	13110.89	13110.89	13082.84	13068.99	13066.84
L15	15782.02	15782.02	15782.02	15782.02	15797.03	15797.03	15811.10	15796.09	15797.08	15797.08	15797.08	15743.01	15601.77
S11	13655.65	13655.65	13655.65	13655.65	13655.65	13655.65	13655.65	13655.65	13674.66	13674.66	13683.67	13591.66[a]	13591.67[b]
S9+Ac	14283.40	14359.50	14359.50	14359.50	14359.50	14359.50	14359.50	14359.50	14302.45	14283.40	14302.45	14372.55	14330.55
L31 type B	9259.36	9259.36	9259.36	9259.36	9259.36	9259.36	9290.37	9290.37	9290.37	9290.37	9290.37	9271.08	9327.44
S16	10234.94	10234.94	10234.94	10234.94	10252.96	10252.96	10234.94	10234.94	10234.94	10234.94	10234.94	10003.54	10230.88

Figure 6.7 The specific set of *m/z* values for eight biomarkers in proteotyping of *Listeria monocytogenes* and other *Listeria* species [72]. [a]*m/z* value of $[M+H]^+$ plus 17 was shown based on the observed MS peak. [b]*m/z* value of $[M+H]^+$ plus *m/z* 14 as methylation. *Source:* Ojima-Kato et al. [72]/PLOS/ CC BY-4.0.

Table 6.4 Results of *Listeria* species from fingerprinting analysis with SARAMIS [72].

Sample	Identity (%)	Family	Genus	Species
L. innocua ATCC33090^T	91.8	Family IV *Listeriaceae*	*Listeria*	sp.
L. innocua GTC02960	99.9	Family IV *Listeriaceae*	*Listeria*	sp.
L. ivanovii GTC02961	84.4	Family IV *Listeriaceae*	*Listeria*	*monocytogenes*
L. ivanovii ATCC49954^T	99.9	Family IV *Listeriaceae*	*Listeria*	sp.
L. seeligeri ATCC35967	91.8	Family IV *Listeriaceae*	*Listeria*	sp.
L. welshimeri GTC02963^T	97.2	Family IV *Listeriaceae*	*Listeria*	sp.
L. rocouritiae GTC16429^T	99.9[a]		*Listeria*	*rocouritiae*
L. grayi ATCC19120	99.9	Family IV *Listeriaceae*	*Listeria*	*grayi*
L. seeligeri JCM7679	81.1	Family IV *Listeriaceae*	*Listeria*	*monocytogenes*
L. seeligeri JCM7683	81.1	Family IV *Listeriaceae*	*Listeria*	*monocytogenes*

[a] The parental database for SuperSpectra™ in SARAMIS.
Source: Ojima-Kato et al. [72]/PLOS/CC BY-4.0.

Taken together, the specific set of eight biomarkers (ribosomal proteins L24, L6, L18, L15, S11, S9, L31 type B, and S16) yielded characteristic MS patterns for the lineages of *L. monocytogenes* and the different species of *Listeria*, and led to the construction of a MS database that was successful in discriminating between these organisms in MALDI-TOF MS fingerprinting analysis followed by advanced proteotyping software Strain Solution analysis.

Although in this study, the standardized database consisted of eight biomarkers that could successfully discriminate *L. monocytogenes* at the serotype level, and different species of *Listeria* that were difficult to identify by conventional fingerprinting methods (Figure 6.7), this database used only naturally isolated strains available from public collections limited to Japan. Therefore, the evaluation of the constructed database still remains to be done. The collaborations with institutions possessing *Listeria* isolates will open a new window to discriminate bacteria in clinical and diagnostic laboratories, and also in food-related industries.

6.4 Discrimination of the *Bacillus cereus* Group and Identification of Cereulide

The genus *Bacillus* comprises Gram-positive and catalase-positive, facultatively anaerobic spore-forming, and rod-shaped bacteria that are ubiquitous in nature, ranging from soil and plants to the human gut. Some, such as *B. subtilis*, *B. amyloliquefaciens*, *B. pumilus*, *B. licheniformis*, and *B. thuringiensis*, are beneficial in food processing of fermented soybean foods [76] and bioinsecticides [77, 78]. On the other hand, others, like the *B. cereus* group, have the conflicting issues of being human pathogens and causing food spoilage [78–80]. Moreover, as the use of RTE foods has increased around the world, foodborne illnesses caused by the *B. cereus* group have become of greater concern, as it is associated with two types of gastrointestinal disease, diarrhea due to the enterotoxin and vomiting due to the emetic toxin. The emetic toxin, cereulide, is 1.2-kDa cyclic dodecadepsipeptide and a potent potassium ionophore [81–83]. As cereulide is heat- and acid-stable, it might not be removed during the process of food production, although emetic bacteria may be. Moreover, sustainable agriculture requires a shift away from chemical pesticides to biopesticides such as *B. thuringiensis*, whose 16S rRNA gene sequence similarity to *B. cereus* is more than 99%, suggesting it is indistinguishable from *B. cereus* [84]. Moreover, a potential risk of foodborne illness associated with *B. thuringiensis* has been reported [85–87]. Thus, from the human health and food safety perspective, the discrimination of bacteria in the *B. cereus* group remains a pressing global issue. Therefore, it has been important to establish a rapid and simple identification method for the genus *Bacillus*. Although the MALDI-TOF MS fingerprint method has become a standard tool for rapid, simple, and reliable identification in clinical and food safety microbiology laboratories, especially in the case of the *B. cereus* group, the MALDI fingerprint method has not yet discriminated them because of their similar fingerprint spectral patterns in principle [7]. Thus, proteotyping using the *S10*-GERMS method had a formidable challenge to overcome the limitations of MALDI-TOF MS analysis. In our first study [37], the best signal quality of MALDI-TOF MS analysis was observed in the 30% (v/v) formic acid (FA) treatment rather than the TFA treatment and, in particular, the high molecular weights of ribosomal proteins with more than the mass value of *m/z* 11 000, such as L18, L22, and S10, coded in *S10* and *spc* operons, were detected with high sensitivity. Accordingly, eight ribosomal subunit proteins (i.e. L18, L22, L24, L29, L30, S10, S14, and S19) were selected as reliable and reproducible biomarkers for rapid bacterial classification of *Bacillus* strains at the species and strain levels by the *S10*-GERMS method. However, they had a particularly limited role in the discrimination of psychrotolerant species of the *B. cereus* group, which cause foodborne illness as well as spoilage. For example, some *B. weihenstephanensis* strains have genes encoding diarrheal toxins, and other strains produce cereulide as an emetic toxin [88–90]. Therefore, Takahashi et al., [91] focused on finding new biomarkers to discriminate the *B. cereus* group, including psychrotolerant species. Among the theoretically calculated masses of ribosomal subunit proteins based on actual sequencing data, the set of four biomarkers (i.e. S10, S16, S20, and L30), of which S16 and S20 were novel and present outside the *S10* and *spc* operons, were selected as promising biomarkers and registered on Strain Solution ver. 2 to discriminate psychrotolerant species, e.g. *B. mycoides* and *B. weihenstephanensis*, of the *B. cereus* group after evaluation of the MALDI-TOF MS analysis. The *m/z* values were as follows (Table 6.5a): *B. cereus* at 6425.6 (L30), 9211.6 (S20), 9987.6 (S16), 11 553.5 (S10); *B. thuringiensis* at 6439.6 (L30), 9227.6 (S20), 9987.6 (S16), 11 567.5 (S10); *B. mycoides* at 6425.6 (L30), 9271.6 (S20), 9972.6 (S16), 11 567.5 (S10); *B. weihenstephanensis* at 6425.6 (L30), 9271.6 (S20), 9987.6 (S16), 11 567.5 (S10). According to the *m/z* values, L30, S20, S16, and S10 have two, three, two and two proteotypes, respectively (Table 6.5b). Based on the phylogenetic tree constructed using a combination of the *m/z* values of four selected biomarkers (Figure 6.8), the *S10*-GERMS method could discriminate the psychrotolerant strain *B. weihenstephanensis* from other mesophilic species, such as *B. mycoides*, which were taxonomically identical [93], as well as highlighting the difference between *B. mycoides* and *B. pseudomycoides* in the *B. cereus* group. Moreover, the three selected biomarkers, S16, S20, and L30, also worked well in discriminating the psychrotolerant strain such as *B. weihenstephanensis* in the *B. cereus* group. In addition, those biomarkers also showed the possibility for identification of *B. thuringiensis* from the *B. cereus* group, suggesting that by

Table 6.5 Theoretical m/z values of ribosomal subunit proteins for proteotyping of *Bacillus cereus* group strains (a) and their proteotype (b) [91].

(a)

Biomarkers	*B. cereus* NBRC 15305	*B. thuringiensis* NBRC 101235	*B. mycoides* NBRC 101228	*B. weihenstephanensis* NBRC 101238
L30	6425.6	6439.6	6425.6	6425.6
S20	9211.6	9227.6	9271.6	9271.6
S16	9987.6	9987.6	9972.6	9987.6
S10	11553.5	11567.5	11567.5	11567.5

(b)

Proteotype	L30	S20	S16	S10
1	6425.6	9211.6	9987.6	11553.5
2	6439.6	9227.6	9972.6	11567.5
3		9271.6		

Source: Takahashi [91]/With permission of Elsevier.

narrowing down the target bacteria, they can be discriminated more specifically by a few characteristic biomarkers. Manzulli et al. [94] analyzed the *B. cereus* group using the species characteristic ion peaks, which were selected by the statistical method processed in the ClinPro Tools software. They identified the specific signals as follows: 4637, 7324, and 9272 Da for *B. weihenstephanensi*; 5422 Da for *B. mycoides*; and 2956, 2968, and 3411 Da for *B. thuringiensis*. In comparison with the biomarkers selected by the *S10*-GERMS method, 9272 Da for *B. weihenstephanensis*, which was selected by the statistical method using ClinPro Tools software (Bruker Microbiology & Diagnostics), may be assigned to the S20 ribosomal subunit protein. Theoretically, *B. weihenstephanensis* and *B. mycoides* have identical mass values of S20. Therefore, caution is required in using characteristic ion peaks selected by the statistical method for the identification of microorganisms by MALDI-TOF MS analysis.

The emetic *B. cereus* group, including psychrotolerant *B. weihenstephanensis*, produces the emetic toxin cereulide synthesized by non-ribosomal enzyme encoded by the *ces* gene, which is a cyclic dodecadepsipeptide composed of three repeats of tetrapeptide units, D-Oxy-Leu-D-Ala-L-Oxy-Val-L-Val amino acids sequence, and acts as a potassium ionophore [88, 95]. Recently, 18 cereulide variants were identified and have different toxicities in severity [95]. Therefore, while many advanced detection methods of the *B. cereus* group and its toxins have been developed [96], the discrimination of a pathogenic strain from other non-pathogenic strains in the *B. cereus* group still needs a reliable, time-saving method for in situ analysis in clinical and food safety microbial laboratories. Because one advantage of MALDI-TOF MS analysis is that it allows rapid and simple detection of cereulide compared with conventional methods, MALDI-TOF MS analysis was used to detect cereulide, and the peaks of its sodium (m/z 1171) and potassium (m/z 1187) adducts were observed [97]. Moreover, taking up the challenge for simultaneous detection of cereulide

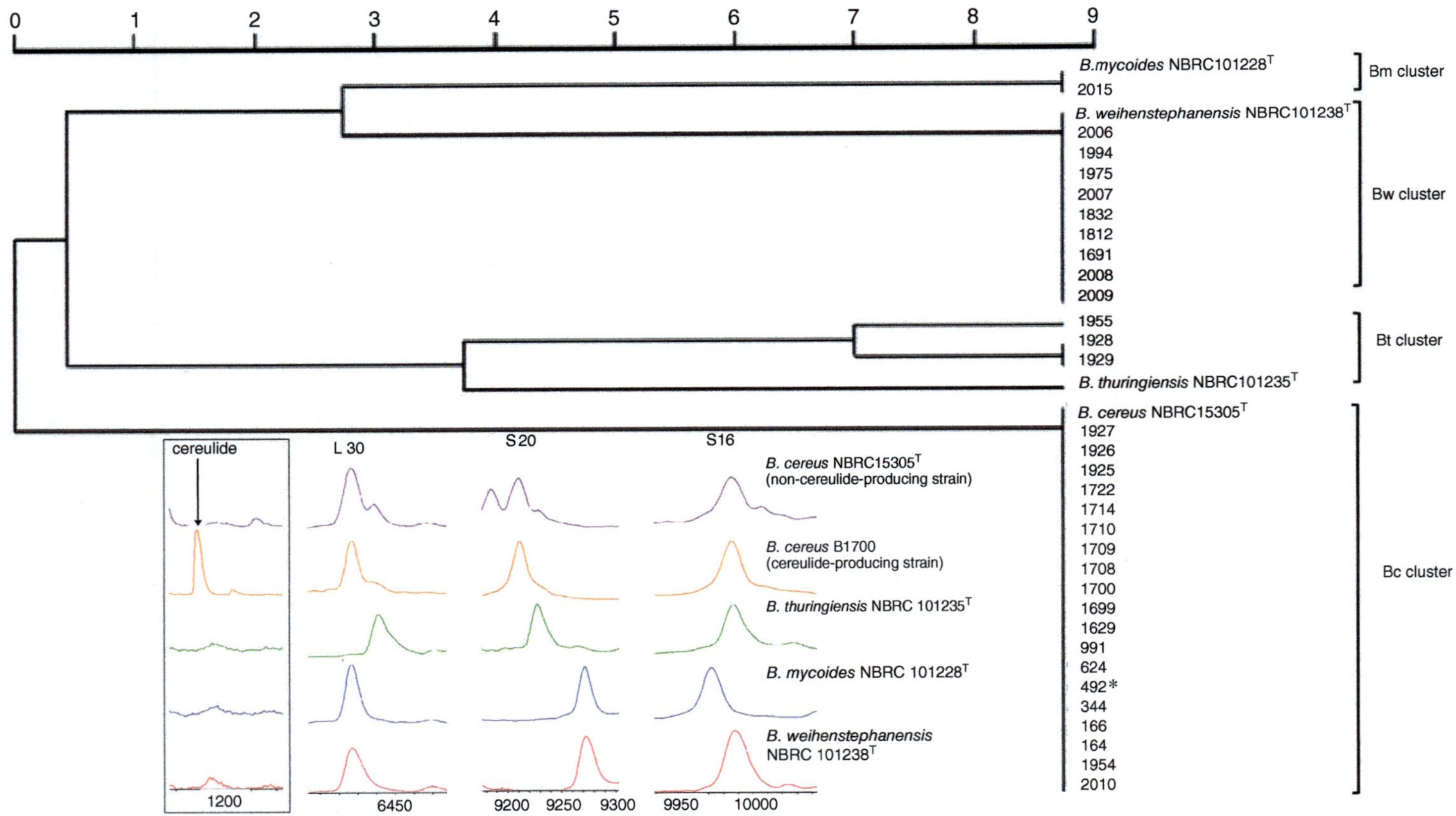

Figure 6.8 Phylogenetic tree based on a set of four biomarkers [91, 92]. The inserts in the figure indicate the detection of cereulide (*m/z* 1190) by simultaneous MALDI-TOF MS analysis of strains of the *B. cereus* group. *A spectrum of cereulide simultaneously detected in the production strain, 492. *Source:* Takahashi [91]/With permission of Elsevier.

and a cereulide-producing strain in the *B. cereus* group [98], bacterial species were first identified by comparing the observed MALDI-TOF MS spectra with the conventional spectral database; if the sample was identified as *B. cereus*, the sample fraction was further analyzed by MALDI-TOF MS using the linear mode in the *m/z* region between 700 and 2000 Da. As a reliable, time-saving, and one-step detection method, our proteotyping method was also used in an attempt to simultaneously discriminate both the cereulide-producing strain in the *B. cereus* group and a psychrotolerant strain such as *B. weihenstephanensis* in the *B. cereus* group. The peak of potassium adducts at *m/z* 1191[99] was only observed by the MALDI-TOF MS whole-cell detection method with high resolution using SA as matrix reagent under our cultivation condition. Accordingly, our proteopyping method combining cereulide detection and the selected biomarkers allowed a simultaneous discrimination of cereulide-producing *B. cereus* group strains from other non-cereulide-producing *B. cereus* group strains using bacterial strains cultured on standard agar plates (Figure 6.8) [92]. This simple, rapid, reliable, and one-step MALDI-TOF MS analysis thus has major potential as a valuable tool for ensuring human health and food safety.

6.5 Identification of Alkylphenol Polyethoxylate-degrading Bacteria in the Environment

Although pesticides are useful materials for crop protection, as long as they are used in agricultural land, they change into various compounds in the process of their degradation and mineralization by environmental microorganisms, and the resultant chemicals pass through groundwater and rivers to the sea. When they enter the hydrosphere, this may raise concerns from the viewpoint of ensuring safe drinking water and protecting the biological systems of rivers and oceans. According to our research [100], the non-toxic pesticide adjuvant, non-ionic surfactant $OPEO_n$, degraded to octylphenols (OP), octylphenol monoethoxylate ($OPEO_1$) and octylphenol diethoxylate ($OPEO_2$), which act as estrogen agonists and androgen antagonists, i.e. metabolic toxicants, and the corresponding octylphenol carboxylates ($OPEC_{1,2}$) through metabolism and/or biodegradation by environmental microorganisms [101, 102]. This biodegradation process was confirmed using MALDI-TOF MS analysis and an oxygen-stable isotope as follows: exo-type shortening of the EO chain accompanied by oxidation of the EO moiety as oxidative biodegradation [103]. Moreover, environmental elements, such as Mg^{2+}, Ca^{2+}, and Fe^{3+}, significantly influence the final degradation metabolites [102, 104]. Sixteen $OPEO_n$-degrading strains isolated from soils in various regions of Japan were classified into four biodegradation patterns based on the final metabolites of $OPEO_n$ as follows: pattern 1, OP as a final product and $OPEO_1$ and $OPEC_1$ as main biodegradation products; pattern 2, $OPEO_2$ and $OPEC_2$ as main biodegradation products; pattern 3, $OPEO_2$ as the main biodegradation product; and pattern 4, $OPEC_n$ as the main biodegradation products. This suggested that the environmental toxicity was strongly dependent on the resultant degradation products by microorganisms in the environment. As the rapid and simple identification of degrading microorganisms is important from the viewpoint of protection from environmental pollution, the *S10*-GERMS method was therefore applied for the identification of 16 $OPEO_n$-degrading isolates.

Table 6.6 Ribosomal protein profile matching for phylogenetic analysis of *P. putida* [105].

	Selected biomarkers[a]													
Strain name	**L36**	**L30**	**L29**	**S17**	**S19**	**L23**	**S14**	**L24**	**S10**	**L22**	**L18**	**S13**	**S11**	**S08**
NBRC 14164^T	1[b]	1	1	1	1	1	1	1	1	1	1	1	1	1
KT2440	1	0[c]	1	0	1	1	0	1	1	1	1	1	1	1
NBRC 15366	1	0	1	0	1	1	0	0	1	1	0	0	1	1
FMP1	1	0	1	0	1	1	0	1	1	1	1	1	1	1
EC31	1	0	1	0	1	1	0	1	1	1	1	1	1	1
S5	1	0	1	0	1	1	0	0	1	1	0	0	1	1
NBRC 3738	1	0	1	0	0	0	0	0	1	1	0	0	0	0

[a] Ribosomal proteins selected as biomarker.
[b] Peaks are observed by MALDI-TOF MS analysis.
[c] Peaks are not observed by MALDI-TOF MS analysis.
Source: Hotta et al. [105]/With permission of Pesticide Science Society of Japan.

Using the 14 biomarkers previously reported for the genus *Pseudomonas* based on the type strain *P. putida* NBRC 14164^T [36], $OPEO_n$-degrading isolates FMP1, EC 31, and S5 were matched for the number of peaks (11, 11, and 9 out of 14 biomarkers, respectively) (Table 6.6), and FMP1 and EC31 had the highest homology with *P. putida* KT2440, which has no $OPEO_n$-degrading ability, while the biomarker profile of the S5 strain was identical to that of *P. putida* NBRC 15366. The *S10*-GERMS method could discriminate $OPEO_n$-degrading isolates clearly at the strain level (Figure 6.9) [105]. Among other $OPEO_n$-degrading isolates, the results of 16Sr RNA gene sequencing suggested the presence of *Chelatococcus thiogangeticum* and *Mesorhizobium asaccharovorans* at the highest homology, with 98.5% and 97.1%, respectively, which have not been reported as OPEOn-degrading bacteria. However, those obtained values suggested that they may belong to new species. To confirm whether the isolated strains were different from *M. thiogangeticum* and *C. asaccharovorans*, their biomarkers were compared with those of type strains *M. thiogangeticum* DSM 17097^T and *C. asaccharovorans* DSM 6462^T, respectively, using nine ribosomal proteins of *M. thiogangeticum* and eight ribosomal proteins of *C. asaccharovorans* encoded in the *S10* and *spc* operons by the *S10*-GERMS method (Table 6.7). The biomarker profiles of *M. thiogangeticum* DSM 17097^T and isolate BNS58 revealed that only one ribosomal subunit protein, S8, was identical among the selected nine biomarkers between them. Moreover, biomarker proteins between *C. asaccharovorans* DSM 6462^T and isolate BSN60 had no similarity at all. The results suggested that they may belong to new species, although further study is required to confirm them as such [105]. This means that proteotyping by the *S10*-GERMS method provides a more accurate and rapid discrimination of bacteria than do 16S rRNA gene sequencing identification techniques. In future, the *S10*-GERMS method will play a useful tool in the discrimination and monitoring of human-made, chemical-degrading bacteria isolated from the environment and thus contribute to the life-cycle impact assessment of human-made chemicals.

Table 6.7 Masses of ribosomal subunit proteins for proteotyping of *M. thiogangeticum* and *C. asaccharovorans* strains [105].

Reference	*M. thiogangeticum*	*M. thiogangeticum*	*C. asaccharovorans*	*C. asaccharovorans*
Peaks	DSM 17097^T	BSN58	DSM 6462^T	BSN60
L30	7120.34	7105.32	7568.81	6935.06
L29	7429.64	7397.6	8331.68	8149.36
S17	9147.55	9103.57	9069.5	8911.3
L23	10 264.93	10 363	10 992.86	11 151.97
S19	10 384.99	10 356.94	10 323.94	10 298.97
S14	11 579.6	11 551.59	11 549.42	11 598.58
L18	12 687.57	12 766.72	12 125.97	12 582.27
L22	14 014.24	14 000.22	14 136.37	13 480.61
S08	14 683.84	14 683.84	14 569.68	14 493.58

Source: Hotta et al. [105]/With permission of Pesticide Science Society of Japan.

6.6 Conclusions and Future Perspectives

Because MALDI-TOF MS instruments are simple and quick to use, MALDI-TOF MS fingerprint analysis has become a standard tool for routine identification of bacterial isolates in clinical and food safety microbiology laboratories. At the same time, along with widespread use of RTE foods, foodborne pathogens, which are ubiquitous in agricultural and food-processing environments, are a considerable and global concern for public health, regulatory authorities, and food industries because they cause not only severe life-threatening diseases but also significant economic losses, due to product recalls and/or withdrawals of suspected food in the distribution chain.

Furthermore, in terms of SDGs, sustainable agriculture prefers biopesticides and organic fertilizers rather than chemical pesticides and fertilizer. However, with the increasing use of biopesticides and organic fertilizers, the hazardous potential of foodborne bacteria such as foodborne illness and/or spoilage will increase because foodborne bacteria can use plants as alternative hosts temporarily [106–108]. When the causative bacteria are toxigenic or spoilage species or serovars, microbial contamination levels of foods derived from biopesticides and organic fertilizers may become associated with the risk of foodborne illness and/or spoilage.

Therefore, the immediate need is to find effective biomarkers to construct accurate biomarker databases which play a pivotal role in discrimination of microorganisms in biomarker-assignment approaches using MALDI-TOF MS proteotyping analysis. There are two methods that can help us to meet this need: one is a statistical method using principal component analysis to determine useful biomarkers from characteristic ion peaks in observed mass spectra, e.g. ClinProTools software [109]; and the other is the *S10*-GERMS method which requires no statistical analysis because the selected biomarkers have a genetically theoretical background. Although the big difference between two methods is

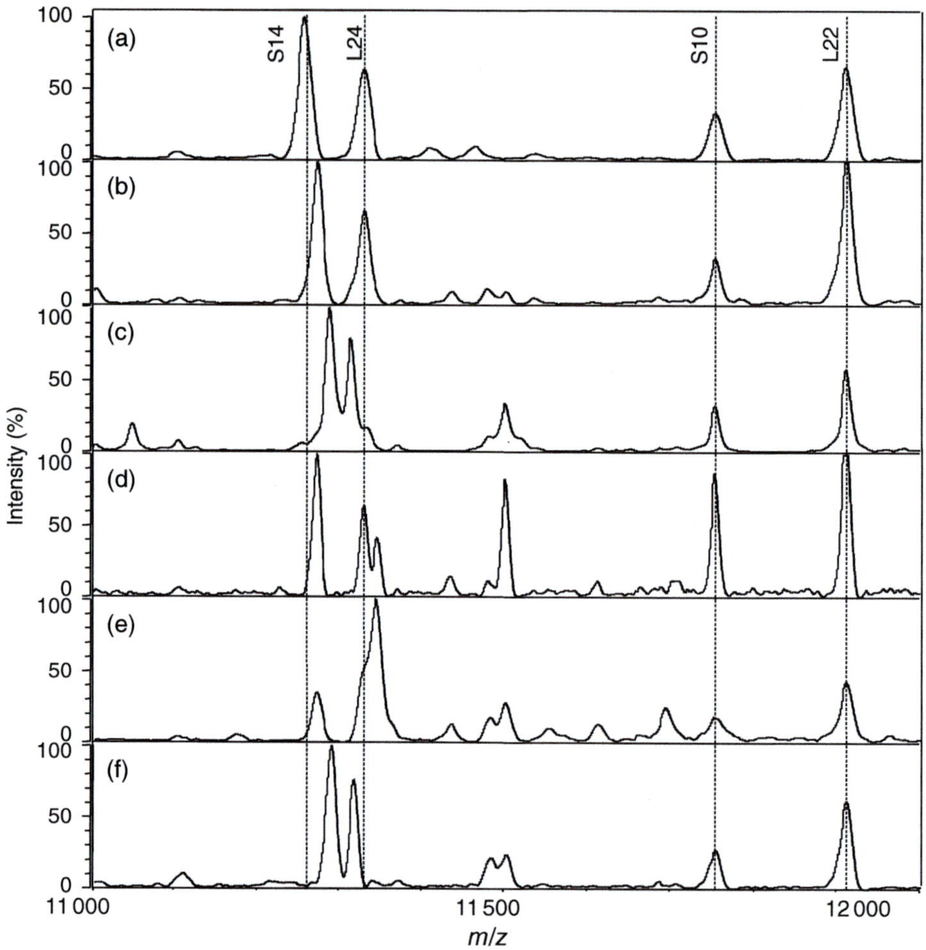

Figure 6.9 MALDI mass spectra of representative observed biomarkers for $OPEO_n$-degrading *P. putida* [105]. (a) *P. putida* NBRC14164^T; (b) KT2440; (c) NBRC 15366; (d) FMP1; (e) EC31; (f) S5. *Source:* Hotta et al. [105]/With permission of Pesticide Science Society of Japan.

whether biomarkers have a theoretical background or not, the methods will contribute to a rapid and simple identification of closely related bacteria with the following impacts: (i) appropriate risk management of public health and food hygiene by rapid and simple proteotyping; (ii) decreasing food waste in association with maintaining resources of food supply and avoiding unnecessary recalls, and reducing the risk of starvation; and (iii) microbiome research for clinical, agricultural, and environmental microbiology.

Currently, WGS has developed rapidly because of the prevailing technology of next-generation sequencing. By enriching the genetic information in the web-accessible WGS database, WGS has proved effective for *in silico* prediction of foodborne pathogens at the serovar level [4, 110, 111] and is becoming the gold standard not only for bacterial discrimination but also for microbiome research in humans, animals, plants, and soil [112, 113]. The enrichment of the web-accessible WGS database will also make possible the creation of a sophisticated method for establishing a database constructed with genetically theoretical biomarkers, leading to proteotyping of bacterial isolates at species, strain, and serovar levels (Figure 6.1). In other words, by using the web-accessible WGS database, this advanced MALDI-TOF MS proteotyping method, which utilizes target-specific biomarkers especially in combination with selection medium for the target microorganism, will pave the way for transformation of the MALDI analysis from a qualitative identification method to one of the possible candidates for quantitative identification of microorganisms in the near future.

References

1 Shank, F., Elliot, E., Wachsmuth, L., and Losikoff, M. (1996). US position on *Listeria monocytogenes* in foods. *Food Control* 7: 229–234.

2 Chen, J.Q., Regan, P., Laksanalamai, P. et al. (2017). Prevalence and methodologies for detection, characterization and subtyping of *Listeria monocytogenes* and *L. ivanovii* in food and environmental sources. *Food Sci. Hum. Wellness* 6: 97–120.

3 Rajapaksha, P., Elbourne, A., Gangadoo, S. et al. (2019). A review of methods for the detection of pathogenic microorganisms. *Analyst* 144: 396–411.

4 Tang, S., Orsi, R.H., Luo, H. et al. (2019). Assessment and comparison of molecular subtyping and characterization methods for *Salmonella*. *Front. Microbiol.* 10: 1591.

5 Anhalt, J.P. and Fenselau, C. (1975). Identification of bacteria using mass spectrometry. *Anal. Chem.* 47: 219–225.

6 Shah, H.N. and Collins, M.D. (1980). Fatty acid and isoprenoid quinone composition in the classification of *Bacteroides melaninogenicus* and related taxa. *J. Appl. Bacteriol.* 48: 75–87.

7 Croxatto, A., Prod'hom, G., and Greub, G. (2012). Applications of MALDI-TOF mass spectrometry in clinical diagnostic microbiology. *FEMS Microbiol. Rev.* 36: 380–407.

8 Fenselau, C. and Demirev, P.A. (2001). Characterization of intact microorganisms by MALDI mass spectrometry. *Mass Spectrom. Rev.* 20: 157–171.

9 Wieser, A., Schneider, L., Jung, J., and Schubert, S. (2012). MALDI-TOF MS in microbiological diagnostics-identification of microorganisms and beyond. *Appl. Microbiol. Biotechnol.* 93: 965–974.

10 Cherkaoui, A., Hibbs, J., Emonet, S. et al. (2010). Comparison of two matrix-assisted laser desorption ionization-time of flight mass spectrometry methods with conventional phenotypic identification for routine identification of bacteria to the species level. *J. Clin. Microbiol.* 48: 1169–1117.

11 Grosse-Herrenthey, A., Maier, T., Gessler, F. et al. (2008). Challenging the problem of clostridial identification with matrix-assisted laser desorption and ionization-time-of-flight mass spectrometry (MALDI-TOF MS). *Anaerobe* 14: 242–249.

12 Idelevich, E.A., Schüle, I., Grünastel, B. et al. (2014). Rapid identification of microorganisms from positive blood cultures by MALDI-TOF mass spectrometry subsequent to very short-term incubation on solid medium. *Clin. Microbiol. Infect.* 20: 1–6.

13 Machen, A., Drake, T., and Wang, Y.F. (2014). Same day identification and full panel antimicrobial susceptibility testing of bacteria from positive blood culture bottles made possible by a combined lysis-filtration method with MALDI-TOF VITEK mass spectrometry and the VITEK2 system. *PLoS One* 9: e87870.

14 Nagy, E., Maier, T., Urban, E. et al. (2009). Species identification of clinical isolates of *Bacteroides* by matrix-assisted laser-desorption/ionization time-of-flight mass spectrometry. *Clin. Microbiol. Infect.* 15: 796–802.

15 Welker, M. and Moore, E.R. (2011). Applications of whole-cell matrix-assisted laser-desorption/ionization time-of-flight mass spectrometry in systematic microbiology (mini review). *Syst. Appl. Microbiol.* 34: 2–11.

16 Branquinho, R., Sousa, C., Lopes, J. et al. (2014). Differentiation of *Bacillus pumilus* and *Bacillus safensis* using MALDI-TOF-MS. *PLoS One* 9: e110127.

17 Khot, P.D. and Fisher, M.A. (2013). Novel approach for differentiating *Shigella* species and *Escherichia coli* by matrix-assisted laser desorption ionization-time of flight mass spectrometry. *J. Clin. Microbiol.* 51: 3711–3716.

18 Mahé, P., Arsac, M., Chatellier, S. et al. (2014). Automatic identification of mixed bacterial species fingerprints in a MALDI-TOF mass-spectrum. *Bioinformatics* 30: 1280–1286.

19 Zhang, L., Borror, C.M., and Sandrin, T.R. (2014). A designed experiments approach to optimization of automated data acquisition during characterization of bacteria with MALDI-TOF mass spectrometry. *PLoS One* 9: e92720.

20 Arnold, R.J. and Reilly, J.P. (1999). Observation of *Escherichia coli* ribosomal proteins and their posttranslational modifications by mass spectrometry. *Anal. Biochem.* 269: 105–112.

21 Pineda, F.J., Antoine, M.D., Demirev, P.A. et al. (2003). Microorganism identification by matrix-assisted laser/desorption ionization mass spectrometry and model-derived ribosomal protein biomarkers. *Anal. Chem.* 75: 3817–3822.

22 Teramoto, K., Sato, H., Sun, L. et al. (2007). Phylogenetic classification of *Pseudomonas putida* strains by MALDI-MS using ribosomal subunit proteins as biomarkers. *Anal Chem.* 79: 8712–8719.

23 Demirev, P.A., Ho, Y.P., Ryzhov, V., and Fenselau, C. (1999). Microorganism identification by mass spectrometry and protein database searches. *Anal. Chem.* 71: 2732–2738.

24 Demirev, P.A., Lin, J.S., Peneda, F.J., and Fenselau, C. (2001). Bioinformatics and mass spectrometry for microorganism identification: proteome-wide post-translational modifications and database search algorithms for characterization of intact *H. pylori*. *Anal. Chem.* 73: 4566–4573.

25 Pineda, F.J., Lin, J.S., Fenselau, C., and Demirev, P.A. (2000). Testing the significance of microorganism identification by mass spectrometry and proteome database search. *Anal. Chem.* 72: 3739–3744.

26 Coenye, T. and Vandamme, P. (2005). Organisation of the *S10*, *spc* and *alpha* ribosomal protein gene clusters in prokaryotic genomes. *FEMS Microbiol. Lett.* 242: 117–126.

27 Victoria, B., Ahmed, A., Zuerner, R.L. et al. (2008). Conservation of the *S10-spc-alpha* locus within otherwise highly plastic genomes provides phylogenetic insight into the genus *Leptospira*. *PLoS One* 3: e2752.

28 Amiri-Eliasi, B. and Fenselau, C. (2001). Characterization of protein biomarkers desorbed by MALDI from whole fungal cells. *Anal. Chem.* 73: 5228–5231.

29 Teramoto, K., Sato, H., Sun, L. et al. (2007b). A simple intact protein analysis by MALDI-MS for characterization of ribosomal proteins of two genome-sequenced lactic acid bacteria and verification of their amino acid sequences. *J. Proteome Res.* 6: 3899–3907.

30 Valentine, N.B., Wahl, J.H., Kingsley, M.T., and Wahl, K.L. (2002). Direct surface analysis of fungal species by matrix-assisted laser desorption/ionization mass spectrometry. *Rapid Commun. Mass Spectrom.* 16: 1352–1357.

31 Running, W.E., Ravipaty, S., Karty, J.A., and Reilly, J.P. (2007). A top-down/bottom-up study of the ribosomal proteins of *Caulobacter crescentus*. *J. Proteome Res.* 6: 337–347.

32 Sato, H., Teramoto, K., Ishii, Y. et al. (2011). Ribosomal protein profiling by matrix-assisted laser desorption/ionization time-of-flight mass spectrometry for phylogenety-based subspecies resolution of *Bifidobacterium longum*. *Syst. Appl. Microbiol.* 34: 76–80.

33 Suh, M.J., Hamburg, D.M., Gregory, S.T. et al. (2005). Extending ribosomal protein identifications to unsequenced bacterial strains using matrix-assisted laser desorption/ionization mass spectrometry. *Proteomics* 5: 4818–4831.

34 Tamura, H., Hotta, Y., and Sato, H. (2013). Novel accurate bacterial discrimination by MALDI-time-of-flight MS based on ribosomal proteins coding in *S10-spc-alpha* operon at strain level *S10*-GERMS. *J. Am. Soc. Mass Spectrom.* 24: 1185–1193.

35 Tamura, H. (2017). *MALDI-TOF MS Based on Ribosomal Protein Coding in S10-Spc-Alpha Operons for Proteotyping in MALDI-TOF and TANDEM MS for Clinical Microbiology* (ed. H.N. Sha and S.E. Gharbia). John Wiley & Sons Ltd.

36 Hotta, Y., Teramoto, K., Sato, H. et al. (2010a). Classification of genus *Pseudomonas* by MALDI-TOF MS based on ribosomal protein coding in *S10-spc-alpha* operon at strain level. *J. Proteome Res.* 9: 6722–6728.

37 Hotta, Y., Sato, J., Sato, H. et al. (2011). Classification of genus *Bacillus* based on MALDI-TOF MS analysis of ribosomal protein coded in *S10* and *spc* operons. *J. Agric. Food. Chem.* 59: 5222–5230.

38 Hotta, Y., Sato, H., Hosoda, A., and Tamura, H. (2012). MALDI-TOF MS analysis of ribosomal proteins coded in *S10* and *spc* operons rapidly classified *Sphingomonaceae* as alkylphenol polyethoxylate-degrading bacteria from the environment. *FEMS Microbiol. Lett.* 330: 23–29.

39 Sato, H., Torimura, M., Kitahara, M. et al. (2012). Characterization of the *Lactobacillus casei* group based on the profiling of ribosomal proteins coded in *S10-spc-alpha* operons as observed by MALDI-TOF MS. *System. Appl. Microbiol.* 35: 447–454.

40 Ojima-Kato, T., Yamamoto, N., Suzuki, M. et al. (2014). Discrimination of *Escherichia coli* O157, O26 and O111 from other serovars by MALDI-TOF MS based on the *S10*-GERMS method. *PLoS One* 9: e113458.

41 Ojima-Kato, T., Yamamoto, N., Iijima, Y., and Tamura, H. (2015). Assessing the performance of novel software strain solution on automated discrimination of *Escherichia coli* serotypes and their mixtures using matrix-assisted laser desorption ionization-time of flight mass spectrometry. *J. Microbiol. Methods* 119: 233–238.

42 CDC. (2015). Serotypes and the Importance of Serotyping *Salmonella*. https://www.cdc.gov/salmonella/reportspubs/salmonella-atlas/serotyping-importance.html#:~:text=Laboratory%20experts%20serotype%20the%20Salmonella,infections%20for%20over%2050%20years. (accessed 27 September 2022).

43 Boore, A., Hoekstra, R., Iwamoto, M. et al. (2015). *Salmonella enterica* infections in the United States and assessment of coefficients of variation: a novel approach to identify epidemiologic characteristics of individual serotypes. *PLoS One* 10: e0145416. https://doi.org/10.1371/journal.pone.0145416.

44 Grimont, P. and Weill, F.X. (2007). *Antigenic Formulae of the Salmonella Serovars*, 9e. Institut Pasteur, Paris, France: WHO Collaborating Centre for Reference and Research on *Salmonella*.

45 Dieckmann, R. and Malorny, B. (2011). Rapid screening of epidemiologically important *Salmonella enterica* subsp. *enterica* serovars by whole-cell matrix-assisted laser desorption ionization–time of flight mass spectrometry. *Appl. Environ. Microb.* 77: 4136–4146.

46 Ojima-Kato, T., Yamamoto, N., Nagai, S. et al. (2017). Application of Proteotyping strain solution™ Ver. 2 software and theoretically calculated mass database in MALDI-TOF MS typing of *Salmonella* serotype. *Appl. Microbiol. Biothecnol.* 101: 8557–8569.

47 Fukuyama, Y., Ojima-Kato, T., Nagai, S. et al. (2019). Improved MALDI-MS method for highly sensitive and reproducible detection of the biomarker peaks for proteotyping of *Salmonella* serotype. *J. Mass Spectrom.* 54: 966–975.

48 Petersen, A., Aarestrup, F.M., Angulo, F.J. et al. (2002). WHO global salm-surv external quality assurance system (EQAS): an important step toward improving the quality of *Salmonella* serotyping and antimicrobial susceptibility testing worldwide. *Microb. Drug Resist.* 8: 345–353.

49 Ashton, P.M., Nair, S., Peters, T.M. et al. (2016). Identification of *Salmonella* for public health surveillance using whole genome sequencing. *PeerJ* 4: e1752.

50 McLauchlin, J., Rees, C., and Genus, I. (2009). *Listeria* Pirie 1940a 383AL. In: *Bergey's Manual of Systematic Bacteriology*, 2e, vol. 3 (ed. P. De Vos, G. Garrity, D. Jones, et al.), 244–257. New York: Springer.

51 US FDA. Draft Compliance Policy Guide Sec. 555.320-*Listeria monocytogenes*. (2008). https://www.federalregister.gov/documents/2008/02/07/08-547/draft-compliance-policy-guide-sec-555320-listeria-monocytogenes (accessed 27 September 2022).

52 Favaro, M., Sarmati, L., Sancesario, G., and Fontana, C. (2014). First case of *Listeria innocua* meningitis in a patient on steroids and eternecept. *JMM Case Rep.* 1: https://doi.org/10.1099/jmmcr.0.003103.

53 Swaminathan, B. and Gerner-Smidt, P. (2007). The epidemiology of human listeriosis. *Microbes Infect.* 9: 1236–1243.

54 Burnett, J., Wu, S.T., den Bakker, H.C. et al. (2020). *Listeria monocytogenes* is prevalent in retail produce environments but *Salmonella enterica* is rare. *Food Control* 113: 107173.

55 Forauer, E., Wu, S.T., and Etter, A.J. (2021). *Listeria monocytogenes* in the retail deli environment: a review. *Food Control* 119: 107443.

56 Farber, J.M., Zwietering, M., Wiedmann, M. et al. (2021). Alternative approaches to the risk management of *Listeria monocytogenes* in low risk foods. *Food Control* 123: 107601.

57 Ottaviani, F., Ottaviani, M. and Agosti, M. (1997) Differential agar medium for *Listeria monocytogenes*. Quimper Froid Symposium Proceedings, Quimper, France (16–18 June 1997), p. 6, ADRIA.

58 Seeliger, H.P.R. and Höhne, K. (1979). Serotyping of *Listeria monocytogenes* and related species. *Methods Microbiol.* 13: 31–49.

59 Graves, L.M. and Swaminathan, B. (2001). PulseNet standardized protocol for subtyping *Listeria monocytogenes* by macrorestriction and pulsed-field gel electrophoresis. *Int. J. Food Microbiol.* 65: 55–62.

60 Rasmussen, O.F., Skouboe, P., Dons, L. et al. (1995). *Listeria monocytogenes* exists in at least three evolutionary lines: evidence from flagellin, invasive associated protein and listeriolysin O genes. *Microbiology* 141: 2053–2061.

61 Ward, T.J., Ducey, T.F., Usgaard, T. et al. (2008). Multilocus genotyping assays for single nucleotide polymorphism-based subtyping of *Listeria monocytogenes* isolates. *Appl. Environ. Microbiol.* 74: 7629–7642.

62 Piffaretti, J.C., Kressebuch, H., Aeschbacher, M. et al. (1989). Genetic characterization of clones of the bacterium *Listeria monocytogenes* causing epidemic disease. *Proc. Natl. Acad. Sci. U S A.* 86: 3818–3822.

63 Roberts, A.J. and Wiedmann, M. (2006). Allelic exchange and site-directed mutagenesis probe the contribution of ActA amino-acid variability to phosphorylation and virulence-associated phenotypes among *Listeria monocytogenes* strains. *FEMS Microbiol. Lett.* 254: 300–307.

64 Jadhav, S., Bhave, M., and Palombo, A.E. (2012). Methods used for the detection and subtyping of *Listeria monocytogenes*. *J. Microbiol. Methods* 88: 327–341.

65 Ragon, M., Wirth, T., Hollandt, F. et al. (2008). A new perspective on *Listeria monocytogenes* evolution. *PLoS Pathog.* 4: e1000146.

66 Salcedo, C., Arreaza, L., Alcalá, B. et al. (2003). Development of a multilocus sequence typing method for analysis of *Listeria monocytogenes* clones. *J. Clin. Microbiol.* 41: 757–762.

67 Borucki, M.K., Krug, M.J., Muraoka, W.T., and Call, D.R. (2003). Discrimination among *Listeria monocytogenes* isolates using a mixed genome DNA microarray. *Vet. Microbiol.* 92: 351–362.

68 Barbuddhe, S.B., Maier, T., Schwarz, G. et al. (2008). Rapid identification and typing of *Listeria* species by matrix-assisted laser desorption ionization-time of flight mass spectrometry. *Appl. Environ. Microbiol.* 74: 5402–5407.

69 Farfour, E., Leto, J., Barritault, M. et al. (2012). Evaluation of the Andromas matrix-assisted laser desorption ionization-time of flight mass spectrometry system for identification of aerobically growing gram-positive bacilli. *J. Clin. Microbiol.* 50: 2702–2807.

70 Hsueh, P.R., Lee, T.F., Du, S.H. et al. (2014). Bruker biotyper matrix-assisted laser desorption ionization-time of flight mass spectrometry system for identification of *Nocardia*, *Rhodococcus*, *Kocuria*, *Gordonia*, *Tsukamurella*, and *Listeria* species. *J. Clin. Microbiol.* 52: 2371–2379.

71 Jadhav, S., Gulati, V., Fox, E.M. et al. (2015). Rapid identification and source-tracking of *Listeria monocytogenes* using MALDI-TOF mass spectrometry. *Int. J. Food Microbiol.* 202: 1–9.

72 Ojima-Kato, T., Yamamoto, N., Takahashi, H., and Tamura, H. (2016). Matrix-assisted laser desorption ionization- time of flight mass spectrometry (MALDI- TOF MS) can precisely discriminate the lineages of *Listeria monocytogenes* and species of *Listeria*. *PLoS ONE* 11: e0159730.

73 McKellar, R.C. (1994). Use of the CAMP test for identification of Listeria monocytogenes. *Appl. Environ. Microbiol.* 60: 4219–4225.

74 Volokhov, D., Rasooly, A., Chumakov, K., and Chizhikov, V. (2002). Identification of *Listeria* species by microarray-based assay. *J. Clin. Microbiol.* 40: 4720–4728.

75 Wieme, A.D., Spitaels, F., Aerts, M. et al. (2014). Effects of growth medium on matrix-assisted laser desorption–ionization time of flight mass spectra: a case study of acetic acid bacteria. *Appl. Environ. Microbiol.* 80: 1528–1538.

76 Meerak, J., Yukphan, P., Miyashita, M. et al. (2008). Phylogeny of γ-polyglutamic acid-producing *Bacillus* strains isolated from a fermented locust bean product manufactured in West Africa. *J. Gen. Appl. Microbiol.* 54: 159–166.

77 Schnepf, E., Crickmore, N., VanRie, J. et al. (1998). *Bacillus thuringiensis* and its pesticidal crystal proteins. *Microbiol. Mol. Biol. Rev.* 62: 775–806.

78 Yamada, S., Ohashi, E., Agata, N., and Venkateswaran, K. (1999). Cloning and nucleotide sequence analysis of gyrB of *Bacillus cereus*, *B. thuringiensis*, *B. mycoides*, and *B. anthracis* and their application to the detection of *B. cereus* in rice. *Appl. Environ. Microbiol.* 65: 1483–1490.

79 Granum, P.E. and Lund, T. (1997). *Bacillus cereus* and its food poisoning toxins. *FEMS Microbiol. Lett.* 157: 223–228.

80 Priest, F.G., Barker, M., Baillie, L.W. et al. (2004). Population structure and evolution of the *Bacillus cereus* group. *J. Bacteriol.* 186: 7959–7970.

81 AgaTa, N., Mori, M., Ohta, M. et al. (1994). A novel dodecadepsipeptide, cereulide, isolated from *Bacillus cereus* causes vacuole formation in Hep-2 cells. *FEMS Microbiol. Lett.* 121: 31–34.

82 Andersson, M.A., Hakulinen, P., Honkalampi-Hämäläinen, U. et al. (2007). Toxicological profile of cereulide, the *Bacillus cereus* emetic toxin, in functional assays with human, animal and bacterial cells. *Toxicon* 49: 351–367.

83 Mikkola, R., Saris, N.E., Grigoriev, P.A. et al. (1999). Ionophoretic properties and mitochondrial effects of cereulide: the emetic toxin of *B. cereus*. *Eur. J. Biochem.* 263: 112–117.

84 Sacch, C.T., Whitney, A.M., Mayer, L.W. et al. (2002). Sequencing of 16S rRNA gene: a rapid tool for identification of *Bacillus anthracis*. *Emerg. Infect. Dis.* 8: 1117–1123.

85 Ehling-Schulz, M., Lereclus, D., and Koehler, T.M. (2019). The *Bacillus cereus* group: *Bacillus* species with pathogenic potential. *Microbiol. Spectr.* 7: 1–60.

86 Johler, S., Kalbhenn, E.M., Heini, N. et al. (2018). Enterotoxin production of *Bacillus thuringiensis* isolates from biopesticides, foods, and outbreaks. *Front. Microbiol.* 23:https://doi.org/10.3389/fmicb.2018.01915.

87 Kim, J.B., Choi, O.K., Kwon, S.M. et al. (2017). Prevalence and toxin characteristics of *Bacillus thuringiensis* isolated from organic vegetables. *J. Microbiol. Biotechnol.* 27: 1449–1456.

88 Guérin, A., Rønning, H.T., Dargaignaratz, C. et al. (2017). Cereulide production by *Bacillus weihen- stephanensis* strains during growth at different pH values and temperatures. *Food Microbiol.* 65: 130–135.

89 Stenfors, L.P., Mayr, R., Scherer, S., and Granum, P.E. (2002). Pathogenic potential of fifty *Bacillus weihenstephanensis* strains. *FEMS Microbiol. Lett.* 215: 47–51.

90 Thorsen, L., Hansen, B.M., Nielsen, K.F. et al. (2006). Characterization of emetic *Bacillus weihenstephanensis*, a new cereulide-producing bacterium. *Appl. Environ. Microbiol.* 72: 5118–5121.

91 Takahashi, N., Nagai, S., Fujita, A. et al. (2020). Discrimination of psychrotolerant *Bacillus cereus* group based on MALDI-TOF MS analysis of ribosomal subunit proteins. *Food Microbiol.* 91: 103542.

92 Takahashi, N., Nagai, S., Fujita, A., Ido, Y., Kato, K., Saito, A., Moriya, Y., Tomimatsu, Y., Kaneta, N., Tsujimoto, Y. and Tamura, H. (2019). MALDI-TOF MS Analysis for simultaneous discrimination of cereulide producing *Bacillus cereus* and psychrotolerant *Bacillus cereus* group from other *B. cereus* group. IUPAC 2019 Ghent Crop Protection Chemistry, Belgique (19–24 May).

93 Liu, Y., Lai, Q., and Shao, Z. (2018). Genome analysis-based reclassification of *Bacillus wei- henstephanensis* as a later heterotypic synonym of *Bacillus mycoides*. *Int. J. Syst. Evol. Microbiol.* 68: 106–112.

94 Manzulli, V., Rondinone, V., Buchicchio, A. et al. (2021). Discrimination of *Bacillus cereus* group members by MALDI-TOF mass spectrometry. *Microorganisms* 9: 1202.

95 Rouzeau-Szynalski, K., Stollewerk, K., Messelhäusser, U., and Ehling-Schulz, M. (2020). Why be serious about emetic *Bacillus cereus*: Cereulide production and industrial challenges. *Food Microbiol.* 85: 103279.

96 Ramarao, N., Tran, S.L., Marin, M., and Vidic, J. (2020). Advanced methods for detection of *Bacillus cereus* and its pathogenic factors. *Sensors* 20: 2667.

97 Ulrich, S., Gottschalk, C., Dietrich, R. et al. (2019). Identification of cereulide producing *Bacillus cereus* by MALDI-TOF MS. *Food Microbiol.* 82: 75–81.

98 Doellinger, J., Schneider, A., Stark, T.D. et al. (2020). Evaluation of MALDI-ToF mass spectrometry for rapid detection of cereulide from *Bacillus cereus* cultures. *Front. Microbiol.* 11: 511674.

99 Ducrest, P.J., Pfammatter, S., Stephan, D. et al. (2019). Rapid detection of *Bacillus* ionophore cereulide in food products. *Sci. Rep.* 9: 1–8.

100 Nishio, E., Yoshikawa, H., Wakayama, M. et al. (2002). Isolation and identification of *Sphingomonas* sp. that yields *tert*-octylphenol monoethoxylate under aerobic conditions. *Biosci. Biotechnol. Biochem.* 66: 1792–1798.

101 Laws, S.C., Carey, S.A., Ferrell, J.M. et al. (2000). Estrogenic activity of octylphenol, nonylphenol, bisphenol A and methoxychlor in rats. *Toxicol. Sci.* 54: 154–167.

102 Shibata, A., Ishimoto, Y., Nishizaki, Y. et al. (2007). The effect of calcium ion on the biodegradation of octylphenol polyethoxylates, and the antiandrogenic activity of their biodegradates. *Appl. Microbiol. Biotechnol.* 77: 195–201.

103 Sato, H., Shibata, A., Wang, Y. et al. (2001). Characterization of biodegradation intermediates of non-ionic surfactants by matrix-assisted laser desorption/ionization–mass spectrometry 1. Bacterial biodegradation of octylphenol polyethoxylate under aerobic conditions. *Polym. Degrad. Stabil.* 74: 69–75.

104 Hotta, Y., Hosoda, A., Sano, F. et al. (2010b). Ecotoxicity by the biodegradation of alkylphenol polyethoxylates depends on the effect of trace elements. *J. Agric. Food Chem.* 58: 1062–1067.

105 Hotta, Y., Wakayama, M., Niwa, K. et al. (2012b). Application of matrix-assisted laser-desorption/ionization time-of-flight mass spectrometry for the identification of alkylphenol polyethoxylate-degrading bacteria in the environment. *J. Pestic. Sci.* 37: 305–311.

106 Arnesen, L.P.S., Fagerlund, A., and Granum, P.E. (2008). From soil to gut: *Bacillus cereus* and its food poisoning toxins. *FEMS Microbiol. Rev.* 32: 579–606.

107 Johnson, N., Litt, P.K., Kniel, K.E., and Bais, H. (2020). Evasion of plant innate defense response by *Salmonella* on lettuce. *Front. Microbiol.* 11: 500.

108 Tuipulotu, D.E., Mathur, A., Ngo, C., and Man, S.M. (2021). *Bacillus cereus*: epidemiology, virulence factors, and host-pathogen interaction. *Trends Microbiol.* 29: 458–471.

109 Wang, H.Y., Lien, F., Liu, T.P. et al. (2018). Application of a MALDI-TOF analysis platform (ClinProTools) for rapid and preliminary report of MRSA sequence types in Taiwan. *PeerJ* 6: e5784.

110 Fricker, A.M., Podlesny, D., and Fricke, W.F. (2019). What is new and relevant for sequencing-based microbiome research? A mini-review. *J. Adv. Res.* 19: 105–112.

111 Pornsukarom, S., van Vliet, A.H.M., and Thakur, S. (2018). Whole genome sequencing analysis of multiple *Salmonella* serovars provides insights into phylogenetic relatedness, antimicrobial resistance, and virulence markers across humans, food animals and agriculture environmental sources. *BMC Genomics* 19: 801.

112 Compant, S., Samad, A., Faist, H., and Sessitsch, A. (2019). A review on the plant microbiome: ecology, functions, and emerging trends in microbial application. *J. Adv. Res.* 19: 29–37.

113 Jansson, J.K. and Hofmockel, K.S. (2018). The soil microbiome – from metagenomics to metaphenomics. *Curr. Opin. Microbiol.* 43: 162–168.

7

Diversity, Transmission and Selective Pressure on the Proteome of *Pseudomonas aeruginosa*

Louise Duncan[1], Ajit J. Shah[2], Malcolm Ward[2], Radhey S. Gupta[3], Bashudev Rudra[3], Alvin Han[3], Ken Bruce[1], and Haroun N. Shah[2]

[1] *School of Cancer and Pharmaceutical Sciences, King's College London, London, UK*
[2] *Department of Natural Sciences, Middlesex University, London, UK*
[3] *Department of Biochemistry and Biomedical Sciences, McMaster University, Hamilton, Canada*

7.1 Introduction: Diversity

7.1.1 *P. aeruginosa*: from 'Atypical' to Diverse

In 1956, Williamson [1] contested the singular morphological description of *Pseudomonas aeruginosa* presented in Bergey's *Manual of Determinative Bacteriology*, Such a limited description was proposed as inadequate at describing this diverse bacterial species. Traditional phenotypical methods for bacterial identification would reveal the phenotypic diversity in *P. aeruginosa* isolates, with the isolation of at least six colony types from different samples [2, 3]. Further variability would be unveiled in the morphologies, pigments, and substrate use, with reports of 'atypical' isolates. Nonetheless, these atypical traits were detected more and more frequently and instead pointed towards an extremely diverse species.

This chapter examines the diversity of *P. aeruginosa* and its close relationship with the proteome. We discuss the cutting-edge mass spectrometry (MS) techniques that are continuing to reveal the extent and impact of proteomic diversity with respect to transmission and human infections of *P. aeruginosa*. In a parallel study, we undertake phylogenomic analysis of the species based upon its core protein sequences against eight other *Pseudomonas* species that may help to elucidate molecular markers that distinguish subtypes.

7.1.2 Phenotypical Diversity in Isolates from Different Environments

7.1.2.1 Clinical Isolates

Pseudomonas aeruginosa represents a significant opportunistic pathogen to humans, remaining one of the leading causes of hospital-acquired pneumonia [4]. It causes a range

Microbiological Identification using MALDI-TOF and Tandem Mass Spectrometry: Industrial and Environmental Applications, First edition. Edited by Haroun N. Shah, Saheer E. Gharbia, Ajit J. Shah, Erika Y. Tranfield, and K. Clive Thompson.

of infection types, including wounds, urinary, ocular, and respiratory. It also frequently causes infections by contaminating contact lens fluid and medical devices [5, 6].

Pseudomonas aeruginosa is a hugely concerning species for those with the genetic lung disease, cystic fibrosis (CF). Due to the high levels of mucus in the CF lung, *P. aeruginosa* infections are often acquired by the time individuals reach adulthood, causing significant lung damage and subsequent failure [7]. Such infections are often not eradicated due to the high levels of innate and acquired resistance.

Given the severity of these infections, CF isolates have been characterized for decades and form much of our understanding of *P. aeruginosa* diversity. Characteristic features include a conversion to mucoidy, loss of flagella-based motility, and loss of pyoverdine production [8–10]. CF isolates also convert to specialised metabolic pathways showing a preference for using the breakdown products of surfactant as a carbon source [11, 12]. Within the CF lung, *P. aeruginosa* can also demonstrate regional phenotypical diversity, evolving as separate populations in different parts of the lung [13]. Additionally, extensive phenotypic diversity is observed within isolates of the same patient [14–16]. Overall, this process of diversification is suggested to be critical to the development of chronic CF infections.

7.1.2.2 Environmental Isolates

Pseudomonas aeruginosa is isolated from a wide range of natural environments such as soils, rivers, and oceans [17–21]. Such isolates have demonstrated an ability to survive across a range of saline and contaminant concentrations. In addition, strains are frequently isolated from hydrocarbon-containing environments, including jet fuel tanks [22–24].

Given the diversity of these natural environments, it seems unlikely that a singular encompassing environmental phenotype of *P. aeruginosa* can be described, as compared with its CF counterpart. Environmental isolates have shown variation in their pyoverdine type, biofilm formation, motility, and virulence in for example *Galleria mellonella* models [20, 25, 26]. The phenotypic diversity observed in these natural environments has reflected the diversity observed within the overall population of *P. aeruginosa*.

7.1.2.3 Veterinary Isolates

Pseudomonas aeruginosa also represents an important veterinary opportunistic pathogen, causing fatal haemorrhagic pneumonia in mink, endometriosis in horses, and mastitis in cows [27–29]. It remains one of the leading causes of otitis externa in canines, often showing high levels of antibiotic resistance [30, 31]. Canine otitis isolates have previously displayed a unique elastase-deficient phenotype that has been demonstrated in both others' and our own investigations [32].

Additionally, *P. aeruginosa* is isolated from the faeces of healthy animal hosts [33]. Isolates were shown to produce many virulence factors, which is concerning given the potential role of healthy animals as transmission reservoirs. Despite causing a range of infection types across a wide array of animal hosts, veterinary isolates remain less well characterized than their clinical and environmental counterparts.

7.1.2.4 Comparing *P. aeruginosa* Phenotypical Profiles from Different Environments

We compared the phenotypical profiles of isolates from three sources: the environment (isolated from River Thames, London), CF infections and veterinary infections (Source: Louise Duncan). The colony morphology, production of virulence factors, substrate use, chemical tolerances, and antibiotic susceptibilities were investigated. There was high variation in the production of all virulence factors within each source. In addition, identical substrate and chemical tolerance profiles were found across all sources. The only notable phenotypes were observed in a select number of CF isolates that reflected many of the known metabolic and functional changes that occur from adaptation to the CF lung. Importantly, this revealed that many environmental and veterinary samples can use the same substrates for growth and produce virulence factors to the same extent as CF isolates. For those phenotypes that could be reliably quantified, the phenotypical profiles of all isolates were compared (see Figure 7.1). Overall, clustering of environmental isolates with those from CF and veterinary sources was observed, suggesting that phenotypical profiles were indistinguishable irrespective of their environments.

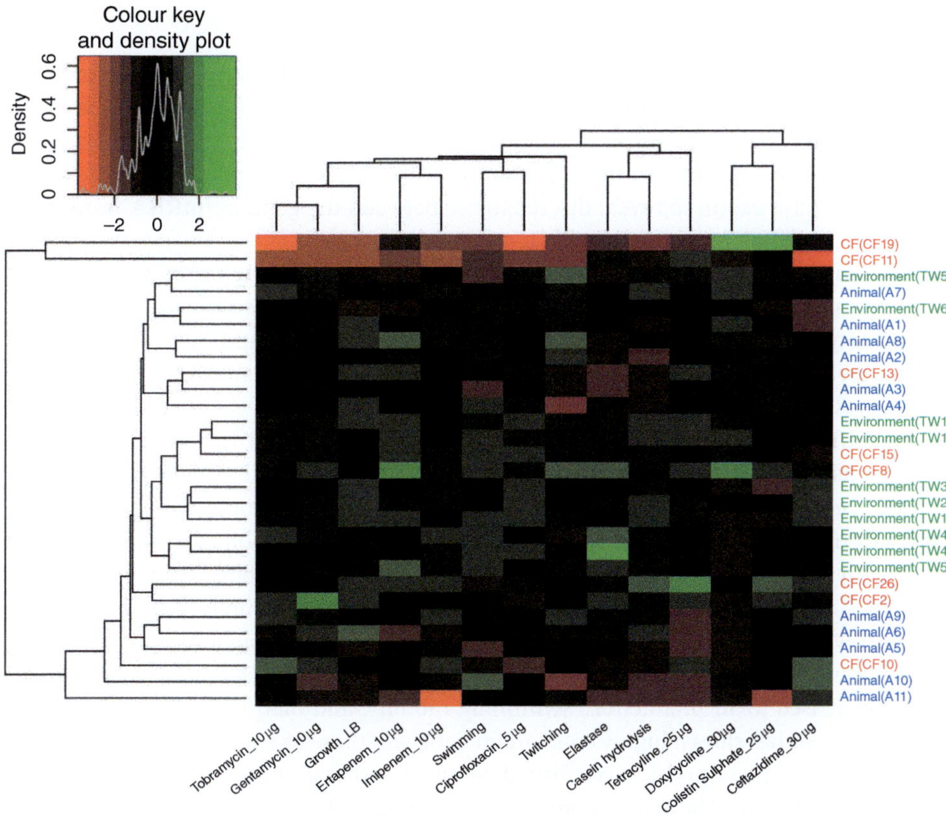

Figure 7.1 Hierarchical cluster analysis of phenotypical profiles of *P. aeruginosa* isolates from CF, environmental, and veterinary sources. All isolates were subject to phenotypical analysis, including characteristics (growth in LB, swimming, and twitching motilities), production of virulence factors (elastase, casein hydrolysis) and antibiotic susceptibility (tobramycin, gentamycin, ertapenem, imipenem, ciprofloxacin, tetracycline, doxycycline, colistin sulphate, ceftazidime). Data were normalized by mean centring and scaled, with green representing values above the mean and red below. Profiles were compared by hierarchical cluster analysis with average linkage. Row and column cluster dendrograms show clustering of isolates and phenotypes, respectively. The colour of isolates represents their *source:* CF (red), environmental (green) and veterinary (blue) – with isolate name in brackets.

7.1.2.5 Antibiotic Resistance in *P. aeruginosa* from Different Environments

Clinical strains of *P. aeruginosa* have displayed global resistance to all classes of antibiotics used in the clinic [34–36]. By contrast, environmental isolates often demonstrate greater susceptibility although multi-drug resistant strains have previously been isolated [20, 24, 37, 38]. We investigated the carriage of antibiotic resistance in isolates from the environment (isolated from River Thames, London) compared with those from CF and veterinary infections (see Figure 7.2). Environmental isolates were susceptible to ciprofloxacin, imipenem, and tobramycin, as compared with CF isolates that showed resistance. Nonetheless, many of the environmental and veterinary isolates were either resistant or close to the breakpoint value for gentamycin, compared with only half of the CF isolates that were resistant. Although we recognize the low number of isolates that were analysed from these environments, this may suggest that household pets and river systems may be reservoirs for gentamycin-resistant isolates.

7.1.3 The Relationship Between Phenotypical and Proteomic Diversity

Many of the described phenotypes are the result of mutations in common genes such as those involved in alginate biosynthesis and quorum sensing [39, 40]. Nonetheless, the complex regulation of gene expression introduces diversity in bacterial gene products, and focus has turned increasingly to the study at other system levels such as the proteome. These analyses have begun to reveal discrepancies between the genome/mRNA transcript information and the proteins produced [41, 42].

The relationship between the proteome and phenotype is particularly well observed in the small colony variants (SCV) of *P. aeruginosa* that are frequently isolated from CF infections. These phenotypical variants display increased twitching mobility, and biofilm formation among an atypical colony morphology [43]. Early proteomic studies revealed differential protein expression between the SCV and the wild-type, in particular, the overexpression of type I and III secreted proteins [44]. Proteomic analysis unveiled the diversity within secreted proteins that can lead to these uniquely adapted morphological phenotypes.

The close relationship of the proteome and phenotype is also evident in the case of adaptive resistance. The proteome exhibits more dynamicity over the genome, allowing cells to respond both rapidly and in a reversible manner, while the genome content remains unchanged. Adaptive resistance can develop from exposure to sub-inhibitory antibiotic concentrations, which induces the formation of rapid and reversible tolerant phenotypes. In both planktonic and biofilm populations of *P. aeruginosa*, adaptive changes are described, including increased toxin production, swimming motility, and biofilm formation [45–47]. Proteomic analysis has shown highly concordant data with these phenotypical changes, with the role of many of these proteins confirmed by knock-out strains and biochemical analyses [48]. Proteomic investigations not only validate changes observed at other system levels but can reveal new insights into function that may be more applicable as therapeutic targets.

7.1.4 Techniques and Practical Considerations for Studying Proteomic Diversity

Early proteomic methods using gel-based separation began to uncover the variation between strains of *P. aeruginosa* [44, 49, 50]. As MS techniques developed, their application in bacterial proteomics would substantially increase the coverage of the proteome and unveil further diversity.

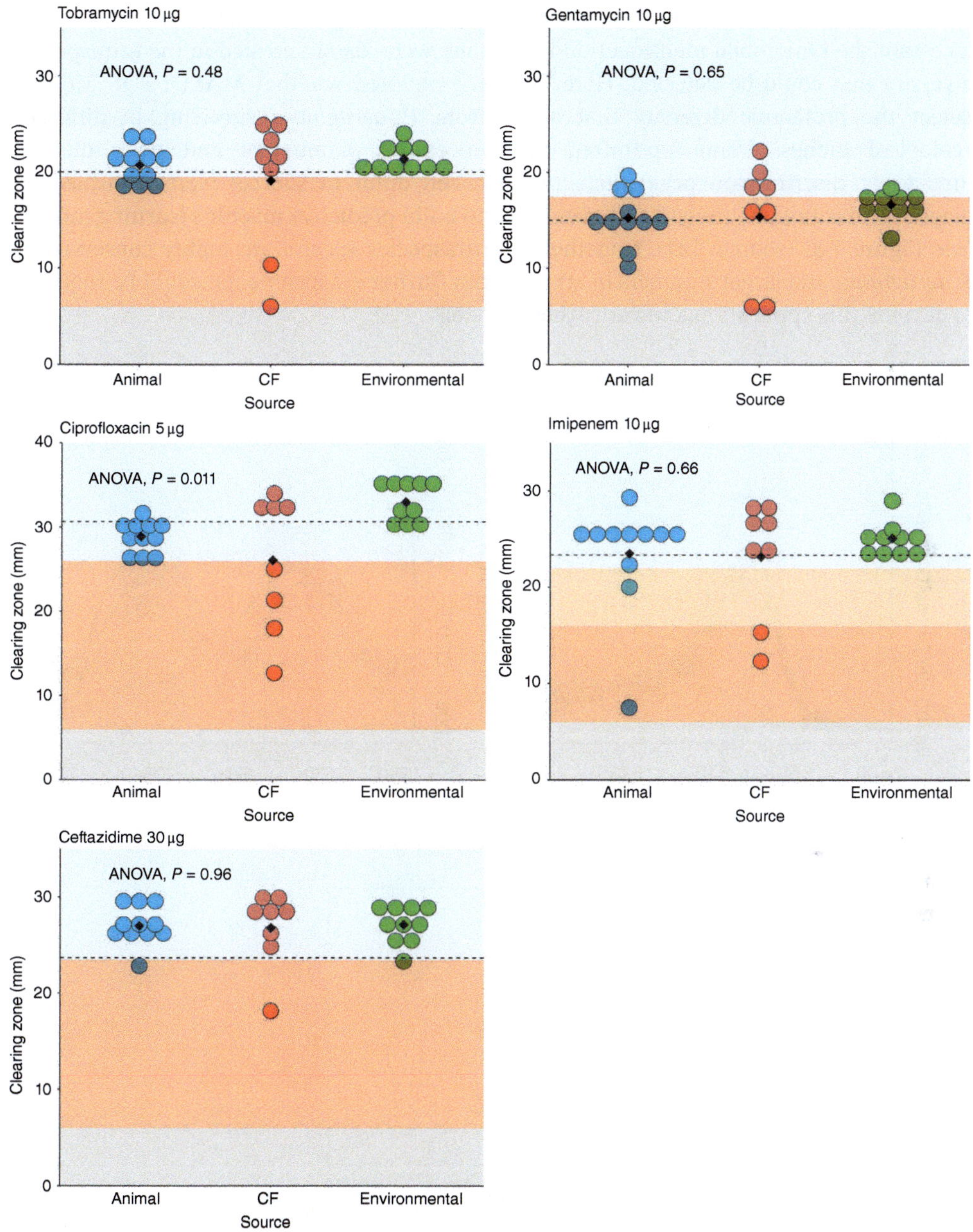

Figure 7.2 Antibiotic resistance in *P. aeruginosa* isolates from CF, environmental, and veterinary sources. Each point indicates clearing zone diameters for isolates from three different source groups: veterinary (blue), CF (red) and the environment (green). Antibiotic breakpoints from published EUCAST breakpoints with the following area colour codes: susceptible (blue), intermediate (yellow), resistant (red), limit of detection (grey). Diamonds represent mean clearing zone diameter for each source group. Dashed line represents clearing zone diameter for reference strain PAO1. The *P*-value from Kruskal–Wallis test is displayed. Data are representative of two independent replicates, performed in triplicate.

As matrix-assisted laser desorption/ionization time-of-flight (MALDI-TOF) MS became an established microbial identification tool, many were also interested in the intraspecies diversity that could be detected. Here, we have explored whether MALDI-TOF MS can detect the proteomic diversity that arises from *P. aeruginosa* surviving in different ecological niches. Using optimized post-processing parameters and peak picking thresholds, discriminant peak intensities between different sources were identified by diagonal discriminant analysis that were used to build predictive machine learning models (see Figure 7.3) (source: LD). Nonetheless, intraspecies spectra are highly conserved in *P. aeruginosa* and larger numbers of strains from further environments would be required to develop this application and verify these models.

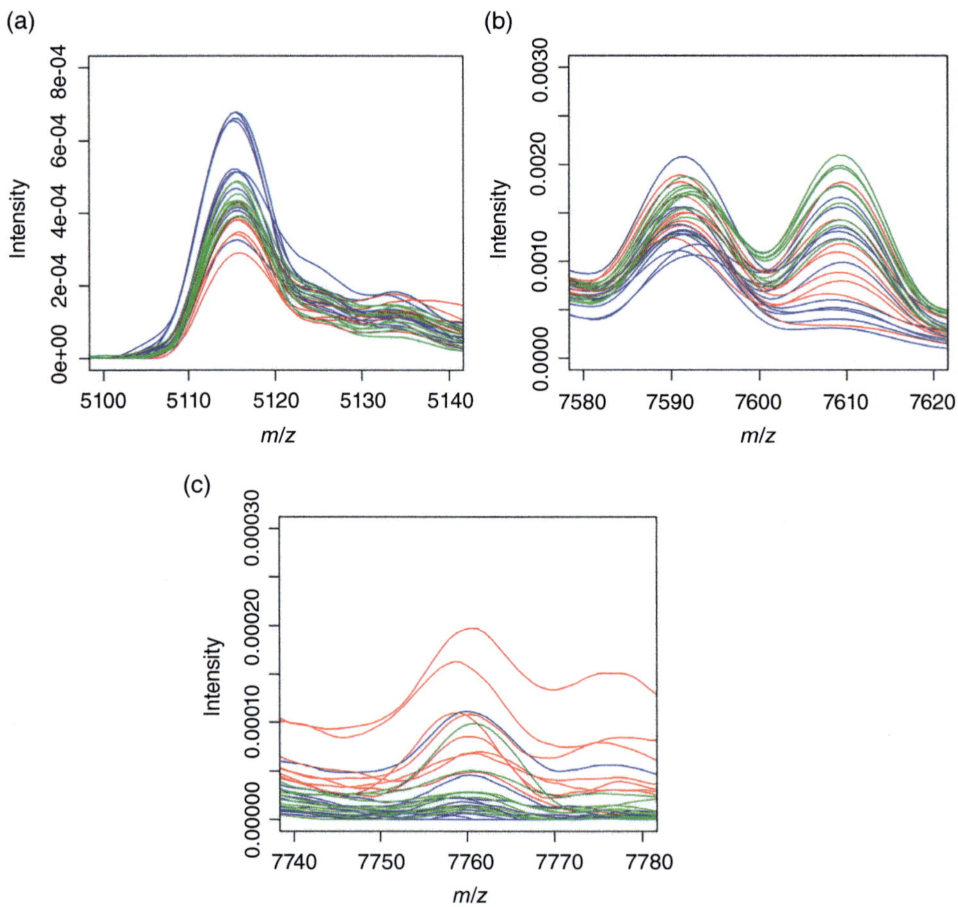

Figure 7.3 Top ranking peaks for source classification of *P. aeruginosa* by MALDI-TOF MS. Twenty-nine isolates of *P. aeruginosa* from CF, environmental, and veterinary isolates were analysed by MALDI-TOF MS and the top-ranking features for each source were identified by diagonal discriminant analysis. (a, b) Top ranking feature for classification of animal isolates at 5115 *m/z* (a), environmental isolates at 7609 *m/z* (b) and CF isolates at 7760 *m/z* (c). Spectra represent the average of three biological replicates with two technical replicates each, acquired on LEXI instrument (Ascend Diagnostics Ltd, Manchester, UK). *Source:* Ascend Diagnostics Ltd, Manchester.

MALDI-TOF MS can also detect differences between antibiotic-susceptible and -resistant strains. This includes the detection of methicillin-resistant *Staphylococcus aureus*, carbapenem resistance in *B. fragilis*, *K. pneumoniae*, and other Enterobacteriaceae [51, 52]. There are still ongoing efforts to validate biomarkers for resistance in *P. aeruginosa*. Nonetheless, MALDI-TOF MS offers high-throughput analysis of strains to examine diversity and identify clinically relevant biomarkers in combination with routine bacterial identification.

With proteomic diversity detected by MALDI-TOF MS, it is of no doubt that MS/MS systems would provide an enhanced view of the proteomic diversity in *P. aeruginosa*. As described previously, *P. aeruginosa* can show distinct phenotypes such as the SCV that have been comparatively profiled by MS/MS (see Section 7.1.3). Previous investigations have revealed differences in the virulence and pathogenic factor production in the proteome of isolates from different infection types, environments, and those that show differences in transmission capacity [53–55].

Overall, MS techniques are well suited to these investigations into diversity. While MS/MS offers greater coverage and resolution of the bacterial proteome over MALDI-TOF MS, it cannot achieve the same throughput and requires extensive data analysis. This therefore limits the number of diverse strains that can be profiled. Nonetheless, as MS/MS systems develop and become integrated into clinical diagnostics, determining these clinically relevant biomarkers will be extremely important.

7.1.5 Proteomic Diversity and MS Applications

Traditional biochemical assays such as the analytical profile index (API) demonstrated the importance of including the intraspecies diversity in microbial identification. As is the case with *P. aeruginosa* and many other species, isolates showed diverse substrate-use profiles that these identification tools were developed on. As microorganism identification technologies developed, it was equally important that this diversity was accounted for. Perhaps the best example of this is MALDI-TOF MS, the success of which is attributed to the all-important databases containing reference spectra from large numbers of geographically diverse isolates from a range of infection types and environments.

As MALDI-TOF MS applications in antibiotic resistance detection and strain typing develop, it is even more important that the intraspecies diversity is incorporated. Whereas microbial identification may use from 60 to 100 biomarkers, these unique applications are often based on as few as one [52, 56, 57]. Therefore, in validating these biomarkers, the whole diversity of a species needs to be included.

7.2 Transmission

7.2.1 The History of *P. aeruginosa* Transmission

The topic of transmission is perhaps more relevant than ever following the COVID-19 pandemic (see Chapter 5). Nonetheless, the devastating impact of human-to-human transmission has long been a reality for individuals with CF. The transmission of *P. aeruginosa* in

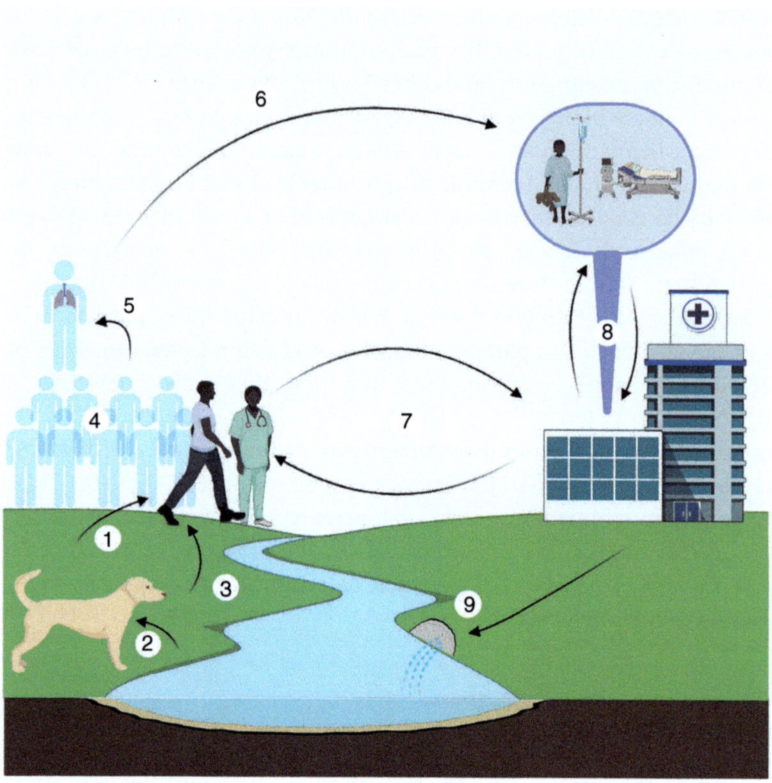

Figure 7.4 Transmission routes of *P. aeruginosa*. *P. aeruginosa* can be transmitted from the environment to humans (1), from the environment to animals including canines (2), and through animal carriage to humans (3). Human-to-human transmission can also occur through individuals carrying *P. aeruginosa* (4). This can be transmitted to individuals with CF (5) who may require hospitalization and transmit *P. aeruginosa* into the clinical environment (6). Healthy individuals and hospital staff may also transmit *P. aeruginosa* into the clinical environment (7), and vice versa. The clinical environment can harbour *P. aeruginosa* and be passed on to patients who can also contaminate the environment (8). Hospital waste can also re-enter the natural environment into water systems (9). Created with Biorender.com.

these communities, particularly within summer camps and shared CF clinics, led to social distancing restrictions and increased surveillance of highly transmissible isolates [58–60]. In these communities, there are many epidemic clones that have caused large outbreaks in hospitals [61–63].

Given that *P. aeruginosa* is isolated from a range of environments, determining other non-human reservoirs has been critical to protecting these vulnerable individuals. There may exist multiple transmission routes by which *P. aeruginosa* can enter the human ecosystem (see Figure 7.4). Many of the epidemic clones have been isolated from both natural and hospital environments [20, 63, 64]. In addition, the household environment and pets are potential reservoirs of concern, although only human-to-animal transmission is reported in the literature [65, 66].

7.2.2 Proteomics and *P. aeruginosa* Transmission

The COVID-19 pandemic revealed the importance of whole-genome sequencing (WGS) in identifying the transmission of variants around the world. WGS and other genotyping methods were also critical in identifying the key transmission events responsible for *P. aeruginosa* outbreaks [67–69]. The impact of both these outbreaks highlights the importance of high-throughput epidemiological tools to identify the source of transmission quickly and prevent further spread of disease.

Given its speed and throughput, many have been interested in the application of MALDI-TOF MS as an epidemiological tool for real-time outbreak investigations. MALDI-TOF MS successfully identified *Escherichia coli* strains responsible for outbreaks and detected distinct peak biomarkers in five high-risk clones of *P. aeruginosa* [70–72]. Alternatively, we have investigated whether MALDI-TOF MS can detect proteomic fingerprints from isolates of different ecological niches to determine transmission events in *P. aeruginosa,* although spectral profiles proved highly conserved (source: LD).

While these approaches are promising, validating both requires examining potential biomarkers against the full spectrum of *P. aeruginosa* diversity. As both clinical laboratories and other industries adopt MALDI-TOF MS for routine bacterial identification and the use of open-source databases increases, there is enormous potential to uncover this diversity to help develop MALDI-TOF MS as a powerful epidemiological tool.

7.2.3 The Impact of Proteomic Diversity on Transmission

By developing MS technologies to identify the source of isolates, we may begin to uncover the transmission pathways of *P. aeruginosa*. However, as much as diversity underpins the use of these tools, it also impacts transmission processes. Transmission describes the process whereby a microorganism leaves one environment to enter a new one, overcoming the barriers involved in transient movement and invading the new host. Even when growing on three different surface types, the proteome of *P. aeruginosa* has demonstrated extensive versatility [73]. As the human body offers a complex niche with multiple threats, the required proteomic response is likely to be extensive and thus offers an opportunity for strains to show variability in their capacity to adapt to the new environment.

Nonetheless, as much as the proteome adapts to a new environment, it can equally be shaped by its previous environment, introducing diversity in the transmission fitness of isolates. We are limited in our understanding of this given that we often cannot identify where an isolate has previously arrived from. Nonetheless, it may be possible to speculate on the potential fitness of isolates that have arrived from different environments based on their phenotypical and proteomic features.

For example, strains that show altered expression of virulence factors and carbon source specificity may have a differential capacity to invade the host and adapt to the unique nutrient profile of the CF lung. By contrast, CF isolates that convert to specialized metabolic pathways may not be able to use different carbon sources in other environments. Nonetheless, we must recognize that the *in vivo* fitness inside the host is difficult to measure, and with the flexibility of the proteome, adapted phenotypes may equally be reversed.

Additionally, analysing the proteome of highly transmissible CF strains can help researchers to understand the key proteins that may promote transmission fitness. This has revealed potential roles for proteins involved in the acquisition of iron, lipopolysaccharide biosynthesis, and specialized metabolic adaptation strategies [53].

7.3 Selective Pressures on the Proteome

7.3.1 Tandem MS Systems for Studying Selected Proteomes

Proteome selection may manifest as an absence or presence of a protein, a change in protein abundance, or a change in the protein form such as a post-translational modification. MS techniques have remained the gold standard in proteomics, with technological advances allowing deeper investigation into all these selective changes.

Earlier in this chapter, the benefits of MS techniques were discussed in examining the diversity that exists within the *P. aeruginosa* proteome (see Section 7.1.4). Here we highlight how these same technologies provide their own advantages in examining the proteome under selection. Tandem MS instruments such as Thermo's Orbitrap offer extensive coverage of the proteome, with reports of 930 to over 2000 proteins detected in *P. aeruginosa* growing in various conditions [41, 48, 73]. In addition, these systems detect proteins of extremely low abundance, detect peptide changes, and quantify changes in protein abundance through data-dependent and -independent acquisition modes [48].

7.3.2 Microenvironment Selection

The proteome of *P. aeruginosa* has been studied under a variety of conditions, from growing on different surfaces to space flight [73–76]. Here, we investigate the proteome of *P. aeruginosa* in environments that may be most relevant to transmission processes.

7.3.2.1 The Human Body and CF Lung

The human host possesses many barriers to bacterial invasion, including but not limited to, temperature, mucosal membranes, the immune system, and antimicrobial threats. Even more so, the CF lung represents a uniquely complex environment. As a result of the defective cystic fibrosis transmembrane conductance regulator (CFTR) gene, thick sticky mucus fills the lungs, in addition to pulmonary surfactant dysfunction and inflammation. CF-adapted isolates show a unique capacity to use the carbon breakdown products of surfactant and a conversion to a mucoid phenotype to resist host responses [11, 77]. Both hypoxic conditions and the presence of sputum alone have been shown to impact the *P. aeruginosa* proteome [76, 78].

Nonetheless, the extent of selection provided by the CF lung is still unknown. Many have suggested that the extensive phenotypic diversity that is present in CF isolates implies there is no single preferred adaptation pathway to the CF lung [79]. By contrast, it is suggested that the nutritional complexity of the CF lung drives *P. aeruginosa* diversification [80].

7.3.2.2 The Natural Environment

The natural microenvironment contains a myriad microbial communities and is influenced by a variety of chemical and physical factors. Although many of these features may be like those of the complex CF lung environment, the microbial communities, nutrient availability, and threats may be substantially more diverse in nature. Consequently, the environmental proteome may be less distinct, maintaining a flexibility in the face of diverse threats. Nonetheless, isolates from different sources reveal potential environment-adapted proteomes, including increased production of chitin-binding proteins and those involved in responding to osmotic pressure changes [54]. The influence of the environment on the proteome is also observed in the presence of unique hydrocarbon-degrading enzymes and altered proteomes from those isolated from contaminated environments [81, 82].

7.3.3 Antimicrobial Selection

Exposure to antibiotics can occur both in the human body and in the natural environment, with frequent reports of contamination through clinical effluents [83–85]. Therefore, understanding the proteomic response of *P. aeruginosa* to antimicrobials is important given its prevalence in many environments.

The process of adaptive resistance arises by sub-inhibitory concentrations of antibiotics that induce a resistant phenotype. As discussed previously, proteomics is ideally placed to understanding this dynamic response (see Section 7.1.3). Previously reported responses to sub-minimum inhibitory concentrations (sub-MIC) include increased production of heat shock proteins, proteases, and quorum sensing regulated proteins [48, 86]. In the current work, we were interested in how this response might differ in isolates from natural environments (source: LD). We therefore compared the proteomic response of an environmental river isolate, TW5A, to the reference strain PAO1 to sub-MIC tobramycin using MS/MS. Overall, 23% of the proteins produced in response to tobramycin were exclusively produced by only one of the strains. Furthermore, we observed differences between the strains in their proteins that showed the largest fold increases in abundance following tobramycin exposure (see Table 7.1). In strain PAO1, the largest fold-change was observed in protein HupB, which has a role in stabilizing DNA under stress conditions. This demonstrated interactions with molecular chaperones and other proteins involved in protein stability. Overall, this suggested an induction of the RpoS stress response in PAO1. By contrast, in strain TW5A, the largest fold increases in proteins were in SpeD and a probable iron trafficking protein. As we did not observe the same induction of proteins produced in response to DNA damage, it is proposed that these proteins of increased abundance may confer a protective role. Overall, 31% of the proteins that showed a notable increase in one strain showed a reduced fold-change in abundance in the other strain. While further analysis of environmental stains is required, this study emphasized the importance of including the full spectrum of *P. aeruginosa* proteomic diversity to understand adaptive resistance processes.

7.4 Conclusions on Studies of the Proteome

Pseudomonas aeruginosa has remained a fascinating species to study due to its inherent diversity and adaptive mechanisms which have been revealed at multiple levels of regulation. Yet while these studies detail such versatility, it is evident that we are only just

Table 7.1 Proteins of notable fold increase in abundance in response to tobramycin in strains PAO1 and TW5A. Ratio values > 1.5 were considered a notable increase in abundance.

		Abundance ratio (tobramycin/control)		
Gene name	**Protein identity**	**PAO1**	**TW5A**	**Function**
hupB	DNA-binding protein HU-beta	2.176	0.728	DNA replication, recombination, modification, and repair
trxA	Thioredoxin	2.037	0.813	Nucleotide biosynthesis and metabolism, energy metabolism Translation, post-translational modification, degradation
algR	Positive alginate biosynthesis regulatory protein AlgR	1.77	0.807	Two-component regulatory system secreted factors (toxins, enzymes, alginate), transcriptional regulators
fur	Ferric uptake protein	1.72	1.399	Transcriptional regulators
rpsB	30S ribosomal protein S2	1.679	1.011	Translation, post-translational modification, degradation
groS/groES	10 kDa chaperonin	1.676	0.865	Chaperones and heat shock proteins
dnaK	Chaperone protein DnaK	1.658	1.283	Adaptation, protection, chaperones, and heat shock proteins, DNA replication, recombination, modification, and repair
etfB	Electron transfer flavoprotein subunit beta	1.646	0.772	Energy metabolism
ppnP	Pyrimidine/purine nucleoside phosphorylase	1.504	0.574	Not assigned
speD	S-adenosylmethionine decarboxylase proenzyme SpeD	2.067	0.553	Central intermediary metabolism
rpmE	50S ribosomal protein L31	1.956	0.735	Translation, post-translational modification, degradation
PSPA7_5885	Probable Fe(2+)-trafficking protein	1.507	1.204	Iron ion binding

All proteins were detected in at least four of the six biological replicates. Protein functions were obtained from the Pseudocap Database (https://www.pseudomonas.com/pseudocap).

beginning to understand its significance in infections. Although previous studies have been largely confined to clinical isolates, increased detection of isolates from environmental sites has prompted investigations into the diversity that exists in non-human niches. Nonetheless, populations of *P. aeruginosa*, including those from veterinary samples, remain notably less well characterized. This is particularly concerning given that household pets have long been considered risk factors for individuals with CF due to the potential of

P. eruginosa transmission. We predict that even further diversity will be revealed in environmental and animal isolates as their roles as transmission reservoirs prompt further study.

To understand the risk of these potential transmission reservoirs, proteomic studies have proved invaluable. These have revealed not only differences in the proteome, but also how this diversity manifests in their response to the threats and challenges of the next environment. Such studies have already demonstrated variation in strain virulence and stress response, with the latter being important in our understanding of how antibiotic resistance develops.

This chapter touches on just a few of the vast number of selective pressures that may shape the proteome and the MS techniques that bring their own advantages and disadvantages to studying the impact of these. MALDI-TOF MS has offered a unique platform to capture diversity in *P. aeruginosa* through its increasing application in routine microbial identification, in addition to being able to detect several clinically relevant biomarkers. MS/MS systems provide greater resolution of the proteome and reliable quantification of protein abundance changes, which are characteristic of differentially selected proteomes. Nonetheless, to capture and determine the full spectrum of proteomic diversity, it will be necessary to analyse large numbers of isolates from varied sources. The increased use of open-access MALDI databases and implementation of MS/MS in diagnostics will reveal regions of the proteome that display the greatest versatility. In turn, this can be used to determine the transmission threat of different ecological niches and in the design of new therapeutic agents that target all *P. aeruginosa* strains.

7.5 Genomic Studies on *Pseudomonas aeruginosa* Strains Revealing the Presence of Two Distinct Clades

The dynamic and diverse genetic composition of *P. aeruginosa* as discussed in Section 7.1 enables this bacterium to colonize diverse environments, including as an opportunistic human pathogen commonly associated with CF and traumatic burn conditions [87]. The strains of *P. aeruginosa* with variable genotypic and phenotypic properties are increasing rapidly worldwide [88, 89] and whole-genome sequences are now available for >6000 *P. aeruginosa* strains in public databases [90, 91]. A number of recent studies based on genomic and biochemical properties of *P. aeruginosa* strains indicate that the available strains form multiple clades in phylogenetic trees [92–99]. Among these clades, one group of strains, which include the type strain for *P. aeruginosa* DSM 50071, has been referred to as the 'classical strains' by some investigators [97, 99–101]. The strains from this group are more pathogenic and commonly associated with nosocomial infections (ventilator-associated pneumonia), CF, and various sepsis syndromes [102]. In comparison to the 'classical strains', strains from another group referred to as the 'clonal outliers' [97, 99, 103, 104] are less pathogenic [100]. This latter clade is identified in a number of studies by the strain PA7, whereas the 'classical' clade is represented by strain PAO1 and PA14 [97, 104, 105]. Although the virulence pattern and pathogenicity of *P. aeruginosa* are multifactorial, the strains from two clades differ in terms of how they exert their cytolytic activity on human cells [98, 106].

The degree of pathogenicity of the strains from 'classical' clade (or clade 1 as it will be referred to in this study) depends largely on the presence of type III secretion system (T3SS), which enables the injection of four main *P. aeruginosa* effectors (ExoS, ExoT, ExoY, and ExoU) [107–110] and several virulence factors (viz. proteases, exotoxin A, pili, flagella, quorum sensing proteins) [111] directly into the host cell cytoplasm. By contrast, the strains from less pathogenic 'outlier clade' (will be referred to as clade 2 in this study) not only lacks the entire T3SS-encoding locus but also the genes encoding type III secreted exoenzymes [100, 106]. In addition, [92] have identified a novel virulence mechanism using exolysin toxins A and B (viz., ExlA and ExlB) which is present in strains lacking T3SS (i.e. clade 2). The pathogenic variation among the *P. aeruginosa* strains is also indicated by several comparative genomic studies [95–97, 99]. These studies indicate that most of the *P. aeruginosa* strains share > 90% of their gene content irrespective of the source or the type of infection. Despite the striking genomic similarity, the strain-to-strain virulence variation is vast, which is due to the presence of virulence factors along with genes involved in metabolic pathways that are important for pathogen adaptation to host-associated lifestyles [112].

7.5.1 Phylogenomic Analysis Reveals the Presence of Two Distinct Clades Within *P. aeruginosa*

The phylogenetic studies provide an important tool for understanding of evolutionary relationships among biological organisms and for differentiating groups that differ from each other [113]. Phylogenetic studies on *P. aeruginosa* strains have been reported by several groups based on single or concatenated sequences of multiple genes and these studies reveal the grouping of these strains in two or more distinct clades referred to by various names including the 'classical' and 'outlier' clades [97, 114, 115]. It has also been noted by several groups that the members of the identified clades differ in terms of their virulence characteristics [96, 99, 116]. In a detailed study in this regard by Sood et al. [97], a phylogenetic tree was constructed for 78 *P. aeruginosa* strains based on sequences of 400 conserved genes. This tree showed the grouping of *P. aeruginosa* strains into two distinct clusters designated as the 'classical' and 'outlier' lineages. The members of these two clades could also be differentiated from each other based on average nucleic acid identity (ANI) analysis. Members of the 'outlier' clade exhibited only 94% ANI to the 'classical' clade, whereas the intragroup ANI values for these two clades were in the range 98–99%. The existence of two distinct clades within *P. aeruginosa* is also indicated in the Genome Taxonomy Database (GDTB) [117], which is based on phylogenetic analyses of 120 ubiquitous single-copy proteins and provides an important resource for phylogenetic-taxonomic inferences. The GTDB places the *P. aeruginosa* strains into two clades, designated as the *P. aeruginosa* clade and *P. aeruginosa*_A clade, corresponding to clade 1 and clade 2 in our designation Figure 7.5.

We have also conducted comprehensive phylogenetic analysis of *P. aeruginosa* strains by constructing a maximum-likelihood phylogenetic tree based on core protein sequences of representative *P. aeruginosa* strains, including the type strain DSM 50071 and with eight other representative *Pseudomonas* species. Although there are > 6000 *P. aeruginosa* strains available in the NCBI database [90], most of them are highly similar or identical

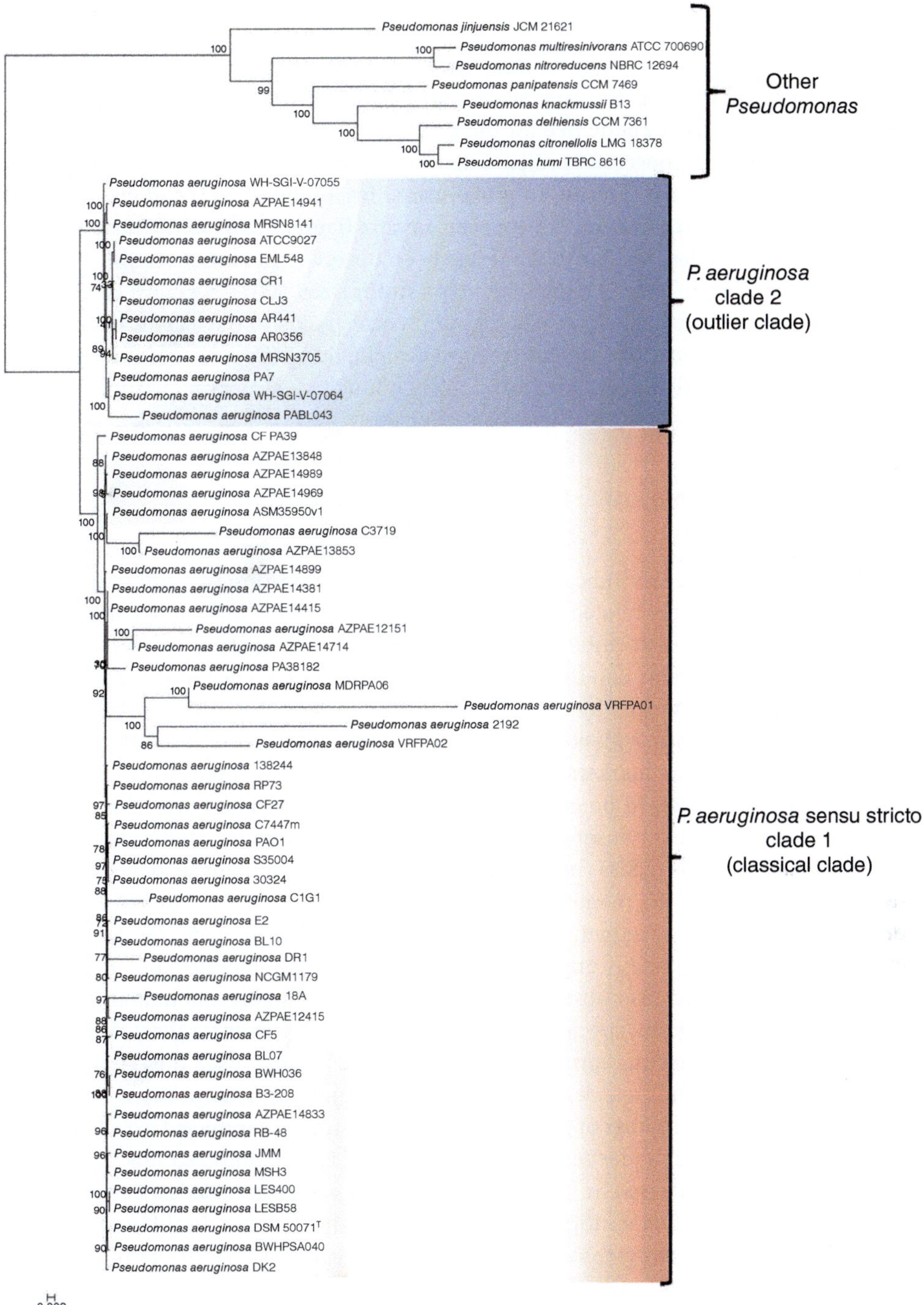

Figure 7.5 A bootstrapped maximum likelihood tree for 57 genome-sequenced *P. aeruginosa* strains representative of different groups/clades of *P. aeruginosa* strains, based on concatenated sequences for 118 conserved proteins. The strains corresponding to the two clades are identified by XX and YY at the end of their names. The two main clusters of *P. aeruginosa* strains that are observed in the tree as *P. aeruginosa* sensu stricto clade 1 (containing the type strain DSM50071) and *P. aeruginosa* clade 2. These two clades have also been referred to in earlier work as 'classical' clade and 'outlier' clade, respectively, in several studies as described in the text.

to each other. Hence, in our analysis we have included only sequences for 57 non-identical strains of *P. aeruginosa* from the two clades described in the GTDB website [117]. A maximum-likelihood phylogenetic tree for these strains was constructed based on concatenated sequences of 118 conserved proteins that comprise the phyloeco set for the class *Gammaproteobacteria* [118]. The resulting tree, based on a total of 39 224 aligned amino acid positions, is shown in Figure 7.5. As seen in this figure, the *P. aeruginosa* strains formed two distinct clades, similar to what has been observed in earlier studies [97, 99, 104]. One of these clades, which contains the type strain DSM 50071, we have labelled as *P. aeruginosa* sensu stricto clade (or clade 1) and it corresponds to the 'classical' clade in earlier studies. The second clade, which we have designated as the *P. aeruginosa* clade 2, corresponds to the 'outlier' clade. Thus, our phylogenomic tree also strongly supports the grouping of *P. aeruginosa* strains into two distinct clades.

7.5.2 Identification of Molecular Markers Distinguishing the Two *P. aeruginosa* Clades

Although phylogenetic studies indicate that *P. aeruginosa* strains form two different clades, these clades are very closely related to each other. Hence, the distinction between these two groups is not readily discernible by means of phylogenetic analysis on 16S rRNA or other single genes or proteins commonly used for such studies. Hence, it is important to identify other molecular means by which members of these two clades can be reliably distinguished from each other. Genome sequences provide an important resource for the identification of molecular markers that can prove useful for distinguishing these two groups from each other. Based on genome sequences, one important class of molecular markers that have proven very useful for evolutionary and taxonomic studies consists of conserved signature indels (CSIs) (inserts or deletions) in gene/protein sequences, which are specifically shared by the members of a given clade [119–121]. Due to the exclusive presence of these molecular signatures in the members of a given clade, they represent synapomorphic characteristics, originating from a common ancestor of the clade. Furthermore, as the presence or absence of these CSIs can be readily and unambiguously ascertained in different strains/species, they provide important means for the demarcation of prokaryotic taxa in molecular terms [122, 123]. We have recently used these markers to reclassify >20 *Pseudomonas* species into two novel genera (viz. *Halopseudomonas* and *Atopomonas*) as well as a number of other genera [124]. Thus, we conducted detailed comparative studies on protein sequences from the genomes of *P. aeruginosa* strains to identify CSIs, which are uniquely shared by the members of two *P. aeruginosa* clades. Results from these analyses have identified multiple CSIs that are exclusively shared by different strains belonging to either the *P. aeruginosa* sensu stricto clade (i.e. clade 1) or the 'outlier' clade, i.e. *P. aeruginosa* clade 2.

In Figure 7.6a, we show partial sequence information for one of the identified CSIs, which is specifically shared by different strains from *P. aeruginosa* clade 1 (i.e. *P. aeruginosa* sensu stricto clade). In the example shown, a one-amino acid (1 aa) insertion (highlighted in pink) in the protein LysR family transcriptional regulator is commonly shared by all strains from clade 1 (marked in the sequence alignment by XX), but it is not found in any of the strains from clade 2 (identified by YY at the end of the

(a)

Group	Strain	Accession	251		292
P. aeruginosa senu stricto clade 1 (classical clade)	*P. aeruginosa* X24509XX	WP_023128735	DIAAGRLIPVLEAYNPGDLEEVHAVY	V	GQGGYLPLRVRAFLD
	P. aeruginosa 3579XX	WP_048520948	--------------------------	-	---------------
	P. aeruginosa DSM50071XX^{T}	EOQ78484	--------------------------	-	---------------
	P. aeruginosa PAO1	WP_003104212	--------------------------	-	---------------
	P. aeruginosa PA14XX	WP_003138496	--------------------------	-	---------------
	P. aeruginosa PA38182XX	CDI91531	--------------------------	-	---------------
	P. aeruginosa MH38XX	WP_003104212	--------------------------	-	---------------
	P. aeruginosa M9A.1XX	WP_003122263	--------------------------	-	---------------
	P. aeruginosa F22031XX	WP_003104212	--------------------------	-	---------------
	P. aeruginosa DR1XX	WP_003104212	--------------------------	-	---------------
	P. aeruginosa BL10XX	WP_003104212	--------------------------	-	---------------
	P. aeruginosa PA1XX	WP_003111185	--------------------------	-	---------------
	P. aeruginosa N002XX	WP_003111185	--------------------------	-	---------------
	P. aeruginosa MSH3XX	WP_003111185	--------------------------	-	---------------
	P. aeruginosa M18XX	WP_003111185	--------------------------	-	---------------
	P. aeruginosa Env16XX	WP_003111185	--------------------------	-	---------------
	P. aeruginosa E2XX	WP_003111185	--------------------------	-	---------------
	P. aeruginosa DK2XX	WP_003111185	--------------------------	-	---------------
	P. aeruginosa CF614XX	WP_003111185	--------------------------	-	---------------
	P. aeruginosa CF18XX	WP_003111185	--------------------------	-	---------------
	P. aeruginosa BL07XX	WP_003111185	--------------------------	-	---------------
	P. aeruginosa YL84XX	WP_010793228	--------------------------	-	---------------
	P. aeruginosa WS136XX	WP_010793228	--------------------------	-	---------------
	P. aeruginosa SG17MXX	WP_010793228	--------------------------	-	---------------
	P. aeruginosa UDLXX	WP_003115019	--------------------------	-	---------------
P. aeruginosa clade 2 (outlier clade)	*P. aeruginosa* PA7YY	WP_004367886	--LE---V-LFAEQMSEYRQPI----		YRNSQ-A--IQC---
	P. aeruginosa CR1	WP_004367886	--LE---V-LFAEQMSEYRQPI----		YRNSQ-A--IQC---
	P. aeruginosa 515477YY	WP_004367886	--LE---V-LFAEQMSEYRQPI----		YRNSQ-A--IQC---
	P. aeruginosa AR0356YY	WP_058145398	--LE---V-LFAEQMSEYRQPI----		YRNSQ-A--IQC---
	P. aeruginosa AR441YY	WP_058145398	--LE---V-LFAEQMSEYRQPI----		YRNSQ-A--IQC---
	P. aeruginosa ATCC33359YY	WP_053814092	--LE---V-LFAEQMSEYRQPI----		YRNSQ-A--IQC---
	P. aeruginosa AUS217YY	WP_004367886	--LE---V-LFAEQMSEYRQPI----		YRNSQ-A--IQC---
	P. aeruginosa AUS483YY	WP_053814092	--LE---V-LFAEQMSEYRQPI----		YRNSQ-A--IQC---
	P. aeruginosa CLJ1YY	WP_004367886	--LE---V-LFAEQMSEYRQPI----		YRNSQ-A--IQC---
	P. aeruginosa EML528YY	WP_004367886	--LE---V-LFAEQMSEYRQPI----		YRNSQ-A--IQC---
	P. aeruginosa T36994	WP_029886084	-L-S---L-L------D-R-PI---F		VG-PSM-A---V-V-
	P. aeruginosa 147	WP_054881876	-L-S---L-L------D-R-PI---F		VG-PSM-A---M-V-
Other *Pseudomonas* (0/>200)	*P. putida* NCTC 912	WP_115294191	-L-S---L-L------D-R-PI---F		VG-PSM-A---V-V-
	P. monteilii SB3078	WP_024086760	-L-N-Q-V-L--------R-PI---F		VG-PAM-A---V-V-
	P. oryzihabitans ATCC 43272	WP_082696389	-LRS-A-V-L---W------PI---F		VG-TTM-A---V-V-
	P. stutzeri ATCC 17588	WP_181102918	-LQR-D-V-L--V-----R-PI---F		VG-SAM-A---V-V-
	P. mendocina ATCC 25411	WP_143506699	-LQR-D-V-L-QAF--D-R-PI---F		IG-PAM-A---V-V-
	P. fluorescens ATCC 13525	WP_038857385	-LQR-D---L--AW----R-PI---F		VG-SAM-A---L-V-

(b)

Group	Strain	Accession	21		61
P. aeruginosa clade 2 (outlier clade)	*P. aeruginosa* NCTC12924YY	VTL98846	FAGALHLAWHRRAERF	SRKDG	RGGRSTGLKALDLADPQAP
	P. aeruginosa PA7YY	WP_012077987	----------------	-----	-------------------
	P. aeruginosa CR1YY	WP_043102528	----------------	-----	-------------------
	P. aeruginosa MRSN3705YY	WP_126632751	----------------	-----	-------------------
	P. aeruginosa LMG5031YY	WP_058136307	----------------	-----	-------------------
	P. aeruginosa EML545YY	WP_053814820	----------------	-----	-------------------
	P. aeruginosa ATCC33359YY	WP_053814820	----------------	-----	-------------------
	P. aeruginosa AR441YY	WP_053814820	----------------	-----	-------------------
	P. aeruginosa AR0356YY	WP_053814820	----------------	-----	-------------------
	P. aeruginosa AUS483YY	WP_043102528	----------------	-----	-------------------
	P. aeruginosa AUS217YY	WP_043102528	----------------	-----	-------------------
	P. aeruginosa 515477YY	WP_058163085	----------------	-----	-------------------
	P. aeruginosa MRSN8141YY	WP_003153474	----------------	-----	-------------------
	P. aeruginosa EML528YY	WP_003153474	----------------	-----	-------------------
P. aeruginosa senu stricto clade 1 (classical clade)	*P. aeruginosa* PAO1XX	NP_254086	----------------		G------------------
	P. aeruginosa DSM50071XX^{T}	WP_003107277	----------------		G------------------
	P. aeruginosa CF27XX	WP_023131408	----------------		G------------------
	P. aeruginosa MSH3XX	WP_016264026	----------------		G------------------
	P. aeruginosa Z61XX	WP_003114447	----------------		G------------------
	P. aeruginosa WC55XX	WP_004351806	----------------		G------------------
	P. aeruginosa S35004XX	WP_003114447	----------------		G------------------
	P. aeruginosa RB-48XX	WP_003114447	----------------		G------------------
	P. aeruginosa PSE305XX	WP_003114447	----------------		G------------------
	P. aeruginosa PGPR2XX	EOQ76595.1	----------------		G------------------
	P. aeruginosa PAO581XX	WP_003114447	----------------		G------------------
	P. aeruginosa MH38XX	WP_015503865	----------------		G------------------
	P. aeruginosa M10XX	WP_003114447	----------------		G------------------
	P. aeruginosa FRD1XX	WP_015503865	----------------		G------------------
	P. aeruginosa CIG1XX	WP_003114447	----------------		G------------------
	P. aeruginosa BWH055XX	WP_003114447	----------------		G------------------
	P. aeruginosa BL24XX	WP_003114447	----------------		G------------------
	P. aeruginosa BL23XX	WP_003114447	----------------		G------------------
	P. aeruginosa BL10XX	WP_023103603	----------------		G------------------
Other *Pseudomonas* (0/>200)	*P. monteilii* SB3078	WP_013970509	----------------		G-------P---E--N--
	P. putida T2-2ZZ	WP_009682905	----------------		G-------P---E--N--
	P. soli LMG 27941	WP_094011564	----------------		G-------P---E--N--
	P. knackmussii B13	WP_043256590	----------------		G-----------T--T--
	P. panipatensis CCM 7	WP_090268356	----------------		G-----------N--M--
	P. jinjuensis JCM 216	WP_084312076	------------S---		G-------P------M--

Figure 7.6 (a) Partial sequence alignment of the LysR family transcriptional regulator protein showing a 1-aa insertion within a conserved region (highlighted) that is exclusively shared by different strains which are a part of the *P. aeruginosa* sensu stricto clade (clade 1) but not found in any other bacterial species or strains from *P. aeruginosa* clade 2. (b) Partial sequence alignment of the DgcB showing a 5-aa insert that is specific for *P. aeruginosa* clade 2 strains but not found in any other bacterial species from strains from *P. aeruginosa* sensu stricto (clade 1). The dashes (-) in the alignment indicate identity with the amino acids on the top line. Accession numbers for different sequences are indicated in the second column and the number on the top indicates the position of the indicated sequence fragment within the protein.

strain name). The CSI shown is present in a conserved region of the protein with dashes (-) in the sequence alignment showing identity with the amino acid present on the top line. As this CSI is also not present in any other *Pseudomonas* species, it represents an insert which is specific for the clade 1 strains, which probably occurred in a common ancestor of this lineage. It should be noted that sequence information in Figure 7.6a,b is shown for only a limited number of strains from clades 1 and 2. However, the other strains from these two clades also show similar sequence characteristics as shown here for the representative strains from these clades. Besides the CSI shown in Figure 7.6a, we have identified eight other CSIs in diverse proteins, which are also specific for the clade 1 strains (unpublished results).

Similar to the CSIs that are specific for the clade 1 strains, our analyses have also identified multiple CSIs that are specific for the *P. aeruginosa* clade 2. In Figure 7.6b, we present sequence information for one of the CSIs, which is exclusively found in members of the *P. aeruginosa* clade 2 ('outlier' clade). This CSI consists of a 5-aa insertion within a conserved region of the dimethylglycine demethylation protein (DgcB) and it is specifically shared by all the strains from *P. aeruginosa* clade 2. This CSI is not present in any strains from *P. aeruginosa* sensu stricto clade 1 and also by other *Pseudomonas* species, indicating that it is a highly specific characteristic of the clade 2 strains. In addition to this CSI, we have identified 14 other CSIs in proteins involved in diverse functions, which are also exclusively present in the *P. aeruginosa* clade 2 strains. Because of the discrete and highly specific molecular nature of the identified CSIs, they provide novel and reliable means for readily distinguishing the two observed clades of *P. aeruginosa* strains from each and for the demarcation of these clades in molecular terms.

In addition to the CSIs that are exclusively present in the two groups of *P. aeruginosa* strains, we have also carried out comparative genomic studies to identify conserved signature proteins (CSPs) [121, 125, 126], whose homologues are uniquely found in the members of these two clades. These analyses have identified numerous CSPs, for which homologues showing significant sequence similarity are exclusively found in either the clade 1 or clade 2 strains (results not shown). The identified CSPs represent unique sets of genes that are only found in two groups of *P. aeruginosa* strains, and they again serve to indicate the distinctness of these clades and provide novel and reliable means for distinguishing them from each other.

In addition to the novel genomic characteristics identified by our work, the members of the two *P. aeruginosa* clades also differ from each other in a number of other phenotypic and genotypic characteristics, including presence or absence of T3SS-encoding loci responsible for their avirulent or less pathogenic characteristic [104], and significant differences in the numbers of pan-genes in the two groups [97]. In addition, Sood et al. [97], and other investigators [95, 96, 99, 101, 103] have noted many other differences that appear specific to the strains from these two clades, including presence or absence of gene clusters required for survival in stress conditions, phenazines or siderophore productions, lipase evolution, as well as the presence of unique biosynthetic pathways (viz., mono-RLs and PYO) and the major protein–protein interaction hubs that are found within these two groups. All of these studies make a strong case that the strains from clade 2 should be recognized as a novel species that is distinct from the clade 1 strains and which corresponds to the species *P. aeruginosa* sensu stricto.

7.6 Final Conclusions

The studies reflected on in this chapter emphasize our continued interest in measuring the impact of the extensive diversity observed in *P. aeruginosa*. From early typing studies examining the distribution of type III secretion systems and type IV pili allele, classification of *P. aeruginosa* would also reveal critical information on strain pathogenicity. Phenotypical analyses of diverse strains also set out to understand clinically relevant variation. Nonetheless, inclusion of diverse strains in population studies continued to demonstrate a non-clonal structure by which isolates from all sources were indistinguishable by their clade and their phenotypes.

Our recent phylogenomic analysis confirms the presence of two clades and identifies molecular markers of CSIs and CSPs. Their identification presents an exciting opportunity to develop proteomic tools for *P. aeruginosa* classification. As discussed in this chapter, MS/MS offers a powerful proteomic tool to investigate the production and/or modification of proteins that manifest from the variable sequence information between the two clades.

In this chapter we have described the proteomic response of *P. aeruginosa* strains to various conditions. This includes the production of stress response proteins and phenazines in the response to sub-MIC tobramycin that may also promote the development of adaptive resistance. Phylogenomic analyses have also described differences in the presence of gene clusters for survival in stress conditions and phenazine production between the two clades. As we continue to characterize the proteomic response of *P. aeruginosa* to different conditions, it will be interesting to examine the contribution of these unique clade features to the proteome and, consequently, how pathogenicity and resistance may develop differentially between strains of these clades.

To conclude, this chapter addresses the benefit of studying the proteome of *P. aeruginosa* at two levels of diversity: (i) examining proteomes that are differentially shaped from the selective pressures of a strain's ecological niche; and (ii) examining how phylogenetic events have shaped the proteomes of strains from different clades. Collectively, these analyses, with the techniques available, currently offer the most comprehensive insight into the impact of *P. aeruginosa* diversity on transmission and human infections.

References

1 Williamson, C.K. (1956). Morphological and physiological considerations of colonial variants of *Pseudomonas aeruginosa*. *J. Bacteriol.* 71: 617–622.

2 Phillips, I. (1969). Identification of *Pseudomonas aeruginosa* in the clinical laboratory. *J. Med. Microbiol.* 2: 9–16.

3 Wahba, A.H. and Darrell, J.H. (1965). The identification of atypical strains of *Pseudomonas aeruginosa*. *Microbiology* 38: 329–342.

4 Kaier, K., Heister, T., Götting, T. et al. (2019). Measuring the in-hospital costs of *Pseudomonas aeruginosa* pneumonia: methodology and results from a german teaching hospital. *BMC Infect Dis* 19: 1028.

5 Koenig, S.M. and Truwit, J.D. (2006). Ventilator-associated pneumonia: diagnosis, treatment, and prevention. *Clin. Microbiol. Rev.* 19: 637–657.

6 Willcox, M.D. (2012). Management and treatment of contact lens-related *Pseudomonas keratitis*. *Clin. Ophthalmol.*, (Auckland, N.Z.) 6: 919–924.

7 Registry, C. F. 2020. *Cystic Fibrosis Strength in Numbers*. UK Cystic Fibrosis Registry Annual Data Report 2019. London: Cystic Fibrosis Trust.

8 De Vos, D., De Chial, M., Cochez, C. et al. (2001). Study of pyoverdine type and production by *Pseudomonas aeruginosa* isolated from cystic fibrosis patients: prevalence of type II pyoverdine isolates and accumulation of pyoverdine-negative mutations. *Arch. Microbiol.* 175: 384–388.

9 Govan, J.R. and Deretic, V. (1996). Microbial pathogenesis in cystic fibrosis: mucoid *Pseudomonas aeruginosa* and *Burkholderia cepacia*. *Microbiol. Mol. Biol. Rev.* 60: 539–574.

10 Mahenthiralingam, E., Campbell, M.E., and Speert, D.P. (1994). Nonmotility and phagocytic resistance of *Pseudomonas aeruginosa* isolates from chronically colonized patients with cystic fibrosis. *Infect. Immun.* 62: 596.

11 Dolan, S.K., Kohlstedt, M., Trigg, S. et al. (2020). Contextual flexibility in *Pseudomonas aeruginosa* central carbon metabolism during growth in single carbon sources. *mBio* 11: e02684–e02619.

12 La Rosa, R., Johansen, H.K., and Molin, S. (2018). Convergent metabolic specialization through distinct evolutionary paths in *Pseudomonas aeruginosa*. *mBio* 9 (2): e00269-18. https://doi.org/10.1128/mBio.00269-18.

13 Jorth, P., Staudinger, B.J., Wu, X. et al. (2015). Regional isolation drives bacterial diversification within cystic fibrosis lungs. *Cell Host Microbe* 18: 307–319.

14 Clark, S.T., Diaz Caballero, J., Cheang, M. et al. (2015). Phenotypic diversity within a *Pseudomonas aeruginosa* population infecting an adult with cystic fibrosis. *Sci. Rep.* 5: 10932–10932.

15 Wilder Cara, N., Allada, G., and Schuster, M. (2009). Instantaneous within-patient diversity of *Pseudomonas aeruginosa* quorum-sensing populations from cystic fibrosis lung infections. *Infect. Immun.* 77: 5631–5639.

16 Workentine, M.L., Harrison, J.J., Weljie, A.M. et al. (2010). Phenotypic and metabolic profiling of colony morphology variants evolved from *Pseudomonas fluorescens* biofilms. *Environ. Microbiol.* 12: 1565–1577.

17 Khan, N.H., Ishii, Y., Kimata-Kino, N. et al. (2007). Isolation of *Pseudomonas aeruginosa* from open ocean and comparison with freshwater, clinical, and animal isolates. *Microb. Ecol.* 53: 173–186.

18 Kimata, N., Nishino, T., Suzuki, S., and Kogure, K. (2004). *Pseudomonas aeruginosa* isolated from marine environments in Tokyo Bay. *Microb. Ecol.* 47: 41–47.

19 Nair, A.V., Joseph, N., Krishna, K. et al. (2015). A comparative study of coastal and clinical isolates of *Pseudomonas aeruginosa*. *Braz. J. Microbiol.* 46: 725–734.

20 Pirnay, J.P., Matthijs, S., Colak, H. et al. (2005). Global *Pseudomonas aeruginosa* biodiversity as reflected in a Belgian river. *Environ. Microbiol.* 7: 969–980.

21 Selezska, K., Kazmierczak, M., Müsken, M. et al. (2012). *Pseudomonas aeruginosa* population structure revisited under environmental focus: impact of water quality and phage pressure. *Environ. Microbiol.* 14: 1952–1967.

22 Gunasekera, T.S., Striebich, R.C., Mueller, S.S. et al. (2013). Transcriptional profiling suggests that multiple metabolic adaptations are required for effective proliferation of *Pseudomonas aeruginosa* in jet fuel. *Environ. Sci. Technol.* 47: 13449–13458.

23 Muriel-Millán, L.F., Rodríguez-Mejía, J.L., Godoy-Lozano, E.E. et al. (2019). Functional and genomic characterization of a *Pseudomonas aeruginosa* strain isolated from the southwestern Gulf of Mexico reveals an enhanced adaptation for long-chain alkane degradation. *Front. Mar. Sci.* 6: https://doi.org/10.3389/fmars.2019.00572.

24 Rado, J., Kaszab, E., Petrovics, T. et al. (2017). Characterization of environmental *Pseudomonas aeruginosa* using multilocus sequence typing scheme. *J. Med. Microbiol.* 66: 1457–1466.

25 Head, N.E. and Yu, H. (2004). Cross-sectional analysis of clinical and environmental isolates of *Pseudomonas aeruginosa*: biofilm formation, virulence, and genome diversity. *Infect. Immun.* 72: 133.

26 Kaszab, E., Radó, J., Kriszt, B. et al. (2021). Groundwater, soil and compost, as possible sources of virulent and antibiotic-resistant *Pseudomonas aeruginosa*. *Int. J. Environ. Health Res.* 31: 848–860.

27 Bai, X., Liu, S., Zhao, J. et al. (2019). Epidemiology and molecular characterization of the antimicrobial resistance of *Pseudomonas aeruginosa* in Chinese mink infected by hemorrhagic pneumonia. *Can. J. Vet. Res.* 83: 122–132.

28 Kidd, T.J., Gibson, J.S., Moss, S. et al. (2011). Clonal complex *Pseudomonas aeruginosa* in horses. *Vet. Microbiol.* 149: 508–512.

29 Park, H.R., Hong, M.K., Hwang, S.Y. et al. (2014). Characterisation of *Pseudomonas aeruginosa* related to bovine mastitis. *Acta Vet. Hung.* 62: 1–12.

30 Penna, B., Thomé, S., Martins, R. et al. (2011). in vitro antimicrobial resistance of *Pseudomonas aeruginosa* isolated from canine otitis externa in Rio de Janeiro, Brazil. *Braz. J. Microbiol.* 42: 1434–1436.

31 Pye, C. (2018). *Pseudomonas otitis* externa in dogs. *Can. Vet. J.* 59: 1231–1234.

32 Petermann, S.R., Doetkott, C., and Rust, L. (2001). Elastase deficiency phenotype of *Pseudomonas aeruginosa* canine otitis externa isolates. *Clin. Diagn. Lab. Immunol.* 8: 632.

33 Ruiz-Roldán, L., Rojo-Bezares, B., De Toro, M. et al. (2020). Antimicrobial resistance and virulence of *Pseudomonas* spp. among healthy animals: concern about exolysin ExlA detection. *Sci. Rep.* 10: 11667–11667.

34 Brzozowski, M., Krukowska, Ż., Galant, K. et al. (2020). Genotypic characterisation and antimicrobial resistance of *Pseudomonas aeruginosa* strains isolated from patients of different hospitals and medical centres in poland. *BMC Infect. Dis.* 20: 693.

35 Peña, C., Cabot, G., Gómez-Zorrilla, S. et al. For the Spanish Network for Research in Infectious, D.(2015). Influence of virulence genotype and resistance profile in the mortality of *Pseudomonas aeruginosa* bloodstream infections. *Clin. Infect. Dis.* 60: 539–548.

36 Saleem, S. and Bokhari, H. (2020). Resistance profile of genetically distinct clinical *Pseudomonas aeruginosa* isolates from public hospitals in central Pakistan. *J. Infect. Public Health* 13: 598–605.

37 Ramsay, K.A., Wardell, S.J.T., Patrick, W.M. et al. (2019). Genomic and phenotypic comparison of environmental and patient-derived isolates of *Pseudomonas aeruginosa* suggest that antimicrobial resistance is rare within the environment. *J. Med. Microbiol.* 68 (11): 1591–1595.

38 Suzuki, Y., Kajii, S., Nishiyama, M., and Iguchi, A. (2013). Susceptibility of *Pseudomonas aeruginosa* isolates collected from river water in Japan to antipseudomonal agents. *Sci. Total Environ.* 450-451: 148–154.

39 Ahmed, S.A.K.S., Rudden, M., Elias, S.M. et al. (2021). *Pseudomonas aeruginosa* pa80 is a cystic fibrosis isolate deficient in rhlri quorum sensing. *Sci. Rep.* 11: 5729.

40 Boucher, J.C., Yu, H., Mudd, M.H., and Deretic, V. (1997). Mucoid *Pseudomonas aeruginosa* in cystic fibrosis: characterization of muc mutations in clinical isolates and analysis of clearance in a mouse model of respiratory infection. *Infect. Immun.* 65: 3838–3846.

41 Erdmann, J., Thöming, J.G., Pohl, S. et al. (2019). The core proteome of biofilm-grown clinical *Pseudomonas aeruginosa* isolates. *Cells* 8: 1129.

42 Rao, J., Damron, F., Basler, M. et al. (2011). Comparisons of two proteomic analyses of non-mucoid and mucoid *Pseudomonas aeruginosa* clinical isolates from a cystic fibrosis patient. *Front. Microbiol.* 2: https://doi.org/10.3389/fmicb.2011.00162.

43 Häußler, S., Ziegler, I., Löttel, A. et al. (2003). Highly adherent small-colony variants of *Pseudomonas aeruginosa* in cystic fibrosis lung infection. *J. Med. Microbiol.* 52: 295–301.

44 Wehmhöner, D., Häussler, S., Tümmler, B. et al. (2003). Inter- and intraclonal diversity of the *Pseudomonas aeruginosa* proteome manifests within the secretome. *J. Bacteriol.* 185: 5807–5814.

45 Bagge, N., Schuster, M., Hentzer, M. et al. (2004). *Pseudomonas aeruginosa* biofilms exposed to imipenem exhibit changes in global gene expression and beta-lactamase and alginate production. *Antimicrob. Agents Chemother.* 48: 1175–1187.

46 Mojsoska, B., Ghoul, M., Perron, G.G. et al. (2021). Changes in toxin production of environmental *Pseudomonas aeruginosa* isolates exposed to sub-inhibitory concentrations of three common antibiotics. *PLoS One* 16: e0248014.

47 Shi, N., Gao, Y., Yin, D. et al. (2019). The effect of the sub-minimal inhibitory concentration and the concentrations within resistant mutation window of ciprofloxacin on mic, swimming motility and biofilm formation of *Pseudomonas aeruginosa*. *Microb. Pathogen.* 137: 103765.

48 Wu, X., Held, K., Zheng, C. et al. (2015). Dynamic proteome response of *Pseudomonas aeruginosa* to tobramycin antibiotic treatment. *Mol. Cell. Proteomics* 14: 2126–2137.

49 Hanna, S.L., Sherman, N.E., Kinter, M.T., and Goldberg, J.B. (2000). Comparison of proteins expressed by *Pseudomonas aeruginosa* strains representing initial and chronic isolates from a cystic fibrosis patient: an analysis by 2-D gel electrophoresis and capillary column liquid chromatography-tandem mass spectrometry. *Microbiology (Reading)* 146 (Pt 10): 2495–2508.

50 Nouwens, A.S., Willcox, M.D.P., Walsh, B.J., and Cordwell, S.J. (2002). Proteomic comparison of membrane and extracellular proteins from invasive (PAO1) and cytotoxic (6206) strains of *Pseudomonas aeruginosa*. *Proteomics* 2: 1325–1346.

51 Nix, I.D., Idelevich, E.A., Storck, L.M. et al. (2020). Detection of methicillin resistance in *Staphylococcus aureus* from agar cultures and directly from positive blood cultures using MALDI-TOF mass spectrometry-based direct-on-target microdroplet growth assay. *Front. Microbiol.* 11: 232.

52 Tang, W., Ranganathan, N., Shahrezaei, V., and Larrouy-Maumus, G. (2019). MALDI-TOF mass spectrometry on intact bacteria combined with a refined analysis framework allows accurate classification of MSSA and MRSA. *PLoS One* 14: e0218951.

53 Hare, N.J., Solis, N., Harmer, C. et al. (2012). Proteomic profiling of *Pseudomonas aeruginosa* AES-1R, PAO1 and PA14 reveals potential virulence determinants associated with a transmissible cystic fibrosis-associated strain. *BMC Microbiol.* 12: 16–16.

54 Liew, S.M., Puthucheary, S.D., Rajasekaram, G. et al. (2021). Proteomic profiling of clinical and environmental strains of *Pseudomonas aeruginosa*. *Mol. Biol. Rep.* 48: 2325–2333.

55 Sewell, A., Dunmire, J., Wehmann, M. et al. (2014). Proteomic analysis of keratitis-associated *Pseudomonas aeruginosa*. *Mol. Vision* 20: 1182–1191.

56 Lau, A.F., Wang, H., Weingarten, R.A. et al. (2014). A rapid matrix-assisted laser desorption ionization-time of flight mass spectrometry-based method for single-plasmid tracking in an outbreak of carbapenem-resistant enterobacteriaceae. *J. Clin. Microbiol.* 52: 2804–2812.

57 Sindt, N.M., Robison, F., Brick, M.A. et al. (2018). MALDI-TOF-MS with PLS modeling enables strain typing of the bacterial plant pathogen *Xanthomonas axonopodis*. *J. Am. Soc. Mass Spectrom.* 29: 413–421.

58 Brimicombe, R.W., Dijkshoorn, L., Van Der Reijden, T.J. et al. (2008). Transmission of *Pseudomonas aeruginosa* in children with cystic fibrosis attending summer camps in the netherlands. *J. Cyst. Fibros.* 7: 30–36.

59 Kidd, T.J., Ramsay, K.A., Hu, H. et al. The ACPinCF Investigator Group(2013). Shared *Pseudomonas aeruginosa* genotypes are common in Australian cystic fibrosis centres. *Eur. Respir. J.* 41: 1091–1100.

60 Ojeniyi, B., Frederiksen, B., and Hoiby, N. (2000). *Pseudomonas aeruginosa* cross-infection among patients with cystic fibrosis during a winter camp. *Pediatr. Pulmonol.* 29: 177–181.

61 Deplano, A., Denis, O., Poirel, L. et al. (2005). Molecular characterization of an epidemic clone of panantibiotic-resistant *Pseudomonas aeruginosa*. *J. Clin. Microbiol.* 43: 1198–1204.

62 Koutsogiannou, M., Drougka, E., Liakopoulos, A. et al. (2013). Spread of multidrug-resistant *Pseudomonas aeruginosa* clones in a university hospital. *J. Clin. Microbiol.* 51: 665–668.

63 Quick, J., Cumley, N., Wearn, C.M. et al. (2014). Seeking the source of *Pseudomonas aeruginosa* infections in a recently opened hospital: an observational study using whole-genome sequencing. *BMJ Open* 4: e006278.

64 Römling, U., Wingender, J., Müller, H., and Tümmler, B. (1994). A major *Pseudomonas aeruginosa* clone common to patients and aquatic habitats. *Appl. Environ. Microbiol.* 60: 1734–1738.

65 Mohan, K., Fothergill, J.L., Storrar, J. et al. (2008). Transmission of *Pseudomonas aeruginosa* epidemic strain from a patient with cystic fibrosis to a pet cat. *Thorax* 63: 839.

66 Schelstraete, P., Van Daele, S., De Boeck, K. et al. (2008). *Pseudomonas aeruginosa* in the home environment of newly infected cystic fibrosis patients. *Eur. Respir. J.* 31: 822.

67 Iversen, B.G., Jacobsen, T., Eriksen, H.-M. et al. (2007). An outbreak of *Pseudomonas aeruginosa* infection caused by contaminated mouth swabs. *Clin. Infect. Dis.* 44: 794–801.

68 Parcell, B.J., Oravcova, K., Pinheiro, M. et al. (2018). *Pseudomonas aeruginosa* intensive care unit outbreak: winnowing of transmissions with molecular and genomic typing. *J. Hosp. Infect.* 98: 282–288.

69 Salm, F., Deja, M., Gastmeier, P. et al. (2016). Prolonged outbreak of clonal MDR *Pseudomonas aeruginosa* on an intensive care unit: contaminated sinks and contamination of ultra-filtrate bags as possible route of transmission? *Antimicrob. Resist. Infect. Control* 5: 53.

70 Cabrolier, N., Sauget, M., Bertrand, X., and Hocquet, D. (2015). Matrix-assisted laser desorption ionization–time of flight mass spectrometry identifies *Pseudomonas aeruginosa* high-risk clones. *J. Clin. Microbiol.* 53: 1395–1398.

71 Christner, M., Trusch, M., Rohde, H. et al. (2014). Rapid MALDI-TOF mass spectrometry strain typing during a large outbreak of Shiga-toxigenic *Escherichia coli*. *PLoS One* 9: e101924–e101924.

72 Egli, A., Tschudin-Sutter, S., Oberle, M. et al. (2015). Matrix-assisted laser desorption/ionization time of flight mass-spectrometry (MALDI-TOF MS) based typing of extended-spectrum β-lactamase producing *E. coli* – a novel tool for real-time outbreak investigation. *PLoS One* 10: e0120624.

73 Guilbaud, M., Bruzaud, J., Bouffartigues, E. et al. (2017). Proteomic response of *Pseudomonas aeruginosa* PAO1 adhering to solid surfaces. *Front. Microbiol.* 8: 1465.

74 Crabbé, A., Schurr Michael, J., Monsieurs, P. et al. (2011). Transcriptional and proteomic responses of *Pseudomonas aeruginosa* PAO1 to spaceflight conditions involve Hfq regulation and reveal a role for oxygen. *Appl. Environ. Microbiol.* 77: 1221–1230.

75 Patrauchan, M.A., Sarkisova, S.A., and Franklin, M.J. (2007). Strain-specific proteome responses of *Pseudomonas aeruginosa* to biofilm-associated growth and to calcium. *Microbiology* 153: 3838–3851.

76 Wu, X., Siehnel, R.J., Garudathri, J. et al. (2019). in vivo proteome of *Pseudomonas aeruginosa* in airways of cystic fibrosis patients. *J. Proteome Res.* 18: 2601–2612.

77 Bjarnsholt, T., Jensen, P.Ø., Fiandaca, M.J. et al. (2009). *Pseudomonas aeruginosa* biofilms in the respiratory tract of cystic fibrosis patients. *Pediatr. Pulmonol.* 44: 547–558.

78 Kamath, K.S., Krisp, C., Chick, J. et al. (2017). *Pseudomonas aeruginosa* proteome under hypoxic stress conditions mimicking the cystic fibrosis lung. *J. Proteome Res.* 16: 3917–3928.

79 Penesyan, A., Kumar, S.S., Kamath, K. et al. (2015). Genetically and phenotypically distinct *Pseudomonas aeruginosa* cystic fibrosis isolates share a core proteomic signature. *PLoS One* 10: e0138527.

80 Schick, A. and Kassen, R. (2018). Rapid diversification of *Pseudomonas aeruginosa* in cystic fibrosis lung-like conditions. *Proc. Natl. Acad. Sci.* 115: 10714–10719.

81 Grady, S.L., Malfatti, S.A., Gunasekera, T.S. et al. (2017). A comprehensive multi-omics approach uncovers adaptations for growth and survival of *Pseudomonas aeruginosa* on n-alkanes. *BMC Genomics* 18: 334–334.

82 Liu, H., Sun, W.-B., Liang, R.-B. et al. (2015). iTRAQ-based quantitative proteomic analysis of *Pseudomonas aeruginosa* SJTD-1: a global response to n-octadecane induced stress. *J. Proteomics* 123: 14–28.

83 Diwan, V., Tamhankar, A.J., Aggarwal, M. et al. (2009). Detection of antibiotics in hospital effluents in India. *Curr. Sci.* 97: 1752–1755.

84 Ekwanzala, M.D., Lehutso, R.F., Kasonga, T.K. et al. (2020). Environmental dissemination of selected antibiotics from hospital wastewater to the aquatic environment. *Antibiotics* 9: 431.

85 Kraupner, N., Hutinel, M., Schumacher, K. et al. (2021). Evidence for selection of multi-resistant *E. coli* by hospital effluent. *Environ. Int.* 150: 106436.

86 Swatton, J.E., Davenport, P.W., Maunders, E.A. et al. (2016). Impact of azithromycin on the quorum sensing-controlled proteome of *Pseudomonas aeruginosa*. *PLoS One* 11: e0147698.

87 Stover, C.K., Pham, X.Q., Erwin, A.L. et al. (2000). Complete genome sequence of *Pseudomonas aeruginosa* PAO1, an opportunistic pathogen. *Nature* 406: 959–964.

88 Gad, G.F., El-Domany, R.A., Zaki, S., and Ashour, H.M. (2007). Characterization of *Pseudomonas aeruginosa* isolated from clinical and environmental samples in Minia, Egypt: prevalence, antibiogram and resistance mechanisms. *J. Antimicrob. Chemother.* 60: 1010–1017.

89 Kouda, S., Ohara, M., Onodera, M. et al. (2009). Increased prevalence and clonal dissemination of multidrug-resistant *Pseudomonas aeruginosa* with the blaIMP-1 gene cassette in Hiroshima. *J. Antimicrob. Chemother.* 64: 46–51.

90 Sayers, E.W., Agarwala, R., Bolton, E.E. et al. (2019). Database resources of the national center for biotechnology information. *Nucleic Acids Res.* 47: D23–D28.

91 Winsor, G., Griffiths, E., Lo, R. et al. (2016). Enhanced annotations and features for comparing thousands of *Pseudomonas* genomes in the *Pseudomonas* genome database. *Nucleic Acids Res.* 44: D646–D653.

92 Basso, P., Ragno, M., Elsen, S. et al. (2017). *Pseudomonas aeruginosa* pore-forming exolysin and Type IV pili cooperate to induce host cell lysis. *mBio* 8 (1): e02250-16.

93 Brzozowski, M., Jursa-Kulesza, J. B. and Kosik-Bogacka, D. (2019). Virulence analysis of 81 of *Pseudomonas aeruginosa* genomes available in public sequence databases. https://assets.researchsquare.com/files/rs-1224/v1/2776dd41-a868-4818-9b0a-17082b591a14.pdf?c=1631825698 (accessed 27 September 2022).

94 Freschi, L., Bertelli, C., Jeukens, J. et al. (2018). Genomic characterisation of an international *Pseudomonas aeruginosa* reference panel indicates that the two major groups draw upon distinct mobile gene pools. *FEMS Microbiol. Lett.* 365 (14): https://doi.org/10.1093/femsle/fny120. PMID: 29897457.

95 García-Reyes, S., Soberón-Chávez, G., and Cocotl-Yanez, M. (2020a). The third quorum-sensing system of *Pseudomonas aeruginosa*: *Pseudomonas quinolone* signal and the enigmatic PqsE protein. *J. Med. Microbiol.* 69: 25–34.

96 García-Reyes, S., Soto-Aceves, M.P., Cocotl-Yañez, M. et al. (2020b). The outlier *Pseudomonas aeruginosa* strain ATCC 9027 harbors a defective LasR quorum-sensing transcriptional regulator. *FEMS Microbiol. Lett.* 367: https://doi.org/10.1093/femsle/fnaa123.

97 Sood, U., Hira, P., Kumar, R. et al. (2019). Comparative genomic analyses reveal core-genome-wide genes under positive selection and major regulatory hubs in outlier strains of *Pseudomonas aeruginosa*. *Front. Microbiol.* 10: 53.

98 Toska, J., Sun, Y., Carbonell, D.A. et al. (2014). Diversity of virulence phenotypes among type III secretion negative *Pseudomonas aeruginosa* clinical isolates. *PLoS One* 9: e86829.

99 Zhang, Z. and Zhang, X. (2021). Evolution of subfamily yy1 lipases in *Pseudomonas aeruginosa*. *Curr. Microbiol.* 78: 3494–3504.

100 Elsen, S., Huber, P., Bouillot, S. et al. (2014). A type III secretion negative clinical strain of *Pseudomonas aeruginosa* employs a two-partner secreted exolysin to induce hemorrhagic pneumonia. *Cell Host Microbe* 15: 164–176.

101 Sood, U., Singh, D.N., Hira, P. et al. (2020). Rapid and solitary production of mono-rhamnolipid biosurfactant and biofilm inhibiting pyocyanin by a taxonomic outlier *Pseudomonas aeruginosa* strain CR1. *J. Biotechnol.* 307: 98–106.

102 Planquette, B., Timsit, J.F., Misset, B.Y. et al. Group, O. S.(2013). *Pseudomonas aeruginosa* ventilator-associated pneumonia. Predictive factors of treatment failure. *Am. J. Respir. Crit. Care Med.* 188: 69–76.

103 Garcia-Ulloa, M., Ponce-Soto, G.Y., Gonzalez-Valdez, A. et al. (2019). Two *Pseudomonas aeruginosa* clonal groups belonging to the PA14 clade are indigenous to the Churince system in Cuatro Cienegas Coahuila, Mexico. *Environ. Microbiol.* 21: 2964–2976.

104 Roy, P.H., Tetu, S.G., Larouche, A. et al. (2010). Complete genome sequence of the multiresistant taxonomic outlier *Pseudomonas aeruginosa* PA7. *PLoS One* 5: e8842.

105 Freschi, L., Jeukens, J., Kukavica-Ibrulj, I. et al. (2015). Clinical utilization of genomics data produced by the international *Pseudomonas aeruginosa* consortium. *Front. Microbiol.* 6: 1036.

106 Huber, P., Basso, P., Reboud, E., and Attree, I. (2016). *Pseudomonas aeruginosa* renews its virulence factors. *Environ. Microbiol. Rep.* 8: 564–571.

107 Engel, J. and Balachandran, P. (2009). Role of *Pseudomonas aeruginosa* type III effectors in disease. *Curr. Opin. Microbiol.* 12: 61–66.

108 Hauser, A.R. (2009). The type III secretion system of *Pseudomonas aeruginosa*: infection by injection. *Nat. Rev. Microbiol.* 7: 654–665.

109 Shaver, C.M. and Hauser, A.R. (2004). Relative contributions of *Pseudomonas aeruginosa* ExoU, ExoS, and ExoT to virulence in the lung. *Infect. Immun.* 72: 6969–6977.

110 Vance, R.E., Rietsch, A., and Mekalanos, J.J. (2005). Role of the type III secreted exoenzymes S, T, and Y in systemic spread of *Pseudomonas aeruginosa* PAO1 in vivo. *Infect. Immun.* 73: 1706–1713.

111 Lyczak, J.B., Cannon, C.L., and Pier, G.B. (2000). Establishment of *Pseudomonas aeruginosa* infection: lessons from a versatile opportunist. *Microbes Infect.* 2: 1051–1060.

112 Panayidou, S., Georgiades, K., Christofi, T. et al. (2020). *Pseudomonas aeruginosa* core metabolism exerts a widespread growth-independent control on virulence. *Sci. Rep.* 10: 9505.

113 Hillis, D.M. (1997). Phylogenetic analysis. *Curr. Biol.* 7: R129–R131.

114 Ozer, E.A., Nnah, E., Didelot, X. et al. (2019). The population structure of *Pseudomonas aeruginosa* is characterized by genetic isolation of exou+ and exos+ lineages. *Genome Biol. Evol.* 11: 1780–1796.

115 Subedi, D., Vijay, A.K., Kohli, G.S. et al. (2018). Comparative genomics of clinical strains of *Pseudomonas aeruginosa* strains isolated from different geographic sites. *Sci. Rep.* 8: 15668.

116 Depke, T., Thoming, J.G., Kordes, A. et al. (2020). Untargeted LC-MS metabolomics differentiates between virulent and avirulent clinical strains of *Pseudomonas aeruginosa*. *Biomolecules* 10 (7): –1041.

117 Parks, D.H., Chuvochina, M., Waite, D.W. et al. (2018). A standardized bacterial taxonomy based on genome phylogeny substantially revises the tree of life. *Nat. Biotechnol.*.

118 Wang, Z. and Wu, M. (2013). A phylum-level bacterial phylogenetic marker database. *Mol. Biol. Evol.* 30: 1258–1262.

119 Gupta, R.S. (1998). Protein phylogenies and signature sequences: a reappraisal of evolutionary relationships among archaebacteria, eubacteria, and eukaryotes. *Microbiol. Mol. Biol. Rev.* 62: 1435–1491.

120 Gupta, R.S. (2014). Identification of conserved indels that are useful for classification and evolutionary studies. *Methods. Microbiol.* 41: 153–182.

121 Gupta, R.S. (2016). Impact of genomics on the understanding of microbial evolution and classification: the importance of Darwin's views on classification. *FEMS Microbiol. Rev.* 40: 520–553.

122 Adeolu, M., Alnajar, S., Naushad, S., and Gupta, R.S. (2016). Genome-based phylogeny and taxonomy of the '*Enterobacteriales*': proposal for *Enterobacterales* ord. nov. divided into the families *Enterobacteriaceae, Erwiniaceae* fam. nov., *Pectobacteriaceae* fam. nov., *Yersiniaceae* fam. nov., *Hafniaceae* fam. nov., *Morganellaceae* fam. nov., and *Budviciaceae* fam. nov. *Int. J. Syst. Evol. Microbiol.* 66: 5575–5599.

123 Patel, S. and Gupta, R.S. (2020). A phylogenomic and comparative genomic framework for resolving the polyphyly of the genus *bacillus*: Proposal for six new genera of *Bacillus* species, *peribacillus* gen. nov., *Cytobacillus* gen. nov., *Mesobacillus* gen. nov., *Neobacillus* gen. nov., *Metabacillus* gen. nov. and *Alkalihalobacillus* gen. nov. *Int. J. Syst. Evol. Microbiol.* 70: 406–438.

124 Rudra, B. and Gupta, R.S. (2021). Phylogenomic and comparative genomic analyses of species of the family *Pseudomonadaceae*: proposals for the genera *Halopseudomonas* gen. nov. and *Atopomonas* gen. nov., merger of the genus *Oblitimonas* with the genus *Thiopseudomonas*, and transfer of some misclassified species of the genus *Pseudomonas* into other genera. *Int. J. Syst. Evol. Microbiol.* 71 (9): https://doi.org/10.1099/ijsem.0.005011.

125 Adeolu, M. and Gupta, R.S. (2014). A phylogenomic and molecular marker based proposal for the division of the genus *Borrelia* into two genera: the emended genus *Borrelia* containing only the members of the relapsing fever *Borrelia*, and the genus *Borreliella* gen. nov. containing the members of the Lyme disease *Borrelia* (*Borrelia burgdorferi* sensu lato complex). *Antonie van Leeuwenhoek* 105: 1049–1072.

126 Gupta, R.S., Naushad, S., and Baker, S. (2015). Phylogenomic analyses and molecular signatures for the class *halobacteria* and its two major clades: a proposal for division of the class *halobacteria* into an emended order *halobacteriales* and two new orders, *haloferacales* ord. Nov. and *Natrialbales* ord. Nov., containing the novel families *Haloferacaceae* fam. Nov. and *natrialbaceae* fam. Nov. *Int. J. Syst. Evol. Microbiol.* 65: 1050–1069.

8

Characterization of Biodegradable Polymers by MALDI-TOF MS

Hiroaki Sato

Research Institute for Sustainable Chemistry, National Institute of Advanced Industrial Science and Technology (AIST), Highahihirosima, Hiroshima, Japan

8.1 Introduction

Polymer materials have highly useful physical characteristics, and as such the quantities of polymers in use have been increasing with the growth of the global economy and rising standards of living. However, the fact that polymers are stable and resist decomposition becomes a drawback when they are discharged into the environment after use. The microplastics produced by the disintegration of waste plastics that find their way into the ocean have become a serious global problem. Biodegradable polymeric materials, which are decomposed by the action of microorganisms present in the environment, have received a great deal of attention as potential environmentally friendly materials.

The material properties of a polymer are evaluated by physical properties such as mechanical strength and melt flow index. The biodegradability of polymers has been generally evaluated in terms of macroscopic changes such as in weight and shape, physical changes such as mechanical strength, and the amount of CO_2 produced during the decomposition process.

Designing or controlling the quality of biodegradable polymers requires detailed molecular characterization, using repeat unit structures, end groups and their combinations, and copolymer composition distribution. If the degradation behavior of biodegradable polymers can be characterized at the molecular level, it should be possible to design and develop environmentally friendly materials that have both the required physical properties and biodegradability suited to the environments in which it accumulates. However, it should be noted here that biodegradation in the environment does not always give optimal results. As described in this chapter, alkylphenol-based surfactants used as an adjunct to pesticides can be degraded by microorganisms to produce alkylphenols that are suspected endocrine disruptors. To understand the fate of polymers in the environment, it is important to develop techniques for molecular characterization.

Microbiological Identification using MALDI-TOF and Tandem Mass Spectrometry: Industrial and Environmental Applications, First edition. Edited by Haroun N. Shah, Saheer E. Gharbia, Ajit J. Shah, Erika Y. Tranfield, and K. Clive Thompson.

Matrix-assisted laser desorption/ionization time-of-flight mass spectrometry (MALDI-TOF MS) has been utilized as an effective technique for analyzing the chemical structures of polymers at the individual molecular level [1–6]. Numerous types of biodegradable polymers have been characterized using this technique (see review by [7]). This chapter deals with the basics and strategies for analyzing the chemical structure of biodegradable polymers by MALDI-TOF MS, using cases reported by the author. We begin by describing MS-based characterization for analyzing changes in the chemical structure of a biodegradable polymer and its molecular structure during the process of biodegradation. We also discuss the effectiveness of high-resolution MALDI-TOF MS, which enables accurate MS of polymers together with a data analysis method, i.e. Kendrick mass defect (KMD) analysis, for revealing the distribution of end groups and copolymerization composition from a very complicated mass spectrum.

8.2 Structural Characterization of Poly(ε-caprolactone) Using MALDI-TOF MS

Polymers are a mixture of macromolecules of various chain lengths that are formed by the polymerization of a single or multiple monomers. Both ends of each polymer chain are terminated with end groups (cyclic polymers without chain ends are also possible), reflecting the conditions under which they were synthesized. The end groups probably play an important role in biodegradation or enzymatic degradation. Degradation that proceeds sequentially from chain terminals is known as "exo-type" degradation. In this case, the structure of the end groups will have a major influence on biodegradability. Analysis of end-group structures is therefore key to understanding the properties of biodegradable polymers.

An overview of the structural characterization by MALDI-TOF MS is given here using the determination of the molecular structure of poly (ε-caprolactone) (PCL) [8]. PCL is a typical biodegradable aliphatic polyester that is used in various applications such as the raw materials for polyurethane, packaging materials, and coating agents. PCL is normally synthesized by ring-opening polymerization of ε-caprolactone using alcohol as an initiator, so the initiator residue is usually incorporated at the α-terminal side of the PCL chain. Trace water in the polymerization system also plays a role as an initiator, resulting in PCL having a carboxyl group at the α-terminal. In the process of synthesizing PCL by ring-opening polymerization catalyzed by metal alkoxide, an intra- or inter-molecular ester exchange reaction may produce cyclic polymer chains or chains with end-groups that differ from the initiator residue. The formation of these various end groups is thought to have a significant impact on the physical and biodegradable properties of PCL.

Figure 8.1 shows whole and expanded mass spectra of the PCL sample. The observed ions here were $[M+Na]^+$, which are generally produced using sodium salts added as a cationization reagent or which are present in the background. All the peaks of the mass spectra shown hereinafter are $[M+Na]^+$. In the observed mass spectrum, the peak intensity reaches a maximum at m/z 1000 on the lowest mass side of the measurement range, and gradually decreases as the mass increases up to m/z 12 000. This phenomenon is common in the mass spectrum of polydisperse polymers and is due to the "mass discrimination

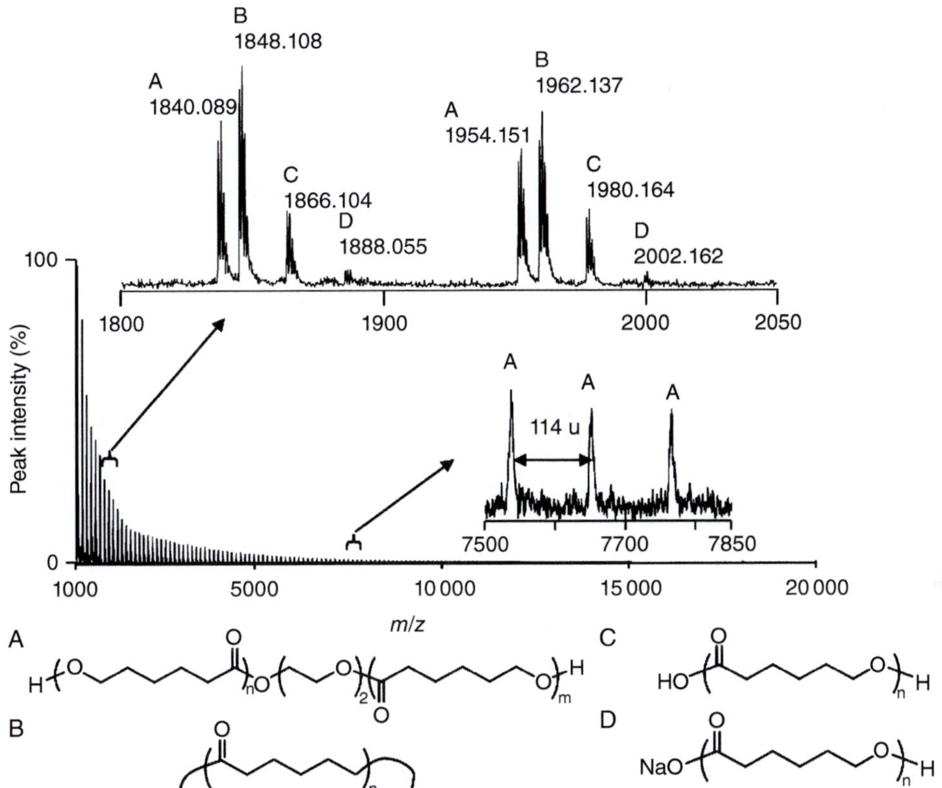

Figure 8.1 Mass spectra of PCL and assigned chemical structures.

effect" in which components on the low-molecular-weight side are predominantly observed in polymers with a wide molecular weight distribution. In response to this, as described later, offline combination with size exclusion chromatography (SEC) is required to observe the accurate molecular weight distribution of polydisperse polymers. In the expanded mass spectrum around *m/z* 2000, there are four series (A–D) with a period of 114-Da intervals. Of these, only the **A** series extends to the higher mass range.

Polymer samples with a simple structure will give a series of peaks with regular intervals on their mass spectrum. The *m/z* values of polymers observed on the mass spectra can be calculated according to Eq. (8.1):

$$m/z = M_{\text{monomer}} \times n + M_{\text{cation}} + M_{\text{residue}} \tag{8.1}$$

where M_{monomer} and M_{cation} are the mass of a repeating unit (monomer) and an adduct cation, n is the number of repeating units, and M_{residue} is the mass of residues such as end groups. In the case of PCL, 116.068 is the mass of the repeating unit and 22.990 is the mass of Na^+, giving

$$m/z = 114.068 \times n + 22.990 + M_{\text{residue}} \tag{8.2}$$

The exact mass of $M_{residue}$ is estimated from the monoisotopic mass of each peak actually observed on the mass spectrum. In brief, the masses of the PCL main chain and sodium cations were excluded from the *m/z* value, taking into account the fact that the value of *n* was maximized under the condition of $0 \leq M_{residue} < 114$. In this way, the residue of the A series, which is the main component on the high-molecular-weight side, has an elemental composition of $C_4H_{10}O_3$, and can be attributed to a PCL diol that has hydroxyl groups at both ends. As the residual mass of the B series is 0, it is a cyclic polymer that has no ends. Cyclic oligomers are commonly observed in industrial polyesters. The residue of the C series is H_2O: these are PCLs with a hydroxyl group and a carboxyl group at each end. In general, compounds with free carboxyl groups give not only the sodium-cationized molecules $[M + Na]^+$ on the mass spectrum but also the peaks of the sodium-cationized molecules of sodium salts $[M - nH + (n + 1) Na]^+$ (*n* is the number of carboxyl groups in the molecule). The D series is the sodium carboxylic acid salt of the C series. The absence of peaks for sodium salts corresponding to the A series is evidence that both their termini end with hydroxyl groups. The terminal structure of the polymers can thus be characterized from the mass and peak intervals observed in the mass spectrum.

The peak distribution of the mass spectrum in Figure 8.1 does not, in fact, accurately reflect the molecular weight distribution. This is due to a phenomenon known as the "mass discrimination effect," which causes a considerable difference between the intensity distribution of the peaks of the sample components observed in the mass spectrum and the true molecular weight distribution as revealed by MALDI-TOF MS. This is because the ionization efficiency and the sensitivity of the detector change a great deal according to the molecular weight of the sample component. In general, if the polydispersity of the sample polymer exceeds about 1.1, the effect is not negligible [9]. It is therefore difficult, in general, to correctly reveal the molecular weight dependence of complex chemical structures, such as the end groups of polymer samples that have a wide molecular weight distribution, using only MALDI-TOF MS. Offline combination of SEC and MALDI-TOF MS is useful for the analysis of such samples. Polydisperse polymer samples are fractionated by SEC into a polydispersity of less than 1.1, after which each fraction is measured using MALDI-TOF MS [10, 11].

Figure 8.2 shows the SEC chromatogram and the MALDI mass spectra of typical fractions of the PCL sample. In SEC, polymer chains are eluted in order of molecular size, so the peak distribution of the mass spectrum of each fraction gradually shifts to a lower molecular weight with fractionation time. Even for the higher masses, around *m/z* 20 000, where almost no peak was observed in pristine PCL samples, a peak was clearly observed for the fraction collected at around 19.75 minutes. The peak-top molecular weight (M_p) of each fraction was then obtained from the mass spectrum, and the logarithm of the M_p value was plotted against the elution time to create a calibration curve and determine average molecular weight. (As an alternative to M_p, number-average molecular weight [M_n] or weight-average molecular weight [M_w] are also possible.) The resulting average molecular weight by SEC/MALDI-TOF MS was almost half that using calibrated using polystyrene (PS) standards. This is due to the difference in molecular size between the PS molecule and the PCL molecule in chloroform, which is used as the mobile phase for SEC. The average molecular weight obtained by the standard PS calibration method consists of "relative" values and is

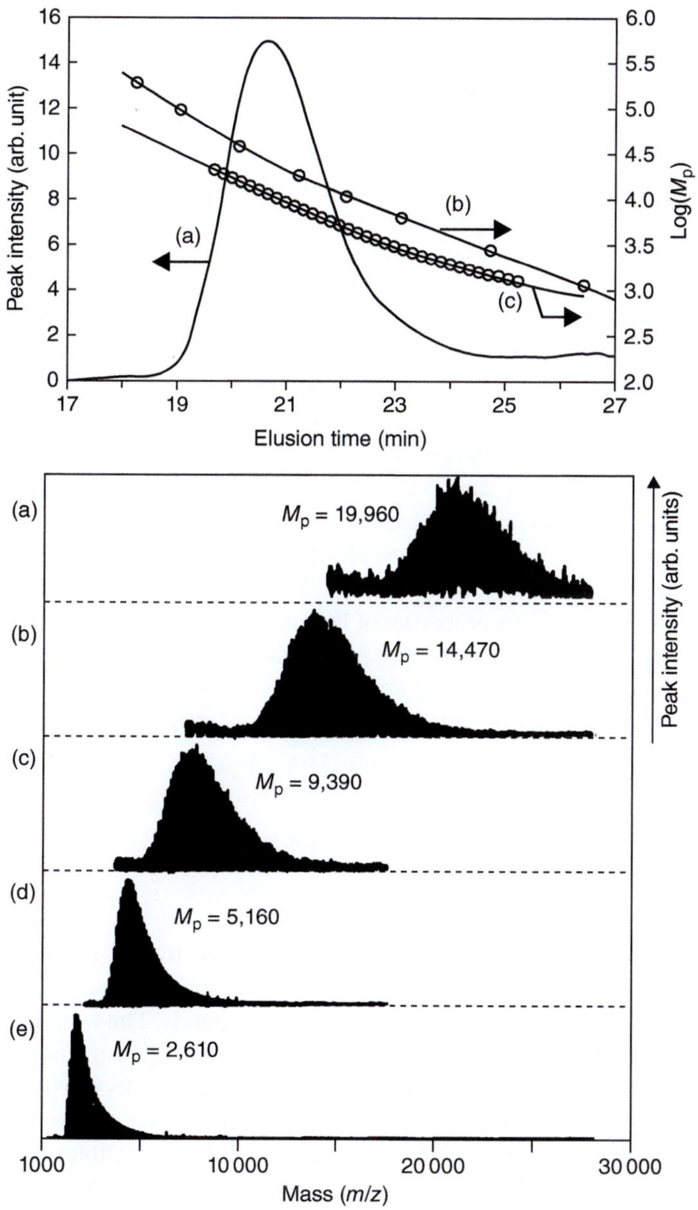

Figure 8.2 Size exclusion chromatography (SEC)/MALDI-TOF MS analysis of PCL. Top: SEC chromatogram of PCL (a) together with calibration curves using polystyrene standards (b) and by SEC/MALDI-MS method (c). Bottom: Typical MALDI mass spectra of PCL fractions eluted at around 19.75 (a), 20.25 (b), 20.92 (c), 21.92 (d), and 23.25 minutes (e), respectively. *Source:* Sato et al. [8]/ with permission of The Society of Polymer Science, Japan.

less reliable. In the SEC/MALDI-TOF MS method, the molecular weight of each fraction is determined and used for calibration, so the "absolute" value can be obtained.

Another advantage of using MALDI-TOF MS as an SEC detector is that elution behavior can be observed for each component with a different molecular structure such as a terminal group [12]. The total peak intensity of the mass spectrum of each fraction is corrected with reference to the peak intensity of the chromatogram determined by the refractive index (RI) detector. The composition of the PCL diol component (the A series in Figure 8.1) increased with rising degree of polymerization, showing a maximum distribution at about 70 mer (molecular weight of about 8000). As the PCL diol had been synthesized using diethylene glycol as an initiator, the molecular chain extends in two directions in the polymerization reaction, resulting in a high-molecular-weight component. By contrast, cyclic PCL (B series) was characteristically observed in the low-molecular-weight region with a molecular weight of $< \sim 4000$. As the cyclic molecule is generally produced by a backbiting reaction or an intramolecular transesterification reaction, the result is low-molecular-weight components. PCL with a hydroxyl group and a carboxyl end (C and D series) tended toward having a higher average molecular weight than did cyclic PCL, but no higher than 100 mer (molecular weight $\sim 11\,000$). This component is probably produced using water, which was present as an impurity, as an initiator. These cyclic PCL and carboxyl-terminal PCL types are by-products, and their presence can adversely affect the various properties of PCL used as a raw material.

As described earlier, MALDI-TOF MS is an ideal tool for characterizing polymer structures. The PCL characterization shown here is just an example. This method can apply essentially the same procedure to different polymers, provided the mass spectrum can be observed with MALDI-TOF MS.

8.3 Biodegradation Profiles of a Terminal-modified PCL Observed by MALDI-TOF MS

The decomposition patterns of biodegradable polymers can be broadly divided into two types: the end type, which randomly cleaves the molecular chain, and the exo type, which undergoes sequential decomposition reaction from the polymer chain ends. The terminal structure of the polymer chain of an exo-type will greatly affect its biodegradation characteristics. Here, an example of identifying the biodegradation pattern using MALDI-TOF MS is demonstrated [13]. A film of PCL that had been terminal-modified with a benzyl group was enzymatically degraded in phosphate buffer solution containing cholesterol esterase from *Pseudomonas* sp. for up to 36 hours. The resultant changes in the terminal structure and molecular weight distribution of the recovered sample were observed.

Figure 8.3 compares the mass spectra of the PCL samples before and after the enzymatic degradation test. Reflecting the fact that this sample was synthesized by controlling the molecular weight distribution, the mass spectrum of the pristine PCL sample before the enzymatic degradation test (Figure 8.3a) has a Gaussian peak distribution with a maximum m/z of 1500. All of the observed peaks are assigned as PCL terminal-modified with a benzyl group, indicated as Bz-PCL. In the mass spectrum of the residual film recovered after enzymatic degradation testing of the PCL sample (Figure 8.3b), a new peak series appears with a maximum around m/z 1000. From the mass of this peak series, they could

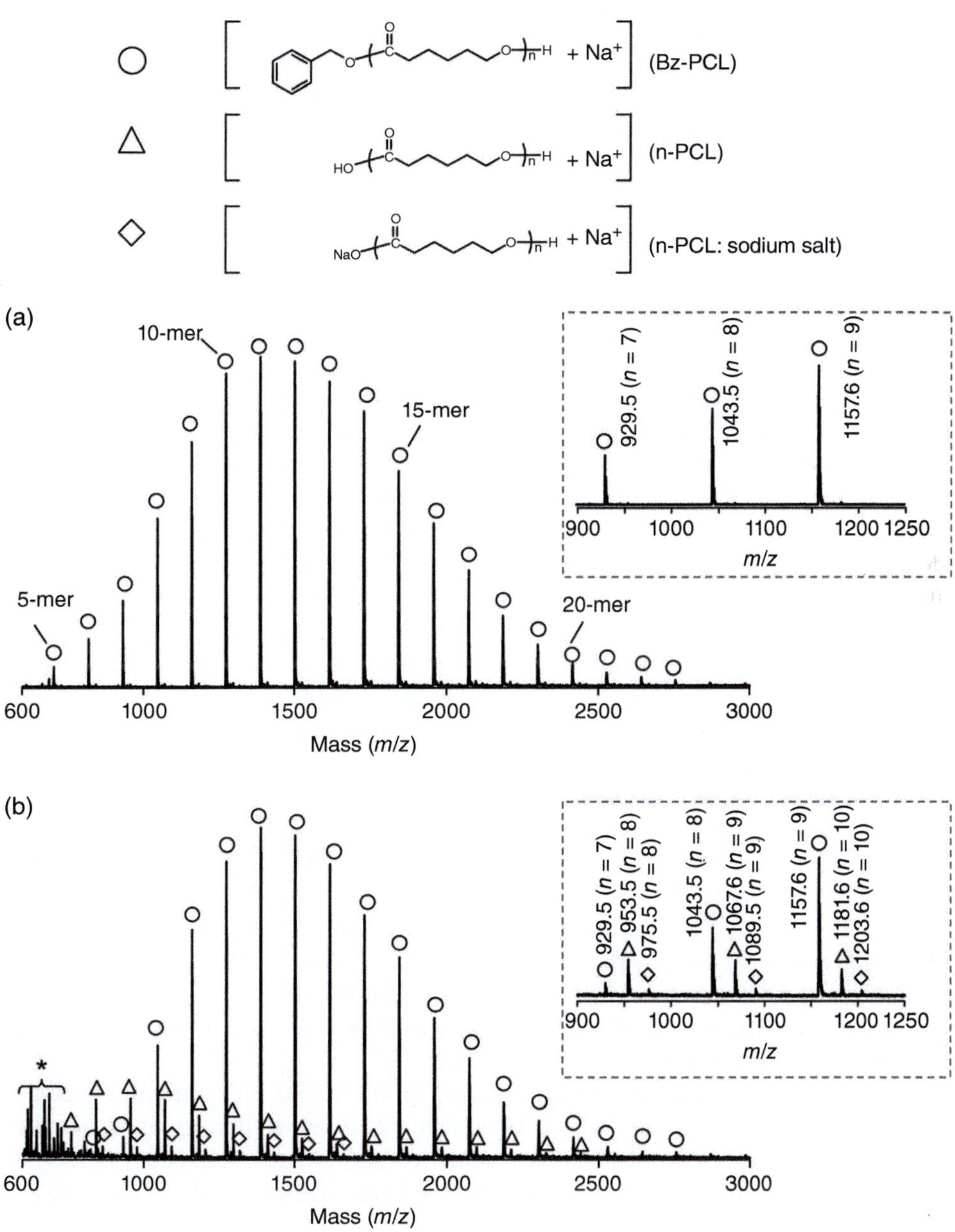

Figure 8.3 Typical MALDI mass spectra of original PCL sample (a) and PCL sample (b) recovered after the enzymatic degradation for 36 hours. Mass numbers in the expanded spectra indicate for monoisotopic mass. *Source:* Reproduced from Sato et al. [13]/With permission of Elsevier.

be assigned as PCL molecules with a hydroxyl group and a carboxyl terminus (n-PCL). If random decomposition (endo-type biodegradation) occurs in the polymer chain during the progress of enzymatic degradation, the peak distribution of the Bz-PCL component in the residual sample should shift to the low-molecular-weight side. However, the molecular weight distribution of the Bz-PCL component is almost the same as that before the biodegradation test. These results reveal that the "exo-type" enzymatic decomposition reaction, which sequentially decomposes from the benzyl terminal side, predominated under the given experimental conditions. As mentioned earlier, MALDI-TOF MS is a powerful method for elucidating the degradation mechanism by using changes in the terminal group and molecular weight distribution as a pointer.

8.4 Bacterial Biodegradation Mechanisms of Non-ionic Surfactants

Alkylphenol polyethoxylates (APEOs), including octylphenol polyethoxylate (OPEO) and nonylphenol polyethoxylate (NPEO), are non-ionic surfactants in which hydrophobic alkylphenol and hydrophilic polyethylene oxide (PEO) are combined. These compounds are used in large quantities as an important raw material for industrial detergents and pesticide additives. When APEOs are dispersed in the environment, PEO chains are selectively biodegraded by microorganisms, and octylphenols, short-chain APEOs, and their oxides accumulate in the environment. These compounds are strongly suspected of having endocrine-disrupting effects.

As an outline of the biodegradation mechanism of APEO, an exo-type biodegradation model in which the biodegradation reaction proceeds from the terminal hydroxyl group side of the PEO chain has been proposed and is currently widely accepted. However, a simple exo-type degradation model does not entirely explain why many components, such as the terminal carboxylic acid component of the PEO chain and the number of two or three EO units, are detected in the environment. To understand the biodegradation mechanism in detail, MALDI-TOF MS is useful for characterizing the biodegradation products at the molecular structure level.

MALDI-TOF MS investigations have been performed to observe changes in the chemical structure and molecular weight distribution of degradation intermediates produced during the OPEO biodegradation process by microorganisms [14, 15]. The results of the molecular characterization of the degradation products observed by MALDI-TOF MS can provide some insight into the mechanism of biodegradation.

Octylphenol polyethoxylate (commercial name: Triton X-100) with a number-average molecular weight (M_n) of 625 was subjected to a bacterial biodegradation test under aerobic conditions for 48 hours using *Pseudomonas putida* S-5 [16] isolated from paddy field soil. After designated periods, the biodegradation intermediates were extracted and analyzed by MALDI-TOF MS.

Figure 8.4 shows the mass spectra of the OPEO sample extracts after the biodegradation test with a soil microorganism (*P. putida*) for 24 and 48 hours. The molecular weight distribution of the pristine sample showed a Gaussian peak distribution with a maximum around *m/z* 670 (corresponding to 10 mer of PEO chain). The mass spectrum of the sample subjected to biodegradation testing shows that the molecular weight distribution shifted to the low mass side, indicating that the degradation of the main chain has considerably progressed. Further new peak sequences have emerged, which have been attributed to the hydroxyl group-oxidized component (OPEC; indicated by "∇" in the figure) and its sodium

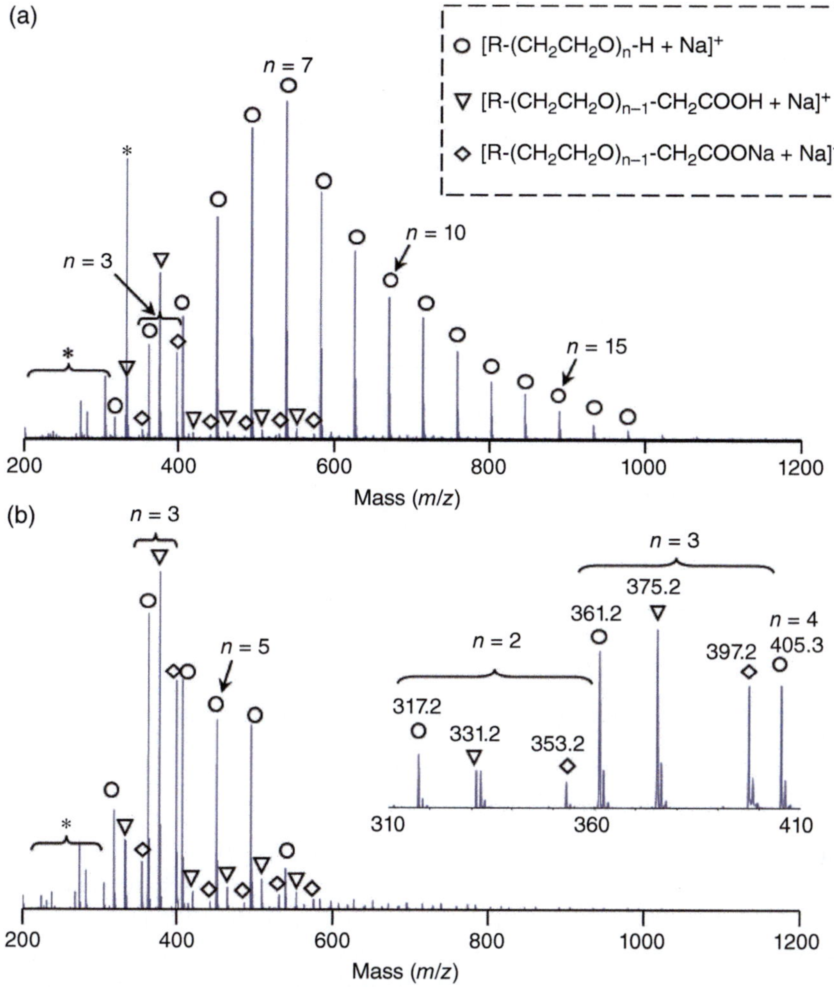

Figure 8.4 Typical MALDI mass spectra of the OPEO extracts after incubation for 24 (a) and 48 hours (b). The asterisk (*) corresponds to impurities and/or matrix peaks. R means *p*-octylphenoxy group. *Source:* Reproduced from Sato et al. [14]/With permission of Elsevier.

salt (OPECNa, indicated by "◇" in the figure) on the hydroxyl group end side of the EO chain of the sample molecule. The formation of this terminal oxide provided strong evidence of the "exo-type" biodegradation mechanism, in which oxidation on the terminal hydroxyl group side and subsequent hydrolysis of the ether bond are repeated. The mechanism of OPEC formation was investigated in greater detail by molecular characterization of the degradation intermediates produced in biodegradation tests using ^{18}O-labeled water [15]. To minimize the use of costly heavy oxygen water, a test system using only 200 μl of culture medium has been established [17]. As shown in Figure 8.5, when incubation was performed in ^{18}O-labeled water, the mass of OPEO did not change, but the mass of OPEC and its sodium salt OPECNa increased by 2 and 4 Da, respectively, indicating that one or two ^{18}O atoms had been introduced into the carboxyl group of OPECs. This result demonstrates that

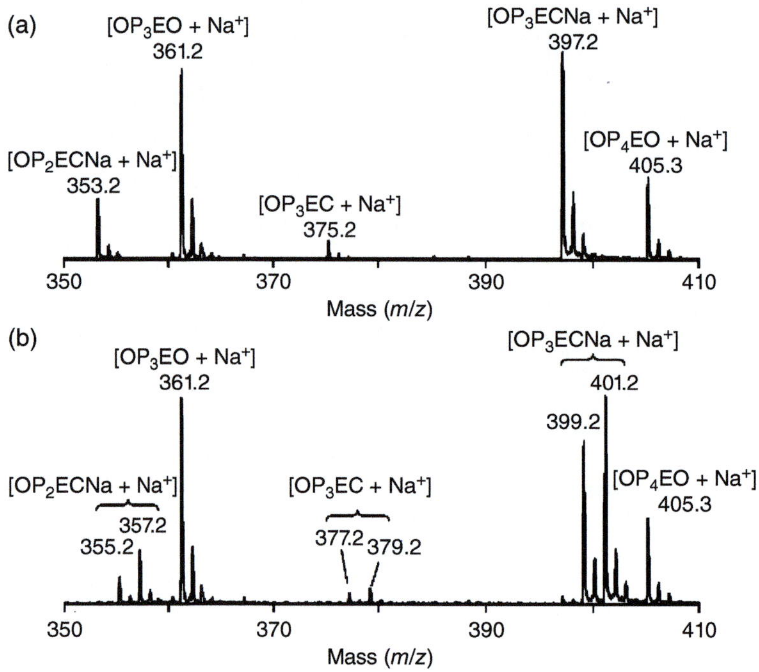

Figure 8.5 Enlarged MALDI mass spectra (trimer region) of the biodegraded OPEO recovered after 48 hours' incubation using normal water (a) and ^{18}O-labeled water (b). *Source:* Reproduced from Sato et al. [15]/With permission of American Chemical Society.

the bacteria utilize water, not air, to oxidize the hydroxyl end side of OPEO. The formation of the OPEC with not only one but also two ^{18}O suggests that an oxygen exchange reaction of the ester has taken place during biodegradation. The oxygen atom in the free carboxyl group does not exchange with the oxygen in water molecules, but the carbonyl oxygen of an ester compound can undergo an exchange reaction with the oxygen in water molecules. During the degradation process of the OPEO molecule, a bacterial enzyme (or coenzyme) and an ester intermediate (enzyme-substrate complex) may have been formed. MALDI-TOF MS investigation with isotope-labeled reagents can provide insights into more detailed biodegradation mechanisms at the molecular structure level.

Studies using MALDI-TOF MS have been further extended to investigate the effects on biodegradability of the chemical structures of surfactants' hydrophobic groups [18]. Because the bacteria used in this study (*P. putida* S-5) only slightly biodegraded hydrophilic PEO ($M_n \sim 1500$), it appears that the hydrophobic moiety plays an important role in the biodegradation of surfactants. The biodegradability of several non-ionic surfactants with hydrophobic groups that have different chemical structures was compared. Here, an OPEO, NPEO, two linear alkylphenol PEOs (AEOs) with different aliphatic chain lengths of C_{12} and C_{14} (A12EO and A14EO), and a PEO with branched aliphatic chains (branched AEO) were investigated. The octamers of all surfactants were separated and purified by silica gel column chromatography to compare only the effects of hydrophobic groups. Each purified sample was subjected to a bacterial biodegradation test for 96 hours and then analyzed by MALDI-TOF MS.

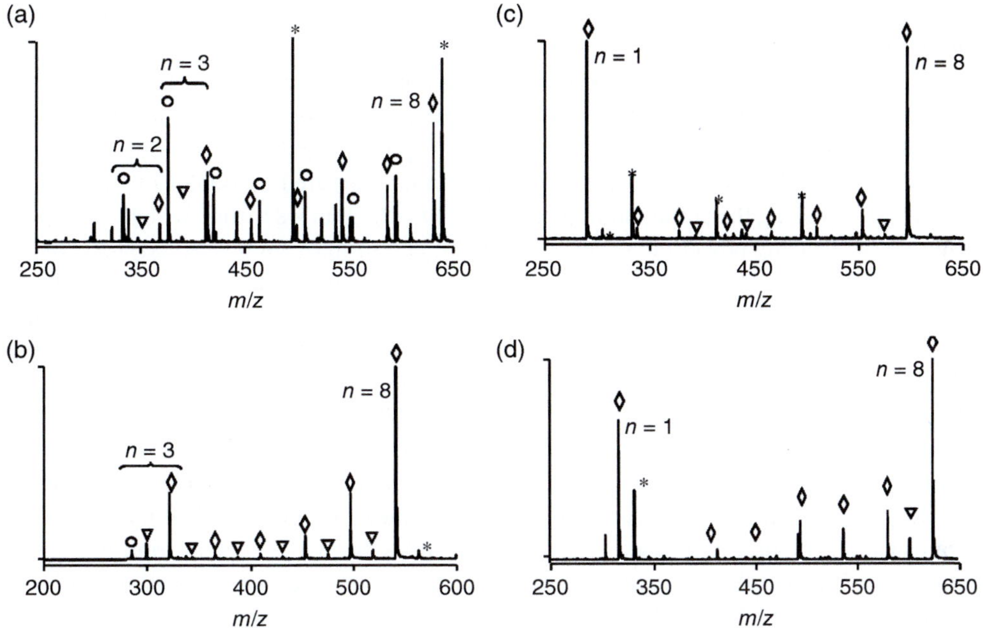

Figure 8.6 Typical MALDI mass spectra of the biodegraded surfactants. (a) NPEO, (b) branched A_8EO, (c) $A_{12}EO$, (d) $A_{14}EO$. *Source:* Reproduced from Sato et al. [18]/With permission of Mass Spectrometry Society of Japan.

The mass spectrum of each biodegraded sample (Figure 8.6) showed that all the samples were chain-shortened and oxidized to form carboxylic acid terminals. Interestingly, the EO chain shortening of APEOs (OPEO and NPEO) and branch AEO stopped at $n = 2$ or 3, while that of linear AEO (A12EO and A14EO) went to $n = 1$. This result suggests that the chemical structure of the surfactant's hydrophobic group has a strong influence on the progress of biodegradation of the PEO chain. As the degradation reaction was stopped at $n = 3$ not only in APEO but also in the branched A8EO, the factor that inhibits the degradation of the PEO chain is probably the steric hindrance of the hydrophobic group rather than the presence of aromatic rings.

As described earlier, MALDI-TOF MS can elucidate the biodegradation mechanism of nonionic surfactants at the molecular structure level. As this method can be measured using only a small amount of sample, it is a great advantage to be able to analyze the details of the degradation mechanism using stable isotopes or samples with controlled chemical structures.

8.5 Advanced Molecular Characterization by High-resolution MALDI-TOF MS Combined with KMD Analysis

The chemical structures of a simple homopolymer can easily be determined by MALDI-TOF MS, as described earlier. However, most industrial polymers are complex materials such as copolymers, polymer blends, and composites. The variety of chemical structures in the complicated polymer samples would give extremely intricate mass spectra if observed using conventional MALDI-TOF MS, resulting in increased likelihood of isobaric interference.

Peak separation of isobaric components within complicated polymer samples can be achieved using high-resolution MS. Figure 8.7 shows a schematic of the ion optical system of a new type of MALDI-TOF MS instrument with a spiral ion trajectory [19–21], the commercial name of which is JMS-S3000 SpiralTOF™, manufactured by JEOL Ltd. The flight path of the ions along the spiral ion trajectory is ~17 m (2.093 m × 8 turns), which results in a high mass-resolving power of 80 000 at full width at half maximum (FWHM) for a peptide sample with an *m/z* of ~2564. This resolving power should in many cases be sufficient to separate the peaks of isobaric components within copolymers. The advantage of spiral-TOF MS is that high-resolution MS can be readily performed using basically the same procedure as with conventional MALDI-TOF MS instruments.

The power of high-resolution MALDI spiral-TOF MS has been demonstrated for the molecular characterization of a commercial PEO "monostearate," which is a mixture of mono-esterified PEO with stearic acid (the main component) or palmitic acid, di-esterified PEO at both ends with two types of fatty acids, and unreacted PEO [22]. Commercial PEO "monostearate" is commonly used as a raw material for cosmetics such as topical lotions and creams. For the sake of quality and safety of the end product, a combination of high-resolution MS and a KMD analysis has been proposed as a powerful analytical strategy to gain unprecedented insights into the composition of pristine non-ionic surfactants as well as their degradation products under normal use.

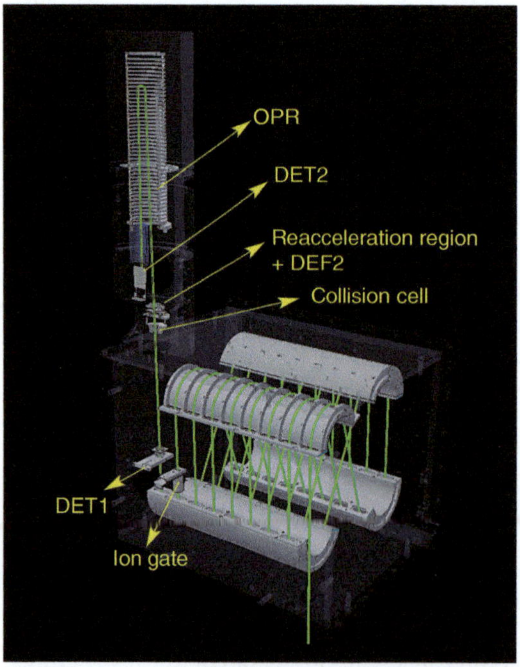

Figure 8.7 Spiral ion trajectory system of JMS-S3000 Spiral TOF. *Source:* Sato [19]/Reproduced with permission from JEOL Ltd.

Figure 8.8a shows the mass spectra of the PEO "monostearate" sample. At first glance, mainly two peak distributions with maxima at *m/z* of ~1700 and ~2000 an be observed in the mass range 1000–3000. This sample was labeled as PEO monostearate, but on the mass spectrum, PEO (denoted as EO_n) without hydrophobic parts was observed to the same extent as PEO monostearate (C_{18}-EO_n), together with PEO monopalmitate (C_{16}-EO_n). The high resolving power of spiral TOF MS reveals components with hydrophobic groups at both ends (denoted as $C_{16/18}$-EO_n-$C_{16/18}$), in which the isobaric peaks of $C_{16/18}$-EO_n-$C_{16/18}$ and isotope of $C_{16/18}$-EO_n are clearly separated. Careful analysis of the high-resolution mass spectra allowed us to identify six types of components. However, it takes a great deal of patience to process the vast collection of peak data observed in high-resolution MS.

Kendrick mass defect analysis is an effective data-processing method that does not rely on peak assignments [23, 24]. It has been proposed to discriminate homologous series that differ only by a number of base units based on the "mass defect." KMD is defined as the

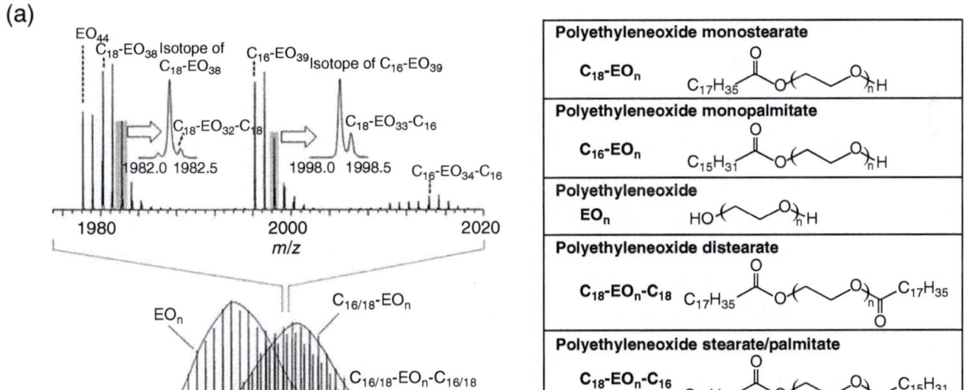

Figure 8.8 Strategy for the compositional characterization of the PEO "monostearate" sample and understanding their decomposition characteristics for a hydrolytic degradation test. (a) High-resolution mass spectrum of the PEO "monostearate" sample and peak assignments. (b) Comparison of KMD plots of the pristine sample and the aged sample after 15 days.

exact Kendrick mass (KM) subtracted from the nominal KM. The KM can be converted from the *m/z* values based on International Union of Pure and Applied Chemistry (IUPAC) mass according to Eq. (8.3),

$$\mathrm{KM} = m/z \times \frac{round\left(R\right)}{R} \tag{8.3}$$

where *R* is the IUPAC mass of a base unit. Here, we apply the IUPAC definition of mass, in which that of ^{12}C is exactly 12. According to this definition, the monoisotopes of the main elements that make up organic compounds are $^{12}C = 12.0000$ u (exactly), $^{1}H = 1.0078$ u, and $^{16}O = 15.9949$ u. In short, components with the same integer mass but different elemental compositions have different masses at the decimal point level. KM is the mass of the base unit, *R*, converted to an integer. As KMD analysis was originally developed in the field of petrochemistry, the methylene unit (-CH_2-) of which IUPAC mass = 14.015 65 has been set as the base unit *R* to convert 14 [25]. KMD values are defined as the differences (i.e. mass defects) between nominal KM (NKM) and KM according to

$$\mathrm{KMD} = \mathrm{NKM} - \mathrm{KM} \tag{8.4}$$

A KMD plot is a two-dimensional bubble chart of NKM versus KMD. For example, when saturated hydrocarbons with different chain lengths are plotted, they are horizontally distributed in the *x*-axis direction at intervals of 14, and components with unsaturated bonds and other functional groups are plotted at positions that deviate from the horizontal plot in the *y*-axis direction. The presence of various structures, such as oxygen-containing compounds and unsaturated hydrocarbon components, can be easily detected simply by inspecting the distribution pattern.

For polymer characterization, *R* sets the IUPAC mass of the repeating unit of a given polymer. An actual KMD plot of the PEO "monostearate" sample converted from the high-resolution mass spectrum (Figure 8.8a) is shown on the left of Figure 8.8b, where $R = 44.0262$ (IUPAC mass of C_2H_4O) was substituted into Eq. (8.3). Congeners with a varying number of EO units line up horizontally, while the components carrying different hydrophobic parts shift along the vertical axis. Different types of components buried in complicated mass spectra clearly appear as separated horizontally aligned lines.

Comparing the KMD plots of the samples before and after degradation testing makes it possible to understand at a glance which components have changed due to the decomposition process. The right-hand part of Figure 8.8b compares the KMD plot of the degradation products recovered after a 15-day hydrolytic degradation test, consisting of immersion in phosphate buffer at pH 7. The composition of the aged surfactants drastically differs from that of the pristine samples. The main degradation products of PEO "monostearate" are shortened and oxidized PEO chains detected at lower mass. Such compositional changes are visually highlighted using the KMD plots from the pristine and aged samples as a reliable way of unraveling the hydrolysis routes. In the case of bacterial biodegradation of OPEO with the "simple" chemical structures described in Section 8.4, the peak assignments of the degradation products could be manually performed. Most industrial polymer materials are, however, a mixture of compounds with various chemical structures, including by-products and impurities. The more complex the molecular structure and composition of the sample, the more effective the KMD analysis.

8.6 Structural Characterization of High-molecular-weight Biocopolyesters by High-resolution MALDI-TOF MS Combined with KMD Analysis

Bacterial polyhydroxyalkanoates (PHAs) are one of the most promising candidates for renewable materials. PHAs are often copolymers and the type of comonomer and copolymer composition of PHA can be controlled by the carbon source fed to the bioreactor [26]. Their physical properties are strongly influenced by the composition and distribution of comonomers, so analysis of the copolymerization composition of PHA is important to be able to understand the characteristics of PHA and for controlling its quality. This chapter describes the usefulness of structural characterization of polymers using MALDI-TOF MS. However, higher molecular weights of bacterial PHAs, which are generally over the hundreds of thousands, represent a major challenge in performing MALDI-TOF MS characterization of intact PHAs [27]. Attempts have therefore been made to gently decompose the polymer chains into sizes that can be analyzed by MS. Partial pyrolysis has been applied for the characterization of bacterial copolyesters [28]. Acid or base catalyzed transesterification is another potential approach to decomposing polyesters into oligomers. Prior to MALDI-TOF MS measurements, hydrochloride in methanol [27] and sodium methoxide in methanol [29] were used as reagents for partial transesterification of PHAs.

An alternative to these processes on the flask scale has also been proposed: simple "on-plate" alkaline degradation of high-molecular-weight polyesters as a sample pre-treatment technique prior to MALDI-TOF MS measurements [30]. The main feature of this method is that sample pretreatment can be performed on a MALDI target plate within a few minutes, where a polymer sample, either by deposition of a polymer solution or by directly attaching a film-like sample, is partially degraded by applying an alkaline solution via simple pipetting. As the resulting oligomer components are always concentrated on the surface layer of the partially degraded sample, it is not necessary to optimize the reaction conditions by trial and error.

Regardless of the decomposition method, the resulting products of oligomerization are a complex mixture of components. Data processing of complicated but informative mass spectra involving end group combinations and copolymer compositions of the degradation products were assisted by a series of advanced KMD analyses to generate visually understandable two-dimensional plots. In this section, the strategy of the combination of three parts, i.e. the on-plate degradation pretreatment, high-resolution MALDI-TOF MS measurements, and advanced KMD analysis as data-processing tools, is demonstrated for the compositional characterization of bacterial poly(3-hydroxybutyric acid-co-3-hydroxyvaleric acid) [P(HB-co-HV)] as an example [30].

The P(HB-co-HV) sample (HV content ~ 12 mol%, $M_n = 1.4 \times 10^5$, $M_w = 3.5 \times 10^5$) was dissolved in tetrahydrofuran (THF) at ~1 mg/ml. About 1 μl of the sample solution was applied to a disposable MALDI target plate to form a polymer thin layer, on which about 1 μl of the alkaline reagent (NaOH in methanol) was spread over the sample layer and left for several minutes until the methanol had evaporated. After removing any excess of alkaline reagent with distilled water by pipetting, the matrix solution was applied on the sample spot. A problem arises here, in that the mass difference of 14 Da between HB and HV units is the same as that caused by esterification of the carboxyl group. To avoid this problem, a stable isotope-labeled reagent, methanol-d_4, was used as the degradation solvent to make a change of 17 Da (+CD_3 -H) for the methyl esters.

The sample processed by on-plate alkaline degradation shows a complicated mass spectrum (Figure 8.9), indicating successful oligomerization. As a result of the scission of ester bonds by transesterification, the observed peaks are a mixture of oligomers with six different types of end group combinations (shown at the bottom of Figure 8.9) and comonomer combinations. Manually analyzing these peaks is not practical, so KMD analysis should be applied.

Mass spectral data were converted to a KMD plot by setting HB as the base unit ($C_4H_6O_2$, $R = 86.0368$) according to Eq. (8.3). Unfortunately, in the resulting KMD plot (Figure 8.10a) all data points cluster in a narrow range along the y-axis, making it difficult to find information about end-group combinations and copolymer composition distributions. The y-axis of the KMD plot roughly reflects the fractional part of the precision mass. For some combinations of end groups, there is no significant difference in the values of the fractional part of the residual mass, resulting in lower resolution in the y-axis direction. The distribution

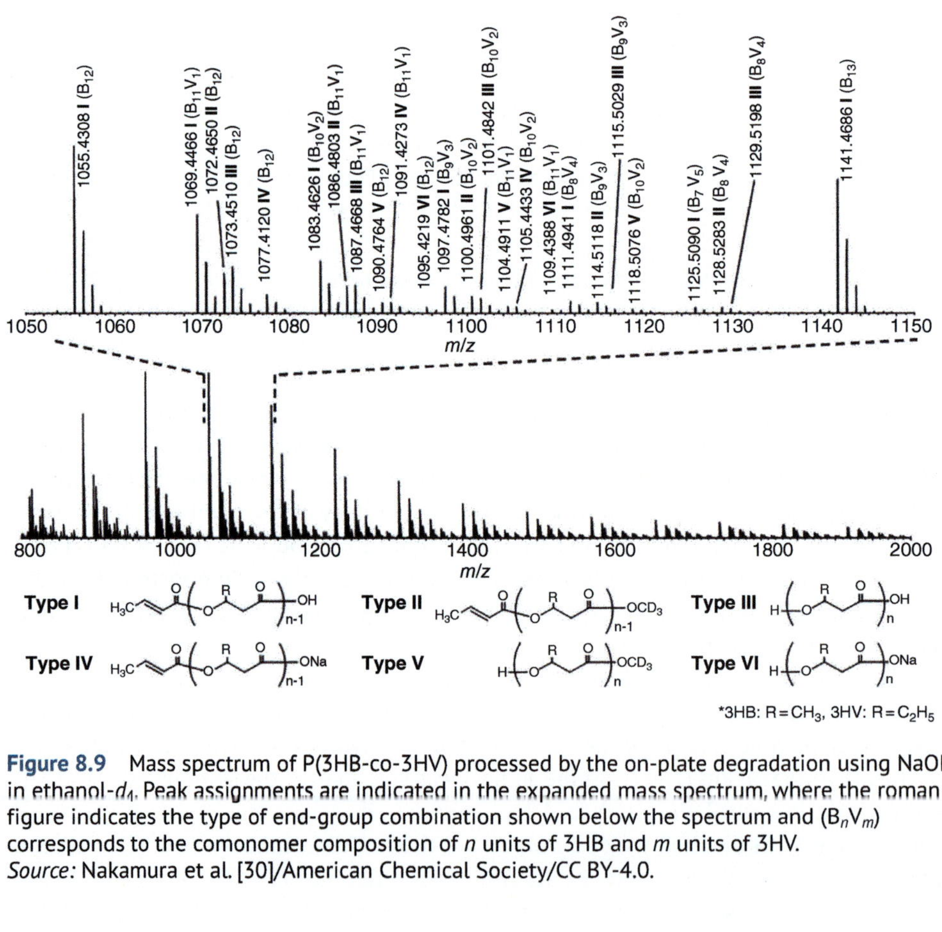

Figure 8.9 Mass spectrum of P(3HB-co-3HV) processed by the on-plate degradation using NaOH in ethanol-d_4. Peak assignments are indicated in the expanded mass spectrum, where the roman figure indicates the type of end-group combination shown below the spectrum and (B_nV_m) corresponds to the comonomer composition of n units of 3HB and m units of 3HV.
Source: Nakamura et al. [30]/American Chemical Society/CC BY-4.0.

in the y-axis direction due to the difference in the terminal group may also overlap with the distribution due to the difference in copolymer composition.

Figure 8.10b is an advanced KMD plot processed by "resolution-enhanced (RE)-KMD, which was proposed to improve the resolution of "regular" KMD plots [31]. In this case, copolymers are separated based on their end groups (types I–VI) to a greater extent than by their comonomer content.

Here, the principle of RE-KMD is briefly explained. RE-KMD differs from regular KMD in that the mass of the base unit is divided by the integer X (the divisor), that is,

$$\mathrm{KM} = m/z \cdot \frac{\mathrm{round}\,(R/X)}{R/X} \tag{8.5}$$

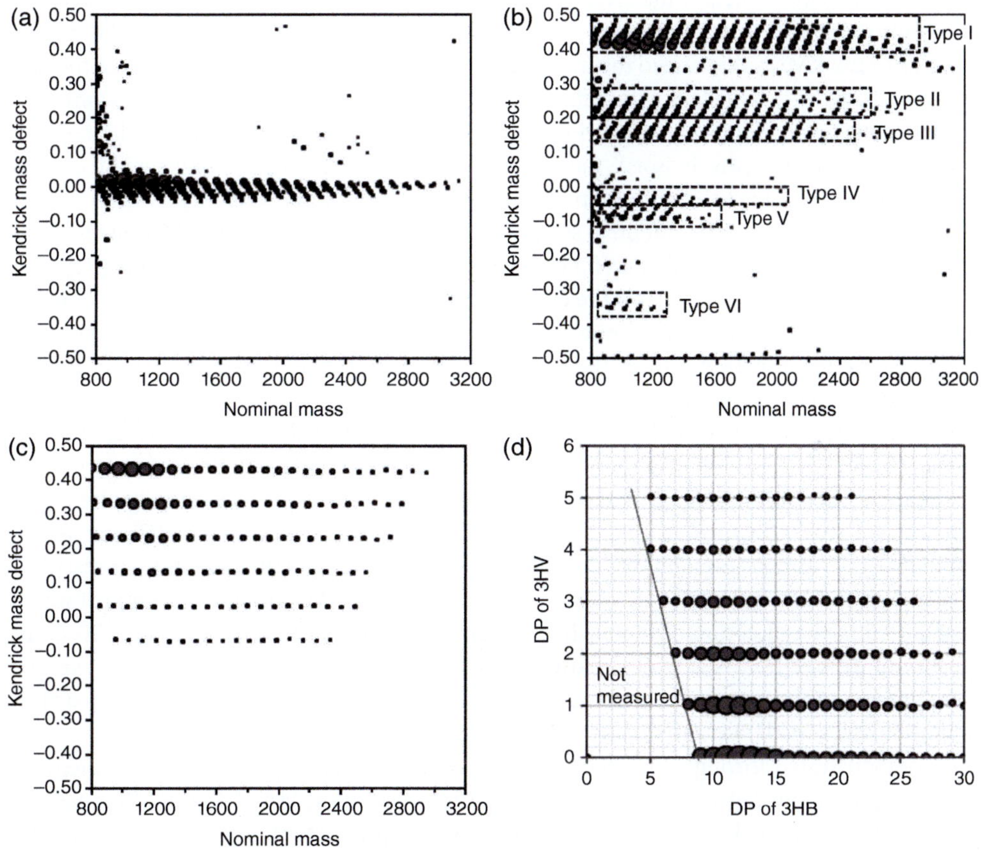

Figure 8.10 KMD plot and DP plot of P(3HB-co-3 HV) processed by the on-plate degradation using NaOH in methanol-d_4. (a) Regular KMD plot (repeat unit: $C_4H_6O_2$, 86.036 23); (b) resolution-enhanced KMD plot (repeat unit: 86.036 23, divisor: X = 92); (c) resolution-enhanced KMD plot for the extracted type I cloud (divisor: X = 62); and (d) DP plot. *Source:* Nakamura et al. [30]/American Chemical Society/CC BY-4.0.

Here, if the divisor X is selected such that $0.5 \leq \text{round}\ (R/X) < 1.5$, then round $(R/X) = 1$, so Eq. (8.5) becomes as follows:

$$\text{KM} = m/z \times \frac{X}{R} = \frac{m/z}{R} X \tag{8.6}$$

The mass of the ion of the polymer is the sum of the main chain, the residue, and the cation, as given in Eq. (8.1). As the KMD values are constant regardless of the difference in degree of polymerization, $n = 0$ may be used. Eq. (8.6) can therefore be transformed as

$$\text{KM} = \frac{\left(M_{\text{residue}} + M_{\text{cation}}\right)}{R} \cdot X \tag{8.7}$$

The fractional part of the sum of the masses of the residue and the cation divided by the monomer mass multiplied by the divisor X will be reflected in the y-direction of the KMD plot. The improvement in resolution with RE-KMD takes advantage of the fact that the rate of increase of the KMD value due to the increase in the divisor X depends on the mass of the end group, and the KMD value falls within the range of ± 0.5 from the definition. In other words, the change in the KMD value when the divisor X is changed is characteristic of a particular residue, so by suitably changing the divisor X, the KMD value can make the best use of the width of ± 0.5 on the y-axis. This allows the resolution of the KMD plot to be improved. Figure 8.10b depicts when $X = 92$ as an optimal situation to show the maximum separation of the differences in the end-group combinations. Guidelines for choosing the best X are described elsewhere [32].

To treat only the information on the copolymerization composition distribution, components with a specific combination of end groups are extracted from the RE-KMD plot using a data extraction function within the data-processing software. Here, the dataset type I, which has higher signal intensities, was extracted.

The extracted dataset was then rearranged by RE-KMD analysis using the alternative divisor $X = 62$ to expand the comonomer distribution (Figure 8.9c). The type I components are neatly placed in the grid, primarily based on the contents of the HV. Finally, the type I distribution was transformed into a degree of polymerization (DP) plot (Figure 8.9d), which is a two-dimensional bubble chart that visually represents the comonomer distribution of HB and HV units (dot size reflects peak intensity). The copolymer composition is distributed in the direction of increasing HV units from the HB homopolymer. This compositional distribution is a typical pattern observed in random copolymers. The average copolymer composition can be estimated from the centroid of the compositional distribution and shows moderate agreement with the value claimed by the supplier.

As demonstrated through the characterization of bacterial copolyester, the strategy of combining on-plate degradation pretreatment, high-resolution MALDI-TOF MS measurements, and advanced KMD analysis as data processing is potentially very useful for the structural characterization of high-molecular-weight PHAs.

References

1 Drzezdzon, J., Jacewicz, D., Sielicka, A., and Chmurzynski, L. (2019). MALDI-MS for polymer characterization – recent developments and future prospects. *TrAC-Trends Anal. Chem.* 115: 121–128.

2 Gruendling, T., Weidner, S., Falkenhagen, J., and Barner-Kowollik, C. (2010). Mass spectrometry in polymer chemistry: a state-of-the-art up-date. *Polym. Chem.* 1: 599–617.

3 Hart-Smith, G. and Barner-Kowollik, C. (2010). Contemporary mass spectrometry and the analysis of synthetic polymers: trends, techniques and untapped potential. *Macromol. Chem. Phys.* 211: 1507–1529.

4 Montaudo, G., Samperi, F., and Montaudo, M.S. (2006). Characterization of synthetic polymers by MALDI-MS. *Prog. Polym. Sci.* 31: 277–357.

5 Soeriyadi, A.H., Whittaker, M.R., Boyer, C., and Davis, T.P. (2013). Soft ionization mass spectroscopy: insights into the polymerization mechanism. *J. Polym. Sci. A-Polym. Chem.* 51: 1475–1505.

6 Weidner, S.M. and Trimpin, S. (2010). Mass spectrometry of synthetic polymers. *Anal. Chem.* 82: 4811–4829.

7 Rizzarelli, P. and Carroccio, S. (2014). Modern mass spectrometry in the characterization and degradation of biodegradable polymers. *Anal. Chim. Acta* 808: 18–43.

8 Sato, H., Tao, H., Ohtani, H., and Aoi, K. (2003a). Characterization of poly(e-caprolactone) by size exclusion chromatography/matrix-assisted laser desorption ionization-mass spectrometry (SEC/MALDI-MS) using MALDI-TOF MS. *Kobunshi Ronbunshu* 60: 305–311.

9 Montaudo, G., Montaudo, M.S., Puglisi, C., and Samperi, F. (1995a). Characterization of polymers by matrix-assisted laser desorption/ionization time-of-flight mass spectrometry: molecular weight estimates in samples of varying polydispersity. *Rapid Commun. Mass Spectrom.* 9: 453–460.

10 Montaudo, G., Garozzo, D., Montaudo, M.S. et al. (1995b). Molecular and structural characterization of polydisperse polymers and copolymers by combining MALDI-TOF mass spectrometry with GPC fractionation. *Macromolecules* 28: 7983–7989.

11 Montaudo, M.S., Puglisi, C., Samperi, F., and Montaudo, G. (1998). Application of size exclusion chromatography matrix-assisted laser desorption/ionization time-of-flight to the determination of molecular masses in polydisperse polymers. *Rapid Commun. Mass Spectrom.* 12: 519–528.

12 Sato, H., Ichieda, N., Tao, H., and Ohtani, H. (2004). Data processing method for the determination of accurate molecular weight distribution of polymers by SEC/MALDI-MS. *Anal. Sci.* 20: 1289–1294.

13 Sato, H., Kiyono, Y., Ohtani, H. et al. (2003b). Evaluation of biodegradation behavior of poly(epsilon-caprolactone) with controlled terminal structure by pyrolysis-gas chromatography and matrix-assisted laser desorption/ionization mass spectrometry. *J. Anal. Appl. Pyrolysis* 68-69: 37–49.

14 Sato, H., Shibata, A., Wang, Y. et al. (2001). Characterization of biodegradation intermediates of non-ionic surfactants by matrix-assisted laser desorption/ionization-mass spectrometry 1. Bacterial biodegradation of octylphenol polyethoxylate under aerobic conditions. *Polym. Degrad. Stab.* 74: 69–75.

15 Sato, H., Shibata, A., Wang, Y. et al. (2003c). Characterization of biodegradation intermediates of nonionic surfactants by MALDI-MS. 2. Oxidative biodegradation profiles of uniform octylphenol polyethoxylate in ^{18}O-labeled water. *Biomacromolecules* 4: 46–51.

16 Nishio, E., Ichiki, Y., Tamura, H. et al. (2002). Isolation of bacterial strains that produce the endocrine disruptor, octylphenol diethoxylates, in paddy fields. *Biosci. Biotechnol. Biochem.* 66: 1792–1798.

17 Shibata, A., Sato, H., Wang, Y. et al. (2003). Development of a novel bacterial biodegradation test method of non-ionic surfactants using ^{18}O-labelled water. *J. Mass Spectrom. Soc. Jpn.* 51: 256–259.

18 Sato, H., Shitbata, A., Yoshikawa, H., and Tamura, H. (2003d). Biodegradation mechanisms of non-ionic surfactants evaluated by MALDI-MS. *J. Mass Spectrom. Jpn.* 51: 415–420.

19 Satoh, T. (2010). Development of JMS-S3000: MALDI-TOF/TOF utilizating a spiral ion trajectory. *JEOL News* 45: 34–37.

20 Satoh, T., Tsuno, H., Iwanaga, M., and Kammei, Y. (2005). The design and characteristic features of a new time-of-flight mass spectrometer with a spiral ion trajectory. *J. Am. Soc. Mass. Spectrom.* 16: 1969–1975.

21 Satoh, T., Sato, T., and Tamura, J. (2007). Development of a high-performance MALDI-TOF mass spectrometer utilizing a spiral ion trajectory. *J. Am. Soc. Mass. Spectrom.* 18: 1318–1323.

22 Fouquet, T., Shimada, H., Maeno, K. et al. (2017). High-resolution Kendrick mass defect analysis of poly(ethylene oxide)-based non-ionic surfactants and their degradation products. *J. Oleo Sci.* 66: 1061–1072.

23 Fouquet, T.N.J. (2019). The Kendrick analysis for polymer mass spectrometry. *J. Mass Spectrom.* 54: 933–947.

24 Sato, H., Nakamura, S., Teramoto, K., and Sato, T. (2014). Structural characterization of polymers by MALDI spiral-TOF mass spectrometry combined with Kendrick mass defect analysis. *J. Am. Soc. Mass. Spectrom.* 25: 1346–1355.

25 Hughey, C.A., Hendrickson, C.L., Rodgers, R.P. et al. (2001). Kendrick mass defect spectrum: a compact visual analysis for ultrahigh-resolution broadband mass spectra. *Anal. Chem.* 73: 4676–4681.

26 Doi, Y., Kunioka, M., Tamaki, A. et al. (1988). Nuclear magnetic-resonance studies on bacterial copolyesters of 3-hydroxybutyric acid and 3-hydroxyvaleric acid. *Macromol. Chem. Phys.* 189: 1077–1086.

27 Abate, R., Ballistreri, A., Montaudo, G. et al. (1993). Quantitative applications of matrix-assisted laser-desorption ionization with time-of-flight mass-spectrometry – determination of copolymer composition in bacterial copolyesters. *Rapid Commun. Mass Spectrom.* 7: 1033–1036.

28 Impallomeni, G., Ballistreri, A., Carnemolla, G.M. et al. (2015). Matrix-assisted laser desorption/ionization time-of-flight vs. fast-atom bombardment and electrospray ionization mass spectrometry in the structural characterization of bacterial poly(3-hydroxyalkanoates). *Rapid Commun. Mass Spectrom.* 29: 811–820.

29 Saeed, K.A., Ayorinde, F.O., Eribo, B.E. et al. (1999). Characterization of partially transesterified poly(beta-hydroxyalkanoate)s using matrix-assisted laser desorption/ionization time-of-flight mass spectrometry. *Rapid Commun. Mass Spectrom.* 13: 1951–1957.

30 Nakamura, S., Fouquet, T., and Sato, H. (2019a). Molecular characterization of high molecular weight polyesters by matrix-assisted laser desorption/ionization high-resolution time-of-flight mass spectrometry combined with on-plate alkaline degradation and mass defect analysis. *J. Am. Soc. Mass. Spectrom.* 30: 355–367.

31 Fouquet, T. and Sato, H. (2017). Extension of the Kendrick mass defect analysis of homopolymers to low resolution and high mass range mass spectra using fractional base units. *Anal. Chem.* 89: 2682–2686.

32 Nakamura, S., Cody, R.B., and Sato, H. (2019b). Graphical ranking of divisors to get the most out of a resolution-enhanced Kendrick mass defect plot. *Anal. Chem.* 91: 2004–2012.

9

Phytoconstituents and Antimicrobiological Activity

Philip L. Poole and Giulia T.M. Getti

School of Science, Faculty of Engineering and Science, University of Greenwich, Kent, UK

9.1 Introduction to Phytochemicals

As new strains of bacteria and virus emerge with resistance to current pharmaceutical drugs as well as parasitic infections, new treatments are needed. Phytoconstituents can be broadly divided into two groups: phytonutrients and phytochemicals. Although both of these groups are important for maintaining good health and the microbiota in this chapter, we will concentrate on phytochemicals that may be used to treat bacterial and parasite infections and how various mass spectroscopy (MS) techniques can give insights into the mechanism of action.

Phytochemicals are, by definition, extracted from tree barks, leaves, bulbs, and cloves. Figure 9.1 places phytochemicals into six common groups with different chemical properties arising from side-chains or reactive groups that are attached to a similar parent chemical group. Many phytochemicals are found to have clinical use in treating medical and physiology disorders, because of their antibacterial and antifungal properties, their ability to treat blood pressure (hypertension), cardiovascular diseases, diabetes, arthritis, and abnormal cellular function, their function as an anticoagulation agent, and their benefit in treating anti-inflammatory hypocholesterolaemia (reduces cholesterol) activity, and cancers. Most show anti-oxidation and anti-inflammatory activities. Further information on phytochemicals can be found in the recent publication by Pal and Nayak [1].

The polyphenols, including the flavonoids, are among the largest groups. Polyphenols that have received promising clinical attention include cicerfuran, a flavonoid, and the stilbenes which have been shown to have antimicrobial (bacterial and parasite) and antifungal activity [2–5]. Similarly allicin (diallythiosulphinate) from garlic (*Allium sativum*) is a volatile organic sulphur compound that has gained much attention for its antibacterial, antiviral, antifungal, and anti-parasite properties [6, 7]. Phytochemicals such as taxol, a cell anti-microtubule agent originally derived from the Pacific yew tree bark, is now a common

Microbiological Identification using MALDI-TOF and Tandem Mass Spectrometry: Industrial and Environmental Applications, First edition. Edited by Haroun N. Shah, Saheer E. Gharbia, Ajit J. Shah, Erika Y. Tranfield, and K. Clive Thompson.

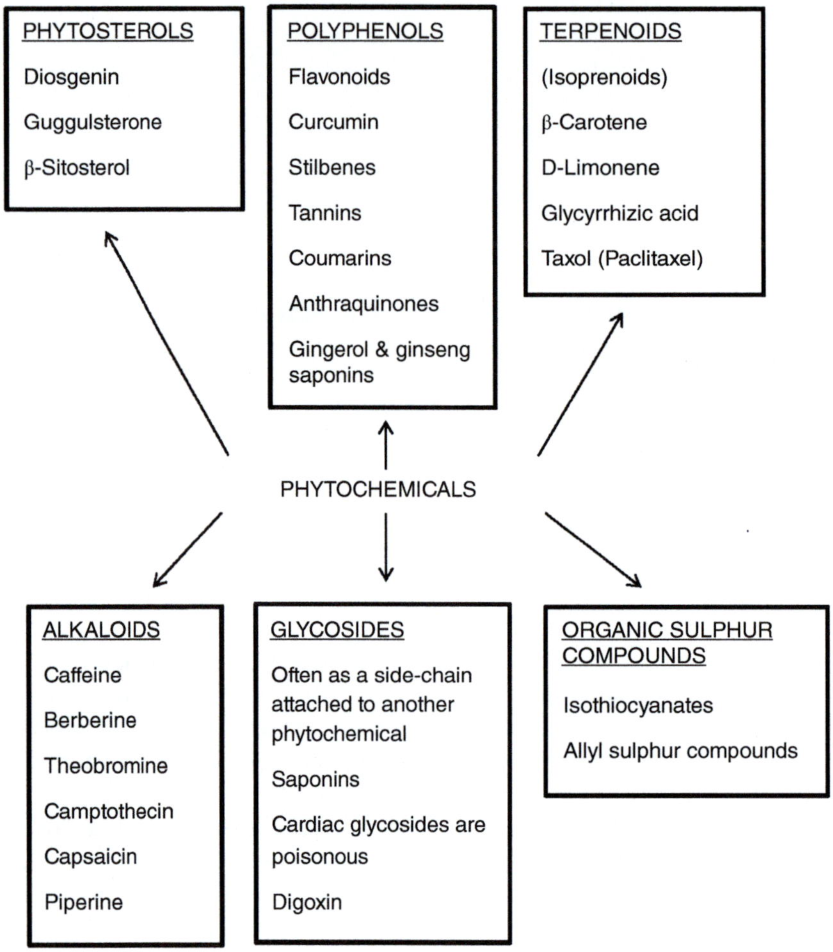

Figure 9.1 Phytochemicals.

treatment for a range of cancers. Recently, monoclonal antibodies conjugated with exatecan or deruxtecan, which are structural analogues or derivatives of camptothecin (CPT) (a quinoline alkaloid derived from the bark of *Camptotheca acuminate* that is a topoisomerase 1 inhibitor), are finding success in cancer treatments. Further CPT derivatives are being investigated as potential pharmaceuticals.

So why do some phytochemicals have these medicinal properties? This is where MS can play its part. But first a phytochemical must be isolated and tested for stability and purity. An initial extraction is generally carried out in an organic solvent such as an alcohol; however, aqueous extraction is preferred in some cases, an example being that of allicin, which is extracted in an aqueous medium [8]. Further purification is generally by high-performance liquid chromatography but care must be taken to ensure stability in the solvents employed. As an example, allicin in aqueous solution at 23°C has a biological half-life of 6.5 days while in 20% ethanol and 70% ethanol it has half-lives of 17.6 and 8.7 days, respectively [9].

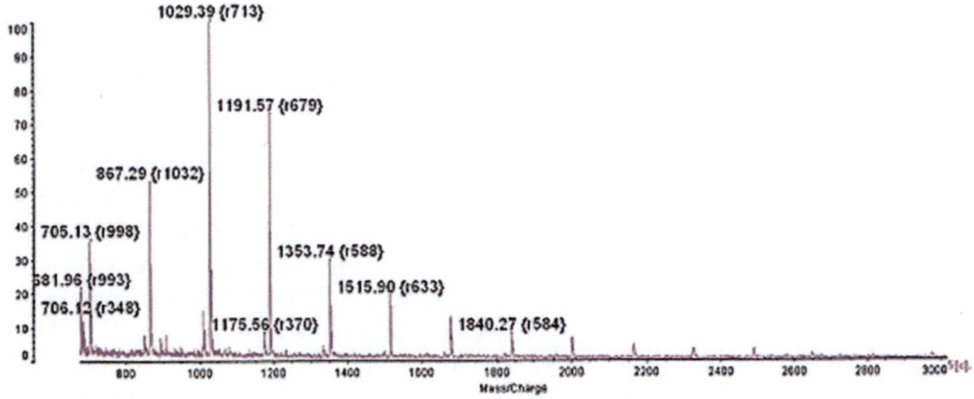

Figure 9.2 MALDI spectrum of pure allicin. *Source:* Getti and Poole [10]/Microbiology Society.

Testing for purity is essential as their isolation is fairly crude. For pure allicin (from Allicin International Ltd). Matrix-assisted laser desorption/ionization time-of-flight (MALDI-TOF) MS was employed by drying down an aqueous solution containing 2.4 μg (15 nmol) of allicin into matrix-prepared wells [14% α-cyano-4-hydroxycinnamic acid (CHCA) in 70% acetonitrile +0.015% trifluoroacetic acid]. Further matrix solution was now placed onto the dried sample and allowed to dry down. The samples were now placed into an AXIMA (Shimadzu Scientific Instruments, Inc.) CRF+ mass spectrometer equipped with a N_2 laser (λ337 nm) and configured for positive mode. The laser power was set to 65 kV. Calibration was achieved with standard peptides (Sigma, UK). Figure 9.2 shows the resultant MALDI-TOF allicin spectrum.

Allicin has a literature value of 162.02 Da, but a mass peak was not found in the mass spectrum at this m/z. However, several peaks were found in the m/z range 700–2500, with mean separations of m/z 162.17 ± 0.009, suggesting that allicin has formed highly concentrated aggregates on drying down. A close examination of Figure 9.2 revealed some satellite peaks with a smaller mass of 16 m/z in front of the allicin aggregation peaks. These may be assigned as diallyl disulphide (which has one less oxygen) that has co-precipitated with the allicin aggregates and is a small contaminant in the allicin purification and demonstrates that even tiny amounts of contaminants may be detected with MALDI-TOF using dried-down samples.

9.2 An Application to Bacteriology

The ability of MALDI-TOF to detect a variety of proteins and cellular components simultaneously makes it a very effective and widespread tool for identification of bacteria and fungi, as well for antimicrobial susceptibility testing [11–17]. However, MALDI-TOF can also be an important tool for determining the mechanism of pharmaceutical inhibition of bacterial protein synthesis, inhibition of cell wall synthesis, direct inhibition of enzymatic activity, alteration of cell membrane permeability, and blockade of specific biochemical pathways.

9.2.1 Allicin Leads to a Breakdown of the Cell Wall of *Staphylococcus aureus*

Cutler and Wilson [8] previously showed allicin to be active against *S. aureus* strains including methicillin-resistant *S. aureus*. To understand this effect allicin has on *S. aureus* we grew two strains of *S. aureus* in separate liquid cultures for 24 hours: Oxford (NCTC 6571), a reference strain, and a clinical isolate strain donated from Royal London and St Bartholomew's Hospital (RLSB). Cells were then treated to 63 μg/mL (final concentration) of allicin for a further 24 hours and then prepared for MALDI analysis as described previously [10]. The mass spectrometer was set up as in Section 9.1 and up to 100 laser shots were collected around each well and averaged with the AXIMA Smooth Gauss program. MALDI-TOF spectra were recorded in the *m/z* range 1000–12 000.

Results showed a clear difference between allicin-treated and non-treated *S. aureus* in both strains (Figure 9.3) and being most evident in the *m/z* region 1000–2000. In the

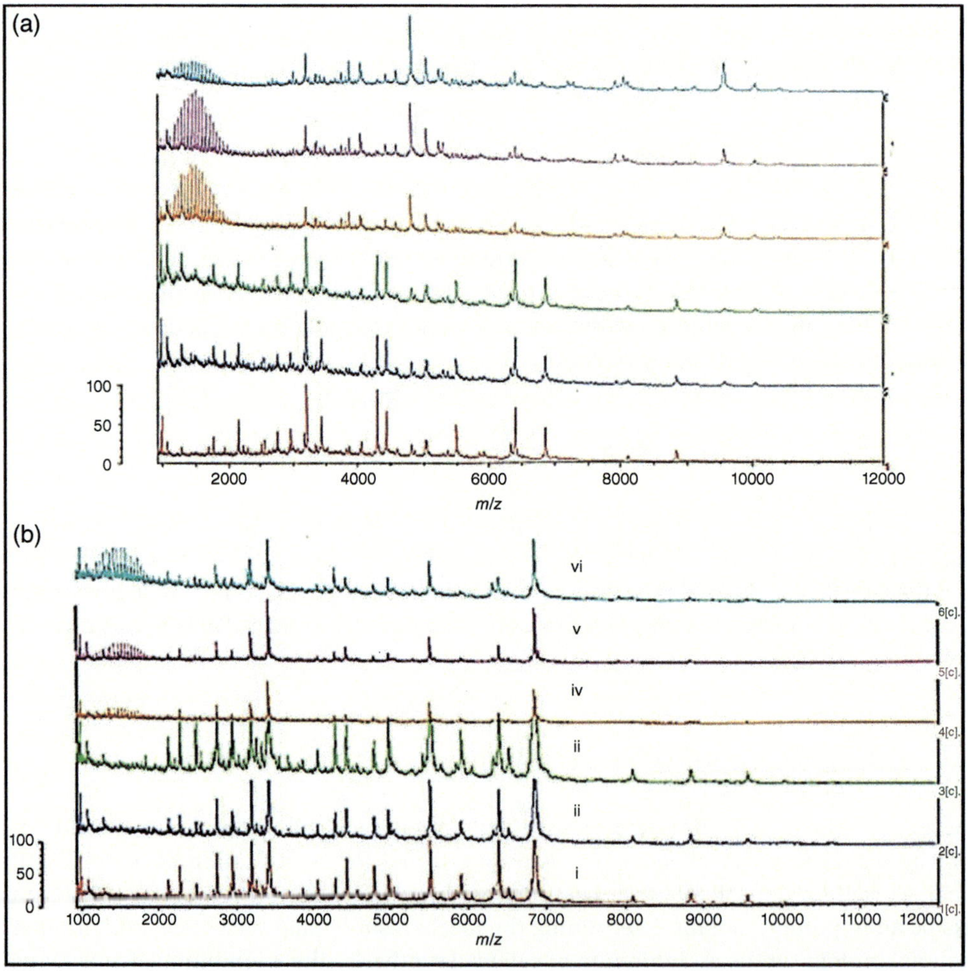

Figure 9.3 MALDI mass spectra recorded as described in the text: (a) *S. aureus* Oxford strain; (b) the RLSB strain. In both of the mass spectra, the top three spectra show those cells that have been treated with 63 μg/mL of allicin for 24 hours while the bottom three spectra are the untreated controls. *Source:* Getti and Poole [10]/Microbiology Society.

untreated cells, seven peaks in the m/z region 2200–3400 and the two peaks in m/z 4300–4500 as well as the single peaks around m/z 5500, 6400, and 6800 are seen in both strains. However, following allicin treatment, both strains showed a significant reduction of the two peaks in the m/z region 4300–4500 and a reduction in the peak at m/z 6400. In the Oxford strain the peaks at m/z 5500 and 6800 were reduced and a new strong peak at m/z 4750 appeared. There were also changes at other mass regions such as the seven peaks at m/z 2200–3400.

Returning to the mass region m/z 1100–2000 for both allicin-treated strains, there was a total of 18 major peaks of an average spacing of ~44 m/z (Figure 9.4) with each major peak followed by a higher mass (up by 15–16 m/z) satellite peak. This peak pattern has characteristics of a sequence of substituted oligosaccharide [18] and as the cells were prepared in saline, the major peaks would be sodium adducts (+23) while the smaller satellite peaks would be potassium adducts (+39). Moreover, the substitutions are commonly those of O-acetylation (OAc) at position C6 on MurNAc applied by the key cytoplasmic membrane enzyme O-acetyltransferase OatA [19] which gives protection against "invading" glycosidase. As these peak patterns are not found in the untreated cells, we may exclude any extraneous attachment of agents arising from the culture medium or from a matrix-induced fragmentation process or by pulse source decay (PSD). Further, as the cells have undergone extensive washing this suggests that these oligosaccharides must remain buried under or in the cell wall during growth and are only released when undergoing MALDI-TOF preparation.

Under normal circumstances, the peptidoglycan layers in the cell wall undergo digestion by hydrolases during cell growth as well as autolysis [20]. To finally confirm if the peptidoglycan fragments may have arisen from peptidoglycan-processing enzymes, it was necessary to examine enzyme specificities as well as their products and susceptibility to allicin [10]. If differences between the calculated and measured mass in some of the

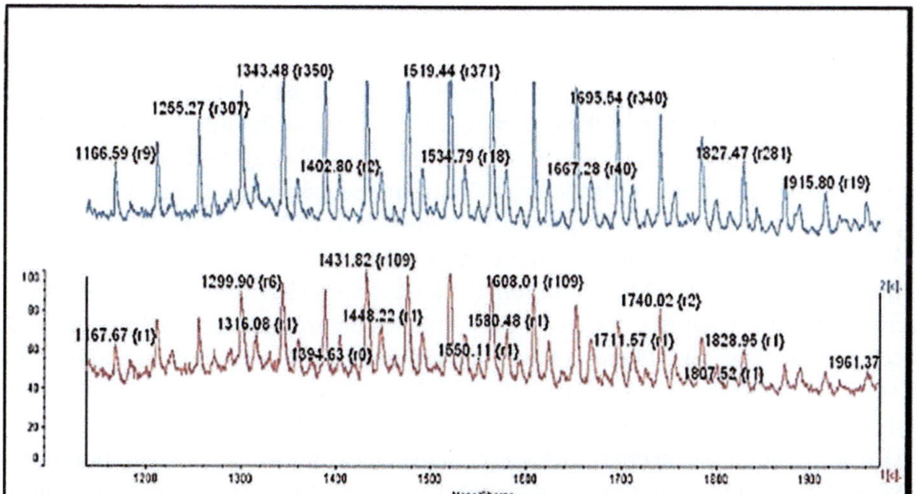

Figure 9.4 MALDI mass spectra. Expanded region 1100–2000 m/z of the treated two strains: the top trace is Oxford and the bottom trace is the RLSB cell line. The values of {r} are a measure of the mass resolution. *Source:* Getti and Poole [10]/Microbiology Society.

assignments are considerably different from expected, this might be attributed to underlining peaks, drifting of the standardization, or baseline variation. We therefore disregarded an assignment with differences $> \pm 1.5\, m/z$ (average relative error of 0.10%) if found in both of the *S. aureus* strains.

One of the key enzymes is a peptidoglycan aminohydrolase, that of N-acetylmuramoyl-L-alanine amidase. This enzyme cleaves the bonds between that of N-acetylmuramoyl→L-alanine and is located on the outer side of the cytoplasmic membrane. Given the vast amount of information of allicin targeting cysteine thiol groups [21, 22], we undertook a search of the NCBI protein sequence data bank (http://www.ncbi.nlm.nih.gov) with the BlastP alignment program. No cysteines likely to interfere with this enzyme activity were found.

Hydrolysis further occurs with endo-β-N-acetylglucosaminidase [23]. This enzyme will hydrolyse bonds between N-acetylglucosamine→N-acetylmuramic acid. We thus calculated peaks in the m/z range 1100–2000 using the sequence:

$$\text{MurNAc}\left(\text{GlcNAc MurNAc}\right)n\ \text{GlcNAc}\left(\text{OAc}\right)y + 23$$

where n is number of disaccharides, y is total number of (OAc) groups and 23 for Na^+.

Mass fragments that may be attributed to this enzyme in the m/z range 1475–1741 with the number of disaccharides (n) at 2 were identified (Table 9.1). As fragments were continuously increasing in O-acetylation, we referred to this as epi-acetylation and this may be explained by the increased spatial flexibility of the released fragments. A BlastP alignment of this enzyme revealed no cysteine residues found in any of 71 matched enzymes entries.

The exolytic transglycosylases, such as the soluble lytic murein transglycosylases, hydrolyze β-1,4 glycosidic bonds between N-acetylmuramic acid→N-acetylglucosamine. This enzyme cleaves off an N-acetylglucosamine and generates an anhMurNAc terminal glycan that is commonly found at the reducing end in peptidoglycans. As the lytic transglycosylases are inhibited by O-acetylation, anhMurNAc(OAc) is not found [24]. Given the mass range of m/z 1100–2000, we thus calculated peaks using:

$$\left(\text{MurNAc GlcNAc}\right)\text{n}\ \left(\text{OAc}\right)\text{y anhMurNAc} + 23$$

Five corresponding masses (Table 9.1) that might be attributed to the enzyme products were found in our mass spectra in the m/z range1255–1432 with the number of disaccharides (n) again at 2. Unfortunately no *S. aureus* soluble lytic transglycosylases (SLTs) have been sequenced. However, other Firmicutes SLT enzyme sequences or domains have been sequenced and, after an examination of 14 SLTs in the UniProtKB (http://ExPASy.org) data bank, no cysteines were found in any of these SLT's [10].

According to Vollmer et al. [20], the minimal oligosaccharide length required for hydrolysis decreases rapidly when less than five or six. This would explain the lengths of cleaved oligosaccharides that are found here. A lower mass value at m/z 1167.59 may be assigned as $(\text{MurNAc})_5(\text{GlcNAc})_4(\text{OAc})_2(\text{Na})_2$ and with increasing values of m/z by two additional O acetylations at each stage. However, the addition of O-acetylation can only continue until all C6 positions on the saccharides are acetylated, then only three other addition steps are possible. Although we cannot rule this out for m/z 1167.59–1299.59, it would suggest that there would be continuous acetylation beyond these values, and with all the oligosaccharides assigned as disodium adducts of the same length, this does not seem likely.

Table 9.1 Best fit of glycan fragments.

	Calculated *m/z*	Oxford *m/z*	RLSB *m/z*	% errors
$(MurNAcGlcNAc)_2$ anhMurNAc	1255.27*	1255.27	1255.93	0.000, −0.052
$(MurNAcGlcNAc)_2$ (OAc) anhMurNAc	1299.29	1299.30	1299.90	−0.001, −0.047
$(MurNAcGlcNAc)_2$ $(OAc)_2$ anhMurNAc	1343.39	1343.48	1343.80	−0.007, −0.031
$(MurNAcGlcNAc)_2$ $(OAc)_3$ anhMurNAc	1387.33	1387.33	1387.84	0.000, −0.037
$(MurNAcGlcNAc)_2$ $(OAc)_4$ anhMurNAc	1431.35	1431.33	1431.82	0.001, −0.033
$MurNAc(GlcNAcMurNAc)_2GlcNAc$	1476.53	1475.33	1475.84	0.081, 0.047
$MurNAc(GlcNAcMurNAc)_2GlcNAc$ (OAc)	1520.55	1519.44	1519.91	0.073, 0.042
$MurNAc(GlcNAcMurNAc)_2GlcNAc$ $(OAc)_2$	1564.57	1563.46	1563.93	0.071, 0.041
$MurNAc(GlcNAcMurNAc)_2GlcNAc$ $(OAc)_3$	1608.59	1607.59	1608.01	0.062, 0.036
Epi-acetylation fragments				
$MurNAc(GlcNAcMurNAc)_2GlcNAc$ $(OAc)_4$	1652.61	1651.72	1652.09	0.057, 0.031
$MurNAc(GlcNAcMurNAc)_2GlcNAc$ $(OAc)_5$	1696.63	1695.54	1696.06	0.064, 0.033
$MurNAc(GlcNAcMurNAc)_2GlcNAc$ $(OAc)_6$	1740.65	1739.36	1740.02	0.074, 0.036

Major peaks (sodium adducts).
Masses (in Da) are taken as follows: GlcNAc 203.24 ($-H_2O$), MurNAc 293.27 ($+H_2O$), anhMurNAc 275.25 ($+H_2O$), GlcN 179.17 ($+H_2O$), OAc 42.02 ($-H_2O$), i.e. $MurNAc(GlcNAcMurNAc)_2GlcNAc$:
$293.27 - 18 + (203.24 + 293.27) - 18 + (203.24 + 293.27) + 203.24 + 23 = 1476.53\,m/z$.
Calculated mass increased by m/z 2.0 for carbonyl reduction (O-acetyl only) (refer to [10]). Fragments marked with an asterisk (*) are not increased by m/z 2.0; these may act as internal standards.
Correlation coefficients of 0.999 are recorded for the difference between both sets of the experimental results and for the calculated values versus Oxford strain and the calculated values versus RLSB 102 strain experimental results. The mean and standard deviation (SD) are calculated from the differences between the two sets of experimental results, giving a value of m/z 0.50 + 0.10, and the differences between the calculated values and the experimental results for the Oxford strain, giving a value of m/z 0.65 + 0.54, and for the RLSB 102 strain, giving a value of m/z 0.58 + 0.08.
Source: Adapted from Getti and Poole [10].

However, it does not alter the interpretation of these results, but one should be aware of ambiguities in assigning exact structures, particularly when dealing with oligosaccharides, but MS/MS fragmentation would aid an exact assignment.

Due of its small size (~162 Da), allicin can easily pass though the peptidoglycan layer and through cell membranes [25] but allicin is unlikely to inhibit the cysteine-free hydrolases, as discussed earlier. However, cell wall synthesis enzymes are some of the first to come into contact with allicin while the cells are still in the culture medium. The peptidoglycan synthesis pathway in *S. aureus* has been well documented [26]. Briefly the enzyme MraY adds Phospho-N-acetylmuramoyl-pentapeptide to a phosphate attached to a lipid membrane component (the undecaprenyl phosphate also known as bactoprenol) to form lipid I with a pentapeptide stem. MurG then adds UMP-GlcNAc to MurNAc of lipid I to form lipid II. This is then followed by the Fem group of enzymes which commences the formation of a pentaglycine bridge by adding five glycines to L-lysine in the stem peptide. After translocation to the outer membrane by a lipid II 'flippase', transglycosylases/transpeptidase adds further MurNAcs and GlcNAcs to form a polysaccharide chain and completing the pentaglycine cross-link. All of these enzymes are free of cysteines or have a cysteine buried within a transmembrane helix (and probably less accessible) with the exception of the Fem enzymes.

The building of the pentaglycyl bridge starts with FemX adding the first glycine to the lipid II intermediate, followed by FemA adding two glycines, followed by FemB adding the final two glycines. The crystal structure of FemA from *S. aureus* has been determined [27]. It was shown to contain two cysteine residues at position 56 and 102 and found in >100 matching FemA sequences. However, these two cysteines are ~12.6 Å apart and are unlikely to form a disulphide bridge, thus both remaining as free thiol groups. Further examination of the crystal structure revealed that cysteine 102 is accessible through a cavity at the protein surface with this cysteine at the floor of this cavity. As allicin is approximately the size of an apolar amino acid, on folding through hydrophobic interaction of the two allyl chains, it could potentially enter or partially enter this cavity and cause a disruption to the protein conformation. Finally, Fem B and Fem X do show some sequence similarity to Fem A, but again the two cysteines in Fem B and the one in Fem X, though not in the same location as in FemA, would remain as free thiol groups.

Sharif et al. [28] reported peptidoglycan fragmentation in *S. aureus* with Fem-deleted mutants containing truncated pentaglycine bridging. Specifically, with FemX-deleted mutants, glycine bridging was totally absent, whereas in FemA-deleted mutants they found that for those with a single bridging glycine, cross-linking was decreased by ~20%, leading to a high concentration of open N-terminus monoglycyl segments. Kim et al. [29] reported the architecture of peptidoglycans in wild-type *S. aureus* was that of peptidoglycan chains forming a vertical scaffold-type model [30], with an average number of disaccharide units of only six. This is in agreement with our findings of oligosaccharides of lengths of five and six saccharides which, when joined together, will have formed peptidoglycan chains of six disaccharides or with a shortened length of one saccharide by the enzymatic removal of a terminal GlcNAc and forming an anhMurNAc terminal end.

Kim et al. concluded that there would be stem peptide "crowding" when forming short bridges and presumably open gap segments between those stem peptides when a bridge is not formed. Applying this same reasoning to the proposed inactivation of FemA by allicin and the possible inactivation of other Fem enzymes, we might therefore expect to find open gap segments, with the weakened cell wall breaking down under turgor pressure and releasing oligosaccharides as identified in our MALDI spectrum.

9.3 Applications to Parasitology

The ability of MALDI-TOF to detect a variety of proteins and cellular components simultaneously makes it a very effective and widespread tool for identification of bacteria and fungi, as well for antimicrobial susceptibility testing [12, 14, 17]. Recently, MALDI-TOF has shown potential for identification of protozoa parasites such as *Leishmania*, *Giardia*, *Plasmodium*, *Cryptosporidium* and *Entamoeba* isolates [31, 32]. Trypanosomatidis identification was also achieved by coupling MALDI-TOF with molecular techniques [33]. Current standards of identification of parasites require labour-intensive microscopic analysis or expensive-to-run molecular methods. MALDI-TOF offers a fast and simple method that requires low-cost consumables. The ability to identify parasites at species level, on top of the previously reported advantages, makes it a suitable identification method in clinical settings, despite the high initial investment [34–37]. The availability of a reference mass spectral library (MSL) for *Leishmania* isolates, accessible through a free Web-based application [mass spectral identification (MSI)] since 2017 has allowed for fast and reliable identification at least to the complex level [38], with examples of successful identification of species, strains, and even of *Leishmania* hybrids [36, 39]. Similarly, MALDI-TOF has shown promise in fast and reliable identification of *Plasmodium* parasite infections of *Anopheles* mosquitoes. The analysis of mosquitoes' cephalothoraxes (head and thorax) allowed the correct identification of 79 out of 80 mosquitoes [31]. Moreover, the success of MALDI-TOF is not limited to the identification of protozoa. This approach has been developed over the past 10 years to support identification of invertebrate pests as well as parasites vectors [40].

9.3.1 Drug Discovery

MALDI-TOF has been used to identify and characterize parasites that developed resistance to mainstream treatments [41–44] which are becoming more prevalent, requiring identification of alternative active ingredients. Phytochemical studies on parasites have underlined that natural products, from both terrestrial and aquatic environments, show promise as novel therapeutics against protozoa infection ([45, 46] [47, 48]). Identification of active products can be based on the detection of activity against a selected drug target or testing against viable microorganisms [49, 50]. Because of their complex life cycle, biological assays have been developed to identify parasites' sensitivity to natural products at various stages of the *Leishmania* life cycle [3, 51–53]. Proteomics analysis has had many successful applications in drug discovery which can be further investigated to support the study and identification of anti-parasitic natural products [54–57]. For example, MALDI-TOF has been used to identify active elements from natural sources [58, 59]. Moreover, it can reliably identify their biological distribution within simple and complex organisms, such as cyanobacteria, marine sponges, and plants [60–62]. This is particularly useful during extraction, as it offers a fast and reliable method for identification of fractions that might contain active ingredients [63].

The use of flavonoids, terpenoides, alkaloids, and glycosides combined with nanotechnology to improve efficacy has great potential for the clinical treatment of leishmaniasis and malaria, as recently reviewed [64, 65]. In general, polyphenols, such as the flavonoids, have been very successful against parasites [66–70]. Polyphenols, phenylethanoid glycosides, and others have shown several biological activities that make them relevant therapeutic options in a variety of diseases with anti-inflammatory, antioxidant, cytotoxic, antibacterial, antivirus, immunomodulatory, enzyme inhibition, and pharmacokinetic properties [71]. Allicin remains of particular interest within the wide range of natural products due to its broad spectrum of antimicrobial properties, including antibacterial, antiparasitic, antiviral. and antifungal [72].

9.3.2 Parasite Characterization

Protozoa parasites undergo significant changes during their life cycle both from a biochemical and a structural point of view. Their ability to change their gene expression and, consequently, their proteome ensures their survival between the very different environments in which they proliferate. For example, *Leishmania* promastigotes survive extracellularly within the digestive system of phlebotomine sandflies as well as binding and entering mammalian cells during infection. Once inside macrophages, parasites need to survive the host cell's oxidative and proteolytic action. Proteomics investigations of different parasite stages identified factors involved in survival and infectivity; for example, proteins involved in resistance to oxidative stress are more abundant at the last stages of growth and differentiation [73–79]. Such insight has supported the identification of potential drug targets as well as a deeper understanding of the link between infective species/strains and disease development [73, 74, 76, 78, 80]. Further insights have been obtained by looking at secreted proteins, which were identified as involved in immune modulation, signal transduction, and early infection [81–83].

9.4 A Proteomic Approach: *Leishmania* Invasion of Macrophages

Proteomics analysis has had many successful applications in drug discovery, which can be further used to support the study and identification of anti-parasitic natural and phytochemical products. Comparative proteomics has proved a powerful tool in providing a better understanding of the changes that take place during protozoa infection of host cells. For example, the use of two-dimensional proteomic analysis in the study of *Leishmania major* and *Leishmania amazonensis* invasion of murine macrophages has revealed changes in the expression of over 160 proteins [84]. Recently a label-free proteomic analysis recorded significant changes in protein expression following infection with *L. major, L. infantum*, and *L. amazonensis* both *in vitro* and *in vivo*, indicating that *Leishmania* infection increased lipid accumulation and apoptosis [85–87]

Similarly, a two-dimensional difference gel electrophoresis-peptide mass fingerprinting (2D-DIGE-PMF) comparative proteomics approach revealed differential expression of over 100 proteins following 24 hours' infection of terminally differentiated THP-1 cells with *L. aethiopica, L. tropica*, and *L. major* [88]. Differential expression was clearly visible following scanning of 2D-DIGE gels containing protein samples extracted from infected THP-1 cells (Cy5-labeled), uninfected control (Cy3-labeled), and internal standard

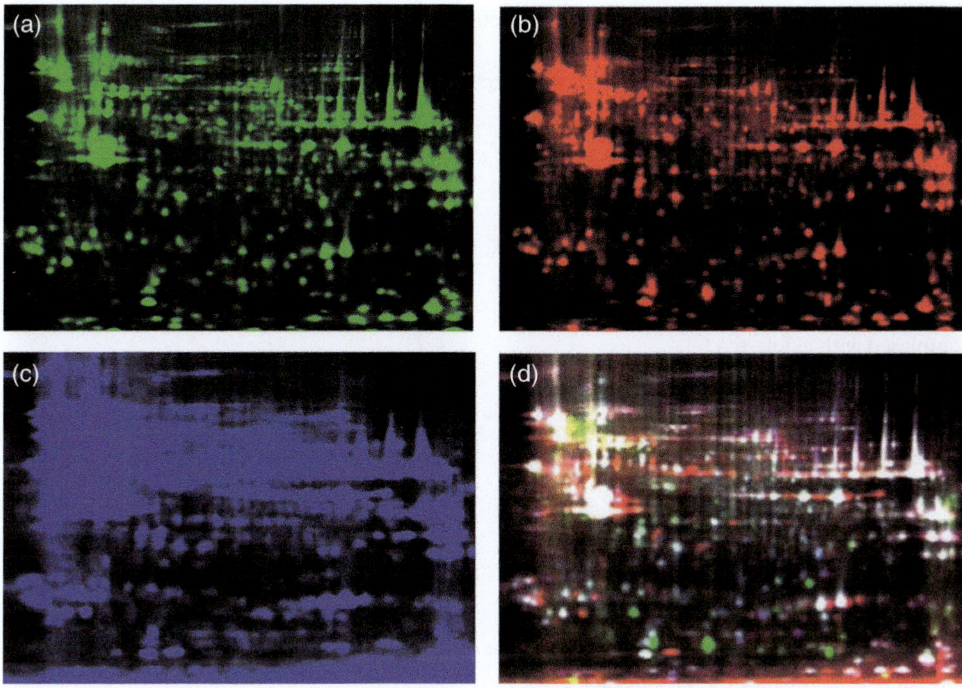

Figure 9.5 2D-DIGE images of the same gel which include the three dyes and three different samples and overlay. Gel containing labelled samples was scanned with a laser scanner Typhoon FLA9500 (GE Healthcare). (a) Proteins derived from uninfected THP-1 samples and labelled with Cy3 are green. (b) Proteins derived from *Leishmania*-infected THP-1 cells and labelled with Cy5 are red. (c) Protein internal standard obtained by pooling the samples together and labelled with Cy2 are blue and used to standardize all gels. (d) Overlay of the spots labelled with both Cy3 and Cy5 showed proteins present in both samples and appearing as bright white spots; unique ones appeared as either red or green-yellow. *Source:* Giulia T.M. Getti (author).

(Cy2-labeled) samples (GE Healthacare, UK) and with DeCyder analysis of differentially expressed spots revealing differences in peak areas (Figure 9.5)

The proteins with the largest relative fold-change (RFC) in gels were excised and identified via PMF using MALDI-TOF MS and Mascot PMF software (Matrix Science). Each of the identified proteins (Table 9.2) had a score equal to or greater than a confidence limit threshold of 95% as estimated by the software and a fold-change of < 0.5 or > 1.5 compared with the uninfected control. Proteins were located on the original 2D gel and their isoelectric point (pI) and molecular weight were confirmed.

The molecular chaperone family Hsp90s are abundant as two major cytoplasmic isoforms, those of alpha and beta (both shown to be elevated in Table 9.2). They are expressed at different times with Hsp90 alpha (HSP90AA1), a stress-induced cytoprotection form, and Hsp90 beta (HSP90AB1) constitutively expressed with Hsp90 alpha proteins, involved in cell cycle regulation and growth promotion, and Hsp90 beta proteins, involved in cytoskeletal stabilization, signal transformation, and cellular transformation. Together they regulate cell proliferation and differentiation [89]. Hsp90 alpha is also important in the regulation of NF-κB, an inducible transcription factor that then regulates immune

Table 9.2 Differentially expressed proteins identified by MALDI-TOF following *Leishmania* infection of terminally differentiated THP-1 cells.

Protein identity	Species	Mass (kDa)	Score	RFC
Heat shock protein HSP 90-alpha 4	*L. aethiopica*	84.6	122	2.43
Heat shock protein 90 kDa protein 1, beta	*L. aethiopica*	83.2	120	2.43
ACTB protein (beta/gamma family)	*L. aethiopica*	40.2	93	1.63
ATPase (15S $Mg_2{}^+$-ATPase p97 subunit)	*L. aethiopica*	89.1	80	−1.75
Human muscle fructose I,6- bisphosphate complexed with aldolase A1	*L. aethiopica*	39.3	69	3.25
Mitochondrial malate dehydrogenase precursor	*L. aethiopica*	35.5	69	−1.72
Annexin A5 chain C1	*L. major*	35.8	114	1.92
Annexin A1	*L. major*	38.7	79	1.55
HNRPHI heterogeneous nuclear ribonucleoprotein H	*L. major*	49	134	1.50
Heterogeneous nuclear ribonucleoprotein A2/Bl isoform A2	*L. major*	35.9	207	1.68
Similar to heterogeneous nuclear ribonucleoprotein A2/B1 isoforms 2 and isoform 4	*L. tropica*	36	79	4.58
Human muscle fructose 1,6- bisphosphate				
Aldolase with fructose 1,6-bisphosphate	*L. tropica*	39.3	88	9.03
Alpha enolase	*L. tropica*	36.3	49	8.35
Triosephosphate isomerase (Tim) chain B complexed with 2-phosphoglycolic acid	*L. tropica*	26.5	63	7.67
Human Glyoxalase I chain D complexed with S-P-nitrobenzyloxycarbonylglutathione	*L. tropica*	20.7	112	4.48

inflammation and cell survival. Moreover, the cleaving of Hsp90s by cathepsin D identified via proteomics induces apoptotic cell death [90], and Wei et al. [91] demonstrated that a series of coumarin pyrazoline derivatives will inhibit Hsp90.

Annexin A1 and A5 are known to be important in apoptotic regulation and counter-inflammatory events and are involved in membrane repair and membrane trafficking [92]. Annexin A5 binds strongly to phosphatidylserine lipids (PS) in a Ca^{2+}-dependent manner on the inner leaf of the cell membrane and stabilizes membrane defects [93]. When annexin A5 is internally bound to the PS, it will inhibit apoptosis and phagocytosis [94]. However, during apoptosis, PS is exposed on the outside and cellular phagocytosis may occur. Heterogeneous nuclear ribonucleoproteins (hnRNPs) associate with nascent pre-mRNAs, packaging them into hnRNP particles as well as many other functions (refer to UniProtKB P22626 at expasy.org).

Fructose 1,6-bisphosphate aldolase was elevated after infection with both *Leishmania aethiopica* and *Leishmania tropica* and found to be complexed with fructose 1,6-bisphosphate. Fructose 1,6-bisphosphate is an allosteric activator of pyruvate kinase and Hsp90 is also known to stabilize the abundance of pyruvate kinase M2 [95].

Triosephosphate isomerase (TIM) and alpha-enolase are also elevated following *L. tropica* infection. The ability of intracellular parasites to regulate host cells' metabolism supports their role in controlling cell survival. It is well established that methylglyoxal formation from conversion of either dihydroxyacetone phosphate or other triose phosphates, lipids and proteins can lead to oxidative stress, cell growth inhibition, and apoptosis [96, 97]. It is worth noting that glyoxalase I was elevated in *L. tropica*.

In summary, the elevation of Hsp90 forms, annexins, and glycolysis enzymes confirms that *Leishmania* parasites can control infected cells' ability to regulate apoptotic processes ([98–101]; [102, 103]).

A similar investigation looked at the changes associated with *L. aethiopica* and *L. mexicana* infection of terminally differentiated THP-1 via gel-free tandem mass tagging (TMT) peptide analysis using liquid chromatography-tandem MS (LC-MS/MS) [104]. Over 200 human proteins were identified as significantly overexpressed following infection with *L. aethiopica* in at least one of three biological replicates. The proteins identified in this study belong to cellular cytoskeletal, tumour suppression, ribosomes, chelators, and immunological factors, giving a further example of the power of proteomics in the understanding of parasite infection processes [104].

9.5 Intracellular *Leishmania* Amastigote Spreading between Macrophages

Leishmaniasis develops after parasites establish themselves as amastigotes inside mammalian cells and start replicating. As relatively few promastigotes survive the innate immune defence [105, 106], the spreading of intracellular amastigotes toward uninfected cells is instrumental to disease progression. Recently, an *in vitro* model of infection spreading was developed and used to identify some of the molecular mechanisms involved in the process [101, 103].

Comparative proteomic investigations based on gel-free TMT LC-MS/MS was used for the first time to gain an overview of the changes associated with *L. aethiopica* spreading [101]. Following analysis of three biological experimental repeats, proteins whose RFC was consistently found in all replicates either as < 0.5 (to indicate downregulation) or > 1.5 (upregulation) were identified. The 19 proteins differentially expressed during *L. aethiopica* spreading were categorized within five functional groups (Figure 9.6).

A closer analysis of the proteins involved with apoptosis pathways confirmed previously published data [103] with a significant overexpression of active caspase 3 (RFC = 1.8) and a two-fold increase of caspase 9 p 35. The latter is a pro-apoptotic protein, the cleaved from of caspase 9, which is involved in the downstream activation of caspase 3 and subsequent induction of apoptosis [107]. Interestingly, proteomic analysis also confirmed downregulation of AKT protein expression (RFC = 0.39), supporting the hypothesis that mitochondrial apoptotic response occurs during the late stages of infection [101].

Overall, a significant downregulation of pro-inflammatory and antimicrobial activity as well as downregulation of anti-inflammatory response was identified. This is consistent with the fact that parasite survival is linked to downregulation of inflammation [108].

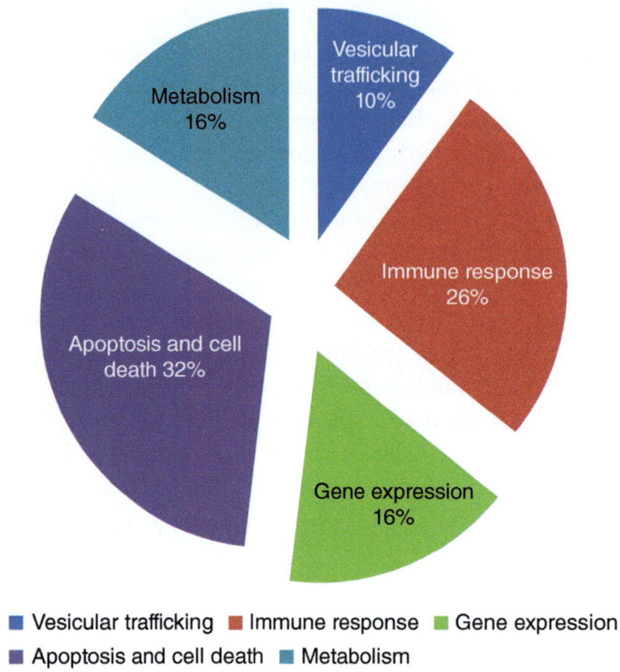

Figure 9.6 Relative distribution of the proteins differentially expressed during *L. aethiopica* spreading. Differentially expressed proteins were identified via TMT LC-MS/MS following comparison of uninfected control samples and *L. aethiopica* infected co-culture.

Similarly, the ability of the parasites to affect expression of a wide range of genes, such as cytokines and chemokines, and NO production is well known and is confirmed by the comparative proteomics results [109–114].

9.6 Potential Virus Applications

Ghildiyal et al. [115] published a recent update on phytochemicals as antiviral agents. The authors highlighted the use of flavonoids, terpenoids, lignins, alkaloids, and coumarins that have antioxidant activity. Studies have been on herpesvirus, human immunodeficiency virus (HIV), influenza, hepatitis virus, and many others. Some human-based treatments as well as *in silico* and *in vitro* studies have also been undertaken. Figure 9.7 shows some of the treatments that have being tried.

SARS-CoV-2 has been suggested as a potential target for phytochemical treatment. Swain et al. [116] summarize a whole spectrum of phytochemicals with reported anti-COVID-19 activity, and analysed them computationally so as to accelerate the development of a drug against SARS-CoV-2. Potential targets would be the proteases that cleave the viral polyprotein. These proteases are cysteine proteases that cleave near the N-terminal end with PL^{pro}, while the C-terminal end of the polyprotein is cleaved by a $3CL^{pro}$ that has a chymotrypsin-like fold. Through the action of $3CL^{pro}$, this releases most of the

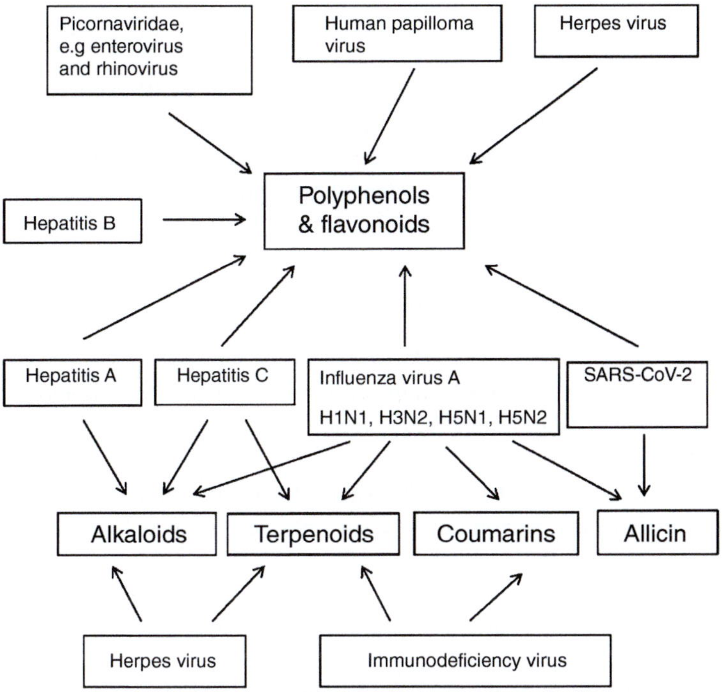

Figure 9.7 Common viruses and tested phytochemical treatments.

fundamental proteins needed for the virus maturation. Importantly, as $3CL^{pro}$ does not have a human analogue, it is a potential target for antiviral drugs and phytochemicals, as discussed in Swain et al. [116] Khubber et al. [117] carried out *in silico* analysis and discussed that a range of flavonoids and allyl sulphur compounds might find a place in the treatment of SARS-CoV-2. Specifically, the flavonoids quercetin and allicin may interact with the main protease M^{pro} (a renamed $3CL^{pro}$), a key enzyme for replication of SARS-CoV-2 and which is responsible for this viral maturation, with structure similarity found in both types 1 and 2 SARS-CoV-2. Recently, Attia et al. [118] and Ben-Shabat et al. [119] discussed the use of phytochemicals for the treatment of SARS-CoV-2 and other viruses, with Attia listing from ligand-receptor computation analysis a number of polyphenols for molecular docking toward the $M^{pro}/3CL^{pro}$ receptor protein.

Panyod et al. [120] discussed COVID-19 and influenza treatments. In particular, they discussed treatments with phytochemicals from garlic, ginger, Chinese mahogany, liquorice, spider lily, and many more. In ginger, the active agents are 6-gingerol, 6-shogaol, and 6-paradol. Phenolic and terpenoids from these plant sources have been tried as potential COVID-19 treatments. Inhibition by ginger was found for avian influenza virus H9N2 and H1N1 invasion of MDCK cells and chick embryo cells. These phytochemicals, according to Panyod et al. [120], could function in a number of ways: by direct action on the virus, by a decrease in virus penetration of cells or by an increased immune response through selective stimulation of cytokines, activation of lymphocytes, increase in natural killer cell counts, and enhanced macrophage action. The actual molecular mode of action was therefore not identified.

Acknowledgements

MALDI-TOF was undertaken at the University of East London School of Health, Sport and Biosciences by PLP. Infection and infection-spreading experiments were carried out and funded by the University of Greenwich, School of Science. TMT LC-MS/MS was outsourced and carried out at the University of Bristol, thanks to internal funding awarded to GTLG by the University of Greenwich.

References

1 Pal, D. and Nayak, A.K. (2021). *Bioactive Natural Products for Pharmaceutical Applications*, Advanced Structured Materials, vol. 140. Springer https://doi.org/10.1007/978-3-030-54027-2.

2 Aslam, S.N., Stevenson, P.C., Kobubun, T. et al. (2009). Antibacterial and antifungal activity of cicerfuran and related 2-arylbenzofurans and stilbenes. *Microbiol. Res.* 164 (2): 191–195.

3 Getti, G.T.M., Aslam, S.N., Humber, D.P. et al. (2006). The effect of Cicerfuran, an Arylbenzofuran from *Cicer bijugum* and related Benzofurans and Stilbenes on Leishmania *aethyopica, L. tropica and L. major*. *Planta Med.* 72 (10): 907–911.

4 Mattio, L.M., Catinella, G., Dallavalle, S. et al. (2020). Stilbenoids: a natural arsenal against bacterial pathogens. *Antibiotics* 9 (6): 336. https://doi.org/10.3390/antibiotics9060336.

5 Mattio, L.M., Catinella, G., Iriti, M. et al. (2021). Inhibitory activity of stilbenes against filamentous fungi. *Ital. J. Food Saf.* 10 (1): 8461. https://doi.org/10.4081/ijfs.2021.8461.

6 Batiha, G.E.L.-S., Beshbishy, A.M., Wasef, L.G. et al. (2020). Chemical constituents and pharmacological activities of garlic (*Allium sativum* L.): a review. *Nutrients* 12 (3): 872. https://doi.org/10.3390/nu12030872.

7 Borlinghaus, J., Foerster, n.R.J., Kappler, U. et al. (2021). Allicin, the odor of freshly crushed garlic: a review of recent progress in understanding Allicin's effects on cells. *Molecules* 26 (1505): 1–22. https://doi.org/10.3390/molecules26061505.

8 Cutler, R.R. and Wilson, P. (2004). Antibacterial activity of a new, stable, aqueous extract of allicin against methicillin-resistant *Staphylococcus aureus*. *Br. J. Biomed. Sci.* 61: 71–74.

9 Fujisawa, H., Suma, K., Origuchi, K. et al. (2008). Biological and chemical stability of garlic-derived allicin. *J. Agric. Food. Chem.* 56: 4229–4235.

10 Getti, G.T.M. and Poole, P.L. (2019). Allicin causes fragmentation of the peptidoglycan coat in *Staphylococcus aureus* by effecting synthesis and aiding hydrolysis: a determination by MALDI-TOF mass spectrometry on whole cells. *J. Med. Microbiol.* 68: 667–677. https://doi.org/10.1099/jmm.0.000950.

11 Clark, A.E., Kaleta, E.J., Arora, A. et al. (2013). Matrix-assisted laser desorption ionization-time of flight mass spectrometry: a fundamental shift in the routine practice of clinical microbiology. *Clin. Microbiol. Rev.* 26: 547–603.

12 Jang, K.S. and Kim, Y.H. (2018). Rapid and robust MALDI-TOF MS techniques for microbial identification: a brief overview of their diverse applications. *J. Microbiol.* 56 (4): 209–216. https://doi.org/10.1007/s12275-018-7457-0.

13 Shah, N.H. and Gharbia, S.E. (2017). *MALDI-TOF and Tandem MS for Clinical Microbiology*. ISBN:9781118960257. John Wiley & Sons Ltd. https://doi.org/10.1002/9781118960226.

14 Singhal, N., Kumar, M., Kanaujia, P.K. et al. (2015). MALDI-TOF mass spectrometry: an emerging technology for microbial identification and diagnosis. *Front. Microbiol.* 6: 791. https://doi.org/10.3389/fmicb.2015.00791.

15 Solntceva, V., Kostrzewa, M., and Larrouy-Maumus, G. (2021). Detection of species-specific lipids by routine MALDI TOF mass spectrometry to unlock the challenges of microbial identification and antimicrobial susceptibility testing. *Front. Cell. Infect. Microbiol.* 10: 621452. https://doi.org/10.3389/fcimb.2020.621452.

16 Van-Belkum, A., Girard, V., Arsac, M. et al. (2016). Matrix-assisted laser desorption ionization-time of flight mass spectrometry for microbial identification in clinical microbiology. In: *Molecular Microbiology Diagnostics Principles and Practice*, *3e*Section 1(8) (ed. D.H. Persing). Wiley ISBN: 978–1–555-81908-8.

17 Weis, C.V., Jutzeler, C.R., and Borgwardt, K. (2020). Machine learning for microbial identification and antimicrobial susceptibility testing on MALDI-TOF mass spectra: a systematic review. *Clin. Microbiol. Infect.* 26 (10): 1310–1317. https://doi.org/10.1016/j.cmi.2020.03.014.

18 Chen, M., Zhu, X., Li, Z. et al. (2010). Application of matrix-assisted laser desorption/ionization time-of-flight mass spectrometry (MALDI-TOF-MS) in preparation of chitosan oligosaccharides (COS) with degree of polymerization (DP) 5–12 containing well-distributed acetyl groups. *Int. J. Mass Spectrom.* 290: 94–99.

19 Bera, A., Herbert, S., Jakob, A. et al. (2005). Why are pathogenic staphylococci so lysozyme resistant? The peptidoglycan O-acetyltransferase OatA is the major determinant for lysozyme resistance of *Staphylococcus aureus*. *Mol. Microbiol.* 55: 778–787.

20 Vollmer, W., Joris, B., Charlier, P. et al. (2008). Bacterial peptidoglycan (murein) hydrolases. *FEMS Microbiol. Rev.* 32: 259–286.

21 Fujisawa, H., Watanabe, K., Suma, K. et al. (2009). Antibacterial potential of garlic-derived allicin and its cancellation by sulfhydryl compounds. *Biosci. Biotechnol. Biochem.* 73: 1948–1955.

22 Loi, V.V., Nguyen, H.N.T.T. et al. (2019). *Staphylococcus aureus* responds to Allicin by global S-Thioallylation – role of the BRX/BSH/YpdA pathway and the Disuphide reductase MerA to overcome Allicin stress. *Free Radic. Biol. Med.* 139: 55–69. https://doi.org/10.1016/j.freeradbiomed.2019.05.018.

23 Boneca, I.G., Huang, Z.H., Gage, D.A. et al. (2000). Characterization of *Staphylococcus aureus* cell wall glycan strands, evidence for a new β-N- acetylglucosaminidase activity. *J. Biol. Chem* 275: 9910–9918.

24 Weadge, J.T., Pfeffer, J.M., and Clarke, A.J. (2005). Identification of a new family of enzymes with potential O-acetylpeptidoglycan esterase activity in both gram-positive and gram-negative bacteria. *BMC Microbiol.* 5: 49.

25 Miron, T., Rabinkov, A., Mirelman, D. et al. (2000). The mode of action of allicin: its ready permeability through phospholipid membranes may contribute to its biological activity. *Biochim. Biophys. Acta* 1463: 20–30.

26 Munch, D. and Sahl, H.-G. (2015). Structural variations of the cell wall precursor lipid II in gram-positive bacteria – impact on binding and efficacy of antimicrobial peptides. *BBA-Membranes* 1848 (11B): 3062–3071. https://doi.org/10.1016/j.bbamem.2015.04.014.

27 Benson, T.E., Prince, D.B., Mutchler, V.T. et al. (2002). X-ray crystal structure of *Staphylococcus aureus* FemA. *Structure* 10: 1107–1115.

28 Sharif, S., Kim, S.J., Labischinski, H. et al. (2009). Characterization of peptidoglycan in fem-deletion mutants of methicillin-resistant *Staphylococcus aureus* by solid-state NMR. *Biochemistry* 48: 3100–3108.

29 Kim, S.J., Chang, J., and Singh, M. (2015). Peptidoglycan architecture of gram- positive bacteria by solid-state NMR. *Biochim. Biophys. Acta* 1848: 350–362.

30 Dmitriev, B.A., Toukach, F.V., Schaper, K.J. et al. (2003). Tertiary structure of bacterial murein: the scaffold model. *J. Bacteriol.* 185: 3458–3468.

31 Laroche, M., Almeras, L., Pecchi, E. et al. (2017). MALDI-TOF MS as an innovative tool for detection of plasmodium parasites in anopheles mosquitoes. *Malar. J.* 16: 5. https://doi.org/10.1186/s12936-016-1657-z.

32 Singhal, N., Kumar, M., and Virdi, J.S. (2016). MALDI-TOF MS in clinical parasitology: applications, constraints and prospects. *Parasitology* 143 (12): 1491–1500. https://doi.org/10.1017/S0031182016001189.

33 Luque-González, A.M., Tabraue-Chavez, M., Lopez-Longerela, B. et al. (2018). Identification of Trypanosomatids by detecting single nucleotide fingerprints using DNA analysis by dynamic chemistry with MALDI-ToF. *Talanta* 176: 299–307. https://doi.org/10.1016/j.talanta.2017.07.059.

34 Cassagne, C., Parlong, F., Jeddi, F. et al. (2014). Identification of *Leishmania* at the species level with matrix-assisted laser desorption ionization time-of-flight mass spectrometry. *Clin. Microbiol. Infect.* 20 (6): 551–557. https://doi.org/10.1111/1469-0691.12387.

35 Evers, D.L., Frye, F.A., and Weina, P.J. (2018). A simple empirical algorithm to distinguish among *Leishmania braziliensis*, major, and tropica species by MALDI-TOF mass spectrometry. *J. Microbiol. Methods* 148: 46–48. https://doi.org/10.1016/j.mimet.2017.12.009.

36 Fernández-Arévalo, A., Ballart, C., Munoz-Basagoiti, J. et al. (2021). Autochthonous and imported tegumentary leishmaniasis in Catalonia (Spain): aetiological evolution in the last four decades and usefulness of different typing approaches based on biochemical, molecular and proteomic markers. *Transboundary Emerging Dis.* https://doi.org/10.1111/tbed.14107.

37 Mouri, O., Morizot, G., Van der Auwera, G. et al. (2014). Easy identification of Leishmania species by mass spectrometry. *PLoS Negl.Trop. Dis.* 8 (6): e2841. https://doi.org/10.1371/journal.pntd.0002841.

38 Lachaud, L., Fernández-Arévalo, A., Normand, A.-C. et al. (2017). Identification of *Leishmania* by matrix-assisted laser desorption ionization–time of flight (MALDI-TOF) mass spectrometry using a free web-based application and a dedicated mass-spectral library. *J. Clin. Microbiol.* 55 (10): 2924–2933. https://doi.org/10.1128/JCM.00845-17.

39 Özbilgin, A., Culha, G., Guray, M.Z. et al. (2021). Diversity of *Leishmania* strains isolated from cutaneous leishmaniasis patients in Turkey and its reflection to clinics in mice model. *Mikrobiyol. Bul.* 54 (3): 429–443. https://doi.org/10.5578/MB.69391.

40 Murugaiyan, J. and Roesler, U. (2017). MALDI-TOF MS profiling-advances in species identification of pests, parasites, and vectors. *Front. Cell. Infect. Microbiol.* 7 (184): 1–9. https://doi.org/10.3389/fcimb.2017.00184.

41 Carnielli, J.B.T., de Andrade, H.M., Pires, S.F. et al. (2014). Proteomic analysis of the soluble proteomes of miltefosine-sensitive and -resistant *Leishmania infantum* chagasi isolates obtained from Brazilian patients with different treatment outcomes. *J. Proteomics* 108: 198–208. https://doi.org/10.1016/j.jprot.2014.05.010.

42 Flores-Treviño, S., Garza-Gonzalez, E., Mendoza-Olazaran, S. et al. (2019). Screening of biomarkers of drug resistance or virulence in ESCAPE pathogens by MALDI-TOF mass spectrometry. *Sci. Rep.* 9 (1): 18945. https://doi.org/10.1038/s41598-019-55430-1.

43 Florio, W., Baldeschi, L., Rizzato, C. et al. (2020). Detection of antibiotic-resistance by MALDI-TOF mass spectrometry: an expanding area. *Front. Cell. Infect. Microbiol.* https://doi.org/10.3389/fcimb.2020.572909.

44 Kumari, S., Kumar, A., Samant, M. et al. (2008). Proteomic approaches for discovery of new targets for vaccine and therapeutics against visceral leishmaniasis. *Proteomics Clin. Appl.* 2 (3): 372–386. https://doi.org/10.1002/prca.200780017.

45 Chowdhury, S.R., Gupta, V.K., and Majumder, H.K. (2021). Targeting DNA topoisomerases in parasitic protozoa by natural products: chemical and biological perspectives. *Stud. Nat. Prod. Chem.* 67: 389–410. https://doi.org/10.1016/B978-0-12-819483-6.00011-4.

46 Orhan, I., Sener, B., Kaiser, M. et al. (2010). Inhibitory activity of marine sponge-derived natural products against parasitic protozoa. *Mar. Drugs* 8 (1): 47–58. https://doi.org/10.3390/md8010047.

47 Fokou, J.B.H., Dize, D., Loe, G.M.E. et al. (2021). Anti-leishmanial and anti-trypanosomal natural products from endophytes. *Parasitol. Res.* 120 (3): 785–796. https://doi.org/10.1007/s00436-020-07035-1.

48 Nweze, J.A., Mbaoji, F.N., Yan-Ming, L. et al. (2021). Potentials of marine natural products against malaria, leishmaniasis, and trypanosomiasis parasites: a review of recent articles. *Infect. Dis. Poverty* 10 (9): 1–19. https://doi.org/10.1186/s40249-021-00796-6.

49 Braun, G.H., Amos, H.P., Candido, A.C.B. et al. (2021). Evaluation of antileishmanial activity of harzialactone a isolated from the marine-derived fungus Paecilomyces sp. *Nat. Prod. Res.* 35 (10): 1644–1647. https://doi.org/10.1080/14786419.2019.1619725.

50 Carter, N.S., Stamper, B.D., Elbarbry, F. et al. (2021). Natural products that target the arginase in *Leishmania* parasites hold therapeutic promise. *Microorganisms* 9 (2): 267. https://doi.org/10.3390/microorganisms9020267.

51 Getti, G.T.M., Durgadoss, P., Domínguez-Carmona, D. et al. (2009). Leishmanicidal activity of Yucatecan medicinal plants on *Leishmania* species responsible for cutaneous leishmaniasis. *J. Parasitol.* 95 (2): 456–460. https://doi.org/10.1645/GE-1675.1.

52 Patel, A.P., Deacon, A., and Getti, G. (2014). Development and validation of four Leishmania species constitutively expressing GFP protein. A model for drug discovery and disease pathogenesis studies. *Parasitology* 141 (04): 501–510. https://doi.org/10.1017/S0031182013001777.

53 Polanco-Hernández, G.M., Gimenez-Turba, A., Salamanca, E. et al. (2021). Leishmanicidal activity and immunomodulatory effect of a mixture of Lupenone and β-Caryophyllene oxide. *Rev. Bras. Farmacogn.* 31 (2): 199–206. https://doi.org/10.1007/s43450-021-00143-8.

54 De Cesare, V., Moran, J., Trayor, R. et al. (2020). High-throughput matrix-assisted laser desorption/ionization time-of-flight (MALDI-TOF) mass spectrometry-based deubiquitylating enzyme assay for drug discovery. *Nat. Protoc.* 15 (12): 4034–4057. https://doi.org/10.1038/s41596-020-00405-0.

55 Abd El-Aziz, T.M., Soares, A.G., and Stockand, J.D. (2020). Advances in venomics: modern separation techniques and mass spectrometry. *J. Chromatogr. B: Anal. Technol. Biomed. Life Sci.* 1160: 122352. https://doi.org/10.1016/j.jchromb.2020.122352.

56 Papac, D.I. and Shahrokh, Z. (2001). Mass spectrometry innovations in drug discovery and development. *Pharm. Res.* 18 (2): 131–145. https://doi.org/10.1023/A:1011049231231.

57 Simon, R.P., Winter, M., Kleiner, C. et al. (2020). MALDI-TOF mass spectrometry-based high-throughput screening for inhibitors of the cytosolic DNA sensor cGAS. *SLAS Discovery* 25 (4): 372–383. https://doi.org/10.1177/2472555219880185.

58 Dolashki, A., Velkova, L., Daskalova, E. et al. (2020). Antimicrobial activities of different fractions from mucus of the garden snail cornu aspersum. *Biomedicines* 8 (9): 315. https://doi.org/10.3390/biomedicines8090315.

59 Liu, R., Khan, R.A.A., Yue, Q. et al. (2020). Discovery of a new antifungal lipopeptaibol from *Purpureocillium lilacinum* using MALDI-TOF-IMS. *Biochem. Biophys. Res. Commun.* 527 (3): 689–695. https://doi.org/10.1016/j.bbrc.2020.05.021.

60 Esquenazi, E., Coates, C., Simmons, L. et al. (2008). Visualizing the spatial distribution of secondary metabolites produced by marine cyanobacteria and sponges via MALDI-TOF imaging. *Mol. Biosyst.* 4 (6): 562–570. https://doi.org/10.1039/b720018h.

61 Rozalén, J., Garcia, M.M., Zalacain, A. et al. (2021). Future trends for the analysis of guayulins in guayule (*Parthenium argentatum* gray) resins. *Ind. Crops Prod.* 159: 113027. https://doi.org/10.1016/j.indcrop.2020.113027.

62 Symma, N. and Hensel, A. (2021). Advanced analysis of oligomeric proanthocyanidins: latest approaches in liquid chromatography and mass spectrometry based analysis. *Phytochem. Rev.* https://doi.org/10.1007/s11101-021-09764-2.

63 Grube, A., Maier, T., Kostrzewa, M. et al. (2007). MS-guided fractionation as a fast way to the identification of new natural products – MALDI-TOF-MS screening of the marine sponge *Stylissa caribica*. *Sect. B J. Chem. Sci.* 62 (4): 600–604. https://doi.org/10.1515/znb-2007-0420.

64 Raj, S., Sasidharan, S., Balaji, S.N. et al. (2020). Review on natural products as an alternative to contemporary anti-leishmanial therapeutics. *J. Proteins Proteomics* 11 (20): 135–158. https://doi.org/10.1007/s42485-020-00035-w.

65 Tariq, A., Adnan, M., Amber, R. et al. (2019). Ethnomedicines and anti-parasitic activities of Pakistani medicinal plants against plasmodia and *Leishmania* parasites. *Ann. Clin. Microbiol. Antimicrob.* 15: 52. https://doi.org/10.1186/s12941-016-0170-0.

66 Chauhan, K., Kaur, G., and Kaur, S. (2018). Activity of rutin, a potent flavonoid against SSG-sensitive and -resistant *Leishmania donovani* parasites in experimental leishmaniasis. *Int. Immunopharmacol.* 64: 372–385. https://doi.org/10.1016/j.intimp.2018.09.026.

67 Chin, S., Behm, C.A., and Mathesius, U. (2018). Functions of flavonoids in plant–nematode interactions. *Plants (Basel)* 7 (4): 85. https://doi.org/10.3390/plants7040085.

68 El-Saber Batiha, G., Beshbishy, A.M., Ikram, M. et al. (2020). The pharmacological activity, biochemical properties, and pharmacokinetics of the major natural polyphenolic flavonoid: quercetin. *Foods* 3: 374. https://doi.org/10.3390/foods9030374.

69 Galvão, B.V.D., Araujo-Lima, C.F., Dos Santos, M.C.P. et al. (2021). *Plinia cauliflora* (Mart.) Kausel (Jaboticaba) leaf extract: *in vitro* anti-*Trypanosoma cruzi* activity, toxicity assessment and phenolic-targeted UPLC-MSE metabolomic analysis. *J. Ethnopharmacol.* 277: 114217. https://doi.org/10.1016/j.jep.2021.114217.

70 Martínez-Castillo, M., Pacheco Yepez, J., Flores, N. et al. (2018). Flavonoids as a natural treatment against *Entamoeba histolytica*. *Front. Cell. Infect. Microbiol.* 8: 209. https://doi.org/10.3389/fcimb.2018.00209.

71 Piwowarczyk, R., Ochmian, I., Lachowicz, S. et al. (2020). Phytochemical parasite-host relations and interactions: a *Cistanche armena* case study. *Sci. Total Environ.* 716: 137071.

72 Choo, S., Chin, V.K., Wong, E.H. et al. (2020). Antimicrobial properties of Allicin used alone or in combination with other medications. *Folia Microbiol.* 65 (3): 451–465. https://doi.org/10.1007/s12223-020-00786-5.

73 Alcolea, P.J., Alonso, A., García-Tabares, F. et al. (2016a). Increased abundance of proteins involved in resistance to oxidative and nitrosative stress at the last stages of growth and development of *Leishmania amazonensis* promastigotes revealed by proteome analysis. *PLoS One* 11 (10): https://doi.org/10.1371/journal.pone.0164344.

74 Alcolea, P.J., Alonso, A., García-Tabares, F. et al. (2016b). Proteome profiling of the growth phases of *Leishmania pifanoi* promastigotes in axenic culture reveals differential abundance of immunostimulatory proteins. *Acta Trop.* 158: 240–247. https://doi.org/10.1016/j.actatropica.2016.03.015.

75 Alcolea, P.J., Alonso, A., Garcia-Tabares, F. et al. (2019). An insight into the constitutive proteome throughout *Leishmania donovani* promastigote growth and differentiation. *Int. Microbiol.* 22 (1): 143–154. https://doi.org/10.1007/s10123-018-00036-2.

76 Cuervo, P., De Jesus, J.B., Junqueira, M. et al. (2007). Proteome analysis of *Leishmania* (Viannia) *braziliensis* by two-dimensional gel electrophoresis and mass spectrometry. *Mol. Biochem. Parasitol.* 154 (1): 6–21. https://doi.org/10.1016/j.molbiopara.2007.03.013.

77 Dea-Ayuela, M.A., Rama-Iñiguez, S., and Bolas-Fernandez, F. (2006). Proteomic analysis of antigens from *Leishmania infantum* promastigotes. *Proteomics* 6 (14): 4187–4194. https://doi.org/10.1002/pmic.200600101.

78 Magalhães, R.D.M., Duarte, M.C., Mattos, E.C. et al. (2014). Identification of differentially expressed proteins from *Leishmania amazonensis* associated with the loss of virulence of the parasites. *PLoS Negl.Trop. Dis.* 8 (4): 1–13. https://doi.org/10.1371/journal.pntd.0002764.

79 Mojtahedi, Z., Clos, J., and Kamali-Sarvestani, E. (2008). *Leishmania major*: identification of developmentally regulated proteins in procyclic and metacyclic promastigotes. *Exp. Parasitol.* 119 (3): 422–429. https://doi.org/10.1016/j.exppara.2008.04.008.

80 Bag, A.K., Saha, S., Sundar, S. et al. (2014). Comparative proteomics and glycoproteomics of plasma proteins in Indian visceral leishmaniasis. *Proteome Sci.* 12 (1): 1–14. https://doi.org/10.1186/s12953-014-0048-z.

81 Cuervo, P., De Jesus, J.B., Saboia-Vahia, L. et al. (2009). Proteomic characterization of the released/secreted proteins of *Leishmania* (Viannia) *braziliensis* promastigotes. *J. Proteomics* 73 (1): 79–92. https://doi.org/10.1016/j.jprot.2009.08.006.

82 Kumar, A., Misra, P., Sisodia, B. et al. (2015). Proteomic analyses of membrane enriched proteins of *Leishmania donovani* Indian clinical isolate by mass spectrometry. *Parasitol. Int.* 64 (4): 36–42. https://doi.org/10.1016/j.parint.2015.01.004.

83 de Oliveira, A.H.C., Ruiz, J.C., Cruz, A.K. et al. (2006). Subproteomic analysis of soluble proteins of the microsomal fraction from two Leishmania species. *Comp. Biochem. Physiol. D: Genomics Proteomics* 1 (3): 300–308. https://doi.org/10.1016/j.cbd.2006.05.003.

84 Menezes, J.P.B., Almeida, T.F., Petersen, A.L.O.A. et al. (2013). Proteomic analysis reveals differentially expressed proteins in macrophages infected with *Leishmania amazonensis* or *Leishmania major*. *Microbes Infect.* 15 (8–9): 579–591. https://doi.org/10.1016/j.micinf.2013.04.005.

85 Negrão, F., Diedrich, J.K., Girgio, S. et al. (2019a). Tandem mass tag proteomic analysis of *in vitro* and *in vivo* models of cutaneous Leishmaniasis reveals parasite-specific and nonspecific modulation of proteins in the host. *ACS Infect. Dis.* 5 (12): 2136–2147. https://doi.org/10.1021/acsinfecdis.9b00275.

86 Negraö, F., Fernandez-Costa, C., Zorgi, N. et al. (2019). Label-free proteomic analysis reveals parasite-specific protein alterations in macrophages following *Leishmania amazonensis, Leishmania major*, or *Leishmania infantum* infection. *ACS Infect. Dis.* 5 (6): 851–862. https://doi.org/10.1021/acsinfecdis.8b00338.

87 Negrão, F., Giorgio, S., Eberlin, M.N. et al. (2019b). Comparative proteomic analysis of murine cutaneous lesions induced by *Leishmania amazonensis* or *Leishmania major*. *ACS Infect. Dis.* 5 (8): 1295–1305. https://doi.org/10.1021/acsinfecdis.8b00370.

88 Getti, G.T.M. (2007) Interaction between *Leishmania* parasites and mammalian macrophages. PhD thesis. University of Greenwich, UK.

89 Sreedhar, A.S., Kalmar, E., Csermely, P. et al. (2004). Hsp90 isoforms: functions, expression and clinical importance. *FEBS Lett.* 562 (1–3): 11–15. https://doi.org/10.1016/S0014-5793(04)00229-7.

90 Fritsch, J., Ficker, R., Klawitter, J. et al. (2016). TNF induced cleavage of HSP90 by cathepsin D potentiates apoptotic cell death. *Oncotarget* 7 (46): 75774–75789.

91 Wei, Q., Ning, J.-Y., Dai, X. et al. (2018). Discovery of novel HSP90 inhibitors that induced apoptosis and impaired autophagic flux in A549 lung cancer cells. *Eur. J. Med. Chem.* 145: 551–558. https://doi.org/10.1016/j.ejmech.2018.01.024.

92 Gerke, V. and Moss, S.E. (2002). Annexins: from structure to function. *Physiol. Rev.* 82 (2): https://doi.org/10.1152/physrev.00030.2001.

93 Lin, Y.C., Chipot, C., and Scheuring, S. (2020). Annexin-V stabilizes membrane defects by inducing lipid phase transition. *Nat. Commun.* 11 (230): 1–13.

94 Kenis, H., van Genderen, H., Deckers, N.M. et al. (2006). Annexin A5 inhibits engulfment through internalization of PS-expressing cell membrane patches. *Exp. Cell Res.* 312 (6): 719–726. https://doi.org/10.1016/j.yexcr.2005.11.023.

95 Xu, Q., Tu, J., Dou, C. et al. (2017). HSP90 promotes cell glycolysis, proliferation and inhibits apoptosis by regulating PKM2 abundance via Thr-328 phosphorylation in hepatocellular carcinoma. *Mol. Cancer* 16: 1–16. https://doi.org/10.1186/s12943-017-0748-y.

96 Cianfruglia, L., Morresi, C., Bacchetti, T. et al. (2020). Protection of polyphenols against glyco-oxidative stress: involvement of glyoxalase pathway. *Antioxidants* 9 (10): 1006. https://doi.org/10.3390/antiox9101006.

97 Distler, M.G. and Palmer, A.A. (2012). Role of glyoxalase 1 (Glo1) and methylglyoxal (MG) in behavior: recent advances and mechanistic insights. *Front. Genet.* 3 (250): 1–10. https://doi.org/10.3389/fgene 2012 00250.

98 DaMata, J.P., Mendes, B.P., Maciel-Lima, K. et al. (2015). Distinct macrophage fates after in vitro infection with different species of *Leishmania*: induction of apoptosis by *Leishmania amazonensis* but not by Leishmania guyanensis. *PLoS One* 10: e0141196.

99 Gupta, P., Srivastav, S., Saha, S. et al. (2016). *Leishmania donovani* inhibits macrophage apoptosis and pro-inflammatory response through AKT-mediated regulation of β-catenin and FOXO-1. *Cell Death & Differ.* 23: 1815–1826.

100 Moshrefi, M., Spotin, A., Kafil, H.S. et al. (2017). Tumor suppressor p53 induces apoptosis of host lymphocytes experimentally infected by *Leishmania major*, by activation of Bax and caspase-3: a possible survival mechanism for the parasite. *Parasitol. Res.* 116 (8): 2159–2166. https://doi.org/10.1007/s00436-017-5517-8.

101 Rai, R., Dyer, P., Richardson, S. et al. (2017). Apoptotic induction induces *Leishmania aethiopic* and *L. mexicana* spreading in terminally differentiated THP-1cells. *Parasitology* 144 (14): 1912–1921. http://dx.doi.org/10.1017/S0031182017001366.

102 Solano-Galvez, S.-G., Alvarez-Hernandez, D.-A., Gutierre-Kobeh, L. et al. (2021). *Leishmania*: manipulation of signaling pathways to inhibit host cell apoptosis. *Ther. Adv. Infect. Dis.* 8: 20499361211014977. https://doi.org/10.1177/20499361211014977.

103 Ranatunga, M., Rai, R., Richardson, S.C.W. et al. (2020). *Leishmania aethiopica* cell-to-cell spreading involves caspase-3, AkT, and NF-jB but not PKC-dactivation and involves uptake of LAMP-1-positive bodies containing parasites. *FEBS J.* 287 (8): https://doi.org/10.1111/febs.15166.

104 Deacon A. D. (2017). Expression of green fluorescent protein (GFP) in *Leishmania* spp. to identify survival strategies the parasite imposes on the human host. PhD thesis. University of Greenwich, UK.

105 Kimblin, N., Peters, N., Debrabant, A. et al. (2008). Quantification of the infectious dose of *Leishmania major* transmitted to the skin by single sand flies. *Proc. Natl. Acad. Sci. USA* 105 (29): 10125–10130. https://doi.org/10.1073/pnas.0802331105.

106 Rogers, M.E., Llg, T., Nikolaev, A.V. et al. (2004). Transmission of cutaneous leishmaniasis by sand flies is enhanced by regurgitation of fPPG. *Nature* 430: 463–467.

107 Twiddy, D. and Cain, K. (2007). Caspase-9 cleavage, do you need it? *Biochem. J* 405: e1–e2.

108 Reverte, M., Eren, R.O., Jha, B. et al. (2021). The antioxidant response favors *Leishmania* parasites survival, limits inflammation and reprograms the host cell metabolism. *PLoS Pathog.* https://doi.org/10.1371/journal.ppat.1009422.

109 Bichiou, H., Rabhi, S., Hamda, C.B. et al. (2021). *Leishmania* parasites differently regulate antioxidant genes in macrophages derived from resistant and susceptible mice. *Front. Cell. Infect. Microbiol.* 11: 748738. https://doi.org/10.338/fcimb.2021.748738.

110 Buates, S. and Mattelawski, G. (2001). General suppression of macrophage gene expression during Leishmania infection. *J. Immunol.* 166: 3416–3422.

111 Gregory, D.J., Sladek, R., Olivier, M., and Matlashewski, G. (2008). Comparison of the effects of *Leishmania major* or *Leishmania donovani* infection on macrophage gene expression. *Infect. Immun.* 76: 1186–1192.

112 Guerfali, F.Z., Laouini, D., Guizani-Tabbane, L. et al. (2008). Simultaneous gene expression profiling in human macrophages infected with *Leishmania major* parasites using SAGE. *BMC Genomics* 9: 238. https://doi.org/10.1186/1471-2164-9-238.

113 Kalavi, K., Jorjani, O., Faghihi, M.A., and Mowla, S.J. (2021). Cytokine gene expression alterations in human macrophages infected by *Leishmania major*. *Cell J.* 4: 476–481. https://doi.org/10.22074/cellj.2021.6524.

114 Meier, C.L., Svensson, M., and Kaye, P.M. (2003). *Leishmania*-induced inhibition of macrophage antigen presentation analyzed at the single-cell level. *J. Immunol.* 171 (12): 6706–6713. https://doi.org/10.4049/jimmunol.171.12.6706.

115 Ghildiyal, R., Prakash, V., Chaudhary, V.K. et al. (2020). Phytochemicals as antiviral agents: recent updates. In: *Plant-derived Bioactives* 12, 279–295. https://doi.org/10.1007/978-981-15-1761-7_12.

116 Swain, S.S., Panda, S.K., and Luyten, W. (2020). Phytochemicals against SARS-CoV as potential drug leads. *Biomed. J.* https://doi.org/10.1016/j.bj.2020.12.002.

117 Khubber, S., Hashemifesharaki, R., Mohammadi, M. et al. (2020). Garlic (*Allium sativum* L.): a potential unique therapeutic food rich in organosulfur and flavonoid compounds to fight with COVID-19. *Nutr. J.* 19 (1): 124. https://doi.org/10.1186/s12937-020-00643-8.

118 Attia, Y.A., Alagawany, M.M., Farag, M.R. et al. (2020). Phytogenic products and phytochemicals as a candidate strategy to improve tolerance to coronavirus. *Front. Vet. Sci.* 7: 57315920. https://doi.org/10.3389/fvets.2020.573159.

119 Ben-Shabat, S., Yarmolinsky, L., Porat, D. et al. (2020). Antiviral effect of phytochemicals from medicinal plants: applications and drug delivery strategies. *Drug Delivery Transl. Res.* 10 (2): 354–367. https://doi.org/10.1007/s13346-019-00691-6.

120 Panyod, S., Chi-T, H., and Sheen, L.-Y. (2020). Dietary therapy and herbal medicine for COVID-19 prevention: a review and perspective. *J. Tradit. Complement. Med.* 10 (4): 420–427. https://doi.org/10.1016/jtcme.2020.05.004.

10

Application of MALDI-TOF MS in Bioremediation and Environmental Research

Cristina Russo and Diane Purchase

Department of Natural Sciences, Faculty of Science and Technology, Middlesex University, The Burroughs, London, UK

10.1 Introduction

The characterization of microorganisms is essential for monitoring environmental and pathogenic microbial communities. Molecular tests, such as polymerase chain reaction (PCR)-based DNA amplification of the 16S or 18S ribosomal RNA (rRNA) gene and their sequencing, represented the gold standard for the identification of microorganisms until the recent arrival of whole-genome sequencing (WGS). Although these techniques have high specificity and sensitivity and have been used extensively, they have limitations for routine laboratory because of their lengthy analysis, high costs, specialized equipment, requirement for a dedicated laboratory, and highly skilled personnel.

Matrix-assisted laser desorption/ionization time-of-flight mass spectrometry (MALDI-TOF MS) has attracted considerable interest in microbiological research, having already established its presence in clinical diagnostic laboratories. MALDI-TOF MS offers many advantages compared with traditional molecular techniques, such as its high throughput, robustness, minimal sample preparation, and low cost for routine laboratory microbial identification [1]. With the development of MALDI-TOF MS, more reproducible and accurate data of microbial species at the genus, species and subspecies level could be generated [2].

MALDI-TOF MS analysis of samples for bioremediation and environmental sites is similar to that used in diagnostic laboratories, viz. microorganisms are first covered or mixed with an excess of matrix where it co-crystallizes with the sample, allowed to dry, and inserted into a MALDI-TOF instrument. The admixture is irradiated by intense laser pulses, which induces thermal decomposition of mainly cellular proteins which are converted into gas-phase ions and separated according to their m/z ratio with a time-of-flight (TOF) analyser. The resulting mass spectrum or peptide mass fingerprint (PMF) is obtained in just a few minutes. The most commonly reported matrices used to generate

Microbiological Identification using MALDI-TOF and Tandem Mass Spectrometry: Industrial and Environmental Applications, First edition. Edited by Haroun N. Shah, Saheer E. Gharbia, Ajit J. Shah, Erika Y. Tranfield, and K. Clive Thompson.

good-quality microbial spectra include a-cyano-4-hydroxycinnamic acid (CHCA), 5-chloro-2-mercaptobenzothiazole (CMBT), and 2,4-dihydroxybenzoic acids (DHB) [3, 4].

An overview of the workflow for the MALDI-TOF MS analysis of microorganisms from bioremediation and environmental sites is illustrated in Figure 10.1.

Peptide mass fingerprinting currently represents one of the major approaches for the identification of an unknown microbe, providing there is an appropriate reference spectrum in the database. Based on the uniqueness of PMF for each microbe, microbial characterization is accomplished by comparing the acquired spectra against a reference spectral library [2]. Consequently, a prerequisite for MALDI-TOF MS-based microbe identification is the availability and accessibility of spectral libraries. A significant drive for microbial identification has led MALDI-TOF manufacturers to develop commercially available databases. For this purpose, Bruker Microbiology & Diagnostics and bioMérieux developed MALDI Biotyper and Vitek MS, respectively [2, 5–7]. These are established databases that use robust algorithms and multivariate statistical approaches to search for correlations between experimental and theoretical spectra, ranging from 2 to 20 kDa, to obtain a spectral match [8].

Multiple approaches can be adopted for the sample preparation step depending on the nature of the sample. In general, a single colony of microorganisms can be directly placed onto a MALDI target plate, or cell extraction may be performed prior to analysis with the aid of organic acids, such as trifluoroacetic acid (TFA) or formic acid (FA) [1]. Interestingly, studies show that different sample preparation methods are employed when different platforms are used [9].

Previous publications have reported the application of MALDI-TOF MS for the characterization of microbes derived from ecological samples, such as water, soil, and roots [10–14]. For example, Lovecka et al. [11] characterized seven isolated strains from contaminated soil using

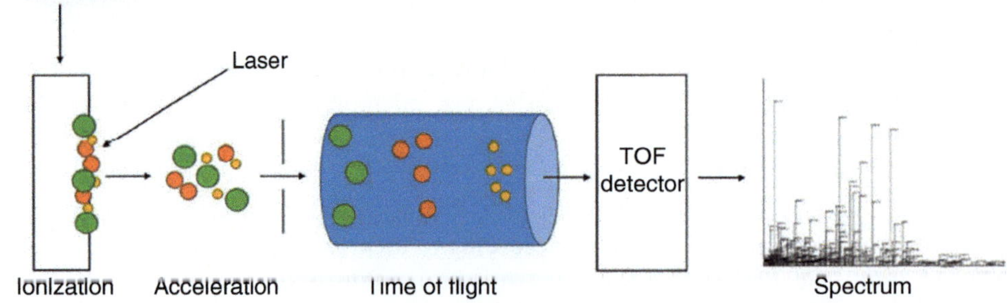

Figure 10.1 A schematic overview of MALDI-TOF MS application for the identification of microorganisms from bioremediation and environmental sites. The laser fires the mixture of sample matrix, producing the desorption and ionization of peptides and proteins; those gas-phase ions are then directed into a mass analyser and separated according to their *m/z*. The measurement of ion relative abundance in the detector results in a mass spectrum. *Source:* From Santos et al. [1] /with permission of Royal Society Chemistry.

MALDI-TOF MS in parallel with 16S rRNA sequencing and showed that *Bacillus* spp. represented the principal species among the isolates. The authors further analysed the pesticide degradation products in media cultures [i.e. g-hexachlorocyclohexane (g-HCH), dichlorodiphenyltrichloroethane (DDT), hexachlorobenzene (HCB)] by these species for their capacity to biodegrade these chemicals with the aim of exploiting their application for bioremediation [11]. A study by Arango et al. [15] utilized MALDI-TOF MS to probe sediments of the Arauca River, Colombia, to initially screen actinobacterial communities such as *Streptomyces* spp. for their bioactive potential, including their potential as antimicrobial agents.

Apart from its use in identification, modification of the MALDI-TOF MS technique is being used in other areas, such as biochemistry and metabolomics, for comprehensive and quantitative analysis of a wide spectrum of metabolites in biological samples. Here MS-based techniques are being used for metabolic fingerprinting to define sample classes. Similar to PMF databases, metabolite libraries and databases are also required to perform these complex tasks [16].

In a more direct biochemical application, Persson et al. [17] used MALDI-TOF MS to detect bacteriochlorophyll a and all the major homologues of bacteriochlorophyll c in the photosynthetic green sulphur bacterium *Chlorobium tepidum*. In a novel application of the technique, the authors reported the detection of pigments and proteins in chlorosomes, as free molecules and in the intact organelles of the bacterium which previously needed to be determined by conventional biochemical and genetic methods.

Recently, the combination of chemical specificity of MS tools with imaging capabilities has allowed the development of mass spectrometry imaging (MSI) techniques. MALDI-TOF MSI is a powerful technique that is able to visualize ion distribution maps of many non-labelled endogenous and exogenous species simultaneously directly on the samples [18] (see also Section 1.6). This technique allows in situ mapping of the peptides or secreted metabolites as well as the study of interactions between microorganisms ([19]; [20]).

Although MALDI-TOF MS has been widely used in clinical microbiology research, its application in environmental microbiology is less well documented. This chapter focuses mainly on the application of MALDI-TOF MS for microbial identification in environmental studies. Other applications where MALDI-TOF MS is proving useful, such as metabolomic studies for investigating microbial activity in relation to the environment and other species as well as bioremediation, will also be discussed. The advantages and disadvantages of this rapid identification technology are compared, and new ways of combining MALDI with other analytical techniques for environmental research are discussed. We envisage that the information will help to broaden its scope in other spheres of science and maximize its application in environmental research.

10.2 Microbial Identification: Molecular Methods and MALDI-TOF MS

Microbial identification in bioremedation and environmental sites can be achieved via phenotypic, proteomic, and genomic methods (Figure 10.2). Phenotypic methods involve the analysis of the biochemical reactions, metabolism, and fermentation activities of the microorganisms. In practice, they are straightforward and not labour-intensive, but the accuracy is lower than the other two methods, especially when samples are derived from non-clinical sites (see Chapter 1). Proteomics methods analyse the ribosomal proteins

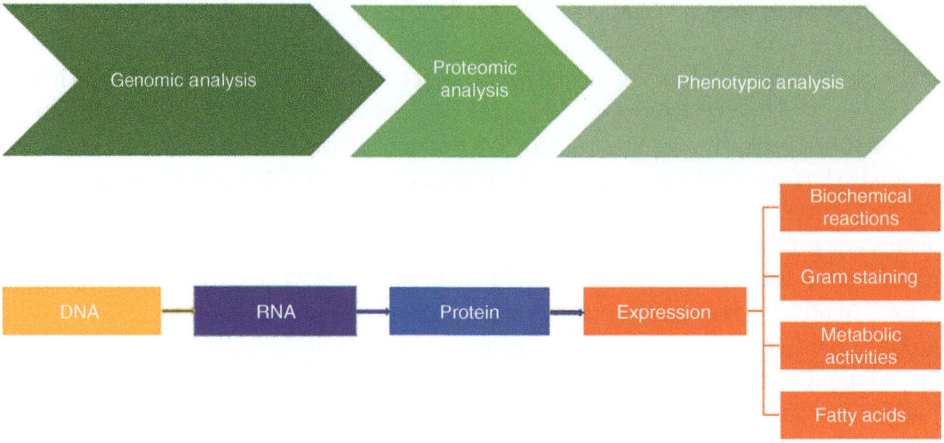

Figure 10.2 Microbial identification methods.

produced by a microorganism to generate a unique profile that can be compared with one in an existing library. Genomic methods compare the sequence of rRNA regions of the microorganism or whole genomes with the existing database and offer the highest degree of accuracy and reproducibility.

Genomic and proteomic analyses are frequently employed to identify microorganisms and find many applications in clinical settings, but each technique has its advantages and limitations. Table 10.1 lists the advantages and limitations of molecular techniques and MALDI-TOF MS analysis for microbial identification.

10.2.1 PCR-based Methods

Molecular methods for microbial identification range from relatively simple DNA amplification-based approaches (PCR, real-time PCR, random amplified polymorphic DNA-PCR) to more complex methods involving restriction fragment analysis, targeted gene, and high-throughput next-generation WGS. PCR-based methods are less time-consuming than conventional culture-based methods and they can be used to identify difficult-to-culture microorganisms. Both the 16S and 18S genes are highly specific within their respective microbial groups and are ideal target for identification. The 16S rRNA gene is ~1600 base pairs long and includes nine hypervariable regions of varying conservation (V1–V9) [21].

Pure bacterial or fungal culture is first cultured on solid or aqueous medium. The microbial cells are collected and lysed, and the DNA is extracted and fragmented. The DNA is then amplified using universal primers and/or species-specific primers, the PCR products are visualized using gel electrophoresis, and this is followed by sequence analysis. Quantitative PCR (qPCR) can also be performed to quantify nucleic acids. The sequence is then compared against known databases for identification using bioinformatics software such as Basic Local Alignment Search Tool (BLAST).

Table 10.1 Comparison of molecular methods and MALDI-TOF MS for microbial identification.

Method	Process	Time	Pros	Cons
DNA sequencing	Sequence-specific amplification of hypervariable regions (16S or 18S rRNA) Sequence-specific amplification of housekeeping genes (MLST)	14–20 h	• Specific, sensitive, rapid and accurate • Can identify fastidious and difficult-to-culture microorganisms • WGS enables multiple identification in a single sample	• Trained laboratory personnel required • Complex workflow • Powerful bioinformatics software required • Expensive
MALDI-TOF	Generate mass spectrum of molecular samples ablated by a laser	1–24 h	• Fast • Accurate • Lower run cost • Trained laboratory personnel not required • A priori knowledge not required	• High initial equipment cost • Limited library • Subspecies identification constrained • Polymicrobial analysis challenging

Source: Adapted from Singhal et al. [4]; Rychert [117]; Rentschler et al. [136].

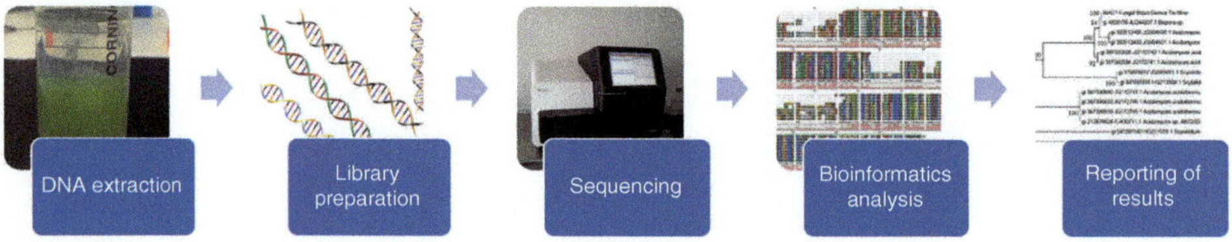

Figure 10.3 General workflow of DNA sequencing for the identification of bacteria and fungi. *Source:* Trinity College Dublin, The University of Dublin.

High-throughput next-generation sequencing is a powerful tool to provide higher resolution and accuracy in identifying microorganisms. It allows culture-independent testing to generate information about microbial community structure and composition of an environment. The process involves obtaining total genomic DNA from the samples, and the DNA is amplified via PCR using appropriate primers (e.g. the 16S V4, V1–3 regions for prokaryotic cells or the ITS2 region for fungal DNA). The DNA library is prepared via barcoding and/or adapter ligation. The sequencing is carried out and the data generated are analysed, the consensus sequence of the gene is aligned and computed, and sequence homologies are determined using a central data repository [e.g. Quantitative Insights Into Microbial Ecology (QIIME)] for similarity search and to select sequence hits to perform phylogenetic analysis and construct a phylogenetic tree (Figure 10.3).

Multilocus sequence typing (MLST) uses nucleotide sequence of fragmented multiple housekeeping genes (commonly between five and seven) along the chromosome to identify the microorganism. The internal gene fragments are specified as allele fragments. The unique combination of polymorphisms observed for each of the loci allows for unambiguous typing of the microorganisms. By sequencing the PCR product of the fragmented housekeeping genes and comparing them against existing databases, it offers a versatile tool for microbial identification and strain typing. MLST databases for more than 50 microorganisms have been established (e.g. www.pubmlst.org). As WGS becomes increasingly available and affordable, MLST information can be extracted from WGS databases using appropriate read mapping-based tools, such as Short Read Sequence Typing for Bacterial Pathogens (SRST2) [22].

10.2.2 MALDI-TOF MS

MALDI-TOF MS offers a faster and more reliable alternative to genomic analysis for the identification of microorganisms [23, 24]. A number of factors have contributed to the success of MALDI-TOF: reduction in time for identification of up to a day [25]; a priori knowledge of the test organisms not required [4]; capability to accurately identify difficult-to-culture anaerobic bacteria in challenging environments [26], and high cost saving in reagents and staffs' time (e.g. up to $2.34 million per annum for patients with bloodstream infection [27]). MALDI-TOF MS has gained huge success in diagnostic clinical laboratories and numerous reports are available (see, e.g., [28–35])

10.3 Combination of MALDI-TOF MS with Other Methods for the Identification of Microorganisms

Over the past two decades, mass spectrometry (MS) has become an established technique for microbial identification. However, the complexity of microbial biomarkers can make MS analysis challenging, reducing the reliability of the results. This challenge derives from the ion suppression effects on biomarkers of interest, which arise due to ionization competition with other compounds within complex biological samples. However, efforts have been made to overcome this drawback by increasing sample purification, desalting, and concentration prior to ionization. Multiple studies investigated the possibility of reducing microbial sample complexity interference by combining MS with other techniques, such as liquid chromatography (LC), gas chromatography (GC), solid-phase extraction, or chip-based microfluid devices and affinity methods [36–44].

A schematic overview of the combination of MS with other techniques is provided in Figure 10.4.

In 2004, Gekenidis et al. [46] developed a novel proteomics workflow focused on microbial identification to the subspecies level. The method included using nano-LC to separate digested proteins extracted from microorganisms and subsequent MALDI-TOF/TOF MS analysis. The combination of nano-LC with MALDI-TOF/TOF MS resulted in MS spectra with a wider mass range and type of proteins identified than classic MALDI biotyping. As proof of concept, specific biomarker peptides for three *Salmonella* subspecies were identified using this approach. In further work, Fagerquist et al. [38] identified a prominent ~10-kDa protein biomarker from MALDI-TOF MS of cell lysates of five thermophilic

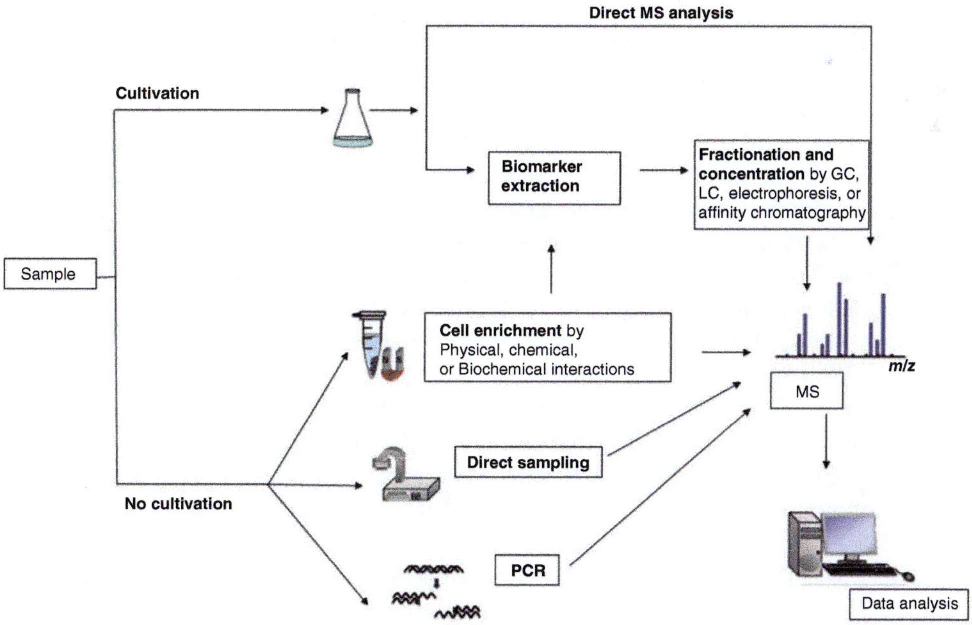

Figure 10.4 Overview of the combination of MS with other methods for microbial enrichment and identification. *Source:* Ho and Reddy [45]/With permission of John Wiley & Sons.

species of *Campylobacter*: *C. jejuni*, *C. coli*, *C. lari*, *C. upsaliensis*, and *C. helveticus*. The DNA-binding protein, designated HU, ionized strongly by MALDI while its molecular weight varied between species and among strains within a species, making it a reliable taxonomic trait. Interestingly, its variation was not due co- or post-translational modifications but instead resulted from changes in the amino acid sequence of the HU protein. Additional studies have demonstrated that MALDI mass spectra of isolated bacterial protein fractions by one- or two-dimensional gel electrophoresis increased the strain-discriminatory power of the technique, leading to higher confidence in the results [47–49].

To increase detection sensitivity, immunomagnetic separation as an affinity-capture technique has also emerged. This technique employs magnetic particles associated with a specific antibody used to isolate a targeted microorganism from a pool of a complex microbes mixture, prior to MALDI-TOF MS analysis. For example, in the work by Madonna et al. [50], magnetic beads coated with polyclonal antibodies raised against serotypes of *Salmonella* were employed to isolate *Salmonella enterica* (formerly *Salmonella choleraesuis*). MALDI-TOF mass spectra of bacteria attached to immunomagnetic beads were acquired, and peaks at 7275, 9525, and 9245 Da and 10.2, 14.4, 17.5, 35.5, 43.3, and 51.7 kDa were attributed to specific biomarkers of the species. The MALDI-TOF MS spectra were not affected by immunomagnetic peaks, and the resulting analysis was rapid and sensitive, enabling the detection of a minimum 1.0×10^7 bacteria cells/ml in less than one hour. These factors point strongly towards the high potential of using immunomagnetic separation with MALDI-TOF MS to select and detect target microorganisms.

Another approach aimed at enhancing spectral protein profiles and quality is based on performing protein characterization while bound to an affinity-capture substrate. Affinity capture is a technique based on the isolation of desired compounds (i.e. biomarkers) by utilizing their binding to a receptor immobilized on a solid support. It is understandable that when the technique is coupled with MALDI, considerations on the compositions of the solid substrate are necessary as it needs to be compatible with MALDI-TOF MS parameters such the presence of matrices and high-vacuum conditions. The use of a polymerized planar lipid bilayer as a substrate has been shown to guarantee the stability required for MALDI-TOF MS analysis [41]. Alternatively, in the work reported by Johnson et al. [51], gold surfaces with immobilized immunoglobulin G and silica surfaces with small peptides were used as platforms to capture biomarkers from *Staphylococcus aureus* samples. Protein A was identified by using MALDI-TOF MS while immobilized to the surfaces. The peptide capture ligands offer the benefits of isolating a biomarker of interest from a mixture of compounds without requiring intact cells, having high stability and small size, which result in a high loading capacity on the surface.

The genotypic method PCR is widely used for microbial identification; however, it has been relatively unsuccessful in the field of microbe classification. Many researchers have shown that the combination of PCR with MS can provide comprehensive microbial genomic information that could not be obtained if the techniques were used individually [52–55]. Indeed, multiple publications have demonstrated how the use of MALDI-TOF MS on bacterial PCR amplicons provided an efficient MS-based screening method for species-level characterization [52, 56–58]. Lefmann et al. [52] showed that MALDI-TOF MS analysis of PCR amplification products was used to differentiate 12 type strains of *Mycobacterium* spp. An unambiguous classification of mycobacteria at the species level was possible as MALDI-TOF MS spectra showed an individual mass signal pattern for each strain. A unique MALDI-TOF MS platform has been available over the last two decades through Sequenom Inc., but the MassArray system has unfortunately not gained widespread usage (see reviews in [59, 60]).

The possibility of combining MALDI-TOF MS with other techniques offers the advantage of providing additional information regarding the identity of microbial samples, enhancing confidence in the results and providing a higher degree of success in strain-level typing.

10.4 Application of MALDI-TOF MS in Environmental and Bioremediation Studies

Table 10.2 cites examples where MALDI-TOF MS has been used to study microbes in air, water, and soil.

10.4.1 The Atmospheric Environment

Rapid identification of airborne microorganisms has garnered a lot of attention in recent years. In addition to identifying airborne pathogens, MALDI-TOF MS have been applied to study the air quality of a variety of outdoor and indoor environments. In outdoor environments, MALDI-TOF MS-based identification found applications in many non-healthcare settings to evaluate biosafety of the air. For example, in a study in outdoor biowaste pretreatment plants, Rasmussen et al. [61] detected 290 bacterial and 37 fungal species at 10^2–10^8 cfu/m^3. Of the bacterial and fungal isolates, 250 and 31 strains were identified to species level, respectively. These organisms have the potential to cause inflammation and to form biofilms as well as being fungicide-resistant, and thus pose a risk to the health of the workers. Similarly, in indoor environments, MALDI-TOF MS has been used to identify microbes in many bioaerosols. Dybwad et al. [68] identified 37 different airborne bacterial genera from a five-month sampling period in a Norwegian underground subway station, demonstrating the suitability of the technology for biological surveillance and public health. The indoor air quality of a glasshouse in a botanical garden was investigated by Kozdrój et al. [69], who found that the concentration of Gram-positive bacteria and fungi did not to exceed the threshold limit. However, concern was raised for the glasshouse workers who were more exposed to the bioaerosol particles. In addition, MALDI-TOF MS was applied to study the impact of tourism in the Škocjan Caves in Slovenia, a UNESCO World Heritage Site. The authors concluded that the tourists were a significant vector and sources of airborne bacteria and the cave system could form a reservoir for the colonization and infection of susceptible hosts [70].

10.4.2 The Aquatic Environment

In aquatic environments, MALDI-TOF MS identification provides a reliable and quick method for monitoring water microbial quality, detecting pathogens in aquaculture, assessing proliferation of antibiotic resistance microorganisms, as well as biodiversity research. For example, MALDI-TOF was used to identify bacteria in routine water analysis in

Table 10.2 Examples of the application of MALDI-TOF MS in the identification of microbes from environmental samples.

Environment	Location	Microorganisms	References
Bioaerosol in biowaste from pre-treatment plants	Denmark	*Aspergillus fumigatus*	Rasmussen et al. [61]
Bioaerosol in waste collection sites	Denmark	180 bacteria and 37 fungal species, including pathogenic *Aspegillus niger* and *Penicillium expansum*	Madsen et al. [62]
Water samples from a petrochemical industry wastewater treatment plant	Brazil	95% agreement with 16S rRNA sequencing, including *Acinetobacter* spp., *Arthrobacter creatinolyticus*, *Bacillus* spp., *Lelliottia amnigena*, *Microbacterium* spp., *Pseudomonas* spp., *Micrococcus luteus*	Antunes et al. [63]
Water samples from private and public water supply wells	USA	*Pseudomonas stutzeri* and *Acinetobacter haemolyticus*	Santos et al. [64]
Water samples from a radioactive material storage pool	France	Bacteria from the phyla Actinobacteria, Bacteroidetes, Firmicutes, and Proteobacteria and fungi from the *Ascomycota* phylum	Hayoun et al. [65]
Soil from a thermal electric power plant	Russia	PAH-degrading bacteria (*Arthrobacter*, *Rhodococcus*, *Oerskovia*, and *Isoptericola*)	Sazykin et al. [66]
Soil from an apple orchard in a cinnabar mining town	Mexico	Mercury-resistant bacteria (*Bacillus muralis* and *Bacillus simplex*) containing *merR* and *merA* genes	Calzada Urquiza et al.[67]

drinking water [71] and the bottling industry [72], for identification of *Legionella* non-pneumophila species with 96.8% sensitivity for species present in the manufacturer's database. Pascale et al. [73] used the method for routine surveillance of environmental *Legionella*, while Vidal et al. [74] employed it for mass screening of marine bacterial symbionts from corals, sponge, fish, and seawater.

A number of MALDI-TOF MS databases are readily available for the most frequently occurring bacterial pathogens of aquatic species such as *Aeromonas* spp., *Carnobacterium piscicola*, *Citrobacter freundii*, *Edwardsiella* spp. *Photoabcteriaum* spp., and *Vibrio parahaemolyticus* [75]. Detecting bacterial pathogens in aquaculture is an essential part of the management process to safeguard stocks as well as protect their human handlers from zoonotic transmission. A MALDI-TOF MS database containing 25 *Virbio* spp. that are potentially pathogenic to marine molluscs has been created to facilitate rapid diagnosis to protect marine aquaculture production [76].

The prevalent use of antibiotics has raised concern about the spread of antimicrobial resistance (AMR). Wastewater effluents from hospitals, veterinary clinics, and farmyards could be a source of antibiotics and their discharge into receiving water enables horizontal gene transfer between resistant and the general population of microorganisms. MALDI-TOF MS provides a useful tool for studying the antibiotic residue and antibiotic resistant genes in aquatic microorganisms. For example, Voigt et al. [77] investigated antibiotic resistance genes and antibiotic-resistant organisms in a river that supplies drinking water reservoir system in Germany using MALDI-TOF MS, LC-MS/MS, and conventional susceptibility testing. The authors concluded that sewage treatment plant effluents are point sources of the antibiotic residue and antibiotic-resistant bacteria. Similarly, the annual changes in antibiotic-resistant coliform bacteria and enterococci in municipal wastewater in Poland was studied using MALDI-TOF MS and 27 antibiotic-resistant species were found, of which 58% were multi-drug-resistant [78].

The rapid and simple operational procedures of MALDI-TOF MS have been pivotal in studies of biodiversity. It has facilitated identification and understanding of the microbial consortia and their ability to survive under stress in diverse external conditions, such as exposure to high levels of industrial effluents. In studying biofilm samples collected from different surfaces entirely submerged in water, Tuohy et al. [79] were able to identify *Deinococcus aquaticus* at subpopulation levels and obtained higher taxonomic resolution than 16S rRNA sequence analysis. In the review by de Vasconcelos et al. [80] on activated sludge microorganisms in textile industry containing azo dyes, MALDI-TOF was singled out as a novel, rapid, and accurate tool to identify bacteria that had the capacity to degrade the dyes [80]. In a different environment setting, Yasir et al. [81] examined the biodiversity of thermophilic microorganisms in hot springs in Saudi Arabia and used them for bioprospecting for thermostable biomolecules. They obtained 536 isolates, of which six species were novel to the niche. In another extreme environment, MALDI-TOF MS identified all 45 bacteria from Arctic Kandalaksha Bay to genus level and 48% to species level [82].

10.4.3 The Terrestrial Environment

In the soil environment, MALDI-TOF MS has been used to identify plant pathogens and for research on sustainable agriculture to investigate the role of microbionts in plants. The rapid detection of plant pathogens is important to prevent the loss of crops and harvests.

For example, MALDI-TOF MS was used to identify the species-specific infection of *Fusarium* spp. on the roots and shoots of asparagus [83]. The relationships of plant growth-promoting bacteria (PGPB) and arbuscular mycorrhizal fungi (AMF) in the rhizosphere with their phytobionts are particularly important, as the microbionts often enhance plant growth, combat environmental stressors, and provide or trigger defence mechanisms against plant pathogens. For example, three PGPB, *Serratia liquefaciens*, *Pseudomonas helleri*, and *Serratia* sp., identified using MALDI-TOF MS and 16S rRNA, were found to be effective against fungal pathogens *Fusarium* spp. while also inducing growth in wheat [84]. They can enhance hormonal balance, maintain nutrient status, and improve plant growth and play an essential role in facilitating nutrient uptake and water flow, especially under abiotic stress conditions [85]. Nineteen AMF species were reliably identified using MALDI-TOF MS to species level [86]. AMF together with plant growth promotion was found to improve growth and defence against infection by *Xanthomonas translucens*, a leaf pathogen [87].

MALDI-TOF MS also found numerous applications in studying microbial biodiversity in terrestrial environments which has helped us to understand the impact of environmental stressors in the distribution and survival of many fungal and bacterial species. Their survival could have a significant impact on the geochemical cycles and the health of the environment. For example, three psychrophilic yeasts (*Rhodotorula muilaginosa*, *Nagnaishia vishniacii*, and *Doioszegia cryoxerica*) have been successfully identified in cold environments by MALDI-TOF MS, helping to provide insights into the effect of temperature on fungal biodiversity [88]. In a thermal electric power plant contaminated with polycyclic aromatic hydrocarbons (PAHs) situated in Novocherkassk, the Russian Federation, MALDI-TOF MS and 16S rRNA sequencing were used to identify 200 isolates and determine their diversity; *R. erythropolis* was found to be the predominant species among the four genera of actinobacteria identified [66].

Water scarcity is a challenge in many regions of the world. To ensure a sustainable water supply, water-reuse and water management concepts are often adopted. Concerns have been raised as to whether reclaimed water for irrigation may contain AMR that could be transferred to terrestrial environments. In addition, the overuse of antibiotics in farming is also a driver for spreading AMR. MALDI-TOF MS has been widely used in the study of AMR of microbes in soil. For example, MALDI-TOF MS provided rapid identification of 40 *Escherichia coli* from 545 samples collected from children's parks area in Ankara, Turkey; these strains were found to be resistant to ciprofloxacin (5%), ampicillin (17%), sulfamethoxazole (15%), streptomycin (12.5%), tobramycin (5%), gentamicin (5%), cefotaxime (2.5%), and ceftazidime (2.5%) [89].

10.4.4 Bioremediation Research Applications

Bioremediation in its broadest sense is a process involving living organisms to treat contamination in an environmental compartment such as air, water, and soil to mitigate the negative impact of the pollutants to an acceptable level. Microorganisms have versatile physical and biochemical capabilities to adapt to very adverse conditions, and consequently their traits can be explored and applied to remediation contamination in a sustainable, environmentally friendly and cost-effective way. The most common approaches for bioremediation involve: (i) the enhancement of naturally occurring biochemical mechanisms in autochthonous microbial populations; (ii) augmentation with exogenous microorganisms that can best degrade the contaminants; (iii) phytoremediation; or (iv) application of cell-free enzymes. Bioremediation treatment can be ex situ or in situ and can be integrated with other remediation processes.

Microbial populations can bioremediate organic pollutants via mineralization where the pollutant is degraded completely to water, CO_2, and simple inorganic compounds, or via partial mineralization to intermediate products. MALDI-TOF MS has been used to identify four halotolerant strains isolated from the Santos Estuary, Brazil, that can effectively degrade bisphenol A. Amongst the isolates, *Shewanella haliois* was the best performing, being able to tolerate up to 15 mg/l bisphenol A and biotransform half of the content in 10 hours [90]. For inorganic pollutants such as metals, microorganisms may detoxify them by accumulation through physical means (e.g. adsorption to cell wall or production of metal-binding molecules), immobilization (e.g. precipitation or complexation) or transformation into less toxic forms (e.g. oxidation or alkylation). For example, a PGPB, *Enterobacter aerogenes*, isolated from a contaminated site near a steel plant in West Bengal, India, was identified using MALDI-TOF MS. The strain was resistant to cadmium (Cd) and was found to reduce oxidative stress, ethylene stress, and Cd uptake in rice seedlings [91].

Phytoremediation is an emerging technology that uses plants and the microbial communities in their rhizosphere for remediation of pollutants. The mechanisms could involve: (i) rhizofiltration, where contaminants are absorbed, concentrated, and precipitated by plant roots; (ii) phytoextraction, where pollutants are extracted and accumulated in the harvestable plant tissues; (iii) phytotransformation, where contaminants are transformed to simple molecules that can be incorporated into the plant tissues; (iv) phytostimulation, where the microbial growth in the rhizosphere is stimulated by the release of chemicals (e.g. auxin); and (v) phytovolatilization, where the unwanted chemicals are converted to volatile forms and released to the atmosphere. PGPB (see, e.g., [92]) and AMF (see, e.g., [93]) also play an important role in phytoremediation. MALDI-TOF MS may be applied to facilitate their identification and understanding of the underlying mechanisms.

The ability of MALDI-TOF MS to rapidly characterize the microbial community in a polluted environment in turn helps in understanding the mechanisms involved and may help to develop useful treatment strategies to manage or remediate the site. MALDI-TOF MS was used to identify the main oxidative stress-induced proteins to Cd and lead (Pb), including subunits alpha, gamma of ATP synthetase, chlorophyll CP26 binding protein, fructose-bisphosphate aldolase and long-chain ribulose bisphosphate carboxylase in *Paspalum fasciulatum* exposed to 30 and 50 mg/kg Cd and Pb, respectively [94]. Copper resistance mechanisms and bioremediation potential of an *Acinetobacter calcoaceticus* strain isolated from a copper (Cu) mine were elucidated using two-dimension MALDI-TOF MS. It was found that the expression of some oxidoreductases such as Cu-resistant protein A were upregulated and the antioxidant defence and Cu efflux pump played a significant role in intracellular Cu detoxification [95]. A *Bacillus* sp. isolated from textile wastewater was reported to produce a biosurfactant to enhance biodegradation of the aromatic amine 4-chloroaniline. MALDI-TOF MS was used to determine the lipopeptide nature of the biosurfactant [96]. An arsenic-resistant fungal strain isolated from a historical tin mine in

Cornwall with the low pH of 1.13 was identified by MALDI-TOF MS as *Acidomyces acidophilus*. The strain was able to resist up to 22 500 and 100 mg/l As(V) and Sb(V), respectively, in Czapex-Dox Agar medium and the strain was able to remove up to 70.30% of 100 mg/l As(V) [97]. MALDI-TOF/TOF was also used to examine the protein responses of *Amynthas acidophilus* when exposed to 100 mg/l of sodium arsenate (Na_2HAsO_4) and potassium antimonate ($KSb(OH)_6$), MALDI-TOF/TOF MS analysis tentatively identified that the following proteins were upregulated in response to exposure to the metal: malate dehydrogenase, phospholipase B, phosphoglycerate kinase, glyceraldehyde-3-phosphate dehydrogenase, and 3-isopropylmalate dehydrogenase (W.K. Chan, personal communication). Thus, MALDI-TOF MS and MALDI-TOF/TOF MS provided powerful tools to understand the mechanisms involved in the resistance exhibited by these organisms and provided in-depth knowledge that is essential to optimize the remediation capability of such organisms.

10.5 Microbial Products and Metabolite Activity

Another facet of MALDI-TOF MS applications in environmental research involves the study of microbial metabolic activities. The isolation and cultivation of microorganisms are essential to investigate their metabolic activity, which represents a starting point for the investigation of their survival in the environment and their function for potential biotechnological and bioremediation applications. However, this calls for large investments in the study of microbial metabolic activities, considering the vast habitats that microorganisms colonize around the Earth.

The advancement of a high-throughput culture technique, known as culturomics, in conjugation with the use of identification techniques, such as MALDI-MS and 16S rRNA gene sequencing, has opened up the opportunity to culture previously uncultured microbes and widen our knowledge and understanding of microbial diversity and activity [98–100]. To rapidly assess bacterial functional traits, Clark et al. [101] developed a novel approach based on the combination of MALDI-TOF MS data from both intact proteins and specialized metabolites. With this approach, metabolic profiles of hundreds of environmental microbial isolates were compared in just a few hours, providing insights into the association between microbial phylogenetic identities with potential environmental functions. Although this work highlighted the discriminatory power of MALDI-TOF MS, it also strongly supported the necessity of developing open sources of MS-based methods to better discriminate microbial subspecies [102]. Also, subtracting interfering medium-related signals from the data before performing sample analysis could be an additional advantage. This latter aspect is particularly relevant for the study of small molecules, as their identification in MALDI spectra could be compromised by matrix peak interferences in the low-mass range. In this regard, Edwards and Kennedy [103] demonstrated that the use of 9-aminoacridine, a matrix with limited interference in the small metabolite range using negative mode, allowing MALDI profiling of over 100 metabolites from *E. coli*.

It is well established that many microbially produced metabolites, known as secondary metabolites, offer a wide variety of biological activities such as antimicrobial, antitumoral, and immunosuppressive biomolecules, etc. For their bioactivity, secondary metabolites have been intensively investigated using MS tools [104]. Fastner et al. [105] employed

MALDI-TOF MS for the first time to examine the diversity of oligopeptide secondary metabolites of *Microcystis* strains. Mass spectra indicated a high interdiversity of oligopeptides such as microcystins, anabaenopeptins, microginins, aeruginosins, and cyanopeptolins and unknown molecules between *Microcystis* colonies that enabled their typing. This approach represents a valuable tool for investigating ecological activities of the genus *Microcystis* [105]. MALDI-TOF MS was used to detect characteristic cyclic lipopeptide biomarkers from *Bacillus* species collected from different soils and seeds in Algeria. These studies led to the identification of surfactin, pumilacidin, lichenysin, kurstakin, and different types of fengycins and revealed high lipopeptide biodiversity in *Bacillus* species [106]. The production of cyclic lipopeptide from *Bacillus* isolates has been widely studied in rice blast disease as they are considered to be responsible for the ability of *Bacillus* species such as *B. altitudinis* and *B. velezensis* to counteract phytopathogens such as *Pyricularia oryzae* by triggering both induced systemic resistance and direct antagonism [107].

However, the production of metabolites is strongly influenced by environmental, temporal, or other external conditions. Antón et al. [108] studied the metabolomic pool between isolates of Extremely Halophilic Bacterium *Salinibacter ruber* (old and newly isolated) from different hypersaline environments worldwide. The results, interestingly, showed a variegated microbial exomatobolome between isolates, despite the high phylogenetic relationship. Another study, by Galleguillos et al. [109], investigated the metabolic activity of different species of acidophilic bacteria in response to osmotic stress. By using ion chromatography and MALDI-TOF MS, it was possible to detect an increase in sugar trehalose synthesis, which was associated with increased osmotic potentials. By investigating the metabolic products using analytical techniques, it was possible to assess and confirm the presence of genes responsible for trehalose synthesis in many chemolithotrophic and heterotrophic acidophilic bacteria, in agreement with data from sequenced genomes. MALDI-TOF MS was also utilized in combination with high-performance LC and nuclear magnetic resonance in order to observe the effects of extracellular signalling in the reprogramming of the secondary metabolome of high-density *Nostoc* strain cultures [110]. The production of secondary metabolites, nostopeptolide, nostamide A, and anabaenopectin, by *Nostoc punctiforme* was detected, supporting the influence of extracellular signalling in the activation of biosynthetic gene clusters (BGCs) responsible for secondary metabolism in *Nostoc* spp. [111].

The development of MS imaging techniques in the last decade has notably attracted interest in metabolomics studies (Chapter 1). The ability to visualize spatial distribution of secreted metabolites directly in a sample has further heightened our knowledge of competitive microbial interactions, which would be undetectable with traditional non-imaging techniques [112]. Chen et al. [20] offered a comprehensive MALDI-TOF MSI analysis of metabolic interactions between the species *Microcystis aeruginosa* and *Pseudomonas grimontii*, bringing further insights into the mechanisms responsible for suppression, increases, and exchange of metabolites. In this particular study, MALDI-TOF MSI data showed that the metabolite phenazine-1-carboxamide secreted by *Pseudomonas grimontii* showed inhibitory activity, hindering the growth of *M. aeruginosa*, whereas microcystins exhibited no suppression activity to *P. grimontii*. This work strongly reinforced the understanding of the molecular mechanisms responsible for these species inter-interactions which were previously investigated by Sandrin and Demirev [113]. Several studies have also reported the combination of MALDI-MSI with LC-MS/MS to investigate factors that can impact metabolite production in *P. aeruginosa*, a bacterium often associated with hospital-acquired infections [114, 115]. MALDI-TOF MSI can also be used to identify metabolic products from intact organisms of differing complexities, as shown in the work of Esquenazi et al. [116], in which several secondary metabolites, such as jamacamide B, curacin A, and curazole, were directly visualized in *Lyngbya majuscula* 3L and JHB, *Oscillatoria nigro-viridis*, *Lyngbya bouillonii*, and a *Phormidium* species, even when they were present as a mixture.

Mass spectrometry is one of the major platforms used for metabolomics studies. Progress in both MS technical advances and pioneering applications has revealed the potential of discovering previously undetectable novel microbial metabolites, as well as investigating alterations in cellular metabolism due to microbial interactions.

10.6 Challenges of Environmental Applications

Successful identification of microorganisms using MALDI-TOF MS is dependent on a number of factors, including the initial investment of the equipment and the availability of a database containing the existing spectra of known organisms. A limited number of spectra in the database can lead to poor discrimination between species, as well as misidentifications. The library must be robust and contain a sufficient number of different species of a group of microorganisms as there are inherent similarities between related species [117]. The lack of data on non-clinical microorganisms has been a crucial limiting factor in applying MALDI-TOF MS in environmental research [118]. Although more spectral profiles have been added to the existing library of environmental samples (e.g. a dedicated MALDI-TOF MS database for identification of drinking water bacteria [71] and a database containing 144 main spectra of *Phytophthora*, an oomycete [119]), there is still a paucity of spectral data compared with the diversity of microorganisms present in the environment.

From a technical point of view, a limitation of MALDI-TOF MS is the low resolving power for microorganism identification, which could compromise the reproducibility of the technique. An important aspect to consider in a MALDI-TOF MS experiment is the sample preparation step, which plays a critical role as it can strongly influence the desorption/ionization process, affecting spectral quality, i.e. peak resolution, sensitivity, and noise [120, 121]. An intense optimization step for the matrix and deposition technique choice is usually performed to obtain good-quality microbial spectra [122, 123].

The reproducibility of MALDI-TOF MS is additionally compromised when the same microorganisms are analysed at different growth stages, as microorganisms generate different amounts of proteins and peptide profiles during their development. This aspect establishes the necessity of recreating the same growth conditions and carrying out comparative inter- and intra-laboratory studies to obtain reproducible results [124]. A comprehensive review of the influence of sample preparation and culture conditions on MALDI-TOF MS results was recently written by Topić Popović et al. [125]. At this time, there is no standardized procedure for sample preparation, but many efforts and optimization strategies have been investigated to enhance the reproducibility of the MALDI-TOF MS technique.

In addition to the choice of matrix, several strategies may be adopted to enhance microbial peptide signals. For example, pre-treatment of cells with ethanol prior to matrix deposition resulted in an improvement in signal/baseline ratio and shot-to-shot reproducibility [126]. Also, the addition of FA to the matrix solution can be used to suppress salt-containing adduct ions derived from culture media and enhance spectral resolution [127]. Organic acids, such as TFA and fluoric acid, are often used to support protein extraction before MALDI analysis in order to generate a more exhaustive PMF and increase accuracy in microbial identification [1, 128]. A number of strategies have been suggested by Zhang and Sandrin [9] to improve the accuracy of identification using MALDI-TOF MS.

Another critical aspect to consider for the successful application of MALDI-TOF MS is the optimal amount of microbial cells. Although the MALDI-TOF MS experiment takes a few minutes to perform, the necessity of microbial cell culturing to obtain sufficient amounts of cells for the analysis should be considered. Commonly, the sensitivity of MALDI-TOF MS is around 10^5–10^7 bacterial cells, depending on the microbial species. The necessity of growing cells could be considered as a limitation; however, it is a common step required for many characterization techniques.

The discriminatory power of MALDI-TOF MS results could be compromised when organisms are inherently similar. For example, at present, MALDI-TOF MS fails to distinguish between *Shigella* and *E. coli*, *Bordetella pertussis* and *B. bronchioseptica*, *Achromobacter xylosoxidans* and *A. ruhlandii*, and *Bacteroides nordii* and *Bacteroides salyersiae*, as well as species within the *Enterobacter cloacae* complex [129–131].

When operating with MALDI-TOF MS instruments, technical issues responsible for instrumental drifts are not rare. These are encountered especially during a long acquisition time and when the sample surface is highly topographically heterogeneous. These aspects, in addition to a varying performance among MALDI-TOF MS instruments, can compromise the reproducibility of microbial peptide mass spectra. Currently, post-processing approaches, including baseline correction and peak alignment followed by peak picking, are employed to enhance MALDI-TOF MS performance by increasing reliable peak identifications and sensitivity [122].

The high instrument cost (around US$200 000) represents another relevant drawback of this technique. The cost is considerably higher than that of other equipment, such as PCR instruments, microscopes, and electrophoresis tanks. However, it needs to be considered that the low cost per experiment in terms of time will compensate for the initial high instrument cost.

Efforts towards instrumental development, upgrading of libraries and software, optimization of sample preparation and processing are tackling the issues associated with the MALDI-TOF MS technique, enhancing its relevance and application in environmental microbiology (Chapters 1, 2).

10.7 Opportunities and Future Outlook

The comparison of PMF of unknown microorganisms against a reference library represents a pivotal aspect for correct microorganism identification. The major constraints against using libraries are the limited availability and associated high costs. In light of these considerations, protein resources that are freely accessible on the internet have recently been developed, such as UniProt (www.uniprot.org [132]). Cheng et al. [133]

developed a robust method for bacterial identification by matching MALDI-TOF spectra to peaks reported in the UniProt spectral library. In this work, the 10 most informative genes were chosen, and through a mathematical algorithm and double cross-validation, it was possible to identify the encoded proteins which were used as protein panels for bacterial identification with an identification accuracy exceeding 80%. This study demonstrates the possibility of relying on freely accessible databases for sample processing, without the need for an expensive MALDI-TOF MS library. The continuous development and validation of in-house libraries have increased the potential of MALDI-TOF MS in microorganism identification, especially in clinical research [134]. However, environmental microorganisms are subjected to stressful conditions that modify their protein profile, leading to misidentification. In the future, improving databases that encounter stress-related peptide modifications should be considered to promote a more reliable environmental microbe characterization by the MALDI-TOF MS technique. In addition, integration and synchronization of spectral libraries in MALDI-TOF MS instruments derived from different manufacturers would enormously enhance the compatibility of the instruments and expandability of the technique by promoting the sharing and processing of data.

The miniaturization of mass spectrometers would also facilitate the application of this technique in environmental monitoring. Portable mass spectrometers offer the benefits of being light and small, and having low power consumption; at the present time they are mainly limited to pollutant monitoring [135]. The continuous development of portable mass spectrometers and their increased versatility represent an exciting future for MS in environmental microbiology research.

With the expansion of MALDI-TOF MS applications, the demand for developing MS instrumentation with enhanced properties (i.e. higher throughput, resolution, and sensitivity) has also increased. Enhancing the limit of detection sensitivity, alongside advances in sample-handling strategies, opens up opportunities for future MALDI analysis directly on a single microbial cell. This would significantly improve the overall speed of analysis, eliminating the time-consuming cell culturing step.

10.8 Conclusions

The identification of the components of microbial communities represents a fundamental aspect of environmental microbiology. MALDI-TOF MS was established as a widespread and powerful tool for detecting clinical and environmental microorganisms by profiling their ribosomal protein pattern. Over recent years, the application of MALDI-TOF MS for microorganism identification, ranging mainly from bacteria to fungi, has been widely investigated. Although yielding considerable success in clinical microbiology, the role of MALDI-TOF MS in environmental research still requires further development. In this regard, the advancement of protein libraries, instrumentation, and optimization of sample preparation is in continuous development in order to permit this technique to be routinely employed for environmental microbiology without requiring validation using gold-standard conventional techniques.

Alongside identification application, the potential use of MALDI-TOF in identifying microbial products and metabolites allows the investigation of microbial activity and the effects on the environment and other microorganisms (Chapter 6). With metabolomics studies, the microbial degradation capacity of pollutants can be investigated, and important additional information on microbial bioremediation capacity can be obtained. Currently, not many studies report the use of MALDI-TOF MS in bioremediation studies, and this could be an avenue for further investigation, increasing the versatility of the technique.

References

1 Santos, I.C., Hildenbrand, Z.L., and Schug, K.A. (2016). Applications of MALDI-TOF MS in environmental microbiology. *Analyst* 141 (10): 2827–2837.

2 Dingle, T.C. and Butler-Wu, S.M. (2013). MALDI-TOF mass spectrometry for microorganism identification. *Clin. Lab. Med.* 33 (3): 589–609.

3 Buszewski, B., Rogowska, A., Pomastowski, P. et al. (2017). Identification of microorganisms by modern analytical techniques. *J. AOAC Int.* 100 (6): 1607–1623.

4 Singhal, N., Kumar, M., Kanaujia, P.K., and Virdi, J.S. (2015). MALDI-TOF mass spectrometry: an emerging technology for microbial identification and diagnosis. *Front. Microbiol.* 6: 791. https://doi.org/10.3389/fmicb.2015.00791.

5 Havlicek, V., Lemr, K., and Schug, K.A. (2013). Current trends in microbial diagnostics based on mass spectrometry. *Anal. Chem.* 85 (2): 790–797.

6 Mather, C.A., Rivera, S.F., and Butler-Wu, S.M. (2014). Comparison of the Bruker Biotyper and Vitek MS matrix-assisted laser desorption ionization-time of flight mass spectrometry systems for identification of mycobacteria using simplified protein extraction protocols. *J. Clin. Microbiol.* 52 (1): 130–138.

7 Pusch, W. (2007). Bruker Daltonics: leading the way from basic research to mass-spectrometry-based clinical applications. *Pharmacogenomics* 8 (6): 663–668.

8 Randell, P. (2014). It's a MALDI but it's a goodie: MALDI-TOF mass spectrometry for microbial identification. *Thorax* 69 (8): 776–778.

9 Zhang, L. and Sandrin, T.R. (2016). Maximizing the taxonomic resolution of MALDI-TOF-MS-based approaches to bacterial characterization: from culture conditions through data analysis. In: *Applications of Mass Spectrometry in Microbiology: From Strain Characterization to Rapid Screening for Antibiotic Resistance*, 147–181. Springer International Publishing https://doi.org/10.1007/978-3-319-26070-9_6.

10 Emami, K., Askari, V., Ullrich, M. et al. (2012). Characterization of bacteria in ballast water using MALDI-TOF mass spectrometry. *PLoS One* 7 (6): https://doi.org/10.1371/journal.pone.0038515.

11 Lovecka, P., Pacovska, I., Stursa, P. et al. (2015). Organochlorinated pesticide degrading microorganisms isolated from contaminated soil. *New Biotechnol.* 32 (1): 26–31.

12 Ruelle, V., el Moualij, B., Zorzi, W. et al. (2004). Rapid identification of environmental bacterial strains by matrix-assisted laser desorption/ionization time-of-flight mass spectrometry. *Rapid Commun. Mass Spectrom.* 18 (18): 2013–2019.

13 Štursa, P., Uhlík, O., Kurzawová, V. et al. (2009). Approaches for cultivable and non cultivable bacteria in real soil. *Plant Soil Environ.* 55 (9): 389–396.

14 Ziegler, D., Mariotti, A., Pflüger, V. et al. (2012). In situ identification of plant-invasive bacteria with MALDI-TOF mass spectrometry. *PLoS One* 7 (5): https://doi.org/10.1371/journal.pone.0037189.

15 Arango, C., Acosta-Gonzalez, A., Parra-Giraldo, C.M. et al. (2018). Characterization of actinobacterial communities from Arauca river sediments (Colombia) reveals antimicrobial potential presented in low abundant isolates. *Open Microbiol. J.* 12 (1): 181–194.

16 Dettmer, K., Aronov, P.A., and Hammock, B.D. (2007). Mass spectrometry-based metabolomics. *Mass Spectrom. Rev.* 26 (1): 51–78.

17 Persson, S., Sönksen, C.P., Frigaard, N.-U. et al. (2000). Pigments and proteins in green bacterial chlorosomes studied by matrix-assisted laser desorption ionization mass spectrometry. *Eur. J. Biochem.* 267 (2): 450–456.

18 Russo, C., Heaton, C., Flint, L. et al. (2020). Emerging applications in mass spectrometry imaging; enablers and roadblocks. *J. Spectral Imaging* 9: 1–21.

19 Aichler, M. and Walch, A. (2015). MALDI imaging mass spectrometry: current frontiers and perspectives in pathology research and practice. *Lab. Invest.* 95 (4): 422–431.

20 Chen, Q., Wang, L., Qi, Y., and Ma, C. (2020). Imaging mass spectrometry of interspecies metabolic exchange revealed the allelopathic interaction between *Microcystis aeruginosa* and its antagonist. *Chemosphere* 259: https://doi.org/10.1016/j.chemosphere.2020.127430.

21 Kim, M., Morrison, M., and Yu, Z. (2011). Evaluation of different partial 16S rRNA gene sequence regions for phylogenetic analysis of microbiomes. *J. Microbiol. Methods* 84 (1): 81–87.

22 Bangayan, N.J., Shi, B., Trinh, J. et al. (2020). MG-MLST: characterising the microbiome at the strain level in metagenomic data. *Microorganisms* (Basel) 8 (5): 684. https://doi.org/10.3390/microorganisms8050684.

23 Han, S., Jeong, Y., and Choi, S. (2021). Current scenario and challenges in the direct identification of microorganisms using MALDI TOF MS. *Microorganisms* (Basel) 9 (9): 1917. https://doi.org/10.3390/microorganisms9091917.

24 Kailasa, S.K., Koduru, J.R., Baek, S.H. et al. (2020). Review on matrix-assisted laser desorption/ionisation time-of-flight mass spectrometry for the rapid screening of microbial species: a promising bioanalytical tool. *Microchem. J.* 159: 105387. https://doi.org/10.1016/j.microc.2020.105387.

25 Ge, M., Kuo, A., Liu, K. et al. (2016). Routine identification of microorganisms by matrix-assisted laser desorption ionisation time-of-flight mass spectrometry: success rate, economic analysis, and clinical outcome. *J. Microbiol. Immunol. Infect.* 50 (5): 662–668.

26 Alcalá, L., Marín, M., Ruiz, A. et al. (2021). Identifying anaerobic bacteria using MALDI-TOF mass spectrometry: a four-year experience. *Front. Cell. Infect. Microbiol.* 11: 521014. https://doi.org/10.3389/fcimb.2021.521014.

27 Patel, T.S., Kaakeh, R., Nagel, J.L. et al. (2017). Cost analysis of implementing matrix-assisted laser desorption ionisation–time of flight mass spectrometry plus real-time antimicrobial stewardship intervention for bloodstream infections. *J. Clin. Microbiol.* 55 (1): 60–67.

28 Branda, J.A., Rychert, J., Burnham, C.D. et al. (2014). Multicenter validation of the VITEK MS v2.0 MALDI-TOF mass spectrometry system for the identification of fastidious gram-negative bacteria. *Diagn. Microbiol. Infect. Dis.* 78 (2): 129–131.

29 Faron, M.L., Buchan, B.W., Hyke, J. et al. (2015). Multicenter evaluation of the Bruker MALDI Biotyper CA system for the identification of clinical aerobic gram-negative bacterial isolates. *PLoS One* 10 (11): e0141350. https://doi.org/10.1371/journal.pone.0141350.

30 Ferreira, L., Sánchez-Juanes, F., González-Ávila, M. et al. (2010). Direct identification of urinary tract pathogens from urine samples by matrix-assisted laser desorption ionization-time of flight mass spectrometry. *J. Clin. Microbiol.* 48 (6): 2110–2115.

31 Garner, O., Mochon, A., Branda, J. et al. (2014). Multi-centre evaluation of mass spectrometric identification of anaerobic bacteria using the VITEK® MS system. *Clin. Microbiol. Infect.* 20 (4): 335–339.

32 Girard, V., Monnin, V., Giraud, D. et al. (2021). Multicenter evaluation of the VITEK MS matrix-assisted laser desorption/ionisation-time of flight mass spectrometry system for identification of bacteria, including *Brucella*, and yeasts. *Eur. J. Clin. Microbiol. Infecti. Dis.* 40 (9): 1909–1917.

33 Uchida-Fujii, E., Niwa, H., Kinoshita, Y., and Nukada, T. (2020). Matrix-assisted laser desorption ionisation–time of flight mass spectrometry (MALDI-TOF MS) for identification of bacterial isolates from horses. *J. Equine Vet. Sci.* 95: 103276. https://doi.org/10.1016/j.jevs.2020.103276.

34 Vatanshenassan, M., Boekhout, T., Meis, J.F. et al. (2019). Candida auris identification and rapid antifungal susceptibility testing against Echinocandins by MALDI-TOF MS. *Front. Cell. Infect. Microbiol.* 9: 20. https://doi.org/10.3389/fcimb.2019.00020.

35 Yuan, Y., Wang, J., Zhang, J. et al. (2020). Evaluation of an optimised method to directly identify bacteria from positive blood cultures using MALDI-TOF mass spectrometry. *J. Clin. Lab. Anal.* 34 (4): e23119. https://doi.org/10.1002/jcla.23119.

36 Afonso, C. and Fenselau, C. (2003). Use of bioactive glass slides for matrix-assisted laser desorption/ionization analysis: application to microorganisms. *Anal. Chem.* 75 (3): 694–697.

37 Bodzon-Kulakowska, A., Bierczynska-Krzysik, A., Dylag, T. et al. (2007). Methods for samples preparation in proteomic research. *J. Chromatogr. B Analyt. Technol. Biomed. Life Sci.* 849 (1–2): 1–31. http://dx.doi.org/10.1016/j.jchromb.2006.10.040.

38 Fagerquist, C.K., Miller, W.G., Harden, L.A. et al. (2005). Genomic and proteomic identification of a DNA-binding protein used in the "fingerprinting" of *Campylobacter* species and strains by MALDI-TOF-MS protein biomarker analysis. *Anal. Chem.* 77 (15): 4897–4907.

39 Ho, Y.P. and Reddy, P.M. (2010). Identification of pathogens by mass spectrometry. *Clin. Chem.* 56 (4): 525–536.

40 Koster, S. and Verpoorte, E. (2007). A decade of microfluidic analysis coupled with electrospray mass spectrometry: an overview. *Lab Chip* 7 (11): 1394–1412.

41 Liang, B., Ju, Y., Joubert, J.R. et al. (2015). Label-free detection and identification of protein ligands captured by receptors in a polymerized planar lipid bilayer using MALDI-TOF MS. *Anal. Bioanal.Chem.* 407 (10): 2777–2789.

42 Moon, D.-C. and Kelley-F, J.A. (1988). A simple desalting procedure for fast atom bombardment mass spectrometry. *Biomed. Environ. Mass Spectrom.* 17: 229–237.

43 Sandrin, T.R., Goldstein, J.E., and Schumaker, S. (2013). MALDI TOF MS profiling of bacteria at the strain level: A review. *Mass Spectrom. Rev.* 32 (3): 188–217.

44 Wang, H. and Hanash, S. (2005). Intact-protein based sample preparation strategies for proteome analysis in combination with mass spectrometry. *Mass Spectrom. Rev.* 24 (3): 413–426.

45 Ho, Y.P. and Reddy, P.M. (2011). Advances in mass spectrometry for the identification of pathogens. *Mass Spectrom. Rev.* 30 (6): 1203–1224.

46 Gekenidis, M.T., Studer, P., Wüthrich, S. et al. (2014). Beyond the matrix-assisted laser desorption ionization (MALDI) biotyping workflow: in search of microorganism-specific tryptic peptides enabling discrimination of subspecies. *Appl. Environ. Microbiol.* 80 (14): 4234–4241.

47 Camara, J.E. and Hays, F.A. (2007). Discrimination between wild-type and ampicillin-resistant *Escherichia coli* by matrix-assisted laser desorption/ionization time-of-flight mass spectrometry. *Anal. Bioanal.Chem.* 389 (5): 1633–1638.

48 Person, M.D., Shen, J., Traner, A. et al. (2006). Protein fragment domains identified using 2D gel electrophoresis/MALDI-TOF. *J. Biomol. Tech.* 17 (2): 145–156.

49 Sun, L., Teramoto, K., Sato, H. et al. (2006). Characterization of ribosomal proteins as biomarkers for matrix-assisted laser desorption/ionization mass spectral identification of *Lactobacillus plantarum*. *Rapid Commun. Mass Spectrom.* 20 (24): 3789–3798.

50 Madonna, A.J., Basile, F., Furlong, E., and Voorhees, K.J. (2001). Detection of bacteria from biological mixtures using immunomagnetic separation combined with matrix-assisted laser desorption/ionization time-of-flight mass spectrometry. *Rapid Commun. Mass Spectrom.* 15 (13): 1068–1074.

51 Johnson, E.M., Ellis, W.R., Powers, L.S., and Wysocki, V.H. (2009). Affinity capture mass spectrometry of biomarker proteins using peptide ligands from biopanning. *Anal. Chem.* 81 (15): 5999–6005.

52 Lefmann, M., Honisch, C., Böcker, S. et al. (2004). Novel mass spectrometry-based tool for genotypic identification of mycobacteria. *J. Clin. Microbiol.* 42 (1): 339–346.

53 Sampath, R., Hall, T.A., Massire, C. et al. (2007a). Rapid identification of emerging infectious agents using PCR and electrospray ionization mass spectrometry. *Ann. N.Y. Acad. Sci.* 1102: 109–120.

54 Sampath, R., Russell, K.L., Massire, C. et al. (2007b). Global surveillance of emerging influenza virus genotypes by mass spectrometry. *PLoS One* 2 (5): https://doi.org/10.1371/journal.pone.0000489.

55 Sauer, S. and Kliem, M. (2010). Mass spectrometry tools for the classification and identification of bacteria. *Nat. Rev. Microbiol.* 8 (1): 74–82.

56 Piao, J., Jiang, J., Xu, B. et al. (2012). Simultaneous detection and identification of enteric viruses by PCR-mass assay. *PLoS One* 7 (8): https://doi.org/10.1371/journal.pone.0042251.

57 Sjöholm, M.I.L., Dillner, J., and Carlson, J. (2008). Multiplex detection of human herpesviruses from archival specimens by using matrix-assisted laser desorption ionization-time of flight mass spectrometry. *J. Clin. Microbiol.* 46 (2): 540–545. 7.

58 Yi, X., Li, J., Yu, S. et al. (2011). A new PCR based mass spectrometry system for high-risk HPV, part I: methods. *Am. J. Clin. Pathol.* 136 (6): 913–919.

59 Honisch, C., Chen, Y., and Hillenkamp, F. (2010). Comparative DNA sequence analysis and typing using mass spectrometry. In: *Mass Spectrometry for Microbial Proteomics* (ed. H.N. Shah and S.E. Gharbia), 441–462. Chichester: John Wiley.

60 Bishop, C., Arnold, C., and Gharbia, S.E. (2010). Transfer of a traditional serotyping system (Kauffmann–White) onto a MALDI-TOF-MS platform for the rapid typing of Salmonella isolates. In: *Mass Spectrometry for Microbial Proteomics* (ed. H.N. Shah and S.E. Gharbia), 463–496. Chichester: Wiley.

61 Rasmussen, P.U., Phan, H.U.T., Frederiksen, M.W., and Madsen, A.M. (2021). A characterisation of bioaerosols in biowaste pre-treatment plants in relation to occupational health. *Waste Manage.*, (Elmsford) 131: 237–248.

62 Madsen, A.M., Frederiksen, M.W., Jacobsen, M.H., and Tendal, K. (2020). Towards a risk evaluation of workers' exposure to handborne and airborne microbial species as exemplified with waste collection workers. *Environ. Res.* 183: 109177. https://doi.org/10.1016/j.envres.2020.109177 ER.

63 Antunes, T.C., Ballarini, A.E., and Van Der Sand, S. (2019). Temporal variation of bacterial population and response to physical and chemical parameters along a petrochemical industry wastewater treatment plant. *An. Acad. Bras. Cienc.* 91 (2): e20180394. https://doi.org/10.1590/0001-3765201920180394 ER.

64 Santos, I.C., Martin, M.S., Carlton, D.D. Jr. et al. (2017). MALDI-TOF MS for the identification of cultivable organic-degrading bacteria in contaminated groundwater near unconventional natural gas extraction sites. *Microorganisms* 5 (3): 47. https://doi.org/10.3390/microorganisms5030047 ER.

65 Hayoun, K., Pible, O., Petit, P. et al. (2020). Proteotyping environmental microorganisms by phylopeptidomics: case study screening water from a radioactive material storage pool. *Microorganisms* (Basel) 8 (10): 1525. https://doi.org/10.3390/microorganisms8101525.

66 Sazykin, I., Minkina, T., Grigoryeva, T. et al. (2019). PAHs distribution and cultivable PAHs degraders' biodiversity in soils and surface sediments of the impact zone of the Novocherkassk thermal electric power plant (Russia). *Environ. Earth Sci.* 78 (19): 1–13.

67 Calzada Urquiza, C., Arvizu Hernández, I., Cruz Medina, J.A. et al. (2017). Identification by MALDI-TOF mass spectrometry of mercury-resistant bacteria associated with the rhizosphere of an apple orchard. *Geomicrobiol. J.* 34 (2): 176–182.

68 Dybwad, M., Granum, P.E., Bruheim, P., and Blatny, J.M. (2012). Characterisation of airborne bacteria at an underground subway station. *Appl. Environ. Microbiol.* 78 (6): 1917–1929.

69 Kozdrój, J., Frączek, K., and Ropek, D. (2019). Assessment of bioaerosols in indoor air of glasshouses located in a botanical garden. *Build. Environ.* 166: 106436. https://doi.org/10.1016/j.buildenv.2019.106436.

70 Mulec, J., Oarga-Mulec, A., Šturm, S. et al. (2017). Spacio-temporal distribution and tourist impact on airborne bacteria in a cave (Škocjan Caves, Slovenia). *Diversity* (Basel) 9 (3): 28. https://doi.org/10.3390/d9030028.

71 Pinar-Méndez, A., Fernández, S., Baquero, D. et al. (2021). Rapid and improved identification of drinking water bacteria using the drinking water library, a dedicated MALDI-TOF MS database. *Water Res.* (Oxford) 203: 117543. https://doi.org/10.1016/j.watres.2021.117543.

72 Sala-Comorera, L., Caudet-Segarra, L., Galofré, B. et al. (2020). Unravelling the composition of tap and mineral water microbiota: divergences between next-generation sequencing techniques and culture-based methods. *Int. J. Food Microbiol.* 334: 108850. https://doi.org/10.1016/j.ijfoodmicro.2020.108850.

73 Pascale, M.R., Mazzotta, M., Salaris, S. et al. (2020). Evaluation of MALDI-TOF mass spectrometry in diagnostic and environmental surveillance of *Legionella* species: A comparison with culture and mip-gene sequencing technique. *Front. Microbiol.* 11: 589369. https://doi.org/10.3389/fmicb.2020.589369.

74 Vidal, L.M.R., Venas, T.M., Gonçalves, A.R.P. et al. (2020). Rapid screening of marine bacterial symbionts using MALDI-TOF MS. *Arch. Microbiol.* 202 (8): 2329–2336.

75 Popovic, N.T., Kazazic, S.P., Strunjak-Perovic, I., and Coz-Rakovac, R. (2017). Differentiation of environmental aquatic bacterial isolates by MALDI-TOF MS. *Environ. Res.* 152: 7–16. https://doi.org/10.1016/j.envres.2016.09.020 ER.

76 Moussa, M., Cauvin, E., Le Piouffle, A. et al. (2021). A MALDI-TOF MS database for fast identification of *Vibrio* spp. potentially pathogenic to marine mollusks. *Appl. Microbiol. Biotechnol.* 105 (6): 2527–2539.

77 Voigt, A.M., Ciorba, P., Döhla, M. et al. (2020). The investigation of antibiotic residues, antibiotic resistance genes and antibiotic-resistant organisms in a drinking water reservoir system in Germany. *Int. J. Hyg. Environ. Health* 224: 113449. https://doi.org/10.1016/j.ijheh.2020.113449.

78 Lepesova, K., Olejnikova, P., Mackuľak, T. et al. (2019). Annual changes in the occurrence of antibiotic-resistant coliform bacteria and enterococci in municipal wastewater. *Environ. Sci. Pollut. Res. Int.* 26 (18): 18470–18483.

79 Tuohy, J.M., Mueller-Spitz, S.R., Albert, C.M. et al. (2018). MALDI-TOF MS affords discrimination of *Deinococcus aquaticus* isolates obtained from diverse biofilm habitats. *Front. Microbiol.* 9: 2442. https://doi.org/10.3389/fmicb.2018.02442.

80 Didier de Vasconcelos, G.M., Mulinari, J., de Souza, S.M.d.A.G.U. et al. (2021). Biodegradation of azo dye-containing wastewater by activated sludge: a critical review. *World J. Microbiol. Biotechnol.* 37 (6): 101. https://doi.org/10.1007/s11274-021-03067-6.

81 Yasir, M., Qureshi, A.K., Khan, I. et al. (2019). Culturomics-based taxonomic diversity of bacterial communities in the hot springs of Saudi Arabia. *Omics* (Larchmont, N.Y.) 23 (1): 17–27.

82 Timperio, A.M., Gorrasi, S., Zolla, L., and Fenice, M. (2017). Evaluation of MALDI-TOF mass spectrometry and MALDI BioTyper in comparison to 16S rDNA sequencing for the identification of bacteria isolated from Arctic Sea water. *PLoS One* 12 (7): e0181860. https://doi.org/10.1371/journal.pone.0181860.

83 Farahani-Kofoet, R.D., Witzel, K., Graefe, J. et al. (2020). Species-specific impact of *Fusarium* infection on the root and shoot characteristics of asparagus. *Pathogens* (Basel) 9 (6): 509. https://doi.org/10.3390/pathogens9060509.

84 Przemieniecki, S.W., Gorczyca, A., Matras, E. et al. (2020). Bacteria isolated from the aeration chamber of wastewater treatment plants used in the biocontrol and promotion of wheat growth. *Agronomy (Basel)* 10 (11): 1792. https://doi.org/10.3390/agronomy10111792.

85 Abdelaal, K., AlKahtani, M., Attia, K. et al. (2021). The role of plant growth-promoting bacteria in alleviating the adverse effects of drought on plants. *Biology* (Basel, Switzerland) 10 (6): 520. https://doi.org/10.3390/biology10060520.

86 Crossay, T., Antheaume, C., Redecker, D. et al. (2017). New method for the identification of arbuscular mycorrhizal fungi by proteomic-based biotyping of spores using MALDI-TOF-MS. *Sci. Rep.* 7 (1): 14306–14316. https://doi.org/10.1038/s41598-017-14487-6.

87 Vannini, C., Domingo, G., Fiorilli, V. et al. (2021). Proteomic analysis reveals how pairing of a mycorrhizal fungus with plant growth-promoting bacteria modulates growth and defense in wheat. *Plant Cell Environ.* 44 (6): 1946–1960.

88 Dalluge, J., Brown, E., and Connell, L. (2019). Toward a rapid method for the study of biodiversity in cold environments: the characterisation of psychrophilic yeasts by MALDI-TOF mass spectrometry. *Extremophiles* Life Under Extreme Conditions 23 (4): 461–466.

89 Caliskan, D., Bakkaloğlu, Z., Numanoglu Cevik, Y. et al. (2021). Maldi-TOF MS identification and antibiotic resistance of *Escherichia coli* isolated from playground. *Microb. Pathogen.* 159: 105155. https://doi.org/10.1016/j.micpath.2021.105155.

90 de Santana, F.S., Gracioso, L.H., Karolski, B. et al. (2019). Isolation of bisphenol A-tolerating/degrading *Shewanella haliotis* strain MH137742 from an estuarine environment. *Appl. Biochem. Biotechnol.* 189 (1): 103–115.

91 Pramanik, K., Mitra, S., Sarkar, A., and Maiti, T.K. (2018). Alleviation of phytotoxic effects of cadmium on rice seedlings by cadmium resistant PGPR strain *Enterobacter aerogenes* MCC 3092. *J. Hazard. Mater.* 351: 317–329.

92 Rani, R., Kumar, V., Gupta, P., and Chandra, A. (2021). Potential use of *Solanum lycopersicum* and plant growth promoting rhizobacterial (PGPR) strains for the phytoremediation of endosulfan stressed soil. *Chemosphere* (Oxford) 279: 130589. https://doi.org/10.1016/j.chemosphere.2021.130589.

93 Zuzolo, D., Guarino, C., Tartaglia, M., and Sciarrillo, R. (2020). Plant-soil-microbiota combination for the removal of total petroleum hydrocarbons (TPH): an in-field experiment. *Front. Microbiol.* 11: 621581. https://doi.org/10.3389/fmicb.2020.621581.

94 Salas-Moreno, M., Contreras-Puentes, N., Rodríguez-Cavallo, E. et al. (2019). Protein carbonylation as a biomarker of heavy metal, Cd and Pb, damage in *Paspalum fasciculatum* Willd. ex Flüggé. *Plants*, (Basel) 8 (11): 513. https://doi.org/10.3390/plants8110513.

95 Kang, W., Zheng, J., Bao, J. et al. (2020). Characterisation of the copper resistance mechanism and bioremediation potential of an *Acinetobacter calcoaceticus* strain isolated from copper mine sludge. *Environ. Sci. Pollut. Res. Int.* 27 (8): 7922–7933.

96 Carolin, C.F., Kumar, P.S., Chitra, B. et al. (2021). Stimulation of *Bacillus* sp. by lipopeptide biosurfactant for the degradation of aromatic amine 4-Chloroaniline. *J. Hazard. Mater.* 415: 125716. https://doi.org/10.1016/j.jhazmat.2021.125716.

97 Chan, W., Wildeboer, D., Garelick, H., and Purchase, D. (2018). Competition of As and other group 15 elements for surface binding sites of an extremophilic *Acidomyces acidophilus* isolated from a historical tin mining site. *Extremophiles*, Life Under Extreme Conditions 22 (5): 795–809.

98 Lagier, J.C., Khelaifia, S., Alou, M.T. et al. (2016). Culture of previously uncultured members of the human gut microbiota by culturomics. *Nat. Microbiol.* 1: http://dx.doi.org/10.1038/nmicrobiol.2016.203.

99 Rahi, P. and Vaishampayan, P. (2020). Editorial: MALDI-TOF MS application in microbial ecology studies. *Front. Microbiol.* 10: https://doi.org/10.3389/fmicb.2019.02954.

100 Sood, U., Kumar, R., and Hira, P. (2021). Expanding culturomics from gut to extreme environmental settings. *MSystems* e0084821. https://doi.org/10.1128/mSystems.00848-21.

101 Clark, C.M., Costa, M.S., Sanchez, L.M., and Murphy, B.T. (2018). Coupling MALDI-TOF mass spectrometry protein and specialized metabolite analyses to rapidly discriminate bacterial function. *Proc. Natl. Acad. Sci. U.S.A.* 115 (19): 4981–4986. https://doi.org/10.1073/pnas.1801247115.

102 Clark, C.M., Costa, M.S., Conley, E. et al. (2019). Using the open-source MALDI TOF-MS IDBac pipeline for analysis of microbial protein and specialized metabolite data. *J. Visualized Exp.* 147: e59219. https://doi.org/10.3791/59219.

103 Edwards, J.L. and Kennedy, R.T. (2005). Metabolomic analysis of eukaryotic tissue and prokaryotes using negative mode MALDI time-of-flight mass spectrometry. *Anal. Chem.* 77 (7): 2201–2209.

104 Covington, B.C., McLean, J.A., and Bachmann, B.O. (2017). Comparative mass spectrometry-based metabolomics strategies for the investigation of microbial secondary metabolites. *Nat. Prod. Rep.* 34 (1): 6–24.

105 Fastner, J., Erhard, M., and von Döhren, H. (2001). Determination of oligopeptide diversity within a natural population of *Microcystis* spp. (cyanobacteria) by typing single colonies by matrix-assisted laser desorption ionization-time of flight mass spectrometry. *Appl. Environ. Microbiol.* 67 (3–12): 5069–5076.

106 Abdellaziz, L., Chollet, M., Abderrahmani, A. et al. (2018). Lipopeptide biodiversity in antifungal *Bacillus* strains isolated from Algeria. *Arch. Microbiol.* 200 (8): 1205–1216.

107 Lam, V.B., Meyer, T., Arias, A.A. et al. (2021). Bacillus cyclic lipopeptides iturin and fengycin control rice blast caused by *Pyricularia oryzae* in potting and acid sulfate soils by direct antagonism and induced systemic resistance. *Microorganisms* 9 (7): 1441. http://dx.doi.org/10.3390/microorganisms9071441.

108 Antón, J., Lucio, M., Peña, A. et al. (2013). High metabolomic microdiversity within co-occurring isolates of the extremely halophilic bacterium *Salinibacter ruber*. *PLoS One* 8 (5): https://doi.org/10.1371/journal.pone.0064701 https://doi.org/10.1016/j.aca.2014.06.032.

109 Galleguillos, P.A., Grail, B.M., Hallberg, K.B. et al. (2018). Identification of trehalose as a compatible solute in different species of acidophilic bacteria. *J. Microbiol.* 56 (10): 727–733.

110 Guljamow, A., Kreische, M., Ishida, K. et al. (2017). High-density cultivation of terrestrial Nostoc strains leads to reprogramming of secondary metabolome. *Appl. Environ. Microbiol.* 83 (23): http://dx.doi.org/10.1128/AEM.01510-17.

111 Gautam, K., Tripathi, J.K., Pareek, A., and Sharma, D.K. (2019). Growth and secretome analysis of possible synergistic interaction between green algae and cyanobacteria. *J. Biosci. Bioeng.* 127 (2): 213–221.

112 McCann, A., Kune, C., la Rocca, R. et al. (2021). Rapid visualization of lipopeptides and potential bioactive groups of compounds by combining ion mobility and MALDI imaging mass spectrometry. *Drug Discovery Today: Technol.* 39 (12): 81–88.

113 Sandrin, T.R. and Demirev, P.A. (2018). Characterization of microbial mixtures by mass spectrometry. *Mass Spectrom. Rev.* 37 (3): 321–349.

114 Michelsen, C.F., Khademi, S.M.H., Johansen, H.K. et al. (2016). Evolution of metabolic divergence in *Pseudomonas aeruginosa* during long-term infection facilitates a proto-cooperative interspecies interaction. *ISME J.* 10 (6): 1330–1336.

115 Phelan, V.V., Fang, J., and Dorrestein, P.C. (2015). Mass spectrometry analysis of *Pseudomonas aeruginosa* treated with azithromycin. *J. Am. Soc. Mass. Spectrom.* 26 (6): 873–877.

116 Esquenazi, E., Coates, C., Simmons, L. et al. (2008). Visualizing the spatial distribution of secondary metabolites produced by marine cyanobacteria and sponges via MALDI-TOF imaging. *Mol. BioSyst.* 4 (6): 562–570.

117 Rychert, J. (2019). Benefits and limitations of MALDI-TOF mass spectrometry for the identification of microorganisms. *J. Infect.* 2 (4): 1–5.

118 Rahi, P., Prakash, O., and Shouche, Y.S. (2016). Matrix-assisted laser desorption/ionization time-of-flight mass-spectrometry (MALDI-TOF MS) based microbial identifications: challenges and scopes for microbial ecologists. *Front. Microbiol.* 7: 1359. https://doi.org/10.3389/fmicb.2016.01359.

119 Bozik, M., Mrázková, M., Novotná, K. et al. (2021). MALDI-TOF MS as a method for rapid identification of *Phytophthora* de Bary, 1876. *PeerJ* (San Francisco, CA) 9: e11662. https://doi.org/10.7717/peerj.11662.

120 Cuénod, A., Foucault, F., Pflüger, V., and Egli, A. (2021). Factors associated with MALDI-TOF mass spectral quality of species identification in clinical routine diagnostics. *Front Cell. Infect. Microbiol.* 11: https://doi.org/10.3389/fcimb.2021.646648.

121 Williams, T.L., Andrzejewski, D., Lay, J.O., and Musser, S.M. (2003). Experimental factors affecting the quality and reproducibility of MALDI TOF mass spectra obtained from whole bacteria cells. *J. Am. Soc. Mass. Spectrom.* 14 (4): 342–351.

122 Almasoud, N., Xu, Y., Nicolaou, N., and Goodacre, R. (2014a). Optimization of matrix assisted desorption/ionization time of flight mass spectrometry (MALDI-TOF-MS) for the characterization of *Bacillus* and *Brevibacillus* species. *Anal. Chim. Acta* 840: 49–57.

123 Ryzhov, V., Hathout, Y., and Fenselau, C. (2000). Rapid characterization of spores of *Bacillus cereus* group bacteria by matrix-assisted laser desorption-ionization time-of-flight mass spectrometry. *Appl. Environ. Microbiol.* 66 (9): 3828–3834.

124 Freiwald, A. and Sauer, S. (2009). Phylogenetic classification and identification of bacteria by mass spectrometry. *Nat. Protoc.* 4 (5): 732–742.

125 Topić Popović, N., Kazazić, S.P., Bojanić, K. et al. (2021). Sample preparation and culture condition effects on MALDI-TOF MS identification of bacteria: a review. *Mass Spectrom. Rev.* https://doi.org/10.1002/mas.21739.

126 Madonna, A.J., Basile, F., Ferrer, I. et al. (2000). On-probe sample pretreatment for the detection of proteins above 15 KDa from whole cell bacteria by matrix-assisted laser desorption/ionization time-of-flight mass spectrometry. *Rapid Commun. Mass Spectrom.* 14 (23): 2220–2229. http://dx.doi.org/10.1002/1097-0231(20001215)14:23%3c2220:AID-RCM155%3e3.0.CO;2-4.

127 Lay, J.O. (2001). MALDI-TOF mass spectrometry of bacteria. *Mass Spectrom. Rev.* 20 (4): 172–194.

128 Tsuchida, S., Umemura, H., and Nakayama, T. (2020). Current status of matrix-assisted laser desorption/ionization/time-of-flight mass spectrometry (MALDI-TOF MS) in clinical diagnostic microbiology. *Molecules* 25 (20): https://doi.org/10.3390/molecules25204775.

129 Hou, T.Y., Chiang-Ni, C., and Teng, S.H. (2019). Current status of MALDI-TOF mass spectrometry in clinical microbiology. *J. Food Drug Anal.* 27 (2): 404–414.

130 Paauw, A., Caspers, M.P.M., Schuren, F.H.J. et al. (2008). Genomic diversity within the *Enterobacter cloacae* complex. *PLoS One* 3 (8): https://doi.org/10.1371/journal.pone.0003018.

131 Pupo, G.M., Lan, R., and Reeves, P.R. (2000). Multiple independent origins of *Shigella* clones of *Escherichia coli* and convergent evolution of many of their characteristics. *Proc. Natl. Acad. Sci. U.S.A.* 97 (19): 10567–10572.

132 Consortium (2007). The universal protein resource (UniProt). *Nucleic Acids Res.* 35 ((SUPPL. 1)): https://doi.org/10.1093/nar/gkl929.

133 Cheng, D., Qiao, L., and Horvatovich, P. (2018). Toward spectral library-free matrix-assisted laser desorption/ionization time-of-flight mass spectrometry bacterial identification. *J. Proteome Res.* 17 (6): 2124–2130.

134 Ceballos-Garzon, A., Amado, D., Vélez, N. et al. (2020). Development and validation of an in-house library of Colombian *Candida auris* strains with MALDI-TOF MS to improve yeast identification. *J. Fungi* 6 (2): https://doi.org/10.3390/jof6020072.

135 Guo, C., Cheng, Y., Liu, M. et al. (2019). Development of the portable mass spectrometer and its application in environmental monitoring. *E3S Web Conf.* 118: http://dx.doi.org/10.1051/e3sconf/201911803055.

136 Rentschler, S., Kaiser, L., and Deigner, H. (2021). Emerging options for the diagnosis of bacterial infections and the characterisation of antimicrobial resistance. *Int. J. Mol. Sci.* 22 (1): 456. https://doi.org/10.3390/ijms22010456.

11

From Genomics to MALDI-TOF MS: Diagnostic Identification and Typing of Bacteria in Veterinary Clinical Laboratories

John Dustin Loy[1] *and Michael L. Clawson*[2]

[1] *School of Veterinary Medicine and Biomedical Sciences, Institute for Agriculture and Natural Resources, University of Nebraska-Lincoln, Lincoln, NE, USA*
[2] *United States Department of Agriculture, Agricultural Research Service, U.S. Meat Animal Research Center, Clay Center, NE, USA*

11.1 Introduction

We live in a microbial world. Since Antonie van Leeuwenhoek first saw unicellular organisms "animalcules" under his microscope in the 1670s, our push to understand bacteria has been driven in part by an expanding recognition that they are arguably the most successful domain of life on the planet [1]. Bacteria have lived and died on planet Earth for billions of years in diverse niches only now being appreciated and understood [2–4]. Life for many eukaryotes flows through bacteria and their fundamental roles in cycling carbon, nitrogen, oxygen, and sulfur, and producing cellular energy [5, 6]. That bacteria also occupy niches involving death and disease to other organisms is not surprising given their overall success as a life form [7, 8]. The potential in which bacteria can operate as pathogens in humans, animals, and other organisms can vary significantly at the subspecies or strain level [9–11]. Thus, it is important in both human and veterinary diagnostic laboratories to identify bacteria at the highest resolution to help illuminate the scope of that potential in each clinical isolate.

Bacteria can vary significantly at the subspecies level in their gene content [11]. Although bacterial species have a core genome, where every member of the population has the same set of genes, individual bacteria typically have additional genes that are unevenly distributed to others of their species, including those that encode pathogenic elements [11, 12]. This is typically due to horizontal gene transfer and selection, which are major drivers of bacterial evolution, with the code and/or signal driving the expression of many pathogenic elements, such as exoproteins, outer membrane proteins (OMPs), and glycans, present on either chromosomes or extrachromosomal genetic elements [11, 13].

In addition to different gene content, there are microbiome components, host factors, and environmental conditions that can result in bacteria expressing themselves as

Microbiological Identification using MALDI-TOF and Tandem Mass Spectrometry: Industrial and Environmental Applications, First edition. Edited by Haroun N. Shah, Saheer E. Gharbia, Ajit J. Shah, Erika Y. Tranfield, and K. Clive Thompson.

pathogenic or non-pathogenic at the subspecies level [11]. Interestingly, some bacteria have prion-like proteins, and there is evidence that archaea may also have them [14–16]. Thus, it may be that non-Mendelian inheritance of prions or other epigenetic factors is blurring the line between genotype and some phenotypes across and within bacteria at the species level. However, for many bacterial pathogens, there is a direct connection between a genotype, expression of virulence factors, and the potential for a pathogenic phenotype. This dogma is the foundation for moving from genomic discovery to expressed protein profiles using matrix-assisted laser desorption/ionization time-of-flight mass spectrometry (MALDI-TOF MS) for bacteria identification and strain typing.

The focus of this chapter is to discuss how whole-genome DNA sequence and protein profiles of veterinary bacteria can be brought together through genomics and MALDI-TOF MS for high-resolution species, subspecies, and sub-strain identification, including screening for strain types with an increased potential for pathogenesis. The nexus of these two emerging approaches helps to provide insights into one of the most frequent challenges faced by veterinary clinical laboratories, to determine if a microbe isolated from a clinical sample undergoing a disease process is etiological or is more likely to be a commensal in nature.

11.2 Genomics

The golden age of genomics was arguably born in the mid-1970s with the publication of the Sanger and Maxam–Gilbert DNA sequencing techniques [17, 18]. It was more loudly ushered in with whole-genome sequencing and assembly of *Haemophilus influenzae* and *Mycoplasma genitalium* in the mid-1990s [19, 20], and punctuated with joint announcements in 2000 that the Human Genome Project (HGP) and the private company Celera had each completed working drafts of the human genome [21, 22]. Over 20 years later, genomic sequencing of viral, prokaryotic, and eukaryotic genomes has become both routine and commonplace. The National Center for Biotechnology Information (NCBI) held more than 20 000 eukaryote genomes, 370 000 prokaryote genomes, and 46 000 virus genomes as of December 2021 [23], and these numbers should continue to increase sharply into the foreseeable future.

Sequencing technologies are commonly referred to by generations [24]. Both the Sanger and Maxam–Gilbert techniques are regarded as first generation. The Maxam–Gilbert technique incorporates chemical modification of DNA followed by digestion and electrophoresis and, although initially popular, is rarely used today [25]. The Sanger technique involves chain terminating nucleotides, was adapted to capillary electrophoresis, and is still used by some laboratories that do not require large-scale sequencing in terms of read numbers or length [24, 25].

The second generation came on board in 2004 and continues today as an approach that employs relatively short read length sequencing of single molecules of DNA with high throughput and massive amounts of reads generated per run [24]. Ion Torrent and Illumina sequencing platforms are both second generation. While they and other second-generation platforms can generate large amounts of sequence data, it can often be difficult to assemble complete chromosomes from second-generation sequence, due to regions of repetitive sequences, including tandem repeats, segmental duplications, or multiple phage genome insertions [26, 27]. The sequence reads are simply too short to span the repeat regions and thus anchor with confidence in the correct region of the genome.

Third-generation sequencing focused on longer read length to circumvent the shortcomings of the second generation. Two primary platform representatives are Pacific BioSciences (PacBio) single molecule, real-time (SMRT) sequencing and Oxford Nanopore Technologies (ONT) [24, 25]. Both technologies involve moving large DNA molecules through a stationary readout system. The PacBio platform became commercially available in 2011 [28]. Single-stranded DNA is circularized with linkers and sequencing takes place on a SMRT cell chip containing pores called zero-mode wavelengths [24, 29]. Each pore contains an immobilized DNA polymerase, which binds to a linker site and initiates sequencing. The strand can then be iteratively sequenced, with the number of passes through a complete circularized DNA strand dependent on the length of the strand. Depending on the exact PacBio sequencer in use, reads 10–20 kb are frequently generated, although the platform is capable of producing sequencing reads that can exceed 135 kb [24, 29].

The ONT platform is considered by some as a bridge between third- and fourth-generation sequencing due to the long length of DNA reads that can be produced at low cost compared with other sequencing technologies [25, 30]. The first ONT product, the MinION, was made commercially available in 2015 [30]. The platform incorporates linkers and a motor protein onto double-stranded DNA [24, 30]. The modified double-stranded DNA is captured onto pores that are on a flow cell membrane. The membrane is immersed in an electrolyte solution, with high resistance and a current flowing through the pores due to ion movement. Once the DNA is captured onto the pore, a single strand moves through, which disrupts ion movement. Each base that moves through the pore can be characterized by the disruption pattern of ion movement, resulting in a single strand of DNA passing through the pore. As each nucleotide passes through the pore, disruption of the ion current occurs with specificity to each type of nucleotide [24, 30]. ONT reads can commonly range from 10 to 60 kb depending on the exact machine and settings, and the platform has generated reads exceeding 1 MB [31].

Although third-generation sequencing technologies produce significantly longer reads than second-generation ones, third-generation platforms have historically had higher error rates [24]. A solution to managing the shortcomings of second- and third-generation technologies has been to combine them, with long and short reads being used for genome assembly and error correction, respectively, an approach often called "hybrid genome assembly" [12, 24]. Short reads from the Illumina platform are particularly useful in error-correcting PacBio long reads, as PacBio reads can contain homopolymer errors, whereas the Illumina platform is not prone to creating them [32, 33]. For organisms with smaller genomes, such as viruses and prokaryotes, this hybrid sequencing approach is an effective way to produce high-quality closed genomes with low error rates, although it requires considerable investment in sequencing resources. With hybrid sequencing, high-quality genomic sequences of microbes can be produced rapidly, and combined, contrasted, or compared with other typing methods such as MALDI-TOF MS, to characterize bacteria by both genomic and phenomic metrics, and to type them with high resolution. This approach is also well suited to understanding aspects of their biology, including a propensity, or lack thereof, for pathogenesis.

11.3 Defining Bacterial Species Through Genomics

It was Aristotle who provided the early groundwork for taxonomic classification. He focused on classifying animals and used the terms "genus" (*genos*) and/or "species" (*eidos*) in his works *History of Animals*, *Parts of Animals*, and *Categories*. As our understanding of life has grown since his time, so too have our taxonomical categories and definitions. "Genus" in particular and "species" have different meanings today compared with the time of Aristotle. We now recognize the microbial world we live in thanks to the observations of Antonie van Leeuwenhoek and the golden age of microbiology that was to follow [34]. Two of the three domains of life (Archaea, Bacteria, Eukarya) are prokaryotic, and our criteria for defining species, especially microbial species, has changed, particularly in recent years with the advent of whole-genome sequencing [35–38].

Classically, microbial taxonomy has relied on a polyphasic integration of phenotypic, genotypic, and phylogenetic information [36, 37]. A multitude of methods have been used to classify microbes in the absence of whole-genome sequencing, such as DNA–DNA hybridization, ribosomal 16S gene sequence and phylogenic analysis, in addition to traditional phenotypic tests that characterize bacterial morphology, physiology, metabolism, and biochemistry [36–38]. DNA–DNA hybridization is a gold standard genotypic tool for microbial species designation, where members of the same species share 70% or greater DNA–DNA relatedness; however, this method is not commonly used today, as variation can occur between laboratories and their protocols, and a database of profiles cannot be developed [38]. Only a few specialized laboratories perform DNA–DNA hybridization, making it inaccessible and impractical for laboratories to use routinely [38].

The ribosomal 16S gene has become an integral tool for determination of microbial species due to high conservation of the gene between prokaryotes and minimal horizontal gene transfer thought to occur between 16S genes [39, 40]. This high conservation allows for "universal" PCR-based assays that can be broadly used to generate 16S sequences from diverse organisms to provide diagnostic and phylogenetic information for microbial identification to clinical laboratories and researchers [41]. A percentage identity $\leq 98.7\%$ between 16S sequences indicates that, for many organisms, the strains of origin are different species [36]. However, different strains can have 16S sequences that are $> 98.7\%$, which precludes using a universal hard cut-off of 16S nucleotide sequence percentage identity and has generated calls for a more rigorous one [42]. Also, bacteria contain multiple copies of 16S genes within their genomes and there can be within-genome variation of their 16S gene sequences [43]. The utility of 16S in clinical laboratories to discriminate species varies depending on the organism, and for some the sequence of additional housekeeping genes or other genes may be required. Consequently, some genus-specific guidance for clinical laboratories has been published to apply sequencing approaches for identification [44]. One complication is that horizontal gene transfer does occur between at least some species of bacteria at the 16S gene locus without detrimental effects and may be widespread as an ongoing evolutionary strategy of bacteria [45]. If this is indeed the case, 16S rDNA loci would be similar to much of the bacterial genome, by being chimeric representatives of both horizontal and vertical descent.

The value of phenotypic tests that characterize bacterial morphology, physiology, metabolism, and biochemistry is not to be diminished, especially when they are collected en masse, and they are still the foundation for identification in clinical laboratories [46, 47]. Newer phenotypic platforms like MALDI-TOF MS have become invaluable tools for microbial identification but are dependent upon correct identification of reference organisms for library entries. These problematic issues and limitations of DNA–DNA hybridization and 16S sequence/phylogeny have facilitated a need for additional methods and approaches for microbial taxonomy based on genotypes or phylogeny. As whole-genome sequencing of microbial genomes has exploded during the current millennium, it was inevitable that better tools for determining microbial taxonomy that utilized genomics would rise to the fore.

One example is the development of metrics that utilize whole-genome sequence and express genome-relatedness either dependently or independently of sequence alignments [38]. Many of the alignment-based tools are either based on average nucleotide identity (ANI) or are very similar to it [38]. ANI is "a pairwise measure of overall similarity between two genome sequences" [48]. In this approach, a genome is fragmented into pieces which are subjected to a search, alignment, and identity calculation with an intact comparison genome [48]. ANI was developed in 2005 and has undergone improvements and modifications since then to become a replacement for DNA–DNA hybridization and a widely used metric for comparing bacterial genomes and taxonomic placement at the species level [38, 49]. To that end, the NCBI now routinely runs ANI on prokaryotic genomes received for submission with a reference type strain to confirm or correct the proposed species designation accompanying the submission [50, 51]. The NCBI uses a default ANI cut-off of 96% with genome coverage values to assure competent matches, although exceptions to the cut-offs are allowed [51] (NCBI, personal communication).

Although ANI is a preferred metric for species identification using whole-genome sequencing, metrics that do not employ alignments for comparing genomes can be useful in confirming ANI results or resolving genomes that are on the threshold of species inclusion or exclusion. Alignment-free methods search for patterns and frequencies in nucleotide sequence at the level of nucleotides (G/C content), dinucleotides, and tetranucleotides [38]. Differences in these characteristics can coincide at the species level for reasons that are not entirely understood [52, 53]. These methods are not computationally demanding to run and are particularly useful when databases are available for comparisons and analyses of genomes from bacteria of unknown species [54]. In summary, there are a multitude of tools that employ different approaches for whole-genome comparisons and accurate microbial identification.

11.4 MALDI-TOF MS

Proteomics has long been explored as a method for microbial identification. Work using whole-organism protein electrophoresis had demonstrated potential and high levels of agreement with gold-standard identification methods such as DNA–DNA hybridization [55]. The potential for MS to be able to recognize specific biomarkers that enable identification has been studied since the 1970s [56]. However, it took application of soft ionization MS techniques, such as MALDI-TOF MS, applied to unfractionated products of microorganisms to realize the potential of proteomic biomarkers for microbial identification [57, 58].

MALDI-TOF MS has revolutionized veterinary clinical microbiology in much the same way it has in human clinical laboratories. Specifically, the use of MALDI-TOF MS to generate proteomic fingerprints from whole cells grown on a solid medium and prepared using a direct-smear approach in a high-throughput and automated manner has moved this technology into the core of microbiology laboratory workflows [59]. The use of MALDI-TOF MS to generate proteomic fingerprints has significant advantages, as results are not strongly influenced by growth conditions and are consistent when generated among multiple laboratories and users [60, 61]. MALDI-TOF MS-based methods for microbial identification have been shown to agree with biochemical or sequencing methods in numerous comparative studies using large collections of organisms in both human and veterinary clinical laboratories, many with 99% or higher levels of agreement to the species level [62–66]. However, these identifications remain database-driven and are limited by the diversity and robustness of the user library [59].

To overcome this challenge, one solution is the development of in-house libraries that are useful to assist in resolving closely related species or those that are novel to a host species or geographic location [67]. The capacity is extremely important in veterinary clinical laboratories where there is tremendous diversity in host species, animal environments, and pathogen frequencies among laboratories. Another approach is the utilization and application of bioinformatics tools for discovery of peaks associated with specific biotypes, phenotypes, pathotypes, or subspecies of interest, which may provide relevant clinical and epidemiological information to veterinarians and diagnosticians. Several examples and an approach to discovery of these peaks will be described in detail later in the chapter.

Numerous MALDI-TOF MS methods have been developed for applications outside of routine identification of bacterial cells and several show promise for veterinary laboratories. These include the direct detection of bacteria from matrices and clinical samples, including blood culture media [68], cerebrospinal fluid [69], and urine [70]. MALDI-TOF MS has also been shown to detect pathogens directly from bronchioalveolar lavage (BAL) fluid or following culture in specific enrichment media, as BAL is a common sample submitted to veterinary laboratories for bovine respiratory disease (BRD) diagnostics [71, 72]. Potential applications outside of bacterial identification in veterinary laboratories also include respiratory and enteric virus identification in infected cell lines [73, 74]. Also, applications towards species-specific identification of fungal organisms, nematode parasites, and even large exoparasites such as ticks and mosquitoes have been explored, highlighting the diverse potential for MALDI-TOF MS applications across veterinary clinical diagnostics [75–79].

One core function of microbiology laboratories is antimicrobial susceptibility testing (AST). AST remains a labor-intensive and time-consuming process, and MALDI-TOF MS has potential applications in this area as well. Several approaches have been developed, including peak discovery approaches, modified growth procedures, isotope labeling, and detection of enzymatic activity [80–82]. Of particular interest to veterinary laboratories is a MALDI Biotyper antibiotic susceptibility test rapid assay (MBT-ASTRA), which uses a semi-quantitative approach to determine microbial growth in an antimicrobial containing broth, making it flexible to use with different microbe and drug combinations. For example, MBT-ASTRA has been successfully used to determine tetracycline resistance in BRD pathogens rapidly [83]. Other approaches have been developed to detect the hydrolysis of drugs in the beta-lactam class by microbial beta-lactamases [84]. Some have explored using stable isotope-containing media or examined microbial growth in microdroplets of antibiotic-containing media directly on the target plate, such as a MALDI-TOF MS-based direct-on-target microdroplet growth assay (DOT-MGA) [85, 86]. Advancement in this area is needed, as most veterinary laboratories rely primarily on disk diffusion or customized broth microdilution assays for AST, both of which require significant labor, time, and quality control, and require inventory of numerous species-specific antibiotic panels or disks.

The use of MALDI-TOF MS as a tool to investigate the relationships of veterinary-associated microbes in space and time epidemiologically has also been explored. These tools are frequently needed for disease outbreak investigations to establish clonality of outbreak strains and examine potential sources of infection. Additionally, for many endemic or opportunistic bacterial diseases, there is often a need for clinicians or veterinarians to explore potential relatedness amongst circulating strains in host populations and to conduct surveillance for emerging strains. Methods, such as multilocus sequence typing and pulsed-field gel electrophoresis have typically been used to examine these relationships; however, both are cost- and time-prohibitive. Additionally, for many veterinary pathogens, there may not be a previously established typing scheme or method, or a previous one may be limited in use, as many pathogens are only studied by a small number of laboratories. As an alternative, some researchers have used MALDI-TOF MS biotyping to explore these relationships [87, 88]. As veterinary laboratories, at least in the United States, are often involved in the procurement of herd-specific isolates for autogenous veterinary vaccines, these types of comparisons may be useful to evaluate herd-level changes in circulating pathogens which may be useful for veterinarians to implement effective disease prevention and management strategies [89].

New approaches to enable typing of organisms to the subspecies level are also required, given that organisms within the same species can have vastly different clinical outcomes. In addition to examples from our research (described later), *Streptococcus equi* ssp. *equi* is known to cause a highly infectious respiratory disease called strangles in horses, while *S. equi* ssp. *zooepidemicus* are opportunistic pathogens or commensals found on mucous membranes [90]. Specific individual mass spectrum peaks have been shown to be associated with these two subspecies, making these peaks extremely useful for diagnostic laboratories and veterinary clinicians [91]. Likewise, *Staphylococcus aureus* has subspecies that are clinically important. *S. aureus* ssp. *aureus* has a broad range of hosts and potential diseases, where *S. aureus* ssp. *anaerobius* appear more limited in host range and disease type. The presence or absence of five specific mass spectrum peaks has been shown to help laboratories distinguish these two subspecies, providing significant benefit to clinicians and microbiologists [92]. Although MALDI-TOF MS approaches have improved identification of the veterinary staphylococci over biochemical analysis, there have remained some challenges using this approach in other closely related veterinary pathogens like *Staphylococcus pseudintermedius* and other members in the *Staphylcoccus intermedius* group (SIG) of organisms [93].

11.5 Combining Genomics with MALDI-TOF MS to Classify Bacteria at the Subspecies Level

We have focused on discovering bacteria at the subspecies level that vary in their association with some of the most significant diseases of cattle. This includes identifying the genetic determinants responsible for those associations and developing MALDI-TOF MS tests that classify subspecies of importance [9, 12, 94–99]. Our approach is a combination of population genetics, clinical metadata, genomic sequencing and analysis, and MALDI-TOF MS profiling (Figure 11.1). We start by assembling a collection of isolates that represent the bacterial species broadly at the population level [9, 94]. The initial goal is to assemble

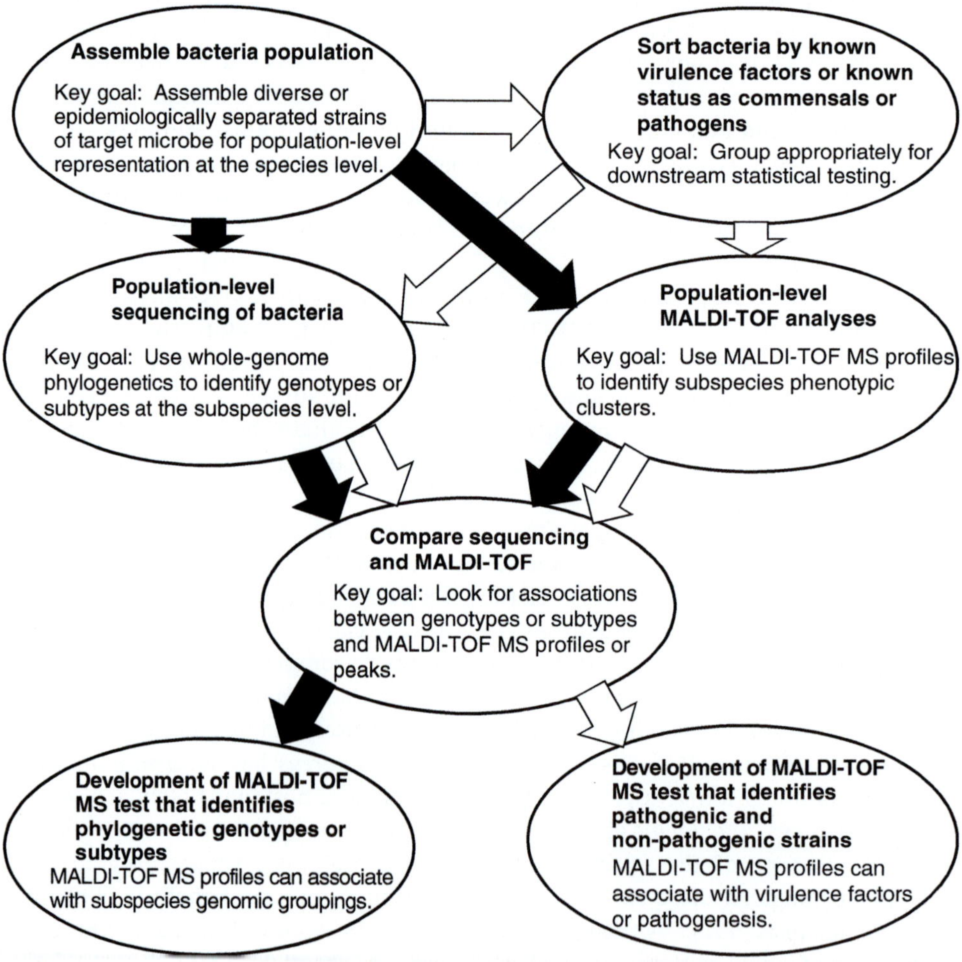

Figure 11.1 Identification of bacteria genotypes or subtypes, virulence, or pathogenesis factors. The black arrows show the pathway to develop MALDI-TOF MS tests for bacteria subspecies genotypes or subtypes identified from whole-genome sequencing. The white arrows show the pathway to develop MALDI-TOF MS tests for virulence or pathogenesis factors harbored by a portion of the bacterial population.

strains that are epidemiologically unlinked in time and space within a host niche of interest, such as cattle, to capture the extent of genomic diversity within the population.

Many of the bacteria we have studied are opportunistic pathogens, such as members of the *Pasteurellaceae* that cause BRD and the *Moraxella* species that cause or associate with infectious bovine keratoconjunctivitis (IBK) or bovine pinkeye, both significant diseases of cattle. Particularly for bacteria characterized as opportunistic pathogens, it is of great importance for population-level strain collections to include strains from animals both with and without signs of clinical disease. Once sequencing and phylogenetic analyses are completed for the strains within a population-level collection, and substrain genotypes or subtypes are identified, associations between those genotypes or subtypes can be merged with metadata components such as disease, or lack thereof, in the host, to potentially reveal associations between them. MALDI-TOF MS tests can be developed and used to identify and classify strains of interest (Figure 11.1).

We first applied this approach to *Mannheimia haemolytica*, which is a major bacterial cause of BRD [100]. A collection of 1131 strains of *M. haemolytica*, all isolated from North American cattle, was sequenced on the Illumina platform [9]. Given that *M. haemolytica* is recognized as an opportunistic pathogen normally found in the upper respiratory tract of cattle and can move down into the lungs during times of stress, the collection consisted of strains isolated from the lungs of cattle with BRD, and the nasopharynx of cattle either with or without BRD. Phylogenetic analysis of the sequences revealed a phylogenetic tree deeply divided into two clades, which represented two distinct genotypes of *M. haemolytica* at the substrain level. Although representatives of both genotypes were found in the diseased lungs of cattle and the nasopharynx of cattle with or without signs of BRD, genotype 2 was predominantly found in diseased lungs over genotype 1. This indicates that genotype 2 is much more of an opportunistic pathogen, and genotype 1 is much more of a commensal to cattle, although experimental work is needed to confirm that. Further sequencing and analyses revealed OMPs that were specific to genotype 2 *M. haemolytica* and not genotype 1, including adhesion proteins that are suspected to be important virulence factors [12]. MALDI-TOF MS profiles of genotype 1 and 2 strains showed peaks specifically associated with the genotypes, and a MALDI-TOF MS test for the two genotypes was developed. Thus, the combined genomics and MALDI-TOF MS approach identified two genotypes of *M. haemolytica* that do not associate equally with BRD, along with DNA sequence, OMPs, and MALDI-TOF MS profiles that distinguish between them. The MALDI-TOF MS-based test that classifies *M. haemolytica* strains into genotypes can provide efficient, cost-effective, and meaningful data to microbiologists and clinicians [97].

We also applied the combined genomics and MALDI-TOF MS approach to *Moraxella bovoculi*, a frequent resident of the bovine eye that is associated with IBK [101]. A collection of 246 strains collected from the eyes of cattle with or without IBK were sequenced on the Illumina platform [96]. Similar to *M. haemolytica*, a phylogenetic tree of the sequences was split into two major clades, representing two major genotypes. Only genotype 1 strains were isolated from IBK eyes; however, future work is needed to show the statistical significance of this due to biases in how genotype 1 and 2 strains were isolated and selected for on mixed plates of field samples. Importantly, genotype 1 strains can have virulence factors such as hemolysins that have not been observed in genotype 2 strains. Two MALDI-TOF MS typing tests, one that distinguishes the two *M. bovoculi* genotypes and one distinguishing strains with hemolysin, were developed for veterinary use [98, 99]. Of note, *M. bovoculi* genotypes 1 and 2 are taxonomically very similar to *Moraxella ovis*, and could be reclassified as substrains of *M. ovis* in the future based on additional whole-genome sequence and ANI analyses [101] (J.D. Loy and M.L. Clawson, unpublished observations).

11.6 Data Exploration with MALDI-TOF MS

Leveraging comparative genomics and clinical metadata can frequently be useful to discover fundamental differences at the genomic level that correlate with more commensal types versus those that may be more associated with pathogenicity. The tagging of single nucleotide polymorphisms (SNPs), which are alleles that tag a particular clade or region of a phylogenetic tree, or other genetic markers identified during genomic analysis can be used to discern genotypes in isolates [12, 102]. However, genetic tests require nucleic acid extraction followed by further processing and analysis. MALDI-TOF MS has long been recognized for its potential to classify numerous organisms at the subspecies level, which is why we have explored developing MALDI-TOF MS tests that accurately score important genotypes at the subspecies level [97, 98, 103].

One way to leverage genomic and MALDI-TOF MS data is to explore mass spectrum for the potential to identify specific peaks associated with genotype or strain differences, by applying techniques initially developed to mine clinical sample mass spectrum data for biomarkers associated with disease states [104]. Once discovered, these peaks can be identified in MALDI-TOF MS spectrum collected during standard MALDI-TOF MS runs for bacterial identification, decreasing the time, labor, and consumable cost for strain typing [105].The first step in this process is to identify a collection of strains to be used as a training set (see Figure 11.2). These strains should represent known diversity within the genotypes to be targeted. High-quality spectra can then be collected from this training set, which, as when creating main spectrum profiles (MSP) files for library addition, should be collected with both technical and biological replication to capture variation. The raw data can then be deposited into libraries for future matching, as well as categorized into genotype or phenotype groups for peak exploration. We have used the software program, ClinProTools, which enables several approaches and machine learning models to explore differences in the training set spectrum and develop classifier algorithms to type unknown isolates [106, 107].

This biomarker-based approach has been used to develop typing strategies at the subspecies level for numerous Gram-negative and Gram-positive human pathogens with some success [108, 109]. There are several software packages, including open source, available that have similar functions and the fundamentals are quite similar [110]. These functions include comparison of spectrum for peak differences between the genotypes or phenotypes, which include statistical analysis and ranking. In some instances, the differences between genotypes in some organisms is represented by a single peak shift that can be readily identified using simple exploration or visualization of raw mass spectrum data. For example, genotypic differences between the two major genotypes of *M. haemolytica* result in a single peak shift that is easily discerned, either with a QuickClassifier model in ClinProTools or by visually inspecting raw mass spectra [97].

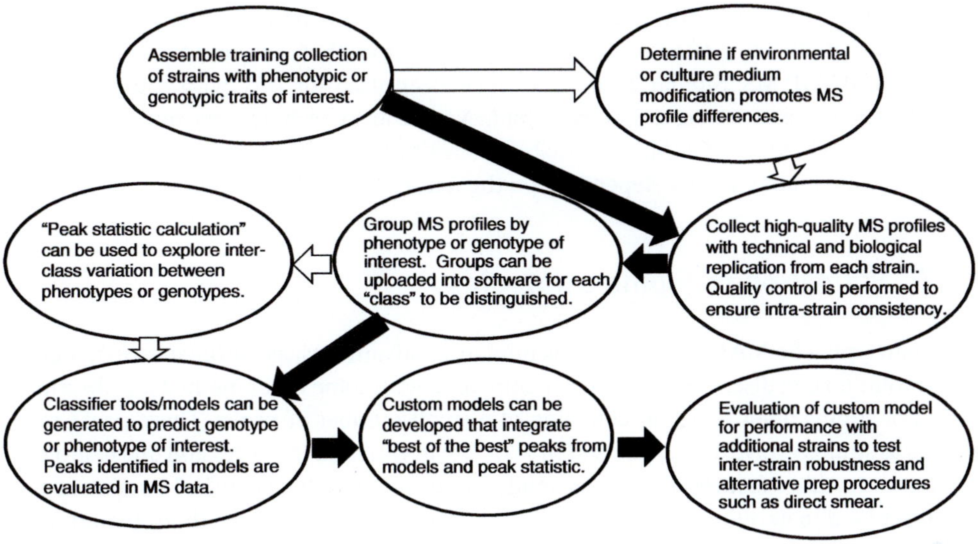

Figure 11.2 Workflow for MALDI-TOF MS biomarker-based typing assays. The black arrows show the suggested required steps for peak discovery. The white arrows show potential steps that may improve results depending on the phenotype or genotype of interest to be typed.

Additionally, if more complex approaches are needed to discover peak differences, classifier models included in the software, such as Genetic Algorithm and Supervised Neural Network, can be applied to find the best combinations of peaks that enable more complex discernment of the genotype of interest. We have found that no single approach or model is optimum in every scenario, and instead a hybrid approach to discover peaks that are consistently included in multiple classifier models is often useful. For example, Hille et al. used several different models within ClinProTools to identify multiple peaks that discriminated genotypic differences in *M. bovoculi* [98]. These peaks were then used to make a custom classifier model within the software that represented the "best of the best" peaks discovered using numerous data exploration methods. Although peaks can be visualized readily in mass spectrum data, we have also found that development and implementation of a model-based classifier is useful to ensure objective classification of unknown strains. Veterinary clinical microbiology and diagnostic laboratories are accredited externally by various means, and the use of objective classifier models with relevant controls make standardizing these procedures for use in quality systems and staff training much more straightforward. The classifier models can also be readily integrated into workflows after identification has been confirmed to ensure models are applied correctly. As the identification algorithms of most instruments already include calibrations and other quality control procedures, if the same spectrum is employed for classification, the use of classifier models can be used as a simple reflex test.

Media formulations can also be used to develop MALDI-TOF MS-based typing approaches that may not be readily apparent using routine non-selective growth media. Hille et al. utilized calcium-supplemented solid media formulations to enhance the toxin expression and/or activity in *M. bovoculi*, which translated into differences in MALDI-TOF MS mass spectrum [99]. The peak differences were detected with model-based approaches. As a result, toxin-producing strains could be distinguished from those without toxin expression with MALDI-TOF MS. Similar approaches have been used for AST with specific media formulations to detect microbial activity [86]. A wide array of media formulations when combined with MALDI-TOF MS may enable many phenotypic or other useful clinical tests to be performed more rapidly, depending on the nature of the phenotype.

11.7 Validation of Typing Strategies

Although typing on MALDI-TOF MS has numerous advantages, as outlined earlier, it does lack resolution and discriminatory power compared with some other methods, as discrimination of strains often relies on detection of a small number of peaks. These peaks may change as strains evolve over space and time or due to variation in expression due to growth conditions or other factors. Therefore, validation of these assays is critical outside of the initial set used to develop the method. One approach recommends five times the number of isolates be used for the validation process that are independent of the training set [105]. However, this provides numerous additional challenges to veterinary diagnostic laboratories, where resources are limited and the number of strains of a certain pathogen is few. It can be challenging to accumulate sufficient strains for robust validation, especially those that represent true diversity across the species. Additionally, the workflow proposed here requires sufficient metadata from the laboratory that isolated the strains to help determine the potential for pathogenicity. Currently, the amount of metadata collected by veterinary clinical laboratories that could be used towards this purpose is tremendous, but it is inconsistent in both depth and format of collected data.

11.8 Future Directions

Both genomics and MALDI-TOF MS will be useful in identifying and characterizing bacteria and other organisms at the subspecies level into the future. Whole-genome sequencing will continue to usher in high-resolution identification of organisms and correct classification of organisms previously misidentified with lower-resolution methods. This process will concurrently improve the robustness of both genomic and MALDI-TOF MS databases, with cross-communication between the platforms. Veterinary diagnostic laboratories are working towards ways of harmonizing the collection of metadata associated with diagnostic and clinical testing, which would greatly enhance the ability to assemble strain collections to further develop and validate new strain typing assays based on both MALDI-TOF MS and genomic sequencing platforms. Additionally, future standardizations on data format and communications systems will allow for aggregation and sharing of data on the thousands of diverse samples tested by veterinary diagnostic laboratories every day [111]. These approaches would help laboratories to collaborate in the validation of new strain typing tools that provide cost-effective and clinically valuable information to veterinarians, researchers, and animal health personnel on animal pathogens, including opportunists and those with complex etiologies. Additionally, as genomic and MALDI-TOF MS platforms continue to improve along with their supporting infrastructure and procedures, and the databases for the two platforms thus become more closely aligned, the capacity for pathogen detection will move closer to real time. Consequently, these technologies should increase our ability to prevent the emergence (or re-emergence) of severe infectious veterinary disease or disease outbreaks, as well as that of newly introduced strains.

References

1 McFall-Ngai, M.J. (2015). Giving microbes their due –animal life in a microbially dominant world. *J. Exp. Biol.* 218 (Pt 12): 1968–1973. https://doi.org/10.1242/jeb.115121.

2 Javaux, E.J. (2019). Challenges in evidencing the earliest traces of life. *Nature* 572 (7770): 451–460. https://doi.org/10.1038/s41586-019-1436-4.

3 Runge, S. and Rosshart, S.P. (2021). The mammalian metaorganism: a holistic view on how microbes of all kingdoms and niches shape local and systemic immunity. *Front. Immunol.* 12: 702378. https://doi.org/10.3389/fimmu.2021.702378.

4 Ando, N., Barquera, B., Bartlett, D.H. et al. (2021). The molecular basis for life in extreme environments. *Annu. Rev. Biophys.* 50: 343–372. https://doi.org/10.1146/annurev-biophys-100120-072804.

5 Todar, K.G. (2006). *Todar's Online Textbook of Bacteriology*. Madison, WI: University of Wisconsin-Madison https://www.textbookofbacteriology.net/ (accessed 27 September 2022).

6 Roger, A.J., Munoz-Gomez, S.A., and Kamikawa, R. (2017). The origin and diversification of mitochondria. *Curr. Biol.* 27 (21): R1177–R1192. https://doi.org/10.1016/j.cub.2017.09.015.

7 Bennet, J.E., Douglas, R., and Blaser, M.J. (2020). *Mandell, Douglas, and Bennett's Principles and Practice of Infectious Diseases*, 9e. Elsevier.

8 Aiello, S.E., Moses, M.A., and Allen, D.G. (2016). *The Merck Veterinary Manual*, 11e. Kenilworth, NJ: Merck & Co.

9 Clawson, M.L., Murray, R.W., Sweeney, M.T. et al. (2016). Genomic signatures of *Mannheimia haemolytica* that associate with the lungs of cattle with respiratory disease, an integrative conjugative element, and antibiotic resistance genes. *BMC Genomics* 17 (1): 982. https://doi.org/10.1186/s12864-016-3316-8.

10 Bono, J.L., Smith, T.P., Keen, J.E. et al. (2012). Phylogeny of Shiga toxin-producing *Escherichia coli* O157 isolated from cattle and clinically ill humans. *Mol. Biol. Evol.* 29 (8): 2047–2062. https://doi.org/10.1093/molbev/mss072.

11 Ehrlich, G.D., Hiller, N.L., and Hu, F.Z. (2008). What makes pathogens pathogenic. *Genome Biol.* 9 (6): 225. https://doi.org/10.1186/gb-2008-9-6-225.

12 Clawson, M.L., Schuller, G., Dickey, A.M. et al. (2020). Differences between predicted outer membrane proteins of genotype 1 and 2 *Mannheimia haemolytica*. *BMC Microbiol.* 20 (1): 250. https://doi.org/10.1186/s12866-020-01932-2.

13 Poole, J., Day, C.J., von Itzstein, M. et al. (2018). Glycointeractions in bacterial pathogenesis. *Nat. Rev. Microbiol.* 16 (7): 440–452. https://doi.org/10.1038/s41579-018-0007-2.

14 Yuan, A.H. and Hochschild, A. (2017). A bacterial global regulator forms a prion. *Science* 355 (6321): 198–201. https://doi.org/10.1126/science.aai7776.

15 Fleming, E., Yuan, A.H., Heller, D.M., and Hochschild, A. (2019). A bacteria-based genetic assay detects prion formation. *Proc. Natl. Acad. Sci. U.S.A.* 116 (10): 4605–4610. https://doi.org/10.1073/pnas.1817711116.

16 Zajkowski, T., Lee, M.D., Mondal, S.S. et al. (2021). The hunt for ancient prions: archaeal prion-like domains form amyloid-based epigenetic elements. *Mol. Biol. Evol.* 38 (5): 2088–2103. https://doi.org/10.1093/molbev/msab010.

17 Sanger, F., Nicklen, S., and Coulson, A.R. (1977). DNA sequencing with chain-terminating inhibitors. *Proc. Natl. Acad. Sci. U. S. A.* 74 (12): 5463–5467. https://doi.org/10.1073/pnas.74.12.5463.

18 Maxam, A.M. and Gilbert, W. (1977). A new method for sequencing DNA. *Proc. Natl. Acad. Sci. U. S. A.* 74 (2): 560–564. https://doi.org/10.1073/pnas.74.2.560.

19 Fleischmann, R.D., Adams, M.D., White, O. et al. (1995). Whole-genome random sequencing and assembly of *Haemophilus influenzae* Rd. *Science* 269 (5223): 496–512. https://doi.org/10.1126/science.7542800.

20 Fraser, C.M., Gocayne, J.D., White, O. et al. (1995). The minimal gene complement of *Mycoplasma genitalium*. *Science* 270 (5235): 397–403. https://doi.org/10.1126/science.270.5235.397.

21 Lander, E.S., Linton, L.M., Birren, B. et al. (2001). Initial sequencing and analysis of the human genome. *Nature* 409 (6822): 860–921. https://doi.org/10.1038/35057062.

22 Venter, J.C., Adams, M.D., Myers, E.W. et al. (2001). The sequence of the human genome. *Science* 291 (5507): 1304–1351. https://doi.org/10.1126/science.1058040.

23 National Center for Biotechnology Information (NCBI). 2021 Genome Information by Organism. https://www.ncbi.nlm.nih.gov/genome/browse#!/overview (accessed 12 December 2021).

24 Hu, T., Chitnis, N., Monos, D., and Dinh, A. (2021). Next-generation sequencing technologies: an overview. *Hum. Immunol.* https://doi.org/10.1016/j.humimm.2021.02.012.

25 Slatko, B.E., Gardner, A.F., and Ausubel, F.M. (2018). Overview of next-generation sequencing technologies. *Curr. Protoc. Mol. Biol.* 122 (1): e59. https://doi.org/10.1002/cpmb.59.

26 Alkan, C., Sajjadian, S., and Eichler, E.E. (2011). Limitations of next-generation genome sequence assembly. *Nat. Methods* 8 (1): 61–65. https://doi.org/10.1038/nmeth.1527.

27 Ho, S.S., Urban, A.E., and Mills, R.E. (2020). Structural variation in the sequencing era. *Nat. Rev. Genet.* 21 (3): 171–189. https://doi.org/10.1038/s41576-019-0180-9.

28 Litchfield N (2011). Pacific Biosciences begins shipments of commercial PacBio RS systems. Press release (27 April). https://www.pacb.com/press_releases/pacific-biosciences-begins-shipments-of-commercial-pacbio-rs-systems (accessed 27 September 2022).

29 Ardui, S., Ameur, A., Vermeesch, J.R., and Hestand, M.S. (2018). Single molecule real-time (SMRT) sequencing comes of age: applications and utilities for medical diagnostics. *Nucleic Acids Res.* 46 (5): 2159–2168. https://doi.org/10.1093/nar/gky066.

30 Lin, B., Hui, J., and Mao, H. (2021). Nanopore technology and its applications in gene sequencing. *Biosensors (Basel)* 11 (7): https://doi.org/10.3390/bios11070214.

31 Miga, K.H., Koren, S., Rhie, A. et al. (2020). Telomere-to-telomere assembly of a complete human X chromosome. *Nature* 585 (7823): 79–84. https://doi.org/10.1038/s41586-020-2547-7.

32 Dohm, J.C., Peters, P., Stralis-Pavese, N., and Himmelbauer, H. (2020). Benchmarking of long-read correction methods. *NAR Genom. Bioinform.* 2 (2): lqaa037. https://doi.org/10.1093/nargab/lqaa037.

33 Besser, J., Carleton, H.A., Gerner-Smidt, P. et al. (2018). Next-generation sequencing technologies and their application to the study and control of bacterial infections. *Clin. Microbiol. Infect.* 24 (4): 335–341. https://doi.org/10.1016/j.cmi.2017.10.013.

34 Kolter, R. (2021). The history of microbiology-a personal interpretation. *Annu. Rev. Microbiol.* 75: 1–17. https://doi.org/10.1146/annurev-micro-033020-020648.

35 Woese, C.R., Kandler, O., and Wheelis, M.L. (1990). Towards a natural system of organisms: proposal for the domains Archaea, Bacteria, and Eucarya. *Proc. Natl. Acad. Sci. U. S. A.* 87 (12): 4576–4579. https://doi.org/10.1073/pnas.87.12.4576.

36 Achtman, M. and Wagner, M. (2008). Microbial diversity and the genetic nature of microbial species. *Nat. Rev. Microbiol.* 6 (6): 431–440. https://doi.org/10.1038/nrmicro1872.

37 Vandamme, P., Pot, B., Gillis, M. et al. (1996). Polyphasic taxonomy, a consensus approach to bacterial systematics. *Microbiol. Rev.* 60 (2): 407–438. https://doi.org/10.1128/mr.60.2.407-438.1996.

38 Sant'Anna, F.H., Bach, E., Porto, R.Z. et al. (2019). Genomic metrics made easy: what to do and where to go in the new era of bacterial taxonomy. *Crit. Rev. Microbiol.* 45: 182–200. https://doi.org/10.1080/1040841X.2019.1569587.

39 Doolittle, W.F. (1999). Phylogenetic classification and the universal tree. *Science* 284: 2124–2128. https://doi.org/10.1126/science.284.5423.2124.

40 Woese, C.R. (2000). Interpreting the universal phylogenetic tree. *Proc. Natl. Acad. Sci. U. S. A.* 97 (15): 8392–8396. https://doi.org/10.1073/pnas.97.15.8392.

41 Weisburg, W.G., Barns, S.M., Pelletier, D.A., and Lane, D.J. (1991). 16S ribosomal DNA amplification for phylogenetic study. *J. Bacteriol.* 173 (2): 697–703.

42 Edgar, R.C. (2018). Updating the 97% identity threshold for 16S ribosomal RNA OTUs. *Bioinformatics* 34 (14): 2371–2375. https://doi.org/10.1093/bioinformatics/bty113.

43 Johnson, J.S., Spakowicz, D.J., Hong, B.Y. et al. (2019). Evaluation of 16S rRNA gene sequencing for species and strain-level microbiome analysis. *Nat. Commun.* 10 (1): 5029. https://doi.org/10.1038/s41467-019-13036-1.

44 Clinical and Laboratory Standards Institute (CLSI) (2008). *Interpretive Criteria for Identification of Bacteria and Fungi by DNA Target Sequencing; Approved Guideline.* CLSI document MM18-A, vol. 12, no. 12. Wayne, PA: CLSI.

45 Miyazaki, K. and Tomariguchi, N. (2019). Occurrence of randomly recombined functional 16S rRNA genes in *Thermus thermophilus* suggests genetic interoperability and promiscuity of bacterial 16S rRNAs. *Sci. Rep.* 9 (1): 11233. https://doi.org/10.1038/s41598-019-47807-z.

46 Bochner, B.R. (2009). Global phenotypic characterization of bacteria. *FEMS Microbiol. Rev.* 33 (1): 191–205. https://doi.org/10.1111/j.1574-6976.2008.00149.x.

47 Carroll, K.C., Pfaller, M.A., Landry, M.L. et al. (2019). *Manual of Clinical Microbiology.* ASM Press.

48 Yoon, S.H., Ha, S.M., Lim, J. et al. (2017). A large-scale evaluation of algorithms to calculate average nucleotide identity. *Antonie Van Leeuwenhoek* 110 (10): 1281–1286. https://doi.org/10.1007/s10482-017-0844-4.

49 Parks, D.H., Chuvochina, M., Chaumeil, P.A. et al. (2020). A complete domain-to-species taxonomy for Bacteria and Archaea. *Nat. Biotechnol.* 38 (9): 1079–1086. https://doi.org/10.1038/s41587-020-0501-8.

50 Federhen, S., Rossello-Mora, R., Klenk, H.-P. et al. (2016). Meeting report: GenBank microbial genomic taxonomy workshop (12–13 May, 2015). *Stand Genomic Sci.* 11: 15. https://doi.org/10.1186/s40793-016-0134-1.

51 Ciufo, S., Kannan, S., Sharma, S. et al. (2018). Using average nucleotide identity to improve taxonomic assignments in prokaryotic genomes at the NCBI. *Int. J. Syst. Evol. Microbiol.* 68 (7): 2386–2392. https://doi.org/10.1099/ijsem.0.002809.

52 van Passel, M.W.J., Kuramae, E.E., Luyf, A.C.M. et al. (2006). The reach of the genome signature in prokaryotes. *BMC Evol. Biol.* 6: 84. https://doi.org/10.1186/1471-2148-6-84.

53 Richter, M. and Rossello-Mora, R. (2009). Shifting the genomic gold standard for the prokaryotic species definition. *Proc. Natl. Acad. Sci. U.S.A.* 106 (45): 19126–19131. https://doi.org/10.1073/pnas.0906412106.

54 Richter, M., Rossello-Mora, R., Oliver Glockner, F., and Peplies, J. (2016). JSpeciesWS: a web server for prokaryotic species circumscription based on pairwise genome comparison. *Bioinformatics* 32 (6): 929–931. https://doi.org/10.1093/bioinformatics/btv681.

55 Vandamme, P., Vancanneyt, M., Van Belkum, A. et al. (1996). Polyphasic analysis of strains of the genus *Capnocytophaga* and Centers for Disease Control group DF-3. *Int. J. Syst. Bacteriol.* 46 (3): 782–791. https://doi.org/10.1099/00207713-46-3-782.

56 Anhalt, J.P. and Fenselau, C. (1975). Identification of bacteria using mass spectrometry. *Anal. Chem.* 47 (2): 219–225. https://doi.org/10.1021/ac60352a007.

57 Lay, J.O. Jr. (2001). MALDI-TOF mass spectrometry of bacteria*. *Mass Spectrom. Rev.* 20 (4): 172–194. https://doi.org/10.1002/mas.10003.

58 Fenselau, C. and Demirev, P.A. (2001). Characterization of intact microorganisms by MALDI mass spectrometry. *Mass Spectrom. Rev.* 20 (4): 157–171. https://doi.org/10.1002/mas.10004.

59 Khot, P.D., Couturier, M.R., Wilson, A. et al. (2012). Optimization of matrix-assisted laser desorption ionization–time of flight mass spectrometry analysis for bacterial identification. *J. Clin. Microbiol.* 50 (12): 3845–3852. https://doi.org/10.1128/jcm.00626-12.

60 Wunschel, S.C., Jarman, K.H., Petersen, C.E. et al. (2005). Bacterial analysis by MALDI-TOF mass spectrometry: an inter-laboratory comparison. *J. Am. Soc. Mass Spectrom.* 16 (4): 456–462. https://doi.org/10.1016/j.jasms.2004.12.004.

61 Jarman, K.H., Daly, D.S., Petersen, C.E. et al. (1999). Extracting and visualizing matrix-assisted laser desorption/ionization time-of-flight mass spectral fingerprints. *Rapid Commun. Mass Spectrom.* 13 (15): 1586–1594. https://doi.org/10.1002/(SICI)1097-0231(19990815)13:15<1586::AID-RCM680>3.0.CO;2-2.

62 Randall, L.P., Lemma, F., Koylass, M. et al. (2015). Evaluation of MALDI-TOF as a method for the identification of bacteria in the veterinary diagnostic laboratory. *Res. Vet. Sci.* 101: 42–49. https://doi.org/10.1016/j.rvsc.2015.05.018.

63 Cherkaoui, A., Hibbs, J., Emonet, S. et al. (2010). Comparison of two matrix-assisted laser desorption ionization-time of flight mass spectrometry methods with conventional phenotypic identification for routine identification of bacteria to the species level. *J. Clin. Microbiol.* 48 (4): 1169–1175. https://doi.org/10.1128/JCM.01881-09.

64 Wilson David, J., Middleton John, R., Adkins Pamela, R.F. et al. (2019). Test agreement among biochemical methods, matrix-assisted laser desorption ionization–time of flight mass spectrometry, and 16S rRNA sequencing for identification of microorganisms isolated from bovine milk. *J. Clin. Microbiol.* 57 (3): e01381–e01318. https://doi.org/10.1128/JCM.01381-18.

65 Wragg, P., Randall, L., and Whatmore, A.M. (2014). Comparison of Biolog GEN III MicroStation semi-automated bacterial identification system with matrix-assisted laser desorption ionization-time of flight mass spectrometry and 16S ribosomal RNA gene sequencing for the identification of bacteria of veterinary interest. *J. Microbiol. Methods* 105: 16–21. https://doi.org/10.1016/j.mimet.2014.07.003.

66 Kuhnert, P., Bisgaard, M., Korczak, B.M. et al. (2012). Identification of animal Pasteurellaceae by MALDI-TOF mass spectrometry. *J. Microbiol. Methods* 89 (1): 1–7. https://doi.org/10.1016/j.mimet.2012.02.001.

67 Robbins, K., Dickey, A.M., Clawson, M.L., and Loy, J.D. (2018). Matrix-assisted laser desorption/ionization time-of-flight mass spectrometry identification of *Moraxella bovoculi* and *Moraxella bovis* isolates from cattle. *J. Vet. Diagn. Invest.* 30 (5): 739–742. https://doi.org/10.1177/1040638718789725.

68 Ulrich, S., Gottschalk, C., Straubinger, R.K. et al. (2020). Acceleration of the identification of sepsis-inducing bacteria in cultures of dog and cat blood. *J. Small Anim. Pract.* 61 (1): 42–45. https://doi.org/10.1111/jsap.13056.

69 Segawa, S., Sawai, S., Murata, S. et al. (2014). Direct application of MALDI-TOF mass spectrometry to cerebrospinal fluid for rapid pathogen identification in a patient with bacterial meningitis. *Clin. Chim. Acta* 435: 59–61. https://doi.org/10.1016/j.cca.2014.04.024.

70 Ferreira, L., Sánchez-Juanes, F., González-Avila, M. et al. (2010). Direct identification of urinary tract pathogens from urine samples by matrix-assisted laser desorption ionization-time of flight mass spectrometry. *J. Clin. Microbiol.* 48 (6): 2110–2115. https://doi.org/10.1128/JCM.02215-09.

71 Van Driessche, L., Bokma, J., Deprez, P. et al. (2019). Rapid identification of respiratory bacterial pathogens from bronchoalveolar lavage fluid in cattle by MALDI-TOF MS. *Sci. Rep.* 9 (1): https://doi.org/10.1038/s41598-019-54599-9.

72 Bokma, J., Van Driessche, L., Deprez, P. et al. (2020). Rapid identification of *Mycoplasma bovis* strains from bovine bronchoalveolar lavage fluid with matrix-assisted laser desorption ionization–time of flight mass spectrometry after enrichment procedure. *J. Clin. Microbiol.* 58 (6): https://doi.org/10.1128/jcm.00004-20.

73 Calderaro, A., Arcangeletti, M.C., Rodighiero, I. et al. (2016). Identification of different respiratory viruses, after a cell culture step, by matrix assisted laser desorption/ionization time of flight mass spectrometry (MALDI-TOF MS). *Sci. Rep.* 6 (1): 36082. https://doi.org/10.1038/srep36082.

74 Sillman, S.J. (2018). *Investigations into the Pathogenesis and Characteristics of Novel Swine Enteric Coronaviruses*, 156. Ann Arbor: The University of Nebraska - Lincoln.

75 Feucherolles, M., Poppert, S., Utzinger, J., and Becker, S.L. (2019). MALDI-TOF mass spectrometry as a diagnostic tool in human and veterinary helminthology: a systematic review. *Parasit. Vectors* 12 (1): 245. https://doi.org/10.1186/s13071-019-3493-9.

76 Becker, P., Normand, A.-C., Vanantwerpen, G. et al. (2019). Identification of fungal isolates by MALDI-TOF mass spectrometry in veterinary practice: validation of a web application. *J. Vet. Diagn. Invest.* 31 (3): 471–474. https://doi.org/10.1177/1040638719835577.

77 Baumbach, C.-M., Müller, S., Reuschel, M. et al. (2021). Identification of zoophilic dermatophytes using MALDI-TOF mass spectrometry. *Front. Cell. Infect. Microbiol.* 11: 365.

78 Yssouf, A., Flaudrops, C., Drali, R. et al. (2013). Matrix-assisted laser desorption ionization–time of flight mass spectrometry for rapid identification of tick vectors. *J. Clin. Microbiol.* 51 (2): 522–528. https://doi.org/10.1128/JCM.02665-12.

79 Rakotonirina, A., Pol, M., Kainiu, M. et al. (2020). MALDI-TOF MS: optimization for future uses in entomological surveillance and identification of mosquitoes from New Caledonia. *Parasit. Vectors* 13 (1): 359. https://doi.org/10.1186/s13071-020-04234-8.

80 Mather Cheryl, A., Werth Brian, J., Sivagnanam, S. et al. (2016). Rapid detection of vancomycin-intermediate *Staphylococcus aureus* by matrix-assisted laser desorption ionization–time of flight mass spectrometry. *J. Clin. Microbiol.* 54 (4): 883–890. https://doi.org/10.1128/JCM.02428-15.

81 Lange, C., Schubert, S., Jung, J. et al. (2014). Quantitative matrix-assisted laser desorption ionization-time of flight mass spectrometry for rapid resistance detection. *J. Clin. Microbiol.* 52 (12): 4155–4162. https://doi.org/10.1128/jcm.01872-14.

82 Florio, W., Baldeschi, L., Rizzato, C. et al. (2020). Detection of antibiotic-resistance by MALDI-TOF mass spectrometry: an expanding area. *Front. Cell. Infect. Microbiol.* 10: 616.

83 Van Driessche, L., Bokma, J., Gille, L. et al. (2018). Rapid detection of tetracycline resistance in bovine *Pasteurella multocida* isolates by MALDI Biotyper antibiotic susceptibility test rapid assay (MBT-ASTRA). *Sci. Rep.* 8 (1): https://doi.org/10.1038/s41598-018-31562-8.

84 Sparbier, K., Schubert, S., Weller, U. et al. (2012). Matrix-assisted laser desorption ionization–time of flight mass spectrometry-based functional assay for rapid detection of resistance against β-lactam antibiotics. *J. Clin. Microbiol.* 50 (3): 927–937. https://doi.org/10.1128/JCM.05737-11.

85 Idelevich, E.A., Sparbier, K., Kostrzewa, M., and Becker, K. (2018). Rapid detection of antibiotic resistance by MALDI-TOF mass spectrometry using a novel direct-on-target microdroplet growth assay. *Clin. Microbiol. Infect.* 24 (7): 738–743. https://doi.org/10.1016/j.cmi.2017.10.016.

86 Demirev, P.A., Hagan, N.S., Antoine, M.D. et al. (2013). Establishing drug resistance in microorganisms by mass spectrometry. *J. Am. Soc. Mass Spectrom.* 24 (8): 1194–1201. https://doi.org/10.1007/s13361-013-0609-x.

87 Berrazeg, M., Diene, S.M., Drissi, M. et al. (2013). Biotyping of multidrug-resistant *Klebsiella pneumoniae* clinical isolates from France and Algeria using MALDI-TOF MS. *PLoS One* 8 (4): e61428. https://doi.org/10.1371/journal.pone.0061428.

88 Mencacci, A., Monari, C., Leli, C. et al. (2013). Typing of nosocomial outbreaks of *Acinetobacter baumannii* by use of matrix-assisted laser desorption ionization-time of flight mass spectrometry. *J. Clin. Microbiol.* 51 (2): 603–606. https://doi.org/10.1128/jcm.01811-12.

89 Schmellik-Sandage, C.S. and Hill, R.E. Jr. (2004). Regulation of autogenous biologicals in the United States. *Dev. Biol.* 117: 9–13.

90 Erol, E., Locke, S.J., Donahoe, J.K. et al. (2012). Beta-hemolytic *Streptococcus* spp. from horses. *J. Vet. Diagn. Invest.* 24 (1): 142–147. https://doi.org/10.1177/1040638711434138.

91 Mani, R.J., Thachil, A.J., and Ramachandran, A. (2017). Discrimination of *Streptococcus equi* subsp. *equi* and *Streptococcus equi* subsp. *zooepidemicus* using matrix-assisted laser desorption/ionization time-of-flight mass spectrometry. *J. Vet. Diagn. Invest.* 29 (5): 622–627. https://doi.org/10.1177/1040638717702687.

92 Pérez-Sancho, M., Vela, A.I., Horcajo, P. et al. (2018). Rapid differentiation of *Staphylococcus aureus* subspecies based on MALDI-TOF MS profiles. *J. Vet. Diagn. Invest.* 30 (6): 813–820. https://doi.org/10.1177/1040638718805537.

93 Silva, M.B., Ferreira, F.A., Garcia, L.N.N. et al. (2015). An evaluation of matrix-assisted laser desorption ionization time-of-flight mass spectrometry for the identification of *Staphylococcus pseudintermedius* isolates from canine infections. *J. Vet. Diagn. Invest.* 27 (2): 231–235. https://doi.org/10.1177/1040638715573297.

94 Loy, J.D. and Brodersen, B.W. (2014). *Moraxella spp.* isolated from field outbreaks of infectious bovine keratoconjunctivitis: a retrospective study of case submissions from 2010 to 2013. *J. Vet. Diagn. Invest.* 26 (6): 761–768. https://doi.org/10.1177/1040638714551403.

95 Dickey, A.M., Loy, J.D., Bono, J.L. et al. (2016). Large genomic differences between *Moraxella bovoculi* isolates acquired from the eyes of cattle with infectious bovine keratoconjunctivitis versus the deep nasopharynx of asymptomatic cattle. *Vet. Res.* 47: 31. https://doi.org/10.1186/s13567-016-0316-2.

96 Dickey, A.M., Schuller, G., Loy, J.D., and Clawson, M.L. (2018). Whole genome sequencing of *Moraxella bovoculi* reveals high genetic diversity and evidence for interspecies recombination at multiple loci. *PLoS One* 13 (12): e0209113. https://doi.org/10.1371/journal.pone.0209113.

97 Loy, J.D. and Clawson, M.L. (2017). Rapid typing of *Mannheimia haemolytica* major genotypes 1 and 2 using MALDI-TOF mass spectrometry. *J. Microbiol. Methods* 136: 30–33. https://doi.org/10.1016/j.mimet.2017.03.002.

98 Hille, M., Dickey, A., Robbins, K. et al. (2020). Rapid differentiation of *Moraxella bovoculi* genotypes 1 and 2 using MALDI-TOF mass spectrometry profiles. *J. Microbiol. Methods* 173: 105942. https://doi.org/10.1016/j.mimet.2020.105942.

99 Hille, M.M., Clawson, M.L., Dickey, A.M. et al. (2021). MALDI-TOF MS biomarker detection models to distinguish RTX toxin phenotypes of *Moraxella bovoculi* strains are enhanced using calcium chloride supplemented agar. *Front. Cell. Infect. Microbiol.* 11: 632647. https://doi.org/10.3389/fcimb.2021.632647.

100 Clawson, M.L. and Murray, R.W. (2014). Pathogen variation across time and space: sequencing to characterize *Mannheimia haemolytica* diversity. *Anim. Health Res. Rev.* 15 (2): 169–171. https://doi.org/10.1017/S1466252314000188.

101 Loy, J.D., Hille, M., Maier, G., and Clawson, M.L. (2021). Component causes of infectious bovine Keratoconjunctivitis – the role of *Moraxella* species in the epidemiology of infectious bovine Keratoconjunctivitis. *Vet. Clin. North Am. Food Anim. Pract.* 37 (2): 279–293. https://doi.org/10.1016/j.cvfa.2021.03.004.

102 Clawson, M.L., Keen, J.E., Smith, T.P. et al. (2009). Phylogenetic classification of *Escherichia coli* O157:H7 strains of human and bovine origin using a novel set of nucleotide polymorphisms. *Genome Biol.* 10 (5): R56. https://doi.org/10.1186/gb-2009-10-5-r56.

103 Lartigue, M.-F. (2013). Matrix-assisted laser desorption ionization time-of-flight mass spectrometry for bacterial strain characterization. *Infect. Genet. Evol.* 13: 230–235. https://doi.org/10.1016/j.meegid.2012.10.012.

104 McDonald, R.A., Skipp, P., Bennell, J. et al. (2009). Mining whole-sample mass spectrometry proteomics data for biomarkers – an overview. *Expert Syst. Appl.* 36 (3, Part 1): 5333–5340. https://doi.org/10.1016/j.eswa.2008.06.133.

105 Sauget, M., Valot, B., Bertrand, X., and Hocquet, D. (2017). Can MALDI-TOF mass spectrometry reasonably type bacteria? *Trends Microbiol.* 25 (6): 447–455. https://doi.org/10.1016/j.tim.2016.12.006.

106 Elssner, T. and Kostrzewa, M. (2006). CLINPROT – a MALDI-TOF MS based system for biomarker discovery and analysis. *Clin Proteomics* 8.

107 Ketterlinus, R., Hsieh, S.-Y., Teng, S.-H. et al. (2005). Fishing for biomarkers: analyzing mass spectrometry data with the new ClinProTools™ software. *Biotechniques* 38 (6S): S37–S40. https://doi.org/10.2144/05386SU07.

108 Boggs, S.R., Cazares, L.H., and Drake, R. (2012). Characterization of a *Staphylococcus aureus* USA300 protein signature using matrix-assisted laser desorption/ionization time-of-flight mass spectrometry. *J. Med. Microbiol.* 61 (5): 640–644. https://doi.org/10.1099/jmm.0.037978-0.

109 Khot, P.D. and Fisher, M.A. (2013). Novel approach for differentiating *Shigella* species and *Escherichia coli* by matrix-assisted laser desorption ionization–time of flight mass spectrometry. *J. Clin. Microbiol.* 51 (11): 3711–3716. https://doi.org/10.1128/jcm.01526-13.

110 López-Fernández, H., Santos, H.M., Capelo, J.L. et al. (2015). Mass-up: an all-in-one open software application for MALDI-TOF mass spectrometry knowledge discovery. *BMC Bioinform.* 16 (1): https://doi.org/10.1186/s12859-015-0752-4.

111 Trevisan, G., Linhares, L.C.M., Schwartz, K.J. et al. (2021). Data standardization implementation and applications within and among diagnostic laboratories: integrating and monitoring enteric coronaviruses. *J. Vet. Diagn. Invest.* 33 (3): 457–468. https://doi.org/10.1177/10406387211002163.

12

MALDI-TOF MS Analysis for Identification of Veterinary Pathogens from Companion Animals and Livestock Species

Dorina Timofte[1], *Gudrun Overesch*[2], *and Joachim Spergser*[3]

[1] *Department of Veterinary Anatomy, Physiology and Pathology, Institute of Infection, Veterinary and Ecological Sciences, University of Liverpool, Leahurst Campus, Neston, UK*
[2] *Institute of Veterinary Bacteriology, Vetsuisse Faculty, University of Bern, Bern, Switzerland*
[3] *Department for Pathobiology, Institute of Microbiology, Vetmeduni – University of Veterinary Medicine Vienna, Vienna, Austria*

12.1 Veterinary Diagnostic Laboratories and the MALDI-TOF Clinical Microbiology Revolution[1]

Matrix-assisted laser desorption/ionization time-of-flight (MALDI-TOF) mass spectrometry (MS) has revolutionized clinical microbiology and has become the reference method for routine bacterial identification in human diagnostic laboratories worldwide. The number of microbial species that can be identified routinely by MALDI-TOF MS continues to grow, owing to increasing numbers of database entries with spectra of well-characterized species.

MALDI-TOF's most important features, such as user-friendliness and cost-efficiency, together with its ability to provide fast and reliable results, has favoured its increasing adoption by veterinary diagnostic laboratories. Several studies have evaluated the performance of MALDI-TOF MS for the identification of certain bacterial genera or species of veterinary importance [1–3]. More recently, commercial providers of MALDI-TOF MS instruments (e.g.Bruker MALDI Biotyper, bioMerieux Vitek MS) have been offering libraries that include reference spectra of an extensive number of veterinary pathogens, allowing faster identification of pathogens from a variety of animal species and specimens [4, 5]. Although only a few studies have investigated the performance of MALDI-TOF MS for the identification of bacteria from animal clinical specimens, Randall et al. [6] have shown that MALDI-TOF MS is an accurate way to identify, at species level, the majority of the routine bacterial isolates encountered in a veterinary diagnostic laboratory [6].

1 Section 12.1 was written by Dorina Timofte.

Microbiological Identification using MALDI-TOF and Tandem Mass Spectrometry: Industrial and Environmental Applications, First edition. Edited by Haroun N. Shah, Saheer E. Gharbia, Ajit J. Shah, Erika Y. Tranfield, and K. Clive Thompson.

12.1.1 MALDI-TOF MS: Reshaping the Workflow in Clinical Microbiology

The clinical microbiology laboratory plays a key role in the diagnosis of infections and a major goal is to provide timely and accurate bacterial identification and antimicrobial susceptibility testing (AST) results. Reducing the time from clinical specimen collection to reporting has a direct impact on patient management and the introduction of MALDI-TOF MS has led to rethinking of the diagnostic workflow in order to achieve this goal. The routine approach to bacterial identification and AST initially involves setting up primary bacterial cultures, which generally takes one day for most fast-growing bacteria, followed by a process of biochemical identification of bacterial isolates and AST, which are usually performed concomitantly and takes another day. However, this is the scenario for processing bacterial isolates obtained as pure cultures, while for polymicrobial cultures, as routinely encountered in clinical samples, additional steps involving colony subculture for isolate purification are required for both identification and AST. All of this, involves a string of processes that is very time-consuming, can take days overall and has several elements of variability where errors can be introduced.

Thus, the potential of MALDI-TOF MS technology for successful rapid, accurate identification of bacterial colonies at species level, without the need for additional overnight incubation (as for biochemical testing), has made it possible to make key changes in the routine workflow for bacterial identification and reporting practice, which may be critically important in enabling effective treatment protocols.

In both human and veterinary medicine, the management of critically ill patients often requires the use of antimicrobial therapy which, without knowing the identity of the pathogen involved, in most cases involves selection of broad-spectrum antimicrobials.

At the Veterinary Microbiology Diagnostic Laboratory (VMDL), University of Liverpool, the introduction of MALDI-TOF MS into our routine diagnostics has led to a reduction of workflow such that, whereas we would previously report both bacterial culture and AST results around 48 hours after sample receipt, we now report bacterial culture results first, at one day (or less) after sample receipt, and then add AST reports after another 24 hours (Figure 12.1). This allows clinicians to tailor initial antibiotic selection based on the

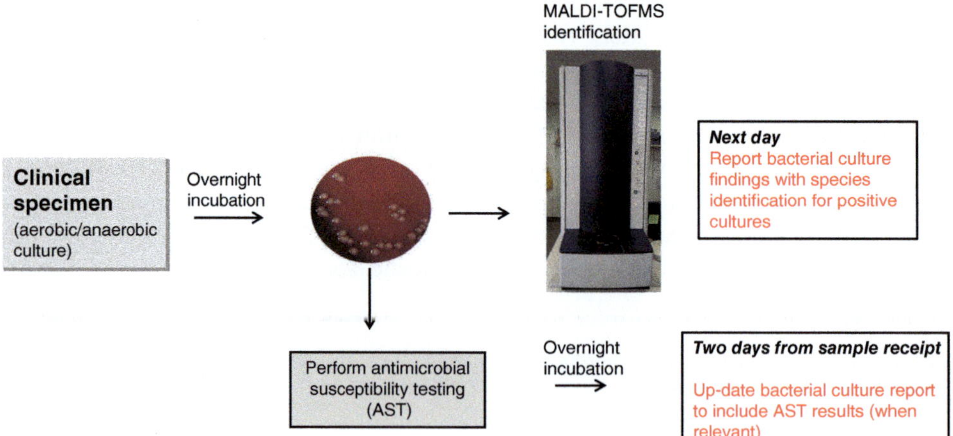

Figure 12.1 Bacterial culture and antimicrobial susceptibility testing workflow at the VMDL, University of Liverpool.

knowledge of: (i) the positive or negative bacterial culture results; and (ii) the general antimicrobial susceptibility profile of the known pathogen in case of positive cultures. This enables a more 'informed' empirical therapy selection, for instance by excluding those antibiotics to which the isolates may exhibit intrinsic resistance. In the VMDL workflow, the AST is reported the next day allowing 'refinement' of initial antibiotic selection, directly impacting on veterinary patient care.

This is one way in which MALDI-TOF MS is enabling an important reduction in the time the results reach the clinic, while at the same time improving the accuracy. In other approaches, shorter incubation times of cultures on solid media have been used to obtain initial inoculums for MALDI-TOF MS identification and also AST, generating results one day earlier than with conventional approaches, therefore facilitating timely clinical decisions for patient management. Consequently, several studies of human infections have demonstrated that the use of MALDI-TOF MS can reduce the time required for organism identification and, hence, decrease the time to selection of effective antibiotic therapy [7–9]. Similar challenges regarding antimicrobial selection and need for rapid use of empirical therapy are faced by the veterinary emergency and critical care (ECC) sector where treatment of critically ill animals cannot be delayed until culture results are available [10, 11]. To highlight these challenges, a recent study investigating antimicrobial prescriptions in a small animal ECC setting found that 45.1% of all cases which had an antimicrobial prescribed had no subsequent evidence of infection [12]. Another study that surveyed UK veterinary practices found that antimicrobials were the second most commonly prescribed medicines (after vaccines) and comprised 17.6% of canine prescriptions and 14.5% of feline prescriptions [13]. In an effort to reduce antibiotic use in companion animals, it is therefore essential to carry out antibiotic-susceptibility testing to guide clinicians in selecting an appropriate treatment regimen for an infection. Therefore, the speed of bacterial (and fungal identification) is critical for both human and animal patients, particularly to inform antimicrobial therapy.

12.1.2 Identification of Bacterial Pathogens Directly from Clinical Specimens

For many years, researchers have attempted to develop systems to identify microorganisms directly from clinical samples. For acute infections such as sepsis and meningitis, early diagnosis is vital for rapid initiation of antimicrobial treatment, and direct detection and identification of microorganisms from clinical specimens [i.e. blood and cerebrospinal fluid (CSF)] have the potential to further reduce the time to diagnosis and improve patient management and outcomes [9]. Initiation of appropriate and timely antibiotic therapy is of paramount importance in increasing the chance of surviving for bacteraemic patients; therefore, MALDI-TOF MS has been widely studied for the identification of microorganisms directly from blood cultures [14, 15]. However, direct pathogen detection requires several steps involving purification and separation of bacterial cells from whole blood or other fluids. Several approaches have been used for this, by applying both (i) in-house laboratory-developed tests (LDTs) and (ii) commercial products. Regardless of the method of choice, most procedures used for purification of bacteria consist of a lysis step followed by centrifugation and washing to remove the matrix proteins or filtration and concentration of bacterial cells [16–19]. To overcome the lack of standardization and validation of these in-house methods, commercial products have been developed, such as the Sepsityper kit (Bruker Microbiology & Diagnostics), the Vitek MS blood culture kit (bioMérieux, Inc., Durham, NC) and the rapid BACpro® II kit (Nittobo Medical Co.,Tokyo, Japan) [20, 21]. Although these systems use different principles for bacterial cell purification, they all have in common the fact that they can be used on any MALDI-TOF MS platform and the process for bacterial identification only takes ~20 minutes. Another slightly different approach is to increase the bacterial load by first performing subculture of blood onto plates of tryptic soy agar (TSA) with 5% sheep blood for a shortened period (e.g. five hours) to allow for growth of microcolonies which are then treated as normal colony growth for MALDI-TOF MS identification [17].

These methods vary greatly in terms of their overall performance. For instance, Prod'hom et al. [19] demonstrated rapid and accurate identification of ~80% of pathogens by obtaining a bacterial pellet from positive blood culture following a simple protocol [19]. On the other hand, other studies have reported successful identification direct from blood cultures, ranging from 60% to 99% concordance to a species level, the variability of which is likely to be linked to the lack of standardization for these methods [16, 22, 23]. Nevertheless, the limitations of this approach have been also highlighted in studies which have shown that although MALDI-TOF MS is an efficient method for direct routine identification of bacterial isolates in blood culture, this methodology may not be successful in the case of polymicrobial samples or infections associated with *viridans* streptococci [24].

There are no data regarding the general capability of MALDI-TOF MS in the identification of bacterial isolates directly from veterinary specimens as compared with the information available in the human clinical practice. However, fast detection and identification of bacterial infection are paramount for optimal use of antimicrobials, particularly in the context of a perceived overuse or misuse of antimicrobials leading to development of antimicrobial resistance, with clear impact on human and animal health. As a result, we also have a need to confidently and rapidly detect and identify the infecting organisms in order to inform the antibiotic selection process for veterinary patients.

Developing protocols for faster detection and bacterial identification with MALDI-TOF MS would be of particular benefit for processing urine samples from companion animals suspected of urinary tract infections (UTIs). These infections are common in dogs with ~14% of all dogs encountering at least one episode of bacterial UTI during their lifetime [25]. Similar to UTIs in humans, among bacteria associated with this condition, *Escherichia coli* has been the most frequently isolated, with up to 30% of UTI cases associated with this species, followed by *Staphylococcus* spp., *Enterococcus* spp., *Proteus* spp., and *Klebsiella* spp. [26–28]. This problem is compounded by the emergence of multidrug-resistant bacteria (MDR) (isolates resistant to three or more antimicrobial categories) in companion animals as both a serious veterinary issue and also the potential threat of zoonotic transmission. A limited number of studies have employed in-house user-friendly protocols for bacterial cell extraction from urine samples by centrifugation at low speed to remove the leukocytes, followed by high-speed centrifugation to collect the bacteria, and achieved 91.8% and 92.7% identification of microorganisms at the species and genus levels, respectively [29]. Another study showed that a concentration of at least 10^3 colony-forming units (CFU)/ml is required to enable MALDI-TOF MS to reliably identify bacteria directly in urine samples [30]. Ferreira et al. [29] have shown that MALDI-TOF MS could enable a UTI diagnosis in minutes, which would be extremely useful for guiding the empirical antimicrobial more efficaciously [29]. Further development of protocols for direct detection of pathogens in urine samples by MALDI-TOF MS would provide an additional and possibly improved approach for informed antibiotic selection for UTI in both human and veterinary patients, thereby impacting on antibiotic stewardship even before performing AST.

12.1.3 Prediction of Antimicrobial Resistance

Given the widespread detection of multi-drug-resistant bacterial strains, particularly in hospital settings, the potential of MALDI-TOF MS for development of rapid AST has also been explored, although with less success in gaining an established place in the clinical laboratory. Instead, various protocols have been trailed for rapid detection of antimicrobial-resistant traits in clinically relevant bacteria [31].

Although several methods have been studied, two of these approaches have the potential for wider adoption in clinical diagnostics. One of these methods is the identification of biomarkers associated with drug resistance or detection of characteristic 'resistance peaks' that aim to identify differences in the mass spectra of susceptible and resistant isolates of a given microorganism. For instance, an extended-spectrum class C beta-lactamase belonging to the *Acinetobacter*-derived cephalosporinases (ADC) family has now been identified in *Acinetobacter baumannii* (m/z 40 279) and evaluated as a possible biomarker for carbapenem resistance [32].

Based on their MALDI-TOF spectra, characteristic 'resistance peaks' have been identified and exploited to discriminate rapidly (minutes) between methicillin-resistant *Staphylococcus aureus* (MRSA) and methicillin-susceptible *Staphylococcus aureus* (MSSA). Several studies have shown that a signal at m/z 2415 is correlated with the expression of a small peptide (PSM-*mec*) which is encoded on three SCCmec cassette types highly associated with MRSA [33, 34]. Despite low sensitivity of the m/z 2415 peak for *mecA* carriage in *S. aureus* and *Staphylococcus epidermidis* (37% and 6%, respectively), the specificity is high (≥98%) [34]. In addition, it has been shown that the PSM-*mec* gene encoding for the PSM-mec peptide is found predominantly among MRSA strains of SCC*mec* types II, III, and VIII, that is a conserved part of the class *mec* gene complex and its expression is highly variable [35].

At the VMDL, we have used the MBT Subtyping Module and software of the MALDI Biotyper to screen for the PSM-*mec* peak in a collection of MRSA isolates (n = 147) from companion animals (cats, dogs, and horses). All isolates were previously confirmed to carry the *mecA* gene by polymerase chain reaction (PCR) (unpublished data), were grown overnight on sheep blood agar plates and then processed by MALDI-TOF following the manufacturer's instructions. All isolates were identified as *S. aureus* with a >2 score but the PSM-mec peak was not detected in any of the companion animal isolates with the subtyping module. However, this does not rule out the use of the MBT Subtyping Module for detection of MRSA in animal isolates as it is likely that the tested isolates were not of SCC*mec* types II, III, and VIII and other factors could account for these results. For instance, it has been demonstrated that detection of antibiotic resistance-associated specific peaks may depend on the local testing conditions (type of culture medium, instruments, and experimental protocols) and that local databases should be built for the accurate detection of resistance profiles [36, 37].

Another approach for detection of antimicrobial resistance is the detection of induced hydrolysis of beta-lactam antibiotics by bacteria producing beta-lactamase, which can be revealed in mass spectra by a decrease of the peak corresponding to the antibiotic and appearance of peaks representing its hydrolysis products [38]. Commercially available products, such as the MBT STAR CEPHA, are designed to work in combination with the MBT Compass STAR-BL software (Bruker Microbiology & Diagnostics) for rapid detection of cephalosporinase activity towards third-generation cephalosporins. Incubation of bacteria producing cephalosporinase with the antibiotic leads to hydrolysis of the antibiotic's beta-lactam ring, converting the antibiotic into an inactive metabolite. Furthermore, a recent study has shown that the test can efficiently distinguish strains resistant to third-generation cephalosporins due to extended-spectrum beta-lactamase (ESBL) production of those that owe their resistance to a combination of narrow spectrum beta-lactamases (e.g. non-ESBL type SHV) and other resistance mechanisms [39]. These types of commercial product are suitable for detection of resistance in veterinary isolates where resistance to expanded spectrum cephalosporins is also common.

12.1.4 Impact in Veterinary Hospital Biosecurity and Epidemiological Surveillance

MALDI-TOF MS has also emerged as a useful diagnostic tool for bacterial typing and subtyping, therefore enabling epidemiological surveillance and detection of nosocomial outbreaks. Similar to management of clinical cases, early identification of emerging pathogens in the hospital environment will impact on hospital infection control and hygiene and ultimately prevention of hospital outbreaks.

In contrast to human medicine, data about veterinary nosocomial or healthcare-associated infections (HAIs) remain limited, although the problem has recently gained increased attention. Surveillance of HAIs in veterinary healthcare settings and infection control still remains in its infancy, despite HAIs posing an increasing threat in modern veterinary practice. This appears to be particularly the case for companion animals, i.e. dogs, cats, and horses, where a growing body of literature has described nosocomial outbreaks of different aetiologies which are often associated with antimicrobial-resistant and sometimes also zoonotic microorganisms. These infections often result in poor patient outcomes and extensive hospital outbreaks as well as posing an additional threat not just to the hospitalized animal patients but also to veterinary staff and animal owners. The most common Gram-negative pathogens implicated in veterinary healthcare-associated outbreaks in small animal and equine hospitals include *Salmonella* serovars, ESBL and pAmpC-producing *E. coli* and other Enterobacterales, and MDR *A. baumannii* [40–42]. Similar to human medicine, UTIs, pneumonia, surgical site infections (SSIs) alongside indwelling device site infections, peripheral and central line-associated bloodstream infections and infectious enteritis account for the vast majority of veterinary HAIs [43].

Strangles is another example of an important infectious equine disease in horses; this is a highly contagious infection of the upper respiratory tract caused by *Streptococcus equi* subspecies *equi*. The disease can occur in horses of any age, and in ponies and donkeys of all types and ages, although young horses typically develop more severe signs. The earlier a diagnosis is made, the sooner the isolation can be implemented, helping to limit disease transmission [44].

Despite these challenges in managing infections and outbreaks, human and veterinary hospitals very rarely have the capacity to investigate the genotype of organisms associated with outbreaks, primarily because conventional strain-typing methods, molecular methods [e.g. macro-restriction pulsed-field gel electrophoresis (PFGE), multilocus sequencing (MLST) and whole-genome sequencing (WGS)] are time-consuming, expensive and require specialized personnel to interpret the results. In addition, results are often not available in a timescale that would be clinically relevant to infection control. Therefore, attempts have been made to exploit the MALDI-TOF MS as a potential tool for strain typing and in an ideal scenario the MALDI-TOF MS data resulting from bacterial identification could potentially be used for isolate typing, allowing for real-time monitoring of outbreak strains. The additional use of MALDI-TOF MS as a typing tool has been demonstrated for an ESBL-producing *Klebsiella pneumoniae* (ESBL-KP) outbreak in the neonatal intensive care unit where the differentiation of the outbreak isolates in two clones was in agreement with cg-MLST results [45]. However, other studies suggested that additional steps need to be taken (e.g. extraction step, designated software tools) in order to increase resolution of MALDI-TOF MS for bacterial strain typing [46]. Nevertheless, the prospect of MALDI-TOF MS as a potential tool for strain typing of human and veterinary isolates, providing real-time surveillance of hospital infections, remains an attractive prospect.

12.2 Identification of *Campylobacter* spp. and *Salmonella* spp. in Routine Clinical Microbiology Laboratories[2]

For several decades, campylobacteriosis and salmonellosis were the first and second most reported foodborne gastrointestinal infections in humans worldwide [47]. Moreover, *Campylobacter* spp. and *Salmonella* spp. are pathogens of notifiable epizootic diseases in veterinary medicine, which are monitored worldwide in surveillance and eradication programmes. Therefore, identification of these two zoonotic pathogens is part of the daily routine in both human and veterinary microbiology laboratories.

12.2.1 General Aspects on the Importance of Species/Subspecies and Serovar Identification of *Campylobacter* spp. and *Salmonella* spp.

In general, identification to the bacterial species level by whole-cell MALDI-TOF MS technology depends upon the representation of a reference spectrum in the databases used. Reference databases should include several strains of the species in question, ideally from different sources, to ensure that mass spectra exhibit a representative phenotypic profile. To date, commercially available databases have been sufficient for identification to genus

2 Section 12.2 was written by Gudrun Overesch.

and species levels for a broad spectrum of microorganisms relevant in human and veterinary medicine. By contrast, identification of more closely related bacteria such as subspecies is more challenging with routine MALDI-TOF MS protocols, as this requires in-depth identification and analysis of subspecies-identifying biomarker ions (SIBIs).

As at 25 March 2022, up to 55 *Campylobacter* species and 16 *Campylobacter* subspecies have been described [48]. With regard to the zoonotic character of *Campylobacter* spp., ~ 20 species are isolated from humans, of which the group of thermophilic *Campylobacter* spp. is of highest relevance. This group includes *C. jejuni* subsp. *jejuni*, *C. jejuni* subsp. *doylei*, *Campylobacter coli*, *Campylobacter lari* subsp. *lari*, *C. lari* subsp. *chonceus* and *C. upsaliensis*. In about 98% of human campylobacteriosis cases reported to the European Food Safety Authority (EFSA), *C. jejuni* (88%) or *C. coli* (10%) are detected [47]. In less than 2% of cases, other *Campylobacter* spp., such as *C. upsaliensis* or *C. lari*, are isolated. For routine clinical laboratories, the identification of thermophilic *Campylobacter* to species level is sufficient, because legal requirements in general refer to *Campylobacter* spp. or *Campylobacter* species, but not to subspecies level. In addition, *C. jejuni* subsp. *doylei* and *C. lari* subsp. *chonceus* are detected only rarely and so far only in human stool samples. By contrast, for veterinary medicine, identification of *Campylobacter fetus* to the subspecies level is mandatory. Currently, *C. fetus* consists of four genetically closely related subspecies or biovars, namely *C. fetus* subsp. *fetus*, *C. fetus* subsp. *venerealis*, *C. fetus* subsp. *venerealis* biovar *intermedius* and *C. fetus* subsp. *testudinum*. *C. fetus* subsp. *fetus* is a commensal in the intestine of ruminants and acts as an opportunistic pathogen causing sporadic abortion in sheep and cattle. Furthermore, the EFSA reported 0.16% *C. fetus* subsp. *fetus* infection in 2020 among all human campylobacteriosis cases [47]. By contrast, *C. fetus* subsp. *venerealis* including biovar *intermedius* is the causative agent of bovine genital campylobacteriosis, a notifiable disease with high economic impact in cattle production that leads to reduced infertility and abortion [49]. *C. fetus* subsp. *testudinum* may be isolated from reptiles and humans and is regarded as opportunistic [50].

The genus *Salmonella* consists of only two species, *S. enterica* and *S. bongori*. Based on the sequences of housekeeping genes and invasion-associated protein genes, *S. enterica* is further divided into seven subspecies: *S. enterica* subsp. *enterica*, *S. enterica* subsp. *salamae*, *S. enterica* subsp. *arizonae*, *S. enterica* subsp. *diarizonae*, *S. enterica* subsp. *houtenae* and *S. enterica* subsp. *indica* [51]. Furthermore, according to the White-Kauffmann-Le Minor scheme, more than 2600 serovars of *Salmonella* are currently defined based on their expression pattern of 46 lipopolysaccharide moieties (O antigens), 114 flagellar proteins (H1/H2 antigens) and capsular components (Vi antigens) [52]. With more than 1500 serovars, *Sa. enterica* subsp. enterica exhibits by far the highest number of serovars. Although fewer than 100 *Salmonella* serovars are reported to cause disease in humans, all *Salmonella* spp. must be considered as pathogenic for humans, and hence legislation for food safety often defines criteria by the presence or absence of *Salmonella* to the genus level. On the other hand, for fresh poultry meat, European food safety criteria are restricted to targeted *S. enterica* serovars, such as *S. Enteritidis* and *S. Typhimurium*, including its monophasic variants [47]. Moreover, national control programmes in poultry in Europe are aimed at reducing the prevalence of specific *S. enterica* serovars that are considered relevant for public health. Currently, *S. Enteritidis* and *S. Typhimurium*, including its monophasic variants, *S. Infantis*, *S.* Virchow, and *S.* Hadar, have been defined as relevant [47]. In veterinary medicine, marked differences in host adaption and clinical outcomes of different *Salmonella* subspecies and serovars exist. Typhoid *S. enterica* serovars [*S. Typhi* and *S. Paratyphi* (A to C)] are host-restricted to humans. Although some *S. enterica* serovars are able to infect a broad range of animal hosts (e.g. *S. Typhimurium*), other serovars are highly restricted to specific host species (e.g. *S. Gallinarum*) [53]. Therefore, identification of *Salmonella* isolates to the subspecies level is needed in routine microbiological diagnostics.

12.2.2 General Aspects on Influence of Media/Culture Environment on Bacterial Species Identification by MALDI-TOF MS

MALDI-TOF MS technology is an excellent tool for fast, high-throughput, and reliable identification of bacterial species in routine diagnostics for both veterinary and human medicine. Nevertheless, the sample preparation and culture conditions affect the outcome of the identification success and have to be adapted to the bacterial species in question [54]. To date, for routine diagnostics, pure cultures of bacteria are grown on solid media to be used for MALDI-TOF MS identification, as identification of bacterial species in mix cultures and biofilms by MALDI-TOF MS requires further research [55]. The matrix [alpha-cyano-4-hydroxycinamic acid (HCCA) in acetonitrile/trifluoroacetic acid] or 2,5-dihydroxybenzoic acid (DHA) in water/ethanol/acetotronitrile are commonly used and result in homogenous molecules with good resolution of mass spectra. Depending on the bacterial cell wall structure, different methods for sample preparation may be used. In general, the direct sample spotting is recommended for a broad variety of Gram-negative, non-mucoid bacteria including *Campylobacter* spp. and *Salmonella* spp. By contrast, many Gram-positive bacteria yield more accurate results when the on-target extraction by overlay with formic acid/ethanol is used. Bacteria with a complex cell wall structure, e.g. mycobacteria, and bacteria belonging to biosafety level 3 need a more laborious full extraction method with simultaneous inactivation of the pathogen [54]. The bacterial cell concentration used for sample spotting also influences the number and intensity of peaks in the main spectra. For *Salmonella* spp., 10^5–10^6 cells turned out to be the optimal cell concentration [56]. Bacterial cultures in liquid media exhibit a higher number and a better quality of peaks in main spectra compared with isolates grown on solid media [57]. However, in routine daily diagnostics, the use of pure culture on solid media was shown to be sufficient for the vast majority of clinically important bacterial species, thereby avoiding laborious techniques for preparation of liquid cultures suitable for MALDI-TOF MS identification procedures [58]. Blood-based media, in particular broth, may affect the protein mass spectra by either higher signals and/or occurrence of extra peaks for bacteria grown in these media. For *Salmonella* spp. media-specific profiles may be detected, but the strain-specific peaks were consistent on all media [59]. For *Campylobacter* spp., isolates grown on modified cefoperazone deoxycholate agar could not be used for direct measurement, as compounds of the agar interfere with ionization of biomolecules, resulting in poor spectra. Identification of isolates grown on *Brucella* or Campylosel agar is feasible [60]. The incubation temperature of the culture before MALDI-TOF MS identification as well as subculturing may also have an effect on the quality of identification. However, reports on the effects of incubation temperature and/or subculturing on the accuracy of MALDI-TOF MS identification vary for different bacteria species. For Gram-negative enteric bacteria, including *Salmonella* spp. and *Campylobacter* spp., no misidentifications were found for various incubation temperatures and several passages [61]. It has been shown that the incubation time of bacteria has an impact on the accuracy of MALDI-TOF MS identification. The best results are achieved when bacteria are in the log or stationary phase of the growth curve, and they should not be analysed in either the lag or the death phase, as this may result in decreased spectral quality and resolution [54]. For *Salmonella* spp., an incubation time of 24–48 hours resulted in similar peak patterns, whereas the intensities were higher with 48 hours' incubation. Incubation for 72 hours showed a loss of specific peaks within the spectra [62]. For *Campylobacter* spp., storage of isolates for up to seven days at room temperature did not affect the success of identification [60].

12.2.3 Possibilities and Limits of Identification of *Campylobacter* spp. by MALDI-TOF MS

12.2.3.1 Thermophilic *Campylobacter* spp.

This group includes *C. jejuni*, *C. coli*, *C. lari*, and *C. upsaliensis*, of which *C. jejuni* and *C. coli* are the by far most important as causative agents of human campylobacteriosis. As early as 2005, Mandrell et al. showed that species identification for *C. jejuni, C. coli, C. lari, C. upsaliensis, C. helveticus*, and *Campylobacter sputorum* from different sources based on SIBIs was possible with MALDI-TOF MS technology [63]. However, for *C. upsaliensis* and *C. helveticus*, identical SIBIs were identified (<10 Da), making the accurate identification by MALDI-TOF MS more challenging between these two species. Another study showed that *C. helveticus* was either misidentified as *Campylobacter hyointestinalis* by one instrument or identified correctly with a low confidence value (50%) by another instrument [64]. Kolínská et al. [65] confirmed successful identification of *C. jejuni* and *C. coli* by MALDI-TOF MS. Bessède et al. [66] compared biochemical and molecular methods for identification of *Campylobacter* spp. with MALDI-TOF MS technology for more than 1000 strains. The accuracy of MALDI-TOF MS was 100% for *C. coli*, *C. fetus*, *C. lari*, *C. sputorum*, *C. upsalienis*, and *Aliarcobacter butzleri*. Only for *C. jejuni* did the accuracy turn out to be 99.4%, as three *C. coli* and one *C. fetus* isolate were misidentified as *C. jejuni* by MALDI-TOF MS [66]. MALDI-TOF MS turned out to be the method of choice for identification of *C. jejuni* subsp. *jejuni, C. jejuni* subsp. *doylei* and *C. coli* with 100% accuracy in the study by Martiny et al. [67]. A comprehensive collaborative study conducted in eight US and Canadian service laboratories with 16 *Campylobacter* spp. and eight non-*Campylobacter* spp. isolates revealed 100% accuracy for the identification of *C. jejuni*, *C. coli,* and *C. lari* when using the Bruker MALDI Biotyper method [68].

Recently, *C. hepaticus*, a new thermophilic *Campylobacter* species, was described, which is the causative agent of spotty liver disease in poultry [69, 70]. This *Campylobacter* species is closely related to *C. jejuni*. Besides difficulties of cultivation, to the best of the authors' knowledge, currently no successful identification by MALDI-TOF MS technology has been reported.

Arcobacter, *Aliarcobacter* spp., and *Helicobacter* spp. are closely related to *Campylobacter* spp. and difficult to differentiate by biochemical testing. As some of these species are also recognized as human and/or animal pathogens, reliable and rapid differentiation of these genera is of utmost importance in clinical diagnostics. Alispahic et al. [60] reported the successful discrimination of *A. butzleri, A. skirrowii, Arcobacter cryaerophilus, Helicobacter pullorum, Helicobacter pametensis*, and *C. jejuni* subsp. *jejuni, C. coli, C. lari, C. hyointestinalis*, and *C. fetus* subsp. *fetus* based on distinctive mass spectra with MALDI-TOF MS [60]. It became evident, particularly in early publications, that uploading reference spectra from a broad variety of isolates was essential for accurate species identification, as this enabled a more representative distribution of the mass spectrum of various phenotypes. This was shown for the identification of *A. butzleri, A. skirrowii*, and *A. cryaerophilus* by van den Abeele et al. [71]. The authors demonstrated that the identification of *A. butzleri* and *A. cryaerophilus* could be significantly improved by enlarging commercial available reference databases with field strains. Furthermore, identification of *A. skirrowii* was significantly improved when a formic acid overlay was used instead of the direct colony transfer protocol [71].

12.2.3.2 Human-hosted *Campylobacter* Species

Campylobacter spp. found in humans include the following: *C. consisus, C. rectus, C. gracilis, C. uroelyticus, C. showae, C. curvus*, and *C. hominis. C. consisus* and *C. gracilis* were first described by Tanner et al. [72], isolated from oral cavity of humans. Its zoonotic and pathogenic potential has not been fully clarified to date. This may be partly due to its genetic diversity, as *C. consisus* belongs to at least two predominant, closely related, but genetically distinct taxa referred to as 'genomospecies'. By conventional MALDI-TOF MS identification procedures, the type strain of *C. consisus* was confidently identified, whereas with other strains of the genomospecies 1, 46% were identified as probable species and another 14% as probable genus. For strains of the genomospecies 2, results were even less definitive, with 5% of isolates not identified to any species [73]. The database contained 14 other strains of *C. consisus* at the time of the study, which demonstrated that the existing strains do not represent the complete phenotypic diversity of this *Campylobacter* species, and further research on both taxonomy and MALDI-TOF MS spectra analysis is needed. *C. rectus* is frequently found in samples from oral cavities and seems to be associated with periodontal diseases. By contrast, the pathogenic role of *C. gracilis* remains controversial [74]. Moreover, although frequently isolated from human genital tract and stool, the pathogenic role of *Campylobacter ureolyticus* is unclear. *C. showae* is also frequently detectable in oral cavities [75]. By contrast, *C. hominis* and *C. curvus* may be isolated from human stool samples. In a study by Martiny et al. [67], one out of six *C. curvus* isolates was not correctly identified by MALDI-TOF MS. Hsieh et al. [64] demonstrated that with the two commercially available MALDI-TOF MS instruments, *C. gracilis, C. rectus, C. hominis, C. curvus*, and *C. showae* were not correctly identified [64]. By contrast, Girard et al. [76] reported correct identification of six *C. rectus* isolates by MALDI-TOF MS. Reference spectra of the type strains *C. rectus* DSM 3260T, *C. gracilis* DSM 19528T, *C. ureolyticus* DSM 20703T, *C. showae* CCUG 30254T, *C. hominis* CCUG 45161T, and *C. curvus* DSM 6644T and several other strains of *Campylobacter* spp. are included in commercially available reference databases of MALDI-TOF MS instruments.

12.2.3.3 *Campylobacter* spp. of Veterinary Importance

For *C. fetus*, identification to the subspecies level is of utmost importance for veterinary medicine. Using Fourier transform infrared and Raman spectroscopies in combination with MALDI-TOF MS, Muhamadali et al. [77] reported that *C. fetus* subsp. *fetus* and *C. fetus* subsp. *venerealis* were recovered in separate clusters. However, Liu et al. [78] demonstrated that *C. fetus* subsp. *fetus* and *C. fetus* subsp. *venerealis* differentiation was not feasible with routine MALDI-TOF MS technology. Whereas two out of 14 *C. fetus* subsp. *fetus* were misidentified with scores > 2. 0, all five *C. fetus* subsp. *venerealis* were misidentified as *C. fetus* subsp. *fetus* [78]. The authors recommended the expansion of commercially available reference databases to solve this problem. Currently, reference spectra of at least four *C. fetus* subsp. *fetus* and *C. fetus* subsp. *venerealis* strains, mainly type strains, are available in commercial databases. By contrast, *C. fetus* subsp. *testudinum* could be distinguished accurately by MALDI-TOF MS from the other two *C. fetus* subspecies [50]. The World Organisation for Animal Health (OIE) *Manual of Diagnostic Tests and Vaccines for Terrestrial Animals* [49] stated that *C. fetus* subsp. *fetus* and *C. fetus* subsp. *venerealis* discrimination was not feasible using MALDI-TOF MS [49].

Campylobacter hyointestinalis consists of two subspecies, *C. hyointestinalis* subsp. *hyointestinalis* and *C. hyointestinalis* subsp. *lawsonii* [79]. Although *C. hyointestinalis* subsp. *lawsonii* could be found as a commensal in the porcine gastrointestinal tract, *C. hyointestinalis* subsp. *hyointestinalis* is more associated ruminants, especially cattle [80]. By contrast, *C. hyointestinalis* subsp. *hyointestinalis* is not restricted to ruminants, but shows a broader mammalian host range, including humans suffering from diarrhoea [81]. To our knowledge, subspecies differentiation of *C. hyointestinalis* by MALDI-TOF MS has not been achieved.

New *Campylobacter* species have been described, especially from shellfish, marine birds, or marine mammals. These include *Campylobacter peloridis, C. subantarticus*, *C. volucris*, and *Campylobacter insulaenigrae*, whereas *C. pinnipediorum* was found in pinnipeds but there appear to be no reports of their delineation using MALDI-TOF MS [82].

12.2.4 Possibilities and Limits of Identification of *Salmonella* spp. by MALDI-TOF MS

Dieckmann et al. [59] published an identification scheme for *Salmonella* spp. at the species and subspecies levels by MALDI-TOF MS technology. The authors showed that for bacterial identification below the species level, the requirements on sample preparation, mass accuracy, quality of spectra, and spectra analysis are much higher than for routine diagnostics on species level by whole-cell MALDI-TOF MS. However, genus-identifying biomarker ions (57 peaks) for identification of *Salmonella* were defined. Despite a high similarity of protein profiles of several *Salmonella* spp. isolates, it was possible to determine SIBIs for discrimination of *S. enterica* (14 peaks) and *S. bongori* (27 peaks). In a further step, sequence variations corresponding to amino acid changes in proteins, which were detectable with MALDI-TOF MS, led to the determination of subspecies-identifying biomarker ions [59]. Of note, the latter were predominantly not linked to ribosomal proteins, showing that for differentiation of very closely related microorganisms, analysis of ribosomal proteins might not be sufficient. The information content of the mass spectra analysed led to a massive increase in data, e.g. more than 300 peaks had to be analysed for subspecies identification, for which the SARAMIS database was used. Identification of *Salmonella* to the serovar level by whole-cell MALDI-TOF MS was described [83]. The authors described a protocol for rapid pre-screening for the successful identification of *S. Enteritidis*, *S. Typhimurium*,

S. Virchow, *S.* Hadar, and *S. Infantis*. The identification procedure was based on a hierarchical decision tree approach consisting of a sequential analysis of mass spectra for the presence or absence of peaks for genus-, species-, subspecies- and serovar-identifying biomarker ions. Although this protocol offered a rapid and cheap alternative to laborious and cost-intensive serovar identification by slide agglutination, the main drawback of this approach is the fact that the vast majority of *Salmonella* serovars still had to rely on the traditional serotyping. Mangmee et al. [84] expanded in a first step the commercial mass spectra database with mass spectra of strains from the six *S. enterica* subspecies and *S. bongori*. With a weighted pattern-matching approach, identification of certain *Salmonella* isolates to subspecies level was achieved. Moreover, expansion of the mass spectra database and developing a machine-learning analysis led to the successful identification of the following five *Salmonella* serovars: *S.* Albany, *S.* Agona, *S.* Altona, *S. Enteritidis*, and *S. Typhimurium*, including its monophasic variant [84]. With a comparable approach, correct identification of *S.* Enteriditis, *S. Typhimurium,* and *S.* Thompson was described [85]. However, when applying routine protocols for sample preparation and commercially available software tools for analysis of mass spectra, *Salmonella* isolates could only be correctly identified to genus level with MALDI-TOF MS technology [86, 87]. A comprehensive collaborative study conducted in 15 European service laboratories with 16 *Salmonella* spp. and eight non-*Salmonella* spp. isolates revealed 100% accuracy for the identification of these *Salmonella* spp. at the genus level when using the Bruker MALDI Biotyper method [88].

Outlook

The sensitivity and specificity of the whole-cell MALDI-TOF MS technology will constantly improve with the expansion of reference databases accompanied by developments in microbial taxonomy and machine-learning peak analysis of main spectra. For the most relevant *Campylobacter* spp., the identification rate at species level of nearly 100% had already been reached 10 years ago [89]. There are a few unsolved problems, of which the inability of *C. fetus* subspecies identification for veterinary microbiology is the most important.

Currently, it is not possible to identify *Salmonella* serotypes with whole-cell MALDI-TOF MS, and thus it is not possible to overcome classical laborious and expensive slide agglutination for *Salmonella* serovar in routine diagnostic laboratories. Perhaps the combination of whole-cell MALDI-TOF MS and liquid chromatography will allow identification of the very closely related *Salmonella* species, subspecies and serovars in future [90]. Moreover, top-down proteomics analysed with the help of computational pipelines is shown to be able to identify several *Salmonella* serovars [91] or proteomic approaches based on traditional two-dimensional electrophoresis and tandem MS [92]. Recently, WGS-based methods for *in silico* prediction of *Salmonella* serovars showed substantially valid results, especially when using SISTR and/or SEQSero bioinformatics tools [93]. As such, WGS methods deliver comprehensive genetic information beyond serovar identification. These are currently broadly installed in routine laboratories and might became a valid alternative to MALDI-TOF MS-based methods, at least for identification of *Salmonella* spp. In a previous volume of this series in 2017, it was shown that the high resolution of WGS data and the rapid analysis by MALDI-TOF MS successfully combined both technologies to identify

Salmonella to the serovar level according to the Kauffman White-Kauffmann-Le Minor scheme [94, 95], utilizing gene sequences of the 'O' and 'H' antigens. Such approaches may be expanded for rapid strain typing using WGS or MLST data that are already available in repositories such as the National Center for Biotechnology Information or the European Molecular Biology Laboratory with a MALDI-TOF MS platform.

12.3 Identification and Differentiation of Mycoplasmas Isolated from Animals[3]

12.3.1 Animal Mycoplasmas at a Glance

Mycoplasmas are the simplest self-replicating prokaryotes that are distinguished from ordinary bacteria by a permanent lack of a cell wall, minute cell size and a small genome with low G+C content [96]. Taxonomically, mycoplasmas are classified in the class *Mollicutes*, which is the only class of the phylum *Mycoplasmatota* [97], commonly referred to as mycoplasmas or mollicutes. The class *Mollicutes* is highly diverse, comprising four orders, five families, nine genera, and more than 200 known species [98]. Mycoplasmas are considered descendants of a common Gram-positive ancestor from which they have evolved by significant genome reduction. As a result of this regressive evolution, mycoplasmas possess limited anabolic and metabolic capabilities and maintain commensal or intimate parasitic lifestyles, relying on nutrients scavenged from their host environment [99]. Mycoplasmas are widespread in the animal kingdom with currently more than 140 validly named and 18 *Candidatus* species detected or isolated in/from vertebrate animals [List of Prokaryotic names with Standing in Nomenclature (LPSN), bacterio.net]. Almost all of these animal mycoplasmas are members of the genera *Mycoplasma*, *Ureaplasma*, and *Acholeplasma*. Some animal mycoplasmas are considered mere commensals, while others are well recognized as opportunists, pathobionts, and primary pathogens causing diseases in different animal species. Pathogenic mycoplasmas usually exhibit a narrow host specificity and predilection for mucous surfaces of the respiratory and urogenital tract, the eyes, the mammary glands, the joints, and serous membranes [100]. A highly specialized group of mycoplasmas, the so-called haemotrophic mycoplasmas or haemoplasmas, only target erythrocytes [101]. Animal mycoplasmas mostly cause mildly to moderately severe, slowly progressive, chronic infections that often display elements of immunopathology. Most important and well recognized mycoplasma diseases of animals are those affecting livestock (cattle, sheep, goats, swine) and poultry, laboratory rodents and wildlife (Table 12.1) [100, 102].

However, several further animal mycoplasmas play an aetiological role as opportunists and pathobionts in diseases of their hosts (including livestock, poultry, companion animals, and wildlife), although reasons and risk factors for disease development and progression are largely unknown [100]. Because of these differences in the clinical relevance of different mycoplasma species, accurate species identification of mycoplasma isolates is essential [103].

3 Section 12.3 was written by Joachim Spergser.

Table 12.1 Examples of important and well-recognized diseases of livestock, poultry, laboratory rodents, and wildlife caused by animal mycoplasmas.

Main host(s)	*Mycoplasma* (*M.*) species	Diseases
Cattle	*M. mycoides* ssp. *mycoides*	Contagious bovine pleuropneumonia
	M. bovis	Bovine respiratory disease, mastitis, arthritis
Goats	*M. capricolum* ssp. *capripneumoniae*	Contagious caprine pleuropneumonia
	M. mycoides ssp. *capri*	Pneumonia, mastitis, arthritis, septicaemia
Sheep, goats, wild ungulates	*M. agalactiae*	Contagious agalactia
	M. ovipneumoniae	Atypical pneumonia
	M. conjunctivae	Infectious keratoconjunctivitis
Swine	*M. hyopneumoniae*	Enzootic pneumonia
	M. hyorhinis	Polyserositis, arthritis
	M. hyosynoviae	Arthritis
	M. suis	Infectious anaemia
Chicken, turkey	*M. gallisepticum*	Chronic respiratory disease, infectious sinusitis
	M. synoviae	Infectious synovitis
Wild songbirds	*M. gallisepticum*	Fatal conjunctivitis
Mice, rats	*M. pulmonis*	Murine respiratory mycoplasmosis
Tortoises	*M. agassizii*	Upper respiratory disease
Alligators	*M. alligatoris*	Fatal septicaemia

Source: Adapted from Rosengarten et al. [100]; Citti and Blanchard [102].

12.3.2 Laboratory Diagnosis of Animal Mycoplasmas

Specific diagnosis of mycoplasma infections in animals is often challenging due to limitations of the current diagnostic tests, including difficulties of culturing mycoplasmas *in vitro*. In general, mycoplasmas are highly fastidious, requiring rich culture media (e.g. Eaton, Friis, Frey, Hayflick, SP-4 medium) for growth, and typically taking a week or more to culture [104]. Isolation and identification of a broad spectrum of animal mycoplasmas are therefore only carried out in specialized laboratories, whereas routine veterinary diagnostic laboratories usually only perform cultural and/or PCR detection of a limited number of *Mycoplasma* species. In specialized laboratories, diagnosis of animal mycoplasmas is widely based on cultural isolation from mucous membranes, secretions, or tissues, and on PCR for the detection of uncultivable (e.g. haemoplasmas) or highly fastidious *Mycoplasma* species (e.g. *Mycoplasma hyopneumoniae*), and mycoplasmas causing diseases of high veterinary importance (e.g. *Mycoplasma bovis*) [105]. By using cultivation procedures, mycoplasmas produce small colonies with diameters of 0.1–1 mm on agar medium, mostly with a characteristic fried egg morphology. Some mycoplasmas, however, are more likely to show centreless, granulated, or mulberry-like colonies, and further differences in size and opacity may occur within a single mycoplasma isolate [106]. Agar plates are incubated at 36–37 °C under 5–7% CO_2 atmosphere for up to 15 days and checked daily for colony formation using a stereomicroscope. Broth cultures are incubated at 36–37 °C in ambient air for 7–15 days or until a colour change of the medium is observed, followed by sub-cultivation on agar plates [106]. Cultivation procedures also include the examination of a limited number of biochemical properties such as glucose fermentation, pyruvate oxidation, and arginine or urea hydrolysis, only allowing the assignment of an isolate to the genus *Ureaplasma* or a metabolic group within the genus *Mycoplasma* [107]. Once cultivated, species identification of an isolate is usually achieved by antigenic or genetic methods. Antigenic identification tests such as colony immunoblotting, membrane filtration-dot immunobinding or epi-immunofluorescence [108, 109], however, depend on specific antisera to each individual mycoplasma, which often vary in their capacity to identify mycoplasmas accurately due to multiple cross-reactions and substantial serological heterogeneity of some *Mycoplasma* species, both leading to erroneous results [110]. Genetic identification of mycoplasma isolates is either based on species-specific PCRs or, more commonly, on universal PCRs targeting the 16S rRNA gene, the 16S-23S intergenic spacer region (ISR) or the *rpoB* gene, followed by sequencing or amplicon analyses [e.g. restriction fragment length polymorphism (RFLP) and denaturing gradient gel electrophoresis (DDGE)] [110–114]. Nevertheless, both serological and genetic identification techniques are time-consuming, labour-intensive, expensive, and not always discriminating. Species-specific PCR systems for the identification of mycoplasma isolates are certainly more rapid and discriminatory; however, with more than 140 animal mycoplasma species currently recognized and with up to 15 mycoplasma species that can be concurrently present in an animal host, it is not feasible to apply multiple individual PCR tests. Hence, there is a pressing need for a single, generic test that can easily and rapidly identify and differentiate mycoplasma isolates. In recent years, MALDI-TOF MS started to be used in veterinary microbiology, enabling rapid and accurate species identification of the vast majority of microorganisms encountered in veterinary diagnostic laboratories [6, 115–117]. The potential and applicability of the novel technique to animal mycoplasma identification were soon evaluated and it became evident that MALDI-TOF MS represents a serious alternative to the cumbersome diagnostics practised so far.

12.3.3 MALDI-TOF MS for the Identification of Animal Mycoplasmas

In general, the performance of MALDI-TOF MS-based identification of animal mycoplasmas is influenced by pre-analytical steps, the composition and quality of the reference database, and the identification criteria selected for species and genus identification.

As most mycoplasmas produce tiny colonies with centres deeply embedded in the agar, a single colony spotted on target often produces poor mass spectra [118]. Transfer of several colonies may work for pure cultures but increases the risk of spotting different *Mycoplasma* species on the target well as mixed cultures are frequently obtained from clinical samples. Furthermore, misidentification was shown to occur when agar is co-transferred with colony material using the direct transfer method, because of interference of spectral mass ions derived from the agar medium components that closely match the *Mycoplasma arginini* and *Mycoplasma alkalescens* reference spectra included in the commercial database [119, 120]. Most studies therefore agree that colony-derived broth cultures provide the best results. Moreover, full protein extraction of broth cultures by formic acid and acetonitrile is preferred because on-target extraction resulted in weaker scores than when the complete extraction procedures were applied [118]. However, a slightly modified protein extraction protocol that omits ethanol precipitation has also been shown to reveal high-quality spectra, concurrently optimizing the laboratory diagnostic workflow [103]. Several studies agree that the composition of the different culture media used to isolate mycoplasmas has no significant influence on MALDI-TOF MS-based mycoplasma identification [103, 118, 121]. Yet, as proteins detected with the current method are predominantly ribosomal proteins [122], the quality of a spectrum may only be optimal when mycoplasma cells are in the exponential phase of growth. In general, $\geq 10^6$ colony-forming units (CFU) spotted onto the target plate have been shown to reveal good- to high-quality spectra from animal mycoplasmas [103, 118, 121]. The incubation time and culture volume required to obtain an optimal number of mycoplasma cells are largely dependent on the species to be identified. For most *Mycoplasma* and *Acholeplasma* species, cells harvested from 1 to 5 ml broth cultures incubated for 48 hours up to 120 hours were appropriate to obtain qualified and interpretable spectra [103, 118, 121, 123, 124]. However, more fastidious mycoplasmas (e.g. *M. hyopneumoniae*, *Mycoplasma dispar*) may require 7–14 days of incubation to harvest sufficient amounts of cells from culture volumes of 1–5 ml [103]. By contrast, identification of *Ureaplasma* species requires culture volumes of ~100 ml, which precludes MALDI-TOF MS for the identification of ureaplasmas in routine diagnostic settings [103, 118].

To date, only a limited number of animal mycoplasmas are represented in the latest commercially available databases, with the few reference spectra apparently being of questionable quality [103, 119, 120]. When reference spectra of 11 animal mycoplasmas were compared with corresponding reference spectra in a commercial database, species identification was limited to 38% of the strains tested, whereas 32% were only identified to the genus level and 30% remained undiagnosed. As culture media and conditions had no significant influence on identification in this study, misidentification using the commercial database could only be explained by poor quality or a much lower number of reference spectra in the commercial database [103]. Several previous reports have emphasized the importance of supplementing reference databases with appropriate reference spectra of multiple strains in order to cover the natural diversity of the species and improve identification success [116, 125, 126]. Consequently, in studies exploring the applicability of MALDI-TOF MS for mycoplasma identification, either in-house reference libraries comprising type and field strains of species missing in the commercial reference database were compiled or the commercial reference databases were extended by including reference spectra from several clinical isolates of the *Mycoplasma* species in question. Such in-house or expanded databases were established to investigate a broad spectrum of animal mycoplasmas, ranging from a single species (i.e. *M. bovis*) [121, 127] to three rodent mycoplasmas [128], 13 ruminant mycoplasmas [118], 23 avian mycoplasmas [123] up to a highly diverse set of animal mycoplasmas, comprising 114 known species and 23 undescribed species, representing three genera (*Mycoplasma*, *Acholeplasma*, and *Ureaplasma*) and 13 phylogenetic groups or clusters [103]. In most of these studies, in-house reference libraries were further validated by testing an appropriate number of epidemiologically independent clinical isolates, which in all cases confirmed the robustness of the established databases [103, 118, 123]. Overall, in all studies, MALDI-TOF MS proved to be an excellent tool for the identification and differentiation of animal mycoplasmas, unequivocally distinguishing phylogenetically closely related *Mycoplasma* species such as *Mycoplasma agalactiae* and *M. bovis* (Figure 12.2) and even subspecies [103, 118]. However, there was a single example where MALDI-TOF MS failed to yield species-level identification (i.e. in the case of *Mycoplasma cottewii* and *Mycoplasma yeatsii*, both members of the so-called *Mycoplasma mycoides* cluster), which was surprising as the remaining members of the cluster could be discerned by their MALDI-TOF mass spectra even at the subspecies level. In fact, a former phylogenetic study demonstrated that *M. cottewii* and *M. yeatsii* exhibited the highest similarity values of phylogenetic marker genes ever observed between two *Mycoplasma* species [110], indicating that further investigations are required to evaluate whether they are different or the same species [103].

Although manufacturers recommend pre-defined cut-off values for genus- and species-level identification (e.g. Bruker Microbiology & Diagnostics advocates log scores of 1.70 and 2.00 for genus- and species-level identification), almost all studies on MALDI-TOF MS-based mycoplasma identification accepted lower species-level identification thresholds ($\geq$1.70 to $\geq$1.80), which allowed 100% species identification rates without any misidentification [103, 118, 121, 123]. However, it has also been recommended to integrate well-defined isolates that produce log scores at the genus-level identification range into the in-house reference database to cover the diversity of mass spectral patterns of a given species and, in consequence, to adjust the identification criteria selected to those provided by the manufacturer [103]. High intraspecific heterogeneity of mass spectral fingerprints may also refer to the subtyping capability of MALDI-TOF MS, as dendrograms generated from mass spectra achieved epidemiologically relevant or subtype-specific clustering for certain *Mycoplasma* species such as *Mycoplasma pneumoniae* (a human *Mycoplasma* species), *M. bovis*, and *M. agalactiae* [118, 129–131].

The taxonomy of *Mycoplasma* or, correctly, Mollicutes species, i.e. the establishment and definition of a species within the taxonomic species category, still relies on polyphasic characterization of phenotypic, genetic, and genomic properties and variations [132]. As the currently used phenotypic tests are very limited and barely discriminating, the development and evaluation of further phenotypic test approaches for the identification and characterization of *Mycoplasma* species are required [110]. Indeed, MALDI-TOF MS has been shown to contribute to the taxonomy of Mollicutes species by unambiguous identification of as-yet-undescribed mycoplasmas obviously occurring in a wide range of different animal species [103, 133]. By employing MALDI-TOF MS as an additive phenotypic test for the description of new species within the genus *Mycoplasma*, such as applied for *M. marinum* sp. nov. and *M. todarodis* sp. nov. [134], *M. struthionis* sp. nov. and *M. nasistruthionis* sp. nov. [135], as well as *M. tauri* sp. nov. [136] would add immense value to the description of these poorly circumscribed species. Altogether, MALDI-TOF MS has been shown to be a powerful and supportive tool for the taxonomic resolution of animal mycoplasmas and will continue to increase in resolution as the technology advances, as described in Chapters 1 and 2 of this volume.

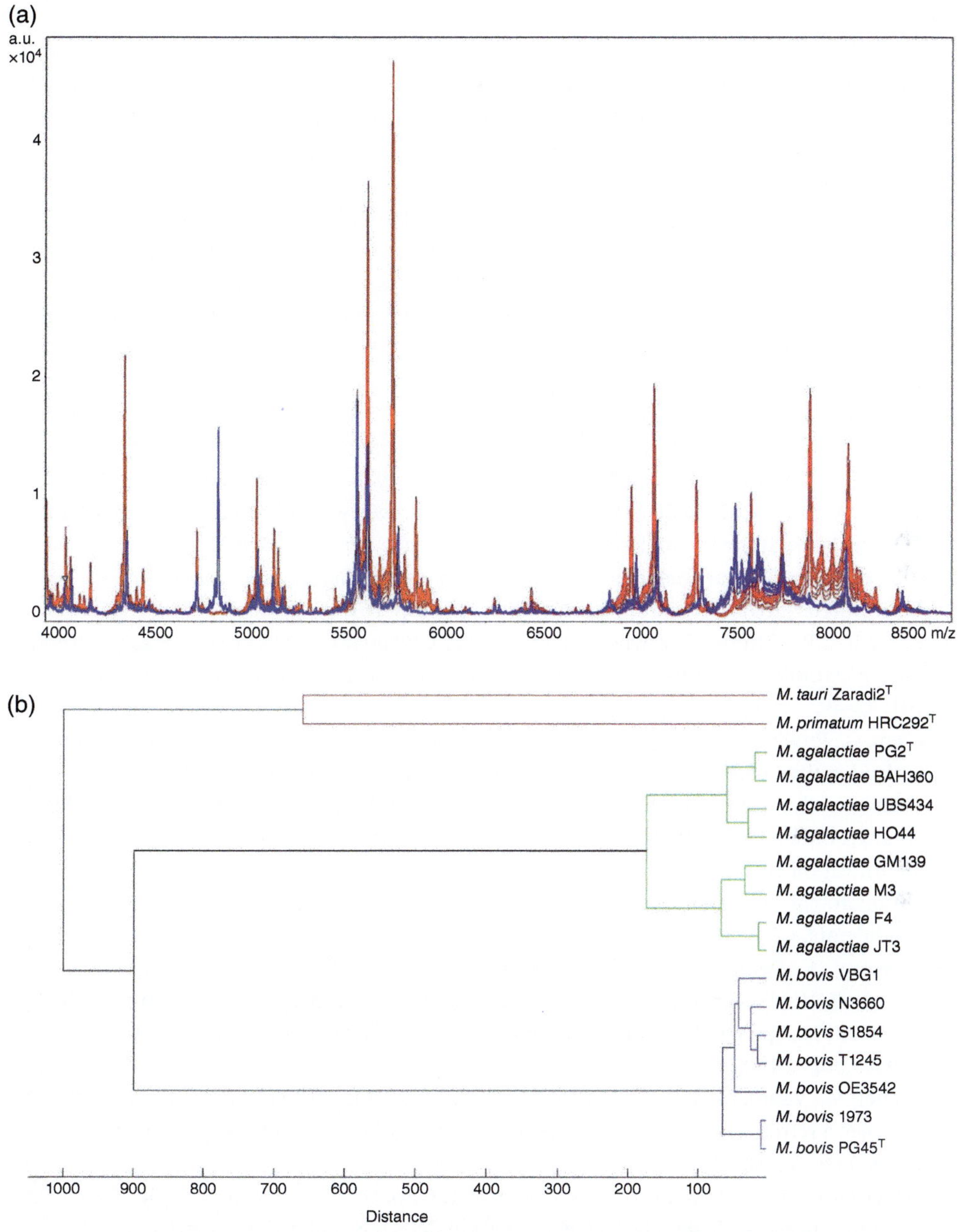

Figure 12.2 (a) Overlays of MALDI-TOF mass spectra using FlexAnalysis 3.4 software (Bruker Daltonics), generated from *Mycoplasma agalactiae* (strain PG2^T, in blue), displaying substantial differences from mass spectra of the closely related species *Mycoplasma bovis* (strain PG45^T, in red). *y*-axis, intensity a.u. = arbitrary unit; *x*-axis, *m/z* = mass-to-charge ratio. (b) MALDI-TOF MS score-oriented dendrogram (Bruker Daltonics) based on distances between reference spectra generated from *M. bovis* and *M. agalactia* strains forming clusters clearly separated from each other and from the closely related species *Mycoplasma tauri* and *Mycoplasma primatum*.

References

1 Hijazin, M., Ulbegi-Mohyla, H., Alber, J. et al. (2012). Identification of *Arcanobacterium* (Trueperella) *abortisuis*, a novel species of veterinary importance, by matrix-assisted laser desorption ionization-time of flight mass spectrometry (MALDI-TOF MS). *Berl. Munch. Tierarztl. Wochenschr.* 125: 32–37.

2 Lorente-Leal, V., Liandris, E., Bezos, J. et al. (2022). Maldi-TOF mass spectrometry as a rapid screening alternative for non-tuberculous mycobacterial species identification in the veterinary laboratory. *Front. Vet. Sci.* https://doi.org/10.3389/fvets.2022.827702.

3 Pavlovic, M., Wudy, C., Zeller-Peronnet, V. et al. (2015). Identification of bacteria isolated from veterinary clinical specimens using MALDI-TOF MS. *Berl. Munch. Tierarztl. Wochenschr.* 128: 24–30.

4 Mani, R.J., Thachil, A.J., and Ramachandran, A. (2017). Discrimination of *Streptococcus equi* subsp. equi and *Streptococcus equi* subsp. *zooepidemicus* using matrix-assisted laser desorption/ionization time-of-flight mass spectrometry. *J. Vet. Diagn. Invest.* 29: 622–627.

5 Werner, G., Fleige, C., Feßler, A.T. et al. (2012). Improved identification including MALDI-TOF mass spectrometry analysis of group D streptococci from bovine mastitis and subsequent molecular characterization of corresponding *Enterococcus faecalis* and *Enterococcus faecium* isolates. *Vet. Microbiol.* 160: 162–169.

6 Randall, L.P., Lemma, F., Koylass, M. et al. (2015). Evaluation of MALDI-ToF as a method for the identification of bacteria in the veterinary diagnostic laboratory. *Res. Vet. Sci.* 101: 42–49.

7 Clerc, O., Prod'hom, G., Vogne, C. et al. (2013). Impact of matrix-assisted laser desorption ionization time-of-flight mass spectrometry on the clinical management of patients with Gram-negative bacteremia: a prospective observational study. *Clin. Infect. Dis.* 56: 1101–1107.

8 Huang, A.M., Newton, D., Kunapuli, A. et al. (2013). Impact of rapid organism identification via matrix-assisted laser desorption/ionization time-of-flight combined with antimicrobial stewardship team intervention in adult patients with bacteremia and candidemia. *Clin. Infect. Dis.* 57: 1237–1245.

9 Verroken, A., Defourny, L., Le Polain De Waroux, O. et al. (2016). Clinical impact of MALDI-TOF MS identification and rapid susceptibility testing on adequate antimicrobial treatment in sepsis with positive blood cultures. *PLoS One* 11: https://doi.org/10.1371/journal.pone.0160537.

10 Escher, M., Vanni, M., Intorre, L. et al. (2011). Use of antimicrobials in companion animal practice: a retrospective study in a veterinary teaching hospital in Italy. *J. Antimicrob. Chemother.* 66: 920–927.

11 Ogeer-Gyles, J., Mathews, K.A., Sears, W. et al. (2006). Development of antimicrobial drug resistance in rectal *Escherichia coli* isolates from dogs hospitalized in an intensive care unit. *J. Am. Vet. Med. Assoc.* 229: 694–699.

12 Robbins, S.N., Goggs, R., Lhermie, G. et al. (2020). Antimicrobial prescribing practices in small animal emergency and critical care. *Front. Vet. Sci.* 7: https://doi.org/10.3389/fvets.2020.00110.

13 Singleton, D.A., Sánchez-Vizcaíno, F., Arsevska, E. et al. (2018). New approaches to pharmacosurveillance for monitoring prescription frequency, diversity, and co-prescription in a large sentinel network of companion animal veterinary practices in the United Kingdom, 2014–2016. *Prev. Vet. Med.* 159: 153–161.

14 Micek, S.T., Welch, E.C., Khan, J. et al. (2010). Empiric combination antibiotic therapy is associated with improved outcome against sepsis due to Gram-negative bacteria: a retrospective analysis. *Antimicrob. Agents Chemother.* 54: 1742–1748.

15 Seifert, H. (2009). The clinical importance of microbiological findings in the diagnosis and management of bloodstream infections. *Clin. Infect. Dis.* 48 (Suppl 4): S238–S245.

16 Bazzi, A.M., Rabaan, A.A., El Edaily, Z. et al. (2017). Comparison among four proposed direct blood culture microbial identification methods using MALDI-TOF MS. *J. Infect. Public Health* 10: 308–315.

17 Faron, M.L., Buchan, B.W., and Ledeboer, N.A. (2017). Matrix-assisted laser desorption ionization-time of flight mass spectrometry for use with positive blood cultures: methodology, performance, and optimization. *J. Clin. Microbiol.* 55: 3328–3338.

18 Moussaoui, W., Jaulhac, B., Hoffmann, A.M. et al. (2010). Matrix-assisted laser desorption ionization time-of-flight mass spectrometry identifies 90% of bacteria directly from blood culture vials. *Clin. Microbiol. Infect.* 16: 1631–1638.

19 Prod'hom, G., Bizzini, A., Durussel, C. et al. (2010). Matrix-assisted laser desorption ionization-time of flight mass spectrometry for direct bacterial identification from positive blood culture pellets. *J. Clin. Microbiol.* 48: 1481–1483.

20 Nomura, F., Tsuchida, S., Murata, S. et al. (2020). Mass spectrometry-based microbiological testing for blood stream infection. *Clin. Proteomics* 17: https://doi.org/10.1186/s12014-020-09278-7.

21 Tsuchida, S., Murata, S., Miyabe, A. et al. (2018). Application of the biocopolymer preparation system, rapid Bacpro® ii kit, for mass-spectrometry-based bacterial identification from positive blood culture bottles by the MALDI Biotyper system. *J. Microbiol. Methods* 152: 86–91.

22 Christner, M., Rohde, H., Wolters, M. et al. (2010). Rapid identification of bacteria from positive blood culture bottles by use of matrix-assisted laser desorption-ionization time of flight mass spectrometry fingerprinting. *J. Clin. Microbiol.* 48: 1584–1591.

23 Ferreira, L., Sánchez-Juanes, F., Porras-Guerra, I. et al. (2011). Microorganisms direct identification from blood culture by matrix-assisted laser desorption/ionization time-of-flight mass spectrometry. *Clin. Microbiol. Infect.* 17: 546–551.

24 La Scola, B. and Raoult, D. (2009). Direct identification of bacteria in positive blood culture bottles by matrix-assisted laser desorption ionisation time-of-flight mass spectrometry. *PLoS One* 4: https://doi.org/10.1371/journal.pone.0008041.

25 Ling, G.V. (1984). Therapeutic strategies involving antimicrobial treatment of the canine urinary tract. *J. Am. Vet. Med. Assoc.* 185: 1162–1164.

26 Hall, J.L., Holmes, M.A., and Baines, S.J. (2013). Prevalence and antimicrobial resistance of canine urinary tract pathogens. *Vet. Rec.* 173: 549.

27 Litster, A., Moss, S.M., Honnery, M. et al. (2007). Prevalence of bacterial species in cats with clinical signs of lower urinary tract disease: recognition of *Staphylococcus felis* as a possible feline urinary tract pathogen. *Vet. Microbiol.* 121: 182–188.

28 Wong, C., Epstein, S.E., and Westropp, J.L. (2015). Antimicrobial susceptibility patterns in urinary tract infections in dogs (2010–2013). *J. Vet. Intern. Med.* 29: 1045–1052.

29 Ferreira, L., Sánchez-Juanes, F., González-Avila, M. et al. (2010). Direct identification of urinary tract pathogens from urine samples by matrix-assisted laser desorption ionization-time of flight mass spectrometry. *J. Clin. Microbiol.* 48: 2110–2115.

30 Köhling, H.L., Bittner, A., Müller, K.D. et al. (2012). Direct identification of bacteria in urine samples by matrix-assisted laser desorption/ionization time-of-flight mass spectrometry and relevance of defensins as interfering factors. *J. Med. Microbiol.* 61: 339–344.

31 Florio, W., Baldeschi, L., Rizzato, C. et al. (2020). Detection of antibiotic-resistance by MALDI-TOF mass spectrometry: an expanding area. *Front. Cell. Infect. Microbiol.* https://doi.org/10.3389/fcimb.2020.572909.

32 Chang, K.-C., Chung, C.-Y., Yeh, C.-H. et al. (2018). Direct detection of carbapenemase-associated proteins of *Acinetobacter baumannii* using nanodiamonds coupled with matrix-assisted laser desorption/ionization time-of-flight mass spectrometry. *J. Microbiol. Methods* 147: 36–42.

33 Josten, M., Dischinger, J., Szekat, C. et al. (2014). Identification of agr-positive methicillin-resistant *Staphylococcus aureus* harbouring the class A mec complex by MALDI-TOF mass spectrometry. *Int. J. Med. Microbiol.* 304: 1018–1023.

34 Rhoads, D.D., Wang, H., Karichu, J., and Richter, S.S. (2016). The presence of a single MALDI-TOF mass spectral peak predicts methicillin resistance in staphylococci. *Diagn. Microbiol. Infect. Dis.* 86: 257–261.

35 Chatterjee, S.S., Chen, L., Joo, H.-S. et al. (2011). Distribution and regulation of the mobile genetic element-encoded phenol-soluble Modulin PSM-mec in methicillin-resistant *Staphylococcus aureus*. *PLoS One* 6: https://doi.org/10.1371/journal.pone.0028781.

36 Jackson, K.A., Edwards-Jones, V., Sutton, C.W., and Fox, A.J. (2005). Optimisation of intact cell MalDI method for fingerprinting of methicillin-resistant *Staphylococcus aureus*. *J. Microbiol. Methods* 62: 273–284.

37 Walker, J., Fox, A.J., Edwards-Jones, V., and Gordon, D.B. (2002). Intact cell mass spectrometry (ICMS) used to type methicillin-resistant *Staphylococcus aureus*: media effects and inter-laboratory reproducibility. *J. Microbiol. Methods* 48: 117–126.

38 Burckhardt, I. and Zimmermann, S. (2011). Using matrix-assisted laser desorption ionization-time of flight mass spectrometry to detect carbapenem resistance within 1 to 2.5 hours. *J. Clin. Microbiol.* 49: 3321–3324.

39 Ota, Y., Furuhashi, K., Hirai, N. et al. (2021). Evaluation of MBT STAR-Cepha and MBT STAR-Carba kits for the detection of extended-spectrum β-lactamases and carbapenemase producing microorganisms using matrix-assisted laser desorption/ionization time-of-flight mass spectrometry. *J. Microbiol. Methods* 183: 106166.

40 Bortolami, A., Zendri, F., Maciuca, E.I. et al. (2019). Diversity, virulence, and clinical significance of extended-spectrum β-lactamase- and pAmpC-producing *Escherichia coli* from companion animals. *Front. Microbiol.* 10: https://doi.org/10.3389/fmicb.2019.01260.

41 Schott, H.C. 2nd, Ewart, S.L., Walker, R.D. et al. (2001). An outbreak of salmonellosis among horses at a veterinary teaching hospital. *J. Am. Vet. Med. Assoc.* 218: 1152–9.

42 Van Der Kolk, J.H., Endimiani, A., Graubner, C. et al. (2019). Acinetobacter in veterinary medicine, with an emphasis on *Acinetobacter baumannii*. *J. Global Antimicrobial Resist.* 16: 59–71.

43 Milton, A.A.P., Priya, G.B., Aravind, M. et al. (2015). Nosocomial infections and their surveillance in veterinary hospitals. *Adv. Anim. Vet. Sci.* 3 (2s): 1–24.

44 Boyle, A.G., Timoney, J.F., Newton, J.R. et al. (2018). *Streptococcus equi* infections in horses: guidelines for treatment, control, and prevention of strangles-revised consensus statement. *J. Vet. Intern. Med.* 32: 633–647.

45 Bar-Meir, M., Berliner, E., Kashat, L. et al. (2020). The utility of MALDI-TOF MS for outbreak investigation in the neonatal intensive care unit. *Eur. J. Pediatr.* 179: 1843–1849.

46 Holzknecht, B.J., Dargis, R., Pedersen, M. et al. (2018). Typing of vancomycin-resistant enterococci with MALDI-TOF mass spectrometry in a nosocomial outbreak setting. *Clin. Microbiol. Infect.* 24: https://doi.org/10.1016/j.cmi.2018.03.020.

47 Authority, E.F.S., Prevention, E.C.f.D., Control (2021). The European Union One Health 2020 Zoonoses report. *EFSA J.* 19: e06971.

48 Parte, A.C., Sardà Carbasse, J., Meier-Kolthoff, J.P. et al. (2020). List of prokaryotic names with standing in nomenclature (LPSN) moves to the DSMZ. *Int. J. Syst. Evol. Microbiol.* 70: 5607–5612.

49 World Organisation for Animal Health (OIE), 2021 *Manual of Diagnostic Tests and Vaccines for Terrestrial Animals.* https://www.oie.int (accessed 28 March 2022) 2021.

50 Fitzgerald, C., Tu, Z.C., Patrick, M. et al. (2014). *Campylobacter fetus* subsp. *testudinum* subsp. nov., isolated from humans and reptiles. *Int. J. Syst. Evol. Microbiol.* 64: 2944–2948.

51 Tindall, B.J., Grimont, P.A.D., Garrity, G.M., and Euzéby, J.P. (2005). Nomenclature and taxonomy of the genus *Salmonella*. *Int. J. Syst. Evol. Microbiol.* 55: 521–524.

52 Grimont, P.A.D. and Weill, F.X. (2007). *Antigenic Formulae of the Salmonella Serovars*, 9e. Paris: WHO Collaborating Centre for Reference and Research on Salmonella, Institut Pasteur.

53 Stevens, M.P. and Kingsley, R.A. (2021). *Salmonella* pathogenesis and host-adaptation in farmed animals. *Curr. Opin. Microbiol.* 63: 52–58.

54 Topić Popović, N., Kazazić, S.P., Bojanić, K. et al. (2021). Sample preparation and culture condition effects on MALDI-TOF MS identification of bacteria: A review. *Mass Spectrom. Rev.* https://doi.org/10.1002/mas.21739.

55 Reeve, M.A. and Bachmann, D. (2019). MALDI-TOF MS protein fingerprinting of mixed samples. *Biol. Methods Protocols* 4: bpz013.

56 Mazzeo, M.F., Sorrentino, A., Gaita, M. et al. (2006). Matrix-assisted laser desorption ionization-time of flight mass spectrometry for the discrimination of food-borne microorganisms. *Appl. Environ. Microbiol.* 72: 1180–1189.

57 Goldstein, J.E., Zhang, L., Borror, C.M. et al. (2013). Culture conditions and sample preparation methods affect spectrum quality and reproducibility during profiling of *Staphylococcus aureus* with matrix-assisted laser desorption/ionization time-of-flight mass spectrometry. *Lett. Appl. Microbiol.* 57: 144–150.

58 Tsuchida, S., Umemura, H., and Nakayama, T. (2020). Current status of matrix-assisted laser desorption/ionization–time-of-flight mass spectrometry (MALDI-TOF MS) in clinical diagnostic microbiology. *Molecules* 25: 4775.

59 Dieckmann, R., Helmuth, R., Erhard, M., and Malorny, B. (2008). Rapid classification and identification of *Salmonellae* at the species and subspecies levels by whole-cell matrix-assisted laser desorption ionization-time of flight mass spectrometry. *Appl. Environ. Microbiol.* 74: 7767–7778.

60 Alispahic, M., Hummel, K., Jandreski-Cvetkovic, D. et al. (2010). Species-specific identification and differentiation of *Arcobacter, Helicobacter* and *Campylobacter* by full-spectral matrix-associated laser desorption/ionization time of flight mass spectrometry analysis. *J. Med. Microbiol.* 59: 295–301.

61 Ford, B.A. and Burnham, C.-A.D. (2013). Optimization of routine identification of clinically relevant Gram-negative bacteria by use of matrix-assisted laser desorption ionization-time of flight mass spectrometry and the Bruker Biotyper. *J. Clin. Microbiol.* 51: 1412–1420.

62 Ruelle, V., El Moualij, B., Zorzi, W. et al. (2004). Rapid identification of environmental bacterial strains by matrix-assisted laser desorption/ionization time-of-flight mass spectrometry. *Rapid Commun. Mass Spectrometry: RCM* 18: 2013–2019.

63 Mandrell, R.E., Harden, L.A., Bates, A. et al. (2005). Speciation of *Campylobacter coli, C. jejuni, C. helveticus, C. lari, C. sputorum*, and *C. upsaliensis* by matrix-assisted laser desorption ionization-time of flight mass spectrometry. *Appl. Environ. Microbiol.* 71: 6292–6307.

64 Hsieh, Y.H., Wang, Y.F., Moura, H. et al. (2018). Application of MALDI-TOF MS systems in the rapid identification of *Campylobacter* spp. of public health importance. *J. AOAC Int.* 101: 761–768.

65 Kolínská, R., Drevínek, M., Jakubů, V., and Zemlicková, H. (2008). Species identification of *Campylobacter jejuni* ssp. *jejuni* and *C. coli* by matrix-assisted laser desorption/ionization time-of-flight mass spectrometry and PCR. *Folia Microbiol.* 53: 403–409.

66 Bessède, E., Solecki, O., Sifré, E. et al. (2011). Identification of *Campylobacter* species and related organisms by matrix assisted laser desorption ionization–time of flight (MALDI-TOF) mass spectrometry. *Clin. Microbiol. Infect.* 17: 1735–1739.

67 Martiny, D., Dediste, A., Debruyne, L. et al. (2011). Accuracy of the API Campy system, the Vitek 2 Neisseria-Haemophilus card and matrix-assisted laser desorption ionization time-of-flight mass spectrometry for the identification of *Campylobacter* and related organisms. *Clin. Microbiol. Infect.* 17: 1001–1006.

68 Bastin, B., Bird, P., Benzinger, M.J. et al. (2019). Confirmation and identification of *Salmonella* spp., *Cronobacter* spp., and other Gram-negative organisms by the Bruker MALDI Biotyper method: collaborative study method extension to include *Campylobacter* species, revised first action 2017.09. *J. AOAC Int.* 102: 1595–1616.

69 Crawshaw, T.R., Chanter, J.I., Young, S.C. et al. (2015). Isolation of a novel thermophilic *Campylobacter* from cases of spotty liver disease in laying hens and experimental reproduction of infection and microscopic pathology. *Vet. Microbiol.* 179: 315–321.

70 Van, T.T.H., Elshagmani, E., Gor, M.C. et al. (2016). *Campylobacter hepaticus* sp. nov., isolated from chickens with spotty liver disease. *Int. J. Syst. Evol. Microbiol.* 66: 4518–4524.

71 Van den Abeele, A.M., Vogelaers, D., Vandamme, P. et al. (2018). Filling the gaps in clinical proteomics: a do-it-yourself guide for the identification of the emerging pathogen *Arcobacter* by matrix-assisted laser desorption ionization-time of flight mass spectrometry. *J. Microbiol. Methods* 152: 92–97.

72 Tanner, A.C.R., Badger, S., Lai, C.H. et al. (1981). *Wolinella* gen. nov., *Wolinella succinogenes* (*Vibrio succinogenes* Wolin et al.) comb. nov., and description of *Bacteroides gracilis* sp. nov., *Wolinella recta* sp. nov., *Campylobacter concisus* sp. nov., and *Eikenella corrodens* from humans with periodontal disease. *Int. J. Syst. Bacteriol.* 31: 432–445.

73 On, S.L.W., Zhang, J., Cornelius, A.J., and Anderson, T.P. (2021). Markers for discriminating *Campylobacter concisus* genomospecies using MALDI-TOF analysis. *Curr. Res. Microbial Sci.* 2: 100019.

74 Liu, F., Ma, R., Wang, Y., and Zhang, L. (2018). The clinical importance of *Campylobacter concisus* and other human hosted *Campylobacter* species. *Front. Cell. Infect. Microbiol.* 8: 243–243.

75 Macuch, P.J. and Tanner, A.C. (2000). *Campylobacter* species in health, gingivitis, and periodontitis. *J. Dent. Res.* 79: 785–792.

76 Girard, V., Monnin, V., Giraud, D. et al. (2021). Multicenter evaluation of the VITEK MS matrix-assisted laser desorption/ionization-time of flight mass spectrometry system for identification of bacteria, including *Brucella,* and yeasts. *Eur. J. Clin. Microbiol. Infect. Dis.* 40: 1909–1917.

77 Muhamadali, H., Weaver, D., Subaihi, A. et al. (2016). Chicken, beams, and *Campylobacter*: rapid differentiation of foodborne bacteria via vibrational spectroscopy and MALDI-mass spectrometry. *Analyst* 141: 111–122.

78 Liu, Y.H., Yamazaki, W., Huang, Y.T. et al. (2019). Clinical and microbiological characteristics of patients with bacteremia caused by *Campylobacter* species with an emphasis on the subspecies of *C. fetus*. *J. Microbiol. Immunol. Infect.* 52: 122–131.

79 On, S.L.W., Bloch, B., Holmes, B. et al. (1995). *Campylobacter hyointestinalis* subsp. *lawsonii* subsp. nov., isolated from the porcine stomach, and an emended description of *Campylobacter hyointestinalis*. *Int. J. Syst. Evol. Microbiol.* 45: 767–774.

80 Miller, W.G., Yee, E., and Chapman, M.H. (2016). Complete genome sequences of *Campylobacter hyointestinalis* subsp. *hyointestinalis* strain LMG 9260 and *C. hyointestinalis* subsp. *lawsonii* strain LMG 15993. *Genome Announc.* 4: e00665–e00616.

81 Costa, D., Lévesque, S., Kumar, N. et al. (2019). Pangenome analysis reveals genetic isolation in *Campylobacter hyointestinalis* subspecies adapted to different mammalian hosts. *Sci. Rep.* 11: 3431.

82 Gilbert, M.J., Miller, W.G., Leger, J.S. et al. (2017). *Campylobacter pinnipediorum* sp. nov., isolated from pinnipeds, comprising *Campylobacter pinnipediorum* subsp. *pinnipediorum* subsp. nov. and *Campylobacter pinnipediorum* subsp. *caledonicus* subsp. nov. *Int. J. Syst. Evol. Microbiol.* 67: 1961–1968.

83 Dieckmann, R. and Malorny, B. (2011). Rapid screening of epidemiologically important *Salmonella enterica* subsp. *enterica* serovars by whole-cell matrix-assisted laser desorption ionization-time of flight mass spectrometry. *Appl. Environ. Microbiol.* 77: 4136–4146.

84 Mangmee, S., Reamtong, O., Kalambaheti, T. et al. (2020). MALDI-TOF mass spectrometry typing for predominant serovars of non-typhoidal *Salmonella* in a Thai broiler industry. *Food Control* 113: 107188.

85 Yang, S.M., Kim, E., Kim, D. et al. (2021). Rapid detection of *Salmonella* Enteritidis, Typhimurium, and Thompson by specific peak analysis using matrix-assisted laser desorption ionization time-of-flight mass spectrometry. *Foods* 10 (5): 933. https://doi.org/10.3390/foods10050933.

86 Kang, L., Li, N., Li, P. et al. (2017). MALDI-TOF mass spectrometry provides high accuracy in identification of *Salmonella* at species level but is limited to type or subtype *Salmonella* serovars. *Eur. J. Mass Spectrometry* 23: 70–82.

87 Martiny, D., Busson, L., Wybo, I. et al. (2012). Comparison of the Microflex LT and Vitek MS systems for routine identification of bacteria by matrix-assisted laser desorption ionization-time of flight mass spectrometry. *J. Clin. Microbiol.* 50: 1313–1325.

88 Bastin, B., Bird, P., Benzinger, M.J. et al. (2018). Confirmation and identification of *Salmonella* spp., *Cronobacter* spp., and other Gram-negative organisms by the Bruker MALDI Biotyper method: collaborative study, first action 2017.09. *J. AOAC Int.* 101: 1593–1609.

89 De Carolis, E., Vella, A., Vaccaro, L. et al. (2014). Application of MALDI-TOF mass spectrometry in clinical diagnostic microbiology. *J. Infect. Dev. Ctries.* 8: 1081–1088.

90 Chen, S.H., Parker, C.H., Croley, T.R., and McFarland, M.A. (2019). Identification of *Salmonella* taxon-specific peptide markers to the serovar level by mass spectrometry. *Anal. Chem.* 91: 4388–4395.

91 Borges Lima, D., Dupré, M., Mariano Santos, M.D. et al. (2021). DiagnoTop: a computational pipeline for discriminating bacterial pathogens without database search. *J. Am. Soc. Mass Spectrom.* 32: 1295–1299.

92 Encheva, V., Wait, R., Begum, S. et al. (2007). Protein expression diversity amongst serovars of *Salmonella enterica*. *Microbiology* 153: 4183–4193.

93 Cooper, A.L., Low, A.J., Koziol, A.G. et al. (2020). Systematic evaluation of whole genome sequence-based predictions of *Salmonella* serotype and antimicrobial resistance. *Front. Microbiol.* 11: 549.

94 Bishop, C., Arnold, C., and Gharbia, S.E. (2010). Transfer of a traditional serotyping system (Kauffmann-White) onto a MALDI-TOF-MS platform for the rapid typing of *Salmonella* isolates. In: *Mass Spectrometry for Microbial Proteomics* (ed. H.N. Shah and S.E. Gharbia), ch. 20. Wiley http://dx.doi.org/10.1002/9780470665497.ch20.

95 Bishop, C., Honisch, C., Goldman, T. et al. (2012). Combined genomarkers approach to *Salmonella* characterization reveals that nucleotide sequence differences in the phase 1 flagellin gene fliC are markers for variation within serotypes. *J. Med. Microbiol.* 61: 1517–1524. https://doi.org/10.1099/jmm.0.047431-0.

96 Razin, S., Yogev, D., and Naot, Y. (1998). Molecular biology and pathogenicity of mycoplasmas. *Microbiol. Mol. Biol. Rev.* 62: 1094–1156.

97 Oren, A. and Gerrity, G.M. (2021). Valid publication of the names of forty-two phyla of prokaryotes. *Int. J. Syst. Evol. Microbiol.* 71: 5056.

98 Brown, D.R., May, M., Bradbury, J.M., and Johansson, K.E. (2018). Mollicutes. In: *Bergey's Manual of Systematics of Archaea and Bacteria* (ed. W.B. Whitman, F. Rainey, P. Kämpfer, et al.). Wiley, in association with Bergey's Manual Trust.

99 Sirand-Pugnet, P., Citti, C., Barre, A., and Blanchard, A. (2007). Evolution of mollicutes: down a bumpy road with twists and turns. *Res. Microbiol.* 158: 754–766.

100 Rosengarten, R., Citti, C., Glew, M. et al. (2000). Host-pathogen interactions in mycoplasma pathogenesis: virulence and survival strategies of minimalist prokaryotes. *Int. J. Med. Microbiol.* 290: 15–25.

101 Messick, J.B. (2004). Hemotrophic mycoplasmas (hemoplasmas): a review and new insights into pathogenic potential. *Vet. Clin. Pathol.* 33: 2–13.

102 Citti, C. and Blanchard, A. (2013). Mycoplasmas and their host: emerging and re-emerging minimal pathogens. *Trends Microbiol.* 21: 196–203.

103 Spergser, J., Hess, C., Loncaric, I., and Ramírez, A.S. (2019). Matrix-assisted laser desorption ionization-time of flight mass spectrometry is a superior diagnostic tool for the identification and differentiation of mycoplasmas isolated from animals. *J. Clin. Microbiol.* 57: 00316–00319.

104 Freundt, E.A. (1983). Culture media for classical mycoplasmas. In: *Methods in Mycoplasmology* (ed. S. Razin and J.G. Tully), 127–136. New York, USA: Academic Press.

105 Markham, P.F. and Noormohammadi, A.H. (2005). Diagnosis of mycoplasmosis in animals. In: *Mycoplasmas: Molecular Biology Pathogenicity and Strategies for Control* (ed. A. Blanchard and G. Browning), 355–382. Norfolk, United Kingdom: Horizon Bioscience.

106 Tully, J.G. (1995). Culture medium formulation for primary isolation and maintenance of mollicutes. In: *Molecular and Diagnostic Procedures in Mycoplasmology* (ed. S. In Razin and J.G. Tully), 33–39. San Diego, USA: Academic Press.

107 Poveda, J.B. (1998). Biochemical characteristics in mycoplasma identification. In: *Methods in Molecular Biology: Mycoplasma Protocols* (ed. R.J. In Miles and R.A.J. Nicholas), 69–70. Totowa, New Jersey, USA: Humana Press, Inc.

108 Poumarat, F., Perrin, B., and Longchambon, D. (1991). Identification of ruminant mycoplasmas by dot immunobinding on membrane filtration (MF dot). *Vet. Microbiol.* 29: 329–338.

109 Poveda, J.B. and Nicholas, R. (1998). Serological identification of mycoplasmas by growth and metabolic inhibition tests. *Methods Mol. Biol.* 104: 105–111.

110 Volokhov, D.V., Simonyan, V., Davidson, M.K., and Chizhikov, V.E. (2012). RNA polymerase beta subunit (*rpoB*) gene and the 16S-23S rRNA intergenic transcribed spacer region (ITS) as complementary molecular marker in addition to the 16S rRNA gene for phylogenetic analysis and identification of the species of the family *Mycoplasmataceae*. *Mol. Phylogenet. Evol.* 62: 515–528.

111 Johansson, K.E., Heldtander, M.U., and Pettersson, B. (1998). Characterization of mycoplasmas by PCR and sequence analysis with universal 16S rDNA primers. *Methods Mol. Biol.* 104: 145–165.

112 McAuliffe, L., Ellis, R.J., Lawes, J.R. et al. (2005). 16S rDNA PCR and denaturing gradient gel electrophoresis; a single generic test for detecting and differentiating *Mycoplasma* species. *J. Med. Microbiol.* 54: 731–739.

113 Ramírez, A.S., Naylor, C.J., Pitcher, D.G., and Bradbury, J.M. (2008). High inter-species and low intra-species variation in 16S-23S rDNA spacer sequences of pathogenic avian mycoplamas offers potential use as a diagnostic tool. *Vet. Microbiol.* 128: 279–287.

114 Spergser, J. and Rosengarten, R. (2007). Identification and differentiation of canine *Mycoplasma* isolates by 16S-23S rDNA PCR-RFLP. *Vet. Microbiol.* 125: 170–174.

115 Bizzini, A. and Greub, G. (2010). Matrix-assisted laser desorption ionization time-of-flight mass spectrometry, a revolution in clinical microbial identification. *Clin. Microbiol. Infect.* 16: 1614–1619.

116 Hess, C., Alispahic, M., and Hess, M. (2016). Application of MALDI-TOF MS in veterinary and food microbiology. In: *MALDI-TOF Mass Spectrometry in Microbiology* (ed. M. Kostrzewa and S. Schubert), 109–125. Norfolk, United Kingdom: Caister Academic Press.

117 Kuhnert, P., Bisgaard, M., Korczak, B.M. et al. (2012). Identification of animal *Pasteurellaceae* by MALDI-TOF mass spectrometry. *J. Microbiol. Methods* 89: 1–7.

118 Pereyre, S., Tardy, F., Renaudin, H. et al. (2013). Identification and subtyping of clinically relevant human and ruminant mycoplasmas by use of matrix-assisted laser desorption ionization–time of flight mass spectrometry. *J. Clin. Microbiol.* 51: 3314–3323.

119 Bokma, J., Pardon, B., Deprez, P. et al. (2020a). Non-specific, agar medium-related peaks can result in false positive *Mycoplasma alkalescens* and *Mycoplasma arginini* identification by MALDI-TOF MS. *Res. Vet. Sci.* 130: 139–143.

120 Lagacé-Wiens, P.R.S., Abbott, A.A., and Karlowsky, J.A. (2019). CHROMagar™ orientation urine culture medium produces matrix-assisted laser desorption ionization-time-of- flight mass spectrometry spectra misidentified as *Mycoplasma arginini* and *Mycoplasma alkalescens*. *Diagn. Microbiol. Infect. Dis.* 94: 113–115.

121 Bokma, J., Pardon, B., Van Driessche, L. et al. (2019). Optimizing identification of *Mycoplasma bovis* by MALDI-TOF MS. *Res. Vet. Sci.* 125: 185–188.

122 Arnold, R.J. and Reilly, J.P. (1999). Observation of *Escherichia coli* ribosomal proteins and their posttranslational modifications by mass spectrometry. *Anal. Biochem.* 269: 105–112.

123 Baudler, L., Scheufen, S., Ziegler, L. et al. (2019). Identification and differentiation of avian *Mycoplasma* species using MALDI-TOF MS. *J. Vet. Diagn. Invest.* 31: 620–624.

124 Bokma, J., Van Driessche, L., Deprez, P. et al. (2020b). Rapid identification of *Mycoplasma bovis* strains from bovine bronchoalveolar lavage fluid with matrix-assisted laser desorption ionization-time of flight mass spectrometry after enrichment procedure. *J. Clin. Microbiol.* 58: e00004–e00020.

125 Alispahic, M., Christensen, H., Bisgaard, M. et al. (2014). MALDI-TOF mass spectrometry confirms difficulties in separating species of the *Avibacterium* genus. *Avian Pathol.* 43: 258–263.

126 Van Veen, S.Q., Claas, E.C., and Kuijper, E.J. (2010). High-throughput identification of bacteria and yeast by matrix-assisted laser desorption ionization-time of flight mass spectrometry in conventional medical microbiology laboratories. *J. Clin. Microbiol.* 48: 900–907.

127 McDaniel, A.J. and Derscheid, R.J. (2021). MALDI-TOF mass spectrometry and high-resolution melting PCR for the identification of *Mycoplasma bovis* isolates. *BMC Vet. Res.* 17: 170.

128 Goto, K., Yamamoto, M., Asahara, M. et al. (2012). Rapid identification of *Mycoplasma pulmonis* isolated from laboratory mice and rats using matrix-assisted laser desorption ionization time-of-flight mass spectrometry. *J. Vet. Med. Sci.* 74: 1083–1086.

129 Barbosa, M.S., Spergser, J., Marques, L.M. et al. (2022). Predominant single stable VpmaV expression in strain GM139 and major differences with *Mycoplasma agalactiae* type strain PG2. *Animals* 12: 265.

130 Becker, C.A., Thibault, F.M., Arcangioli, M.A., and Tardy, F. (2015). Loss of diversity within *Mycoplasma bovis* isolates collected in France from bovines with respiratory diseases over the last 35 years. *Infect. Genet. Evol.* 33: 118–126.

131 Xiao, D., Zhao, F., Zhang, H. et al. (2014). Novel strategy for typing *Mycoplasma pneumoniae* isolates by use of matrix-assisted laser desorption ionization-time of flight mass spectrometry coupled with ClinProTools. *J. Clin. Microbiol.* 52: 3038–3043.

132 Brown, D.R., Whitcomb, R.F., and Bradbury, J.M. (2007). Revised minimal standards for description of new species of the class Mollicutes (division *Tenericutes*). *Int. J. Syst. Evol. Microbiol.* 57: 2703–2719.

133 Hennig-Pauka, I., Sudendey, C., Kleinschmidt, S. et al. (2020). Swine conjunctivitis associated with a novel *Mycoplasma* species closely related to *Mycoplasma hyorhinis*. *Pathogens* 10: 13.

134 Ramírez, A.S., Vega-Orellana, O.M., Viver, T. et al. (2019). First description of two moderately halophilic and psychrotolerant *Mycoplasma* species isolated from cephalopods and proposal of *Mycoplasma marinum* sp. nov. and *Mycoplasma todarodis* sp. nov. *Syst. Appl. Microbiol.* 42: 457–467.

135 Spergser, J., Botes, A., Nel, T. et al. (2020). *Mycoplasma nasistruthionis* sp. nov. and *Mycoplasma struthionis* sp. nov. isolated from ostriches with respiratory disease. *Syst. Appl. Microbiol.* 43: 126047.

136 Spergser, J., DeSoye, P., Ruppitsch, W. et al. (2022). *Mycoplasma tauri* sp. nov. isolated from the bovine genital tract. *Syst. Appl. Microbiol.* 45: 126292.

13

MALDI-TOF MS: from Microbiology to Drug Discovery

Ruth Walker[1]*, Maria E. Dueñas*[1]*, Alan Ward*[2]*, and Kaveh Emami*[3]

[1] *Laboratory for Biological Mass Spectrometry, Biosciences Institute, Newcastle University, Newcastle upon Tyne, UK*
[2] *School of Biology, Newcastle University, Newcastle upon Tyne, UK*
[3] *FUJIFILM Diosynth Biotechnologies, Billingham, UK*

13.1 Introduction

Matrix-assisted laser desorption/ionization time-of-flight (MALDI-TOF) mass spectrometry (MS) is a highly versatile, powerful technique that has been used in recent decades as an exploratory technique in proteomics, metabolomics, analysis of tissue samples, and for characterizing microorganisms. The use of mass spectrometers to characterize microorganisms dates from 1975 [1]. Methods were then rapidly optimized and expanded to include clinical and environmental species. However, these methods were too labour-intensive, expensive, and inefficient to be widely applied, but the invention of MALDI-TOF MS in 1985 [2] introduced a rapid, reliable, and robust MS method for microbial identification.

The method for microbial identification is now well established in clinical and environmental contexts with numerous publications in which MALDI-TOF MS has been used for characterization of bacteria and viruses [3–5], archaea [6], yeast [7, 8], filamentous fungi [9, 10], nematodes [11], insects [12], scallops [13], fish [14, 15], and algae [16]. In addition, MALDI-TOF MS biotyping of microorganisms has been applied in defence (e.g. identifying anthrax [17, 18]) and food safety (e.g. for strain identification), food fraud, and for inspection of the animal and plant trades such as meat typing [19].

As the molecular basis for the differences in MALDI-TOF MS spectra is generally unknown, the differences cannot be assumed to reflect phylogeny. However, there is evidence showing that the majority of the peaks in bacterial MALDI-TOF mass spectral profiles are from ribosomal and other highly abundant housekeeping proteins, which are generally independent of growth conditions [20, 21]. Nonetheless, MALDI-TOF MS has the capability to discriminate between closely related organisms, exploring their epidemiology and taxonomy, enabling understanding of disease patterns and their transmission (e.g. necrotizing fasciitis [22]), identifying factors that increase risk of a disease (e.g. temperature) and exploring the best prevention and treatment strategies [23].

Microbiological Identification using MALDI-TOF and Tandem Mass Spectrometry: Industrial and Environmental Applications, First edition. Edited by Haroun N. Shah, Saheer E. Gharbia, Ajit J. Shah, Erika Y. Tranfield, and K. Clive Thompson.

Furthermore, MALDI-TOF MS has been utilized to provide insights into the underlying human molecular mechanisms and pathways of non-microbial diseases by applying it to mammalian cells. Extensive application of MALDI-TOF MS has also been applied to imaging as it can be used to determine the spatial distribution of identified/detected molecules while maintaining sample integrity. This allows for post-imaging histological staining, such as haematoxylin and eosin staining [24, 25]. In 2006, MALDI-TOF MS spectral fingerprinting was first used to differentiate between three mammalian cell lines [26], K562, BHK21, and GM15226, which are human myelo-monocytic, rodent kidney fibroblast and human lymphoblast cell lines, respectively. Since then, technological and experimental developments have vastly improved the ability of researchers, enabling them to leverage the technology to produce novel breakthroughs.

In drug discovery, discovering a 'hit' compound that has desired activity is achieved by performing a high-throughput screen. One of the key outcomes in the screen is to identify and quantify biomarkers in a readout [27]. Current early screens in the drug development pathway tend to be performed using label-based, high-throughput screening assays such as fluorescence [28–30] or antibody-based [31, 32] assays. Although there have been many improvements over the years, these indirect measurements have many shortcomings as they require the expensive, time-consuming development of specific probes and involve extensive orthogonal confirmation as they are susceptible to false positives or negatives. Therefore, confirmation assays are required to validate hits [33–35]. MALDI-TOF MS provides scientists with a label-free, robust, reliable alternative.

All these methods have recently been transformed by an expanding application of statistical analysis tools such as R [36–39] and various machine-learning strategies. This has allowed scientists to conduct more detailed data mining and analysis [40, 41].

However, there are some limitations to MALDI-TOF MS, such as its application to small molecules due to matrix ion interference with the analyte ions and detector saturation in the low-mass range [42]. Nevertheless, there have been attempts to use the method for metabolites [43].

Our laboratories have explored the uses of MALDI-TOF MS to investigate environmental microbiology, mammalian cell fingerprinting, drug discovery, and tissue imaging. Our work has been developed using Bruker Ultraflex II and Bruker rapifleX. We have observed how technological improvements have aided faster and more accurate results, but MALDI-TOF MS is also capable of providing information that is not available from other approaches. In the following sections, we will highlight recent advances in environmental microbiology, mammalian cell fingerprinting, drug discovery, and data analysis. We will also discuss the limitations encountered by scientists when investigating low-mass MALDI-TOF MS applications.

13.2 Microbial Fingerprinting

Although MALDI-TOF MS assays have been used since its infancy to study molecules, the main and most rapidly growing use of MALDI-TOF MS is in the identification of bacteria, fungi, and other microorganisms, also known as biotyping or microorganism fingerprinting. MALDI-TOF MS provides speed and cost-effective advantages over previous methods, as well as providing improved data acquisition, as it is a highly discriminatory technique and allows the introduction of additional phenotypic markers [44, 45]. MALDI-TOF MS can be used not only for microorganism identification and differentiation, but also for detection of biomarkers that can be used as indicators of specific bacterial strains. This method can be complementary to the acquisition of data obtained from sequencing rRNA or other reference genes and can generate unique biochemical fingerprints for the subtyping of species. As this topic is addressed further in other chapters of the book, we will briefly mention some applications in environmental and food microbiology and processing.

13.2.1 Environmental

13.2.1.1 Actinobacteria

There are many studies using MALDI-TOF MS for successful characterization of actinobacteria. Approximately two-thirds of natural antibiotics are produced by actinomycetes [46] and, of these, about 75% are antibiotics from members of the genus *Streptomyces* [47–49]. Therefore, successful identification and characterization of *Streptomyces* strains present an area of significant interest, which has been dependent on established reference spectral databases. Several groups have supplemented these commercial databases with data from laboratory-developed libraries, enhancing identification by adding local strains. For example, in Buckwalter et al.'s study [50], the initial identification and validation of the protein fingerprinting results were validated by sequencing of 16S rRNA gene or other housekeeping genes. The investigators added MALDI-TOF MS fingerprints of the manufacturer (Bruker Microbiology & Diagnostics) to their in-house-created spectra of the strains obtained from the German Collection of Microorganisms and Cell Cultures (Deutsche Sammlung von Mikroorganismen und Zellkulturen, DSMZ), the American Type Culture Collection (ATCC), and the Culture Collection, University of Göteborg (CCUG). Using the combined library, they successfully identified several, previously unidentifiable *Mycobacterium*, *Nocardia*, and actinobacterial species. In another study, Loucif et al. combined their established laboratory specific *Streptomyces* mass spectral library with the original Bruker Biotyper 3.0 library and successfully identified 20 previously unidentifiable aquatic strains, isolated from two Algerian lakes, in less than 30 minutes [51].

13.2.1.2 Aquatic Microorganisms

13.2.1.2.1 Ballast Water

Shipping transfers billions of tons of ballast water around the globe every year. This ballast water transports many non-indigenous living organisms, including microorganisms. Therefore, the microbiology of ballast water is of growing environmental significance because discharged ballast water may contain infectious organisms and lead to their global dissemination [52].

MALDI-TOF MS with the Biotyper software 2.0 (Bruker Microbiology and Diagnostics) has been used to characterize more than 30 marine bacterial species in ballast water [53]. In the study, seawater samples collected from the North Sea were incubated in 55 000-litre steel ballast tanks. All the isolated strains were successfully identified at the genus level and the results were compared to the 16S rRNA gene sequencing results. Part of the research results are summarized in Table 13.1 in which a score of ≥ 2 indicates identity of

Table 13.1 Identification of marine bacteria from North Sea ballast water.

Biotyper software V2 ID	Biotyper score	16S rRNA gene ID	Nucleotide matching (%)	Accession number
Bacillus mycoides	2.359	*B. mycoides*	99.6	AB607163
Enterococcus faecalis	2.360	*Enterococcus hirae*	99.6	AB607142
Enterococcus faecium	2.502	*E. faecium*	99.9	AB607143
E. hirae	2.410	*E. hirae*	99.9	AB607166
Halomonas aquamarina	1.568	*Halomonas venusta*	99.7	AB607152
Lactobacillus paraplantarum 10641_DSM	1.750	*Lactobacillus pentosus*	99.8	AB607167
Proteus vulgaris	2.024	*Pseudomonas stutzeri*	99.2	AB607149
Pseudoalteromonas sp.	2.519	*Pseudoalteromonas atlantica*	98.0	AB607158
Pseudomonas aeruginosa	2.270	*P. aeruginosa*	99.8	AB607169
Pseudomonas brenneri	2.113	*Pseudomonas synxantha*	99.7	AB607159
Pseudomonas gessardii	2.128	*P. synxantha*	99.7	AB607132
Pseudomonas monteilii	2.288	*Pseudomonas fulva*	98.9	AB607147
P. stutzeri	2.386	*P. stutzeri*	99.6	AB607133
Pseudomonas tolaasii	2.110	*P. tolaasii*	99.4	AB607148
Serratia liquefaciens	2.062	*Serratia plymuthica*	99.8	AB607146
S. plymuthica	2.185	*S. plymuthica*	99.9	AB607156
Tenacibaculum discolor	2.048	*Tenacibaculum sp.*	99.5	AB607162
Vibrio alginolyticus	2.167	*Vibrio rotiferianus*	99.0	AB607144
Vibrio cyclitrophicus	2.168	*V. cyclitrophicus*	98.0	AB607154
Vibrio fortis	1.543	*Vibrio lentus*	99.4	AB607134
Vibrio gigantis	2.104	*V. cyclitrophicus*	99.5	AB607141
Vibrio tasmaniensis	2.087	*Vibrio splendidus*	99.7	AB607140

Identification of bacteria in ballast water using MALDI-TOF MS and 16S rRNA gene sequencing. The score by Biotyper software (V2) ranged from 0 to 3. Scores > 1.7 indicate a positive genus ID and the scores < 1.7 mean no reliable identification. Nucleotide matching indicates the percentage of rRNA gene nucleotides matching the query and publicly available sequences at NCBI by BLAST search. Accession numbers are DNA DataBank of Japan (DDBJ) deposition,
Source: Table adapted from Emami et al. [53].

the strains to the species level with high confidence, scores around 1.7 indicate only correct identification of genus and scores < 1.7 suggest no reliable identification even at the genus level. Although the Biotyper database lacked many environmental species, successful identification by 16S rRNA sequencing of one *Pseudoalteromonas tetraodonis* sample allowed it to be added to the database. This enabled successful identification of seven unknown strains as *Pseudoalteromonas* by MALDI-TOF MS. Furthermore, MALDI-TOF MS successfully distinguished between isolates of *Enterococcus hirae* and *Enterococcus faecalis*, whereas both species were identified as *E. hirae* by 16S rRNA gene sequencing.

13.2.1.2.2 *Discrimination of* Pseudoalteromonas *Strains*

Members of the genus *Pseudoalteromonas* belong to a large and widely distributed marine bacteria, many of which are of ecological and biotechnological interest [54]. A large percentage of members of this genus are associated with marine eukaryotes [55] and some strains are pathogens of other organisms [56]. In addition to their ecological importance, some members of the genus *Psudoalteromonas* produce cold-adapted industrially valuable enzymes [57, 58] and other strains have the potential to produce antibiotics [59].

We have tested the power of MALDI-TOF MS to differentiate *Pseudoalteromonas* strains by creating an enhanced library combining reference spectra for all 31 existing strains at the DSMZ, and the Japan Collection of Microorganisms (JCM). MALDI-TOF analysis of 13 *Pseudoalteromonas* isolates from UK waters successfully identified them to genus level [60]. Additionally, some biomarker peaks were identified that could distinguish some closely related stains such as *Pseudoalteromonas tunicata*, *Pseudoalteromonas ulvae*, *Pseudoalteromonas aurantia,* and *Pseudoalteromonas citrea*. As expected, a greater resolution by MALDI-TOF MS was obtained than by 16S rRNA gene sequencing.

13.2.1.2.3 *Algae*

MALDI-TOF MS has also been used in clinical and environmental contexts to identify and differentiate between algal strains. We evaluated the capacity of MALDI-TOF MS to function as both a complementary and a standalone tool to distinguish between closely related strains of the commercially important microalgae *Dunaliella* and some other microalgae strains [16]. In all, 36 microalgae whole-cell isolates from the Culture Collection of Algae and Protozoa (CCAP; www.ccap.ac.uk), were clearly classified by MALDI-TOF MS. The results were compared with the internal transcribed spacer (ITS) sequences using ITS4 and ITS5 primer pairs. Analysis of the data revealed that MALDI-TOF MS has more resolution power for discriminating *Dunalliela* strains (Figure 13.1), highlighting the high level of identification possible with this technique.

Furthermore, Baumeister et al. optimized a live-cell matrix-free laser desorption/ionization MS for the purposes of taxonomic identification of environmental marine plankton [54]. Using mass spectral pattern matching, they successfully identified the microalga strains to the species level, most belonging to *Guinardia*, *Coscinodiscus*, and *Chaetoceros* genera. Unlike MALDI-TOF MS, their method also recognizes metabolites including pigments such as chlorophyll, carotene, xanthophyll, lipids, and Zwitterions.

13.2.2 Terrestrial Microbiology

Like the other microorganisms, characterization of soil-associated microorganisms by MALDI-TOF MS has been achieved by many laboratories worldwide, both in the past and in recent years [61–63]. For example, MALDI-TOF MS successfully differentiated individual strains of the nitrogen-fixing actinobacteria, *Frankia*, clustering 37 isolates into their three respective host plant groups [64].

In addition to the identification of soil actinobacterial species, their metabolites have also been subjected to investigation. For example, Traxler et al. used combined nanospray desorption electrospray ionization coupled with MALDI-TOF imaging to identify over 600 metabolites produced by *Streptomyces coelicolor* [65]. This included 12 different

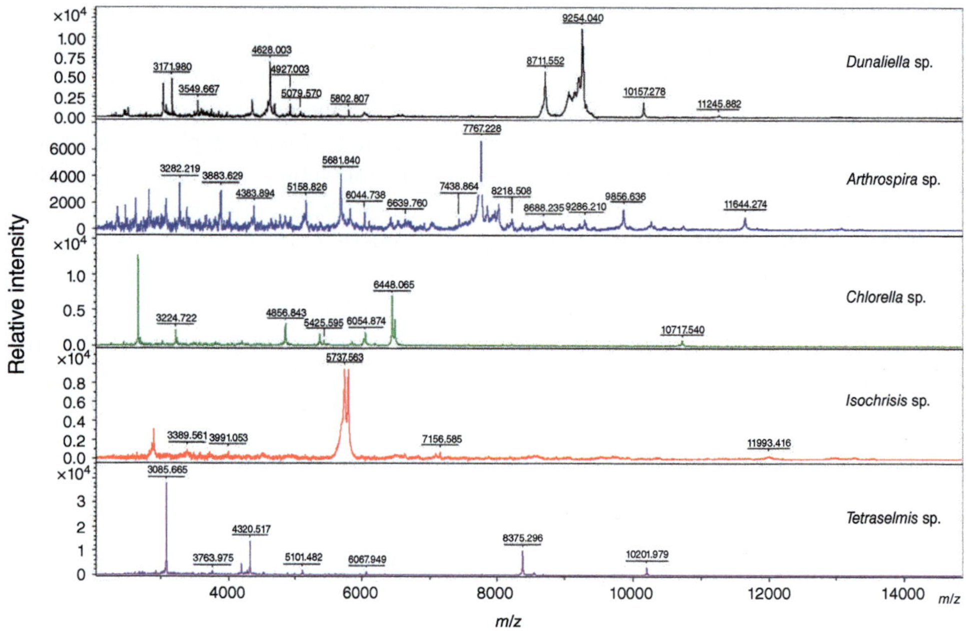

Figure 13.1 Comparison of different microalgae MALDI-TOF mass spectral patterns. The spectra of the different genera are clearly distinguishable from each other. *Source:* Adapted from Emami et al. [16].

siderophores, some of which had not previously been reported as being produced by *S. coelicolor* and many of which were potentially novel compounds which could be used to treat various conditions.

13.2.3 Food and Food Safety

A natural extension of microorganism detection and identification from clinical and environmental strains, MALDI-TOF MS has been used to detect microorganism contamination of food. Furthermore, over the last decade, the advantages of this method have led to the expansion of its use in food storage, safety, and identification [66].

13.2.3.1 Food Storage Effect on Identification

Rau et al. constructed a MALDI-TOF MS database of muscle meat from several animals, including mammals, reptiles, and bird species [67]. They examined over 1000 meat samples against their database. Based on their study, the common species of the major meat categories in the market could be reliably identified by MALDI-TOF MS. Not surprisingly, in some cases the differentiation was unsuccessful, for example when comparing wild boar with pork meat. This could be because species samples were lacking in the database, or it could be due to the absence of a highly capable validation method. They also concluded that: (i) the influence of long-term frozen storage on the species identification by

MALDI-TOF MS is negligible, and (ii) protein extracted from raw or heat-treated meat is still suitable for animal species identification by MALDI-TOF MS.

13.2.3.2 Insects

In an interesting study, Ulrich et al. [68] studied identification of edible Proteinaceous insects as human food sources using MALDI-TOF MS. They purchased freeze-dried snacks such as worms (larvae of *Alphitobius diapernius, Tenebrio molitor*) or the whole insects (*Acheta domesticus, Locusta migratoria*) from commercial suppliers, ground them into powder and extracted the protein contents as material for MS analysis and were able to successfully identify the insects using Biotyper 3.0 software [68].

13.3 Mammalian Cell Fingerprinting

Mammalian fingerprinting was first studied as an extension of microorganism fingerprinting because, although microorganism fingerprinting provides the opportunity to identify microbes, their mechanisms of resistance, and improved treatment, identification and quantification of mammalian fingerprints can provide insights into the underlying human molecular mechanisms and pathways, aiding the understanding of disease to a greater extent than just that caused by microorganisms, as well as helping to identify potential drugs, as depicted in the mammalian workflow shown in Figure 13.2. However, compared with microbial cells, eukaryotic cells have a much higher complexity, with intricate cellular networks and cell cycle states influenced by their spatial anatomy. They are composed of a wide variety of molecules from peptides to nucleotides to lipids, which make them challenging to analyse by MS [69].

When trying to investigate disease and discover novel human therapeutics, the most phenotypically relevant screening method is used to investigate the properties and responses of the potential drugs. Therefore, drug discovery scientists use cellular assays at the earliest point in the drug discovery pipeline as possible, because they provide information that is not available from biochemical assays. Cellular MALDI-TOF MS assays provide a unique opportunity to fill this gap. Since the first application of MALDI-TOF MS to generate a mammalian spectral fingerprint, significant technology advances and research breakthroughs have vastly improved the speed, reliability, and ease with which this technology can be utilized, enabling numerous and various novel uses for MALDI-TOF cellular assays both in primary research and in drug discovery.

13.3.1 Differentiation of Cell Lines and Response to Stimuli

MALDI-TOF MS assays have aided basic research by providing a rapid, reliable, and definitive method to authenticate cell types and explore basic cellular processes. It has become evident that to truly understand the complex interplay between healthy and disease states, it is extremely important to identifying the composition and profiles of cell types. However, traditional methods for cell line identification are expensive and labour-intensive and are thus not routinely applied. Misidentification, false authentication, or lack of recognition of cross-contamination can lead to false conclusions being drawn and published, which can

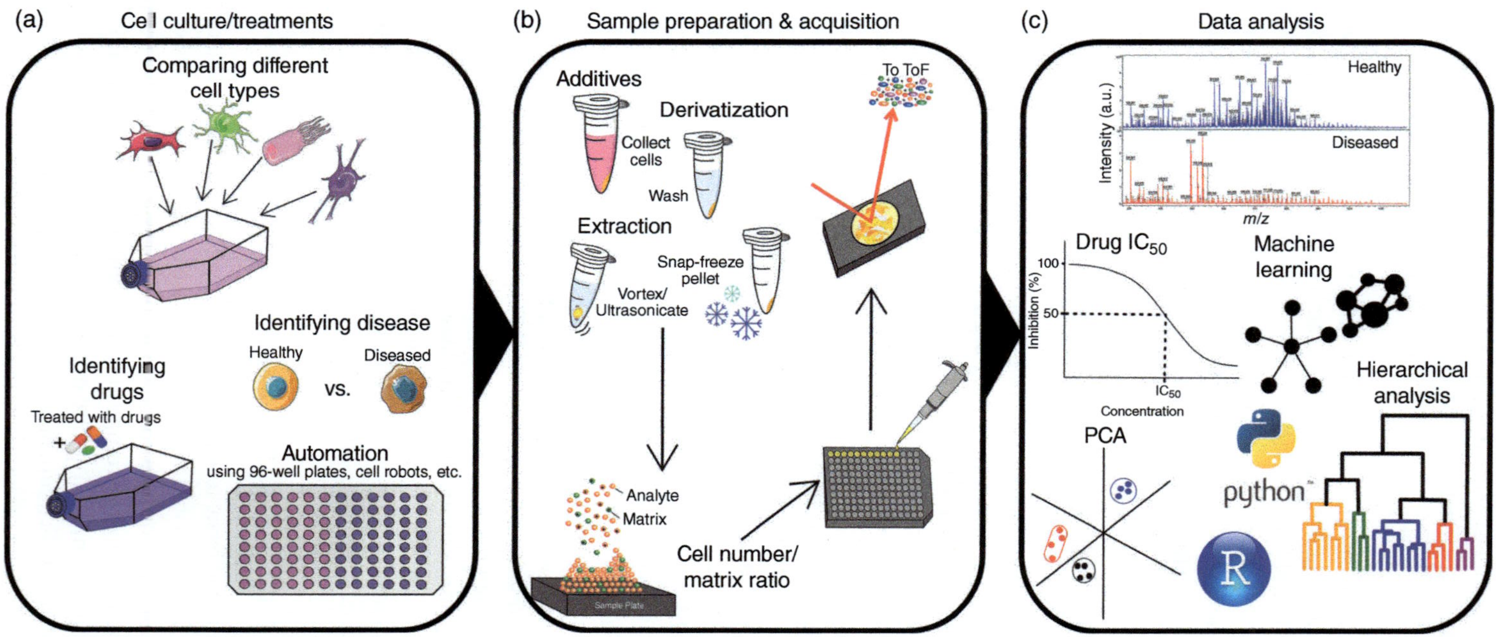

Figure 13.2 Mammalian MALDI-TOF MS assays workflow. (a) Cell culture/treatments. Different cell types can be compared, identified, and phenotyped. For example, MALDI-TOF MS can also be used to distinguish between healthy and cancerous cells and to identify inhibitors. These cell treatments can be automated, making them more high-throughput by culturing cells in 96-well plates, utilizing cell handling robots, etc. (b) Sample preparation. Additives, extraction and derivatization depend upon the goal, e.g. lipids, proteins, etc. Matrix choice, method of spotting, and cell number/matrix ratio are key to being able to successfully ionize the cells to achieve the experimental aim. Sample and matrix are spotted onto the target plate, which is then inserted into the MALDI-TOF MS, and spectra are collected. (c) Data analysis. Spectral data undergoes informatic analysis such as baseline subtraction, spectral alignment, peak detection, normalization, etc. Then, statistical analysis such as principal component analysis (PCA), analysis of variance (ANOVA), and hierarchical clustering are carried out. Machine-learning methods can also be implemented to discriminate between different conditions without bias. These analysis methods provide the ability to characterize drugs such as determining their IC_{50}.

have a significant impact on the whole scientific community when further studies are based on these results. Although mainly developed for microorganism biotyping, leveraging reference spectral database technology for whole-cell MALDI-TOF MS allowed Karger et al. to identify 66 cell culture lines from 34 species [70]. Building on this work, Vogel et al. [71] developed a MALDI-TOF MS protocol and reference spectral database for cell line identification, introducing a rapid, cost-effective method to ensure true cell line identification. Later, Povey et al. expanded this by phenotyping large bioreactor-grown recombinant mammalian cell lines, providing a method for the wider scientific community to leverage this technology to benefit society as whole, as large bioreactors are essential for the production of many biopharmaceuticals, such as recombinant therapeutic proteins [72].

Exploring the response of cells to stimuli is also important in determining the underlying mechanisms of action. Recently, Daumas et al. analysed peripheral blood mononuclear cells (PBMC) from patients with severe sepsis and healthy donors using MALDI-TOF MS and was able to monitor PBMC activation and determine whether sepsis was present [73], which, when applied clinically, allowed patients to be treated much sooner, providing better clinical outcomes, minimizing hospital stays, and reducing treatment cost.

In addition to studying cell populations, investigating single mammalian cells, often from heterogeneous populations, offers an invaluable opportunity to reduce biological noise, gain more comprehensive insights into key changes such as mechanisms of cellular functionality, and characterize subtypes and rare cells. For example, Ong et al. used single-cell MALDI-TOF MS to classify neurone and endocrine cell populations based on their peptide profiles [74]. Despite these different and diverse uses for mammalian cell fingerprinting, no robust sample preparation method had been developed for mammalian cell assays, but recently Heap et al. used MALDI-TOF MS to profile the differentiation of embryonic stem cells, developing a robust sample workflow [75], thus making possible the transition from primary research to high-throughput drug discovery screening methods.

13.3.2 Cancer Diagnostics

One of the major cellular areas to which MALDI-TOF MS has been applied is cancer diagnostics, because it relies on the correct identification of cancer type, species, and metastatic potential (see Chapter 15). The main reason that biopsies and subsequent testing are performed is to determine whether abnormal growths are cancerous and, if so, whether they are invasive or non-invasive. Serafim et al. definitively classified six cancer cell lines according to cancer types, species, and metastatic potential by MALDI-TOF MS – differentiating non-invasive from invasive cells [76]. By speeding up this identification process using MALDI-TOF MS, it enables earlier intervention, increasing the likelihood of a positive outcome.

Furthermore, traditional methods of identifying cancers that have mixed cell types is difficult and expensive, and patients often have a low survival chance because it is often identified too late to enable specific targeting of the cancer cell types involved. However, MALDI-TOF MS has been shown by Petukhova et al. [77] to have effective detection capability of cancer cells in mixed-cell populations, as they successfully profiled murine fallopian tube and high-grade ovarian carcinoma cells and two-component cell mixtures using MALDI-TOF MS.

Brain tumours are also among the most difficult types of cancer to characterize; however, Petre et al. showed the ability of MALDI-TOF MS to distinguish between glioblastoma cell lines [78], providing precise grading of brain tumours not currently available and therefore allowing physicians to adopt the best treatment plan. Moreover, Heap et al. combined this ability to utilize cell line differentiation to identify nilotinib as an inhibitor of inflammation in acute myeloid leukaemia and multiple myeloma [79], highlighting the use of MALDI-TOF MS as a tool both for cancer diagnostics and for potentially discovering the cure.

13.3.3 Biomarkers

Diseases such as cancer are often detected or classified by subtypes defined by biomarkers, which can be proteins associated with subtypes or particular prognostic stages. Discovering these biomarkers for different cancer or different disease types speeds up diagnosis and therefore patient treatment, which is key for patient outcome. There have been several attempts using laborious proteomics tools such as MS coupled with two-dimensional gel electrophoresis for cancer biomarker discoveries [80, 81]. However, gel-free methods are now the common approach to biomarker discovery. Recently, Zhao et al. identified glutathione S-transferase π 1 as a prognostic biomarker for multiple myeloma using MALDI-TOF MS [82].

Treatment plans for some cancers/diseases can be dependent on biomarker identification because some treatments have also been found to be more effective for patients who have certain biomarkers or target specific enzyme isoforms only present in a specific subpopulation. Treatment efficacy can also be quantified by using biomarkers. For example, Weigt et al. successfully analysed the abundance of lipid drug-response markers such as polyacetylated histones for cellular drug response by monitoring BCR-ABL tyrosine kinase inhibitors in leukaemia cells and identified the potassium adduct of phosphatidylcholine PC(36:1) and haem B as biomarkers for response [83]. Before these drugs become treatments, these biomarkers of response are often used as a key outcome to determine whether the drugs will progress through the drug development process.

13.4 Drug Discovery Using MALDI-TOF

The aim of the drug discovery pipeline is to use the fastest, most reliable, cost-efficient, unbiased, and comprehensive phenotypically relevant screening method possible to identify and discover potential new drugs. These technologies are often used as key methods to identify hits within lead identification in the early stages of the pharmaceutical drug discovery pipeline. Other established biophysical reference techniques, such as nuclear magnetic resonance spectroscopy, isothermal titration calorimetry, and cryogenic electron microscopy, can provide label-free valuable information on kinetic, thermodynamic, and structural molecule–ligand interactions but do not yet achieve the speeds required for high-throughput screening (HTS) of libraries > 10 000 compounds. MS-based technologies, and in particular MALDI-TOF MS, offers a sensitive, robust, reproducible, and label-free direct readout producing greater analytical insights into aforementioned cellular-based assays or enzyme assays described in the following section.

13.4.1 Enzymatic Assays

Since 2006, MALDI-TOF MS has been used more and more widely to undertake HTS enzyme assays, generating IC_{50} curves by measuring substrate-to-product ratios and producing hits that can be taken further into the drug discovery pipeline. These assays have focused on *in vitro* assays with simple readouts such as identifying inhibitors of post-translational modifications like phosphorylation, methylation, and ubiquitination [84–86]. The sample analysis speeds in this study were still relatively slow for HTS speeds, with a 10-second-per-spot analysis required to acquire good spectra. The development of the rapifleX mass spectrometer overcame this and enables a one-second sample analysis speed that rivals more conventional HTS assays, expanding the field MALDI-TOF MS assays rapidly.

The first study utilizing the high speed of these new-generation MALDI-TOF MS instruments for drug discovery was the development of an assay by Ritorto et al. [85] to investigate the specificity and drugability of deubiquitylases (DUBs). Following this, another ubiquitin-based assay was developed by De Cesare et al. to identify compounds that inhibit E3 ligase activity [86]. These studies presented a novel screening assay that was fast, sensitive, and highly reproducible. More recently, Heap et al. identified inhibitors of salt-inducible kinases interesting targets in inflammatory diseases [87], Guitot et al. used MALDI-TOF MS to characterize inhibitors of protein lysine methyltransferases [84], and Machálková et al. developed a label-free MALDI-TOF MS method to study the kinetics and undertake inhibitor screening of the Alzheimer's disease drug target β-secretase [88]. Since then, further advances have been made, focusing on increasing the speed of MALDI-TOF assays to enable ultra-high-throughput inhibitor screens, often through automation.

13.4.1.1 Targeting Antibiotic Resistance Using MALDI-TOF MS Enzymatic Assays

MALDI-TOF MS methods have also been established for monitoring enzymatic activities of biomolecules. With the continued rise in antibiotic resistance, finding antibiotics with novel modes of action has become an increasing priority and, in bacteria, cell walls are often successfully targeted. When screening different growth inhibitor compounds for a mutant *Bacillus subtilis* lacking all four class A penicillin-binding proteins (Δ4), a small molecule, isoquinoline-1-carboxilic acid (IQC), was identified to inhibit growth with relatively high activity in the mutant compared with the wild-type [89]. We also discovered that this molecule utilized a novel method and targets cytidine triphosphate synthetase (CTPS) encoded by *pyrG* in *B. subtilis*, which is an important protein target for key infectious diseases such as tuberculosis [90] and African sleeping sickness [91]. CTPS is involved in pyrimidine biosynthesis, catalysing the following reaction:

$$\mathrm{ATP} + \mathrm{H_2O} + \mathrm{L-glutamine} + \mathrm{UTP} \rightarrow \mathrm{ADP} + \mathrm{CTP} + 2\mathrm{H}^+ + \mathrm{L-glutamate} + \mathrm{phosphate}$$

To assess the activity of the enzyme in the presence of the inhibitor, we created an IQC-resistant strain, and compared this with the mutant using MALDI-TOF MS by measuring the substrate abundance (UTP, *m/z* 484) and the reaction product (CTP, *m/z* 483) in the reaction. Other methods, such as spectrophotometric assays, could not be used because the compound absorbed strongly at the wavelength used to measure CTP production.

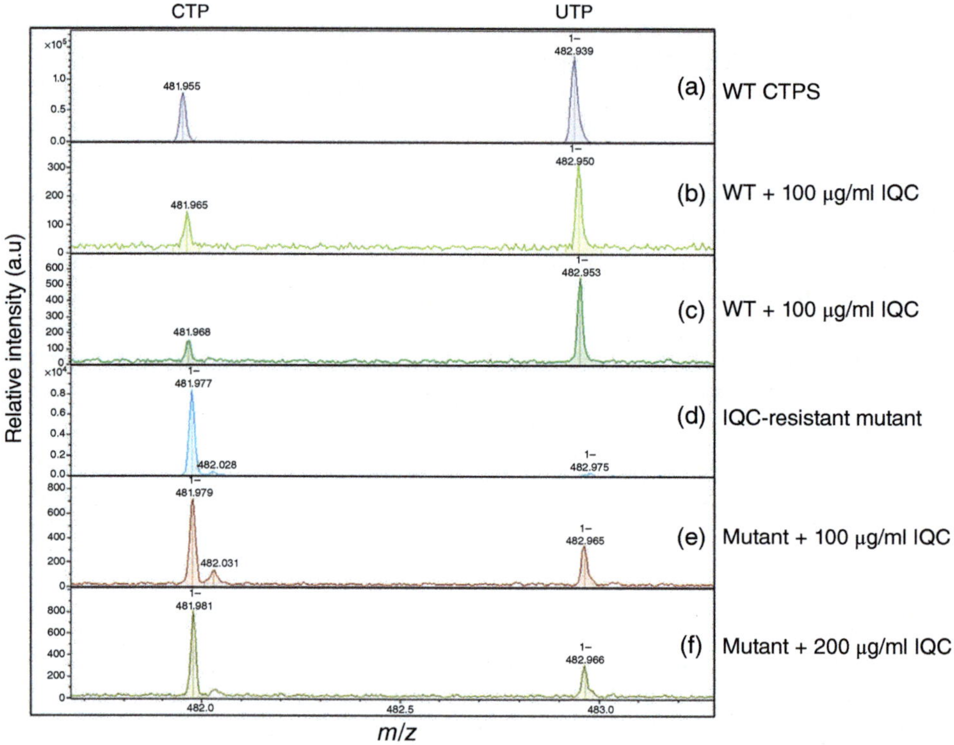

Figure 13.3 CTP synthetase assay using MALDI-TOF MS. Wild-type (WT) CTPs (A) and the inhibitor (IQC)-resistant mutant (D) convert UTP (*m/z* 484) to CTP (*m/z* 483) in the reaction mix. CTP production is decreased upon addition of IQC to the WT [100 μg/ml (B), 200 μg/ml (C)], but has no effect on the IQC-resistant mutants [(E) and (F), respectively]. MS was performed in the negative mode. *Source:* Adapted from Emami et al. [89].

After incubation of the reactions at 37 °C for 30 minutes, it was revealed that: (i) the reaction can be followed using mass spectra (nearly all UTP in the reaction mixture was converted to CTP); (ii) the addition of the inhibitor (IQC) to the reaction indeed inhibited the catalytic activity of the wild-type CTPS; and (iii) the resistant mutant is active in making CTP from UTP even after addition of relatively high concentrations of the inhibitor into the reaction [89]. An example of the assay results is shown in Figure 13.3.

13.4.2 Cellular-based Assays for Drug Discovery

Most of the MALDI-TOF MS-based HTS approaches conducted so far have focused on *in vitro* assays with simple readouts and have been limited to peptide/protein-centric activity assays. Applying this technology for cellular assays remains a challenge, mostly due to the interference from matrix peaks and the matrix-dependent analyte selectivity. Nevertheless, MALDI-TOF MS campaigns have been used to track individual metabolites such as trimethylamine [92], acetylcholine [93], and cyclic GMP-AMP [94] in a cell-based assay. In recent years, affinity-based screening, i.e. affinity selection MS (AS-MS) has been used to circumvent the limitations of current MALDI-MS techniques, such as ion suppression and interference. Using AS-MS, Simon et al. [94] presented a validation screen, demonstrating the general applicability of the platform for the HTS-compatible assessment of protein–ligand interactions.

13.4.3 Automation in Drug Discovery

Additional challenges come with using a MALDI-MS assay as part of HTS, such as increasing throughput speed and automation, which tend to be linked. Liquid handling robots have been linked to mass spectrometers in various forms to automate MALDI assays such as the multichannel pipettor CyBio well [95], the Mosquito liquid handler robot (TTP Labtech) [87, 96], and the acoustic liquid handler called the 'Echo' [97]. The end goal is to develop ultra-HTS MALDI-TOF, using low-sample volumes and fast analysis speed (less than one second per sample).

In a recent screening campaign, the Mosquito liquid handler robot was used to uniformly deposit sample and matrix, from nanolitres to microlitres, onto the target [87, 96]. As a proof of concept, Haslam et al. described and validated an ultra-HTS-compatible enzyme assay for small molecules and peptide analytes [96]. The overall spotting process in this setup required an hour, which is great but not ideal for ultra-HTS which requires screening of ≥100 000 compounds per day, 2–3 s/sample. For MALDI-TOF MS to be implemented in automation, there is a need for HTS-compatible sample preparation. To address this limitation, Boehringer Ingelheim reported a fully automated, in-house liquid handler MALDI spotting station which enabled them to achieve 0.6 s/sample for spotting and 0.4 s/sample for MS analysis [94, 98], matching the requirements of ultra-HTS compatibility. Here, a CyBio liquid handling platform was equipped with a 1536-channel ceramic tip head to transfer the 1536-well plate within one step. Since then, these automated platforms have been leveraged to perform inhibitor screens for protein–ligand interactions such as human cyclic GMP-AMP synthase inhibition [94], protein tyrosine phosphatase 1B [99], and microbial trimethylamine-lyase inhibitors [92].

Another liquid handler compatible with MALDI-TOF MS instrumentation, the acoustic liquid handler, offers an alternative to tip dispensing. Here, the samples are deposited directly onto a MALDI target using acoustic deposition, which enabled spotting of a complete 1536 target plate in under eight minutes with no drying time required [100]. This was significantly faster than the available CyBio system; however, there are drawbacks with respect to a volume cut-off in the source plate as well as potential instrument bottlenecks.

13.4.4 Assay Multiplexing

Although other MS techniques such as liquid chromatography tandem MS (LC–MS/MS) have been used to multiplex assays in the past [101–103], for example assessing multiple enzymes for deficiency in dried blood spots of newborns [103], they only evaluated the presence of specific enzyme activity or particular drug inhibition of a targeted enzyme in a set. By pre-balancing multiple reaction monitoring transitions during method development, a direct readout of enzyme activity can be achieved by monitoring the substrate : product ratio and if they have different substrate : product ratios, multi-substrate assays are also possible [104]. Recently, Xu et al. quantitatively multiplexed an orthogonal MALDI-TOF MS assay for multiple protein kinases to simultaneously assess activity of protein kinase A, Abl, Src, and inhibition induced by H-89. Excellent readout sensitivity was shown by the simultaneous tracing and characterization of the competitive binding behaviour between Abl and Src inactivation with the closely related drugs imatinib, dasatinib, and ponatinib [105]. Cell viability/apoptosis markers can also be used to identify whether drugs are toxic, allowing the multiplexing also to be used for orthogonal assays [30, 106]. This emphasizes the capability of MALDI-MS to potentially screen multiple enzymes or measure multiple physiological components in the same screen.

13.4.5 MS Imaging in Drug Discovery

MALDI-TOF mass spectrometers can also be used to acquire two-dimensional arrays of hundreds of thousands of mass spectra over a tissue sample to study numerous diseases and therapeutics. Known as imaging mass spectrometry (IMS), this allows the visualization and localization of chemical species within a sample in a two-dimensional, and sometimes three-dimensional, map. Typically, IMS is performed using high-resolution instruments, such as the MALDI Fourier-transform ion cyclotron resonance (FTICR) MS, as certain metabolites need the high resolving power to confidently identify them. Most of these instruments are not fast or well-suited for HTS, therefore MALDI-TOF instruments are still needed for these approaches. In a proof-of-concept study, Weigt et al. [83] combined the power of both MALDI-FTICR and MALDI-TOF MS instruments to develop a cellular drug discovery assay. In this study, MALDI-FTICR was first used to identify lipids of interest before being transferred to a MALDI-TOF MS platform that rapidly profiled compounds and showed good IC_{50} correlation of hits with a biochemical assay.

The goal of using IMS is to preserve the spatial information inherent in the sample. Therefore, this technology has been used to study the localization of therapeutics and to study numerous diseases. Evaluating the drug distribution directly on tissue is essential in understanding how the drug is transferred, if the drug has reached its target, and the drug's efficacy. Even though MS imaging is mainly used to visualize the localization of compounds in tissue sections, the instrumentation also lends itself well to microarray format, which is often used to increase sample throughput. These microarray formats can be used for tissue or organ sections as these grid-like applications generate defined coordinates for systematic sampling across a surface [107].

13.4.6 MALDI-2

Recently, the major technological advance in MALDI technology with the introduction of the MALDI-2, a MALDI source coupled to a timsTOF flex instrument, has expanded the application range of the MALDI even further by increasing the small molecule/lipid detection by one to two orders of magnitude. To achieve this, the MALDI-2 fires a second orthogonal laser into the material desorbed by the first laser, ionizing neutral matrix and allowing a post-ionized charge transfer to the analyte, thereby increasing the amount of charged analyte particles to be detected. The polarization of this second laser significantly increases analyte sensitivity for molecules such as lipids, hormones, glycans, etc. However, this technology has mainly been leveraged for spatialOMx studies incorporating MALDI imaging; for example, Bien et al. [108] were able to develop a 'molecular histology' approach for human colorectal cancer tissue which could potentially be used for stratifying patients according to subtypes caused by distinct neutral glycosphinogolipids, leading to more specific, rapid treatments which are required for cures [108, 109]. Moreover, Soltwisch et al. produced a sizable increase in detected features of rat brain and testes tissue and Niehaus et al. achieved a pixel size of 600 nm when visualizing brain tissue [107, 110]. These investigations represent only the start of the utilization of this technique, therefore demonstrating a valuable new tool for biomedical, cell biology, and drug discovery research.

13.5 Limitations/Challenges, Future Outlook, and Conclusions

13.5.1 Sample Preparation Limitations

When developing a MALDI-TOF MS assay, sample preparation is crucial in detecting the target molecules with high-quality spectra while maintaining spectral reproducibility. However, sample preparation methods are inherently limited by several challenges involved, such as: matrix, interference from low-molecular-mass matrix clusters, buffer compatibility, TOF mass resolution limitations, and mixed cultures.

13.5.1.1 Matrix

Matrix choice plays a crucial role in the detection of the desired molecules. For example, in early MALDI research, Karas et al. [111] were analysing biologically active human platelets and cell fractions; however, it was not until they switched matrix from 2,5-dihydroxybenzoic acid (DHB) to 3-hydroxypicolinic acid (HPA), which has been found to be good for detecting nucleotides, that they saw many signals on the spectra. These signals were later identified as nucleotides which are not visible with DHB [112]. Matrix choice can be affected by the molecular weight range and most often the mode (positive or negative) in which the assay analytes can be found. For example, sinapinic acid (SA), α-cyano-4-hydroxycinnamic acid (HCCA), and 2′,4′,6′-trihydroxyacetophenone monohydrate (THAP) have been found to be good matrices for high-molecular-weight proteins > 2 kDa mainly in positive mode, whereas 9-aminoacridine (9AA) is basic, so it works better in the negative mode [113–116]. However, recently matrices have begun to be employed that can be used well in both positive and negative modes (dual-mode matrices), such as 2-hydrazinoquinoline (2-HQ), which can be used to sensitively detect neutral, sialic, and low-molecular-weight carbohydrates. 2-HQ has also been shown to improve glycan sensitivity 100-fold compared with using DHB [117]. These many influencing factors are emphasized by the fact that different cell types can be visualized better using different matrices [69].

MALDI-MS normally produces singly charged molecules $[M+H]^+$ if the samples are totally desalted; however, MALDI-MS is tolerant to salt, and adducts of sodium $[M+Na]^+$ and potassium $[M+K]^+$ are generally found in the MALDI-MS spectrum [118]. Determining the best sample: matrix ratio is also crucial because too much of either can lead to a lack of or supressed signal. If the matrix is too concentrated, the wrong sample: matrix concentration ratio is used, or the matrix and sample are incompatible, this can lead to non-homogeneous spots on the target plate with either crystal clumps in different areas over the target spot or the 'coffee-ring' effect, where a concentrated ring forms on the outside of the spot.

13.5.1.2 Interference from Low-molecular-mass Matrix Clusters

When working in the lower mass region, researchers often use different matrices that do not produce peaks in this region to avoid ion suppression and interference from matrix peaks, such as nanomaterials [119] or graphite/graphene oxide nanoribbons/graphene [120, 121]. Some of these alternative doped carbon-graphene matrices are also dual-mode matrices [122]. Matrix-related ions can also be supressed, e.g. with cyclodextrin, to produce high-quality MALDI spectra with negligible matrix background peaks and strong analyte signals [123–125].

13.5.1.3 Buffer Compatibility

Although many other MS techniques, including electrospray ionization, are heavily influenced by sample purity, MALDI-TOF MS has a relatively high tolerance for biological contaminants. However, MALDI-TOF MS analyte ion suppression has been found to be induced by many of the biochemical components incorporated into assays to preserve physiological conditions, e.g. salts, cofactors, enzymes, etc. Chandler et al. undertook a systematic screening of many of the common additives using acetylcholine as a mimic substrate assay [126], providing guidance for designing assay conditions. However, the influence of buffers on specific assays requires optimization due to the interaction between the biology and pharmacological enzyme kinetics.

13.5.1.4 TOF Mass Resolution Limitations

MALDI-MS, which in theory has the ability to study an unlimited range of molecules, is most often coupled to a time-of-flight (TOF) analyser. TOF instruments represent a compromise between mass and useful resolution because they often have linear and reflector modes, which measure the mass-dependent time it takes for the ions to move from the ion source to the detector in a field-free vacuum. Linear mode allows detection of <100 to 100 kDa, while reflectron mode allows detection of smaller molecules (≤ 2 kDa). However, despite being able to detect a large range of molecules from m/z 100 to 100 000, MALDI-TOF MS instruments have a relatively low resolving power [127], which affects their utility. For example, true identification of lipid species is not possible from a mixed lipid sample or when studying cell contents; identification must be carried out on a higher resolution instrument such as a FTICR mass spectrometer.

13.5.2 Data Analysis and Application of Machine Learning

Over the last decade, the development of new MS instrumentation and applications has resulted in an increase in complexity and dimensionality of MS data. Therefore, several strategies need to be implemented to extract meaningful information from the datasets generated. Data processing usually comprises data pre-processing, data compression, and post-processing. One of the main problems with MALDI-TOF MS data is the high variation, as spectra can be different in terms of length of data, wave shape, peak position, and alignment. Therefore, the initial key step in MALDI-TOF MS data processing is the pre-processing pipeline.

A complete analysis pipeline for MALDI-TOF pre-processing can be accomplished in an automated fashion using software platforms, such as R (i.e. MALDIquant R package) or Matlab (i.e. spectrum and signal analysis functions). In a typical pre-processing pipeline, quality controls are implemented after importing the raw data. Then, smoothing filters and variance-stabilizing transformation are applied, followed by baseline correction and normalization. Lastly, the spectra are aligned, peaks are detected, and a feature matrix is created. Overall, pre-processing can help to reduce experimental variance within the dataset and extract relevant information from large datasets.

Pre-processing is typically followed by supervised and unsupervised clustering of the data. A wide array of statistical and machine-learning methods [i.e. principal component analysis (PCA), clustering, heat map, K-means, supervised classification-random forests, etc.] can be used to aid the data analysis. Typically, component analysis is used in MALDI-TOF MS data analysis, which requires a statistical or machine-learning algorithm to cluster the data. For example, PCA is used to reduce the dimensionality of the dataset by converting complex data into a set of linearly uncorrelated values, called principal components. This provides an overview of complex data by decreasing the dimensionality based on feature. Supervised clustering can also be used when the specified set of classes is known, and the goal of the analysis is to classify the new dataset into one of those classes. Partial least-squares (PLS) regression is a supervised classification method, where classes of data are annotated with known labels. Both supervised and unsupervised classification methods reduce data down to the most important features (i.e. m/z values).

Machine learning is starting to play a larger role in the analysis of MS data, by developing algorithms to find relationships in MALDI-TOF MS data and using these to identify data to make predictions for new datasets. The utility of combining MALDI-TOF MS with machine learning in the field of medicinal chemistry was demonstrated in a study by van Oosten et al. [40]. In this study, antibiotic compound screening was conducted, and the assay offered a straightforward, high-throughput, label-free readout for ligand binding to highly relevant antibiotic target sites. This study used quadratic support vector machine classification models to reliably classify specific proteomic signature, showing that machine learning can extend the MALD-TOF MS toolbox towards phenotypic screening of compounds.

13.6 Future Outlook/Conclusions

Overall, MALDI-TOF MS is a highly versatile and powerful technique which has been applied in many different areas and has become a key technology to identify novel candidates for enzyme and cellular assays, highlighting its importance in drug discovery. With data analysis and machine-learning pipelines constantly being improved, the potential applications of MALDI-TOF MS continue to expand, and in the future we expect further utilization of the MALDI-2 platform and continued method and technology development, leveraging research to overcome present limitations and expand the scope of what is possible for MALDI MS, especially in the drug discovery field.

References

1 Anhalt, J.P. and Fenselau, C. (1975). Identification of bacteria using mass spectrometry. *Anal. Chem.* 47 (2): 219–225.

2 Tanaka, K., Waki, H., Ido, Y. et al. (1988). Protein and polymer analyses up to m/z 100 000 by laser ionization time-of-flight mass spectrometry. *Rapid Commun. Mass Spectrometry* 2 (8): 151–153.

3 Tran, N.K., Howard, T., Walsh, R. et al. (2021). Novel application of automated machine learning with MALDI-TOF-MS for rapid high-throughput screening of COVID-19: a proof of concept. *Sci. Rep.* 11 (1): 8219.

4 Deulofeu, M., García-Cuesta, E., Peña-Méndez, E.M. et al. (2021). Detection of SARS-CoV-2 infection in human nasopharyngeal samples by combining MALDI-TOF MS and artificial intelligence. *Front. Med. (Lausanne)* 8: 661358.

5 Sjöholm, M.I.L., Dillner, J., and Carlson, J. (2008). Multiplex detection of human herpesviruses from archival specimens by using matrix-assisted laser desorption ionization-time of flight mass spectrometry. *J. Clin. Microbiol.* 46 (2): 540–545.

6 Dridi, B., Raoult, D., and Drancourt, M. (2012). Matrix-assisted laser desorption/ionization time-of-flight mass spectrometry identification of Archaea: towards the universal identification of living organisms. *APMIS* 120 (2): 85–91.

7 Qian, J., Cutler, J.E., Cole, R.B., and Cai, Y. (2008). MALDI-TOF mass signatures for differentiation of yeast species, strain grouping and monitoring of morphogenesis markers. *Anal. Bioanal. Chem.* 392 (3): 439–449.

8 Normand, A.C., Gabriel, F., Riat, A. et al. (2020). Optimization of MALDI-ToF mass spectrometry for yeast identification: a multicenter study. *Med. Mycol.* 58 (5): 639–649.

9 Becker, P.T., Stubbe, D., Claessens, J. et al. (2015). Quality control in culture collections: confirming identity of filamentous fungi by MALDI-TOF MS. *Mycoscience* 56 (3): 273–279.

10 Zvezdanova, M.E., Escribano, P., Ruiz, A. et al. (2018). Increased species-assignment of filamentous fungi using MALDI-TOF MS coupled with a simplified sample processing and an in-house library. *Med. Mycol.* 57 (1): 63–70.

11 Nagorny, S.A., Aleshukina, A.V., Aleshukina, I.S. et al. (2019). The application of proteomic methods (MALDI-toff MS) for studying protein profiles of some nematodes (dirofilaria and ascaris) for differentiating species. *Int. J. Infect. Dis.* 82: 61–65.

12 Kaufmann, C., Steinmann, I.C., Hegglin, D. et al. (2012). Spatio-temporal occurrence of *Culicoides* biting midges in the climatic regions of Switzerland, along with large scale species identification by MALDI-TOF mass spectrometry. *Parasit. Vectors* 5 (1): 246.

13 Stephan, R., Johler, S., Oesterle, N. et al. (2014). Rapid and reliable species identification of scallops by MALDI-TOF mass spectrometry. *Food Control* 46: 6–9.

14 Mazzeo, M.F., Giulio, B.D., Guerriero, G. et al. (2008). Fish authentication by MALDI-TOF mass spectrometry. *J. Agric. Food Chem.* 56 (23): 11071–11076.

15 Volta, P., Riccardi, N., Lauceri, R., and Tonolla, M. (2012). Discrimination of freshwater fish species by matrix-assisted laser desorption/ionization-time of flight mass spectrometry (MALDI-TOF MS): a pilot study. *J. Limnol.* 71 (1): e17.

16 Emami, K., Hack, E., Nelson, A. et al. (2015). Proteomic-based biotyping reveals hidden diversity within a microalgae culture collection: an example using *Dunaliella*. *Sci. Rep.* 5 (1): 10036.

17 Gallegos-Candela, M., Boyer, A.E., Woolfitt, A.R. et al. (2018). Validated MALDI-TOF-MS method for anthrax lethal factor provides early diagnosis and evaluation of therapeutics. *Anal. Biochem.* 543: 97–107.

18 Kuklenyik, Z., Boyer, A.E., Lins, R. et al. (2011). Comparison of MALDI-TOF-MS and HPLC-ESI-MS/MS for endopeptidase activity-based quantification of anthrax lethal factor in serum. *Anal. Chem.* 83 (5): 1760–1765.

19 Flaudrops, C., Armstrong, N., Raoult, D., and Chabrière, E. (2015). Determination of the animal origin of meat and gelatin by MALDI-TOF-MS. *J. Food Compos. Anal.* 41: 104–112.

20 Ryzhov, V. and Fenselau, C. (2001). Characterization of the protein subset desorbed by MALDI from whole bacterial cells. *Anal. Chem.* 73 (4): 746–750.

21 Pranada, A.B., Schwarz, G., and Kostrzewa, M. (2016). MALDI biotyping for microorganism identification in clinical microbiology. In: *Advances in MALDI and Laser-Induced Soft Ionization Mass Spectrometry* (ed. R. Cramer), 197–225. Cham: Springer International Publishing.

22 Grier, J.T., Arivett, B.A., Ramírez, M.S. et al. (2021). Two *Acinetobacter baumannii* isolates obtained from a fatal necrotizing fasciitis infection display distinct genomic and phenotypic characteristics in comparison to type strains. *Front. Cell. Infect. Microbiol.* 11: 635673.

23 Fang, L., Xu, H., Ren, X. et al. (2020). Epidemiology and risk factors for Carbapenem-resistant *Klebsiella pneumoniae* and subsequent MALDI-TOF MS as a tool to cluster KPC-2-producing *Klebsiella pneumoniae*, a retrospective study. *Front. Cell. Infect. Microbiol.* 10: 462.

24 Wang, J., Qiu, S., Chen, S. et al. (2015). MALDI-TOF MS imaging of metabolites with a N-(1-Naphthyl) ethylenediamine dihydrochloride matrix and its application to colorectal cancer liver metastasis. *Anal. Chem.* 87 (1): 422–430.

25 Yalcin, E.B. and de la Monte, S.M. (2015). Review of matrix-assisted laser desorption ionization-imaging mass spectrometry for lipid biochemical histopathology. *J. Histochem. Cytochem.* 63 (10): 762–771.

26 Zhang, X., Scalf, M., Berggren, T.W. et al. (2006). Identification of mammalian cell lines using MALDI-TOF and LC-ESI-MS/MS mass spectrometry. *J. Am. Soc. Mass Spectrom.* 17 (4): 490–499.

27 Dueñas, M.E., Peltier-Heap, R.E., Leveridge, M. et al. (2022). Advances in high-throughput mass spectrometry in drug discovery. *EMBO Mol. Med.* e14850. https://doi.org/10.15252/emmm.202114850.

28 Elsadek, B.E.M. and Hassan, M.H. (2021). Screening for new peptide substrates for the development of albumin binding anticancer pro-drugs that are cleaved by prostate-specific antigen (PSA) to improve the anti tumor efficacy. *Biochem. Biophys. Rep.* 26: 100966.

29 Holdgate, G.A. and Hemsley, P.E. (2021). Ligand discovery: high-throughput binding: fluorescence polarization (anisotropy). *Methods Mol. Biol.* 2263: 231–246.

30 Dong, G., Yasgar, A., Peterson, D.L. et al. (2020). Optimization of high-throughput methyltransferase assays for the discovery of small molecule inhibitors. *ACS Comb. Sci.* 22 (8): 422–432.

31 Bowen, W. (2010). High-throughput antibody screening using high-sensitivity microplate cytometry. *Nat. Methods* 7 (12): 8–9.

32 Hofmann, T., Schmidt, J., Ciesielski, E. et al. (2020). Intein mediated high throughput screening for bispecific antibodies. *MAbs* 12 (1): 1731938.

33 Wigle, T.J., Swinger, K.K., Campbell, J.E. et al. (2015). A high-throughput mass spectrometry assay coupled with redox activity testing reduces artifacts and false positives in lysine demethylase screening. *J. Biomol. Screen.* 20 (6): 810–820.

34 Yang, Z.Y., He, J.H., Lu, A.P. et al. (2020). Frequent hitters: nuisance artifacts in high-throughput screening. *Drug Discov. Today* 25 (4): 657–667.

35 McLaren, D.G., Shah, V., Wisniewski, T. et al. (2021). High-throughput mass spectrometry for hit identification: current landscape and future perspectives. *SLAS Discov.: Adv. Sci. Drug Discov.* 26 (2): 168–191.

36 RStudioTeam (2021). *RStudio: Integrated Development for R*. Boston, MA: RStudio, Inc.

37 Palarea-Albaladejo, J., McLean, K., Wright, F., and Smith, D.G.E. (2018). MALDIrppa: quality control and robust analysis for mass spectrometry data. *Bioinformatics (Oxford, England)* 34 (3): 522–523.

38 S G. MALDIquantForeign: Import/Export routines for MALDIquant2019, https://cran.biodisk.org/web/packages/MALDIquantForeign/vignettes/MALDIquantForeign-intro.pdf (accessed 3 October 2022).

39 Gibb, S. and Strimmer, K. (2012). MALDIquant: a versatile R package for the analysis of mass spectrometry data. *Bioinformatics (Oxford, England)* 28 (17): 2270–2271.

40 van Oosten, L.N. and Klein, C.D. (2020). Machine learning in mass spectrometry: a MALDI-TOF MS approach to phenotypic antibacterial screening. *J. Med. Chem.* 63 (16): 8849–8856.

41 Xie, Y.R., Castro, D.C., Bell, S.E. et al. (2020). Single-cell classification using mass spectrometry through interpretable machine learning. *Anal. Chem.* 92 (13): 9338–9347.

42 Santos, I.C., Hildenbrand, Z.L., and Schug, K.A. (2016). Applications of MALDI-TOF MS in environmental microbiology. *Analyst* 141 (10): 2827–2837.

43 Edwards, J.L. and Kennedy, R.T. (2005). Metabolomic analysis of eukaryotic tissue and prokaryotes using negative mode MALDI time-of-flight mass spectrometry. *Anal. Chem.* 77 (7): 2201–2209.

44 Ng, H.J., Webb, H.K., Crawford, R.J. et al. (2013). Updating the taxonomic toolbox: classification of *Alteromonas* spp. using multilocus phylogenetic analysis and MALDI-TOF mass spectrometry. *Antonie Van Leeuwenhoek* 103 (2): 265–275.

45 Dieckmann, R., Graeber, I., Kaesler, I. et al. (2005). Rapid screening and dereplication of bacterial isolates from marine sponges of the Sula ridge by intact-cell-MALDI-TOF mass spectrometry (ICM-MS). *Appl. Microbiol. Biotechnol.* 67 (4): 539–548.

46 Challis, G.L. and Hopwood, D.A. (2003). Synergy and contingency as driving forces for the evolution of multiple secondary metabolite production by *Streptomyces* species. *Proc. Natl. Acad. Sci.* 100 (suppl 2): 14555–14561.

47 Aouiche, A., Bijani, C., Zitouni, A. et al. (2014). Antimicrobial activity of saquayamycins produced by *Streptomyces* spp. PAL114 isolated from a Saharan soil. *J. Mycol. Méd.* 24 (2): e17–e23.

48 Smaoui, S., Mathieu, F., Elleuch, L. et al. (2012). Taxonomy, purification and chemical characterization of four bioactive compounds from new *Streptomyces* sp. TN256 strain. *World J. Microbiol. Biotechnol.* 28 (3): 793–804.

49 Linares, D.M., Fernández, M., Del-Río, B. et al. (2012). The tyrosyl-tRNA synthetase like gene located in the tyramine biosynthesis cluster of *Enterococcus duransis* transcriptionally regulated by tyrosine concentration and extracellular pH. *BMC Microbiol.* 12 (1): 23.

50 Buckwalter, S.P., Olson, S.L., Connelly, B.J. et al. (2016). Evaluation of matrix-assisted laser desorption ionization-time of flight mass spectrometry for identification of mycobacterium species, *Nocardia* species, and other aerobic Actinomycetes. *J. Clin. Microbiol.* 54 (2): 376–384.

51 Loucif, L., Bendjama, E., Gacemi-Kirane, D., and Rolain, J.-M. (2014). Rapid identification of *Streptomyces* isolates by MALDI-TOF MS. *Microbiol. Res.* 169 (12): 940–947.

52 Ruiz, G.M., Rawlings, T.K., Dobbs, F.C. et al. (2000). Global spread of microorganisms by ships. *Nature* 408 (6808): 49–50.

53 Emami, K., Askari, V., Ullrich, M. et al. (2012). Characterization of bacteria in ballast water using MALDI-TOF mass spectrometry. *PLoS One* 7 (6): e38515.

54 Ivanova, E.P., Gorshkova, N.M., Zhukova, N.V. et al. (2004). Characterization of *Pseudoalteromonas distincta*-like sea-water isolates and description of *Pseudoalteromonas aliena* sp. nov. *Int. J. Syst. Evol. Microbiol.* 54 (Pt 5): 1431–1437.

55 Holmström, C. and Kjelleberg, S. (1999). Marine *Pseudoalteromonas* species are associated with higher organisms and produce biologically active extracellular agents. *FEMS Microbiol. Ecol.* 30 (4): 285–293.

56 Choudhury, J.D., Pramanik, A., Webster, N.S. et al. (2015). The pathogen of the great barrier reef sponge *Rhopaloeides odorabile* is a new strain of *Pseudoalteromonas agarivorans* containing abundant and diverse virulence-related genes. *Marine Biotechnol.* 17 (4): 463–478.

57 Furhan, J. (2020). Adaptation, production, and biotechnological potential of cold-adapted proteases from psychrophiles and psychrotrophs: recent overview. *J. Genet. Eng. Biotechnol.* 18 (1): 36.

58 Ivanova, E.P., Bakunina, I.Y., Nedashkovskaya, O.I. et al. (2003). Ecophysiological variabilities in ectohydrolytic enzyme activities of some *Pseudoalteromonas* species, *P. citrea*, *P. issachenkonii*, and *P. nigrifaciens*. *Curr. Microbiol.* 46 (1): 0006–0010.

59 Bowman, J.P. (2007). Bioactive compound synthetic capacity and ecological significance of marine bacterial genus *Pseudoalteromonas*. *Mar. Drugs* 5 (4): 220–241.

60 Emami, K., Nelson, A., Hack, E. et al. (2016). MALDI-TOF mass spectrometry discriminates known species and marine environmental isolates of *Pseudoalteromonas*. *Front. Microbiol.* 7 (104): https://doi.org/10.3389/fmicb.2016.00104.

61 Uhlik, O., Strejcek, M., Junkova, P. et al. (2011). Matrix-assisted laser desorption ionization (MALDI)-time of flight mass spectrometry- and MALDI Biotyper-based identification of cultured biphenyl-metabolizing bacteria from contaminated horseradish rhizosphere soil. *Appl. Environ. Microbiol.* 77 (19): 6858–6866.

62 Çevik, Y.N. and Ogutcu, H. (2020). Identification of bacteria in soil by MALDI-TOF MS and analysis of *Bacillus* spp., *Paenibacillus* spp. and *Pseudomonas* spp. with PCA. *Anal. Chem. Lett.* 10 (6): 784–797.

63 Calzada Urquiza, C., Arvizu Hernández, I., Cruz Medina, J.A. et al. (2017). Identification by MALDI-TOF mass spectrometry of mercury-resistant bacteria associated with the rhizosphere of an apple orchard. *Geomicrobiol. J.* 34 (2): 176–182.

64 Hahn, D., Mirza, B., Benagli, C. et al. (2011). Typing of nitrogen-fixing *Frankia* strains by matrix-assisted laser desorption ionization-time-of-flight (MALDI-TOF) mass spectrometry. *Syst. Appl. Microbiol.* 34 (1): 63–68.

65 Traxler, M.F., Watrous, J.D., Alexandrov, T. et al. (2013). Interspecies interactions stimulate diversification of the *Streptomyces coelicolor* secreted metabolome. *MBio* 4 (4): e00459–e00413.

66 Akimowicz, M. and Bucka-Kolendo, J. (2020). MALDI-TOF MS – application in food microbiology. *Acta Biochim. Pol.* 67 (3): 327–332.

67 Rau J, Hiller E, Männig A, Dyk M, Wenninger O, Stoll P, et al. Animal Species Identification of Meat Using MALDI-TOF Mass Spectrometry. ChemRxiv. Cambridge: Cambridge Open Engage. 2021.

68 Ulrich, S., Kühn, U., Biermaier, B. et al. (2017). Direct identification of edible insects by MALDI-TOF mass spectrometry. *Food Control* 76: 96–101.

69 Munteanu, B. and Hopf, C. (2013). Emergence of whole-cell MALDI-MS biotyping for high-throughput bioanalysis of mammalian cells? *Bioanalysis* 5 (8): 885–893.

70 Karger, A., Bettin, B., Lenk, M., and Mettenleiter, T.C. (2010). Rapid characterisation of cell cultures by matrix-assisted laser desorption/ionisation mass spectrometric typing. *J. Virol. Methods* 164 (1): 116–121.

71 Vogel, G., Strauss, A., Jenni, B. et al. (2011). Development and validation of a protocol for cell line identification by MALDI-TOF MS. *BMC Proc.* 5 (Suppl 8): 45.

72 Povey, J.F., O'Malley, C.J., Root, T. et al. (2014). Rapid high-throughput characterisation, classification and selection of recombinant mammalian cell line phenotypes using intact cell MALDI-ToF mass spectrometry fingerprinting and PLS-DA modelling. *J. Biotechnol.* 184: 84–93.

73 Daumas, A., Alingrin, J., Ouedraogo, R. et al. (2018). MALDI-TOF MS monitoring of PBMC activation status in sepsis. *BMC Infect. Dis.* 18 (1): 355.

74 Ong, T.H., Kissick, D.J., Jansson, E.T. et al. (2015). Classification of large cellular populations and discovery of rare cells using single cell matrix-assisted laser desorption/ionization time-of-flight mass spectrometry. *Anal. Chem.* 87 (14): 7036–7042.

75 Heap, R.E., Segarra-Fas, A., Blain, A.P. et al. (2019). Profiling embryonic stem cell differentiation by MALDI TOF mass spectrometry: development of a reproducible and robust sample preparation workflow. *Analyst* 144: 6371–6381.

76 Serafim, V., Shah, A., Puiu, M. et al. (2017). Classification of cancer cell lines using matrix-assisted laser desorption/ionization time-of-flight mass spectrometry and statistical analysis. *Int. J. Mol. Med.* 40 (4): 1096–1104.

77 Petukhova, V.Z., Young, A.N., Wang, J. et al. (2019). Whole cell MALDI fingerprinting is a robust tool for differential profiling of two-component mammalian cell mixtures. *J. Am. Soc. Mass Spectrom.* 30 (2): 344–354.

78 Petre, G., Durand, H., Pelletier, L. et al. (2020). Rapid proteomic profiling by MALDI-TOF mass spectrometry for better brain tumor classification. *Proteomics Clin. Appl.* 14 (5): 1900116.

79 Heap RE, Marín-Rubio JL, Heunis T, Dueñas ME, Inns J, Saxty B, et al. (2021). A MALDI-TOF assay identifies nilotinib as an inhibitor of inflammation in acute myeloid leukemia and multiple myeloma. *bioRxiv*. https://doi.org/10.1101/2021.03.29.437557.

80 Vasudev, N.S., Ferguson, R.E., Cairns, D.A. et al. (2008). Serum biomarker discovery in renal cancer using 2-DE and prefractionation by immunodepletion and isoelectric focusing; increasing coverage or more of the same? *Proteomics* 8 (23–24): 5074–5085.

81 Thierolf, M., Hagmann, M.-L., Pfeffer, M. et al. (2008). Towards a comprehensive proteome of normal and malignant human colon tissue by 2-D-LC-ESI-MS and 2-DE proteomics and identification of S100A12 as potential cancer biomarker. *Proteomics Clin. Appl.* 2 (1): 11–22.

82 Zhao, J., Wang, M., He, P. et al. (2020). Identification of glutathione S-transferase π 1 as a prognostic proteomic biomarker for multiple myeloma using proteomic profiling. *Oncol. Lett.* 19 (3): 2153–2162.

83 Weigt, D., Sammour, D.A., Ulrich, T. et al. (2018). Automated analysis of lipid drug-response markers by combined fast and high-resolution whole cell MALDI mass spectrometry biotyping. *Sci. Rep.* 8 (1): 11260.

84 Guitot, K., Drujon, T., Burlina, F. et al. (2017). A direct label-free MALDI-TOF mass spectrometry based assay for the characterization of inhibitors of protein lysine methyltransferases. *Anal. Bioanal. Chem.* 409 (15): 3767–3777.

85 Ritorto, M.S., Ewan, R., Perez-Oliva, A.B. et al. (2014). Screening of DUB activity and specificity by MALDI-TOF mass spectrometry. *Nat. Commun.* 5: 4763.

86 De Cesare, V., Johnson, C., Barlow, V. et al. (2018). The MALDI-TOF E2/E3 ligase assay as universal tool for drug discovery in the ubiquitin pathway. *Cell Chem. Biol.* 25 (9): 1117–27.e4.

87 Heap, R.E., Hope, A.G., Pearson, L.A. et al. (2017). Identifying inhibitors of inflammation: a novel high-throughput MALDI-TOF screening assay for salt-inducible kinases (SIKs). *SLAS Discov.* 22 (10): 1193–1202.

88 Machálková, M., Schejbal, J., Glatz, Z., and Preisler, J. (2018). A label-free MALDI TOF MS-based method for studying the kinetics and inhibitor screening of the Alzheimer's disease drug target β-secretase. *Anal. Bioanal. Chem.* 410 (28): 7441–7448.

89 Emami, K., Wu, L.J., and Errington, J. (2020). A small molecule inhibitor of CTP synthetase identified by differential activity on a *Bacillus subtilis* mutant deficient in class A penicillin-binding proteins. *Front. Microbiol.* 11 (2001).

90 Chiarelli, L.R., Mori, G., Orena, B.S. et al. (2018). A multitarget approach to drug discovery inhibiting *Mycobacterium tuberculosis* PyrG and PanK. *Sci. Rep.* 8 (1): 3187.

91 Fijolek, A., Hofer, A., and Thelander, L. (2007). Expression, purification, characterization, and in vivo targeting of trypanosome CTP synthetase for treatment of African sleeping sickness*. *J. Biol. Chem.* 282 (16): 11858–11865.

92 Winter, M., Bretschneider, T., Thamm, S. et al. (2019). Chemical derivatization enables MALDI-TOF-based high-throughput screening for microbial trimethylamine (TMA)-lyase inhibitors. *SLAS Discov.: Adv. Sci. Drug Discov.* 24 (7): 766–777.

93 Chandler, J., Haslam, C., Hardy, N. et al. (2016). A systematic investigation of the best buffers for use in screening by MALDI–mass spectrometry. *SLAS Discov.: Adv. Life Sci. R&D* 22 (10): 1262–1269.

94 Simon, R.P., Winter, M., Kleiner, C. et al. (2020). MALDI-TOF mass spectrometry-based high-throughput screening for inhibitors of the cytosolic DNA sensor cGAS. *SLAS Discov.: Adv. Sci. Drug Discov.* 25 (4): 372–383.

95 Winter, M., Bretschneider, T., Kleiner, C. et al. (2018). Establishing MALDI-TOF as versatile drug discovery readout to dissect the PTP1B enzymatic reaction. *SLAS Discov.: Adv. Sci. Drug Discov.* 23 (6): 561–573.

96 Haslam, C., Hellicar, J., Dunn, A. et al. (2016). The evolution of MALDI-TOF mass spectrometry toward ultra-high-throughput screening: 1536-well format and beyond. *J. Biomol. Screen.* 21 (2): 176–186.

97 Beeman, K., Baumgärtner, J., Laubenheimer, M. et al. (2017). Integration of an in situ MALDI-based high-throughput screening process: a case study with receptor tyrosine kinase c-MET. *SLAS Discov.: Adv. Life Sci. R&D* 22 (10): 1203–1210.

98 Winter, M., Ries, R., Kleiner, C. et al. (2019). Automated MALDI target preparation concept: providing ultra-high-throughput mass spectrometry–based screening for drug discovery. *SLAS Technol.: Transl. Life Sci. Innovat.* 24 (2): 209–221.

99 Simon, R.P., Winter, M., Kleiner, C. et al. (2021). MALDI-TOF-based affinity selection mass spectrometry for automated screening of protein–ligand interactions at high throughput. *SLAS Discov.: Adv. Sci. Drug Discov.* 26 (1): 44–57.

100 Chin, J., Wood, E., Peters, G.S., and Drexler, D.M. (2016). Acoustic sample deposition MALDI-MS (ASD-MALDI-MS):a novel process flow for quality control screening of compound libraries. *J. Lab. Autom.* 21 (1): 204–207.

101 Roddy, T.P., Horvath, C.R., Stout, S.J. et al. (2007). Mass spectrometric techniques for label-free high-throughput screening in drug discovery. *Anal. Chem.* 79 (21): 8207–8213.

102 Krenkel, H., Hartmane, E., Piras, C. et al. (2020). Advancing liquid atmospheric pressure matrix-assisted laser desorption/ionization mass spectrometry toward ultrahigh-throughput analysis. *Anal. Chem.* 92 (4): 2931–2936.

103 Gelb, M.H., Turecek, F., Scott, C.R., and Chamoles, N.A. (2006). Direct multiplex assay of enzymes in dried blood spots by tandem mass spectrometry for the newborn screening of lysosomal storage disorders. *J. Inherit. Metab. Dis.* 29 (2–3): 397–404.

104 Rathore, R., Corr, J.J., Lebre, D.T. et al. (2009). Extending matrix-assisted laser desorption/ionization triple quadrupole mass spectrometry enzyme screening assays to targets with small molecule substrates. *Rapid Commun. Mass Spectrometry* 23 (20): 3293–3300.

105 Xu, H., Liu, M., Huang, X. et al. (2018). Multiplexed quantitative MALDI MS approach for assessing activity and inhibition of protein kinases based on postenrichment dephosphorylation of phosphopeptides by metal-organic framework-templated porous CeO(2). *Anal. Chem.* 90 (16): 9859–9867.

106 Dong, H., Shen, W., Cheung, M.T.W. et al. (2011). Rapid detection of apoptosis in mammalian cells by using intact cell MALDI mass spectrometry. *Analyst* 136 (24): 5181–5189.

107 Niehaus, M., Soltwisch, J., Belov, M.E., and Dreisewerd, K. (2019). Transmission-mode MALDI-2 mass spectrometry imaging of cells and tissues at subcellular resolution. *Nat. Methods* 16 (9): 925–931.

108 Bien, T., Perl, M., Machmüller, A.C. et al. (2020). MALDI-2 mass spectrometry and immunohistochemistry imaging of Gb3Cer, Gb4Cer, and further glycosphingolipids in human colorectal cancer tissue. *Anal. Chem.* 92 (10): 7096–7105.

109 Heijs, B., Potthoff, A., Soltwisch, J., and Dreisewerd, K. (2020). MALDI-2 for the enhanced analysis of N-linked glycans by mass spectrometry imaging. *Anal. Chem.* 92 (20): 13904–13911.

110 Soltwisch, J., Heijs, B., Koch, A. et al. (2020). MALDI-2 on a trapped ion mobility quadrupole time-of-flight instrument for rapid mass spectrometry imaging and ion mobility separation of complex lipid profiles. *Anal. Chem.* 92 (13): 8697–8703.

111 Karas, M., Bachmann, D., Bahr, U., and Hillenkamp, F. (1987). Matrix-assisted ultraviolet laser desorption of non-volatile compounds. *Int. J. Mass Spectrometry and Ion Processes* 78: 53–68.

112 Schlüter, H., Offers, E., Brüggemann, G. et al. (1994). Diadenosine phosphates and the physiological control of blood pressure. *Nature* 367 (6459): 186–188.

113 Vermillion-Salsbury, R.L. and Hercules, D.M. (2002). 9-Aminoacridine as a matrix for negative mode matrix-assisted laser desorption/ionization. *Rapid Commun. Mass Spectrometry* 16 (16): 1575–1581.

114 Liu, H., Du, Z., Wang, J., and Yang, R. (2007). Universal sample preparation method for characterization of bacteria by matrix-assisted laser desorption ionization-time of flight mass spectrometry. *Appl. Environ. Microbiol.* 73 (6): 1899–1907.

115 Conway, G.C., Smole, S.C., Sarracino, D.A. et al. (2001). Phyloproteomics: species identification of Enterobacteriaceae using matrix-assisted laser desorption/ionization time-of-flight mass spectrometry. *J. Mol. Microbiol. Biotechnol.* 3 (1): 103–112.

116 Vargha, M., Takáts, Z., Konopka, A., and Nakatsu, C.H. (2006). Optimization of MALDI-TOF MS for strain level differentiation of *Arthrobacter* isolates. *J. Microbiol. Methods* 66 (3): 399–409.

117 Lin, X., Xiao, C., Ling, L. et al. (2021). A dual-mode reactive matrix for sensitive and quantitative analysis of carbohydrates by MALDI-TOF MS. *Talanta* 235: 122792.

118 Leite, J.F., Hajivandi, M.R., Diller, T., and Pope, R.M. (2004). Removal of sodium and potassium adducts using a matrix additive during matrix-associated laser desorption/ionization time-of-flight mass spectrometric analysis of peptides. *Rapid Commun. Mass Spectrometry* 18 (23): 2953–2959.

119 Lu, M., Yang, X., Yang, Y. et al. (2017). Nanomaterials as assisted matrix of laser desorption/ionization time-of-flight mass spectrometry for the analysis of small molecules. *Nanomaterials (Basel)* 7 (4): 87.

120 Lee, D., Kim, Y., Jalaludin, I. et al. (2021). MALDI-MS analysis of disaccharide isomers using graphene oxide as MALDI matrix. *Food Chem.* 342: 128356.

121 Huang, X., Liu, Q., and Jiang, G. (2019). Tuning the performance of graphene as a dual-ion-mode MALDI matrix by chemical functionalization and sample incubation. *Talanta* 199: 532–540.

122 Zhao, H., Li, Y., Wang, J. et al. (2018). Dual-ion-mode MALDI MS detection of small molecules with the O–P,N-doped carbon/graphene matrix. *ACS Appl. Mater. Interfaces* 10 (43): 37732–37742.

123 McCombie, G. and Knochenmuss, R. (2004). Small-molecule MALDI using the matrix suppression effect to reduce or eliminate matrix background interferences. *Anal. Chem.* 76 (17): 4990–4997.

124 Smirnov, I.P., Zhu, X., Taylor, T. et al. (2004). Suppression of α-cyano-4-hydroxycinnamic acid matrix clusters and reduction of chemical noise in MALDI-TOF mass spectrometry. *Anal. Chem.* 76 (10): 2958–2965.

125 Yamaguchi, S., Fujita, T., Fujino, T., and Korenaga, T. (2008). Suppression of matrix-related ions using cyclodextrin in MALDI mass spectrometry. *Anal. Sci.* 24 (11): 1497–1500.

126 Chandler, J., Haslam, C., Hardy, N. et al. (2017). A systematic investigation of the best buffers for use in screening by MALDI–mass spectrometry. *SLAS Discovery: Adv. Sci. Drug Discov.* 22 (10): 1262–1269.

127 Jang, K.-S. and Kim, Y.H. (2018). Rapid and robust MALDI-TOF MS techniques for microbial identification: a brief overview of their diverse applications. *J. Microbiol.* 56 (4): 209–216.

14

Rapid Pathogen Identification in a Routine Food Laboratory Using High-throughput MALDI-TOF Mass Spectrometry

Andrew Tomlin

ALS Laboratories (UK) Ltd, Rotherham, UK

14.1 Introduction

In 2015, the contract food microbiology laboratory ALS Rotherham, formerly ALcontrol, acquired a matrix-assisted laser desorption/ionization time-of-flight (MALDI-TOF) mass spectrometry (MS) Biotyper (Bruker Microbiology & Diagnostics) instrument and incorporated the technique into routine microbiological testing standard operating procedures (SOPs) to deliver rapid and accurate pathogen confirmation results in a drastically reduced time-frame. ALS Rotherham is a high-throughput food and waters microbiology laboratory operating under ISO 17025 that tests ~15 000 samples per week and, as such, the impact of integrating such a technology has been significant. This chapter is written from this perspective and aims to provide industry insight into how this technique has been implemented, how it has been utilized, and how it can continue to transform food testing in the future. As ALS Rotherham currently operates a Bruker MALDI Biotyper (MBT), all research and testing referred to herein therefore used this system and its relevant computer software as required.

14.2 MALDI-TOF MS in Food Microbiology

MALDI-TOF MS has been utilized widely in food microbiology and increasingly so in recent years, and consequently the vices and virtues of this technique have been extensively reviewed in the literature [1, 2]; hence only a brief contextual summary of the technology will be provided in this chapter.

When analysing a bacterial, yeast, or mould isolate, a sample is prepared on a MALDI target plate using a very simple procedure. The preparation involves addition of the organism in question via smearing a colony onto a target spot using a sterile cocktail stick or similar, then subsequent covering with an energy-absorbing matrix, α-cyano-4-hydroxycinnamic acid (CHCA) matrix. This dries on each sample spot and co-crystallizes

Microbiological Identification using MALDI-TOF and Tandem Mass Spectrometry: Industrial and Environmental Applications, First edition. Edited by Haroun N. Shah, Saheer E. Gharbia, Ajit J. Shah, Erika Y. Tranfield, and K. Clive Thompson.

with the analyte. A laser is then fired at each target spot, causing desorption and ionizing the sample. The matrix absorbs the laser energy and allows the samples to retain their structure, which would otherwise be destroyed by direct irradiation from the laser. The addition of matrix separates this technology from other 'hard' ionization techniques such as gas chromatography mass spectrometry (GC-MS) and thus MALDI-TOF MS is sometimes referred to as a soft ionization technique. Once ionized, the protonated ions are accelerated along the flight tube and separated based on size or mass : charge ratio (m/z). In MALDI-TOF MS, the samples are then detected and measured using a TOF analyser [3]. From this TOF information, a spectrum is generated for each sample, as in Figure 14.1.

At ALS Laboratories, MALDI-TOF MS is principally used as a confirmation technique for pathogens from food, water, and swab samples. Samples are processed using an appropriate test methodology to isolate a desired target organism and, if a presumptive isolate of this target is acquired, MALDI-TOF MS analysis can then be carried out using a single colony. As an example, the timeline for *Salmonella* spp. detection is shown in Figure 14.2, using both the traditional ISO method and MALDI-TOF MS in the confirmation. Once an isolated colony on an agar plate has been acquired, the MALDI-TOF MS confirmation can be carried out directly. No subculture of the organism is required, as good-quality, reproducible results are easily obtainable from selective media and a range of different agars can be used with this technique. The ability to test without any further growth stages is a significant advantage of MALDI-TOF MS analysis and enables results to be attained rapidly; in the case of *Salmonella* detection this removes two 24-hour growth periods and so results are available two days faster. In food microbiology, the impact of two extra days

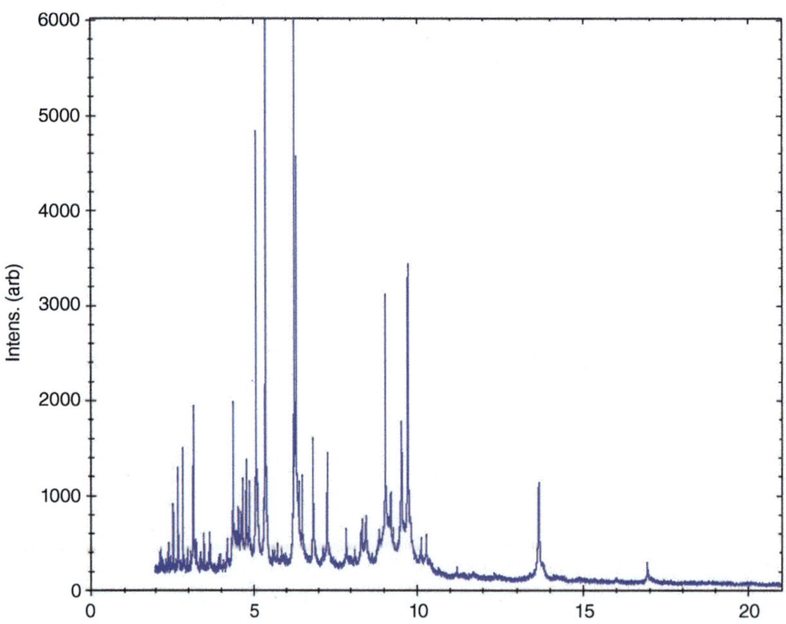

Figure 14.1 MALDI-TOF MS spectra for a bacterial test standard sample.

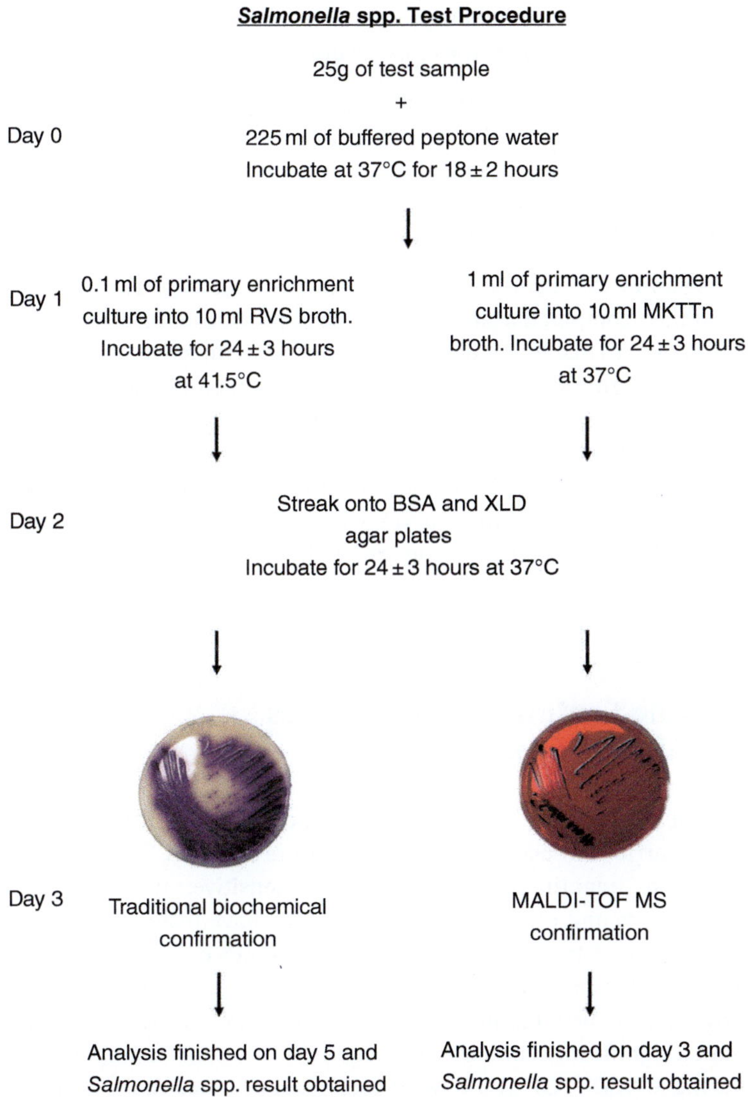

Figure 14.2 Example workflow for *Salmonella* detection using conventional culture media methods. RVS, Rappaport–Vassiliadis soya peptone broth; MKTTn, Muller–Kauffmann Tetrathionate-Novobiocin broth.

cannot be understated; this can prevent or reduce the likelihood of serious illness and save significant sums of money.

Several methods of sample preparation for MALDI-TOF MS analysis exist, such as direct transfer, extended direct transfer, and the full extraction method. After an internal validation was carried out by ALS evaluating each extraction method [1], it was determined that the extended direct transfer constituted the optimal extraction technique when factoring in efficacy and the time taken to carry out the process. This extended direct transfer essentially provides a partial extraction of the sample but does not require the same degree of hands-on time as a full extraction. The validation process is described in detail in Cook et al. [1]. An example of the workflow for this type of extraction is as follows:

Smear an isolated colony as a thin film directly onto a sample position on a MALDI target plate.
Overlay the material with 1 µl of 70% formic acid and dry at room temperature.
Overlay the material with 1 µl of CHCA matrix.
Run the target plate on the MALDI-TOF MS instrument.

Using the Bruker Compass and FlexControl software programmes, spectra were obtained from each sample spot, and then compared against a database to provide an identification of the organism spotted onto the MALDI target plate. At the time of writing, ALS currently operates MALDI-TOF MS instruments at the majority of its UK food microbiology sites. At ALS Rotherham specifically, the above technique is used for the UKAS accredited confirmation of *Salmonella* spp., *Listeria* spp., and *Campylobacter* spp. with more organisms planned to be added to the accredited scope. Using the Bruker systems, automation of the sample preparation is possible along with integration with a Laboratory Information Management Systems (LIMS), both of which are being explored by ALS specifically for use in a high-throughput laboratory. Specialist identification work is occasionally also carried out by request, using the latest databases available for bacteria, yeast, and mould isolates.

14.3 Review of Existing Confirmation Techniques and Comparison to MALDI-TOF MS

Before the implementation of MALDI-TOF MS into the food microbiology setting, confirmation testing was generally carried out using conventional culture methods. These methods can give qualitative and/or quantitative results, are very sensitive, and are often inexpensive [4]. For these reasons and others, these test methods have historically been regarded as the gold standard [5] and are the ISO reference test methods in food microbiology. These ISO methods are used as part of an ISO 16140 validation when validating any new rapid method to show that the new method is valid and provides at least equivalent results. At ALS Rotherham a significant number of pathogen confirmation testing is currently being carried out using MALDI-TOF MS; however, it should be noted that a proportion of confirmation testing is still carried out using traditional biochemical methods, for a variety of reasons. Most frequently, this is due to UKAS accredited MALDI-TOF MS confirmation not yet being available for more niche pathogens or Food Business Operators (FBOs) opting to remain using existing reference methods where the turnaround time is less critical and lower testing costs are the primary focus.

The majority of the reference methods involve some form of biochemical test gallery, such as the Analytical Profile Index (API) 20E. For a test such as this, a colony acquired from a selective agar must first be purified onto a non-selective media before use. The API20E (the biochemical gallery used for the confirmation of *Salmonella* spp.) comes in the form of a test strip with various dehydrated substrates present in each well which react based on enzymatic activity. Addition of a bacterial suspension rehydrates each well and then the strip is incubated. After incubation, colour changes are observed either immediately or upon the addition of reagents. The combination of reactions (or lack thereof) are used to build a score, which is entered into a database to return a result. From the point of having a presumptive isolate on an agar plate, this testing takes around 48 hours before a confirmed species result is acquired. Although they are established, well verified, and cost-effective, these tests are limited by database size, require some user interpretation to observe reactions and, crucially, take a considerable amount of time to carry out in terms of both hands-on and incubation time.

The food testing industry often requires results as quickly as possible, particularly as many products are 'positive release' samples, meaning that products are in quarantine until the microbiological testing can be completed and results are satisfactory. The shortest possible turnaround time for testing is therefore desirable and, in some cases, critical, especially for short shelf life products. To address the primary issue with biochemical tests – speed – a great number of other techniques for effective bacterial identification exist, such as the polymerase chain reaction (PCR), restriction fragment length polymorphism (RFLP), Fourier transform infrared spectroscopy (FTIR) and more [6]. Many of these are expensive, laborious, require specialist knowledge, lack sufficient validation data, and are time-consuming to carry out.

Using MALDI-TOF MS in lieu of the traditional confirmation techniques addresses the issue of time, being able to provide a confirmed result from an isolate on an agar plate, while also not suffering from the issues listed earlier relating to other rapid methods. In particular, the time required to analyse each sample is low with MALDI-TOF MS, a huge benefit for a contract laboratory processing a large number of samples daily. When integrated into the test methods for *Salmonella* spp. or *Listeria* spp. in use at ALS, MALDI-TOF MS pathogen confirmation provides a confirmed result after three and four days for *Salmonella* and *Listeria*, respectively, which is a drastic reduction in time compared with the reference methods, which take five or six days from the start of testing.

In food microbiology, targeted confirmation testing is often carried out in which the target organism is known and determining the presence (or absence) of the specific species or genus is the goal. *Listeria monocytogenes* is a hugely important foodborne pathogen capable of causing serious illness and even death amongst people in certain risk groups. However, the *Listeria* genus is a particularly challenging to speciate due to the highly homologous nature of species at the genetic and proteomic level. Testing for *Listeria* spp. (and *L. monocytogenes*) makes up a substantial amount of routine microbiological testing for FBOs and so reliable differentiation between the different species of the *Listeria* genus is critical. Using MALDI-TOF MS and a specific subtyping software module, a rapid and accurate result can be obtained for these organisms. In addition to challenging *Listeria* species identification, MALDI-TOF MS can also be used to distinguish between several other poorly differentiated taxa, such as *Mycobacterium chimaera* from *Mycobacterium intracellulare* and *Elizabeth/Kingia* species. MALDI-TOF MS can also be used to identify antibiotic resistance markers in key pathogens such as methicillin-resistant *Staphylococcus aureus* (MRSA) [7] and carbapenem-resistant *Klebsiella pneumoniae* (CRKP) [8]. This is an emerging area of development with MALDI-TOF MS, particularly in the clinical context, but as resistance testing is at present not a key factor in food microbiology, it will not be discussed further here.

Another key advantage of MALDI-TOF MS is that an unknown sample can be analysed and high-quality, accurate results can be obtained. If the organism present is in the extensive database (as of May 2022 there are 12000+ entries), then an identification can be obtained. By contrast, biochemical galleries can only contain a limited number of reactions by their nature and so they generally test for a specified family of bacteria (such as the API20E, which tests for members of the *Enterobacteriaceae* family). A series of traditional microbiology tests must therefore first be carried out (such as oxidase, catalase, and Gram staining) in order to inform which biochemical gallery should be used. Unless a screening process has been carried out on an isolate first in the form of a selective culture media or enzyme-linked immunosorbent assay (ELISA), it may not be possible to know which gallery to use without carrying these tests out, which adds further time and costs to the identification process.

Despite the plethora of benefits, the main disadvantage of MALDI-TOF MS technology must be discussed: the cost. A six-figure investment (of GB pounds) for a new MALDI-TOF MS Biotyper machine is likely to price smaller laboratories out of the technology if grants are not available. It should be considered, however, that as this technology is designed for high throughput, this may not be an issue in practice. In contrast to the initial instrument cost, operational costs for running MALDI-TOF MS instruments are exceptionally low, with only a handful of low-cost reagents being required for testing [2]. When factoring in labour savings and this minimal requirement for consumables, this initial cost has the potential to be offset over the course of the instrument's life and an eventual reduction in testing costs could be seen depending on the volume of samples processed. Biochemical tests have a clear advantage in terms of costs initially as they are cheap to purchase and carry out. If the target organism for confirmation is known from the outset of testing, then a specific test can be selected and purchased en masse. To confirm a broad range of organisms, however, a stock of various test galleries must be held to enable the identification of these. In such a scenario, utilization of MALDI-TOF MS would have clear benefits when compared with the biochemical tests, as only one simple stock of reagents is needed.

Overall, implementation of MALDI-TOF MS as a confirmation technique for routine bacterial pathogens identification has been extremely advantageous to ALS and its clients, providing a rapid and accurate pathogen confirmation solution that can be operated easily, has a large database of microorganisms, is validated against the reference methods, and has very low running costs.

14.4 Strain Typing Using MALDI-TOF MS

Often in food microbiology, pathogen detections (such as *L. monocytogenes* or *Escherichia coli*) occur in groups, for example where a series of detections are seen across a number of products with common ingredients, or swabs from a variety of locations across a production process. The ability to link pathogen detections, or to be able to say that they are not the same isolate, is key in discerning the root cause of an issue, and therefore in addressing the problem. When such an event transpires, it is typically required that further analysis be carried out on several bacterial isolates as part of an investigation process, with the aim of discerning whether the isolates detected at each point are similar or clearly different. In the case of ALS, this analysis was previously subcontracted to a laboratory capable of carrying out advanced testing such as ribotyping or even whole-genome sequencing (WGS), which by their nature would incur significant costs and long turnaround times. The high cost of this further analysis was in some cases prohibitive to carrying out the analysis, and the long turnaround times often meant that the important information obtained from this type of analysis lost its meaning and value as more time elapsed. As such, a faster alternative to these techniques was sought.

Beyond routine confirmation of bacterial isolates from food and water samples, MALDI-TOF MS can also be used to get some resolution beyond species [9–11]. Broadly speaking this is achieved by carrying out a full biochemical extraction on a bacterial sample, generating many high-quality spectra, processing the spectra, and carrying out a subsequent comparison between different isolates. MALDI-TOF MS strain typing is most effectively used to screen batches of samples for further investigative analysis. In practice, this essentially means utilizing this more cost-effective technique to screen high volumes of samples to reduce the number of samples needing further, more expensive analysis, or even removing the need for this further analysis completely. It has therefore been proposed that strain typing by MALDI-TOF MS constitutes a reliable, cost-effective, and fit-for-purpose method to gain some resolution between isolates of pathogens from certain genera detected in routine food and water microbiology. Removing the need to subcontract such analysis and the possibility for quick turnaround times using this technique enables this testing to be carried out more rapidly than previously used techniques. In light of this potential, ALS has implemented strain typing by MALDI-TOF MS at the Rotherham food laboratory for this purpose. After an advanced training course with the MALDI-TOF MS manufacturer, a series of small-scale verifications were carried out at the laboratory to assess the efficacy of the technique and the laboratory's ability to carry it out.

14.5 Verification Trial

Two key bacterial organisms were selected for these trials based on previous client requirements: *L. monocytogenes* and *E. coli*. For both of the trials, a series of isolates of the same species but from different origins were tested, including isolates that were known to be different [National Collection of Type Cultures (NCTC) strains] and unknown wild-types. The isolates used were as below in Table 14.1.

All samples were held in MicroBank™ cryovials from initial acquisition until use in the trial. Wild-type samples were recovered through routine testing at the laboratory, and NCTC samples were purchased from Public Health England (now UK Health Security Agency). Isolates were cultured on Columbia Blood Agar (CBA) plates for 24 ± 3 hours at $37 \pm 1\,°C$ for use in the trial. All samples were processed for MALDI-TOF MS analysis using the full extraction protocol to ensure the highest quality results. All analyses were carried out using a Bruker MALDI-TOF MS Biotyper and counterpart FLexControl, FlexAnalysis, and Compass Explorer software, all licensed by the manufacturer. The procedure followed is outlined below:

- Transfer 300 μl of high-performance liquid chromatography (HPLC) grade water into an Eppendorf tube.
- Using a 1 μl loop, transfer isolated colonies from the culture plate into the water and mix thoroughly until the material is completely in suspension.
- Add 900 μl absolute ethanol and mix thoroughly.
- Spin down the microbial material in a centrifuge for two minutes at 13 500 rpm.
- Decant the supernatant, avoiding contact with the pellet.

Table 14.1 Isolates used in the verification trial and their details.

Bacterial organism	Strain details
Listeria monocytogenes	NCTC 10527
L. monocytogenes	NCTC 7973
L. monocytogenes	Wild-type swab
L. monocytogenes	Wild-type butter
L. monocytogenes	Wild-type olives
L. monocytogenes	Wild-type spray
Escherichia coli	NCTC 10418
E. coli	NCTC 12241
E. coli	NCTC 12923
E. coli	NCTC 13216
E. coli	Wild-type 1
E. coli	Wild-type 2

- Spin down the microbial material in a centrifuge for two minutes at 13 500 rpm.
- Remove any remaining supernatant.
- Air-dry the pellet for at least five minutes at room temperature.
- Add 25 μl of 70% formic acid and mix by pipetting the solution up and down.
- Add 25 μl of acetonitrile (ACN) to the tube and mix by pipetting up and down two to three times.
- Centrifuge the tube for two minutes at 13 500 rpm.
- Deposit 1 μl of the supernatant onto a vacant sample position of a MALDI target plate, then overlay with 1 μl of CHCA matrix once dried.

Each sample was deposited onto the target plate eight times and run three times each in order to generate ample data for the analysis. A Bacterial Test Standard (BTS) was also run to calibrate the process. Before spectra were generated, the MALDI-TOF MS instrument was appropriately checked and calibrated to ensure optimal performance. Once sufficient spectra were acquired, a quality control check of the isolates was run to ensure species identity of the isolates and to check that the spectra were of a high enough quality for the analysis. Following this, a BTS calibration check was carried out in the FlexAnalysis software to ensure that all of the peaks in the BTS spectra were with acceptable limits (±300 ppm). After this, the test isolates' spectra were processed to ensure that no outliers or flat-line spectra were present, that baseline subtraction and smoothing was carried out, and that subsequently the peak shift between all spectra generated for each isolate was within the acceptable tolerance (500 ppm) across the mass range to be used for analysis. Isolates that met these criteria were saved and could be used in the final strain typing analysis. All isolates tested in these two verification trials met the criteria, meaning that all 12 samples in total (six for each species) were acceptable for use in the final comparative

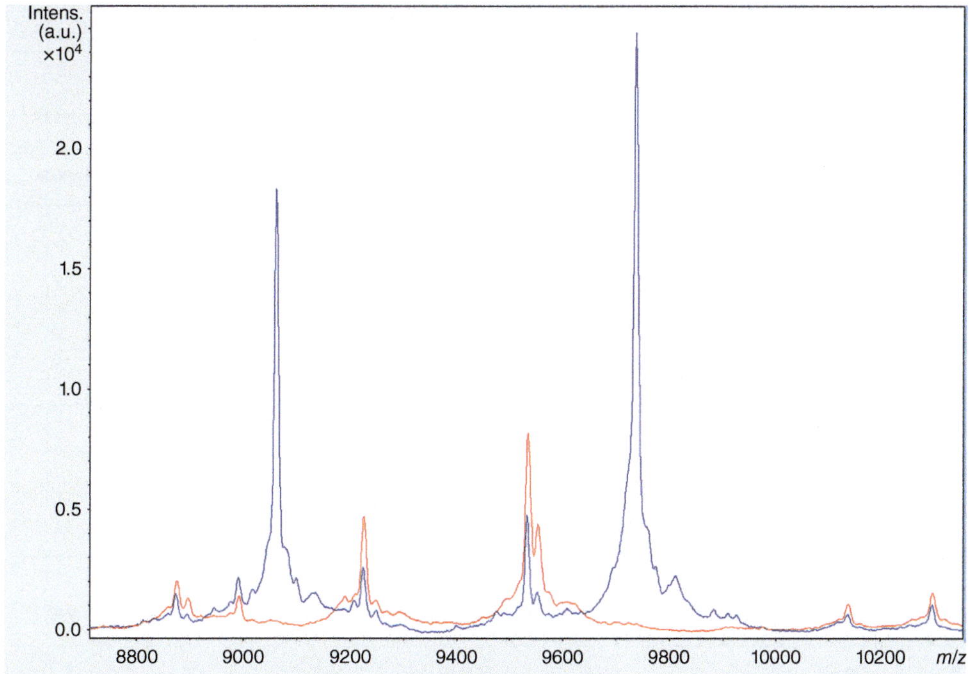

Figure 14.3 Overlaid spectra of two wild-type *E. coli* isolates.

analysis. The final processed spectra were then compared to each other and viewed in several different fashions to facilitate comparison, such as the overlaid style seen in Figure 14.3. With spectra overlaid, significant differences, present or absent peaks and mass shifts could easily be observed. All data were recorded in a spreadsheet and tabulated in Tables 14.2 and 14.3.

Based on these data, different combinations of peaks present or absent at the various m/z ratios were assigned arbitrary strain references to indicate differences from other isolates in the trial. Figure 14.3 shows an example of the overlaid spectra of two *E. coli* isolates from the trial. A clear difference between two isolates can be seen based on the presence of peaks in one, but not the other. By carefully analysing each dataset in this way, differences such as this (if present) could be discerned between the isolates.

Tables 14.4 and 14.5 display the overall results of the two strain typing trials and translate the information from Tables 14.2 and 14.3 into a more usable and client-facing format in which differences are made clear by the differing strain references assigned to each isolate.

In the *L. monocytogenes* trial, all but the NCTC 7973 and the wild-type butter isolates could be distinguished as different strains. The NCTC 7973 and the wild-type butter *L. monocytogenes* samples were indistinguishable at this level of investigation. All isolates in the *E. coli* portion of the trial were allocated different strain references and were thus demonstrated to be different from each other using this technique. This trial outlines the capability and potential of MALDI-TOF MS strain typing as a technique to distinguish between food pathogen isolates of the same species.

Table 14.2 *Listeria monocytogenes* strain typing peak comparison.

Listeria monocytogenes peak presence/absence							
Peak no.	***m/z***	**NCTC 7973**	**NCTC 10527**	**Wild-type (WT) butter**	**WT olives**	**WT spray**	**WT swab**
1	3970	X		X		X	X
2	4321	X	X	X	X	X	X
3	7142	X		X	X	X	
4	5588	X		X	X		
5	7940	X		X		X	X

Table 14.3 *Escherichia coli* strain typing peak comparison.

Escherichia coli peak presence/absence							
Peak no.	***m/z***	**NCTC 10418**	**NCTC 12241**	**NCTC 12923**	**NCTC 13216**	**Wild-type (WT) 1**	**WT 2**
1	4165	X		X	X	X	X
2	4535	X			X	X	
3	4872	X			X	X	
4	6592	X					
5	9739	X			X	X	
6	5384	X	X	X	X	X	X
7	9067	X			X	X	
8	3470					X	
9	3355					X	
10	3206	X	X	X	X		

Table 14.4 Results for the *Listeria monocytogenes* strain typing trial.

Listeria trial results	
Isolate details	**Assigned strain reference**
L. monocytogenes NCTC 10527	LM001
L. monocytogenes NCTC 7973	LM002
L. monocytogenes wild-type swab	LM003
L. monocytogenes wild-type butter	LM002
L. monocytogenes wild-type olives	LM004
L. monocytogenes wild-type spray	LM005

Table 14.5 Results for the *Escherichia coli* strain typing trial.

***E. coli* trial results**	
Isolate details	**Assigned strain reference**
E. coli NCTC 10418	EC001
E. coli NCTC 12241	EC002
E. coli NCTC 12923	EC003
E. coli NCTC 13216	EC004
E. coli Wild Type 1	EC005
E. coli Wild Type 2	EC006

14.6 Limitations of MALDI-TOF MS Strain Typing and Future Studies

Although this technique has many obvious benefits, there are also several limitations and comparisons to other technologies that must be considered. This technique can be used effectively with a range of organisms, but there are some food-critical organisms that MALDI-TOF MS cannot distinguish with high accuracy beyond species level, such as *Salmonella* spp. [12], due to its low discriminatory ability between subtypes and closely related serovars. In order to distinguish *Salmonella* isolates, traditional serological confirmation testing must be carried out or other equivalent validated testing methods [13]. While the efficacy of MALDI-TOF MS strain typing is clear both here and in the literature, it is not as widely recognized or characterized as other techniques used for this purpose, such as WGS. The principal drawback of other techniques such as WGS is the cost, which at the time of writing can leave it prohibitively expensive to FBOs when carrying out further testing and, in this respect, MALDI-TOF MS strain typing compares very favourably. Another frequently used technique for distinguishing between isolates of the same species, and perhaps a more comparable technique, is ribotyping [14]. A molecular technique capable of typing bacteria to the strain level, ribotyping is widely recognized in the food industry as a go-to technique for beyond species analysis. Very limited data are available comparing efficacy of these two techniques, particularly with food-relevant microorganisms; however, some research has been published using both techniques to strain type *Clostridioides difficile* [15], which suggests that MALDI-TOF MS strain typing could be a viable alternative to ribotyping. Some limited internal unpublished data in which several wild-type *Listeria welshimeri* isolates were analysed also suggest that this may be the case; however, further work must be carried out to draw any reliable conclusions. As such, it can only be said that these two techniques appear similar in efficacy at this stage.

Although the limited data presented in this trial show good efficacy of this technique for two different bacterial species, it is a small-scale trial and so further trials are planned in which the number of *L. monocytogenes* and *E. coli* isolates tested will be expanded, and different genera of organisms will be used to test the robustness of the technique in other areas. Emerging research indicates this is possible for a number of key organisms in food microbiology [16–19].

Recognizing the limitations discussed, the scope of MALDI-TOF MS strain typing must be clearly defined and, as such, this technique is proposed to be a more economical fit-for-purpose test for the food industry as a first step in beyond species analysis for certain food pathogens. As before with MALDI-TOF MS, the running costs of this strain typing analysis are low, meaning that strain typing is possible at a reasonable cost for a laboratory that already has MALDI-TOF MS capabilities. The technique can be carried out readily by experienced users, and advanced training courses focusing on the technique are available, requiring only minimal investment.

Utilization of this technique (and thus internalization of a previously subcontracted form of analysis) has brought a previously costly and lengthy analysis into reach for a routine food microbiology laboratory. This, in turn, has helped enable FBOs to act rapidly in a crisis and ultimately helped to maintain and improve food safety.

14.7 *Listeria* Detection by MALDI-TOF MS

In food microbiology, the ability to rapidly detect the presence of a pathogen is of critical importance, primarily to public health, but also to the FBO, as the presence of pathogens in products can trigger expensive product recalls and food wastage. MALDI-TOF MS is currently used to this end in food microbiology, to drastically reduce the length of confirmation testing in the pathogen detection process, as discussed previously; however, the technique also has the potential to be used more directly for the detection of pathogens [1, 20].

Based on the constant demands of the industry to obtain results faster and the impact that MALDI-TOF MS has already made in this area, the study below has been carried out to test the potential of a MALDI-TOF MS protocol in which the technique is used to detect *Listeria* spp. directly. This procedure aims to deliver results faster than the current gold standard ISO detection methods (against which all novel methods are tested according to ISO 16140) and even the latest industry-leading techniques. The *Listeria* ISO detection method (ISO 11290-1) takes five to six days to acquire a confirmed *Listeria* species result [21]. A plethora of alternative methods are available for *Listeria* spp. detection and many of these can acquire results in two to three days [22]. This trial aimed to explore the potential of detecting *Listeria* spp. in 24–48 hours in food and swab samples, acquiring a species-level result using a MALDI-TOF MS-based protocol.

14.8 Trial Sample Preparation Procedure

This method combines the use of a conventional food sample preparation, dilution in a selective media, and overnight incubation with a subsequent modified MALDI-TOF MS full extraction procedure. The MALDI-TOF MS procedure described below was followed for all samples, with minor adjustments where required at the initial preparation stage depending on the sample matrix.

The samples in this trial were inoculated using cultures of *Listeria* from the NCTC frozen on microbank beads, grown for 24 hours in Brain Heart Infusion (BHI) broth, and diluted down to appropriate concentrations using maximum recovery diluent (MRD). Approximate concentrations were discerned by plating out a dilution series on CBA plates, which enabled the number of colony-forming units (cfu) per ml of an overnight *Listeria* BHI culture grown from a bead to be estimated. A fresh culture was then set up using a new bead, grown in BHI for 24 hours, which was then used to inoculate the samples based on the concentration calculated from the previous culture. This inoculating procedure was used throughout all the trials described in this study.

All samples were tested with the proposed *Listeria* detection method and the *Listeria* spp. ISO detection method in tandem. Testing alongside the ISO detection method ensured the viability of the organisms throughout the test procedure and allowed a comparison between the two methods.

Industry-standard sample preparation was utilized in this trial, and all possible preparation methods, incubation times, and temperatures were aligned with the current ISO *Listeria* detection method. Three sample matrices were chosen for this trial: a wheat biscuit, chicken sandwiches, and a swab. Two iterations of the chicken sandwich trial were run separately. Enrichment broth-only samples were also run to assess the effects of the sample matrix and to test the procedure at a fundamental level. The sample matrices were selected under the rationale that these sample types constitute a reasonable cross-section of those typically tested for *Listeria* spp. detection and enumeration. These sample types would also present a challenge to the test method through the debris produced in the sample homogenate. Food products were inoculated at 10 and 100 cfu/g, with the swabs being inoculated at 10 and 100 cfu/area swabbed. These inoculation levels were chosen as the 10 cfu/g would challenge the limit of detection (LOD) of the test and is also the LOD of enumerating *Listeria* spp. tests, whereas 100 cfu/g is the legal limit for *L. monocytogenes* in ready-to-eat foods [23]. As food products, these samples also provide a significant quantity of debris in the sample matrix (when compared with a swab sample) and so challenge the robustness of the proposed test method. To help remove some of the debris downstream, separator bags were be used in the primary enrichment stage.

All samples were incubated for 25 ± 1 hours at 30 ± 1 °C. After this, 0.2-ml aliquots of the sample were taken and processed as per the ISO detection method by subculturing into Full Fraser (FF) broth and incubating at 37 ± 1 °C for 24 ± 2 hours. One-millilitre aliquots were also taken at this stage for the 24-hour testing of the primary enrichment broth using the proposed MALDI-TOF MS detection protocol described earlier. A BTS was run alongside each MALDI-TOF run to verify the integrity of the analysis. The BTS sample constitutes a modified *E. coli* isolate and covers the mass range used for this analysis. The MALDI-TOF MS analysis procedure described was carried out on day 1 using the primary enrichment broth, and then again on day 2 using the secondary enrichment broth (FF). In this way the viability of a 24 and 48-hour test method was evaluated. This preliminary trial consisted of five configurations for each matrix tested: a *L. monocytogenes* sample at a concentration of 10 cfu/g, a *Listeria innocua* sample at 10 cfu/g, *L. monocytogenes* sample at 100 cfu/g, a *L. innocua* sample at 100 cfu/g, and an uninoculated blank sample. Results for the analysis of this portion of the trial are displayed in Tables 14.6 and 14.7. These two

Table 14.6 Twenty-four- and forty-eight-hour results of detection by MALDI-TOF MS.

Sample matrix	Inoculation organism	Inoculation level	24-hour results	48-hour results
Sponge swab	None	Blank	No peaks found	No peaks found
Sponge swab	*Listeria monocytogenes*	10 cfu/area swabbed	No peaks found	*L. monocytogenes*
Sponge swab	*L. monocytogenes*	100 cfu/area swabbed	No peaks found	*L. monocytogenes*
Sponge swab	*Listeria innocua*	10 cfu/area swabbed	No peaks found	*L. innocua*
Sponge swab	*L. innocua*	100 cfu/area swabbed	No peaks found	*L. innocua*
Sandwich 1	None	Blank	No peaks found	No peaks found
Sandwich 1	*L. monocytogenes*	10 cfu/g	*L. monocytogenes*	*L. monocytogenes*
Sandwich 1	*L. monocytogenes*	100 cfu/g	*L. monocytogenes*	*L. monocytogenes*
Sandwich 1	*L. innocua*	10 cfu/g	No peaks found	*L. innocua*
Sandwich 1	*L. innocua*	100 cfu/g	No peaks found	*L. innocua*
Wheat biscuits	None	BLANK	No peaks found	No peaks found
Wheat biscuits	*L. monocytogenes*	10 cfu/g	No peaks found	*L. monocytogenes*
Wheat biscuits	*L. monocytogenes*	100 cfu/g	No peaks found	*L. monocytogenes*
Wheat biscuits	*L. innocua*	10 cfu/g	No peaks found	*L. innocua*
Wheat biscuits	*L. innocua*	100 cfu/g	No peaks found	*L. innocua*
Sandwich 2	None	BLANK	*L. citreum*[a]	No peaks found
Sandwich 2	*L. monocytogenes*	10 cfu/g	*L. citreum*	*L. monocytogenes*
Sandwich 2	*L. monocytogenes*	100 cfu/g	*L. citreum*	*L. monocytogenes*
Sandwich 2	*L. innocua*	10 cfu/g	*P. extremorientalis*[b]	*L. innocua*
Sandwich 2	*L. innocua*	100 cfu/g	*P. fluorescens*[b]	*L. innocua*
Media only	None	Blank	No peaks found	No peaks found
Media only	*L. monocytogenes*	10 cfu/g	No peaks found	*L. monocytogenes*
Media only	*L. monocytogenes*	100 cfu/g	No peaks found	*L. monocytogenes*
Media only	*L. innocua*	10 cfu/g	No peaks found	*L. innocua*
Media only	*L. innocua*	100 cfu/g	*L. innocua*	*L. innocua*

[a] *Leuconostoc.*
[b] *Pseudomonas.*

Table 14.7 ISO results for *Listeria* spp. detection.

Sample matrix	Inoculation organism	Inoculation level	Final result
Sponge swab	None	n/a	Not detected
Sponge swab	*Listeria monocytogenes*	10 cfu/area swabbed	Detected
Sponge swab	*L. monocytogenes*	100 cfu/area swabbed	Detected
Sponge swab	*Listeria innocua*	10 cfu/area swabbed	Detected
Sponge swab	*L. innocua*	100 cfu/area swabbed	Detected
Sandwich 1	None	n/a	Not detected
Sandwich 1	*L. monocytogenes*	10 cfu/g	Detected
Sandwich 1	*L. monocytogenes*	100 cfu/g	Detected
Sandwich 1	*L. innocua*	10 cfu/g	Detected
Sandwich 1	*L. innocua*	100 cfu/g	Detected
Wheat biscuits	None	n/a	Not detected
Wheat biscuits	*L. monocytogenes*	10 cfu/g	Detected
Wheat biscuits	*L. monocytogenes*	100 cfu/g	Detected
Wheat biscuits	*L. innocua*	10 cfu/g	Detected
Wheat biscuits	*L. innocua*	100 cfu/g	Detected
Sandwich 2	None	n/a	Not detected
Sandwich 2	*L. monocytogenes*	10 cfu/g	Detected
Sandwich 2	*L. monocytogenes*	100 cfu/g	Detected
Sandwich 2	*L. innocua*	10 cfu/g	Detected
Sandwich 2	*L. innocua*	100 cfu/g	Detected
Media only	None	n/a	Not detected
Media only	*L. monocytogenes*	10 cfu/g	Detected
Media only	*L. monocytogenes*	100 cfu/g	Detected
Media only	*L. innocua*	10 cfu/g	Detected
Media only	*L. innocua*	100 cfu/g	Detected

species were selected due to their being the two most commonly isolated species of *Listeria* from food testing at ALS Rotherham. The MALDI-TOF MS section of the protocol, carried out after 24 and 48 hours on the primary and secondary enrichment broths, is as follows:

- After incubation, samples were gently agitated and the suspension was evaluated for the level of debris.
- 1 ml of culture was aliquoted into a 1.5-ml Eppendorf tube, taking care to avoid debris where possible.
- Samples were centrifuged for two minutes at 13 500 rpm to pellet, and supernatant was removed.
- 1 ml of ultrapure HPLC-grade water added and pellet resuspend.

- Samples centrifuged again for two minutes at 13 500 rpm to pellet, and supernatant removed.
- Pellet was resuspended in 300 µl of ultrapure HPLC-grade water, then 900 µl of 100% ethanol (EtOH) was added. Samples were subsequently vortexed.
- Samples were centrifuged for two minutes at 13 500 rpm to pellet, and supernatant was removed. Samples were centrifuged again at 13 500 rpm for two minutes, and then the residual supernatant was removed and samples were allowed to air-dry for 15 minutes.
- Pellets were then resuspended into 25 µl of 70% formic acid, and then the same volume of ACN was added and mixed by pipetting up and down.
- Samples were centrifuged for two minutes at 13 500 rpm.
- 1 µl of supernatant was pipetted onto a MALDI-TOF target plate and dried at room temperature.
- 1 µl of CHCA solution was pipetted onto the same MALDI-TOF target spot and dried at room temperature.
- Samples were run on a MALDI-TOF MS Biotyper using the standard procedures to reflect current routine identification practices.
- Appropriate controls were run, which constituted an uninoculated control for the entire process and a BTS run with each MALDI-TOF MS run.

14.9 Initial Trial

Negative control samples run in this trial performed as expect, with no *Listeria* being detected in any uninoculated samples. The ISO testing results were congruent with the MALDI-TOF MS procedure and so all *Listeria* spp. were detected where possible. BTS samples run with each MALDI-TOF MS run passed the quality control check and validated that the MALDI-TOF MS was running correctly and was detecting the mass range analysed effectively.

After 24 hours, *Listeria* spp. were detected in 3 out of 20 inoculated samples analysed. Two of these were high-confidence identifications, with one being a low-confidence identification. All 'sandwich 2' (including the uninoculated control samples) returned non *Listeria* bacterial identifications, indicating high levels of background flora. This presents a challenge to this method as this clearly shows that several organisms can grow in the Half Fraser enrichment broth and compete with the *Listeria* spp. when background flora of this type are present. This issue would probably be seen with a number of sample matrices typically tested for *Listeria*. Despite the use of separator bags, a significant amount of sample debris was present throughout the process, particularly in the wheat biscuits samples, which is thought to have interfered with the process and is probably a key factor in the lack of detections observed after 24 hours. Due to the small volume subcultured from the primary enrichment broth in the secondary broth, the presence of the debris is hugely diluted down and becomes relatively insignificant, meaning that when the aliquot is removed from the secondary enrichment broth (after 48 hours) it is relatively free from debris. This enabled a much purer and more clearly defined pellet free from debris to be obtained in the centrifugation step in samples from the 48-hour segment. Overall, the results from this 24-hour test section of the trial (in particular the lack of an identification result from the enrichment media-only samples) showed that in its current format, this technique cannot perform *Listeria* spp. detection reliably and consistently after 24 hours.

After 48 hours, *Listeria* spp. were detected in all sample matrices and identified to species correctly for both *L. innocua* and *L. monocytogenes*, at all concentrations tested. High-confidence identification was obtained for all samples. These results show the potential for this method to detect *Listeria* spp. after 48 hours at a concentration of at least 10 cfu/g for a food product and 10 cfu/swab. This suggests that the LOD is comparable to that of a standard ISO *Listeria* spp. enumeration (10 cfu/g), which has the added value of being a quantitative test method and, when combined with MALDI-TOF MS, could also give a species result within 48 hours. As the ISO enumeration method is less labour-intensive, a LOD of 10 cfu/g for the *Listeria* spp. detection by MALDI-TOF MS does not constitute a competitive level of resolution when compared with other readily available tests. It was noted, however, that the lower inoculation level of 10 cfu/swab used in the swab testing returned accurate, high-confidence results after 48 hours, which suggests that a lower LOD (than 10 cfu/g) could be possible after 48 hours of incubation with products. The neutralizing buffer sponge swabs used in this trial weighed ~10 g and so it is hypothesized that given the detected result for the 10 cfu/area swabbed, a LOD of at least 1 cfu/g for a product could be possible. As the theoretical LOD of the ISO detection method is 1 cfu/25 g swab, this MALDI-TOF MS detection method should be challenged to this same limit to explore the true LOD of this method and whether it is possible to match the industry standard ISO detection method.

14.10 Limit of Detection Trial

Based on these data, a lower LOD of 1 cfu/25 g was tested using enrichment media-only samples to explore whether the test procedure could detect *Listeria* spp. without the interference of a sample matrix. In this trial, samples were subcultured after 24 hours as before, but the primary enrichment broth was then reincubated for a further 4 hours and tested by the MALDI-TOF MS procedure after 28 hours (not 24 hours) to explore the possibility of a same day test with an extended incubation. The 48-hour test procedure followed the same primary and secondary enrichment processes as the initial trial. Each bacterial species was run in duplicate due to the difficultly of inoculating at such a low level, with an ISO detection test again being run in tandem to verify the presence/viability of the organisms inoculated. Results for this testing are in Table 14.8. The negative control and BTS samples performed as expected. Despite the increased incubation time, it was still not possible to detect any *Listeria* spp. after 28 hours using this procedure. All results from this trial after 48 hours were concordant with the ISO method results, although no growth was observed in one sample for both tests. This is considered to be a failed inoculation, due to the low inoculation level attempted.

It is clear from all these data that the proposed MALDI-TOF MS method in its current guise cannot reliably detect *Listeria* spp. after 24–28 hours, but has the potential to be explored further for obtaining an accurate qualitative *Listeria* species result within 48 hours. *Listeria* spp. are present at the 24–28 hour stage in the primary enrichment broth at levels sufficient to be detected by both the ISO detection process and the 48-hour MALDI-TOF MS methods. As no peaks were found in the MALDI-TOF MS analysis section after

Table 14.8 MALDI-TOF MS detection and ISO detection results for 1 cfu/25 g LOD.

Sample matrix	Inoculation organism	Inoculation level	28-hour results	48-hour results	Final ISO result
Media only	None	Blank	No peaks found	No peaks found	Not detected
Media only	*Listeria monocytogenes*	1 cfu/25 g	No peaks found	No peaks found	Not detected
Media only	*L. monocytogenes*	1 cfu/25 g	No peaks found	*L. monocytogenes*	Detected
Media only	*Listeria innocua*	1 cfu/25 g	No peaks found	*L. innocua*	Detected
Media only	*L. innocua*	1 cfu/25 g	No peaks found	*L. innocua*	Detected

28 hours, the final MALDI target plate did not therefore have sufficient genomic material with which to acquire an identification and so adjustments to the protocol must be made in order for the method to be effective after 24–28 hours, with a LOD that is in line with the most effective and rapid existing detection methods.

14.11 Method Optimization, Further Prospects, and Conclusions

A MALDI-TOF MS-based *Listeria* spp. detection protocol has shown promise in this trial and others [20], but there are several key areas for optimization. Ideally, the process should be reliable and repeatable after 24–48 hours for this methodology to be relevant in modern-day food microbiology, as many existing techniques are capable of obtaining results within 48 hours. Although most of these alternative methods require expensive hardware and reagents, it has to be noted that this MALDI-TOF MS does too, and so, practically speaking, the protocol may only be viable and economical for a laboratory that already has a MALDI-TOF MS instrument in place. For the purpose of improving this technique, three key areas have been highlighted:

controlling sample debris;
selectivity;
sensitivity.

As discussed earlier, utilization of separator bags can help to reduce the levels of sample debris; however they cannot remove the issue entirely, particularly after 24 hours. In order to improve this method, the level of debris must be further reduced/removed. One way this could be achieved is using a filtration system in which aliquots of the primary enrichment broth are filtered through a membrane that allows the bacteria through but not the debris. This, however, would require a good deal of optimization itself to ensure that *Listeria* cells are not trapped inside food debris on the filter and would also require purchase of further equipment. Such a filtration system would also by its nature require a high volume of single-use consumables, which would be costly, and the use of these is increasingly becoming a concern in modern times. As such, a simpler solution would be preferable.

The presence of background flora in one of the samples caused non-target organism identifications after 24 hours, which could mask the presence of low levels of *Listeria* spp. and cause erroneous results. One approach to addressing this, taken by Yalcin-Karasu et al. [20], would be to utilize a more selective enrichment broth, such as 'ONE Broth *Listeria*'. This particular broth is a supplemented broth that has been demonstrated to be comparable (and, in some cases, superior) to the two-stage Half Fraser, Full Fraser enrichment process of 11290-1 [24], but is designed for a 24 ± 2-hour incubation period. This alternative medium could increase the selectivity of the process, but in the trial described here, the standard ISO enrichment process was intentionally chosen to make the proposed method as cost-effective as possible by utilizing a well-established, highly available and low-cost culture medium. This protocol was designed to be as economical as possible, based on the needs of the UK food microbiology industry. As such, while utilizing an alternative growth medium such as this would probably improve the method, it would address this issue only and incur further cost which would need to be evaluated carefully. Despite this added cost, this is a promising area for exploration.

In order to reliably achieve results after 24 hours, the sensitivity of this method must be improved. To do this, simply put, (more) *Listeria* needs to be present on the MALDI target plate at the point of the MALDI-TOF MS analysis. Thus, a larger aliquot of primary enrichment medium could be taken at the initial stage of the process and more wash steps incorporated. This would require a larger centrifuge to house 15-ml tubes (or larger) and the pellet obtained from such a centrifugation step would therefore be 10-fold larger than those obtained in the described protocol. This was not explored in this trial because a centrifuge of this size was unavailable, as they are not regular a requirement in routine food microbiology laboratories. As this technique is designed with a view to applying it in the food microbiology industry, efforts have been focused on utilizing existing equipment where possible. Using a centrifuge of this size would incur further cost and so this was not explored. It should also be noted that using a larger initial aliquot would increase the volume of debris present in the sample and so this adjustment could, in effect, simply move the problem further downstream.

Despite these adjustments having the potential to improve the process, each comes with several issues and principally addresses each problem only. As such, a solution that can impact all of these areas with only one procedural modification would be preferable. Utilizing an immunomagnetic separation (IMS) process on an aliquot from the primary enrichment sample may constitute such a solution. IMS is currently utilized within the industry very successfully in other applications [25] to provide concentrated material from primary enrichment broths to be used for further analysis. This step would concentrate any *Listeria* spp. present in the sample by attaching any *Listeria* spp. present to *Listeria*-specific paramagnetic beads through a *Listeria*-specific antibody–antigen reaction. The magnetic bead antibody–antigen complex could then be extracted through magnetism and deposited into a separate aliquot of ultrapure water, which would provide a concentrated, specific sample free from debris. The paramagnetic beads would then need to be removed from the microorganisms by washing the concentrated sample using EDTA or similar to prevent damage to the MALDI-TOF MS machine [26], thus releasing the *Listeria* spp. into solution and allowing the unattached magnetic beads to be removed and the concentrated *Listeria* spp. to be purified and ready for the MALDI-TOF MS extraction process.

A combination of these improvements will probably improve the efficacy of this procedure and so further work should aim at discerning the adjustments that provide maximum improvement to the method's efficacy, while incurring the minimum expense. It is hypothesized that a combination of using a more selective enrichment broth and an IMS will constitute the most effective test methodology. Once the optimal configuration has been discerned, a wide variety of sample matrices typically tested in routine food microbiology should be tested along with the remaining four key *Listeria* species not discussed here that the ISO detection method identifies (*L. ivanovii*, *L. seeligeri*, *L. grayi*, and *L. welshimeri*). As new species continue to emerge, it may be prudent to undertake limited study on various further species, such as *L. aquatica* and *L. weihenstephanensis*, to ensure all species are detectable using this method. The effects of mixed species should also be explored, as in the current configuration it is possible that a species could be masked by the greater presence of another. One way in which this could be addressed would be to use species (rather that genus) specific beads in the IMS step; however, this would of course mean that the test would only detect one species of *Listeria*, such as *L. monocytogenes*. In the context of food microbiology, this could be acceptable in most cases as *L. monocytogenes* is the primary organism of concern in this genus and a significant body of legislation pertains specifically to this pathogen. If the test were to use *Listeria* spp. beads only, the result reported after 24–48 hours could only confidently be described as a *Listeria* genus result and so purification of the post-IMS aliquot onto a culture medium similar to the confirmation stage of the ISO method which could differentiate the different species would be required (this would, however, add a further 24–48 hours to the testing timeline). Operationally speaking, the difference between these two options would be small and so it would be a possibility to carry out both versions in parallel depending on client requirements.

Both the strain typing and *Listeria* detection protocols outline the potential of MALDI-TOF MS to continue contributing to food microbiology, reducing testing turnaround times, increasing resolution and, overall, having a positive impact on the industry and food safety standards. Given the wide scope of applications, it is likely that these niche areas will expand and even cross over between current applications may emerge as the scientific landscape changes. One potential area of this could be the already existing market for *S. aureus* detection in food, blending with antimicrobial resistance detection using MALDI-TOF MS from the clinical setting as this issue continues to rise in significance [27].

A further emerging area for MALDI-TOF MS use is mould identification. As the databases available continue to expand, further scope for expansion in this field may be possible for food microbiology, as currently available databases are focused primarily on clinical isolates. Generally, in food microbiology, mould plate counts are assessed to provide cfu/g levels, but rarely is further work carried out, because traditional culture methods for mould identification are very complex and require expert interpretation as the field of mycology is incredibly intricate [28]. As such, this further work is very costly. MALDI-TOF MS could simplify this process hugely [29] and make a vast amount of information more readily accessible to FBOs. Identifying a mould and thus understanding it can be instrumental in finding the root cause of an issue and enable an FBO to effectively address a problem. Further work is needed in this area, but it is clear that MALDI-TOF MS has huge potential to make a significant impact in this area once the databases are curated and effective and comprehensive SOPs are produced.

References

1 Cook, N., D'Agostino, M., and Thompson, K.C. (2016). *Molecular Diagnostic Methods*, 1e. Academic Press.

2 Pavlovic, M., Huber, I., Konrad, R., and Busch, U. (2013). Application of MALDI-TOF MS for the identification of food borne bacteria. *Open Microbiol. J.* 7: 135–141.

3 Singhal, N., Kumar, M., Kanaujia, P.K., and Virdi, J.S. (2015). MALDI-TOF mass spectrometry: an emerging technology for microbial identification and diagnosis. *Front. Microbiol.* 6 (791).

4 Doyle, M.P. (2001). *Food Microbiology: Fundamentals and Frontiers*, 2e. Washington: ASM Press.

5 Mandal, P.K., Biswas, A.K., Choi, K., and Pal, U.K. (2011). Methods for rapid detection of foodborne pathogens: an overview. *Am. J. Food Technol.* 6: 87–102.

6 Akimowicz, M. and Bucka-Kolendo, J. (2020). MALDI-TOF MS – application in food microbiology. *Acta Biochim. Pol.* 67 (3): 327–332.

7 Nix, I.D., Idelevich, E.A., Storck, L.M. et al. (2020). Detection of methicillin resistance in *Staphylococcus aureus* from agar cultures and directly from positive blood cultures using MALDI-TOF mass spectrometry-based direct-on-target microdroplet growth assay. *Front. Microbiol.* 11 (1664-302X): 232.

8 Fang, L., Xu, H., Ren, X. et al. (2020). Epidemiology and risk factors for Carbapenem-resistant *Klebsiella pneumoniae* and subsequent MALDI-TOF MS as a tool to cluster KPC-2-producing Klebsiella Pneumoniae, a retrospective study. *Front. Cell. Infect. Microbiol.* 10 (2235–2988): 462.

9 Barbuddhe, S.B., Maier, T., Schwarz, G. et al. (2020). Rapid identification and typing of *Listeria* species by matrix-assisted laser desorption ionization-time of flight mass spectrometry. *Appl. Environ. Microbiol.* 74 (17).

10 Christner, M., Trusch, M., Rohde, H. et al. (2014). Rapid MALDI-TOF mass spectrometry strain typing during a large outbreak of Shiga-toxigenic *Escherichia coli*. *PLoS One* 9 (7).

11 Wang, H.Y., Lee, T.Y., Tseng, Y. et al. (2018). A new scheme for strain typing of methicillin-resistant *Staphylococcus aureus* on the basis of matrix-assisted laser desorption ionization time-of-flight mass spectrometry by using machine learning approach. *PLoS One* 13 (3).

12 Kang, L., Li, N., Li, P. et al. (2017). MALDI-TOF mass spectrometry provides high accuracy in identification of *Salmonella* at species level but is limited to type or subtype *Salmonella* serovars. *Eur. J. Mass Spectrom.* 23 (2): 70–82.

13 Diep, B., Barretto, C., Portmann, A.C. et al. (2019). Salmonella serotyping; comparison of the traditional method to a microarray-based method and an in silico platform using whole genome sequencing data. *Front. Microbiol.* 10 (2554).

14 Matloob, M. and Griffiths, M. (2014). *Ribotyping and Automated Ribotyping of Listeria monocytogenes*, 85–93. New York: Humana Press.

15 Calderaro, A., Buttrini, M., Martinelli, M. et al. (2021). Rapid classification of *Clostridioides difficile* strains using MALDI-TOF MS peak-based assay in comparison with PCR-ribotyping. *Microorganisms* 9 (661).

16 Horváth, B., Peles, F., Szél, A. et al. (2020). Molecular typing of foodborne coagulase-positive *Staphylococcus* isolates identified by MALDI-TOF MS. *Acta Alimentaria* 49 (3): 307–313.

17 On, S.L., Zhang, J., Cornelius, A.J., and Anderson, T.P. (2021). Markers for discriminating *Campylobacter concisus* genomospecies using MALDI-TOF analysis. *Curr. Res. Microbiol. Sci.* 2 (100019).

18 Quéro, L., Courault, P., Cellière, B. et al. (2020). Application of MALDI-TOF MS to species complex differentiation and strain typing of food related fungi: case studies with *Aspergillus* section *Flavi* species and *Penicillium roqueforti* isolates. *Food Microbiol.* 86 (103311).

19 Zhang, J., Plowman, J.E., Tian, B. et al. (2020). An improved method for MALDI-TOF analysis of wine-associated yeasts. *J. Microbiol. Methods* 172 (105904).

20 Yalcin-Karasu, S., Soylemez-Milli, N., Eren, O., and Eryasar-Orer, K. (2021). Reducing time in detection of *Listeria monocytogenes* from food by MALDI-TOF mass spectrometry. *J. Food Sci. Technol.* 58 (11): 4102–4109.

21 Gasanov, U., Hughes, D., and Hansbro, P.M. (2005). Methods for the isolation and identification of *Listeria* spp. and *Listeria monocytogenes*: a review. *FEMS Microbiol. Rev.* 29 (5): 851–875.

22 Välimaa, A.-L., Tilsala-Timisjärvi, A., and Virtanen, E. (2015). Rapid detection and identification methods for *Listeria monocytogenes* in the food chain – a review. *Food Control* 55: 103–114.

23 EFSA BIOHAZ Panel (EFSA Panel on Biological Hazards), Ricci, A., Allende, A. et al. (2018). *Listeria monocytogenes* contamination of ready-to-eat foods and the risk for human health in the EU. *EFSA J.* e05134.

24 Oxoid, 2004. *Oxoid Poster, Folio No. 1033*. Basingstoke: Oxoid Ltd.

25 Althaus, D., Zweifel, C., Johler, S., and Stephan, R. (2016). Performance of the assurance GDS® assay for the detection of *L. monocytogenes* in pure cultures and spiked food samples. *J. Food Microbiol. Saf. Hyg.* 1 (1).

26 Sun, J., Shi, H., Xue, Y. et al. (2021). Releasing bacteria from functional magnetic beads is beneficial to MALDI-TOF MS based identification. *Talanta* 225 (121968).

27 Výrostková, J., Regecová, I., Zigo, F. et al. (2022). Antimicrobial resistance of *Staphylococcus* sp. isolated from cheeses. *Animals* 12 (36).

28 Pitt, J.I. and Hocking, A.D. (2009). *Fungi and Food Spoilage*, 3e. New York: Springer.

29 Sacheli, R., Henri, A.S., Seidel, L. et al. (2020). Evaluation of the new id-fungi plates from Conidia for MALDI-TOF MS identification of filamentous fungi and comparison with conventional methods as identification tool for dermatophytes from nails, hair and skin samples. *Mycoses* 63: 1115–1127.

15

Detection of Lipids in the MALDI Negative Ion Mode for Diagnostics, Food Quality Control, and Antimicrobial Resistance

Yi Liu, Jade Pizzato, and Gerald Larrouy-Maumus

Faculty of Natural Sciences, Department of Life Sciences, MRC Centre for Molecular Bacteriology & Infection, Imperial College, London, UK

15.1 Introduction

Matrix-assisted laser desorption/ionization-mass spectrometry (MALDI-MS) has become a popular technique in clinical and industrial laboratories for rapid diagnosis, toxicological studies, and the food industry, although most such studies have been based on positive ion mode. The rationale behind such phenomena was a focus of studies on cytosolic proteins that are preferentially ionized in protonated form and effectively detected by positive ion mode [1–4] compared with other molecules, such as sugars, nucleic acids, and lipids. Moreover, in addition to positive ion mode, analyses in negative ion mode could offer greater molecular coverage, leading to an increase in the sensitivity and selectivity of the detection of specimens of interest by the ability to detect negatively charged molecules such as phospholipids and species-specific lipids [5]. In MALDI-MS, samples are co-crystallized with a UV-absorbing substance called a matrix, which efficiently absorbs energy from ultraviolet or infrared laser pulses and transfers this energy to samples, allowing so-called 'soft ionization'. Ions with different polarities can form, but the ionizing nature of different molecules may lead to the preferential formation of either positive or negative ions. In negative ion mode, a reversed accelerating voltage and analytical fields need to be applied to allow the accurate detection of negative ions. Therefore, the differences in ion source and detection methods between positive and negative ion modes also limit the use of negative ion mode as, until now, the majority of commercially available routine mass spectrometers could not be operated in negative ion mode, limiting the scope of the applications of routine MALDI-MS. Currently, for routine MALDI mass spectrometer analyses, the MALDI Biotyper sirius® offers the versatility to detect markers in both positive and negative ion modes.

Based on its potential, the detection of molecules in negative ion mode needs to be explored to improve the detection of various biological molecules that are preferentially

Microbiological Identification using MALDI-TOF and Tandem Mass Spectrometry: Industrial and Environmental Applications, First edition. Edited by Haroun N. Shah, Saheer E. Gharbia, Ajit J. Shah, Erika Y. Tranfield, and K. Clive Thompson.

ionized into negative ions and to address the challenges of diagnostics that cannot be approached by using positive ion mode alone. For example, various common phospholipid species, including phosphatidylinositol (PI), phosphatidylglycerol (PG), phosphatidic acid (PA), and cardiolipins (CLs), can be detected better in negative ion mode, and these are the major phospholipids with roles in membrane construction, signalling, and immunoregulation; hence, studies on their abundances, localization, and relative metabolism in both healthy and diseased samples are important [6]. In this chapter, we aim to review the current use of negative ion mode MALDI-MS and discuss its potential in different fields, including infection and cancer diagnostics, toxicology, and food safety. We hope this chapter will show the readers a broad range of uses for negative ion mode in research and diagnostics and illustrate its complementarity to the current positive ion mode MALDI-MS to address the challenges of diagnostics, toxicology, and food fraud using lipids as markers.

15.2 Applications of Lipids in Clinical Microbiology Diagnostics

15.2.1 Use of Cell Envelope Lipids for Bacterial Identification

The bacterial cell envelope is extremely rich in molecules that could be used as biomarkers for identification. For example, in Gram-negative bacteria, the cell envelope is composed of a plasma membrane, a thin layer of peptidoglycan, and an outer membrane. Within the outer membrane, endotoxin or lipopolysaccharide (LPS) is found in the outer leaflet [7]. It is a glycolipid composed of three parts: lipid A, a hydrophobic section of the molecule responsible for some aspects of the toxicity of Gram-negative bacteria; the hydrophilic core oligosaccharide; and the O antigen, a polysaccharide comprising the outermost domain of LPS. The structure of LPS varies significantly among different species, with the O antigen being the most diverse part of the molecule [8]. The structure and composition of the O antigen can vary between strains of one species and have been used to assign serotypes of *Salmonella* and *Escherichia coli* [9, 10]. Casabuono et al. reported the structural characterization of lipid A from *Shigella flexneri* variant X, and discrete variants could serve to discriminate *Shigella* species from *E. coli* [11].

In addition, modifications to the lipid A portion of LPS can be specific to different bacterial species. The lipid A saccharide backbone of the majority of Gram-negative bacteria consists of two glucosamine units in beta $(1 \rightarrow 6)$ glycosidic linkages. Amino groups (at positions 2 and 2′) and hydroxy groups (at positions 3 and 3′) of glucosamines are commonly substituted by 3-hydroxy fatty acids, most often by 3-hydroxytetradecanoic acid. Other fatty acids (usually saturated, unbranched) are ester-linked to hydroxy acids by their hydroxy group. Phosphate groups can be observed at positions 1 and 5′ of the glucosamine units. Other common substituents of lipid A besides phosphate groups are commonly ethanolamine, beta-mannopyranose, 4-aminoarabinose, galacturonic acid, and glycine [7].

Additionally, for mycobacterial species, which include the human pathogen *Mycobacterium tuberculosis* and harbour a complex cell envelope architecture [12–14], surface-exposed lipids and glycolipids have been found to be useful markers for mycobacterial identification and discrimination without the need for specific sample preparation [15–17].

These molecules represent an interesting target for identification as they are detectable on the outer membrane and thus provide an easier extraction route than existing protein-based methods, which can involve mechanical lysis and protein enrichment steps [18]. The negative ion mode in MALDI time-of-flight (TOF) MS becomes a crucial tool for the development of novel bacterial identification protocols, given that many classes of lipids, such as cardiolipins, phospholipids, and glycolipids, ionize preferentially in the negative mode [5]. For example, the acylation and phosphorylation pattern of lipid A allows the study of its composition, role in toxicity, and even ability to distinguish between closely related bacterial species.

In this context, a range of different matrices have been investigated for their use in negative-mode MALDI-TOF. The strong base 1,8-bis(dimethylamino)naphthalene, which acts as a proton sponge, has proved to be a good matrix for lipid analysis, producing clear spectra with low background for a range of fatty acids from Gram-positive lactobacilli [19]. Another matrix, 5-chloro-2-mercaptobenzothiazole, gives a high-quality spectrum of pyrophosphate anions fragmented from diphosphorylated lipid A for many pathogenic Gram-negative bacteria, such as *Yersinia pestis*, *Pseudomonas aeruginosa*, and *Salmonella enterica* [20]. A matrix widely used for analysis in positive ion mode, 2,5-dihydroxybenzoic acid (DHB), also works in negative mode: MALDI-MS revealed hexaacylated, pentaacylated, and tetraacylated molecular species in the lipid A profile of Gram-negative *Bordetella hinzii*, consistent with data from plasma desorption MS and nuclear magnetic resonance analyses [21]. DHB has also been used to detect and discriminate mycobacteria [15–17]. Norharmane is a dual-polarity matrix ideal for hydrophobic molecules and, as such, has emerged as a potent tool in bacterial lipidomics research. MALDI profiling of *Francisella novicida* monophosphoryl lipid A using norharmane has shown a limit of detection at the picogram scale and even the ability to co-profile the host–pathogen interaction in infected tissue [22].

With the growing use of MALDI-TOF MS in the field of microbiology, profiles of lipids and glycolipids, such as lipid A and species-specific lipids, have already been established for a number of bacteria. While these continue to be explored, a focus has arisen on building databases and programs for identification based on existing data. The interest in publicly accessible databases come from the need for quick and reliable clinical diagnosis. Accordingly, researchers have developed a fast lipid analysis technique (FLAT) and a Python program for computational analysis that can shorten the bacterial identification time to less than an hour [23]. Their method uses norharmane as a matrix for negative-mode MALDI analysis, and the data obtained are comparable to those from microextraction, producing similar spectra, but FLAT is better at conserving phospholipids in the relevant mass range. Different pH conditions are needed for optimal lipid extraction from Gram-negative bacteria, and as it is not always possible to know which group the isolate belongs to, one preparation with and one without buffer are recommended. The Python script measures similarity between the isolate and known bacterial spectra by cosine similarity [23]. This technique has the potential to become a clinical assay delivered at the point of care and provides an additional benefit of being able to identify colistin resistance. Another research group has created an approach called the Lipid A Spectral Library (LASL) to identify bacterial species based on a comparison model to a library of known glycolipid mass spectra [24]. Compared with other identification approaches, such as Biotyper, cosine correlation, and intensity-weighted Euclidean distance, LASL correctly identified a higher number of bacteria at false discovery rates lower than 0.01. Additionally, the machine-learning model can be trained to pick out key ion peaks from the spectra. This means that the more MALDI-TOF data that are entered into the library, the better LASL will become at assigning lipid profiles to specific bacterial species [24]. A comprehensive library would compile spectra covering a range of possible lipid A modifications found within a species, providing an asset in the clinical diagnosis of human pathogens.

An important facet of bacterial identification based on MALDI-TOF lipid analysis is lipid extraction. Bacteria generally require different methods of lipid extraction due to their cell wall structure, which can delay the diagnosis time from sample collection to the confirmation of species ID. This can be remedied by a novel method that uses a short incubation with aqueous sodium acetate buffer for quicker extraction that retains a high amount of intact lipid A while preserving labile moieties [25]. The treated samples can then be analysed in negative ion mode MALDI-TOF using a norharmane matrix, resulting in an optimized identification process that takes less than an hour and works for fungal species as well as Gram-positive and Gram-negative bacteria. Taking the concept even further, a protocol to detect lipid A directly on the surface of Gram-negative bacteria has been developed. Overnight cultures of bacteria are simply heat-inactivated and washed in double-distilled water before being overlaid with DHB matrix for MALDI-MS in negative ion mode, thus eliminating the need for chemical extraction and purification steps [26]. This idea of lipid detection directly from inactivated or live intact cells is also useful for mycobacterial species. Their identification can be particularly difficult due to the extensive protocols necessary to extract and purify intracellular proteins [18, 27]. By contrast, the diverse lipid composition of the mycobacterial membrane presents a number of targets for species-specific identification. In fact, whole mycobacteria (from either solid or liquid culture) that are heat-killed, mixed with DHB, and analysed in both positive- and negative-mode MALDI-TOF produce spectra that enable discrimination between bacteria in the *M. tuberculosis* complex [17] and in the *Mycobacterium abscessus* complex [16]. Such discrimination is not possible using protein profiling in positive ion mode. A method that does not rely on growing bacterial culture would further accelerate the process from isolation to diagnosis for clinical microbiology applications. An approach using metal oxide laser ionization mass spectrometry imaging (MOLI-MSI) for detecting pathogens directly from tissue samples has been shown as a proof-of-concept [28]. MOLI is a variation of MALDI that uses a catalytic metal oxide to cleave fatty acid ions on surface lipids in the tissue sample being analysed. This method is capable of identifying microbes using fatty acid profiling and, coupled with the imaging aspect, even allows visualization of the spatial distribution of these microbes on tissue sections. Along with bacterial identification, determination of the drug susceptibility pattern of bacterial pathogens could provide added value to MALDI-MS in negative ion mode and could aid in speeding up diagnostics.

15.2.2 Detection of Cell Envelope Lipids and their Modifications to Determine Bacterial Drug Susceptibility

Among the cell envelope lipids and their modifications that can illuminate antimicrobial susceptibility, lipid A and its modifications can also be indicative of antibiotic resistance. The MALDI is a good diagnostic instrument to help determine which treatment options would be effective without the need for conventional minimum inhibitory concentration

(MIC) tests. Quick detection of bacterial susceptibility to polymyxins is of interest due to the rising number of strains that are resistant to this class of last-resort antibiotics [29]. Colistin, or polymyxin E, is used to treat infections caused by multidrug-resistant (MDR) Gram-negative bacteria, including species in the ESKAPEE group of pathogens (*Enterococcus faecium*, *Staphylococcus aureus*, *Klebsiella pneumoniae*, *Acinetobacter baumannii*, *P. aeruginosa*, ***E**nterobacter* species, and *E. coli*) [30, 31]. The ESKAPEE bacteria have been placed on a global watchlist because together they are currently the leading cause of hospital-acquired infections, and their MDR strains are responsible for life-threatening disease.

Resistance to colistin most commonly arises from chemical modifications of the lipid A portion of LPS and can be mediated through chromosomal mutations or by the activity of plasmid-encoded mobile colistin resistance (MCR) genes. These modifications include the addition of cationic groups such as 4-amino-4-deoxy-L-arabinose (L-Ara4N) or phosphoethanolamine (pETN), which reduce the binding of colistin to the positively charged LPS due to electrostatic repulsion [32]. Analysis of bacterial lipids using MALDI-TOF MS represents a valuable tool for detecting colistin resistance, as it can directly identify lipid modifications.

To address the need for quick clinical methods to detect whether bacteria are susceptible to colistin, researchers have developed the MALDIxin test. For this test, bacteria are cultured on Mueller–Hinton agar for 18–24 hours, after which a single colony is resuspended in distilled water, washed, incubated with 1% acetic acid for 10 minutes to 2 hours at 98 °C, and washed again to remove protein and salt contaminants before being mixed on the MALDI target plate with either DHB or norharmane matrix for analysis using negative ion mode ([33–35] [36, 37]). Colistin resistance is determined from the spectra by observing and quantifying the shift in peaks caused by lipid A modifications. The acquired mass spectra enable the differentiation of colistin-susceptible and colistin-resistant strains on the basis of the mass-to-charge ratio (m/z) of the major lipid A peaks. The addition of pETN to a phosphate group at position 1 of lipid A results in an m/z shift of +123 of the native lipid A-related peak, while L-Ara4N modification causes an m/z shift of +131. In contrast to *E. coli* and *K. pneumoniae,* for example, *A. baumannii* acquires resistance solely by the addition of pETN to native lipid A as this species does not encode any of the genes necessary for L-Ara4N biosynthesis [34].

The MALDIxin test can be optimized for a range of Gram-negative pathogens by changing the duration of the mild-acid hydrolysis step. In *P. aeruginosa*, resistance is identified through mass shifts of a set of peaks due to the addition of one or two L-Ara4N residues to the phosphate groups of native lipid A [38]. Colistin resistance due to plasmid-encoded *mcr* (mobile colistin resistance) has not been widely reported in *P. aeruginosa*, nor has the addition of pETN to lipid A. Resistant *K. pneumoniae* isolates produce spectra showing sets of peaks corresponding to the addition of pETN or L-Ara4N, or both [37]. Interestingly, the MALDIxin method was also able to distinguish between plasmid- and chromosome-encoded resistance as, in *K. pneumoniae*, resistance due to chromosomal determinants leads to the addition of L-Ara4N, whereas resistance due to *mcr* leads to the addition of pETN. The same conclusion is valid for *S. enterica* [36]. Similar to *K. pneumoniae* and *S. enterica*, the profiles for resistant *E. coli* show pETN-modified lipid A for strains where colistin resistance is mediated by MCR protein, and L-Ara4N-modified lipid A for strains where resistance is chromosomally encoded [35]. However, some chromosomal mutations in *E. coli* can also lead to the addition of pETN [39]. The MALDIxin test can identify whether colistin resistance is encoded on the *mcr-1* plasmid or on the bacterial chromosome, even when examining only pETN modifications [33]. Chromosome-encoded resistance shows mass spectra where the peak shift corresponds to a pETN residue added to the phosphate group at position 1 of native lipid A, and *mcr*-encoded resistance shows spectra with a smaller shift, indicating a pETN residue on the phosphate group at position 4′ of native lipid A together with a loss of phosphate group at position 1.

In addition to the applications to polymyxin resistance cited earlier, more generally, lipid profiling by MS systems in negative ion mode has the potential to be used for broad antibiotic susceptibility testing. For example, using MALDI coupled to high-resolution MS [Fourier transform ion cyclotron resonance (FTICR) MS], Schenk et al. reported on the analogous distribution of altered fatty acids and glycerophospholipids from *E. coli* strains upon exposure to sublethal concentrations of norfloxacin, a member of the class of fluoroquinolone antibiotics [40]. They concluded that variations in lipid content in the strain correspond to the extent of antibiotic resistance. Additionally, several papers report that antibiotic-resistant strains, such as rifampin-resistant *M. tuberculosis*, and daptomycin-resistant *Enterococcus faecalis*, display altered lipid compositions [41, 42]. By detecting these alterations and their relative abundances in negative ion mode, lipid profiling by MALDI-MS may be able to identify resistant microorganisms. However, this would require a large-scale analysis of the lipid composition, which is a highly complex task if done manually, and machine-learning techniques could support those investigations [43]. Taken together, these studies show that MALDI-TOF MS in negative ion mode has the potential to revolutionize the diagnosis of drug-resistant bacteria (Figure 15.1).

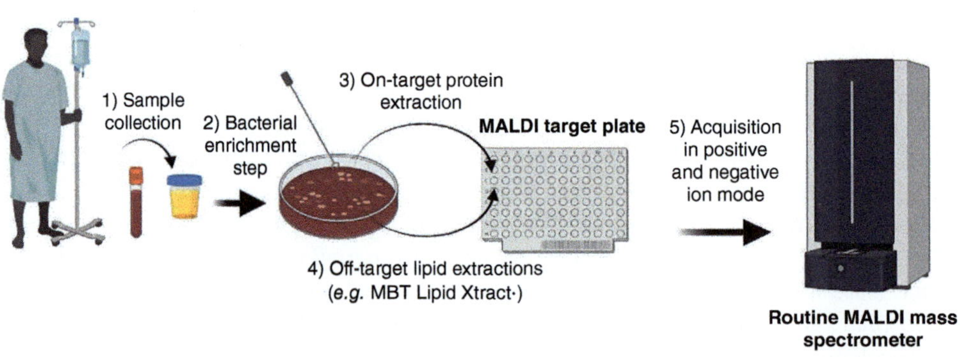

Figure 15.1 Proposed workflow using MALDI in positive and negative ion modes for bacterial samples in the clinic. Biological fluids or tissues are collected (1), and bacteria are enriched using dedicated agar Petri dishes (e.g. blood agar, Mueller–Hinton agar) (2). After overnight incubation at a defined temperature, one colony is directly deposited onto the MALDI target plate and overlaid with a suitable matrix for protein extraction and co-crystallization (3). In parallel, one colony is treated for lipid extraction and enriched using, for example, the MBT Lipid Xtract® kit from Bruker (4). The MALDI target plate is loaded in the routine MALDI Biotyper sirius system, and the data are acquired in positive ion mode for protein profiling and bacterial identification and in negative ion mode for bacterial identification if the positive ion mode data are inconclusive with regard to antimicrobial susceptibility (e.g. polymyxin resistance).

15.2.3 Detection of Lipids in MALDI Negative Ion Mode for Fungal Identification

Lipid profiling via MALDI-TOF MS can help to characterize yeast and fungal species, as well as distinguish between them. A study showed that membrane phospholipid composition was better at differentiating between closely related strains of yeast than protein biotyping alone [44]. Researchers analysed a number of yeast and fungi isolates using MALDI-TOF in positive ion mode with 6-aza-2-thiothymine (ATT) matrix and in negative ion mode with 9-aminoacridine (9-AA) matrix. The spectra from positive ion mode were generally of higher quality, particularly for filamentous fungi, but the negative mode spectra still demonstrated good intrastrain reproducibility. Yeast strains with complex phylogeny could be clustered into groups according to the different phosphatidylcholine (PC) and lyso-PC (LPC) species detected by MALDI-MS. Lipid phenotyping in negative ion mode could become a powerful tool for quickly identifying and differentiating among yeast strains and fungal species that are genetically very closely related.

Additionally, glycosyl-inositol-phosphoceramide (GIPC), which is a type of lipid found on the plasma membrane of fungi, is another good candidate for identification. The molecule has high polarity and is hard to extract using conventional methods, so a study developed a one-step sample preparation protocol for the detection of GIPCs for four fungal plant pathogens using MALDI-TOF in negative ion mode [45]. Prior to analysis, the filamentous fungi were heat-inactivated, washed, and mixed with a super-DHB matrix consisting of a 9 : 1 mixture of 2,5-dihydroxybenzoic acid and 2-hydroxy-5-methoxybenzoic acid solubilized at 10 mg/ml in chloroform–methanol (9 : 1 v/v). The profiles for all four common fungal plant pathogens, *Botrytis cinerea*, *Fusarium graminearum*, *Neurospora crassa*, and *Ustilago maydis*, showed a peak at *m*/*z* 924.6 [M-H]-, corresponding to an inositol-phosphoceramide composed of C18-phytoshingosine and 2-monohydroxylated-C24:0 fatty acids, and groups of peaks at higher masses, representing 1–4 additional hexose residues. The detection of GIPC species by MALDI-MS in negative ion mode and their relative abundances could be used to distinguish among pathogens.

15.2.4 Detection of Lipids in MALDI Negative Ion Mode for Parasite Identification

The spatial distribution of different biomolecules in tissue samples is an emerging field of research. In this context, negative ion mode in MALDI-imaging mass spectrometry (MALDI-IMS) has been used to analyse the full-body phospholipid composition of *Caenorhabditis elegans* [46]. Worms were freeze-cracked with liquid nitrogen to remove their cuticle exoskeleton, dried, and then coated with either DHB or 9-AA matrix for positive or negative ion mode, respectively. The authors used whole-body MALDI-IMS to analyse the fatty acid composition of PC and its spatial distribution in wild-type *C. elegans* compared with *fat-1* mutants, which lack an *n-3* fatty acid desaturase gene. Whereas positive ion mode produced PC profiles consistent with the liquid chromatography-electrospray ionization tandem MS (LC-ESI-MS/MS) data, negative ion mode revealed molecules, probably phosphatidylinositol species, that were detected nearly exclusively in either the wild-type or mutant. This study highlights the importance of analysis using negative ion mode in the whole-body MALDI-IMS technique to differentiate strains with distinct metabolic profiles. Charting the abundance of different biomolecules allows the comparison of lipidomics at the individual level, meaning that MALDI-IMS analysis of organisms with varied gene expression profiles can help to pave the way for an integrated omics approach to nematode research.

15.2.5 Detection of Lipids in MALDI Negative Ion Mode for Virus Identification

Some viruses, such as influenza and SARS-CoV-2, possess a lipid envelope [47, 48]. It is generally agreed that the composition of this membrane differs from that of the host, and it has been hypothesized that specific lipids could be necessary for viral function. This study used viruses known to infect bacterial and archaeal species to validate a novel technique for viral lipid identification [49]. Using a 9-AA matrix and MALDI-TOF in both positive and negative ion modes, the authors were able to generate spectra for the intact virus as well as subviral particles. The lipid profiles acquired in negative ion mode had a lower background, and for archaeal viruses these were the only profiles that contained significant peaks. Phosphatidylethanolamine (PE), PG, and CL were found to be the major viral envelope components for bacteriophages, in agreement with previous studies. Additionally, cardiolipin-like structures were identified for the first time in archaeal viruses. MALDI-TOF MS can become a valuable resource to bridge the gap in knowledge of virus lipidomics, and the method developed in this paper can be used to study the lipid dynamics that occur during infection. As this technique requires only minute amounts of intact virus (1×10^7 particle formation units), it is well suited for work with pathogenic human viruses, for which purification can be difficult and hazardous.

15.3 Applications of the Detection of Lipids in Negative Ion Mode MALDI-MS in Cancer Studies

Cancer is the second leading cause of death globally, and profiles of tumour cells, which can arise from different organs and tissues, show high heterogeneity. Accurate diagnosis of early-stage cancer and classification of cancer subtypes will greatly improve treatment outcomes and reduce therapy costs [50]. Changes in metabolomic and lipidomic profiles have been linked to cell morphology and the development of cancer [51–53], and the roles of phospholipid metabolism in carcinogenesis and tumour progression have been studied in-depth [54, 55].

MALDI-MS has been found to be efficient in profiling tumour samples in proteomics and lipidomics without extensive processing time or a large quantity of samples [56]. The recent development of MSI techniques, which generate mass spectra across the tissue sample section and visualize them as the intensity map of the targeted m/z, has enabled direct lipid profiling with information on the spatial distributions within samples and has become a powerful tool in cancer studies [6, 57, 58]. A general workflow for cancer studies with MALDI-MSI starts with sample collection from humans or animals, which need to be obtained rapidly, flash-frozen, and stored until sectioning. Tissue sections are thaw-mounted onto an indium tin oxide (ITO)-coated glass slide and layered with suitable matrix solution for MALDI-MSI screening. The differential ionization properties of phospholipids necessitate the use of negative ion mode for a better range of detection: e.g. PI, PS, PA, and

PG preferentially form deprotonated ions [6]. For data analysis, a plot of signal intensity versus *m*/*z* is generated and combined with microscopic data, e.g. images of haematoxylin & eosin (H&E) staining of the adjacent tissue session, to obtain the spatial distribution of the peak of interest with histological information. MS/MS is usually performed directly after MALDI-MSI to identify the peaks. Currently, MALDI-MS and MALDI-MSI are used mainly for the differentiation and subtyping of cancers by the detection of specific patterns of lipid profiles, while studies of drug–tumour interactions and therapeutic outcomes represent another potential application of these techniques.

15.3.1 Lipids and MALDI Negative Ion Mode for Diagnosis of Lung Cancer

Lung cancer is the leading cause of death among all cancers worldwide, with more than 2.26 million cases and 1.8 million deaths in 2020 [59]. It can be broadly grouped as non-small cell lung cancer (NSCLC) and small cell lung cancer (SCLC), which account for ~85% and ~15% of total diagnoses, respectively. Amongst NSCLCs, the two most common subtypes are adenocarcinomas and squamous cell carcinomas, which have different mutation profiles and corresponding therapeutic strategies [60]. Correct and rapid subtyping is therefore important for lung cancer diagnosis and the initiation of treatment.

Lee et al. [60] used lipid MALDI profiles in addition to histopathological diagnosis of NSCLC to distinguish between normal tissue and tumour cells and between adenocarcinoma and squamous cell carcinoma. Twenty-one pairs of retrospective surgical tissue samples (lung cancers and adjacent normal tissues) were included in this study, with two serial sections obtained for each sample, one for H&E histological staining and the other for MALDI-MS analysis with a binary matrix of 7 mg/ml each of DHB and α-cyano-4-hydroxycinnamic acid (CHCA) dissolved in 70% methanol with 1% piperidine and 0.1% trifluoroacetic acid (TFA) for maximal identification of phospholipids in both positive and negative ion modes. MS/MS analysis was also performed to confirm the phospholipids identified in MALDI-MS. A total of 144 features (78 in positive ion mode and 66 in negative ion mode) were identified, showing a significant difference between cancer and normal tissues. Cancer tissue samples were found to have 12 overexpressed and 32 underexpressed features compared with normal tissues, with PC 34:1 found in positive ion mode and confirmed by MS/MS. Negative ion mode identified 31 out of 32 underexpressed peaks in cancer samples that were not annotated. When comparing adenocarcinomas and squamous cell carcinomas, 11 peaks were found to be differentially expressed, 9 of which were overexpressed in adenocarcinomas. A sterol (ST-OH 42 : 1 [M-H]-) was identified by negative ion mode and confirmed by MS/MS to be overexpressed in adenocarcinomas.

In another study, He et al. detected cell type-specific lipid profiles in six different lung cancer cell lines, including adenocarcinoma, squamous cell carcinoma and large cell carcinoma [53]. MALDI-FTICR MS was performed in both positive and negative modes with DHB and 1,8-bis(dimethylamino)naphthalene (DMAN) as the matrices, respectively. A total of 43 lipids were identified by partial least-squares discriminant analysis (PLS-DA) with significant contributions to the differentiation of these cell lines, and 15 lipids from 6 categories [PE, PI, PC, sphingophospholipid (SPLs), PA, fatty acid (FA)] were further identified by collision-induced dissociation experiments. In negative ion mode, three PE species, five FA species and one PI were identified with differential abundances between cell lines. Based on the differential patterns of lipid profiles in these cell lines, the authors suggested using lipid profiles, e.g. a cluster of PA 36:2, PA 38:3, and sphingomyelin (SM) 34:1 for lung cancer cell line classifications.

15.3.2 Lipids and MALDI Negative Ion Mode for the Diagnosis of Breast Cancer

Breast cancer is the second most common cancer following lung cancer and accounts for ~30% of cancers in females [61]. It is highly heterogeneous, and clinical subtyping is based mainly on the status of hormone receptors, namely, oestrogen receptor (ER) and progesterone receptor (PR), together with human epidermal growth factor 2 (HER2) [62]. These subgroups of breast cancer show distinct lipid profiles, and the use of MS in lipidomic studies improves our understanding of disease progression and serves as a potential diagnostic tool [63].

Silva et al. [64] used MALDI-MS to compare lipid profiles in breast active carcinoma (AC) and breast cancer-free (CF) samples from 28 patients. To overcome the ion suppression effects caused by the abundant and easily ionized PC species and obtain better resolution of less common phospholipid species, the authors applied two different matrices, DHB (10 mg/ml in methanol) and 9-AA (10 mg/ml in acetone), in both positive and negative modes. Despite showing higher numbers of signals, spectra obtained with DHB, as the matrix did not show differences between AC and CF samples, and the authors used spectra with 9-AA as the matrix for major analyses. Significant differences between the ratios of Lyso-PC 16:0 and PC 16:0/18:2 were identified between AC and CF samples and between samples of different breast cancer stages (II and III). These findings may indicate increases in the activity of phospholipase A2 (PLA2), which cleaves PC to produce LPC and was previously reported to be linked with cancer progression [65]. In negative ion mode, PI species could be detected, and the profiles showed significant differences between the AC and CF samples. For example, PI 16:0/18:1 was detected in AC but not in CF samples. At breast cancer stage I, there was no significant difference in PI profiles between AC and CF samples, while in samples of breast cancer stages II and III, the differences between AC and CF were more profound. Some negatively charged ions at higher m/z were also detected and were suspected to be phosphorylated PI species, but further confirmation is required, together with studies on enzyme activities related to PI metabolism.

In another study, Kang et al. [66] used histology-directed MALDI-MS to obtain proteomic and lipidomic profiles of breast cancer and adjacent normal tissue samples for subtyping. Thirty-four pairs of breast cancer and adjacent normal tissue samples were first classified by histology into three classes (luminal, HER2+, and triple negative), followed by protein and lipid MALDI-MS analysis in both positive and negative ion modes, with sinapinic acid (20 mg/ml in 50 : 50 acetonitrile: 0.1% TFA) and binary CHCA + DHB (7 mg/ml each in 70% methanol with 0.1% TFA and 1% piperidine) as the matrices for protein and lipid MALDI, respectively. Eighty-five proteins and 144 lipid features (78 in positive ion mode and 66 in negative ion mode) were identified for all samples studied, and negative ion mode revealed 15 peaks in lipid MALDI analysis that showed differential distribution with a feature selection $P < 10^{-5}$, although they were not annotated in this study. Class prediction analysis was able to differentiate cancer and normal tissues with 94.1% accuracy at a feature selection $P < 0.001$, suggesting clear differences in protein and lipid profiles between normal and cancer cells. Similar to other cancer types discussed earlier, PC 34:1 was found to be overexpressed in cancer, which can also be linked to changes in the levels of 18:1/16:0 PC species, as discussed by Silva et al. [64]. Breast cancer intrinsic types also have different protein and lipid profiles, as found by this study, and in random test sets, a median prediction accuracy of 80.0–81.3% was achieved in the classification of these subtypes, although the test was limited by small sample numbers ($n = 34$). Another limitation of this study was incomplete annotations of peaks, especially those identified in negative ion mode.

15.3.3 Lipids and MALDI Negative Ion Mode for Diagnosis of Other Cancers

Similar to the studies discussed above, MALDI-MS technologies using negative ion mode can be applied to different cancers for rapid diagnosis and accurate subtyping. This could be especially important for cases of rare cancers, as the limited understanding of specific biomarkers for rare cancers limits the efficiency of conventional diagnostic methods, and quick identification with high accuracy that differentiates these cases from other more common cancers is necessary for timely treatment and prognosis. One example of such studies is discussed in the following.

There is a need for biomarkers for cholangiocarcinoma (bile duct cancer), which accounts for ~10% of all primary liver cancer cases and shows increasing incidence. Moreover, differentiation between primary liver cancer and those originating from other sites, including the lung, pancreas, breast, and gastrointestinal tract, is also important for proper diagnosis. Park et al. [58] used MALDI-MSI to investigate whether cholangiocarcinomas could be differentiated from normal tissue and other types of cancers by lipid profiles. Four pairs of retrospective cholangiocarcinoma samples and adjacent normal tissue samples were used in this study for cholangiocarcinoma-specific lipid profile identification in addition to conventional histopathological staining methods. A binary matrix composed of DHB and CHCA (7 mg/ml each dissolved in 70% methanol plus 0.1% TFA and 1% piperidine) was used for MALDI-TOF analysis in both positive and negative modes, and MALDI-MS/MS analysis was performed as confirmation. Similar to other studies, PC 34:1 was found to be increased in cancer samples compared with normal tissue. Although negative ion mode analysis was applied and five peaks (one overexpressed and four underexpressed in cholangiocarcinoma tissues) were found at a feature selection $P < 0.05$, they were not confirmed by MS/MS. The study also investigated differences in lipid profiles between cholangiocarcinomas and cancers derived from other organs (pancreatic cancer and colorectal cancer). Cholangiocarcinomas could be differentiated from colorectal cancer with a prediction accuracy of 99.7–100%, whereas 48.1–55.8% prediction accuracy was achieved for separation between cholangiocarcinomas and pancreatic cancers, suggesting fewer differences in the lipid profiles of these two cancer types.

Taken together, MALDI-MS in negative ion mode combined with positive ion mode could provide an interesting avenue for the rapid detection of cancer and its stages and could generate hypotheses regarding the role of lipids in cancer development and drug interactions.

15.3.4 Lipids and MALDI Negative Ion Mode for Drug–Cell Interactions and Prognosis

Predicting tumour response to therapeutics can help in choosing the correct treatment regimen for a specific case of cancer for better outcome and can direct the development of new drugs by identifying potential biomarkers and pathways. By identifying phospholipid signatures of extracellular vesicles (EVs) secreted by NSCLC cells by MALDI-MS, Jung et al. aimed to correlate lipid profiles with cellular resistance to gefitinib, an epidermal growth factor receptor (EGFR) tyrosine kinase inhibitor (TKI) and one of the standard treatments for NSCLC [67]. EV phospholipid profiles of gefitinib-resistant cells showed distinctive characteristics compared with sensitive cells, with 20 lipids in classes of PC, SM and LPC identified in positive ion mode and 15 in negative ion mode, including PI, PG and Lyso-PI. The authors suggested that further evaluative studies could be performed on these lipid profiles to identify levels of gefitinib resistance and to predict the treatment outcome of NSCLC patients.

In another study, Chen et al. [68] used a three-dimensional multicellular tumour spheroid (MCTS) model of the A549 lung cancer cell line to mimic tumour microenvironments and studied its response to hydroxychloroquine, an anti-malaria drug that showed effectiveness in various tumour models. A combination of MALDI-MSI in positive and negative ion modes, LC-MS, and H&E histological staining revealed a time-dependent penetration gradient of hydroxychloroquine into MCTS and spatial distribution of lipid species upon drug treatment. Changes in lipidomic profiles after treatment, including upregulation of PC, PE, and PI and downregulation of PG and TG species, were identified. In negative ion mode, differential distributions of PE, PS, and PI species were found, which complemented the data from positive ion mode. This study serves as an example of how MALDI-MSI could be applied to explore cellular drug response in a complex cancer microenvironment.

15.4 Applications of the Detection of Lipids and MALDI-MS in Alzheimer's Disease Studies

Alzheimer's disease (AD) is a progressive neurodegenerative disorder and the leading cause of dementia worldwide [69, 70]. The roles of lipids in the neuropathology and progression of AD have been extensively studied, such as the roles of short-chain fatty acids in AD pathogenesis, including the modulation of neuroinflammation, the maintenance of the blood–brain barrier and the regulation of brain metabolism [71]; the links between the abundances and spatial distribution of CLs and neuronal dysfunction and diseases [72]; and the ways in which sphingolipids and other bioactive lipids affect neuroinflammation [73, 74]. Lipidomic studies of these lipid species and their spatial distributions will enhance our understanding of AD mechanisms and promote the development of therapeutics, and MALDI-MS techniques, especially MALDI-MSI, are widely used for these purposes.

Cholinergic neurotransmission deficiencies, which are caused by loss of basal forebrain cholinergic neurons (BFCNs) and impaired functions of choline metabolism, leading to deficiencies in short-term learning and memory loss, have been described in AD [69].

Llorente-Ovejero et al. [69] studied the involvement of membrane phospholipid modulations in cholinergic neurotransmission deficiencies in a mouse model of AD with loss of BFCNs. Positive and negative MALDI-MSI was performed with sublimed 2-mercaptobenzothiazole (MBT) as a matrix on mouse brain tissue sections to localize changes in lipid species induced by lesions of BFCNs. The result was coupled with acetylcholine staining and autoradiography measuring activities of muscarinic acetylcholine receptor (mAchR). In addition to various PC, PE, and SM species detected in positive ion mode, negative ion mode detected additional SM, PE, and PS species that showed variations in levels, expanding the range of lipid profiling. Altogether, these results indicated a possible direction for future AD therapeutics: interfering with specific lipid metabolism and mAchR activities to potentiate cholinergic neurotransmission.

The roles of neurolipid-based signalling by the endocannabinoids lysophosphatidic acid and sphingosine 1-phosphate in lipid regulation were studied by Gonzalez de San Roman et al. [70], with MALDI-MSI used to quantify the intensities of lipid species in different brain regions from a triple transgenic mouse model that mimics AD development. In this study, SM and PI species were detected in negative ion mode but not in positive mode, and the combination of both positive and negative ion mode data achieved a larger coverage of changes in lipid profiles. In another study by Zhang et al. [75], MALDI-MSI and desorption electrospray ionization (DESI)-MSI were used to study the metabolism and spatial distribution of various brain lipid species in an early-stage AD mouse model. These two MSI techniques complemented each other, resulting in a wider range of peak detection. In conclusion, the integration of MALDI-MS with other methods, such as microscopy, metabolomic pathway studies, genomics, and other imaging techniques, could help to expand the detection range and validation of lipid species and enhance our understanding of the complex biological events in AD.

15.5 Applications of MALDI in Negative Ion Mode and the Detection of Lipids in Toxicology

Understanding the side-effects of drugs and selecting those with less toxicity is an important aspect in drug discovery and validations. The currently available assays for toxicity measurements are based on measurements of cellular energetic status and can be time-consuming and expensive [76]. MALDI-MS technologies have been applied in toxicology studies by probing changes in the cell lipidome as a result of altered energetic status and metabolism, without extensive separation and extraction steps. In this case, the detection of lipid species, including PI, PS, PG, PE, and CL, in negative ion mode could maximize the range of detection and contribute to rapid and accurate profiling of the cellular lipidome. For example, Rebollo-Ramirez et al. [76] studied changes in lipid profiles in HeLa cells and mouse bone marrow-derived macrophage (BMDM) cells after treatment with the aminoglycoside kanamycin. In addition to conventional bioenergetics studies, including the Seahorse XF Cell Mito Stress assay and membrane potential assay, an intact cell lipidomic study using negative MALDI-MS was performed by directly loading the cells treated or untreated with the antibiotics on the plate and adding DHB as a matrix. The authors found an altered ratio between the abundances of PI and CL species in HeLa cells after kanamycin treatment, which could be the result of changes in bioenergetics and fatty acid metabolism in these cells upon exposure to kanamycin. Although later experiments suggested no direct correlation between mitochondrial respiratory capacity and CL : PI ratio, this rapid methodology could be applied to additional cell types and purposes, acting as a screening tool to visualize any changes in the cell lipidome before further comprehensive studies.

Understanding drug–host interactions also helps us to better understand drug mechanisms and exploit new therapeutic potentials. In another study, by Giraud-Gatineau et al., the antimycobacterial antibiotic bedaquiline (BDQ) was found to interact with the host immune system and reprogramme macrophages to exert higher bactericidal activities against bacterial infections [77]. MALDI-MS was used to assess changes in lipid profiles after BDQ treatment, and an increase in PI abundance was identified. Together with RNA-sequencing and bioenergetic studies, these results confirmed metabolomic changes in macrophages in response to BDQ treatment as the basis of bactericidal reprogramming. The authors suggested that BDQ could be used as a host-directed therapy to enhance the host immune response against different bacterial infections. Taken together, MALDI-MS in negative ion mode could provide information on compound toxicity and impact cell bioenergetics and lipid composition. That methodology could, for example, be valuable in drug screening and save time in lead optimization.

15.6 Lipids and MALDI Negative Ion Mode for Food Fraud Detection

Food fraud generally refers to intentional violations of food industry regulations and the deception of customers regarding the quality, ingredients, origin, or production process of food [78]. It is mainly driven by economic profits and global competition amongst manufacturers and can lead to safety issues, including allergies and food-related illnesses [78]. MALDI-MS has been found to be a valuable tool for the analysis of food adulteration because of its high sensitivity and simple protocol for sample processing [79]. Studies using MALDI-MS techniques for food fraud detection have been rising in recent years, including methods to detect the adulteration of dairy products [80–82] and cooking oils [83–85].

Despite the numerous studies in the field with positive ion mode MALDI-MS over recent years, negative ion mode has been used less for this purpose and should be explored. Zhang et al. used MALDI-MSI in both positive and negative ion modes to study phospholipid distributions in rice [86]. Phospholipids are micronutrients of rice but are of great significance in its nutritional value. The hydrolysis of phospholipids into free fatty acids and subsequent oxidation are affected by the storage conditions of rice, leading to changes in flavour and nutritional values. In this study, tissue sections of white rice and brown rice were obtained for analysis, and sublimed DHB, which was deposited onto sample sections by a sublimation spraying procedure (180 °C, 1.5 μm thickness), was selected as the matrix after a series of method optimization trials. MALDI-MSI identified 12 phospholipid species belonging to PC and LPC, which also showed characteristic spatial distributions in different tissues of rice grain (bran, endosperm, and embryo). Differences in phospholipid profiles between white and brown rice also suggested a correlation between phospholipid distribution and degree of processing, which could be used as indicators of potential adulterations when suppliers process aged rice to sell as new rice for a higher profit [86].

Another study by Enomoto et al. [87] analysed the spatial distribution of PI species in different pork tissues by MALDI-MSI. The abundance of PI species in different foods varies from trace (in duck meat) to more than 40% (in sea bass eggs) and is a major source of polyunsaturated fatty acids and animal food-derived inositol, with various health benefits [87, 88]. The spatial distribution of PI species is less well understood. In this study, pork sections from four different tissues were prepared with sublimed 9-AA deposited by a vacuum deposition system as a matrix for MALDI-MSI in negative ion mode, and *m/z* values of 850–920 were measured. PI species were identified by MS/MS and by LC-ESI-MS of the total lipid extract of the same tissue. Twelve PI species were identified by MALDI-MS/MS and LC-ESI-MS/MS, and their concentrations in each tissue were compared to map a spatial distribution in different pork tissues. Compared with the LC-MS method, which requires extensive lipid extraction and sample preparation steps, the MALDI-MSI method showed advantages in the rapid and direct visualization of the spatial distributions of phospholipids in food samples and could illuminate their nutritional values [87]. The addition of negative ion mode studies to the current MALDI-MS workflow for food adulteration detection and quality checks could expand the range of detectable nutrients and potentially facilitate the identification of new markers for food fraud.

15.7 Conclusions and Future Development of Lipids and their Detection in MALDI in Negative Ion Mode

Over recent decades, routine MALDI-MS in positive ion mode has proved its considerable usefulness in a wide range of topics. However, relying solely on this mode of ionization considerably reduces the number of biomarkers that could be identified and therefore reduces the scope of the use of MALDI and its range of applications. That is why, along with positive ion mode, MALDI-MS in negative ion mode has the potential not only to complement the data and information generated by the positive ion mode but also to address some of the challenges that cannot be addressed by the acquisition of positive ion mode spectra alone. In addition to the detection of proteins in positive ion mode, lipids and glycolipids were of particular interest among the molecules that can be detected in the negative ion mode, along with appropriate matrices, as shown by the examples presented in this chapter (Table 15.1). Indeed, lipid profiling in negative ion mode has the potential to be employed alongside protein fingerprinting and could help to address some of the current limits of MALDI-MS, thereby further broadening its utility in diagnostics and other routine applications. Such large-scale datasets should generate new databases and new data analysis workflows that use machine-learning algorithms to facilitate the interpretation and integration of the data for rapid and effective decision-making.

Acknowledgments

This work was supported by the MRC Confidence in Concept Fund and an ISSF Wellcome Trust grant (105603/Z/14/Z, to GL-M).

Table 15.1 MALDI matrices and their applications in negative ion mode.

Matrix	Structure	Solvents	Wavelengths (nm)	Applications	Key References
2,5-Dihydroxybenzoic acid (DHB)	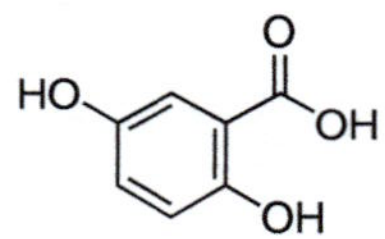	Ethanol 25% Chloroform/methanol 9 : 1 v/v Sublimation	337, 355, 266	Lipid A analysis; toxicology analysis; food fraud analysis	[23, 33, 34, 76, 77]
9-Aminoacridine (9-AA)	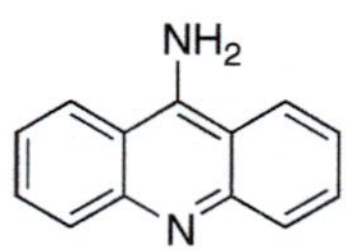	2-Propanol/acetonitrile, 60/40, v/v Acetone Sublimation	337, 355	Analysis of major fungal lipid classes, viral lipid envelopes, and intact mitochondria; cancer classification; tumour-drug interactions; food fraud analysis	[44] [49, 64, 67, 87]
1,8-bis(Dimethylamino) naphthalene (DMAN)	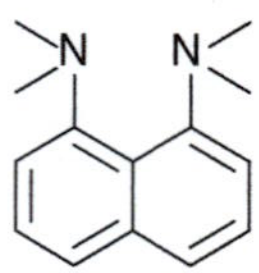	Chloroform/methanol, 1 : 1, v/v Sublimation	337, 355	Analysis of free fatty acids, glycolipids; phosphatidylglycerols, and cardiolipins; lung cancer subtyping	[19, 53]
5-Chloro-2-mercaptobenzothiazole	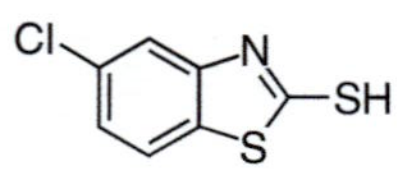	Chloroform/methanol/water, 4 : 4 : 1, v/v/v Sublimation	337	Lipid A analysis; AD studies (brain tissue section MSI)	[20, 69]
Norharmane	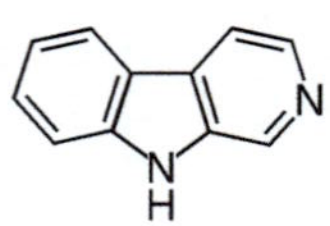	Chloroform/methanol, 9 : 1, v/v Chloroform/methanol, 2 : 1, v/v Chloroform/methanol/water, 12 : 6 : 1, v/v/v	337	Lipid A analysis and cardiolipins	[22, 23, 25, 38] [89]
DHB + CHCA		70% MeOH +1% piperidine + 0.1% TFA	355	Cancer classification	[58, 60, 66]

References

1 Chivte, P., LaCasse, Z., Seethi, V.D.R. et al. (2021). MALDI-ToF protein profiling as a potential rapid diagnostic platform for COVID-19. *J. Mass Spectrom. Adv. Clin. Lab.* 21: 31–41. https://doi.org/10.1016/j.jmsacl.2021.09.001.

2 Croxatto, A., Prod'hom, G., and Greub, G. (2012). Applications of MALDI-TOF mass spectrometry in clinical diagnostic microbiology. *FEMS Microbiol. Rev.* 36 (2): 380–407. https://doi.org/10.1111/j.1574-6976.2011.00298.x.

3 Kostrzewa, M. (2018). Application of the MALDI Biotyper to clinical microbiology: progress and potential. *Expert Rev. Proteomics* 15 (3): 193–202. https://doi.org/10.1080/14789450.2018.1438193.

4 Singhal, N., Kumar, M., Kanaujia, P.K., and Virdi, J.S. (2015). MALDI-TOF mass spectrometry: an emerging technology for microbial identification and diagnosis. *Front. Microbiol.* 6: 791. https://doi.org/10.3389/fmicb.2015.00791.

5 Solntceva, V., Kostrzewa, M., and Larrouy-Maumus, G. (2020). Detection of species-specific lipids by routine MALDI TOF mass spectrometry to unlock the challenges of microbial identification and antimicrobial susceptibility testing. *Front. Cell. Infect. Microbiol.* 10: 621452. https://doi.org/10.3389/fcimb.2020.621452.

6 Mamun, A., Islam, A., Eto, F. et al. (2020). Mass spectrometry-based phospholipid imaging: methods and findings. *Expert Rev. Proteomics* 17 (11–12): 843–854. https://doi.org/10.1080/14789450.2020.1880897.

7 Raetz, C.R. and Whitfield, C. (2002). Lipopolysaccharide endotoxins. *Annu. Rev. Biochem.* 71: 635–700. https://doi.org/10.1146/annurev.biochem.71.110601.135414.

8 Lerouge, I. and Vanderleyden, J. (2002). O-antigen structural variation: mechanisms and possible roles in animal/plant-microbe interactions. *FEMS Microbiol. Rev.* 26 (1): 17–47. https://doi.org/10.1111/j.1574-6976.2002.tb00597.x.

9 Liu, B., Knirel, Y.A., Feng, L. et al. (2014). Structural diversity in *Salmonella* O antigens and its genetic basis. *FEMS Microbiol. Rev.* 38 (1): 56–89. https://doi.org/10.1111/1574-6976.12034.

10 Orskov, F. and Orskov, I. (1992). *Escherichia coli* serotyping and disease in man and animals. *Can. J. Microbiol.* 38 (7): 699–704.

11 Casabuono, A.C., van der Ploeg, C.A., Roge, A.D. et al. (2012). Characterization of lipid A profiles from *Shigella flexneri* variant X lipopolysaccharide. *Rapid Commun. Mass Spectrom.* 26 (17): 2011–2020. https://doi.org/10.1002/rcm.6306.

12 Brennan, P.J. (2003). Structure, function, and biogenesis of the cell wall of *Mycobacterium tuberculosis*. *Tuberculosis (Edinb.)* 83 (1–3): 91–97.

13 Dulberger, C.L., Rubin, E.J., and Boutte, C.C. (2020). The mycobacterial cell envelope – a moving target. *Nat. Rev. Microbiol.* 18 (1): 47–59. https://doi.org/10.1038/s41579-019-0273-7.

14 Vincent, A.T., Nyongesa, S., Morneau, I. et al. (2018). The mycobacterial cell envelope: a relict from the past or the result of recent evolution? *Front. Microbiol.* 9: 2341. https://doi.org/10.3389/fmicb.2018.02341.

15 Gonzalo, X., Broda, A., Drobniewski, F., and Larrouy-Maumus, G. (2020). Performance of lipid fingerprint-based MALDI-ToF for the diagnosis of mycobacterial infections. *Clin. Microbiol. Infect.* https://doi.org/10.1016/j.cmi.2020.08.027.

16 Jia Khor, M., Broda, A., Kostrzewa, M. et al. (2021). An improved method for rapid detection of *Mycobacterium abscessus* complex based on species-specific lipid fingerprint by routine MALDI-TOF. *Front. Chem.* 9: 715890. https://doi.org/10.3389/fchem.2021.715890.

17 Larrouy-Maumus, G. and Puzo, G. (2015). Mycobacterial envelope lipids fingerprint from direct MALDI-TOF MS analysis of intact bacilli. *Tuberculosis (Edinb.)* 95 (1): 75–85. https://doi.org/10.1016/j.tube.2014.11.001.

18 Mather, C.A., Rivera, S.F., and Butler-Wu, S.M. (2014). Comparison of the Bruker Biotyper and Vitek MS matrix-assisted laser desorption ionization-time of flight mass spectrometry systems for identification of mycobacteria using simplified protein extraction protocols. *J. Clin. Microbiol.* 52 (1): 130–138. https://doi.org/10.1128/JCM.01996-13.

19 Calvano, C.D., Zambonin, C.G., and Palmisano, F. (2011). Lipid fingerprinting of Gram-positive lactobacilli by intact-matrix-assisted laser desorption/ionization mass spectrometry using a proton sponge based matrix. *Rapid Commun. Mass Spectrom.* 25 (12): 1757–1764. https://doi.org/10.1002/rcm.5035.

20 Jones, J.W., Shaffer, S.A., Ernst, R.K. et al. (2008). Determination of pyrophosphorylated forms of lipid A in Gram-negative bacteria using a multivaried mass spectrometric approach. *Proc. Natl. Acad. Sci. U. S. A.* 105 (35): 12742–12747. https://doi.org/10.1073/pnas.0800445105.

21 Aussel, L., Brisson, J.R., Perry, M.B., and Caroff, M. (2000). Structure of the lipid A of *Bordetella hinzii* ATCC 51730. *Rapid Commun. Mass Spectrom.* 14 (7): 595–599. https://doi.org/10.1002/(SICI)1097-0231(20000415)14:7<595::AID-RCM919>3.0.CO;2-4.

22 Scott, A.J., Flinders, B., Cappell, J. et al. (2016). Norharmane matrix enhances detection of endotoxin by MALDI-MS for simultaneous profiling of pathogen, host, and vector systems. *Pathog. Dis.* 74 (8): https://doi.org/10.1093/femspd/ftw097.

23 Sorensen, M., Chandler, C.E., Gardner, F.M. et al. (2020). Rapid microbial identification and colistin resistance detection via MALDI-TOF MS using a novel on-target extraction of membrane lipids. *Sci. Rep.* 10 (1): 21536. https://doi.org/10.1038/s41598-020-78401-3.

24 Ryu, S.Y., Wendt, G.A., Chandler, C.E. et al. (2019). Model-based spectral library approach for bacterial identification via membrane glycolipids. *Anal. Chem.* 91 (17): 11482–11487. https://doi.org/10.1021/acs.analchem.9b03340.

25 Liang, T., Leung, L.M., Opene, B. et al. (2019). Rapid microbial identification and antibiotic resistance detection by mass spectrometric analysis of membrane lipids. *Anal. Chem.* 91 (2): 1286–1294. https://doi.org/10.1021/acs.analchem.8b02611.

26 Larrouy-Maumus, G., Clements, A., Filloux, A. et al. (2016). Direct detection of lipid A on intact gram-negative bacteria by MALDI-TOF mass spectrometry. *J. Microbiol. Methods* 120: 68–71. https://doi.org/10.1016/j.mimet.2015.12.004.

27 Kostrzewa, M., Nagy, E., Schrottner, P., and Pranada, A.B. (2019). How MALDI-TOF mass spectrometry can aid the diagnosis of hard-to-identify pathogenic bacteria – the rare and the unknown. *Expert Rev. Mol. Diagn.* 19 (8): 667–682. https://doi.org/10.1080/14737159.2019.1643238.

28 Basu, S.S., McMinn, M.H., Gimenez-Cassina Lopez, B. et al. (2019). Metal oxide laser ionization mass spectrometry imaging (MOLI MSI) using cerium(IV) oxide. *Anal. Chem.* 91 (10): 6800–6807. https://doi.org/10.1021/acs.analchem.9b00894.

29 Nordmann, P., Jayol, A., and Poirel, L. (2016). Rapid detection of polymyxin resistance in Enterobacteriaceae. *Emerg. Infect. Dis.* 22 (6): 1038–1043. https://doi.org/10.3201/eid2206.151840.

30 Furniss, R.C.D., Dortet, L., Bolland, W. et al. (2019). Detection of Colistin resistance in *Escherichia coli* by use of the MALDI Biotyper Sirius mass spectrometry system. *J. Clin. Microbiol.* 57 (12): https://doi.org/10.1128/JCM.01427-19.

31 Poirel, L., Jayol, A., and Nordmann, P. (2017). Polymyxins: antibacterial activity, susceptibility testing, and resistance mechanisms encoded by plasmids or chromosomes. *Clin. Microbiol. Rev.* 30 (2): 557–596. https://doi.org/10.1128/CMR.00064-16.

32 Jeannot, K., Bolard, A., and Plesiat, P. (2017). Resistance to polymyxins in Gram-negative organisms. *Int. J. Antimicrob. Agents* 49 (5): 526–535. https://doi.org/10.1016/j.ijantimicag.2016.11.029.

33 Dortet, L., Bonnin, R.A., Pennisi, I. et al. (2018a). Rapid detection and discrimination of chromosome- and MCR-plasmid-mediated resistance to polymyxins by MALDI-TOF MS in *Escherichia coli*: the MALDIxin test. *J. Antimicrob. Chemother.* 73 (12): 3359–3367. https://doi.org/10.1093/jac/dky330.

34 Dortet, L., Potron, A., Bonnin, R.A. et al. (2018b). Rapid detection of colistin resistance in *Acinetobacter baumannii* using MALDI-TOF-based lipidomics on intact bacteria. *Sci. Rep.* 8 (1): 16910. https://doi.org/10.1038/s41598-018-35041-y.

35 Furniss, R.C.D., Dortet, L., Bolland, W. et al. (2019). Detection of colistin resistance in *Escherichia coli* using the MALDI Biotyper Sirius mass spectrometry system. *J. Clin. Microbiol.* https://doi.org/10.1128/JCM.01427-19.

36 Dortet, L., Bonnin, R.A., Le Hello, S. et al. (2020a). Detection of Colistin resistance in *Salmonella enterica* using MALDIxin test on the routine MALDI Biotyper Sirius mass spectrometer. *Front. Microbiol.* 11: 1141. https://doi.org/10.3389/fmicb.2020.01141.

37 Dortet, L., Broda, A., Bernabeu, S. et al. (2020b). Optimization of the MALDIxin test for the rapid identification of colistin resistance in *Klebsiella pneumoniae* using MALDI-TOF MS. *J. Antimicrob. Chemother.* 75 (1): 110–116. https://doi.org/10.1093/jac/dkz405.

38 Jeannot, K., Hagart, K., Dortet, L. et al. (2021). Detection of Colistin resistance in *Pseudomonas aeruginosa* using the MALDIxin test on the routine MALDI Biotyper Sirius mass spectrometer. *Front. Microbiol.* 12: 725383. https://doi.org/10.3389/fmicb.2021.725383.

39 Janssen, A.B. and van Schaik, W. (2021). Harder, better, faster, stronger: Colistin resistance mechanisms in *Escherichia coli*. *PLoS Genet.* 17 (1): e1009262. https://doi.org/10.1371/journal.pgen.1009262.

40 Schenk, E.R., Nau, F., Thompson, C.J. et al. (2015). Changes in lipid distribution in *E. coli* strains in response to norfloxacin. *J. Mass Spectrom.* 50 (1): 88–94. https://doi.org/10.1002/jms.3500.

41 Lahiri, N., Shah, R.R., Layre, E. et al. (2016). Rifampin resistance mutations are associated with broad chemical remodeling of *Mycobacterium tuberculosis*. *J. Biol. Chem.* 291 (27): 14248–14256. https://doi.org/10.1074/jbc.M116.716704.

42 Mishra, N.N., Bayer, A.S., Tran, T.T. et al. (2012). Daptomycin resistance in enterococci is associated with distinct alterations of cell membrane phospholipid content. *PLoS One* 7 (8): e43958. https://doi.org/10.1371/journal.pone.0043958.

43 Weis, C.V., Jutzeler, C.R., and Borgwardt, K. (2020). Machine learning for microbial identification and antimicrobial susceptibility testing on MALDI-TOF mass spectra: a systematic review. *Clin. Microbiol. Infect.* 26 (10): 1310–1317. https://doi.org/10.1016/j.cmi.2020.03.014.

44 Stubiger, G., Wuczkowski, M., Mancera, L. et al. (2016). Characterization of yeasts and filamentous fungi using MALDI lipid phenotyping. *J. Microbiol. Methods* 130: 27–37. https://doi.org/10.1016/j.mimet.2016.08.010.

45 Saromi, K., England, P., Tang, W. et al. (2020). Rapid glycosyl-inositol-phospho-ceramides fingerprint from filamentous fungi pathogens using the MALDI Biotyper Sirius system. *Rapid Commun. Mass Spectrom.* e8904. https://doi.org/10.1002/rcm.8904.

46 Hameed, S., Ikegami, K., Sugiyama, E. et al. (2015). Direct profiling of the phospholipid composition of adult *Caenorhabditis elegans* using whole-body imaging mass spectrometry. *Anal. Bioanal. Chem.* 407 (25): 7589–7602. https://doi.org/10.1007/s00216-015-8932-7.

47 Ivanova, P.T., Myers, D.S., Milne, S.B. et al. (2015). Lipid composition of viral envelope of three strains of influenza virus – not all viruses are created equal. *ACS Infect. Dis.* 1 (9): 399–452. https://doi.org/10.1021/acsinfecdis.5b00040.

48 Theken, K.N., Tang, S.Y., Sengupta, S., and FitzGerald, G.A. (2021). The roles of lipids in SARS-CoV-2 viral replication and the host immune response. *J. Lipid Res.* 62: 100129. https://doi.org/10.1016/j.jlr.2021.100129.

49 Vitale, R., Roine, E., Bamford, D.H., and Corcelli, A. (2013). Lipid fingerprints of intact viruses by MALDI-TOF/mass spectrometry. *Biochim. Biophys. Acta* 1831 (4): 872–879. https://doi.org/10.1016/j.bbalip.2013.01.011.

50 Bandu, R., Mok, H.J., and Kim, K.P. (2018). Phospholipids as cancer biomarkers: mass spectrometry-based analysis. *Mass Spectrom. Rev.* 37 (2): 107–138. https://doi.org/10.1002/mas.21510.

51 Doria, M.L., Cotrim, C.Z., Simoes, C. et al. (2013). Lipidomic analysis of phospholipids from human mammary epithelial and breast cancer cell lines. *J. Cell. Physiol.* 228 (2): 457–468. https://doi.org/10.1002/jcp.24152.

52 Fu, Y., Zou, T., Shen, X. et al. (2021). Lipid metabolism in cancer progression and therapeutic strategies. *MedComm (2020)* 2 (1): 27–59. https://doi.org/10.1002/mco2.27.

53 He, M., Guo, S., Ren, J., and Li, Z. (2016). In situ characterizing membrane lipid phenotype of human lung cancer cell lines using mass spectrometry profiling. *J. Cancer* 7 (7): 810–816. https://doi.org/10.7150/jca.14310.

54 Cheng, M., Bhujwalla, Z.M., and Glunde, K. (2016). Targeting phospholipid metabolism in cancer. *Front. Oncol.* 6: 266. https://doi.org/10.3389/fonc.2016.00266.

55 Wang, W., Bai, L., Li, W., and Cui, J. (2020). The lipid metabolic landscape of cancers and new therapeutic perspectives. *Front. Oncol.* 10: 605154. https://doi.org/10.3389/fonc.2020.605154.

56 Lee, P.Y., Yeoh, Y., Omar, N. et al. (2021). Molecular tissue profiling by MALDI imaging: recent progress and applications in cancer research. *Crit. Rev. Clin. Lab. Sci.* 58 (7): 513–529. https://doi.org/10.1080/10408363.2021.1942781.

57 Arentz, G., Mittal, P., Zhang, C. et al. (2017). Applications of mass spectrometry imaging to cancer. *Adv. Cancer Res.* 134: 27–66. https://doi.org/10.1016/bs.acr.2016.11.002.

58 Park, Y.S., Yoo, C.W., Lee, S.C. et al. (2011). Lipid profiles for intrahepatic cholangiocarcinoma identified using matrix-assisted laser desorption/ionization mass spectrometry. *Clin. Chim. Acta* 412 (21–22): 1978–1982. https://doi.org/10.1016/j.cca.2011.07.008.

59 Thai, A.A., Solomon, B.J., Sequist, L.V. et al. (2021). Lung cancer. *Lancet* 398 (10299): 535–554. https://doi.org/10.1016/S0140-6736(21)00312-3.

60 Lee, G.K., Lee, H.S., Park, Y.S. et al. (2012). Lipid MALDI profile classifies non-small cell lung cancers according to the histologic type. *Lung Cancer* 76 (2): 197–203. https://doi.org/10.1016/j.lungcan.2011.10.016.

61 Loibl, S., Poortmans, P., Morrow, M. et al. (2021). Breast cancer. *Lancet* 397 (10286): 1750–1769. https://doi.org/10.1016/S0140-6736(20)32381-3.

62 Medina, M.A., Oza, G., Sharma, A. et al. (2020). Triple-negative breast cancer: a review of conventional and advanced therapeutic strategies. *Int. J. Environ. Res. Public Health* 17 (6): https://doi.org/10.3390/ijerph17062078.

63 Honig, P.K. (2022). The "coming of age" of real-world evidence in drug development and regulation. *Clin. Pharmacol. Ther.* 111 (1): 11–14. https://doi.org/10.1002/cpt.2465.

64 Silva, C.L., Perestrelo, R., Sousa-Ferreira, I. et al. (2020). Lipid biosignature of breast cancer tissues by matrix-assisted laser desorption/ionization time-of-flight mass spectrometry. *Breast Cancer Res. Treat.* 182 (1): 9–19. https://doi.org/10.1007/s10549-020-05672-9.

65 Perestrelo, R., Petkovic, M., and Silva, C.L. (2021). Analytical platforms for the determination of phospholipid turnover in breast cancer tissue: role of phospholipase activity in breast cancer development. *Metabolites* 11 (1): https://doi.org/10.3390/metabo11010032.

66 Kang, H.S., Lee, S.C., Park, Y.S. et al. (2011). Protein and lipid MALDI profiles classify breast cancers according to the intrinsic subtype. *BMC Cancer* 11: 465. https://doi.org/10.1186/1471-2407-11-465.

67 Jung, J.H., Lee, M.Y., Choi, D.Y. et al. (2015). Phospholipids of tumor extracellular vesicles stratify gefitinib-resistant nonsmall cell lung cancer cells from gefitinib-sensitive cells. *Proteomics* 15 (4): 824–835. https://doi.org/10.1002/pmic.201400243.

68 Chen, Y., Wang, T., Xie, P. et al. (2021). Mass spectrometry imaging revealed alterations of lipid metabolites in multicellular tumor spheroids in response to hydroxychloroquine. *Anal. Chim. Acta* 1184: 339011. https://doi.org/10.1016/j.aca.2021.339011.

69 Llorente-Ovejero, A., Martinez-Gardeazabal, J., Moreno-Rodriguez, M. et al. (2021). Specific phospholipid modulation by muscarinic signaling in a rat lesion model of Alzheimer's disease. *ACS Chem. Nerosci.* 12 (12): 2167–2181. https://doi.org/10.1021/acschemneuro.1c00169.

70 de San, G., Roman, E., Llorente-Ovejero, A. et al. (2021). Modulation of neurolipid signaling and specific lipid species in the triple transgenic mouse model of Alzheimer's disease. *Int. J. Mol. Sci.* 22 (22): https://doi.org/10.3390/ijms222212256.

71 Chen, H., Meng, L., and Shen, L. (2022). Multiple roles of short-chain fatty acids in Alzheimer disease. *Nutrition* 93: 111499. https://doi.org/10.1016/j.nut.2021.111499.

72 Falabella, M., Vernon, H.J., Hanna, M.G. et al. (2021). Cardiolipin, mitochondria, and neurological disease. *Trends Endocrinol. Metab.* 32 (4): 224–237. https://doi.org/10.1016/j.tem.2021.01.006.

73 Lee, J.Y., Jin, H.K., and Bae, J.S. (2020). Sphingolipids in neuroinflammation: a potential target for diagnosis and therapy. *BMB Rep.* 53 (1): 28–34.

74 de Wit, N.M., Mol, K., Rodriguez-Lorenzo, S. et al. (2020). The role of sphingolipids and specialized pro-resolving mediators in Alzheimer's disease. *Front. Immunol.* 11: 620348. https://doi.org/10.3389/fimmu.2020.620348.

75 Zhang, X., Wu, C., and Tan, W. (2021a). Brain lipid dynamics in amyloid precursor protein/Presenilin 1 mouse model of early Alzheimer's disease by desorption electrospray ionization and matrix assisted laser desorption ionization-mass spectrometry imaging techniques. *J. Proteome Res.* 20 (5): 2643–2650. https://doi.org/10.1021/acs.jproteome.0c01050.

76 Rebollo-Ramirez, S., Krokowski, S., Lobato-Marquez, D. et al. (2018). Intact cell lipidomics reveal changes to the ratio of cardiolipins to phosphatidylinositols in response to kanamycin in HeLa and primary cells. *Chem. Res. Toxicol.* 31 (8): 688–696. https://doi.org/10.1021/acs.chemrestox.8b00038.

77 Giraud-Gatineau, A., Coya, J.M., Maure, A. et al. (2020). The antibiotic bedaquiline activates host macrophage innate immune resistance to bacterial infection. *Elife* 9: https://doi.org/10.7554/eLife.55692.

78 Valletta, M., Ragucci, S., Landi, N. et al. (2021). Mass spectrometry-based protein and peptide profiling for food frauds, traceability and authenticity assessment. *Food Chem.* 365: 130456. https://doi.org/10.1016/j.foodchem.2021.130456.

79 Herrero, M., Simo, C., Garcia-Canas, V. et al. (2012). Foodomics: MS-based strategies in modern food science and nutrition. *Mass Spectrom. Rev.* 31 (1): 49–69. https://doi.org/10.1002/mas.20335.

80 Calvano, C.D., De Ceglie, C., Monopoli, A., and Zambonin, C.G. (2012). Detection of sheep and goat milk adulterations by direct MALDI-TOF MS analysis of milk tryptic digests. *J. Mass Spectrom.* 47 (9): 1141–1149. https://doi.org/10.1002/jms.2995.

81 Calvano, C.D., De Ceglie, C., Aresta, A. et al. (2013). MALDI-TOF mass spectrometric determination of intact phospholipids as markers of illegal bovine milk adulteration of high-quality milk. *Anal. Bioanal. Chem.* 405 (5): 1641–1649. https://doi.org/10.1007/s00216-012-6597-z.

82 England, P., Tang, W., Kostrzewa, M. et al. (2020). Discrimination of bovine milk from non-dairy milk by lipids fingerprinting using routine matrix-assisted laser desorption ionization mass spectrometry. *Sci. Rep.* 10 (1): 5160. https://doi.org/10.1038/s41598-020-62113-9.

83 Calvano, C.D., Aresta, A., and Zambonin, C.G. (2010). Detection of hazelnut oil in extra-virgin olive oil by analysis of polar components by micro-solid phase extraction based on hydrophilic liquid chromatography and MALDI-ToF mass spectrometry. *J. Mass Spectrom.* 45 (9): 981–988. https://doi.org/10.1002/jms.1753.

84 De Ceglie, C., Calvano, C.D., and Zambonin, C.G. (2014). Determination of hidden hazelnut oil proteins in extra virgin olive oil by cold acetone precipitation followed by in-solution tryptic digestion and MALDI-TOF-MS analysis. *J. Agric. Food Chem.* 62 (39): 9401–9409. https://doi.org/10.1021/jf504007d.

85 Di Girolamo, F., Masotti, A., Lante, I. et al. (2015). A simple and effective mass spectrometric approach to identify the adulteration of the Mediterranean diet component extra-virgin olive oil with corn oil. *Int. J. Mol. Sci.* 16 (9): 20896–20912. https://doi.org/10.3390/ijms160920896.

86 Zhang, Y.X., Zhao, X.B., Ha, W. et al. (2021b). Spatial distribution analysis of phospholipids in rice by matrix-assisted laser desorption/ionization time-of-flight mass spectrometry imaging. *J. Chromatogr. A* 1651: 462302. https://doi.org/10.1016/j.chroma.2021.462302.

87 Enomoto, H., Takeda, S., and Hatta, H. (2021). Spatial analysis of phosphatidylinositol molecular species in pork chop tissues using matrix-assisted laser desorption/ionization-mass spectrometry imaging. *J. Oleo Sci.* 70 (7): 979–987. https://doi.org/10.5650/jos.ess21042.

88 Lordan, R., Tsoupras, A., and Zabetakis, I. (2017). Phospholipids of animal and marine origin: structure, function, and anti-inflammatory properties. *Molecules* 22 (11): https://doi.org/10.3390/molecules22111964.

89 Leung, L.M., Fondrie, W.E., Doi, Y. et al. (2017). Identification of the ESKAPE pathogens by mass spectrometric analysis of microbial membrane glycolipids. *Sci. Rep.* 7 (1): 6403.

16

Use of MALDI-TOF MS in Water Testing Laboratories

Matthew Jones[1], Nadia Darwich[1], Rachel Chalmers[2,3], K. Clive Thompson[4], and Bjorn Nielsen[5]

[1] *Dŵr Cymru Welsh Water, Glaslyn Laboratory, Newport, UK*
[2] *Cryptosporidium Reference Unit, Public Health Wales Microbiology and Health Protection, Singleton Hospital, Swansea, UK*
[3] *Swansea University Medical School, Singleton Park, Swansea, UK*
[4] *ALS, Life Sciences, Rotherham, UK*
[5] *ALS Environmental, Wakefield, UK*

16.1 Introduction

The importance of wholesome and clean drinking water has been recognized since the time of the Romans, with major advances being made during the nineteenth and twentieth centuries. In 1854, British scientist John Snow investigated an epidemic of cholera in London which led him to conclude that contaminated water from the Broad Street pump was the root cause of the disease and, consequently, the removal of the handle stopped the epidemic [1]. This work carried out by John Snow reinforced the importance of sanitation and in the following decades led to the development of water purification and sewage drainage systems in cities and towns.

As part of this improvement in water treatment, early quality standards were developed in the late nineteenth century, including the use of faecal indicator organisms, particularly *Escherichia coli* and thermotolerant coliforms [2]. Since these early standards, *Clostridium perfringens* and enterococci have become widely used as indicators of contamination or to monitor system integrity. Enterococci, in particular, have been shown to have a useful relationship with waterborne pathogens and this is why they are used in the monitoring of drinking water for assessing public health risk. In conjunction with chlorination, the indicator approach has been paramount to protecting public health [3].

Indicator bacteria are not typically dangerous to human health themselves; they are used to indicate the presence of a health risk or potential contamination. Faecal material can enter the water cycle through many sources such as effluents from wastewater treatment plants, livestock, poultry manure or direct contamination from animals, including wildlife.

Microbiological Identification using MALDI-TOF and Tandem Mass Spectrometry: Industrial and Environmental Applications, First edition. Edited by Haroun N. Shah, Saheer E. Gharbia, Ajit J. Shah, Erika Y. Tranfield, and K. Clive Thompson.

This faecal matter may contain species of pathogenic bacteria, such as *Salmonella* or *Campylobacter*, parasites such as *Cryptosporidium*, or viruses such as norovirus, associated with gastroenteritis. The variety and low concentrations of these pathogens make them challenging, time-consuming, and costly to test for individually. As such the analysis of indicator bacteria is the most sensitive and effective method for detecting faecal and, therefore, potentially dangerous contamination in drinking water.

Each indicator bacterium has a varying degree of association to faecal matter; for example, some coliforms may originate from non-faecal sources such as soil or decaying vegetation, e.g. *Serratia fonticola*, which is commonly found in water but is of no known health significance, or *Citrobacter*, *Klebsiella*, and *Enterobacter* species which are found both in faeces and in extraintestinal environments such as soil and water [3]. Therefore, when coliform bacteria are isolated from drinking water samples, it is useful to determine which species are present, to understand their possible source and significance. *Escherichia coli* belongs to the coliform group and occurs in the faeces of all mammals, often in high numbers. This abundance, combined with the simple and cost-effective methods available for their detection, is why *E. coli* is regarded as a primary indicator for faecal contamination of both treated and untreated water and has formed the cornerstone of microbiological water quality assessment for over 100 years [2].

Within the *Enterococcus* genus there are a number of species that occur in the faeces of humans and warm-blooded animals. They are considered to be secondary indicators of faecal contamination and are used to assess the significance when coliforms are present in the absence of *E. coli* or to provide further information when assessing the extent of potential faecal contamination. The identification of *Enterococcus* species may provide an indication of the source of contamination, with certain species being present in the faeces of humans and various animals (e.g. *Enterococcus faecalis*) whereas others are not typically found in human faeces (e.g. *Enterococcus avium*) [3].

Clostridium perfringens is a sulphite-reducing bacterium, commonly found in human and animal faeces but at a much lower concentration than other indicator organisms and, as such, is a less sensitive indicator of faecal contamination. *C. perfringens* produces environmentally resistant spores that survive in water for much longer than vegetative cells of other faecal indicators, and so can indicate remote contamination. *Clostridia* are removed from water by coagulation and filtration, but the spores can be resistant to chlorine at concentrations normally used in water treatment [3].

National and international bodies such as the European Committee for Standardization (CEN), British Standards Institution (BSI), and International Organization for Standardization (ISO) oversee the standardization and publication of standard methods for the examination of drinking water. Within the UK, in 1973, the Department of the Environment became responsible for all aspects of the water cycle and established the Standing Committee of Analysts (SCA) to review and keep up to date the methods recommended for water examination. These are now published under the *Microbiology of Drinking Water* series [3] (available online at www.standingcommitteeofanalysts.co.uk).

Over the long period of history that indicator bacteria have been used, reliance has been placed on detection methods being relatively simple and rapid. A wide range of techniques and methods are available for bacterial isolation, confirmation, and identification. These can be culture-based, involving the use of chromogenic media or biochemical testing, or molecular, such as high-throughput sequencing of the 16S rRNA gene. The most widely used methodology in water testing laboratories is membrane filtration analysis, where organisms are concentrated on a membrane filter which is then placed on an agar medium containing specific constituents for the growth of the target organism. Isolation of colonies is followed by confirmatory tests involving subculture using an agar substrate and a series of biochemically based tests. It is essential that these methods are rapid, as a quick response to bacterial detections in drinking water is often required to avoid potential public health risks.

The analysis of faecal indicator bacteria in regulatory drinking water samples is often carried out by water utility companies with their own laboratory testing services. In the UK, however, some water utility companies and local authorities that are responsible for the regulation of private water supplies utilize independent, commercial, environmental testing or public health laboratories as partners for their water analysis. Routine water analysis for various bacterial parameters is, however, also carried out in many other fields and sectors. In-premise water systems, for example, are sampled and tested for water hygiene purposes. This is a large industry in the UK with many service providers carrying out risk assessments and management of these water systems. Typical analysis is for the same faecal indicator organisms as regulatory drinking water testing, but also for other organisms deemed through risk assessment to be hazards, e.g. *Legionella* spp. and *Pseudomonas aeruginosa*. In addition, the analysis of surface waters is routinely conducted due to its use as a source water in drinking water production, as bathing waters and as the repository of treated waste waters. These analyses and many others are often carried out by commercial, environmental testing laboratories. This wider spectrum of customers and sectors creates a broader scope of potential analytical methods

Protein analysis offers rapid and accurate microbial identification of organisms through the application of matrix-assisted laser desorption/ionization time-of-flight mass spectrometry (MALDI-TOF MS), which analyses high-abundance proteins (mainly ribosomal) of a microbial cell. MALDI-TOF MS analysis has been widely used in the clinical field and has been shown to have equivalent resolution for species identification as 16S rRNA gene sequencing [4]. There are several studies demonstrating accurate identification performance of MALDI-TOF MS for a diverse group of microorganisms such as Gram-negative non-fermenting bacteria [5], enteric Gram-negative bacteria [6], and heterotrophic bacteria isolated from drinking water [7].

MALDI-TOF MS has had an immediate, significant impact on water microbiology in its use for confirmatory analysis of faecal indicator organisms in drinking water, and on analysis of environmental waters, waste waters and water in buildings for *Legionella* species and other relevant pathogens such as *Pseudomonas* species in place of traditional confirmation methods. MALDI-TOF MS is an improved confirmatory and identification analysis because it provides an accurate and rapid result that negates the need for subculture and biochemical testing.

This chapter reports work on the introduction of MALDI-TOF MS for drinking water analysis for bacterial identification and subsequent confirmation of indicator organisms at both a drinking water laboratory and a commercial environmental testing laboratory. This work also formed the basis of a recent *Microbiology of Drinking Water* series publication by the SCA which has standardized the MALDI-TOF MS method for water and is supported by comprehensive validation studies, demonstrating equivalent or better performance when compared with the traditional confirmation methods [8]. This chapter also details novel work investigating the possibility of using MALDI-TOF MS for the identification of *Cryptosporidium* species.

16.2 Application in a Drinking Water Laboratory

16.2.1 Introduction

MALDI-TOF MS has been used for the identification of bacteria for the past 30 years. With the development, in the last 10 years, of benchtop, easy-to-use instruments, that have built-in reference databases covering a vast number of microorganisms, MALDI-TOF MS has become a routine identification tool in clinical laboratories. Its use in environmental laboratories is rapidly expanding and this section details the use of MALDI-TOF MS in a drinking water laboratory.

Dŵr Cymru Welsh Water (DCWW) is a not-for-profit water company which supplies drinking water and wastewater services to three million people living in Wales and some parts of England. This involves treating and supplying over 800 million litres of water every day via a network covering 27 000 km [9]. DCWW has its own laboratories which carry out over 300 000 tests each year, including microbiological testing for indicator bacteria. The use of MALDI-TOF MS was ideal for meeting one of DCWW's key drivers of developing and implementing laboratory methods that are rapid, reliable, economical, and accurate. The development of rapid and accurate analytical methods improves the laboratories' capability to provide water quality data that is essential for protecting the public health of its customers.

MALDI-TOF MS analysis is of benefit to drinking water laboratories in terms of protecting public health as it provides a bacterial identification result which can be used for a subsequent confirmation result via a relatively simple and easy-to-use method in terms of preparation and analysis. Traditional methods, e.g. coliform confirmation via subculture onto Tryptone Nutrient Agar (TNA) and then assessing ortho-nitrophenyl-β-galactoside (ONPG), oxidase, and indole reactions [10], delay the reporting of a confirmed result by 18–24 hours due to the subculturing and incubation period. MALDI-TOF MS does not require a subculturing step as a small amount of material can be taken from a single bacterial colony isolated on a primary agar, e.g. Membrane Lactose Glucuronide Agar (MLGA), and transferred to a MS target plate.

The MALDI-TOF MS method also offers benefits in terms of cost savings compared with traditional methods. A previous study documented a saving of 57.1% in reagent cost [11] and internal assessment at DCWW has calculated a cost saving of ~50% in terms of reagent and consumable cost. The initial investment in the equipment is high, but it has the potential to be offset through the reduction in reagent cost and the negating of actions taken following presumptive bacterial detections which later do not confirm, via the reporting of only confirmed results in a rapid time-frame.

The MALDI-TOF mass spectrometer generates a unique proteomic fingerprint. Using the software package provided by the manufacturer (Bruker Microbiology & Diagnostics),

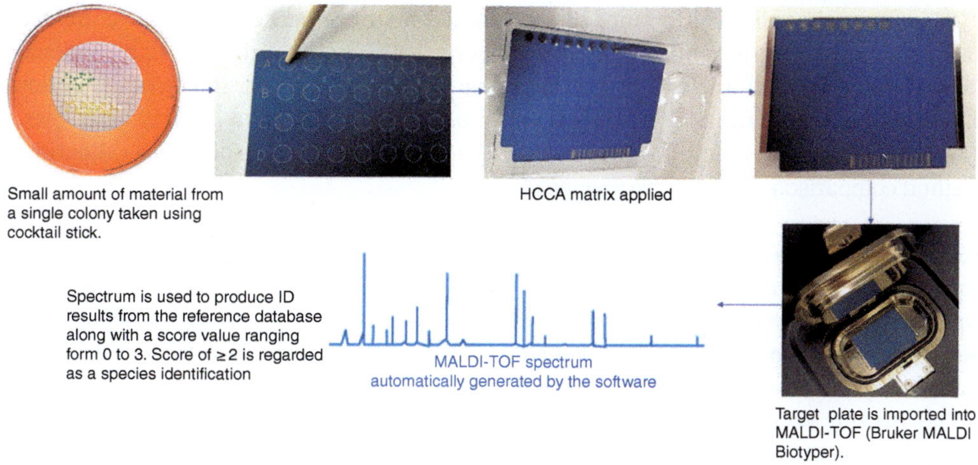

Figure 16.1 Identification process using MALDI-TOF (MALDI Biotyper, Bruker Microbiology and Diagnostics).

the spectrum pattern of the molecular fingerprint is used to identify the bacteria by matching it against spectra in the reference database. A score value is assigned to each identification, where a score value ≥ 2.00 is considered an excellent probability for identification at the species level and a score value between 1.70 and 1.99 is considered an excellent probability for identification at the genus level. This result is used to determine if the bacteria identified is confirmed for the specific indicator bacteria being analysed for, i.e. coliforms, *E. coli*, *C. perfringens*, or *Enterococcus*. An overview of the analytical process is displayed in Figure 16.1.

The MALDI-TOF MS analysis process, including sample preparation, takes around 60 minutes to analyse 96 target spots. For the equivalent number of colonies, this would take around 20 hours using the traditional agar and biochemical-based methods, meaning that the MALD-TOF MS is able to provide more detailed results in terms of species identification in a more rapid time-frame. This enables the laboratory to report confirmed bacterial results from drinking water samples 18–20 hours sooner, ensuring a more rapid response to bacterial detection in drinking water for the protection of public health. It also means that DCWW is reacting to confirmed results rather than presumptive results, so that any response actions taken are due to a genuine indicator bacteria detection. This removes the time, effort, and cost of responding to presumptive indicator bacteria detections which do not confirm 18–20 hours later.

16.2.2 Method Validation

As part of developing the MALDI-TOF MS method at DCWW's laboratories, validation was carried out to determine whether confirmation and identification by MALDI-TOF MS were equivalent to the current United Kingdom Accreditation Service (UKAS) ISO 17025-accredited agar and biochemical-based methods. This validation study also set out to determine the necessary performance characteristics of the MALDI-TOF MS method.

To carry out the validation, a range of typical water matrices were chosen to cover treated and raw surface and ground water. Samples were spiked using traceable [e.g. National Collection of Type Cultures (NCTC) or American Type Culture Collection (ATCC)] strains of bacteria and naturally derived bacteria from untreated surface and ground water sites. The validation of the method was split into several stages: reference database validation, method comparison, and agar assessment.

16.2.2.1 Reference Database Validation

To assess the reference database, a series of known traceable reference material and wild-type strains of bacteria, including coliforms, *Enterococcus* and *C. perfringens* along with non-target organisms such as *Pseudomonas* species, were spiked into treated drinking water samples and colonies identified via MALDI-TOF MS.

The Bruker Microbiology and Diagnostics environmental reference database in the MALDI Biotyper at the time of the validation study (MBT Compass Library v9) contained 8468 strains of bacteria, covering eight phyla. For coliforms there were 596 strains covering 40 genera of coliforms, including the most commonly identified organisms in drinking water. For *Enterococcus* there were 109 strains covering 34 species of *Enterococcus*, including the most commonly identified organisms in drinking water samples. For *C. perfringens* there were nine strains of *C. perfringens*. Identification analysis using the reference and wild-type bacterial strains demonstrated the suitability of the reference database within the Bruker Microbiology and Diagnostics MALDI-TOF MS system for the identification and subsequent confirmation of bacteria isolated from drinking water (Table 16.1).

As demonstrated in Table 16.1 the percentage of correct identifications was 92.2%, which is in line with other studies that have investigated the feasibility of MALDI-TOF MS for bacterial identifications [12, 13]. The misidentified organisms were *Shigella sonnei* and *Clostridium beijerinckii*, which were identified as *E. coli* and *Clostridium diolis*, respectively. *Shigella* species and *E. coli* are closely related and share many common

Table 16.1 Identification of spiked samples for a range of organisms typically isolated and identified in drinking water.

Microorganism group	Number of colonies analysed	Number of correct identifications	Number of incorrect identifications
Enterobacteriaceae	1235	1125 (91.1%)	110 (8.9%)
Enterococci	882	859 (97.4%)	23 (2.6%)
Clostridium perfringens	399	395 (99.0%)	4 (1.0%)
Non-target	360	272 (75.6%)	88 (24.4%)
Total	2876	2651 (92.2%)	225 (7.8%)

Enterobacteriaceae group contained *Citrobacter freundii*, *Enterobacter aerogenes*, *Enterobacter cloacae*, *Hafnia alvei*, *Raoultella planticola*, *Klebsiella oxytoca*, *Pantoea agglomerans*, *E. coli* O157, *Shigella sonnei*, *Klebsiella pneumoniae*, *Escherichia vulneris*, *Leclercia adecarboxylata* and *Serratia liquefaciens*. Enterococci group contained (all genera *Enterococcus*) *E. faecalis*, *E. casseliflavus*, *E. hirae*, *E. avium*, *E. faecium*, *E. mundtii*, *E. gallinarum*. Non-target group contained *Clostridium bifermentans*, *Clostridium beijerinckii*, *Pseudomonas putida*, *Pseudomonas aeruginosa*, *Aeromonas hydrophila* and *Staphylococcus warneri*.

characteristics [14] and, because of this, differentiation of *Shigella* species from *E. coli* cannot be done using MALDI TOF MS at this present time. During this validation study, the *S. sonnei* colonies isolated on MLGA produced green colonies, which would be considered confirmed *E. coli* [10], indicating that the agar-based method is also unable to distinguish between *Shigella* and *E. coli*. As many *E. coli* and *Shigella* are of similar public health significance, this limitation of the MALDI-TOF MS method was not considered a significant issue for confirmatory and identification analysis in drinking water at DCWW laboratories. *C. beijerinckii* and *C. diolis* are very similar in terms of 16S rRNA gene sequences and a reclassification of *C. diolis* as a later heterotypic synonym of *C. beijerinckii* has been proposed [15]. This was not considered a significant issue as neither of these organisms would be confirmed as the relevant target species, *C. perfringens*. Factoring in these know limitations of MALDI-TOF, this would be a correct identification percentage of 98.6%.

16.2.2.2 Method Comparison

To compare the traditional and MALDI-TOF MS methods, a series of known traceable reference materials and wild-type strains of bacteria were spiked into treated drinking water samples and analysed using both the traditional methods [10, 16, 17] and MALDI-TOF MS. Analysis of these spiked, treated water samples demonstrated that MALDI-TOF MS identification analysis and the subsequent confirmatory analysis performed better than the traditional UKAS ISO 17025 methods accredited for each organism (Table 16.2).

For coliform confirmation, the MALDI-TOF MS method correctly confirmed 99.1% of presumptive colonies, as compared with the traditional method [10] which correctly confirmed 91.9%. This discrepancy was due to *Hafnia alvei* not hydrolysing ONPG which has been documented in a previous study where it was reported that 14% of *H. alvei* strains did not hydrolyse ONPG [18]. For *C. perfringens* confirmation, the MALDI-TOF MS method

Table 16.2 Comparison of MALDI-TOF confirmation method and reference methods in the confirmation on spiked treated water samples.

	Traditional methods		MALDI-TOF method	
Microorganism group	Number of colonies analysed	Number of correctly confirmed colonies	Number of colonies analysed	Number of correctly confirmed colonies
Coliforms	1235	1135 (91.9%)	1235	1224 (99.1%)
Enterococci	882	776 (88.0%)	882	859 (97.4%)
Clostridium perfringens	399	310 (77.7%)	399	395 (99.0%)
Non-target	360	360 (100%)	360	360 (100%)

The traditional method for coliforms was subculture onto TNA and then biochemical testing for ONPG, oxidase and indole. The traditional method for enterococci was hydrolysation of aesculin on KAAA and catalase testing. The traditional method for *C. perfringens* was subculture on to CAB and acid phosphatase testing.

Table 16.3 Mean confidence score value results (n = minimum of 4) for a range of NCTC organisms isolated on a range of agars used within a drinking water laboratory.

	Agar – mean score value							
Organism	**MLGA**	**TSC**	**S&B**	**YEA**	**NA**	**TNA**	**MAC**	**CAB**
Klebsiella aerogenes	2.06	N/A	N/A	2.53	2.52	2.52	2.50	N/A
Escherichia coli	2.06	N/A	N/A	2.37	2.44	2.46	2.26	N/A
Pseudomonas aeruginosa	2.32	N/A	N/A	2.46	2.30	2.46	2.32	N/A
Enterococcus faecalis	N/A	N/A	2.19	2.30	2.10	N/A	N/A	N/A
Staphylococcus warnerii	N/A	N/A	2.18	2.15	2.07	N/A	N/A	N/A
Bacillus cereus	N/A	N/A	N/A	N/A	2.01	N/A	N/A	2.16
Clostridium perfringens	N/A	2.24	N/A	2.46	2.45	N/A	N/A	2.40
Clostridium bifermentans	N/A	2.00	N/A	2.10	2.13	N/A	N/A	2.12

N/A, this agar and organism combination was not tested as it is not typically cultured on this agar.

correctly confirmed 99% of presumptive colonies, compared with 77.7% by the traditional method [16]. This discrepancy was due to *C. perfringens* NCTC 8797 obtained from a reference material provider; on contacting the provider they had anecdotal evidence that other laboratories had found that this particular *C. perfringens* was atypically acid phosphatase-negative. For *Enterococcus* confirmation, the MALDI-TOF MS method correctly confirmed 97.4% of presumptive colonies, compared with 88% by the traditional method [17]. This discrepancy was due to *E. avium*, where 61.5% of colonies did not confirm as enterococci due to a failure to hydrolyse aesculin on Kanamycin Aesculin Azide Agar (KAAA), as compared with only 0.9% that failed to be identified as *Enterococcus* using MALDI-TOF MS.

This method comparison highlights the advantages of the MALD-TOF MS method. It provides a more accurate result and reduces the potential for false negatives, as it does not rely on subculture and further biochemical testing. This is of huge benefit to drinking water laboratories, as confirmed results can be reported 18–20 hours sooner and there is also a reduction in false-positive presumptive results and false negatives due to atypical biochemical reactions, e.g. *H. alvei*.

16.2.2.3 Agar Assessment

A range of agars that are typically used in a drinking water laboratory were assessed by subculturing a range of bacteria onto both selective and non-selective agars (Table 16.3). The agars used for assessment were MLGA, Tryptose Sulphite Cycloserine (TSC), Slanetz and Bartley agar (S&B), Yeast Extract Agar (YEA), Nutrient Agar (NA), Tryptone Nutrient Agar (TNA), MacConkey Agar (MAC), and Columbia Agar Base (CAB).

16.2.3 Application Within Drinking Water Laboratory

As detailed in Sections 16.2.2.1 and 16.2.2.2 the MALDI-TOF MS demonstrated better performance for the identification and subsequent confirmation of indicator bacteria isolated from drinking water. The use of MALDI-TOF MS for routine confirmation of indicator bacteria provides a more rapid confirmed result, enabling a quicker response to a potential public health issue. It also prevents responding to unconfirmed presumptive detections, resulting in time and cost savings from conducting unnecessary investigations and actions such as resampling. An added bonus is the detailed information obtained from MALDI-TOF MS in terms of species identification. This allows investigations to be carried out looking for trends or patterns in identifications, with the potential for source tracking of bacterial contaminations in drinking or raw water samples.

Potential future applications of MALDI-TOF MS in drinking water laboratories may be to combine it with other new technologies such as flow cytometry or cell sorting, which has been done in the clinical field [19]. This would enable the heterotrophic bacteria in drinking water and raw water samples to be identified, which would provide more information on complex bacterial communities [7]. This information could be of use, as heterotrophic bacteria might include pathogens and can impact water quality through the formation of biofilms on pipes or storage tanks. A developing future interest within the water industry is antimicrobial resistance, and the use of MALDI-TOF MS for the detection of antibiotic resistance is detailed in other chapters within this book. The identification of *Cryptosporidium* is also a possible future application, and there is more information on this in Section 16.4.

16.3 Application in Water Hygiene and Environmental Laboratory Testing

16.3.1 Introduction

Commercial testing laboratories have a broad range of potential customers, markets, regulatory testing requirements, and test methods where MALDI-TOF MS analysis can be applied. Typical commercial environmental testing laboratories in the UK may be involved in various fields, e.g. regulatory drinking water analysis, water hygiene analysis, wastewater testing, bathing and recreational water analysis, and analysis of water for use in healthcare settings and processes.

ALS Environmental UK are one of the UK's largest providers of analytical services, providing a range of UKAS-, Monitoring Certification Scheme (MCERTS)- and DWTS-accredited analyses for water, land, and waste monitoring. It provides testing services to drinking water and wastewater utility customers, in addition to water treatment clients operating in the water hygiene field. It is part of ALS Global, a global testing services company operating from over 350 locations in 55 countries.

For a commercial laboratory with a focus on service to the customer, the rapid nature of the MALDI-TOF MS technique is an important factor. Commercial Laboratories such as ALS Environmental UK have seen the potential for the application of the MALDI-TOF MS technique, and ALS Environmental UK were pioneers in implementing this as an alternative confirmation method in UK Environmental testing in 2013, utilizing the Bruker Microbiology and Diagnostics system. The approach has since been adopted in other ALS laboratories globally.

There are many benefits of this technique, for both the laboratory and the customer, and it provides a unique selling point to differentiate the service offering from competitor laboratories. The advantage of immediate confirmed results as opposed to a presumptive result notification to be followed by a confirmed result some days later allows the client to act upon the result sooner, which is a benefit to public health. Also, the additional information that MALDI-TOF MS provides in the form of a species identification, where previously the confirmed result might only be to the genus or a group of organisms, gives the client an additional tool for further investigations.

In addition to the applications described in Section 16.2 for drinking water analysis, the following sections describe other applications of the MALDI-TOF MS technique in routine water microbiology testing.

16.3.2 *Legionella* Testing

Legionellae are waterborne bacteria found in natural aquatic environments. Artificial aquatic environments have created a niche for *Legionella* species to survive and proliferate, along with providing the mechanisms that allow their spread and the infection of humans. Monitoring for the presence of *Legionella* in these systems is therefore important for public health.

There are over 60 described species of *Legionella* with at least 26 species reported as responsible for human infections. Artificial aquatic systems such as cooling towers, hot and cold water systems within buildings and spa pools can create the ideal environments for *Legionella* to thrive, due to readily available nutrients, ideal temperatures, and areas of low flow or stagnation. In addition, *Legionella* may form part of biofilms in such systems or may exist intracellularly within amoeba [20, 21], which affords protection from biocide activity and may offer additional nutrient availability.

Although there is no regulatory requirement to test for the presence of *Legionella* in such water systems in the UK, the management of the risk it presents is governed by existing legislation, e.g. the Health and Safety at Work Act 1974. To this end, codes of practice and guidance have been published [22–24] in addition to industry best-practice guidance to assess the risk of *Legionella* in these systems [25]. This will indicate the need for monitoring through laboratory analysis to assess the risk, identify sources, validate control measures, and verify that ongoing controls remain effective. The requirement to test for *Legionella* spp. has, however, been added to the latest EU Drinking Water Directive.

Legionella analysis is typically carried out using a culture method according to the standard reference method [26] and is typically longer than many other water testing parameters, requiring a 10-day incubation period. Suspect or 'presumptive' *Legionella* colonies are identified and confirmatory analysis takes a further two days due to subculture and the relatively slow growth of the organism. The method is illustrated in Figure 16.2.

Alternative, rapid techniques such as polymerase chain reaction (PCR) can be employed, however, the results are expressed in genomic units (GUs) as opposed to colony-forming units (CFU). In the UK, the current guidance on action limits and actions to be taken is based on results from the culture method expressed as CFU. Therefore, water treatment specialists are currently unable to base routine actions on PCR results.

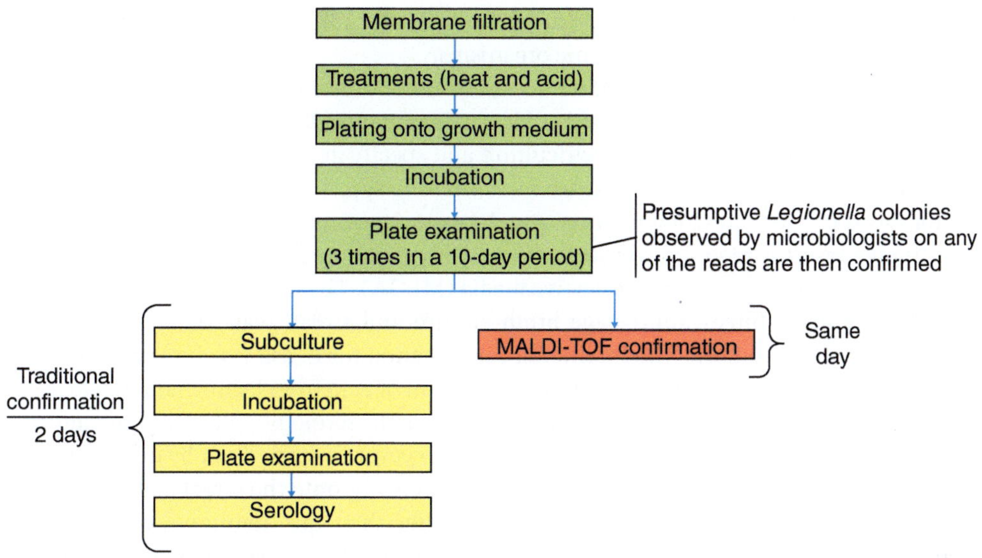

Figure 16.2 *Legionella* analytical process, including confirmatory approaches.

The application of MALDI-TOF MS analysis as an alternative confirmatory technique to the culture method therefore allows the method to be shortened and still provide a result that can be referenced against current UK guidance. Notification of a presumptive *Legionella* result means a decision on whether action should be taken or whether to wait a further two days for the confirmed outcome. The MALDI-TOF MS technique allows an immediate confirmation of presumptive colonies, removing uncertainty and potentially speeding up remedial action.

Legionella identification on MALDI-TOF MS systems also typically gives a species identification. Traditional techniques provide a confirmation that the colony is a species of *Legionella*, and serology using commercially available latex agglutination kits can identify species of concern such as *Legionella pneumophila*; however, internal data reviews have shown this only accounts for ~30% of positive samples.

The reference databases typically provided with MALDI-TOF MS instruments do not contain all species of *Legionella*. This presents options for how the chosen MALDI-TOF MS system can be applied to the analytical method. If no reliable identification is received when using an incomplete *Legionella* library, this may not give a sufficiently reliable outcome to determine that the presumptive colony is not a species of *Legionella*. ALS Environmental took the decision to build an enhanced library for *Legionella* detection containing all available reference strains of *Legionella* species. Following verification exercises (described in Section 16.3.6), this was demonstrated to give a suitably reliable outcome to allow the presence or not of *Legionella* to be determined, without reverting to other confirmatory methods.

16.3.3 Wastewater and Sewage Sludge Microbiology

Treated sewage sludge may be utilized by application to agricultural land as a fertilizer and soil improver [27]. Various processes can be applied during sewage treatment, but in the context of the use of sewage sludge in agriculture the purpose is the control of toxic substances and reduction of pathogenic organisms.

Microbiological analysis is carried out to assess and monitor the effectiveness of sewage treatment processes. National standard guidance documents [28] describe different categories of sewage sludge to be used in agriculture and specify guideline maximum allowable concentrations of *E. coli* and the absence of *Salmonella* spp.

Analysis is carried out using national standard reference methods [29, 30]. For *E. coli* a quantitative result is required so this is typically by membrane filtration of homogenized and serially diluted samples onto selective media (MLGA). The analysis of *Salmonella* spp. is qualitative and utilizes multi-stage broth culture and streak plate on Xylose Lactose Deaminase (XLD) medium. The traditional confirmation for *E. coli* involves subculture of suspect colonies and demonstration of the production of acid from lactose and indole from tryptophan, and *Salmonella* confirmation is through serology following subculture. MALDI-TOF MS identification of the organisms, however, is carried out by directly transferring suspect colonies from MLGA and XLD, respectively, onto the target plate. This generates a confirmed outcome one to two days earlier than the traditional confirmation methods. This is of benefit to the client as the results of the analysis may be required to enable release of large volumes of treated sewage sludge.

Salmonella can only be reliably identified to genus level using the Bruker Microbiology and Diagnostics system, which is suitable for the purpose of the analysis described above.

16.3.4 Healthcare Water Testing

Water used in healthcare environments presents additional risks due to their users being increasingly susceptible, as, for example, in the case immunocompromised individuals. There are additional, lower action limits for *Legionella* spp. in healthcare premises, and guidance on managing the risks of *P. aeruginosa* in healthcare settings is available [31] due to its growing importance as an opportunistic pathogen. Analysis for these parameters in building water systems is therefore common for water hygiene proposes. The rapid identification through MALDI-TOF MS is of benefit when identifying and taking rapid action on public health risks in these high-risk settings.

Testing water from specific healthcare equipment such as washer disinfectors for *P. aeruginosa* is carried out for a similar purpose. The presence of non-tuberculous mycobacteria has also been deemed a risk from these systems as well as heater cooler units used for cardiac bypass surgery [32]. The Bruker Microbiology and Diagnostics Mycobacteria library had been shown in internal studies to be very effective at identifying species of mycobacteria isolated by the standard reference method and on other growth media, which is an area for potential further study.

Emerging pathogens such as the 'ESKAPE' pathogens (*Enterococcus faecium*, *Staphylococcus aureus*, *Klebsiella pneumoniae*, *Acinetobacter baumannii*, *P. aeruginosa*, and *Enterobacter* spp.) [33] and *Stenotrophomonas maltophilia* [34] are of particular significance in healthcare settings. Some of these species are often waterborne and may be isolated using typical water microbiology methods. These species are contained in the Bruker Microbiology and Diagnostics libraries, and the MALDI-TOF MS technique allows their identification, where previously a confirmation to genus or group of organisms would have been available.

16.3.5 Investigative Analysis

Commercial testing laboratories carry out investigative or project-based analysis on water and associated materials. This may be further investigation of water systems where microbial contamination has been observed. The rapid identification of isolated colonies significantly reduces the time taken for often bespoke projects. In addition, fungal analysis and identification may form part of such exercises. MALDI-TOF MS allows rapid identification of certain groups of fungi. This analysis utilizes more in-depth extraction and preparation techniques than is typical for bacteria. However, traditional fungal identification methods utilize microscopy examination, which may take much longer and requires specific training and skills for what may be an infrequent, non-routine analysis.

16.3.6 Method Validation

Traditional confirmatory approaches for the methods described in this chapter are part of national and international standard methods and are approved and specified by regulators and approval bodies for use in these industries. Methods will also have been validated by the laboratory and accredited by the national accreditation body UKAS. To deviate from standard methods requires validation that the alternative technique is suitable. As the first testing laboratory in this industry to utilize the MALDI-TOF MS technique as a confirmatory approach, the validation process required the approval of the accrediting body and also of the UK drinking water regulator. There was also little published guidance available for its use in this manner. ALS adopted a staged approach, and examples of its various aspects are described in the following sections. This, along with the validation described in Section 16.2.2, informed the approach now documented in The Microbiology of Water and Associated Materials series [8].

16.3.6.1 Characterization of Intended Use

The MALDI-TOF MS procedure must first be characterized in terms of its intended use, i.e. bacterial identification and/or confirmatory analysis. For confirmatory analysis, the user should define how the identifications generated constitute a positive or negative confirmation outcome. This is straightforward for a selective method for a single species or genus, where it is clear if the colony has been identified as the target organism(s) or not. Examples of such methods are identification of *P. aeruginosa*, *C. perfringens*, *E. coli*, *Legionella* spp., *Enterococcus* spp., and *Salmonella* spp.

Methods where the parameter is not a defined species or genus, but a group of bacteria usually defined by common biochemical characteristics, require a different definition of a positive or negative confirmation outcome. Coliform bacteria are one of the main faecal indicator organisms for drinking water and fall into this category. These are traditionally regarded as members of the family Enterobacteriaceae and, following the primary isolation step using selective media, they are defined as oxidase-negative and produce acid from lactose or express the enzyme β-galactosidase. To define, therefore, how an identification achieved by MALDI-TOF MS is interpreted as a confirmation outcome must be based on comparison with the traditional confirmation (see Section 16.3.6.4). An approved and validated list of species or genera regarded as 'coliforms' (and those that should not be regarded as such) can then be generated based on this exercise and also on literature reviews.

16.3.6.2 Library Assessment

An initial library assessment identified the most appropriate system and libraries for the intended application. Known limitations of the libraries, i.e. difficulties distinguishing between species, were taken into consideration when carrying out method validation. Where libraries were incomplete, not suitable for the intended use, or required additional organisms, this formed part of further validation or library development.

A practical library assessment carried out using known, traceable, and appropriate reference organisms determined if the library was suitable for the intended application. This assessment also included 'wild-type' strains of target and non-target microorganisms typically isolated by the laboratory, and also interfering organisms that typically grow on the medium.

16.3.6.3 Assessment of Variables

The colony material to be analysed in routine testing will be subjected to many variables. Examples are given in Table 16.4.

In addition to the agars shown to be suitable in Table 16.3, the agars in Table 16.5 were demonstrated to be suitable.

Table 16.4 Variables to be assessed during validation.

Variable	How assessed through validation
Water sample matrix	Ensure analysis of colonies isolated from routine samples received by the laboratory of all typical matrices (Section 16.3.6.4)
Agar	Ensure colonies are isolated on typical agars from reference methods
Colony preparation	Assess scores achieved from different colony preparation techniques – full protein extraction; direct 'smearing' with addition of formic acid; direct 'smearing' without addition of formic acid
Target plate selection	Disposable or reusable. If reusable, validate that the cleaning protocol is effective
Colony age	If colonies are not to be analysed immediately, assess the impact on identification and confidence scores of typical storage times and conditions
Prepared sample stability and storage	If prepared colonies are not to be analysed immediately or transported to other laboratories, assess the impact on identification and confidence scores of typical storage times and conditions

Table 16.5 Additional growth media assessed to be suitable.

Method	Agars
Legionella spp.	Buffered Charcoal Yeast Extract (BCYE) and Glycine Vancomycin Polymyxin Cycloheximide (GVPC)
Pseudomonas aeruginosa	*Pseudomonas* CN (PCN)
Salmonella spp.	Xylose Lactose Deaminase (XLD)

Table 16.6 Repeatability and reproducibility exercise results.

Parameter	Analyst repeatability	Instrument repeatability	Reproducibility
Salmonella species	30 (100%)	54 (100%)	48 (100%)
Legionella species	124 (100%)	71 (100%)	97 (100%)
Enterococci	30 (100%)	103 (100%)	30 (100%)
Clostridium perfringens	33 (100%)	98 (100%)	33 (100%)
Total coliforms	71 (100%)	81 (100%)	32 (100%)
Escherichia coli	57 (100%)	100 (100%)	31 (100%)
Pseudomonas aeruginosa	40 (100%)	60 (100%)	32 (100%)
Total	385 (100%)	567 (100%)	303 (100%)

For colony preparation the 'direct smearing' technique without the addition of formic acid was found to be suitable for the methods carried out for routine drinking water, water hygiene, and wastewater methods. Reliable identifications were demonstrated for at least three days after colony preparation, with storage at ambient or refrigerated temperatures. (However, please note that care must be taken with refrigeration as this can result in moisture being taken up when water-absorbent materials are used.) The cleaning protocols were demonstrated to be effective for reusable target plates and documented into in-house methods.

A large initial contributor to variability in the identifications and associated confidence scores was found to be the colony smearing technique when preparing colonies on the target plates. Robust training and assessment of the repeatability of outcomes from individual analysts and the reproducibility of outcomes from different analysts by repeat preparation of colony isolates were therefore required. In addition, the ability of the instrument to achieve repeatable identifications was also assessed (Table 16.6). If multiple instruments are to be used, measurement of reproducibility between instruments may also be suitable.

16.3.6.4 Comparison Assessment

A statistical comparison of the traditional and MALDI-TOF MS confirmation outcomes is necessary and typically forms the largest part of a validation exercise. Colonies of target and non-target organisms isolated using the laboratory's test methods derived from routine testing, reference cultures, and relevant proficiency testing scheme samples are confirmed. A suitable assessment such as that detailed in BS EN ISO 13843 [35] can then be used to assess the data, presenting the outcomes in a 2×2 diagram such as that in Figure 16.3.

The fraction of the confirmations correctly assigned (in this case, matching the traditional confirmation outcome) can be defined as the method efficiency (E) and calculated as $E = (a + d)/n$. In addition, to assess if one confirmation method performs significantly differently from another, a test based on the Poisson index of dispersion X^2, where

		Traditional confirmation method		
		+	−	
Biotyper confirmation	+	*a*	*b*	*a* + *b*
	−	*c*	*d*	*c* + *d*
		a + *c*	*b* + *d*	*n*

Figure 16.3 2 × 2 diagrammatic representation of colony confirmation outcomes; *n*, total no. of colonies tested; *a*, no. of colonies confirmed by both methods; *b*, no. of colonies confirmed by Biotyper, but not by traditional method; *c*, no. of colonies not confirmed by Biotyper, but confirmed by traditional method; *d*, no. of colonies.

$X^2 = (c - b)^2/(c + b)$ can be carried out. The confirmation results achieved from an exercise such as this are shown in Table 16.7.

An acceptable efficiency value was determined for this exercise to be > 95%, with values of < 95% but > 90% prompting further investigation. In addition, if the value of X^2 was ≥ 4, the methods were deemed to be 'different'. If the value of X^2 was < 4, the methods were deemed to be 'not different' (see Table 16.8).

This identified that the outcomes were suitable for all parameters other than enterococci and *C. perfringens*, prompting further investigation, including independent identification by a third-party laboratory. For *Enterococcus* spp., six colonies from raw water sources identified as *Enterococcus malodoratus* (which is associated with faecal sources) were not confirmed by the traditional method. For *C. perfringens,* six colonies showed a weak confirmation reaction using the traditional method beyond the time limit given. This indicated better performance by the MALDI-TOF MS method and, when considered in the significant difference assessments, gave outcomes of 'not different'.

The seven non-matching results for *Salmonella* were all from heavily contaminated sewage sludge samples, indicating that subculture to create suitable purity plates may be required before identification when single, well-isolated colonies are difficult to identify.

16.3.6.5 Ongoing Verification

As a laboratory accredited to UKAS ISO 17025, the performance of methods and instruments requires ongoing verification. Trending and monitoring of the confidence scores achieved from identification of routine analytical quality control cultures, and also the peak values achieved from routine system calibration (e.g. values achieved when calibrating the Bruker Microbiology and Diagnostics system with Bacterial Test Standard) can be used as system suitability checks for detecting drift; however, in practice this has rarely been observed. Performance is also verified on an ongoing basis by suitable results achieved through participation in proficiency testing schemes.

When updates are made to the installed libraries, particularly through routine updates to manufacturer-supplied libraries, these should be assessed to establish if there is any impact on the identification that may be achieved, and how this might affect the interpretation of that identification as a confirmation outcome of the wider method to which it is applied. This may prompt further verification exercises.

Table 16.7 Results of colony confirmation using traditional and MALDI-TOF techniques.

Salmonella species		Traditional confirmation method		
		+	−	
Biotyper confirmation	+	112	2	114
	−	5	13	18
		117	15	132
Legionella species		Traditional confirmation method		
		+	−	
Biotyper confirmation	+	378	0	318
	−	0	6	6
		318	6	384
Enterococci		Traditional confirmation method		
		+	−	
Biotyper confirmation	+	190	6	196
	−	0	18	18
		190	21	214
Clostridium perfringens		Traditional confirmation method		
		+	−	
Biotyper confirmation	+	179	9	188
	−	0	80	80
		179	89	268
Total coliforms		Traditional confirmation method		
		+	−	
Biotyper confirmation	+	348	10	358
	−	4	156	160
		352	166	518
Escherichia coli		Traditional confirmation method		
		+	−	
Biotyper confirmation	+	175	5	180
	−	6	11	17
		181	16	197
Pseudomonas aeruginosa		Traditional confirmation method		
		+	−	
Biotyper confirmation	+	66	0	66
	−	0	166	232
		66	166	229

Table 16.8 Outcome of efficiency and significant difference assessments.

Parameter	Efficiency (target > 90%)	X^2 result (target ≥ 4)	X^2 outcome
Salmonella species	94.7%	1.29	Not different
Legionella species	100%	0	Not different
Enterococci	97%	6	Different
Clostridium perfringens	96%	9	Different
Total coliforms	97%	2.57	Not different
Escherichia coli	94%	0.09	Not different
Pseudomonas aeruginosa	100%	0	Not different

16.3.7 Conclusion on Suitability for Use in an Environmental Testing Laboratory

The exercises described in Section 16.3.6 demonstrated that bacterial identification by MALDI-TOF MS is suitable for use in a commercial environmental and water hygiene testing laboratory as an alternative confirmation technique to those stated in national and international reference methods.

The MALDI-TOF MS confirmation gave a reliable outcome for *Legionella* spp., which is suitable to meet the requirements of the reference method. Identifications to species level are, however, achieved for the majority of colonies where previously a result of '*Legionella* species' would have been the outcome. This gives the client additional information about water system contaminants that was previously unavailable. An internal review of historical confirmation data has shown that greater than 50% of colonies identify as *Legionella anisa*. This is a potential area for further study to assess the prevalence of species other than *L. pneumophila* and their impact on public health. It is usual practice for laboratories to report the presence of *L. pneumophila* and its serogroups. This is typically achieved through the use of commercially available serology kits. MALDI-TOF MS now provides this species identification, but it has so far proved difficult to identify the species further to serogroup level. Serotyping is still therefore typically carried out using latex agglutination, but this is an area for potential further work.

As described in Section 16.3.6.1, the definition of a positive coliform when using MALDI-TOF MS is based on validation against the outcomes of the traditional confirmation and on published, recognized coliform genera. A reference list of acceptable coliform genera (and also those genera commonly not confirming as coliforms) is therefore required. The additional species information that is now available with MALDI-TOF MS confirmation is of use to the client for investigation and trending of bacterial contamination.

The benefits of MALDI-TOF MS confirmations for a commercial testing laboratory and its clients are many. The lack of subculture and interpretation of biochemical tests ensures that identification can be achieved on the same day. This identification is then interpreted as a confirmation result for the method and reported immediately to the client, rather than a presumptive result pending confirmation. A presumptive result might have required additional, unnecessary sampling, incurring associated costs such as engineers' and samplers' time, additional sample container, and additional analysis, which might also lead to an increased carbon footprint. The MALDI-TOF MS 'real-time' confirmed result could therefore prevent this. For the laboratory, having a unified confirmation technique across the microbiology methods reduces the training requirement for new staff and the MALDI-TOF provides immediate feedback during training. The need for most subculture agars and biochemical reagents is eliminated, giving a consumable cost saving, as is the additional presumptive reporting step. For the client, as the end user of the result, the reduced turnaround time through rapid confirmation allows system management decisions to be taken on confirmed results, removing uncertainty. More rapid intervention can take place, which is of benefit for the water systems being managed, and for public health.

16.4 Potential Application for *Cryptosporidium* Identification

The protozoan parasite *Cryptosporidium* has been a challenge for the water industry for half a century, since the first waterborne outbreaks of cryptosporidiosis were identified in the 1980s [36]. The oocyst stage, shed in faeces of a wide variety of animals including humans, may break through water treatment barriers and is resistant to chlorine disinfection [37]. Ingestion of even single numbers of human infective oocysts can lead to infection with the sporozoites contained within, causing gastroenteritis characterized by watery diarrhoea, abdominal pain, nausea, vomiting, low-grade fever, and anorexia [38]. Symptoms appear, on average, five to seven days after ingestion and can last for around two weeks. For severely immunocompromised patients, e.g. those with untreated HIV/AIDS or congenital immunodeficiencies, or who are on immunosuppressive drug treatments, *Cryptosporidium* can cause severe intractable diarrhoea and may infect the biliary tree and extraintestinal organs [39]. In Africa and South-east Asia, *Cryptosporidium* is one of the leading causes of morbidity and mortality in malnourished young children [40]. Long-term consequences of infection in malnourished children include growth faltering and cognitive defects [41], and, in high-income countries, persistent gastrointestinal upset, irritable bowel syndrome [42], myalgia/arthralgia, and fatigue [43], and an association with colon cancer have been reported [44].

Waterborne cryptosporidiosis is of concern globally, but outbreaks are more evident where surveillance is in place and where water supplies are routinely chlorinated, as *Cryptosporidium* survives this disinfection strategy that controls most other waterborne pathogens. There are significant economic and social impacts of *Cryptosporidium* outbreaks. A large waterborne outbreak in Milwaukee, Wisconsin, in 1993 that affected 403 000 people was estimated to cost US $96.2 million [45]. In Sweden, an outbreak with an attack rate of 45% of 60 000 residents was estimated to result in 50 000 sick leave days [46]. There is also a high impact of the interventions needed to mitigate the detection of *Cryptosporidium* oocysts in public water supplies and control outbreaks, including public notices to boil water and provision of bottled water. In Galway, Ireland, in 2007, notices to boil all water for consumption were in place for 158 days, affecting over 120 432 people and costing around €19 million [47].

Table 16.9 The most recent *Cryptosporidium* outbreaks linked to public water supplies in England and Wales and the information provided by genotyping *Cryptosporidium* spp.

Where	When	Infecting species and gp60 subtype	Epiemiological, environmental, and microbiological evidence (*Cryptosporidium* spp. and gp60 subtype) for association with drinking water
South-east England	2005	*Cryptosporidium hominis*	Analytical epidemiology Oocysts detected in treated water, <1/10 litres Not genotyped
Wales	2005–2006	*Cryptosporidium hominis* IbA10G2	Analytical epidemiology Oocysts detected in source and tap water, <1/10 litres *Cryptosporidium hominis* IbA10G2
East Midlands of England	2008	*Cryptosporidium cuniculus* VaA18	Descriptive epidemiology Oocysts detected at water treatment works and in distribution, <1/10 litres Oocysts in tap water and gut contents of a dead rabbit retrieved from a chlorine contact tank *Cryptosporidium cuniculus* VaA18
South-west England	2013	*Cryptosporidium hominis* IbA10G2	Analytical epidemiology Oocysts detected at works highest levels being 0.021, 0.023, and 0.024/10 litres *Cryptosporidium hominis*

A fault tree analysis of outbreaks in the European Union during 1995–2005 identified chronic filtration failures or livestock and rainfall in the catchment as key faults in protozoal (mainly *Cryptosporidium*) outbreaks [48]. Latterly, *Cryptosporidium* outbreaks in the UK have been attributed to human or wildlife sources, indicated, in part, by the *Cryptosporidium* species detected (Table 16.9). The application of molecular genotyping to *Cryptosporidium* in stools and water samples is beneficial in outbreak investigations to strengthen the evidence of links between cases of cryptosporidiosis with water supplies (Table 16.9). Genotyping is also beneficial to supplement catchment data to track sources of contamination to humans, farmed animals, or wildlife. Sequence analysis of part of the *Cryptosporidium* ssu rRNA gene is most commonly used to identify species; subtyping by sequencing part of the gp60 gene has been used to match isolates from cases of illness with those found in drinking water, source waters, and animals (Table 16.9). *Cryptosporidium* genotyping is currently a post hoc test from the microscope slides used for oocyst monitoring [49], but it could be performed by parallel testing where DNA is extracted from sample concentrates.

The characterization of *Cryptosporidium* spp. has focused on genetic analysis, and improved whole-genome reference sequences are available [50], enabling more accurate understanding of the genetics of the parasite, including new targets for microbial identification and characterization. Alternatively, for bacteria and fungi, the analysis of the proteome by MALDI-TOF MS has provided microbial identification and strain typing, and spectral databases containing peptide mass fingerprints of type and wild strains are well established for clinical and environmental microbiology [51]. Analysis of the cellular proteome additionally provides information derived directly from gene products and metabolism. Unfortunately, for *Cryptosporidium*, a major constraint is the availability of sufficient biomass for such studies. The study of extracellular stages from stools and water samples – oocysts and sporozoites – by MS would require purification of high numbers. Although improved culture systems are emerging, they remain limited in the amount of material generated [52], and oocyst purification methods from stools need to be refined. As demonstrated by Gathercole et al., this is possible for samples containing a sufficient number of parasites, and for the first time, distinct MALDI-TOF MS mass spectra have been reported for *Cryptosporidium parvum* and *Cryptosporidium hominis* derived from clinical stool samples, in addition to those from reference isolates propagated *in vivo* [53]. Stool samples from which $>2\times10^5$ oocysts were purified using salt flotation followed by potassium bromide gradient centrifugation successfully produced mass spectra from a microflex LT/SH instrument with flexControl and flexAnalysis software (Bruker Microbiology & Diagnostics). Although this work showed the potential for detection and characterization of high numbers of oocysts found in stools, it is doubtful that the application to water samples will be realized.

References

1 Tulchinsky, T.H. (2018). John Snow, cholera, the Broad Street pump; waterborne diseases then and now. In: *Case Studies in Public Health* (ed. T.H. Tulchinsky), 77–99. Academic Press.

2 Edberg, S.C., Rice, E.W., Karlin, R.J., and Allen, M.J. (2000). *Escherichia coli*: the best biological drinking water indicator for public health protection. *J. Appl. Microbiol.* 88: 106S–116S.

3 Standing Committee of Analysts. (2002). *The Microbiology of Drinking Water. Part 1 – Water quality and public health*. Methods for the Examination of Waters and Associated Materials. https://secureservercdn.net/160.153.137.128/fkp.e54.myftpupload.com/wp-content/uploads/Blue_Book_Library/176.pdf (accessed 5 October 2022).

4 Sarvari, K.P., Soki, J., Ivan, M. et al. (2018). MALDI-TOF MS versus 16S rRNA sequencing: minor discrepancy between tools in identification of bacteroides isolates. *Acta Microbiol. Immunol. Hung.* 65: 173–181.

5 Mellmann, A., Cloud, J., Maier, T., and Keckevoet, U. (2008). Evaluation of matrix-assisted laser desorption ionization-time-of-flight mass spectrometry in comparison to 16S rRNA gene sequencing for species identification of non-fermenting bacteria. *J. Clin. Microbiol.* 46: 1946–1954.

6 Ford, B.A. and Burnham, C.D. (2013). Optimization of routine identification of clinically relevant Gram-negative bacteria by use of matrix-assisted laser desorption, ionization–time of flight mass spectrometry and the Bruker Biotyper. *J. Clin. Microbiol.* 51 (5): 1412–1420.

7 Pinar-Méndez, A., Fernández, S., Baquero, D. et al. (2021). Rapid and improved identification of drinking water bacteria using the drinking water library, a dedicated MALDI-TOF MS database. *Water Res.* 203: 117543.

8 Standing Committee of Analysts. (2020). *The Microbiology of Drinking Water. Part 4 – The identification of microorganisms using MALDI-TOF mass spectrometry*. Methods for the Examination of Waters and Associated Materials. https://secureservercdn.net/160.153.137.128/fkp.e54.myftpupload.com/wp-content/uploads/Blue_Book_Library/270.pdf (accessed 5 October 2022).

9 Dŵr Cymru Welsh Water. (2021). Providing clean water (website). https://www.dwrcymru.com/en/our-services/water/providing-clean-water (accessed 12 October 2021).

10 Standing Committee of Analysts. (2016). *The Microbiology of Drinking Water. Part 4 – Methods for the isolation and enumeration of coliform bacteria and* Escherichia coli *(including E. coli O157:H7)*. Methods for the Examination of Waters and Associated Materials. https://secureservercdn.net/160.153.137.128/fkp.e54.myftpupload.com/wp-content/uploads/Blue_Book_Library/277.pdf (accessed 5 October 2022).

11 Tan, K.E., Ellis, B.C., Lee, R. et al. (2012). Prospective evaluation of a matrix-assisted laser desorption ionization-time of flight mass spectrometry system in a hospital clinical microbiology laboratory for identification of bacteria and yeasts: a bench-by-bench study for assessing the impact on time to identification and cost-effectiveness. *J. Clin. Microbiol.* 50: 3301–3308.

12 Eigner, U., Holfelder, M., Oberdorfer, K. et al. (2009). Performance of a matrix-assisted laser desorption ionization time-of-flight mass spectrometry system for the identification of bacterial isolates in the clinical routine laboratory. *Clin. Lab.* 55: 289–296.

13 Reading, N., Dadrah, A., Symmonds, A. et al. (2012). *Evaluation of the Vitek MS® MALDI-TOF Mass Spectrometry System in a Routine Clinical Laboratory*. Department of Microbiology, Sandwell & West Birmingham Hospitals NHS Trust.

14 Devanga Ragupathi, N.K., Muthuirulandi Sethuvel, D.P., Inbanathan, F.Y., and Veeraraghavan, B. (2017). Accurate differentiation of *Escherichia coli* and *Shigella* serogroups: challenges and strategies. *New Microbes New Infect.* 21: 58–62.

15 Kobayashi, H., Tanizawa, Y., Sakamoto, M. et al. (2020). Reclassification of *Clostridium diolis* Biebl and Spröer 2003 as a later heterotypic synonym of *Clostridium beijerinckii* Donker 1926 (approved lists 1980) emend. Keis et al. 2001. *Int. J. Syst. Evol. Microbiol.* 70 (4): 2463–2466.

16 Standing Committee of Analysts. (2012). *The Microbiology of Drinking Water. Part 5 – Methods for the isolation and enumeration of enterococci*. Methods for the Examination of Waters and Associated Materials. https://secureservercdn.net/160.153.137.128/fkp.e54.myftpupload.com/wp-content/uploads/Blue_Book_Library/238.pdf (accessed 5 October 2022).

17 Standing Committee of Analysts. (2021). *The Microbiology of Drinking Water. Part 6 – Methods for the isolation and enumeration of sulphite-reducing clostridia and* Clostridium perfringens *by membrane filtration*. Methods for the Examination of Waters and Associated Materials. https://secureservercdn.net/160.153.137.128/fkp.e54.myftpupload.com/wp-content/uploads/Blue_Book_Library/275.pdf (accessed 5 October 2022).

18 Boadi, S., Wren, M., and Morris-Jones, S. (2010). Selective testing of β-galactosidase activity in the laboratory identification of *Salmonella* and *Shigella* species. *J. Clin. Pathol.* 63: 1101–1104.

19 Gu, Y.F., Li, Y., Zhang, X.L. et al. (2019). A new method aimed to quickly identify pathogen and drug susceptibility test based on matrix-assisted laser desorption/ionization time of flight mass spectrometry combined with flow cytometry. *Surg. Infect.* 20: 219–224.

20 Rowbotham, T.J. (1980). Preliminary report on the pathogenicity of *Legionella pneumophila* for freshwater and soil amoebae. *J. Clin. Pathol.* 33: 1179–1183.

21 Logan-Jackson, A. and Rose, J.B. (2021). Cooccurrence of five pathogenic *Legionella* spp. and two free-living amoebae species in a complete drinking water system and cooling towers. *Pathogens* 10: 1407. https://doi.org/10.3390/pathogens10111407.

22 Health and Safety Executive. 2013. Legionnaires' disease. The control of legionella bacteria in water systems. Approved Code of Practice and Guidance L8. www.hse.gov.uk/pubns/books/l8.htm (accessed 5 October 2022).

23 Health and Safety Executive. 2014. Legionnaires' disease. Technical Guidance. HSG 274 (3 parts). www.hse.gov.uk/pubns/books/hsg274.htm (accessed 5 October 2022).

24 Health and Safety Executive. 2017. Control of legionella and other infectious agents in spa-pool systems. HSG282. https://www.hse.gov.uk/pubns/books/hsg282.htm (accessed 5 October 2022).

25 British Standards Institution. 2019. BS 8580-1:2019 Water quality – Risk assessments for Legionella control. Code of Practice. https://shop.bsigroup.com/products/water-quality-risk-assessments-for-legionella-control-code-of-practice-1 (accessed 5 October 2022).

26 International Organization for Standardization. 2017a. BS EN ISO 11731:2017 Water quality. Enumeration of Legionella. https://shop.bsigroup.com/ProductDetail/?pid=000000000030266280 (accessed 5 October 2022).

27 Department for Environment and Rural Affairs. 2018. Guidance. Sewage sludge in agriculture: code of practice for England, Wales and Northern Ireland. https://www.gov.uk/government/publications/sewage-sludge-in-agriculture-code-of-practice/sewage-sludge-in-agriculture-code-of-practice-for-england-wales-and-northern-ireland#producers-test-sewage-sludge (accessed 5 October 2022).

28 The Standing Committee of Analysts. 2003a. *The Microbiology of Sewage Sludge (2003) - Part 1 - An overview of the treatment and use in agriculture of sewage sludge in relation to its impact on the environment and public health.* Methods for the Examination of Waters and Associated Materials. https://secureservercdn.net/160.153.137.128/fkp.e54.myftpupload.com/wp-content/uploads/Blue_Book_Library/188.pdf (accessed 5 October 2022).

29 The Standing Committee of Analysts. 2003b. *The Microbiology of Sewage Sludge (2003) - Part 3 -Methods for the Isolation and Enumeration of* Escherichia coli, *including Verocytotoxigenic* Escherichia coli. Methods for the Examination of Waters and Associated Materials. https://secureservercdn.net/160.153.137.128/fkp.e54.myftpupload.com/wp-content/uploads/Blue_Book_Library/190.pdf (accessed 5 October 2022).

30 The Standing Committee of Analysts. 2004. *The Microbiology of Sewage Sludge (2004) - Part 4 -Methods for the detection, isolation and enumeration of* Salmonellae. Methods for the Examination of Waters and Associated Materials. https://secureservercdn.net/160.153.137.128/fkp.e54.myftpupload.com/wp-content/uploads/Blue_Book_Library/191.pdf (accessed 5 October 2022).

31 Department of Health. 2016. Health Technical Memorandum 04-01: Safe water in healthcare premises. Department of Health. https://www.england.nhs.uk/publication/safe-water-in-healthcare-premises-htm-04-01/ (accessed 5 October 2022).

32 Walker, J., Moore, G., Collins, S. et al. (2017). Microbiological problems and biofilms associated with *Mycobacterium chimaera* in heater-cooler units used for cardiopulmonary bypass. *J. Hosp. Infect.* 96 (3): 209–220. https://doi.org/10.1016/j.jhin.2017.04.014.

33 Mulani, M.S., Kamble, E.E., Kumkar, S.N. et al. (2019). Emerging strategies to combat ESKAPE pathogens in the era of antimicrobial resistance: a review. *Front. Microbiol.* 10: 539. https://doi.org/10.3389/fmicb.2019.00539.

34 Kanamori, H., Weber, D.J., and Rutala, W.A. (2016). Healthcare outbreaks associated with a water reservoir and infection prevention strategies. *Clin. Infect. Dis.* 62 (11): 1423–1435. https://doi.org/10.1093/cid/ciw122.

35 International Organization for Standardization. 2017b. BS EN ISO 13843:2017 Water quality Requirements for establishing performance characteristics of quantitative microbiological methods. https://shop.bsigroup.com/products/water-quality-requirements-for-establishing-performance-characteristics-of-quantitative-microbiological-methods/standard (accessed 5 October 2022).

36 Karanis, P., Kourenti, C., and Smith, H. (2007). Waterborne transmission of protozoan parasites: a worldwide review of outbreaks and lessons learnt. *J. Water Health* 5: 1–38.

37 Betancourt, W.Q. and Rose, J.B. (2004). Drinking water treatment processes for removal of *Cryptosporidium* and *Giardia*. *Vet. Parasitol.* 126: 219–234.

38 Chalmers, R.M. and Davies, A.P. (2010). Minireview: clinical cryptosporidiosis. *Exp. Para.* 124: 138–146.

39 Hunter, P.R. and Nichols, G. (2002). Epidemiology and clinical features of *Cryptosporidium* infection in immunocompromised patients. *Clin. Microbiol. Rev.* 15: 145–154.

40 Kotloff, K.L., Nataro, J.P., Blackwelder, W.C. et al. (2013). Burden and aetiology of diarrhoeal disease in infants and young children in developing countries (the Global Enteric Multicenter Study, GEMS): a prospective, case-control study. *Lancet* 382: 209–222.

41 Khalil, I.A., Troeger, C., Rao, P.C. et al. (2018). Morbidity, mortality, and long-term consequences associated with diarrhoea from *Cryptosporidium* infection in children younger than 5 years: a meta-analyses study. *Lancet Glob. Health* 6: e758–e768.

42 Carter, B.L., Stiff, R.E., Elwin, K. et al. (2019). Health sequelae of human cryptosporidiosis-a 12-month prospective follow-up study. *Eur. J. Clin. Microbiol. Infect. Dis.* 38: 1709–1717.

43 Insulander, M., Silverlås, C., Lebbad, M. et al. (2013). Molecular epidemiology and clinical manifestations of human cryptosporidiosis in Sweden. *Epidemiol. Infect.* 141: 1009–1020.

44 Osman, M., Benamrouz, S., Guyot, K. et al. (2017). High association of *Cryptosporidium* spp. infection with colon adenocarcinoma in Lebanese patients. *PLoS One* 12: e0189422.

45 Corso, P.S., Kramer, M.H., Blair, K.A. et al. (2003). Cost of illness in the 1993 waterborne *Cryptosporidium* outbreak, Milwaukee, Wisconsin. *Emerg. Infect. Dis.* 9: 426–431.

46 Ridderstedt, F., Widerström, M., Lindh, J., and Lilja, M. (2018). Sick leave due to diarrhea caused by contamination of drinking water supply with *Cryptosporidium hominis* in Sweden: a retrospective study. *J. Water Health* 16: 704–710.

47 Chyzheuskaya, A., Cormican, M., Srivinas, R. et al. (2017). Economic assessment of waterborne outbreak of cryptosporidiosis. *Emerg. Infect. Dis.* 23: 1650–1656.

48 Risebro, H.L., Doria, M.F., Andersson, Y. et al. (2007). Fault tree analysis of the causes of waterborne outbreaks. *J. Water Health* 5 (Suppl. 1): 1–18.

49 Di Giovanni, G.D., Rochelle, P.A., Hoffman, R.M., and Akagi, Y. (2015). Maximizing the value of your Lt2 *Cryptosporidium* monitoring. *J. AWWA* 107: 73–75.

50 Baptista, R.P., Li, Y., Sateriale, A., Sanders, M.J., Brooks, K.L., Tracey, A., Ansell, B.R.E., Jex, A.R., Cooper, G.W., Smith, E.D., Xiao, R., Dumaine, J.E., Berriman, M., Striepen, B., Cotton, J.A., Kissinger J.C. (2021). Long-read assembly and comparative evidence-based reanalysis of *Cryptosporidium* genome sequences reveal new biological insights. *bioRxiv*. https://doi.org/10.1101/2021.01.29.428682.

51 Singhal, N., Kumar, M., Kanaujia, P.K., and Virdi, J.S. (2015). MALDI-TOF mass spectrometry: an emerging technology for microbial identification and diagnosis. *Front. Microbiol.* 6: 791.

52 Bhalchandra, S., Cardenas, D., and Ward, H.D. (2018). Recent breakthroughs and ongoing limitations in *Cryptosporidium* research. *F1000Res* 7: Faculty Rev-1380.

53 Gathercole, R., Tranfield, E., Xia, D. et al. (2021). Analysis of *Cryptosporidium* spp. from clinical samples by matrix-assisted laser desorption ionisation-time of flight mass spectrometry. *J. Appl. Microbiol.* 131: 1840–1847.

17

A New MALDI-TOF Database Based on MS Profiles of Isolates in Icelandic Seawaters for Rapid Identification of Marine Strains

Sibylle Lebert[1], Viggó Þór Marteinsson[1,2,3], and Pauline Vannier[1]

[1] *Microbiology Group, Matís, Reykjavík, Iceland*
[2] *Faculty of Food Science and Nutrition, University of Iceland, Reykjavík, Iceland*
[3] *The Agricultural University of Iceland, Reykjavík, Iceland*

17.1 Introduction

Oceans cover about 71% of our planet and thus represent the largest natural reservoir on Earth. Among the numerous ecosystems present in the oceanic environments, microbes play an essential role. They govern most of the large geochemical cycles, such as the carbon, nitrogen, and phosphorus cycles, and supply 98% of the primary production on Earth [1]. This should be viewed in perspective with the fact that marine microorganisms are omnipresent in oceanic waters, with a general distribution of 10^4–10^7 cells/ml [2, 3]. They represent about 70% of the whole marine biomass, whereas the estimate is around 2% for the terrestrial biomass [4, 5]. In total, marine and deep subsurface biomass represent 76 Gt of carbon, i.e. about six-fold less than the terrestrial biomass. As the number of studies on the important role of the oceans in regulating and predicting the climate continues to increase, it is impossible to ignore the extent to which ocean ecosystems are threatened partly because of anthropogenic pressures [6]. In particular, the North Atlantic, which represents most of the Icelandic seawaters, comes under major changes on various scales [7]. However, there is still a lack of studies in the field of marine microbiology on diversity and the effect of climate change on the oceans [8]. The axiom that less than 1% of living bacteria are cultivable in laboratory conditions should be challenged [9, 10]. We should instead focus more on our ability to cultivate the remaining 99% of uncultivated marine microorganisms [11, 12]. This constitutes the most challenging part of studying microorganisms. According to the microbiota cultivation-based studies of Venter et al., the dominant bacterial species found in the ocean are members of *Vibrio*, *Alteromonas*, and *Pseudomonas* genera [13]. Bacterial communities inhabiting polar regions are far less studied, whereas the ongoing global warming has a profound influence on the Arctic. Iceland, located in the northern part of the Atlantic Ocean, is influenced by two main oceanic currents: the North

Microbiological Identification using MALDI-TOF and Tandem Mass Spectrometry: Industrial and Environmental Applications, First edition. Edited by Haroun N. Shah, Saheer E. Gharbia, Ajit J. Shah, Erika Y. Tranfield, and K. Clive Thompson.

Atlantic current and the East Greenland current. So far, very little is known about the cultivated microbial part of the northern hemisphere and Icelandic marine waters are still poorly studied. Cultivating microorganisms can provide important information in terms of identifying the key players in the oceans, their genetic diversity, and their efficient survival mechanisms developed during evolution. Marine microorganisms represent one the main origins of biomolecules for pharmacology purposes [14], as well as cosmetics or bioremediation. In a more general sense, marine microorganisms, which include bacteria, constitute the most substantial potentially valorized source of the blue bioeconomy and biotechnology [15].

The most conventional method for identifying microbial diversity is 16S rRNA gene sequencing of either DNA extracted directly from the environment (uncultivated microbiomes) or DNA from a cultivated strain which can then be tested further. Sequencing is still a costly and time-consuming method that relies on the knowledge of small invariable regions of the 16S rRNA gene. 16S rRNA gene sequencing is often the only option for identification of many marine strains. Nevertheless, a much quicker and more cost-effective method is needed to identify marine isolates. The matrix-assisted laser desorption/ionization time-of-flight mass spectrometry (MALDI-TOF MS) technique appears to be a perfect method for such identification or at least for clustering isolates before further identification with a conventional method such as 16S rRNA gene sequencing. Environmental databases such as the marine microbial database are less accessible compared with those in the fields of clinical and food microbiology. A growing interest in the utilization of the MALDI-TOF MS technique in environmental microbiology has been observed over the past few years. For instance, good results have been observed in environmental studies, mainly on aquatic pathogens [16] but also on marine microbial strains isolated from seawater with a very high rate of identification [17]. Indeed, by using the spectral signature to identify microorganisms, MALDI-TOF MS is quicker, more cost-effective, and proven to be accurate for bacterial identification as compared with classical phenotypic and genotypic methods. However, its application in this field raises many challenges, the main one being the lack of representation of environmental isolates in the Bruker commercial database. Most of the databases of spectra available worldwide are designed for human pathogenic strains. Some MS libraries such as SpectraBank are in open access, but the number of added spectra is drastically slowing and close to stationary [18]. This lack leads to a high rate of misidentification or non-identification. In this context, many researchers cope with this deficiency and enhance their identification rate by creating their own in-house databases [19]. However, as these databases are not available in open access, the problem is only solved on a local scale. The objective of this study was therefore to generate an in-house and extended database to obtain a more accurate identification of Icelandic marine environmental microbial species in comparison with the Bruker database.

17.2 Selection and Cultivation of the Strains

The strains used in this study were isolated from Icelandic seawaters using samples collected during trips around Iceland in May 2012 and May 2013. Briefly, four different media were used: Marine Broth Agar™ (Difco), R2A Agar™ (Difco), Trypticase Soy Yeast Extract

Agar, and Cythophaga marine media. Plates of different media were inoculated with 100 µl of the seawater samples and incubated at 4 and 10°C until growth occurred. Each colony that looked different was picked and streaked at least three times on a new plate to obtain pure culture with the three-sector 'T streak' method [20]. After being isolated, the pure strains were stored in the Icelandic Strain Collection and Records (ISCAR; http://iscar.matis.is) at −140°C in a 20% glycerol stock solution.

Strains were cultivated on an agar plate in the first trial of identification. After streaking them at least three times more, their 16S rRNA genes were partially amplified and sequenced. In total, 132 cultivated strains were identified to the genus level and chosen for screening. The influences of various media, incubation time, and temperature on cell growth and MALDI-TOF MS data acquisition were also tested. In line with the 1999 results of Arnold et al. [21], for the same strain different spectral profiles were found when different cultivation parameters were assessed (Figure 17.1). Of the 132 selected strains, 52 were chosen to be analyzed a second time on the MALDI-TOF Biotyper and their 16S rRNA gene was fully sequenced. The results of the first and second genotypic identifications combined, as well as the sampling locations and IDs, are shown in Table 17.1. These two selected batches (batch of 132 then batch of 52) were both taken as being representative of the diversity of genera found in the Icelandic seawaters to create the in-house database.

According to the results of the second trial, the strains of the third trial were cultivated on four different media according to the salinity needs of the bacteria: Marine Broth Agar™ (Difco), R2A Agar™ (Difco), Trypticase Soy Yeast Extract Agar, and Cythophaga marine media. They were all grown at 17°C for 48–72 hours, depending on the strains, under aerobic conditions. Results were obtained for 51 strains as the strain ISCAR-5352 was lost during the process.

17.3 Genotypic Identification

The V1–V4 region of the 16S rRNA gene was first amplified with the 27F/806R primer set [22, 23]. Although this 16S rRNA gene region has shown good results so far, a recent study highlighted that the taxonomic results obtained with the 16S rRNA gene partially sequenced do not always match the expected taxonomic resolution [24]. Therefore, for creating our database and identifying our cultivated strains to the species level accurately, the whole 16S rRNA gene of all our isolates in this project was sequenced.

DNA extraction was performed with the MasterPure kit (Epicentre Technologies Corp.) on fresh cultivated and isolated strains according to the manufacturer's instructions. DNA was resuspended in Tris buffer (10 mM, pH 8), the quality and concentration were assessed with the NanoDrop (ND-1000 UV-Vis; Thermo Scientific™), and stored at −20°C until further analysis. The whole 16S rRNA gene was amplified by PCR with three different primer sets: 27F [23] and 806R [22], 515F [25] and 1046R [1] and 784F [26] and 1492R [23]. Amplified extracts were cleaned using ExoSAP-IT™ (ThermoFisher Scientific), dyed using BigDye™ Terminator v3.1 (ThermoFisher Scientific) and purified according to the Angecourt CleanSEQ Dye-Terminator Removal protocol (Beckman Coulter). Purified amplicons were then sequenced using the Sanger method (3070 DNA Analyser – Hitachi – Applied Biosystems).

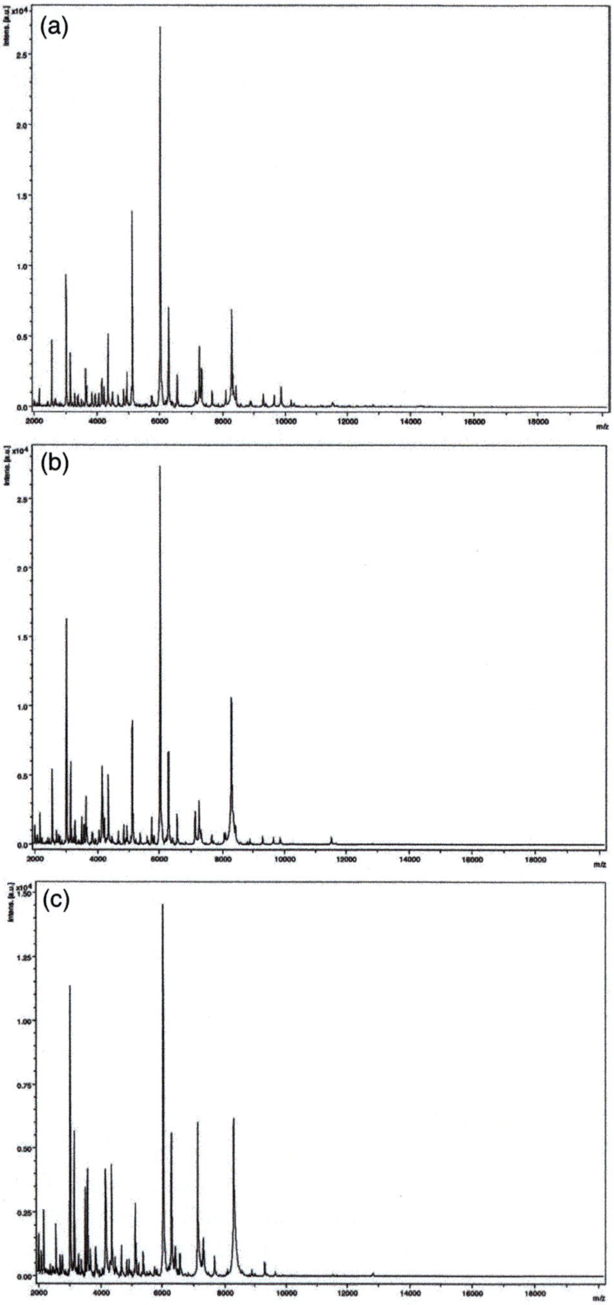

Figure 17.1 Different spectra of strain ISCAR-5297 belonging to *Halomonas*, obtained after cultivation under six different conditions (view from FlexAnalysis 3.1 software after applying MBT_StandardF Method, a baseline subtraction and a smoothing). These spectra were chosen as a "showcase" for all other spectra obtained from the other 51 isolates in this study. (a–f) Spectra obtained after growing strain ISCAR-5297 at 17°C for four days on Cytophaga marine medium (optimal culture conditions) (a), at 17°C for five days on Cytophaga marine medium (b), at 30°C for four days on Cytophaga marine medium (c), at 30°C for five days on Cytophaga marine medium (d), at 17°C for four days on Cytophaga marine medium diluted twice (e) and at 17°C for four days on Marine agar medium (f).

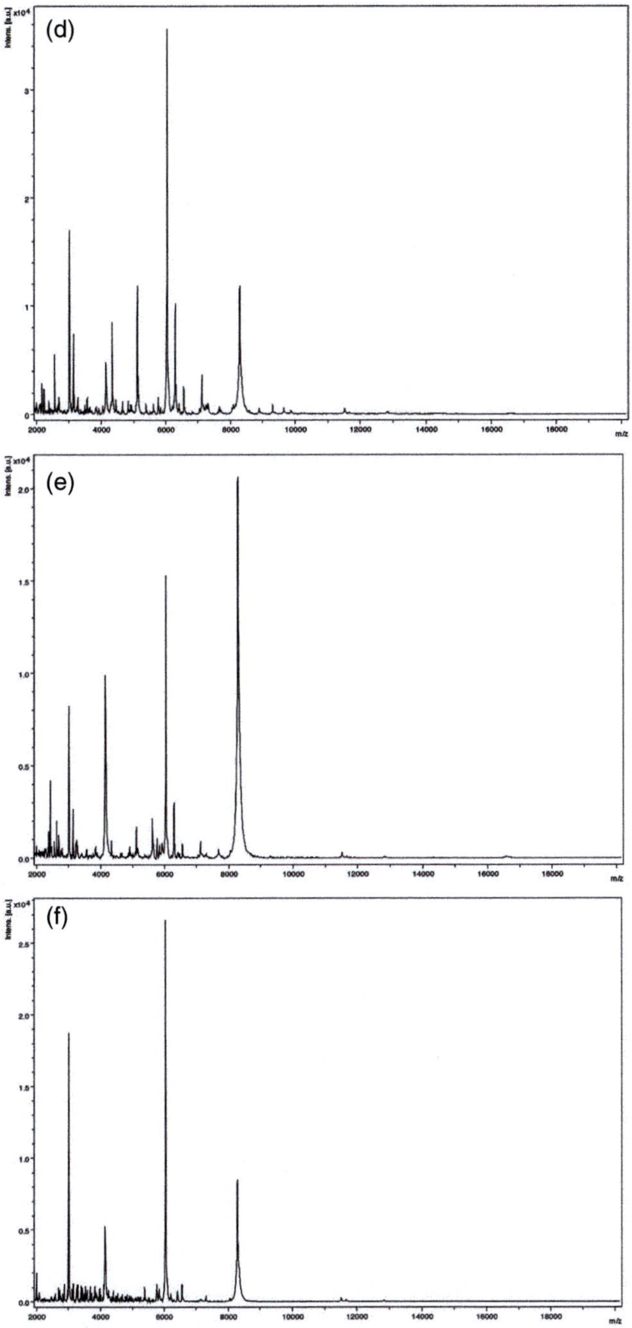

Figure 17.1 (Continued)

Table 17.1 Background information on the strains at the start of the study.

ISCAR number	Original water sample number	Sampling location	Identification from first and second trials
ISCAR-5306	INGOB-001	63°46′0.12″ N, 16°35′58.92″W	*Alteromonas*
ISCAR-5451	FX002-010	64°20′2.4″ N, 23°0′10.8″ W	*Brevundimonas*
ISCAR-5216	SIGLD-001	67°59′58.92″ N, 18°49′47.28″ W	*Croceibacter*
ISCAR-5277	SIGLD-001	67°59′58.92″ N, 18°49′47.28″ W	*Croceibacter*
ISCAR-5191	LANND-001	67°59′59.28″ N, 12°40′10.92″ W	*Dietzia*
ISCAR-5435	LANND-001	67°59′59.28″ N, 12°40′10.92″ W	*Dietzia*
ISCAR-5540	SELVE-001	62°58′54.48″ N, 21°28′57.72″ W	*Colwellia*
ISCAR-5314	ICUNK-001	64°57′46.8″ N, 19°1′14.88″ W	*Pseudoalteromonas*
ISCAR-5377	KOGUC-001	66°41′2.4″ N, 23°9′9″ W	*Dasania*
ISCAR-5297	LANNC-001	66°36′56.52″ N, 14°15′45.72″ W	*Halomonas*
ISCAR-5688	LANNC-001	66°36′56.52″ N, 14°15′45.72″ W	*Halomonas*
ISCAR-5564	SELVE-001	62°58′54.48″ N, 21°28′57.72″ W	*Idiomarina*
ISCAR-5606	HORNB-001	66°59′57.12″ N, 21°34′50.52″ W	*Joostella*
ISCAR-5257	SIGLE-001	67°59′56.4″ N, 18°50′1.68″ W	*Loktanella*
ISCAR-5706	SIGLE-001	67°59′56.4″ N, 18°50′1.68″ W	*Loktanella*
ISCAR-5264	LANND-001	67°59′59.28″ N, 12°40′10.92″ W	*Marinobacter*
ISCAR-5439	ICUNK-001	64°57′46.8″ N, 19°1′14.88″ W	*Marinobacter*
ISCAR-5186	HORNB-001	66°59′57.12″ N, 21°34′50.52″ W	*Marinobacter*
ISCAR-5228	KOGUD-001	67°35′3.48″ N, 23°56′8.88″ W	*Marinobacter*
ISCAR-5319	SIGLE-001	67°59′56.4″ N, 18°50′1.68″ W	*Marinobacter*
ISCAR-5693	SIGLC-001	66°32′3.12″ N, 18°49′57.72″ W	*Nocardioides*
ISCAR-5511	LANND-001	67°59′59.28″ N, 12°40′10.92″ W	*Sphingopyxis*
ISCAR-5523	LATRC-001	65°35′6″ N, 24°55′19.2″ W	*Sphingopyxis*
ISCAR-5328	LATRC-001	65°35′6″ N, 24°55′19.2″ W	*Strenotrophomonas*
ISCAR-5362	KROSC-001	64°59′55.32″ N, 10°6′54″ W	*Croceibacter*
ISCAR-5467	SIGLD-001	67°59′58.92″ N, 18°49′47.28″ W	*Pseudoalteromonas*
ISCAR-5442	SELVC-001	63°28′59.52″ N, 20°54′7.92″ W	*Pseudomonas*
ISCAR-5463	SIGLD-001	67°59′58.92″ N, 18°49′47.28″ W	*Pseudomonas*
ISCAR-5493	SIGLD-001	67°59′58.92″ N, 18°49′47.28″ W	*Sulfitobacter*
ISCAR-5315	FX007-012	64°20′2.4″ N, 26°0′14.4″ W	*Idiomarina*
ISCAR-540	ICUNK-001	64°57′46.8″ N, 19°1′14.88″ W	*Psychrobacter*
ISCAR-5417	KOGUD-001	67°35′3.48″ N, 23°56′8.88″ W	*Psychroserpens*
ISCAR-5524	LANND-001	67°59′59.28″ N, 12°40′10.92″ W	*Sphingopyxis*
ISCAR-5835	LATRC-001	65°35′6″ N, 24°55′19.2″ W	*Colwellia*
ISCAR-5243	SIGLE-001	67°59′56.4″ N, 18°50′1.68″ W	*Rhodococcus*
ISCAR-5202	LANNC-001	66°36′56.52″ N, 14°15′45.72″ W	*Alteromonas*

Table 17.1 (Continued)

ISCAR number	Original water sample number	Sampling location	Identification from first and second trials
ISCAR-5309	LANNC-001	66°36′56.52″ N, 14°15′45.72″ W	*Colwellia*
ISCAR-5490	SIGLC-001	66°32′3.12″ N, 18°49′57.72″ W	*Porphybacter*
ISCAR-5258	SIGLE-001	67°59′56.4″ N, 18°50′1.68″ W	*Reinekea*
ISCAR-5482	FX002-010	64°20′2.4″ N, 23°0′10.8″ W	*Rhodococcus*
ISCAR-5222	FX002-014	64°20′2.4″ N, 23°0′10.8″ W	*Rhodococcus*
ISCAR-5466	LANND-001	67°59′59.28″ N, 12°40′10.92″ W	*Dasania*
ISCAR-5468	LATRC-001	65°35′6″ N, 24°55′19.2″ W	*Rhodococcus*
ISCAR-5647	SIGLD-001	67°59′58.92″ N, 18°49′47.28″ W	*Rhodococcus*
ISCAR-5828	FX007-008	64°20′2.4″ N, 26°0′14.4″ W	*Shimia*
ISCAR-5479	FX007-012	64°20′2.4″ N, 26°0′14.4″ W	*Sneathiella*
ISCAR-5185	HORNB-001	66°59′57.12″ N, 21°34′50.52″ W	*Colwellia*
ISCAR-5382	LANND-001	67°59′59.28″ N, 12°40′10.92″ W	*Strenotrophomonas*
ISCAR-5210	SIGLD-001	67°59′58.92″ N, 18°49′47.28″ W	*Sulfitobacter*
ISCAR-5837	FX003-015	64°20′2.4″ N, 23°15′7.2″ W	*Sulfitobacter*
ISCAR-5259	LANNC-001	66°36′56.52″ N, 14°15′45.72″ W	*Sulfitobacter*
ISCAR-5352	LATRD-001	66°1′5.52″ N, 26°48′10.08″ W	*Bacillus*

The gathered sequences were analyzed and merged into contigs using Sequencher (v.5.2.4; Genes Codes Corporation). Ends were trimmed using the automatic settings, and empty sequences and those presenting a quality rate inferior to 85% were removed. The quality of the sequences was assessed by checking the chromatograms, and sequences were curated manually before being assembled. Sequences at least 1100 bp long were obtained to have a query length long enough for BLAST alignment. An exception was made for the ISCAR-5693 strain, identified as a *Nocardioides*, as only 765 bp of the 16S rRNA gene was successfully amplified. The decision to add it to the batch of candidates for the database creation stemmed from the uniqueness of its genus.

The 16S ribosomal RNA sequences (Bacteria and Archaea) database was selected on the BLAST algorithm and only highly similar sequences (megablast) were considered. The sequences presenting the smaller *e*-value, the highest query coverage and identity percentages were selected as the best matches. All the best matches chosen for the strains presented an *e*-value of 0.0 and a query coverage > 99%, apart from strain ISCAR-5216, which presented a query coverage of 82%. As for strain ISCAR-5693, an exception was made, and the strain was kept in the batch of potential entry to the database. Usually, an identity percentage of at least 97% can be used to differentiate microbial strains to the species level, whereas the limit of 95% can be used for the genus level [27]. Among the 51 sequenced strains, 46 were identified at the species level, 3 at the genus level and 2 presented identity scores under 95% despite a high quality of sequences. These last results could be interesting for further strain characterization studies. Even though they were not clearly identified,

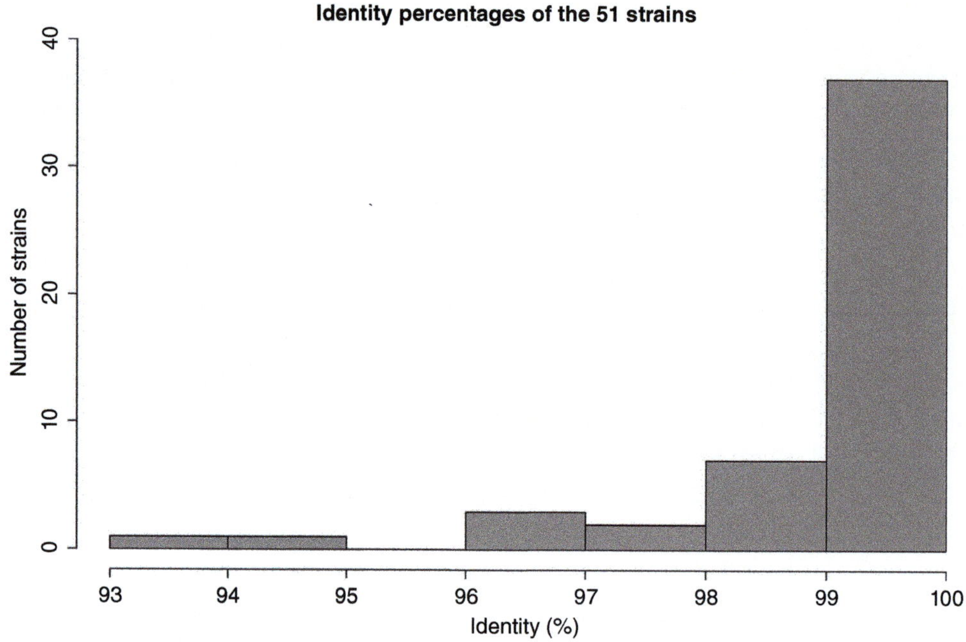

Figure 17.2 Repartition of identity percentages of all strains found with the best-aligned sequence on BLAST.

they are representative of the microbial diversity found in the seawaters around Iceland and thus need to be added to the database. The results of the identity percentages obtained with the best aligned sequence on BLAST for each strain is shown in Figure 17.2.

17.4 MALDI-TOF MS Data Acquisition and Database Creation

In parallel with the genotypic identification, MALDI Biotyper data acquisition was performed. First, protein extraction and purification steps were performed with 70% ethanol, 70% formic acid, and 100% acetonitrile, according to the protocol in Faron et al. [28]. As with genotypic identification, only fresh colonies cultivated under optimal conditions were used. A volume of 1 μl of the supernatant of the purified protein extracts was spotted in eight replicates on a reusable main spectral profile (MSP) 96 polished steel BC target (Bruker Microbiology & Diagnostics). The target was air-dried at room temperature under a fume hood for at least 10 minutes. Formerly, 1 μl of a saturated α-cyano-4-hydroxycinnamic acid (CHCA) matrix solution (Bruker Daltonics) was added. According to the manufacturer's instructions, calibration of each run was assessed by a triplicate of 1 μl of bacterial test standard (BTS) (Bruker Microbiology & Diagnostics) also covered by 1 μl of CHCA matrix [29]. The BTS is a protein extract of *Escherichia coli* DH5 alpha with a specific protein profile in MALDI-TOF mass spectra. It also contains two proteins to extend the overall mass range up to 17 kDa (total range 3.6–17 kDa).

The acquisition of mass spectra was done using a Microflex LT mass spectrometer (Bruker Microbiology & Diagnostics) combined with the MALDI Biotyper software package and the reference database MBT DB-7854 MSP (Bruker Microbiology & Diagnostics). The analyzed spectra were in the m/z range 2000–20 000 Da.

For each tested isolate, the first run was carried out using the Compass software (MBT Compass 4.1; Bruker Microbiology & Diagnostics), which compares the extract to a selected database and calculates a log score of identification. This software is usually used in pathogen screening. The goal of this first run was to assess the ability of the commercial MALDI Biotyper database to identify marine microbial strains. Identification log scores ≥ 2.0 was accepted as a reliable identification at the species level and log scores between ≥ 1.7 and < 2.0 were considered as reliable at the genus level, whereas identification with a log score < 1.7 were considered unreliable [30]. Three isolates were accurately identified: strains ISCAR-5185 and ISCAR-5382 at the species level (*Stenotrophomonas maltophilia*) with log score > 2.090, and strain ISCAR-5451, which was only assigned to the genus level: *Brevundimonas*. A second run carried out using another piece of MALDI Biotyper software, FlexControl (FlexControl 3.4; Bruker Microbiology & Diagnostics) was done. For each new run, a new protein extract and therefore a new target were prepared using the same protocol. This software, although less convenient to use, allows the user to generate multiple spectra per spot and to export the generated spectra more easily to other software for further analysis. Each sample is spotted onto the target plate 8 times and then for each spot, a minimum of three spectra were generated, thus keeping the number of spectra per strain to a minimum of 24 spectra per run.

The generated spectra with both methods (Compass and FlexControl) were imported into the FlexAnalysis software (FlexAnalysis 3.4; Bruker Microbiology & Diagnostics) to control the quality and to sort and to modify them. According to the manufacturer's instructions, the MBT_Standard. FAMS Method was used before performing a baseline subtraction and a smoothing steps. Spectra displaying flatline, outliers, and abnormalities were removed and then each spectrum was checked individually to spot any differences. Bruker recommends a minimum of 10 clean and similar spectra to create one MSP. According to our genotypic results, some of our strains were only identified at the genus level (or less) and thus an additional precaution about the quality of our potential identification via MALDI-TOF MS was taken. A minimum of 12 spectra for one strain was thus performed to formally be used to generate one strain-specific MSP. If this limit could not be reached because of a high rate of suppressed spectra for some strains, another run with these strains, using the same methods, was performed.

Once corrected spectra were collected, they were imported into the Compass Explorer software (MBT Compass Explorer 4.1; Bruker Microbiology & Diagnostics) to proceed to the MSP creation. A unique MSP calculated with at least 12 corrected spectra with high frequency of appearance (> 75%) was generated per strain. We succeeded in obtaining an average of 19 spectra per strain (lower boundary of a confidence interval at level 0.95 on the expectation of a sample of unknown distribution and variance for a sample > 30) (Figure 17.3). Metadata were modified to have the name of the strain identified with BLAST associated with the corresponding MSP. Cultivation conditions, MALDI-TOF MS data acquisition methods, and BLAST results were also added. Later, a dendrogram with only these in-house-created MSPs was automatically generated. A genus was represented by a few isolates of different species, and thus by creating a dendrogram the relationships

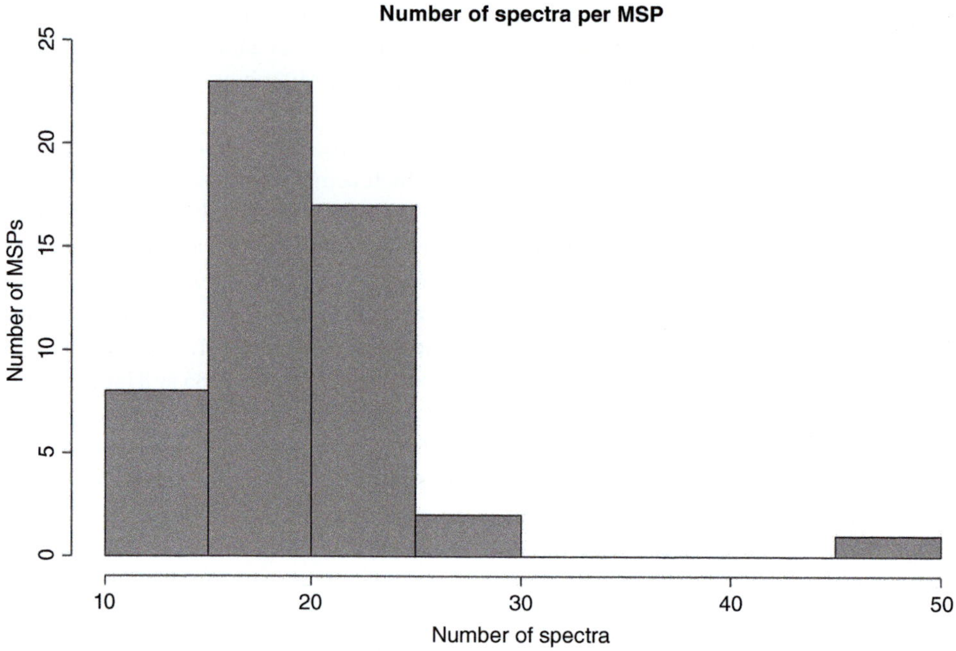

Figure 17.3 Repartition of the number of spectra used per strain-specific MSP.

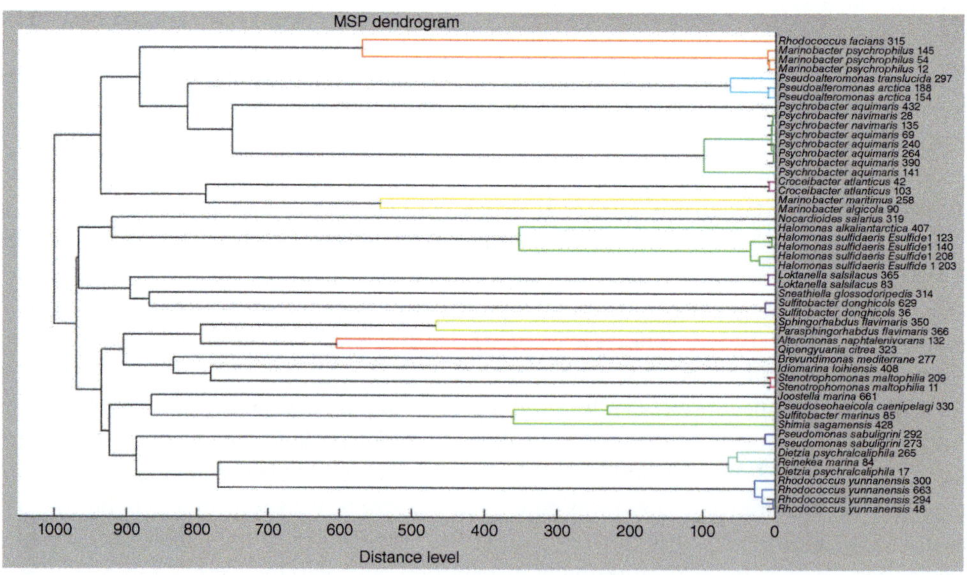

Figure 17.4 Dendrogram with the 51 marine bacterial strains from the in-house database.

between the different isolates could be shown (Figure 17.4). If a MSP of a specific genus, which was represented multiple times in the batch, was not properly clustered on the dendrogram, the quality of the spectra used for the MSP as well as those of the other MSPs from the same genus were checked and compared.

When all these conditions were checked, using the taxonomy tree editor of Compass Explorer, a library containing all the in-house MSPs was created and saved. The in-house library was thus constructed, including 51 isolates of 24 different genera.

17.5 Verification of the Accuracy of the Home-made Database

To verify the pertinence of this home-made database to identify a strain correctly, the MSPs generated for the database creation were tested against all existing bacterial entries of the commercial Bruker databases (7588 entries) on Compass Explorer. A greater identification rate was observed in comparison with the results obtained during the first run. This can be explained by the fact that MSPs are multiple corrected spectra resulting in one high-quality spectrum. Moreover, an important update of the software was done and more entries became available in between the beginning of the database creation and the testing. As seen in the Table 17.2, the commercial database was still not able to properly identify our

Table 17.2 Genotypic and MALDI-TOF MS in comparison with MBT DB-7854 MSP (Bruker Daltonics) database identification results.

ISCAR number	Genotoypic identification from third trial (identity percentage)	MALDI identification (log score)
ISCAR-5306	*Alteromonas naphtalenivorans* (99.71%)	Not reliable identification (score < 1.7)
ISCAR-5451	*Brevundimonas mediterranea* (99.69%)	Misidentified as *Brevundimonas intermedia* (score 2.150)
ISCAR-5216	*Croceibacter (atlanticus)* (100%)	Not reliable identification (score < 1.7)
ISCAR-5277	*Croceibacter atlanticus* (100%)	Not reliable identification (score < 1.7)
ISCAR-5191	*Dietzia psychralcaliphila* (99.62%)	Identified as a *Dietzia* (score 1.780)
ISCAR-5435	*Dietzia psychralcaliphila* (99.55%)	Not reliable identification (score < 1.7)
ISCAR-5540	*Halomonas alkaliantarctica* (99.32%)	Not reliable identification (score < 1.7)
ISCAR-5314	*Halomonas sulfidaeris Esulfide1* (99.19%)	Not reliable identification (score < 1.7)
ISCAR-5377	*Halomonas sulfidaeris Esulfide1* (99.10%)	Not reliable identification (score < 1.7)
ISCAR-5297	*Halomonas sulfidaeris Esulfide1* (99.17%)	Not reliable identification (score < 1.7)
ISCAR-5688	*Halomonas sulfidaeris Esulfide1* (99.10%)	Not reliable identification (score < 1.7)
ISCAR-5564	*Idiomarina loihiensis* (99.93%)	Not reliable identification (score < 1.7)
ISCAR-5606	*Joostella marina* (99.93%)	Not reliable identification (score < 1.7)
ISCAR-5257	*Loktanella salsilacus* (99.92%)	Not reliable identification (score < 1.7)
ISCAR-5706	*Loktanella salsilacus* (100%)	Not reliable identification (score < 1.7)
ISCAR-5264	*Marinobacter algicola* (99.24%)	Not reliable identification (score < 1.7)
ISCAR-5439	*Marinobacter maritimus* (98.22%)	Not reliable identification (score < 1.7)
ISCAR-5186	*Marinobacter psychrophilus* (99.79%)	Not reliable identification (score < 1.7)
ISCAR-5228	*Marinobacter psychrophilus* (99.86%)	Not reliable identification (score < 1.7)
ISCAR-5319	*Marinobacter psychrophilus* (99.72%)	Not reliable identification (score < 1.7)

(Continued)

Table 17.2 (Continued)

ISCAR number	Genotoypic identification from third trial (identity percentage)	MALDI identification (log score)
ISCAR-5693	*Nocardioides* (99.48%)	Not reliable identification (score < 1.7)
ISCAR-5511	*Sphingorhabdus flavimaris* (99.55%)	Not reliable identification (score < 1.7)
ISCAR-5523	*Parasphingorhabdus flavimaris* (99.48%)	Not reliable identification (score < 1.7)
ISCAR-5328	*Pseudoalteromonas* (99.27%)	Not reliable identification (score < 1.7)
ISCAR-5362	*Pseudoalteromonas arctica* (99.34%)	Not reliable identification (score < 1.7)
ISCAR-5467	*Pseudoalteromonas translucida* (94.42%)	Not reliable identification (score < 1.7)
ISCAR-5442	*Pseudomonas sabuligrini* (98.05%)	Not reliable identification (score < 1.7)
ISCAR-5463	*Pseudomonas sabuligrini* (98.22%)	Not reliable identification (score < 1.7)
ISCAR-5493	*Pseudoseohaeicola* (96.11%)	Not reliable identification (score < 1.7)
ISCAR-5315	*Psychrobacter aquimaris* (99.84%)	Not reliable identification (score < 1.7)
ISCAR-5401	*Psychrobacter aquimaris* (99.64%)	Identified as a *Psychrobacter* (score 1.980)
ISCAR-5417	*Psychrobacter aquimaris* (99.51%)	Misidentified as *Psychrobacter namhaensis* (score 2.000)
ISCAR-5524	*Psychrobacter aquimaris* (99.64%)	Identified as a *Psychrobacter* (score 1.990)
ISCAR-5835	*Psychrobacter aquimaris* (99.64%)	Not reliable identification (score < 1.7)
ISCAR-5243	*Psychrobacter aquimaris* (99.57%)	Identified as *Psychrobacter* (score 1.920)
ISCAR-5202	*Psychrobacter nivimaris* (99.42%)	Not reliable identification (score < 1.7)
ISCAR-5309	*Psychrobacter nivimaris* (99.55%)	Identified as a *Psychrobacter* (score 1.960)
ISCAR-5490	*Qipengyuania citrea* (96.72%)	Not reliable identification (score < 1.7)
ISCAR-5258	*Reinekea* (93.90%)	Not reliable identification (score < 1.7)
ISCAR-5482	*Rhodococcus fascians* (99.59%)	Rightly identified as *Rhodococcus fascians* (score 2.350)
ISCAR-5222	*Rhodococcus yunnanensis* (98.99%)	Not reliable identification (score < 1.7)
ISCAR-5466	*Rhodococcus yunnanensis* (98.98%)	Not reliable identification (score < 1.7)
ISCAR-5468	*Rhodococcus yunnanensis* (99.17%)	Not reliable identification (score < 1.7)
ISCAR-5647	*Rhodococcus yunnanensis* (99.06%)	Not reliable identification (score < 1.7)
ISCAR-5828	*Shimia sagamensis* (98.87%)	Not reliable identification (score < 1.7)
ISCAR-5479	*Sneathiella glossodoripedis* (98.64%)	Not reliable identification (score < 1.7)
ISCAR-5185	*Stenotrophomonas maltophilia* (99.64%)	Rightly identified as *Stenotrophomonas maltophilia* (score 2.220)
ISCAR-5382	*Stenotrophomonas maltophilia* (99.71%)	Rightly identified as *Stenotrophomonas maltophilia* (score 2.160)
ISCAR-5210	*Sulfitobacter* (97.64%)	Not reliable identification (score < 1.7)
ISCAR-5837	*Sulfitobacter donghicola* (97.72%)	Not reliable identification (score < 1.7)
ISCAR-5259	*Sulfitobacter marinus* (96.30%)	Not reliable identification (score < 1.7)

environmental strains. Three isolates were correctly identified at the species level: two *S. maltophilia* (ISCAR-5185 and ISCAR-5182) and one *Rhodococcus fascians* (ISCAR-5482). However, another four isolates, genotypically identified as *Rhodococcus* (ISCAR-5222, ISCAR-5466, ISCAR-5468, and ISCAR-5647), were not identified even at the genus level with MALDI-TOF MS despite the high quality of spectra. These isolates were identified as *Rhodococcus yunnanensis*. According to a study published in 2005 [31], this species is known to be phylogenetically further away from other species of the same genus, the closest being *R. fascians* with an identity of 98.8%. This could explain the lack of identification for these strains even if 96 *Rhodococcus* isolates are present in the commercial database. Strain ISCAR-5191 had its genus correctly identified as a *Dietzia* strain. However, strain ISCAR-5435 presented similar genotypic identification results (*Dietzia psychralcaliphila*) but even with a slightly lower identification percentage (0.07 lower) it was not identified at the genus level. This may be because only eight *Dietzia* isolates are present in the database. *Brevundimonas* was not properly assigned at the species level (strain ISCAR-5451) but at the genus level. Finally, among the eight *Psychrobacter* isolates, one has been misidentified at the species level with a score of 2.000 which could be acceptable as a genus or a species level (strain ISCAR-5417). Four others have been identified at the genus level and three were not identified.

Finally, to verify the accuracy of the home-made database, a new batch of the 51 strains were cultivated under optimal conditions and tested. The same protein extraction and purification methods were used. Targets were prepared by depositing each extract in triplicate. The first run was done using Compass as operating software and the home-made database as reference database, whereas the second run was done using FlexControl. The spectra obtained with the second run were processed with FlexAnalysis in the same way as for the database creation, and then MSPs were created on Compass Explorer. These MSPs were only created with two or three spectra and matched against the 51 home-made database entries and against all the bacterial entries of the commercial Bruker database. The results are shown in Table 17.3. The identification results obtained with the home-made database with both runs were similar and only slight variations were observed in the scores.

With the commercial database as reference, only two strains were identified at the species level (strains ISCAR-5689 and ISCAR-5243OW) and four at the genus level (strains ISCAR-5208, ISCAR-5496, ISCAR-5506, and ISCAR-5591). However, using the home-made database, with both methods (Compass only or FlexControl, FlexAnalysis, and CompassExplorer) only nine strains were not successfully identified; four were identified at the genus level and thus 39 were identified at the species level. Every strain that was identified by the commercial database was also identified by the home-made database with a higher score. The genus found by the first one was always corroborated by the results of the second one. For example, strain ISCAR-5496 was identified as a *Psychrobacter* with a score of 1.880 using Bruker's database and as *Psychrobacter nivimaris* with a score of 2.710 using the home-made database. Some isolates were identified at the species level with both databases, although the two databases did not indicate the same species. For example, strain ISCAR-5243OW was identified as a *Psychrobacter namhaensis* by the commercial database and as a *Psychrobacter aquimaris* by the home-made database. In view of the scores (2.070 for the commercial and 2.590 for the home-made) we can state with caution that the genus of this isolate is *Psychrobacter*. Even though the threshold score of 2.000 required to differentiate bacteria into species was reached for both these identifications,

Table 17.3 Background information and MALDI Biotyper identification results obtained for the test batch with both databases.

ISCAR number	Original water sample name	Sampling location	Identification 2014	Results with commercial database	Results with home-made database
ISCAR-5471	FX002-010	64°20′2.4″ N, 23°0′10.8″ W	*Croceibacter*	Not identified (score < 1.7)	Identified as *Pseudoalteromonas arctica* (score 2.580)
ISCAR-5488	FX002-010	64°20′2.4″ N, 23°0′10.8″ W	*Marinobacter*	Not identified (score < 1.7)	Identified as *Psychrobacter aquimaris* (score 2.150)
ISCAR-5268	FX007-008	64°20′2.4″ N, 26°0′14.4″ W	*Sulfitobacter*	Not identified (score < 1.7)	Not identified (score < 1.7)
ISCAR-5324	FX007-012	64°20′2.4″ N, 26°0′14.4″ W	*Colwellia*	Not identified (score < 1.7)	Identified as *Halomonas sulfidaeris* Esulfide1 (score 2.420)
ISCAR-5211	HORNB-001	66°59′57.12″ N, 21°34′50.52″ W	*Sulfitobacter*	Not identified (score < 1.7)	Identified as *Sulfitobacter donghicola* (score 2.260)
ISCAR-5211	HORNB-001	66°59′57.12″ N, 21°34′50.52″ W	*Sulfitobacter*	Not identified (score < 1.7)	Identified as *Halomonas yunnanensis* (score 2.620)
ISCAR-5591	HORNB-001	66°59′57.12″ N, 21°34′50.52″ W	*Hydrogenophaga*	Identified as a *Psychrobacter* (score 1.750)	Identified as *Psychrobacter nivimaris* (2.470)
ISCAR-5308	ICUNK-001	64°57′46.8″ N, 19°1′14.88″ W	*Alteromonas*	Not identified (score < 1.7)	Identified as *Alteromonas naphtalenivorans* (score 2.590)
ISCAR-5310	ICUNK-001	64°57′46.8″ N, 19°1′14.88″ W	*Halomonas*	Not identified (score < 1.7)	Identified as *Halomonas sulfidaeris* Esulfide1 (score 2.380)
ISCAR-5536	ICUNK-001	64°57′46.8″ N, 19°1′14.88″ W	*Halomonas*	Not identified (score < 1.7)	Identified as a *Halomonas* (score 1.910)
ISCAR-5333	INGOB-001	63°46′0.12″ N, 16°35′58.92″ W	*Pseudoalteromonas*	Not identified (score < 1.7)	Identified as a *Pseudoalteromonas* (score 1.860)
ISCAR-5404	INGOB-001	63°46′0.12″ N, 16°35′58.92″ W	*Colwellia*	Not identified (score < 1.7)	Identified as *Dietzia psychralcaliphila* (score 2.460)
ISCAR-5454	INGOB-001	63°46′0.12″ N, 16°35′58.92″ W	*Sphingopyxis*	Not identified (score < 1.7)	Identified as *Marinobacter maritimus* (score 2.600)

ISCAR-5474	INGOB-001	63°46′0.12″ N, 16°35′58.92″ W	*Marinobacter*	Not identified (score < 1.7)	Not identified (score < 1.7)
ISCAR-5584	INGOB-001	63°46′0.12″ N, 16°35′58.92″ W	*Pseudoalteromonas*	Not identified (score < 1.7)	Identified as *Pseudoalteromonas arctica* (score 2.260)
ISCAR-5377O	KOGUC-001	66°41′2.4″ N, 23°9′9″ W	*Dasania*	Not identified (score < 1.7)	Identified as *Halomonas sulfidaeris Esulfide1* (score 2.110)
ISCAR-5377B	KOGUC-001	66°41′2.4″ N, 23°9′9″ W	*Dasania*	Not identified (score < 1.7)	Identified as a *Halomonas* (score 1.860)
ISCAR-5377S	KOGUC-001	66°41′2.4″ N, 23° 9′ 9″ W	*Dasania*	Not identified (score < 1.7)	Not identified (score < 1.7)
ISCAR-5565	KOGUD-001	67°35′3.48″ N, 23°56′8.88″ W	*Psychroserpens*	Not identified (score < 1.7)	Identified as *Halomonas alkaliantarctica* (score 2.400)
ISCAR-5685	KOGUD-001	67°35′3.48″ N, 23°56′8.88″ W	*Rhodococcus*	Not identified (score < 1.7)	Not identified (score < 1.7)
ISCAR-5243O3	KOGUD-001	67°35′3.48″ N, 23°56′8.88″ W	*Psychroserpens*	Not identified (score < 1.7)	Identified as *Halomonas alkaliantarctica* (score 2.130)
ISCAR-5243OW	KOGUD-001	67°35′3.48″ N, 23°56′8.88″ W	*Psychroserpens*	Identified as *Psychrobacter namhaensis* (score 2.070)	Identified as *Psychrobacter aquimaris* (score 2.590)
ISCAR-5286	LANNC-001	66°36′56.52″ N, 14°15′45.72″ W	*Colwellia*	Not identified (score < 1.7)	Identified as *Alteromonas naphtalenivorans* (score 2.180)
ISCAR-5291	LANNC-001	66°36′56.52″ N, 14°15′45.72″ W	*Pseudomas*	Not identified (score < 1.7)	Not identified (score < 1.7)
ISCAR-5384	LANNC-001	66°36′56.52″ N, 14°15′45.72″ W	*Sulfitobacter*	Not identified (score < 1.7)	Identified as *Sulfitobacter donghicola* (score 2.380)
ISCAR-5601	LANNC-001	66°36′56.52″ N, 14°15′45.72″ W	*Halomonas*	Not identified (score < 1.7)	Identified as a *Halomonas* (score 1.980)
ISCAR-5229	LANND-001	67°59′59.28″ N, 12°40′10.92″ W	*Dietzia*	Not identified (score < 1.7)	Identified as *Dietzia psychralcaliphila* (score 2.350)

(*Continued*)

Table 17.3 (Continued)

ISCAR number	Original water sample name	Sampling location	Identification 2014	Results with commercial database	Results with home-made database
ISCAR-5284	LANND-001	67°59′59.28″ N, 12°40′10.92″ W	*Dietzia*	Not identified (score < 1.7)	Identified as *Dietzia psychralcaliphila* (score 2.310)
ISCAR-5414	LANND-001	67°59′59.28″ N, 12°40′10.92″ W	*Dietzia*	Not identified (score < 1.7)	Identified as *Dietzia psychralcaliphila* (score 2.200)
ISCAR-5539	LANND-001	67°59′59.28″ N, 12°40′10.92″ W	*Hyphomonas*	Not identified (score < 1.7)	Identified as *Halomonas alkaliantarctica* (score 2.510)
ISCAR-5435	LANND-001	67°59′59.28″ N, 12°40′10.92″ W	*Dietzia*	Not identified (score < 1.7)	Not identified (score < 1.7)
ISCAR-5496	LATRC-001	65°35′6″ N, 24°55′19.2″ W	*Dietzia*	Identified as a *Psychrobacter* (score 1.880)	Identified as *Psychrobacter nivimaris* (score 2.710)
ISCAR-5504	LATRC-001	65°35′6″ N, 24°55′19.2″ W	*Dietzia*	Not identified (score < 1.7)	Identified as *Dietzia psychralcaliphila* (score 2.440)
ISCAR-5506	LATRD-001	66°1′5.52″ N, 26°48′10.08″ W	*Pelagicola*	Identified as a *Psychrobacter* (score 1.840)	Identified as *Psychobacter nivimaris* (score 2.420)
ISCAR-5464	SELVC-001	63°28′59.52″ N, 20°54′7.92″ W	*Sulfitobacter*	Not identified (score < 1.7)	Identified as *Pseudomonas sabuligrini* (score 2.590)
ISCAR-5322	SELVD-001	62°58′52.68″ N, 21°28′45.12″ W	*Pseudoalteromonas*	Not identified (score < 1.7)	Identified as *Pseudoalteromonas arctica* (score 2.570)
ISCAR-5385	SIGLC-001	66°32′3.12″ N, 18°49′57.72″ W	*Pseudoalteromonas*	Not identified (score < 1.7)	Identified as *Pseudoalteromonas arctica* (score 2.370)
ISCAR-5574	SIGLC-001	66°32′3.12″ N, 18°49′57.72″ W	*Psychroserpens*	Not identified (score < 1.7)	Identified as *Halomonas alkaliantarctica* (score 2.230)
ISCAR-5184	SIGLD-001	67°59′58.92″ N, 18°49′47.28″ W	*Sulfitobacter*	Not identified (score < 1.7)	Identified as *Sulfitobacter donghicola* (score 2.170)

ISCAR-5208	SIGLD-001	67°59′58.92″ N, 18°49′47.28″ W	*Colwellia*	Identified as a *Psychrobacter* (score 1.840)	Identified as *Psychrobacter aquimaris* (score 2.670)
ISCAR-5225	SIGLD-001	67°59′58.92″ N, 18°49′47.28″ W	*Rhodococcus*	Not identified (score <1.7)	Identified as *Rhodococcus yunnanensis* (score 2.420)
ISCAR-5300	SIGLD-001	67°59′58.92″ N, 18°49′47.28″ W	*Alteromonas*	Not identified (score <1.7)	Identified as *Pseudoalteromonas arctica* (score 2.390)
ISCAR-5497	SIGLD-001	67°59′58.92″ N, 18°49′47.28″ W	*Sulfitobacter*	Not identified (score <1.7)	Identified as *Pseudoalteromonas arctica* (score 2.470)
ISCAR-5522	SIGLD-001	67°59′58.92″ N, 18°49′47.28″ W	*Pseudomonas*	Not identified (score <1.7)	Identified as *Pseudomonas sabuligrini* (score 2.540)
ISCAR-5589	SIGLD-001	67°59′58.92″ N, 18°49′47.28″ W	*Pseudoalteromonas*	Not identified (score <1.7)	Identified as *Pseudoalteromonas arctica* (score 2.300)
ISCAR-5600	SIGLD-001	67°59′58.92″ N, 18°49′47.28″ W	*Erythrobacter*	Not identified (score <1.7)	Not identified (score <1.7)
ISCAR-5214	SIGLE-001	67°59′56.4″ N, 18°50′1.68″ W	*Marinobacter*	Not identified (score <1.7)	Identified as *Marinobacter psychrophilus* (score 2.160)
ISCAR-5245	SIGLE-001	67°59′56.4″ N, 18°50′1.68″ W	*Marinobacter*	Not identified (score <1.7)	Not identified (score <1.7)
ISCAR-5267	SIGLE-001	67°59′56.4″ N, 18°50′1.68″ W	*Rhodococcus*	Not identified (score <1.7)	Identified as *Rhodococcus yunnanensis* (score 2.490)
ISCAR-5689	SIGLE-001	67°59′56.4″ N, 18°50′1.68″ W	*Psychrobacter*	Identified as *Psychrobacter alimentarius* (score 2.020)	Identified as *Psychrobacter nivimaris* (score 2.720)
ISCAR-5451	SIGLE-001	67°59′56.4″ N, 18°50′1.68″ W	*Hydrogenophaga*	Not identified (score <1.7)	Identified as *Marinobacter maritimus* (score 2.570)

objectively it would be reasonable to tend towards identification with the highest score. This partition limit could therefore be considered with caution. For both the strains that were identified at the species level with both databases, the difference of log score was > 10%. This result corroborates the addition of a category of misleading identification presented by Papalia et al. in 2020: log score $\geq$2.0 and Δ >10% between two identified species [30]. Here, this condition applies to the results presented by two different databases for the same extract and not for the first two results of the same database. Albeit the results were promising, nine strains were not identified while they belong to a genus already known and entered into the database. For instance, strain ISCAR-5435 known to be a *Dietzia* according to the 16S rRNA gene sequencing result was not identified; likewise, for strains ISCAR-5245 and ISCAR-5474, which were supposed to be assigned to the *Marinobacter* genus. Even though the genotypic identification was only partial, this absence of properly assigned strains demonstrates a lack of representation of the marine diversity in this home-made database. Additionally, as for the creation of the database, a difference between the assumed genus and the identified genus can be noted for many isolates. For the database creation, in accordance with the results of Yarza et al., different genera were found by sequencing the complete 16S rRNA gene compared with the sequencing of the V1–V4 regions that was previously done [32]. Here, compared with the results of V1–V4 sequencing, different genera were found for 19 strains out of 51 by carrying out a MALDI identification with a database specially designed for environmental isolates. The genera *Colwellia*, *Dietzia*, and *Dasania* were found to be not properly identified both times. Further experiments are needed to confirm the hypothesis that MALDI identification performs better than sequencing of the V1–V4 regions of the 16S rRNA gene.

Roughly, 12% of the tested isolates were identified at the genus level with the Bruker database, while 83% of them were correctly identified by the newly generated database. Among the identified strains with the second database, 91% of them were identified at the species level with a score > 2.130.

17.6 Conclusions

A marine database with 51 isolates belonging to 24 different genera was successfully generated. This allowed us to increase the identification rate of marine microbial isolates, at least to the genus level, by 60% in comparison with the Bruker commercial database. As previously stated, this is just a small, non-exhaustive representation of the Icelandic marine microbial diversity and it needs to be extended to improve its accuracy further. However, this marine environmental microbial strain database represents a promising tool for the rapid identification of environmental microbial strains to increase our knowledge of cultivated marine microorganisms in the North Atlantic.

Funding

This work was supported by the EU project MaCuMBA (Marine Microorganisms: Cultivation Methods for Improving their Biotechnological Applications; grant agreement no. 311975) and by the Icelandic research fund (IRF) RANNIS, project MIME (Microbes in the Icelandic Marine Environment, grant no. 163266-051).

References

1 Sogin, M.L., Morrison, H.G., Huber, J.A. et al. (2006). Microbial diversity in the deep sea and the underexplored "rare biosphere.". *Proc. Natl Acad. Sci. U.S.A.* 103 (32): 12115–12120.

2 Overmann, J. and Lepleux, C. (2016). Marine bacteria and archaea: diversity, adaptations, and culturability. In: *The Marine Microbiome: An Untapped Source of Biodiversity and Biotechnological Potential* (ed. L.J. Stal and M.S. Cretoiu), 21–55. Springer International Publishing https://doi.org/10.1007/978-3-319-33000-6_2.

3 Wang, F., Li, M., Huang, L. et al. (2021). Cultivation of uncultured marine microorganisms. *Mar. Life Sci. Technol.* 3: 117–120.

4 Bar-On, Y.M. and Milo, R. (2019). The biomass composition of the oceans: a blueprint of our blue planet. *Cell* 179: 1451–1454. https://doi.org/10.1016/j.cell.2019.11.018.

5 Bar-On, Y.M., Phillips, R., and Milo, R. (2018). The biomass distribution on Earth. *Proc. Natl. Acad. Sci. U. S. A.* 115: 6506–6511. https://doi.org/10.1073/pnas.1711842115.

6 Ramirez-Llodra, E., Tyler, P.A., Baker, M.C. et al. (2011). Man and the last great wilderness: human impact on the deep sea. *PLoS One* 6: https://doi.org/10.1371/journal.pone.0022588.

7 Halpern, B.S., Walbridge, S., Selkoe, K.A. et al. (2008). A global map of human impact on marine ecosystems. *Science* 319: 948–952.

8 Hauser, D.D.W., Tobin, E.D., Feifel, K.M. et al. (2016). Disciplinary reporting affects the interpretation of climate change impacts in global oceans. *Global Change Biol.* 22: 25–43. https://doi.org/10.1111/gcb.12978.

9 Martiny, A.C. (2019). High proportions of bacteria are culturable across major biomes. *ISME J.* 13: 2125–2128. https://doi.org/10.1038/s41396-019-0410-3.

10 Steen, A.D., Crits-Christoph, A., Carini, P. et al. (2019). High proportions of bacteria and archaea across most biomes remain uncultured. *ISME J.* 13: 3126–3130. https://doi.org/10.1038/s41396-019-0484-y.

11 Hofer, U. (2018). The majority is uncultured. *Nat. Rev. Microbiol.* 16: 716–717. https://doi.org/10.1038/s41579-018-0097-x.

12 Jiao, J.Y., Liu, L., Hua, Z.S. et al. (2021). Microbial dark matter coming to light: challenges and opportunities. *Natl. Sci. Rev.* 8: https://doi.org/10.1093/nsr/nwaa280.

13 Venter, J.C., Remington, K., Heidelberg, J.F. et al. (2004). Environmental genome shotgun sequencing of the Sargasso Sea. *Science* 304: 66–74. https://doi.org/10.1126/science.1093857.

14 Youssef, D.T.A. (2021). Marine-derived biomolecules. *Biomolecules* 11: 1–3. https://doi.org/10.3390/biom11010012.

15 Rotter, A., Barbier, M., Bertoni, F. et al. (2021). The essentials of marine biotechnology. *Front. Mar. Sci.* 8: https://doi.org/10.3389/fmars.2021.629629.

16 Popović, N.T., Kazazić, S.P., Strunjak-Perović, I., and Čož-Rakovac, R. (2017). Differentiation of environmental aquatic bacterial isolates by MALDI-TOF MS. *Environ. Res.* 152: 7–16. https://doi.org/10.1016/j.envres.2016.09.020.

17 Timperio, A.M., Gorrasi, S., Zolla, L., and Fenice, M. (2017). Evaluation of MALDI-TOF mass spectrometry and MALDI BioTyper in comparison to 16S rDNA sequencing for the identification of bacteria isolated from Arctic sea water. *PLoS One* 12: https://doi.org/10.1371/journal.pone.0181860.

18 Böhme, K., Fernández-No, I.C., Barros-Velázquez, J. et al. (2012). SpectraBank: an open access tool for rapid microbial identification by MALDI-TOF MS fingerprinting. *Electrophoresis* 33: 2138–2142.

19 Ashfaq, M.Y., Da'na, D.A., and Al-Ghouti, M.A. (2022). Application of MALDI-TOF MS for identification of environmental bacteria: a review. *J. Environ. Manage.* 305: https://doi.org/10.1016/j.jenvman.2021.114359.

20 Cappucino, J.G. and Sherman, N. (2008). *Microbiology: A Laboratory Manual*, 8e (ed. B. Cummings). San Francisco, CA: Pearson International.

21 Arnold, R.J., Karty, J.A., Ellington, A.D., and Reilly, J.P. (1999). Monitoring the growth of a bacteria culture by MALDI-MS of whole cells. *Anal. Chem.* 71: 1990–1996.

22 Takai, K. and Nakamura, K. (2011). Archaeal diversity and community development in deep-sea hydrothermal vents. *Curr. Opin. Microbiol.* 14: 282–291. https://doi.org/10.1016/j.mib.2011.04.013.

23 Weisburg, W.G., Barns, S.M., Pelletier, D.A., and Lane, D.J. (1991). 16S ribosomal DNA amplification for phylogenetic study. *J. Bacteriol.* 173 (2): 697–703.

24 Johnson, J.S., Spakowicz, D.J., Hong, B.Y. et al. (2019). Evaluation of 16S rRNA gene sequencing for species and strain-level microbiome analysis. *Nat. Commun.* 10: https://doi.org/10.1038/s41467-019-13036-1.

25 Caporaso, J.G., Lauber, C.L., Walters, W.A. et al. (2012). Ultra-high-throughput microbial community analysis on the Illumina HiSeq and MiSeq platforms. *ISME J.* 6: 1621–1624. https://doi.org/10.1038/ismej.2012.8.

26 Andersson, A.F., Lindberg, M., Jakobsson, H. et al. (2008). Comparative analysis of human gut microbiota by barcoded pyrosequencing. *PLoS One* 3: https://doi.org/10.1371/journal.pone.0002836.

27 Schloss, P.D. and Handelsman, J. (2005). Introducing DOTUR, a computer program for defining operational taxonomic units and estimating species richness. *Appl. Environ. Microbiol.* 71: 1501–1506.

28 Faron, M.L., Buchan, B.W., and Ledeboer, N.A. (2017). Matrix-assisted laser desorption ionization-time of flight mass spectrometry for use with positive blood cultures: methodology, performance, and optimization. *J. Clin. Microbiol.* 55: https://doi.org/10.1128/JCM.00868-17.

29 Calderaro, A., Martinelli, M., Motta, F. et al. (2014). Comparison of peptide nucleic acid fluorescence in situ hybridization assays with culture-based matrix-assisted laser desorption/ionization-time of flight mass spectrometry for the identification of bacteria and yeasts from blood cultures and cerebrospinal fluid cultures. *Clin. Microbiol. Infect.* 20: https://doi.org/10.1111/1469-0691.12490.

30 Papalia, M., Figueroa Espinosa, R., Steffanowski, C. et al. (2020). Expansion and improvement of MALDI-TOF MS databases for accurate identification of *Achromobacter* species. *J. Microbiol. Methods* 172: https://doi.org/10.1016/j.mimet.2020.105889.

31 Zhang, Y.Q., Li, W.J., Kroppenstedt, R.M. et al. (2005). *Rhodococcus yunnanensis* sp. nov., a mesophilic actinobacterium isolated from forest oil. *Int. J. Syst. Evol. Microbiol.* 55: 1133–1137. https://doi.org/10.1099/ijs.0.63390-0.

32 Yarza, P., Yilmaz, P., Pruesse, E. et al. (2014). Uniting the classification of cultured and uncultured bacteria and archaea using 16S rRNA gene sequences. *Nat. Rev. Microbiol.* 12: 635–645. https://doi.org/10.1038/nrmicro3330.

18

MALDI-TOF MS Implementation Strategy for a Pharma Company Based upon a Network Microbial Identification Perspective

Lynn Johnson[1], *Christoph Hansy*[2], *and Hilary Chan*[3]

[1] *Central Quality - Microbiology and Contamination Control, National Resilience, Inc., USA*
[2] *Global Sterility Assurance and Microbiology, Takeda Pharmaceutical Co. Ltd, Vienna, Austria*
[3] *Global Sterility Assurance and Microbiology, Takeda Pharmaceutical Co. Ltd, Lexington, MA, USA*

18.1 Introduction

18.1.1 Microbial Identifications from a Pharmaceutical Industry Perspective

Microbial identification is an essential part of a comprehensive contamination control strategy for pharmaceutical and biotech companies. A thorough understanding of microbial characteristics (e.g. source, growth and survival capabilities and potential pathogenicity to patients) derived from accurate microbial identifications is necessary to understand the quality of the environment in which the product is manufactured, to investigate the source of microorganisms recovered from product, and to understand potential impact on products and patients. Microbial identification is also necessary for trending programs and identification confirmation (e.g. for microbial production cell lines, prepared cultures, biological indicators [BIs], etc.). Microbiology laboratories supporting pharmaceutical manufacturing need robust, reliable, and accurate methods for microbial identification, with an increasing need for rapid results to enable investigations and support product disposition decisions.

18.1.2 Historical Evolution

Within the last few decades, microbial identification methodologies have evolved by leaps and bounds, from classical (traditional) to novel methods with advanced/rapid technology and increased automation. This has led to a shift from manual, bench-scale tests and culture methods to the use of more analytical, proteomic, biochemical, and genomic methodologies (see Figure 18.1 [1]). Additionally, microbiological laboratories in pharmaceutical companies are increasingly focused on waste reduction/lean/six sigma practices to improve efficiencies and reduce costs. Quality risk management (QRM) per ICH Q9 is also at the

Microbiological Identification using MALDI-TOF and Tandem Mass Spectrometry: Industrial and Environmental Applications, First edition. Edited by Haroun N. Shah, Saheer E. Gharbia, Ajit J. Shah, Erika Y. Tranfield, and K. Clive Thompson.

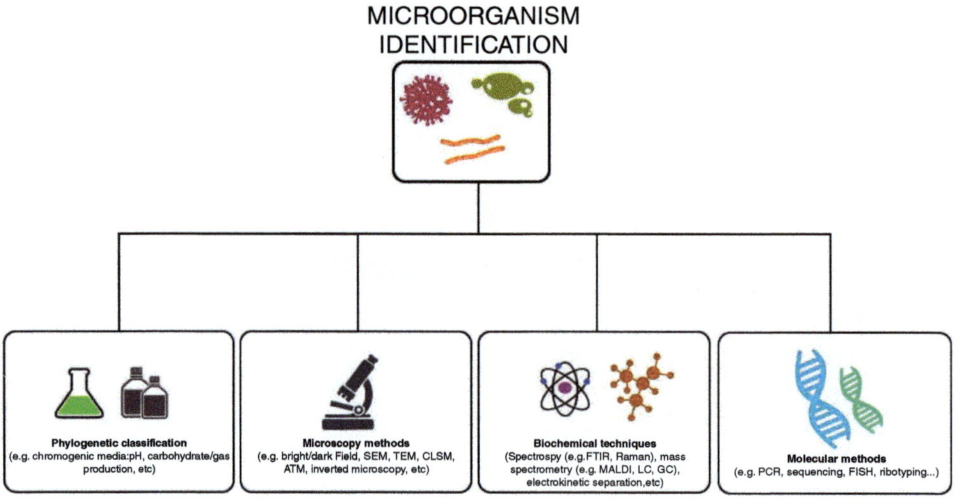

Figure 18.1 Microbial identification techniques.

forefront of laboratory and manufacturing operations. These factors, along with the increasing regulatory body acceptance and prevalence of modern microbial methods (synonymous with alternative/rapid microbial methods [ARMMs/RMMs]) place microbial identifications in a more strategic light.

In a modern microbiology laboratory supporting good manufacturing practice (GMP) pharmaceutical production, considerations for microbial identification technologies typically include priorities such as turnaround time (test preparation plus time to result), specificity, expansive and industry-relevant library, instrument and material cost, data integrity, vendor support, and ease of use (e.g. time/complexity of assay, data analysis). Because of the criticality for product disposition and support of root cause analysis for investigations, acquiring results as close to real time as possible is desirable. Refer to Table 18.1 for a comparison of these attributes between current common microbial identification technologies.

Table 18.1 Comparison of microbiological identification technologies.

Technology	Material cost	Test execution time	Time to result	Specificity
Traditional culture methods	Inexpensive	Moderate	Long	Relatively specific
Single species or multiplex RT-PCR methods	Expensive	Moderate	Short	Highly specific
MALDI-TOF mass spectrometry	Inexpensive	Slight	Short	Limited to database
rRNA base sequencing	Moderately expensive	Moderate	Moderate	Limited by 400–500 base pairs
Whole-genome sequencing	Expensive	Long	Long	Highly specific

RT-PCR, reverse transcription polymerase chain reaction.
Source: Adapted from Cundell [2] and USP Chapter <1113> [3].

Although the microbial identification methods listed in Table 18.1 are beneficial for several reasons, matrix-assisted laser desorption/ionization time-of-flight mass spectrometry (MALDI-TOF MS, or MALDI-TOF) represents a technology that can be an excellent overall fit for pharma/biotech microbiology laboratories. Pharmaceutical laboratories typically perform microbial identification for many types of isolates, such as environmental isolates recovered from cleanrooms where manufacturing production takes place and from critical utilities used to support manufacturing processes, as well as isolates from raw materials, solutions, excipients, and media used in production, and from the product itself (intermediates, drug substance, and drug product). The identification data is critical to trending, testing for the absence of specified microorganisms, investigation of sources of contamination/root cause analysis, and to inform decision-making for batch disposition. MALDI-TOF systems are capable of inexpensive, high throughput, and short turnaround time to results. This chapter will focus on the implementation and use of MALDI-TOF MS in modern pharmaceutical company laboratories and the use of this technology as part of a company network/global microbial identification program.

18.2 Regulatory Requirements/Guidance for Microbial Identification

The importance of microbial identification and characterization in the pharmaceutical/biotech industry is evident in both regulatory body and industry requirements and guidance documents. In general, the use of MALDI-TOF for microbial identification in the pharma industry is deemed appropriate for the type and range of scenarios where identification to the genus and/or species level is required. Table 18.A.1 in the Appendix summarizes examples of relevant regulatory requirements and industry guidance for microbial identification.

18.3 Strategic Approaches to MALDI-TOF Implementation Within the Modern Microbial Methods Framework

18.3.1 Incorporation of MALDI-TOF into a Technical Evaluation Roadmap

Companies are often tasked with identifying a technology roadmap strategy that outlines the evaluation and implementation of technologies over a period of time (e.g. implementation of core technologies across the manufacturing network within five years). This roadmap considers the vision and goals of the company, laboratory and manufacturing capabilities, personnel, and available technologies and outlines a structured approach. Ideally, an aligned approach would be applied across all the sites and would account for specific site variabilities. With identification technologies, these are numerous and variable and when deciding on a technology for implementation, one must consider a number of factors to identify how the technology roadmap will be outlined.

In order to properly lay out the technical roadmap, the steps involved in implementation must be understood first. Numerous resources, in addition to the regulatory guidelines, are available that outline the framework for implementation of modern microbial methods.

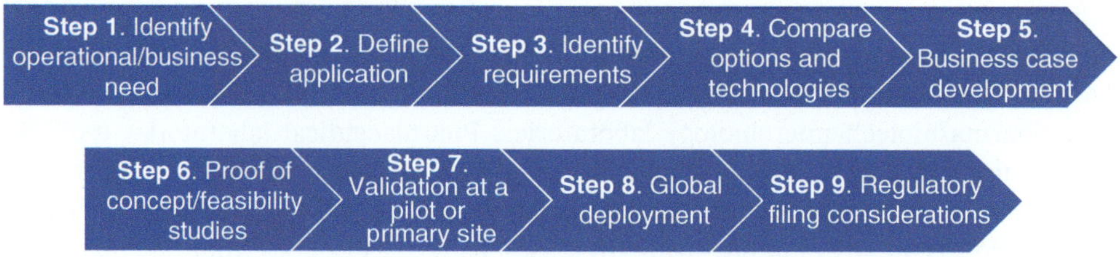

Figure 18.2 Implementation strategy. *Source:* Adapted from BioPhorum [4].

These resources provide a holistic approach to evaluation, validation, and implementation. Specific industry approaches have been published from BioPhorum and the Parenteral Drug Association (PDA), both of which detail the process from initial technology evaluation (defining the business need and application) through regulatory filing considerations after validation. Figure 18.2 provides a general overview of the implementation process derived from the above-mentioned references and company experience [4, 5].

18.3.2 Initial Implementation Planning Stage

Steps 1 through 5 in Figure 18.2 represent the research and up-front analysis taken to gain endorsement to onboard a technology and is considered the initial implementation planning stage.

18.3.2.1 Roles and Responsibilities (Global/Local, Partners/IT, Stakeholders)

Network implementation strategies in global pharmaceutical companies are complex and require thorough planning. It is essential to plan and drive the implementation project with a cross-functional, multidisciplinary team with defined roles and responsibilities.

To ensure proper communication and to define the roles and responsibilities, it is recommended to use tools such as the Responsibility and Relationship Matrix, which is an evolution of the traditionally used RACI Matrix (Responsible, Accountable, Consulted, and Informed). A good example is provided by Tonchia [6], who summarizes the relationship and responsibilities as following:

1) General responsibility (concerning the final/overall deliverables)
2) Operational responsibility (being delegated to perform groups of tasks)
3) Specific responsibility (for single tasks)
4) Necessary consultation
5) Possible consultation
6) Compulsory communication (to a specified person)
7) Necessary approval (by the person involved).

In the setting of a pharmaceutical company, the following roles/departments should be minimally included in a network implementation project:

- A project manager
- Quality assurance and quality control
- Maintenance and metrology

- Validation
- IT and data integrity subject matter experts
- Safety, security, sustainability, health.

Additional consideration should be given to the fact that, before initiating such projects, executive sponsorship might be required, including regular reporting on the achievements/roadblocks. Also, consideration regarding the involvement of local and/or global functions for the mentioned departments may be necessary and might vary, depending on how the company is set up.

It is important to ensure endorsement from senior leadership and, in some cases, across departments. If the benefits can be seen across multiple use cases, traction is gained, and endorsement becomes more likely. This also helps when planning out resources (i.e. for both personnel and budgeting) for implementation. An additional aspect is that endorsement may also need to be secured from both site and global strategy perspectives. Global functions can coordinate the technology roadmap across multiple sites, ensure the technology is the right fit, and help with alignment of procurement, validation, maintenance, and continuous improvement. Site functions provide the resources for execution activities and benefit from leveraging lessons learned and best practices from prior executions. Strong support, and ideally, a project sponsor drive the project and, as the technology is implemented, the success stories serve to promote use throughout the company and create a more detailed roadmap for subsequent implementation opportunities.

18.3.2.2 Considerations When Selecting a Vendor/Model

When the technology to be implemented has been defined (in this case, one for proteomic microbial identification), a crucial next step is to select the proper vendor/model type to fit all the expectations and requirements.

Each company has its own vendor/supplier selection and approval guidelines and their own sections that need to be included in the user requirement specification (URS). However, specifically for modern technologies, it is recommended to include as a minimum the following considerations if they are not covered by the existing selection process:

- Microbial identification systems are also intensively used in the food and beverage industry, as well as in healthcare settings. Vendors often have a historical background in the purpose/industry branch for which their equipment has been initially designed. The background and expertise of the vendor in the field of pharmaceutical quality control environments should be considered for the overall decision-making process, as the experience and maturity of the vendor can vary significantly, especially in terms of software systems, including connectivity and data integrity, training approaches, knowledge of the industry guidelines, and validation support.
- The vendor presence, including reasonable reaction times in different geographical regions, depending on the planned roll-out strategy (including availability of spare parts).
- The lack of an extensive, continuously updated, and validated database to support accurate microbial identification will lead to a high number of unknown or incorrect identification results. It is crucial to include a thorough evaluation of the database provided by the vendor, in particular the accuracy of mold identification as molds are commonly classified as critical organisms for the pharmaceutical industry. Unlike other

traditional microbiological identification systems, such as biochemical cartridge-based instruments, MALDI-TOF can identify not only bacteria, yeast and molds, thus highlighting the importance of selecting an appropriate vendor/model with an extensive and accurate library. Stein et al. [7] showed excellent examples of how successful identification of mold is dependent on the database and how the identifications from different databases have been obtained with varying results, with an overall combined identification rate of 77.4%, although the literature describes a success rate for the libraries used of up to 98.8%.

- Available laboratory space needs to be considered, given the typically large footprint of a MALDI-TOF system. In addition, one must consider the technical specification in terms of power supply, heating, ventilation, and air conditioning requirements, or the overall room classification where the instrument is planned to be housed.
- Recurring expenses for consumables and maintenance are more relevant from a long-term perspective than the initial costs of acquisition. Therefore, a thorough business case, including a return-on-investment calculation, should be available prior to the final acquisition. For a network implementation strategy, it is recommended to negotiate in advance a gradual discount for subsequent systems, but also for the expected increasing number of consumables.
- A classical literature search of available papers and publications related to the systems of interest will additionally help to obtain a holistic overview of the various solutions.
- An evaluation of automation/robotization opportunities, e.g. for the sample preparation steps, and application to the system should be added, if of interest to the company or strategy.

To conclude the selection process and rank the different vendors in a quantitative, traceable manner throughout the decision-making process, tools such as quality function deployment (QFD) and analytic hierarchy process (AHP) [8] can be used to facilitate the outcome.

18.3.2.3 Overall Identification Process Flow and MALDI-TOF as the Defined Application

The process for microbial identification using MALDI-TOF is quite simple, as there is minimal reliance on benchtop analysis. It is recommended to perform Gram stain on isolates prior to identification using MALDI-TOF to ensure the appropriate method is used (direct transfer to target plate/slide or extraction method for mold and some Gram-positive spore-forming bacteria) and to substantiate the identification result (Figure 18.3).

18.3.2.4 Benefits of an In-house System for Pharmaceutical Companies Compared with Outsourcing

Implementation of MALDI-TOF MS in-house affords several benefits over sending out to an external laboratory for identification. This method does not require highly experienced analysts, such as when performing genotypic identifications, but still provides an accurate microbial identification by generating a spectral analysis that is compared to a database of spectra. It allows for a rapid result with very little sample needed, low cost of processing, and higher throughput for most identification, thereby reducing the reliance on contract

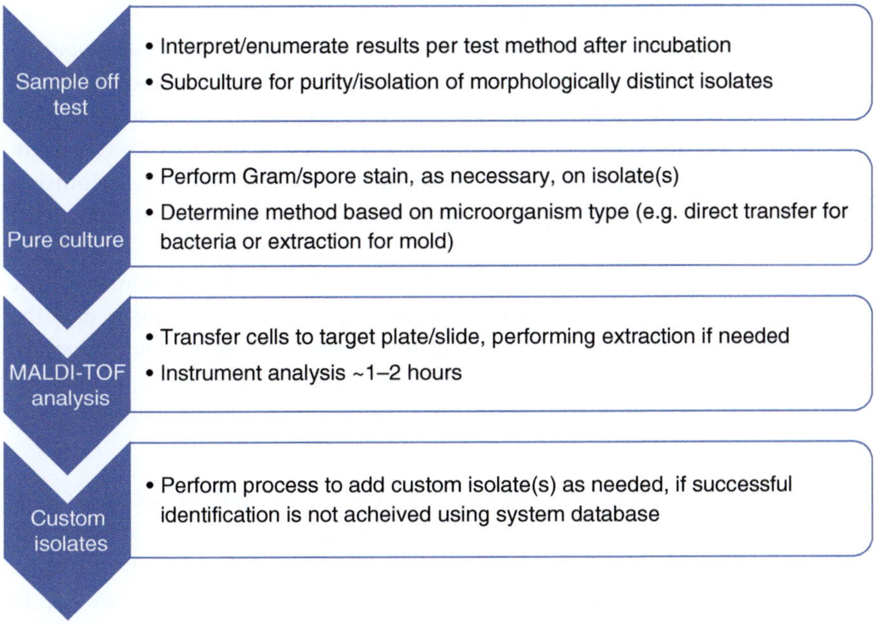

Figure 18.3 MALDI-TOF in the overall identification process flow.

testing laboratories [9]. For those where identification cannot be obtained, there still is the option to send out the sample to be further analyzed by an external laboratory or, if available, to be sequenced on an internal genotypic identification system.

Although some external laboratories may have a larger database for comparison, others may rely on a primarily clinical database or a limited one based on manufacturer-provided information. Having a MALDI-TOF MS in-house provides the opportunity to create a tailored microbial identification program, including creating custom identifications from facility isolates for the laboratory.

When considering companies with several manufacturing sites and laboratories, there can be variation in the site-specific microbial identification strategies, e.g. whether samples are tested at an external laboratory or in-house, the identification capabilities in the laboratory, type of instrumentation, etc. Laboratories tend either to be fully outsourced (no internal identification capabilities) or to use an identification system in combination with external outsourcing. However, even within these two categories, differences can be substantial. For example, one facility might depend on phenotypic identification or only have Gram stain capability, thereby relying more heavily on external testing laboratories, whereas another might have genotypic identification, but struggle with the assay complexity and training and end up still sending a limited subset to an external laboratory. Selection of the method is generally dependent on volume of identification, level of identification, and cost per test. The threshold lies around 20–30 identifications per month, at which point in-house identification using a semi-automated method such as the MALDI-TOF becomes a cost-effective option [10].

18.3.2.5 The Center of Excellence (CoE) Approach

The benefits and strategy described above can be further expanded on a global or regional level through a Center of Excellence (CoE) program. Regional CoEs are created to perform centralized microbial identifications, designed to provide a common set of best practices and work standards and to support establishing new identification technologies. Logistics are related to shipping and submission of isolates, cost responsibilities, sample request and data communication, and the need to use an external laboratory for further identification.

The CoE should generally consist of a MALDI-TOF, to be used as the workhorse of the identification program, and a system capable of performing genetic sequencing in case the MALDI-TOF does not produce an identification. If both these systems do not produce an identification, then the CoE will be responsible for sending the sample out to an external laboratory for further analysis. With this general structure in mind, one can envision a CoE being used in the following ways:

1) As the primary identification hub for the network/region – this will include sites with and without first-order identification capability (e.g. Gram stain, phenotypic methods).
2) As a supplement to the current site identification program – where a site already has a MALDI-TOF or similar system, the CoE can still serve as a supplement or back-up in case the site library cannot produce an identification or if the system is down, e.g. for maintenance or repair.

In practice, the CoE serves the role of a traditional external contract laboratory, so there are some parallel logistics that should be considered. Some find benefit in establishing a general statement of work between the site and CoE to establish guidelines for work performed, timelines, shipment of isolates, and cost center charges. From here, the following workflow is recommended based on company experience:

1) *Isolate shipment.* Determine the shipment logistics, including appropriate packaging and temperature tracking. Note that there may be restrictions when shipping out of the country where costs may become prohibitive and a regional or site-specific strategy may make more sense to implement.
2) *Isolate submission.* Submission of pure cultures and, ideally, a Gram stain should be performed so that the receiving laboratory (CoE) can be assured that the correct isolate was submitted. A central communication site (e.g. SharePoint) should be utilized for ease of transparency for sample chain of custody and transfer of information. The communication site should not be used for formal results reporting. Additionally, an alternative process is needed to ensure business continuity if the network connection is down, and the communication site is unavailable.
3) *Additional microbial identifications.* In cases where the isolate cannot be identified by the CoE, the CoE will send the sample out to a contract laboratory. The statement of work that was generated between the site and CoE should dictate which cost center will be charged for the additional services.
   - Mold identification and strain typing – these are often specialized identification capabilities that not all laboratories have available. If the CoE does not have in-house capabilities, these services should be sent directly to the contract laboratory.

Table 18.2 Pros and cons of implementing a CoE program.

Pros	Cons
Consistency in identification approach	Site variability in identification interpretation without global guidance
Cost savings to perform internally	External laboratory tends to have a larger database due to the varied customer base
Quicker turnaround time internally, as evident with pandemic constraints when using external laboratories	Identifying the appropriate CoE location
Supports lean laboratories efforts – fewer instruments, fewer staff, less training needed, allowing for focused efforts from staff	Logistics can be difficult to sort – shipping costs, testing agreements, results sharing
Internal control over database, option to try to centralize database for network	

4) *Results reporting.* The turnaround times should generally align with those expected from contract laboratories (e.g. 0 days/high priority, 1–2 days, 3–5 days, 5–7 days). The benefit of having a CoE is that there is much more flexibility in communicating urgency of results and prioritization.

The pros and cons of implementing a CoE are outlined in Table 18.2, which can provide supportive information for building a business case.

A benefit of MALDI-TOF worth highlighting is the ability to create a customized library based on the microbial identifications across multiple sites, which can support and strengthen the company's root cause analysis and trending abilities. This allows for a library based on isolates that are specific not only to pharmaceutical manufacturing, but also to the specific company. Having this curated list can aid in trending at single or multiple sites and in identifying whether other sites may have isolated common organisms during investigations. Quality control is maintained to internal standards and oversight within the CoE provides consistency in evaluation and interpretation of results in creating the custom library.

18.3.2.6 Building a Business Case for the MALDI-TOF as a Network Strategy

The MALDI-TOF instrument provides flexibility in use and scale-up options in the pharmaceutical industry, as it is not as complex as the genotypic options but still provides a robust method of identification. By driving toward a global or regional CoE approach, microbial identification processes become aligned and major/immediate benefits can be seen, such as the following:

- *Critical investigations.* When microbial issues arise requiring investigation, such as out-of-level (OOL), out-of-specification (OOS), or presence of indicator organisms, immediate actions may be required to contain the issue and remediate/decontaminate. It is also important to perform an efficient, effective root cause analysis, assess the impact of the issue, determine the appropriate corrective and preventative actions (CAPAs), which address the root cause(s), and evaluate the effectiveness of the CAPAs. Throughout the investigation process, accurate and rapid microbial identification can be essential. MALDI-TOF offers a benefit in providing timely, critical microbial identification results that can be used to support the investigation, such as determination of the source of the microorganism(s) through investigative sampling/identification and facilitation of the evaluation of impact based on the microbial characteristics derived from the identification.

- *Alignment with regulatory requirements/guidance.* United States Pharmacopeia (USP) Chapters <62> and <71> [11, 12] are examples of requirements for microbial identifications as either confirmatory (USP Chapter <62>) or critical for determining validity (USP Chapter <71>). Because the tests for specified microorganisms and sterility are critical in determining the outcome of batch disposition, a rapid time to result and accuracy are important factors for microbial identification. MALDI-TOF MS is a technology that allows for this. Additionally, the recently published EU Annex 1 revision (2022) [13] requires manufacturing facilities to have a Contamination Control Strategy (CCS), whereby having an accurate and timely microbial identification is critical in supporting root cause analysis of microbial contaminations.

Table 18.3 provides considerations for a business case based on the elements and information presented here.

Table 18.3 Business case considerations.

Proposal/ business need	A multifaceted approach is outlined below. The goal is to reduce costs associated with using an external laboratory and gain control over internal data: • Implement a MALDI-TOF system at a pilot site. • Expand use by implementing at additional key sites and identifying a network CoE or regional CoEs.
Financial aspect	A global approach is recommended to best leverage pricing advantages. Different strategies include: • A tiered approach, where the first instrument is purchased, and subsequent instruments are purchased at reduced pricing. • Or, if implementation is all within a short period of time, a bulk discounted purchasing strategy can be negotiated. Within any procurement strategy, consider negotiating the cost for consumables and preventative maintenance activities.
Benefits and rationale	• Can tailor identification program based on sample throughput • Ease of operation (reduced complexity), training • Opportunity to create a custom identification/database • Increased control over turnaround time, flexibility with prioritization • Improved communication • Validated and accepted technology in the industry • Streamlined (internal vs. external) quality agreements
Risks	• Strategy may not make sense for sites with low sample throughput or limited identification needs. • Lack of personnel to perform validation. • Inconsistent identification strategies, if implemented by site instead of through a global strategy, could lead to the inability to obtain an identification or inaccurate identification, and delays in investigations.

18.3.3 Implementation Strategy – From Feasibility Studies to Global Deployment

Referring to Figure 18.2, the implementation strategy encompasses steps 6–9 and represents the execution phase that follows endorsement of the business case.

18.3.3.1 Pilot Trials/Feasibility

The pharmaceutical industry is currently undergoing the so-called fourth industrial revolution (industry 4.0), which includes more flexible project implementation strategies.

A proof-of-concept strategy, often also called a pilot-site strategy, allows an organization to validate and prove the innovative technology thoroughly before roll-out to the whole network and also to reduce unknown risk factors, disseminate lessons learned accordingly, and, overall, to clear the road and speed up subsequent implementations [14].

This is recommended for global network implementation plans for a MALDI-TOF system as well. An organization's capacity to establish needs, prepare for the pilot, determine application duties, design a pilot strategy and plan, and monitor the pilot is critical to the success of such a strategy.

The key aspects to consider are:

- Identification of the pilot site, considering the capacity and knowledge of the respective site.
- Project plan and timelines.
- Monitoring the overall implementation progress, focusing on improvement and risk identification.
- Post-implementation monitoring to ensure the system is running smoothly in the routine setting.
- Sharing the documentation and the lessons learned and, if required, adapt the strategy for subsequent implementations.

18.3.3.2 Risk Assessment/Risk-based Validation Approach

Quality risk management has become an obligatory tool for pharmaceutical businesses in recent years. QRM is an ongoing process of reducing risks to product quality throughout its life cycle to maximize benefits while balancing risk. It is a method for assessing, controlling, communicating, and reviewing risks to the quality of a pharmaceutical product [15].

Risk assessment/risk-based implementation approaches provide benefits if used for technology implementation projects as well. The U.S. Food and Drug Administration (FDA) guidance for industry *Q9 Quality Risk Management* [16] provides an overview of the typical risk management process and refers to some of the most commonly used tools (Figure 18.4).

Risk evaluation in combination with a pilot implementation strategy as mentioned in Section 18.3.3.1 will facilitate the identification and mitigation of the vast majority of potential high and medium risks for the global implementation of modern technologies and can also ensure a smooth subsequent validation.

Examples of commonly used tools for risk assessment are as follows [16]:

- Basic risk management facilitation methods
  (flowcharts, check sheets, etc.)

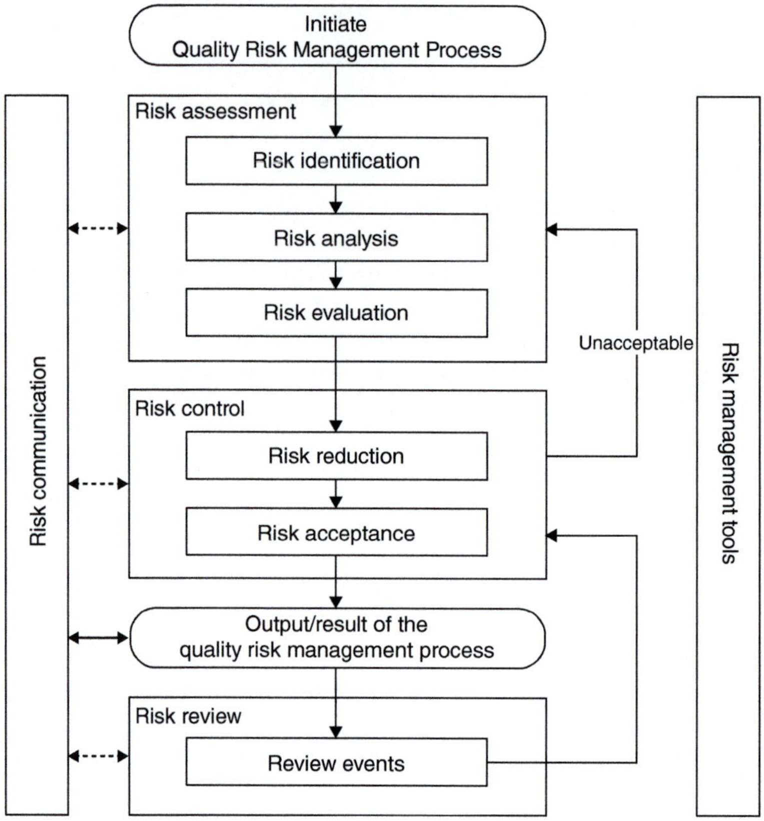

Figure 18.4 Typical quality risk management process [16].

- Failure mode effects analysis (FMEA)
- Failure mode, effects, and criticality analysis (FMECA)
- Fault tree analysis (FTA)
- Hazard analysis and critical control points (HACCP)
- Hazard operability analysis (HAZOP)
- Preliminary hazard analysis (PHA).

18.3.3.3 Network Validation Approach

Having completed the selection of a specific model, the process of purchasing, and risk assessment, the subsequent qualification/validation can start.

Each company will follow their in-house guidelines for performing the qualification/validation of a new piece of equipment, and often the terminologies are interpreted and applied in slightly different ways; for example, qualification, verification, and validation are frequently used terminologies, but can be applied differently at different companies or even in the various pharmacopeias. Figure 18.5 shows the typical core elements contained in qualification/validation pathway.

- Overall Risk Assessment
- User Requirement Specification (URS)
- Design Qualification (DQ)
- Functional Design Specification (FDS)
- Installation Qualification (IQ)
- Operational Qualification (OQ)
- Performance Qualification (PQ)
- Method Validation
- Validation Maintenance

General Qualification activities, typically performed in collaboration with the vendor.

Method Validation, according pharmacopeial and in-house requirements, can be performed as part of the PQ.

Validation Maintenance as post-implementation activity, to grant the validated state.

Figure 18.5 Typical qualification/validation pathway.

The activities that are categorized under general qualification can be diverse based on the individual company guidelines and based on the level of involvement by the vendor or a third party performing some of these qualification activities.

Therefore, we will focus on recommendations for the method validation portion for a MALDI-TOF identification system from the European Pharmacopeia (EP) and the USP. If guidance from these sources is followed, the validation will most likely be accepted around the globe; however, it is important to consider and check the local or regional guidelines as well and to adapt the strategy accordingly.

As indicated in Table 18.A.1 there are a variety of regulatory requirements and guidelines available related to microbial identification. The following compendial chapters have been identified to provide guidance around the validation of microbial identification systems:

- USP Chapter <1113> [3] – Microbial Characterization, Identification and Strain Typing
- USP Chapter <1223> [17] – Validation of Alternative Microbiological Methods
- EP 5.1.6 [18] – Alternative Methods for Control of Microbiological Quality.

Table 18.4, provides a comparison of two of these chapters to highlight the differences in some key parameters for identification methods. Furthermore, Table 18.5 illustrates the differences in the required validation parameters for the different method types in general.

Table 18.4 Comparison of the validation core elements: EP versus USP.

	USP Chapter <1113>	**EP 5.1.6**
General approach	• Parallel testing of ~50 isolates with an existing system • Testing 12–15 known stock cultures • Confirming 2–50 • Identifications with a reference laboratory	• General qualification of equipment (URS, DQ, FDS, IQ, OQ, PQ) • Primary method validation by supplier • Method validation
Validation parameters	• Accuracy • Reproducibility • Sensitivity • Positive and negative predictive value	• Accuracy • Precision • Robustness

Source: Adapted from Sandle [19]

Table 18.5 Overview of the required validation parameters in EP 5.1.6 [18], USP Chapter <1223> [17], and USP Chapter <1113> [3].

	Qualitative test		**Quantitative test**		**Identification test**	
Validation parameter	**EP 5.1.6**	**USP Chapter <1223>**	**EP 5.1.6**	**USP Chapter <1223>**	**EP 5.1.6**	**USP Chapter <1113>**
Accuracy	×		×	×	×	×
Precision	×		×	×	×	
Specificity	×	×	×	×		×
Limit of detection	×	×		×		
Limit of quantification			×	×		
Linearity			×	×		
Range			×	×		
Robustness	×	×	×	×	×	
Repeatability (parameter mentioned in USP Chapter <1223> [17] only)		×		×		
Reproducibility (parameter mentioned in USP Chapter <1223> [17] only)						×
Ruggedness (parameter mentioned in USP Chapter <1113> [3] only)		×		×		
Equivalency	×	×		×		
Suitability (parameter mentioned in EP 5.1.6 [18] only)	×		×			
Sensitivity (parameter mentioned in USP Chapter <1113> [3] only)						×
Positive and negative predictive value (parameter mentioned in USP Chapter <1113> [3] only)						×

For the detailed definition of all the above parameters, it is recommended to consult the referenced pharmacopeial chapters.

Applying the same comparison principle to all three of the referenced chapters and including the requirements for alternative quantitative and qualitative test methods in addition to the identification methods, the level of misalignment becomes even more visible (see Table 18.5).

Considering that the requirements shown in the chapters usually reflect the minimum expected requirements, and take into account that globally operating companies will thrive on a holistic and widely accepted validation approach, the interpretation of Table 18.5 should be to cover, at a minimum, all the seven marked validation parameters in the two far right columns for identification methods.

Combining this interpretation with the discussed pilot-site strategy (Section 18.3.3.1), this stipulates that the selected pilot site needs to perform a holistic full validation package, considering all the differences. However, subsequent sites, implementing the same technology according to the same principles, will have the opportunity to benefit from the generated knowledge and documents in such a way that they can follow a risk-based, reduced validation strategy, leveraging data from: (i) the primary validation package from the vendor, as the key aspects will have been confirmed by the pilot-site validation; and (ii) the available in-house validation package as well as from the initial risk assessment from the pilot site (risk-based validation in this case refers to identifying what needs to be validated "again" on the other sites and what can be considered as covered or low-risk considering the available information from the pilot-site validation package).

From a strategic perspective this will enhance the implementation speed for the whole network, although it is well acknowledged that the primary site will have a significantly higher workload, leading to potential resistance from volunteers. Nonetheless, the overall success and benefit for the network should outweigh the concerns of individual sites.

18.4 Conclusions

As the industry becomes more advanced and companies seek to streamline and reduce costs, the need for fast, accurate, and low-cost methods increases. Because microbial identifications play such an important role for pharmaceutical companies, from both business and quality perspectives, modern, robust, simple, and high-throughput systems are a must. For companies with multiple manufacturing sites, from a regional and/or global perspective, using a network microbial identification strategy with MALDI-TOF MS as the principal technology can be beneficial in terms of reduced cost, CoE, and ability to build an extensive internal supplemental custom identification library, which strengthens the companies' abilities in root cause analysis and trending. By using a robust, risk-based network validation strategy with a well-defined process for isolate submission, testing, and reporting, pharmaceutical companies can achieve these goals.

18.A Appendix

Table 18.A.1 Microbial identification requirements and guidance.

Reference	Relevant information
USP Chapter <1113> – Microbial characterization, identification, and strain typing [3]	Discusses different methods for microbial characterization. Regarding whole-cell composition phenotypic method by MALDI-TOF spectrometry, the system is successfully used in pharmaceutical microbiology testing laboratories by providing "information that enables microbiologists to make informed decisions regarding product risk and to recognize changes in environmental microflora." USP Chapter <1113> states that for investigations, "phenotypic identification alone is sufficient and will enable scientists to conduct a thorough investigation and to recommend appropriate corrective actions as needed."
FDA – Guidance for industry sterile drug products produced by aseptic processing [20]	Highlights the importance of accurate and precise microbial identification for the purposes of monitoring and understanding the state of microbial control. Correlation of isolates from media fill or product sterility test failures and environmental monitoring samples can provide valuable information for investigations. Microbial differentiation can be important in trending programs, revealing migration of microorganisms from areas of lower control to higher control and aseptic processing rooms. The FDA guidance also emphasizes the level of identification, to species (or, where appropriate, genus) level. Finally, although the FDA guidance states that genotypic methods are especially valuable for investigation of failures (e.g. sterility, media fill), appropriate biochemical and phenotypic methods can be used for routine identification. Note: since publication of this guidance document in 2004, MALDI-TOF has become a more widely available and commonly used microbial identification technology in the pharma laboratory for investigation of failures as well as routine identification.
USP Chapter <1223> – Validation of alternative microbiological methods [17]	Provides guidance on the selection, evaluation, and use of microbiological methods as alternatives to compendial methods. This chapter is primarily concerned with methods, and excludes identification systems, although sections on data quality and instrument qualification can be helpful as guidance.
EU – GMP Annex 1 (2022), Manufacture of sterile medicinal products [13]	States that "Microorganisms detected in the grade A and grade B areas should be identified to species level and the potential impact of such microorganisms on product quality (for each batch implicated) and overall state of control should be evaluated. Consideration should also be given to the identification of microorganisms detected in grade C and D areas (for example where action limits or alert levels are exceeded) or following the isolation of organisms that may indicate a loss of control, deterioration in cleanliness or that may be difficult to control such as spore-forming microorganisms and moulds and at a sufficient frequency to maintain a current understanding of the typical flora of these areas."
USP General notices, 6.30 – Alternative and harmonized methods and procedures [21]	States that "Alternative methods and/or procedures may be used if they provide advantages in terms of accuracy, sensitivity, precision, selectivity, or adaptability to automation or computerized data reduction, or in other special circumstances."

Table 18.A.1 (Continued)

Reference	Relevant information
USP Chapter <1225> – Validation of compendial procedures [22]	Contains data elements required for validation of analytical procedures.
USP Chapter <62> [11], EP 2.6.13 [23], PMDA 4.05 [24] (harmonized) – Microbiological examination of nonsterile products: tests for specified microorganisms	Specify the need for microbial identification as a requirement for confirmatory identification tests for microorganisms that grow on selective or diagnostic media and demonstrate defined morphological characteristics.
USP Chapter <71> [12], EP 2.6.1 [25], PMDA 4.06 [26] (harmonized) – Sterility tests	Highlight the importance of microbial identification as it allows for invalidation of the sterility test if the identification of the isolate(s) from the test can be unequivocally ascribed to faults with sterility test material and/or technique.
USP Chapter <1116> – Microbiological control and monitoring of aseptic processing environments [27]	Recommends that microbial isolates be identified at a rate sufficient to support the environmental monitoring program.
EP 5.1.6 – Alternative methods for control of microbiological quality [18]	Provides information on choosing an alternative microbiological method as a supplement or as an alternative to conventional microbiological methods and to provide validation guidance on these methods. For identification tests, EP 5.1.6 highlights the extent of database and customization of the database to allow the user to add microorganisms not already included as factors to consider in the validation.
PDA TR 33 – Valuation, validation and implementation of alternative and rapid microbiological methods [5]	Provides guidance on validation and implementation of alternative/rapid microbiological methods, including criteria for acceptability of the RMM to the compendial/classical method and guidance for proving this to regulatory agencies.
Bain et al. [28] – Microbial monitoring for biological drug substance manufacturing: an industry perspective	Industry best-practices paper provides justification for the level of microbial identification and importance of speciation in contamination investigation. Recommends that all drug substance and drug product microorganisms be identified to the species level.
Griffin and Reber – Microbial identification: the keys to a successful program [29]	Collection of chapters from industry experts containing scientific rationale for the application of current technology, experience/case studies, and literature review for modern pharmaceutical microbiology identification programs.
PIC/S – Recommendations on sterility testing [30]	States that routine identification of environmental microorganisms to at least the genus level should assist in detecting trends and that techniques such as molecular typing will be required for identification of microorganisms if evidence of identity of environmental and test isolates is the sole rationale used to invalidate the original sterility test.
FDA – Guide to inspections of microbiological pharmaceutical quality control laboratories [31]	Reference material for investigators and other FDA personnel which discusses the importance of microbial identification.

Abbreviation: PMDA, Pharmaceuticals and Medical Devices Agency.

References

1 Franco-Duarte, R., Černáková, L., Kadam, S. et al. (2019). Advances in chemical and biological methods to identify microorganisms – from past to present. *Microorganisms* 7 (5): 130. https://www.mdpi.com/2076-2607/7/5/130/htm (accessed 9 February 2022).

2 Cundell, T. (2020). Transition to modern microbiological methods in the pharmaceutical industry. *American Pharmaceutical Review*. https://www.americanpharmaceuticalreview.com/Featured-Articles/571418-Transition-to-Modern-Microbiological-Methods-in-the-Pharmaceutical-Industry/ (accessed 9 February 2022).

3 United States Pharmacopeia (USP) n.d. Microbial characterization, identification, and strain typing <1113>. In: *USP–NF*. Rockville, MD: USP.

4 BioPhorum. 2020 A framework for the evaluation, validation and implementation of alternative and rapid microbiological testing methods. https://www.biophorum.com/download/alternative-and-rapid-micro-methods-armm-a-framework-for-the-evaluation-validation-and-implementation-of-alternative-and-rapid-microbiological-testing-methods (accessed 17 January 2022).

5 Parenteral Drug Association (PDA) (2013). Evaluation, validation and implementation of alternative and rapid microbiological methods. *Technical Report No. 33* (revised). Bethesda, MD: Parenteral Drug Association.

6 Tonchia, S. (2018). Project communication. In: *Industrial Project Management. Management for Professionals*, 143–152. Berlin, Heidelberg: Springer https://doi.org/10.1007/978-3-662-56328-1_12.

7 Stein, M., Tran, V., Nichol, K.A. et al. (2018). Evaluation of three MALDI-TOF mass spectrometry libraries for the identification of filamentous fungi in three clinical microbiology laboratories in Manitoba, Canada. *Mycoses* 61: 743–753. https://doi.org/10.1111/myc.12800.

8 Alinezad, A., Seif, A., and Esfandiari, N. (2013). Supplier evaluation and selection with QFD and FAHP in a pharmaceutical company. *Int. J. Adv. Manuf. Technol.* 68: 355–364. https://doi.org/10.1007/s00170-013-4733-3.

9 Moldenhauer, J. (2011) Proteotypic identification methods – a change in identification methods. *American Pharmaceutical Review*. https://www.americanpharmaceuticalreview.com/Featured-Articles/112514-Proteotypic-Identification-Methods-A-Change-in-Identification-Methods (accessed 2 February 2022).

10 Sandle, T. (2017). Microbial identification strategy for pharmaceutical microbiology. *J. GxP Compliance* 21: 11–20.

11 United States Pharmacopeia (USP) n.d. Microbiological examination of nonsterile products: tests for specified microorganisms <62>. In: *USP–NF*. Rockville, MD: USP.

12 United States Pharmacopeia (USP) n.d. Sterility tests <71>. In: *USP–NF*. Rockville, MD: USP.

13 European Commission Annex 1: Manufacture of sterile medicinal products. Consultation document. In: *The Rules Governing Medicinal Products in the European Union, Volume 4. EU Guidelines for Good Manufacturing Practice for Medicinal Products for Human and Veterinary Use*. 2022

14 Barata, João; Cunha, Paulo Rupino Da; and Coyle, Sharon, (2018) "Guidelines for Using Pilot Projects in the Fourth Industrial Revolution". *ACIS 2018 Proceedings*. 80. https://aisel.aisnet.org/acis2018/80

15 Ismael, O.A. and Ahmed, M.I. (2020). Using quality risk management in pharmaceutical industries: a case study. *Calitatea* 21 (178): 106–113.

16 Food and Drug Administration (FDA) (2006). *Q9 Quality Risk Management*. Guidance for Industry. US Department of Health and Human Services.

17 United States Pharmacopeia (USP) n.d. Validation of alternative microbiological methods <1223>. In: *USP–NF*. Rockville, MD: USP.

18 European Directorate for the Quality of Medicines and HealthCare (EDQM) n.d. Alternative methods for control of microbiological quality – 5.1.6. In: *European Pharmacopoeia 10.3*. Strasbourg, France: EDQM.

19 Sandle, T. (2013). Automated microbial identification: a comparison of USP and EP approaches. *American Pharmaceutical Review*. https://www.americanpharmaceuticalreview.com/Featured-Articles/140520-Automated-Microbial-Identification-A-Comparison-of-USP-and-EP-Approaches/ (accessed 25 January 2022).

20 Food and Drug Administration (FDA) (2004). *Sterile Drug Products Produced by Aseptic Processing – Current Good Manufacturing Practice*. Guidance for Industry. US Department of Health and Human Services.

21 United States Pharmacopeia (USP) General Notices and Requirements applying to standards, tests, assays, and other specifications of the United States Pharmacopeia. 6.30 Alternative and harmonized methods and procedures. In: *USP–NF*. Rockville, MD: USP. 2021

22 United States Pharmacopeia (USP) n.d. Validation of compendial procedures <1225>. In: *USP–NF*. Rockville, MD: USP.

23 European Directorate for the Quality of Medicines and HealthCare (EDQM) n.d. Microbiological examination of non-sterile products: test for specified micro-organisms – 2.6.13. In: *European Pharmacopoeia 10.3*. Strasbourg, France: EDQM.

24 Pharmaceutical Medical Devices Agency (PMDA) n.d. Microbiological examination of non-sterile products – 4.05. In: *Japanese Pharmacopoeia*. Tokyo, Japan: PMDA.

25 European Directorate for the Quality of Medicines and HealthCare (EDQM) n.d. Sterility – 2.6.1. In: *European Pharmacopoeia 10.3*. Strasbourg, France: EDQM.

26 Pharmaceutical Medical Devices Agency (PMDA) n.d. Sterility test – 4.06. In: *Japanese Pharmacopoeia*. Tokyo, Japan: PMDA

27 United States Pharmacopeia (USP) n.d. Microbiological control and monitoring of aseptic processing environments <1116>. In: *USP–NF*. Rockville, MD: USP.

28 Bain, D., Hardy, D., Bell, B.L. et al. (2015). Microbial monitoring for biological drug substance manufacturing: an industry perspective. *PDA J. Pharm. Sci. Technol.* 69 (3): 451–460. https://journal.pda.org/content/69/3/451.

29 Griffin, M. and Reber, D. (ed.) (2012). *Microbial Identification: The Keys to a Successful Program*. Bethesda, MD: PDA.

30 Pharmaceutical Inspection Co-operation Scheme (PIC/S) (2007) *Recommendations on Sterility Testing*. https://www.gmp-compliance.org/files/guidemgr/PI%20012-3%20Recommendation%20on%20Sterility%20Testing.pdf (accessed 15 February 2022).

31 Food and Drug Administration (FDA) (1996). Microbiological Pharmaceutical Quality Control Labs. "Guide to Inspections of Microbiological Pharmaceutical Quality Control Laboratories" https://www.fda.gov/inspections-compliance-enforcement-and-criminal-investigations/inspection-guides/microbiological-pharmaceutical-quality-control-labs-793 (accessed 15 February 2022).

19

MALDI-TOF MS – Microbial Identification as Part of a Contamination Control Strategy for Regulated Industries

Christine E. Farrance and Prasanna D. Khot

Charles River Laboratories, Newark, DE, USA

19.1 Industry Perspective

19.1.1 Introduction to Regulated Industries

Manufacturing industries produce goods with the use of labor, machines, and chemical or biological processing or formulation. Many of these industries – including pharmaceutical, dietary supplements, and medical devices – are controlled by specific governmental regulations. The products manufactured by these regulated industries can be used by people for critical care, treatment of illness, daily personal care, and consumer products. Globally, regulations and guidance documents exist to ensure the quality of manufactured products. A "quality" product consistently meets the expectation for its use; it consistently and reliably delivers the performance and other characteristics stated on the label, and most importantly is not contaminated.

Microbial contaminants in pharmaceutical, food, and cosmetic products can represent a risk to public health and consumers. Release of contaminated products can cause tremendous financial losses to a company because of product recalls, fines, and complex internal investigations, in addition to the significant damage to a company's reputation due to loss of brand confidence by consumers [1, 2]. As a result, many companies implement an environmental monitoring (EM) program to help ensure products from their manufacturing facilities are free of microbial contamination and are safe for the consumer to use. The purpose of an EM program is to demonstrate a state of control by detecting organisms in their facilities as a function of location and time before these microbes can potentially contaminate a product [3–7]. Microbial control will vary and differ based on the product and manufacturing methods.

Pharmaceutical companies understand the responsibility they have to patients to deliver safe therapies and that means adherence to current good manufacturing practice (GMP). Sterile pharma is easy to define as there cannot be any microorganisms or toxins such as

Microbiological Identification using MALDI-TOF and Tandem Mass Spectrometry: Industrial and Environmental Applications, First edition. Edited by Haroun N. Shah, Saheer E. Gharbia, Ajit J. Shah, Erika Y. Tranfield, and K. Clive Thompson.

endotoxins in the final product. These manufacturers design their quality system for sterility assurance. For the non-sterile industries, including pharma and personal care products (PCPs), the product must be free of adulteration. Three levels of testing are required for non-sterile products: microbial enumeration, testing for the absence of specified microorganisms, and screening for objectionable microorganisms. The manufacturer is responsible for establishing specifications for purity of the product or incoming components or ingredients, as well as limits on the types of contaminants that may adulterate the finished product.

Microorganisms are incredibly diverse, and microbial contamination of pharmaceutical, consumer care, or other products represents two hazard types: the health hazard or the risk of infection, and spoilage. The regulations do not designate adulteration or objectionable microorganisms, but they can be broadly defined as microorganisms that, owing to their numbers in the product and/or their pathogenicity, can cause infection in the patient or consumer via the route of administration/use of that pharmaceutical or PCP. Additionally, microorganisms can proliferate in a product, adversely affecting the chemical, functional, and therapeutic properties of that product, and render it less effective or alter the product to a point where a consumer would be discouraged from using it.

19.1.2 Contamination Control Strategy

Given the negative consequences of contamination, manufacturers should develop and be committed to a contamination control strategy (CCS). Previously, a focus was placed on sampling methods for the measurement of potential contamination. This approach has been remodeled to prevent and control potential contamination from reaching the product. A CCS is the integration of many interrelated components which are defined under a system of oversight and regulation. At its foundation are scientific and manufacturing knowledge, quality risk management, personnel awareness, and a culture for quality. Contamination controls are for raw materials and consumables, personnel training and hygiene, equipment cleaning and sanitization, facility and utility design, and process design (e.g. see Figure 19.1). The validation and monitoring of these controls, such as personnel, in-process, pest control, and EM, are critical.

The biggest risk to the process comes from humans, followed by water, production equipment, and process flow. A robust CCS will make use of multiple quality risk management tools to understand both direct opportunities for contamination and how failure of a control may allow contamination. Monitoring limits or levels and interpreting the data will indicate their significance in terms of the facility state of control and the efficacy of a CCS and contribute to making sound data-driven and scientifically justified manufacturing and product release decisions.

19.1.3 Tracking and Trending EM Data

Monitoring the manufacturing environment is one way to control microbial contamination. Microbes can be detected before they become a problem, and the data provide the tools to evaluate and resolve contamination events. EM is standard in the pharmaceutical manufacturing environment [6, 7]. While it is not mandated for consumer products, many

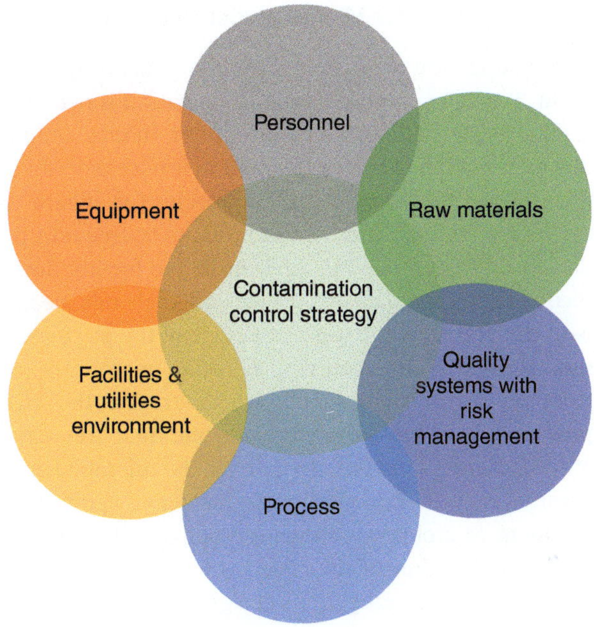

Figure 19.1 Conceptual image of a CCS, which consists of a planned set of controls for microorganisms, pyrogens, and particulates that are derived from current product and process understanding and are to ensure process performance and product quality. The controls can include parameters and attributes related to materials and components, facility and equipment, operating conditions, in process controls, and the associated methods and frequency of monitoring and control of these aspects.

companies are implementing EM to enhance safety standards and manage risk. Different locations and materials are sampled and plated on culture media to look for microorganism growth (personnel, air, surfaces, water, raw materials, equipment). The number of microorganisms is counted and recorded, and in certain cases the organisms are identified.

The data collected during the execution of an EM program must be utilized and evaluated through tracking and trending reports – both recovery rate reports and trending by microorganism name. EM trending can illustrate historical patterns, temporal patterns, and links between shared ingredients, and evaluate the effect of controls on the manufacturing environment. An identification method must produce consistent results to understand repeated trends and reoccurrences. The more accurate and reproducible the identification, the more confidence there will be in the EM data and in the operational processes.

Environmental monitoring provides a baseline profile of what microorganisms are present in the manufacturing environment and defines what is typical for a facility. By routinely sampling and identifying organisms, manufacturers can detect possible contaminants and identify sites where there is risk of contaminating the product. The microbial baseline gives a reference, which facilitates faster and more effective investigations, identifies areas for improvement, responds proactively to changes in the environment, documents consistent quality, and demonstrates a state of control.

An ideal EM program should have the ability to detect and identify all categories of microbes – bacteria, yeasts, and molds (filamentous fungi). The microorganisms in a facility come from soil, water, air, people, and ingredients. The diversity of organisms that are recovered from manufacturing facilities within and across taxonomic domains is immense. Unlike organisms in clinical settings, it is not just pathogenic organisms that are of concern for manufacturing industries, but also microorganisms that can proliferate in a product and adversely affect the critical attributes of that product.

19.1.4 Drivers for Microbial Identification

Characterization of recovered microorganisms provides vital information for the EM program. Typically, some level of identification is necessary on any organism found within or originating from the plant environment. In most industries, microbial identification is usually required when more information is necessary for mitigation and to manage risk. Thus, the EM program should require species (or, where appropriate, genus) identification of microorganisms in these environments at frequent intervals to establish a valid, current database of contaminants present in the facility during processing. The identification of recovered microorganisms is needed to rule out specific pathogens or objectionable organisms as defined by the use and nature of the product. The identity of a microorganism can indicate if it is a reoccurring resident microorganism or a new transient contaminant. Routine identifications can also determine the efficacy of cleaning and sanitization protocols, and the suitability of rotational frequency of disinfectants and sporocides. Manufacturers want both to manage the microbial count and exclude objectionable microorganisms from non-sterile products and to exclude all microorganisms from sterile products. Microbial identification helps to define the microbial contamination risk level, while the location of the source of the microorganisms (water, personnel) is useful in product failure investigations designed to mitigate risk.

19.1.5 Level of Resolution of an Identification

There are many drivers in regulated industries for microbial identifications and these will vary by industry, process, and product. The focus should be on microbial safety and risk assessment. There is a hierarchy of response to the recovery of a microorganism from a manufacturing facility, process, or product: basic microbial characterization, identification, and strain typing. Basic microbial characterization comprises the determination of the Gram stain reaction, cellular and colony morphology, spore formation, and key biochemical parameters (catalase, oxidase, coagulase). These are diagnostic features but not speciation. Identification is the assignment of a taxonomic rank, or genus/species name, to a microorganism. Many times, species-level identifications are required, but this is not always necessary. Strain typing provides detailed information about the genetic differences between isolates of the same species and is often required during investigations. Identification can be performed using different technologies that vary in the ability to differentiate the vast range of microorganisms that can be found in manufacturing facilities, and in the time it takes to get a result, accuracy, and reproducibility.

Depending on the source of the isolate, there can be different requirements for accuracy needed to manage risk, and different requirements for how quickly the identification needs to be obtained. Isolates are often described as "routine" or "critical" and these labels are often correlated with environmental or product-related samples. If the isolate is recovered from a product, its criticality is, in turn, dependent on the type of product. There is significant response if the product is designed to be sterile. If it is non-sterile, requirements for the identification will vary according to the use or application, and whether it is a personal care or household product. Primary screening can be sufficient for risk assessment either to recommend a corrective action or to dictate the need to obtain a species-level identification. Different levels of confidence in the identification are needed when the microbe is recovered from different areas or sources, such as the loading dock, reagent preparation area, carts or trollies, or gowning area. The level of characterization depends on the criticality of the area involved and on whether an investigation warrants further identification.

19.1.6 Global Harmonization

Manufacturing of pharmaceuticals, and even food, consumer, and personal care products, is highly globalized with an international supply chain of raw materials, intermediates, and final products. Many of these companies have microbiologists and quality assurance personnel with responsibilities for global operations or investigations who recognize and advocate for a harmonized microbial identification platform for all plants around the world. Use of different identification technologies, reference libraries, and quality procedures at the various manufacturing sites along the pipeline can create inconsistency, wasted resources, and challenges for collating and trending EM data and during investigations related to product contamination. A harmonized approach has many advantages, including a standardized approach to the level of identification, as well as standardized techniques to perform microbial identification. Using standardized methods and reference libraries results in consistent data capture and repositories, and in the confidence that data interpretation is not based on subjective measures if using clearly defined analytical methods. With the global movement of materials between sites, a harmonized microbial identification platform can aid in investigations, as there will be confidence in those global data. The implementation of new technologies can be made using global datasets and platform approaches.

19.1.7 Validation Requirements for Regulated Industries

A company that adheres to GMP and intends to implement new technologies, including microbial identification systems, will need to plan and execute an instrument and software qualification as well as method validation. This internal validation is generally planned and executed by the end user but can have documentation and support from the vendor as needed. The company's quality system and standard operating procedures should govern the validation plans and required documentation for both the system itself and its reference database. The results of the validation demonstrate the performance, consistency, and reliability of the method [8, 9]. In addition, metrics such as accuracy, precision, specificity, and repeatability are needed for validation [9]. If system verification is sufficient, USP Chapter <1113> provides informative guidance to that end [10]. A well planned and executed validation process can refer to references on the validation of a computerized laboratory system under GAMP 5 [11], 21 CFR Part 11 [12], and the USP Chapter <1058>, "Analytical instrument qualification" [13]. While these references provide guidance on general elements for an instrument validation, specific modifications relevant to microbial identification systems are needed, including confirmation that the analytical procedure used for a specific test is suitable for its intended purpose – a requirement for GMP.

19.1.8 Summary

Documents such as the FDA's *Sterile Drug Products Produced by Aseptic Processing – Current Good Manufacturing Practice* [4] and the Parenteral Drug Association's Technical Report No. 67, *Exclusion of Objectionable Microorganisms from Nonsterile Pharmaceuticals, Medical Devices, and Cosmetics* [14], present requirements and guidance for the manufacture of a wide range of products, including over-the-counter drug products and PCP in addition to what is in the titles. Regulators and manufacturing companies are working together to make these products safe and efficacious.

A CCS helps to assure process performance and product quality. An important control system is EM, the recovery and identification of microorganisms to the species level, and strategies for managing microbial risks associated with the presence of microorganisms. The identification of environmental isolates provides a snapshot of the environmental control of a location and helps in understanding what "typical" looks like, so a drift away from normal can be made apparent. Consistent identification methods allow for sound comparisons and interpretations while appropriate trending can enhance control strategies. Quality and control, and the evaluation of the risk, will ultimately link to the protection of the patient and the consumer.

19.2 Technical Perspective

19.2.1 Identification Technologies

As discussed in Section 19.1, in addition to detecting and enumerating the number of microorganisms in an environmental sample, it is important to identify representative recovered isolates. To detect changes in trends, it is important that the isolates be successfully, consistently, and correctly identified. All characterizations and identifications must begin with core microbiological techniques as pure cultures are essential for all further analyses. A broad characterization can give an indication of crude or large-scale issues such as people, soil, or water sources of an organism [15], but accurate genus and species identification will give a high-resolution map of the microbiota of a facility.

The identification methods need to be relatively inexpensive, fast, yet accurate, with relevant and the appropriate breadth of organisms in the reference database. There are varied levels of operational needs, and these needs dictate different approaches to microbial identification. The criticality of the samples can be prioritized and correlated with the accuracy of the method used for the identification. Three basic categories of identification methods are routinely used for the identification of recovered environmental microbial isolates to the genus/species level: phenotypic, proteotypic, and genotypic. If using phenotypic systems, confidence to the genus level should be considered, as accuracy to the species level is not high, while proteotypic systems provide a very confident species-level identification. The most highly accurate resolution at the species level is achieved through genotypic or sequence-based approaches [4, 16–19].

19.2.2 Phenotypic Systems

In the past, microbial identification of bacteria and fungi was only obtainable using a variety of phenotypic methods, including staining, culturing, and chemical tests. These methods are still used today as they are cost effective, require minimal equipment, and are easy to perform. However, traditional identification techniques usually cannot match the combination of speed to result, reproducibility, and accuracy of modern methods. For these reasons, they are usually performed as a preliminary step to obtain a presumptive identification before a more modern technique is used. Additionally, the commercial phenotypic systems available have outdated and very limited reference databases, leading to incorrect or no identifications. The user must be vigilant about ensuring the purity of the sample being applied to the phenotypic systems as it is not clear from the data if a sample is mixed, and a blended phenotypic description will result. The accuracy and reproducibility of these systems is low. However, for genus-level confidence, they could be sufficient.

Protein-based and sequence-based methods decrease time to result, increase speciation, and decrease errors for microbial identification. This is especially true with fungi which are slow-growing, and conventional methods are often ambiguous [20].

19.2.3 Proteotypic Systems

Matrix-assisted laser desorption/ionization time-of-flight mass spectrometry (MALDI-TOF MS) systems used for identification come with a large capital expenditure, but are simple to use, have an extremely fast time to result, and are highly accurate and reproducible. The turnaround time for a result can be minutes as compared to days with conventional biochemical methods. Testing can be performed on very small samples and has a very low reagent cost [21]. The commercial systems have traditionally had clinical applications and this is reflected in the relevancy of their databases, but there is an extensive publication history in support of the accuracy of the results obtained for many, but not all, groups of organisms that have representation in the reference libraries. Performance can be improved by the creation and addition of reference spectra by the user for under-represented species or to increase the diversity of the representation to reflect isolates found during EM, but this can be beyond the capability for some industrial quality control microbiology laboratories.

19.2.4 Genotypic Systems

Genotypic identifications have been shown, with convincing scientific data, to be the most accurate for bacteria and fungi; they are more discriminatory, robust, and reproducible as the classification is based on conserved nucleic acid sequences and a phylogenetic analysis, and the technology is not dependent on culture or environmental conditions, which is not the case for the phenotypic and proteotypic systems [16–19, 22]. Sequence-based technologies for identification also come with large capital and consumable expenses, are complex to perform, but have a relatively short time to result, and produce data of highest specificity, reliability, and reproducibility.

19.2.5 The Importance of the Reference Database

Every technology is dependent on the robustness and diversity of its database for the correct identification. Whether using a phenotypic, proteotypic, or genotypic system, accuracy is linked to the quality of the underlying data and the associated taxonomic descriptor for those data. Phenotypic systems must have accurate and complete morphological and biochemical descriptors of the type or typical reference strains. Proteotypic systems must have high-quality reference spectra while genotypic systems are dependent on the quality and accuracy of the reference sequences. Additionally, the appropriate and current taxonomic names must be associated with the reference biochemical, spectral, or sequence data.

19.2.6 MALDI-TOF in Regulated Industries

During routine monitoring, accurate and consistent identification methods are important in managing risk and for applying risk-based assessment tools. MALDI-TOF MS is often the first choice for the routine identification of bacterial and yeast isolates in many industries because of its relatively high accuracy and reproducibility to the species level, low costs, easy operations, and shortened time to result as compared with sequence-based analyses. On the other hand, for routine identification of filamentous fungi, the adoption of MALDI-TOF MS has progressed more slowly because of the limitation of the reference libraries and the difficulties of sample preparations, including protein extractions. Recent methodological improvements and expansion of supplemental databases focusing on filamentous fungi have significantly increased the identification rate and use of MALDI-TOF MS for this group of organisms. There are two primary commercially available instrument providers. The identification systems differ in their databases and identification algorithms, and the numerical rankings are reported differently. However, they function similarly.

One area that is driving the adoption of MALDI-TOF MS for microbial identifications is the increased scrutiny of fidelity or integrity of data produced in regulated industries. Data integrity (DI) is mentioned in many global regulations, and warning letters note failures to meet these regulations, which are therefore a violation of GMP. DI reflects the recording, analysis, retention, and accuracy of data. DI helps to assure that records are complete, intact, traceable, and maintained in their original context. These attributes are essential for data used for the establishment of a control strategy. MALDI-TOF MS is well suited to enhance DI in microbial identification, especially as compared with phenotypic systems. Human error and subjectivity are minimized with the move to objective data analysis, and the original data can be saved in a secure location as an electronic record. Improving DI is about improving the quality and the in-process control of the production of our drug supply.

19.2.7 Outsourcing

An additional consideration for manufacturers of regulated products is whether the different identification technologies and expertise will be obtained and run in-house or if the work is outsourced to a contract testing laboratory to take advantage of a wider breadth of experience and systems. There is even a combination of the two, where samples are run in-house but an external reference database with expanded relevant coverage of environmental organisms is used for the final identification.

Consideration for running a system in-house includes the cost and effort to validate a system and the reference database, the cost to run and maintain the system, the cost of the reagents and consumables, the system throughput, the time to result, system resolution, accuracy, and ease of accessing, interpreting, integrating, and tracking the data [23, 24]. Selection of a system to use will depend on particular needs, such as the number of tests per week, the amount of labor, experience, skills, and education required of the technician to perform the test, the depth and breadth of the database, the manufacturing niche, the impact of the organisms on the product or people, and the criticality of the sample to the process. The best system (MALDI-TOF MS or sequencing) for the type of sample or the identification needed must be chosen.

Outsourcing eliminates a number of the major disadvantages of the more modern identification methods, primarily capital and depreciation costs for equipment, resources in the form of highly specialized/trained personnel, and the setup of specialized facilities. The majority of the identification needs occur at the manufacturing facilities, and requires rapid, accurate identifications in the event of an excursion to manage product disposition and mitigation. For some industries, it is difficult to justify the development of internal capabilities and resources for the non-phenotypic approaches, and facilities may not be in the position to make this type of capital expense. Thus, this need is outsourced, with an identification that can be turned around in 24 hours.

19.2.8 Summary

Previously, the identification of microorganisms was a multi-day activity. After initial recovery of microorganisms following growth on nutrient medium, identification required subcultures and other classical microbiological techniques. Newer methods analyze the protein or the genetic profile of the microorganisms. The implementation of MALDI-TOF MS has made significant inroads in the last decade into routine identifications in manufacturing facilities with its high accuracy, short time to result, and low cost per test. In 2012, a book from the Parenteral Drug Association contained a chapter that discussed the validation of a MALDI-TOF MS identification platform in a pharmaceutical company as it represented a new technology in the identification arena [25]. Today it is recognized that pharmaceutical companies should employ "modern microbiological methods to identify representative isolates, e.g. MALDI supplemented with 16S rRNA sequencing" [26]. Accurate and consistent identification methods, such as MALDI-TOF MS, are critical in managing risk by yielding data that allow for comprehensive and reliable tracking and trending during routine monitoring and more confidence in any resulting decisions.

19.3 MALDI-TOF MS Microbial Identification Workflow at a High-throughput Laboratory

19.3.1 MALDI-TOF MS Principles for Microbial Identification

Microbial identification with MALDI-TOF relies on profiling abundant proteins, predominantly ribosomal, within a microorganism [27, 28]. Briefly, an analyte co-crystallized with organic matrix is ionized with a high-intensity laser. The ions desorb from the MALDI-TOF target plate and travel through the flight tube. The time it takes for the ions to reach the detector is proportional to their mass. A profile of mass of the proteins or peptides and the intensity of those ions generates a spectrum that is presumed to be unique to each species and forms the basis for an identification system. The spectrum generated from an unknown isolate is then matched to a database (or library) containing spectra of known strains belonging to specific species. The closeness of the match generates a numerical score value. The magnitude of this score value can be validated to determine the threshold for a species-level identification.

19.3.2 Organism Cultivation for Microbial Identification with MALDI-TOF MS

In the pharmaceutical, biotech, and consumer care industries, the most common media types for EM are nutrient-rich like trypticase soy agar (TSA) for bacteria, and Sabouraud agar (SAB) for fungi. If the organisms are recovered from water systems, then Reasoner's 2A agar (R2A) may be used. Occasionally, other media types, such as TSA blood agar, TSA with Lecithin and Tween, and potato dextrose agar (PDA) are also used. TSA with Lecithin and Tween is specifically designed for recovering microbes from sanitized surfaces and personnel in cleanrooms and isolators. All these media types are compatible with testing the microorganisms directly by MALDI-TOF MS [29–31]. Recently, ID Fungal Plate (IDFP), a specialized agar medium with a membrane, has begun to be used for cultivating filamentous fungi [32–34]. The medium is designed to promote mycelial growth and the membrane enables ease of harvesting mycelia for testing by MALDI-TOF MS. Most EM uses aerobic growth conditions at incubation temperatures in the range of 25–35°C. Thermophiles such as *Geobacillus stearothermophilus*, typically used as control organisms in sterilization processes, may be recovered at 55°C.

19.3.3 Sample Preparation for Microbial Identification with MALDI-TOF MS

The most common sample preparation method for testing bacteria and yeast with MALDI-TOF MS is the extended direct smear method. It is a rapid method, which involves smearing a thin layer of microbial cells onto the MALDI-TOF MS target and then adding formic acid and the organic matrix. Approximately 75% of the species diversity found during EM produces identifications with this method. This method is more successful for Gram-negative species than for Gram-positive ones. A relatively more specialized method, called the extraction method, which utilizes formic acid and acetonitrile, can be used for those organisms that do not produce an identification with the direct smear method [29, 31, 35, 36]. In general, higher-quality spectra characterized by more peaks are produced by the extraction method than by the extended direct smear method. Organisms like fungi have a higher identification rate by MALDI-TOF MS when the extraction method is used [37].

19.3.4 Sample Processing Workflow for Microbial Identification

Quality control microbiology laboratories that support EM programs with medium (40–200 per day) to high sample volumes (> 200 per day) could consider a workflow that utilizes MALDI-TOF MS as the first phase of identification. This process would start with a pure

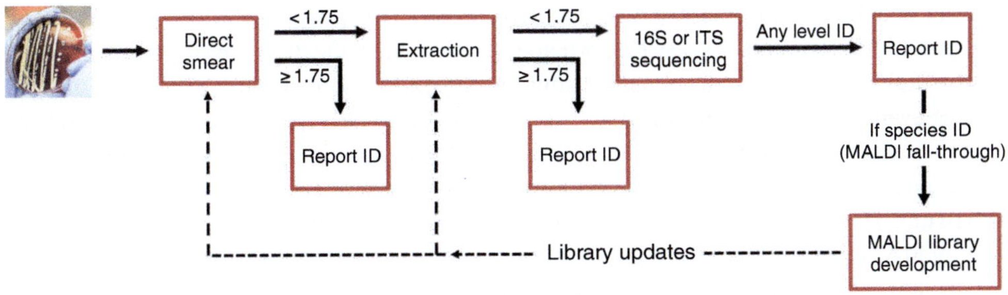

Figure 19.2 Workflow for processing samples for microbial identification by MALDI-TOF MS backed up by sequencing and a feedback loop of targeted library development.

and actively growing subculture of the microorganism, typically 24–72 hours old. When adequate growth is observed, the extended direct smear method can be attempted. If the result is not a reliable identification as determined by pre-defined score thresholds, then the extraction method can be attempted either from cell mass obtained from the same subculture or from another fresh subculture if the cells might have entered the stationary growth phase. The success of identification will also depend on the extent of library coverage. If a spectrum is generated and the "no identification" result is due to not finding a match to the library, then a second identification system is needed, which should have wider library coverage in terms of strain and species diversity. The secondary identification system could be at a contract laboratory or specialized identification library that can be augmented to the current MALDI-TOF MS system. Alternatively, an identification system based on sequencing of ribosomal gene targets, such 16S or ITS, that is linked to a comprehensive library and that uses a phylogenetic approach to find the closest taxonomic placement of the unknown sequence could be used.

An advanced workflow for microbial identification that is completely self-sufficient is shown in Figure 19.2. Contract laboratories or companies that have developed centers of excellence for microbial identification often establish such workflows. These workflows also involve continuous improvement of the MALDI-TOF MS library coverage driven by the no identification rate of testing by MALDI-TOF MS and using sequencing of ribosomal targets such 16S or ITS as the reference identification method. Although such identification workflows require significant investment of specialized resources, they can generate a collective identification rate of > 99%.

19.3.5 Data Interpretation

Data interpretation guidelines to establish species-level confidence directly impact the accuracy and identification rate of a MALDI-TOF MS identification system. Higher accuracy leads to greater confidence in the identifications to track and trend the facilities' microbiota during EM. Use of a microbial identification platform such as MALDI-TOF MS in regulated industries requires that the instrument and methods are validated for their intended use. One commercial instrument, the MALDI Biotyper (Bruker Microbiology and Diagnostics), produces results that range from a log reduced score (or match factor) of 3.0

(perfect match) to 0 (no match). Charles River established custom MALDI-TOF MS data interpretation guidelines for assigning a species-level identification based on two fitness-for-use studies. More than a thousand microbial identifications representing ~350 unique species evaluated during the fitness-for-use studies were also generated using 16S rDNA sequencing. The identification generated by 16S rDNA sequencing was used as a reference to evaluate the score cut-off necessary for an accurate species-level identification generated by MALDI-TOF MS. An optimal cut-off for an accurate species-level identification generated by MALDI-TOF MS needed to fulfill the following criteria: (i) highest rate possible of species-level identifications in agreement with 16S rDNA data; (ii) lowest rate possible of no species-level identification; (iii) lowest rate possible of species-level identifications in disagreement with 16S rDNA data. Detailed analysis of all identification results showed that a score ≥ 1.75 meets the three criteria required for an accurate species-level identification using MALDI-TOF MS, when 16S rDNA sequencing serves as a reference. Developing a binary classification decision based on a single score threshold (≥ 1.75) could reduce the potential subjectivity of high- and low-confidence identifications that can make operational decisions difficult.

19.3.6 Importance of a Sequence-based Secondary (or Fall-through) Identification System

Although the species and strain diversity coverage of MALDI-TOF MS systems has grown significantly in the past decade, there appears to be a consistently growing fraction of isolates recovered during EM from manufacturing facilities across the globe that do not have coverage in a MALDI-TOF MS library. These isolates that do not result in an identification are referred to as "fall-through" isolates. Having a sequence-based secondary identification provides the best-case scenario for fall-through isolates to generate an identification, albeit not always at the species level. This is most effectively understood through two real-world EM datasets from pharmaceutical companies. One of these companies is based in the United States and the other in Europe. Figure 19.3a,b, representing 5246 isolates, and Figure 19.4a,b, representing 5582 isolates, show the identification rate of these cohorts over two years, respectively. As seen in these figures, the MALDI-TOF MS system with a supplemented reference library, which follows the processing workflow shown in Figure 19.2, resulted in identification rates of 88.3% and 89.6%, respectively. When the fall-through isolate cohort was tested by sequencing as the next step in the identification workflow, an additional 10.8% and 9.6% of the isolates were successfully identified, resulting in final identification rates of 99.1% and 99.2%, respectively.

It is interesting to evaluate what isolates result in no identifications by MALDI-TOF MS but have successful identification by sequencing. These fall-through isolates either do not generate spectra by MALDI-TOF MS due to the impact of suboptimal sample processing, data acquisition, or subculture conditions, or, if they do generate spectra, they do not have a spectral match in the MALDI-TOF MS library. However, when tested by sequencing, as long as sequence data are generated, a phylogenetic approach always links the sequence of the unknown isolate to the closest species represented in the library, but only when a phylogenetic tree is used in interpreting the identification, otherwise a percentage difference or similarity to the closest related library entry is indicated. With the custom MALDI-TOF MS

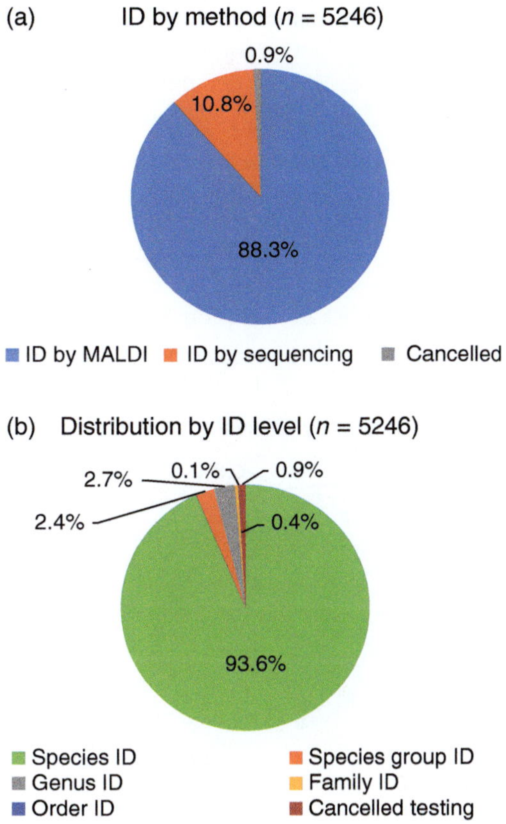

Figure 19.3 (a, b) Identification by method (a) and distribution of these identifications by taxonomic resolution (b) representing a cohort of 5246 EM isolates spanning two years of a pharmaceutical manufacturer in Europe. These isolates were tested by following the processing workflow shown in Figure 19.2. Identifications were generated using MALDI-TOF MS or 16S rRNA gene sequencing, or not successfully identified (cancelled).

data interpretation guidelines presented in Section 19.3.5, Charles River's MALDI-TOF MS identifications are either to the species level or at the species group- or complex-level resolution for closely related species, e.g. *Burkholderia cepacia* complex. However, as seen in Figures 19.3b and 19.4b, a small fraction of the identifications tested by sequencing result in genus-level, family-level, or order-level identification resolution. In most of these cases, it is because the unknown isolate recovered from the manufacturing facility does not match any known published species and is very likely to be a novel organism yet be formally classified and published (Figure 19.5).

19.4 MALDI-TOF MS Library Development and Coverage

19.4.1 Importance of Library Development Under a Quality System

Having access to the most relevant, accurate, and compliant microbial libraries is crucial to obtaining correct species-level identifications. Figure 19.6 illustrates the key reasons why microbial identification libraries matter. Data generated from the MALDI-TOF MS system for microbial identification are intricately linked to its library. First, the success of an identification will depend on whether a match to the spectral profile of the unknown isolate is

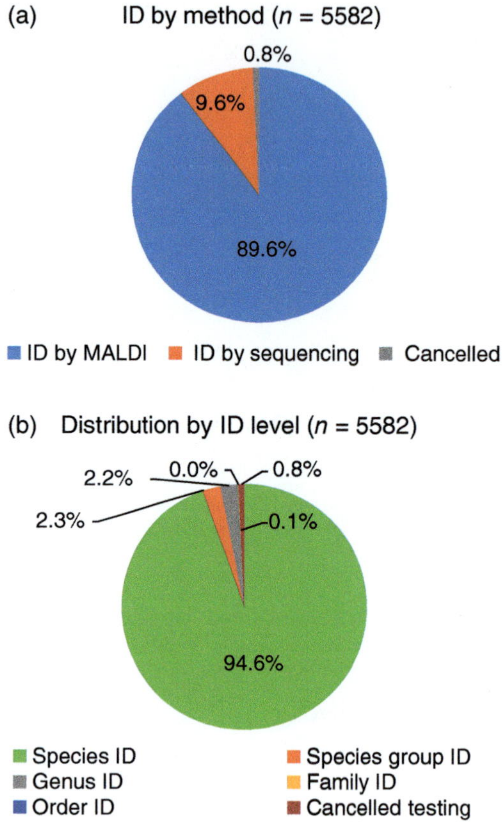

Figure 19.4 (a, b) Identification by method (a) and distribution of these identifications by taxonomic resolution (b) representing a cohort of 5582 EM isolates spanning two years of a pharmaceutical manufacturer in the United States. These isolates were tested by following the processing workflow shown in Figure 19.2. Identifications were generated using MALDI-TOF MS or 16S rRNA gene sequencing, or not successfully identified (cancelled).

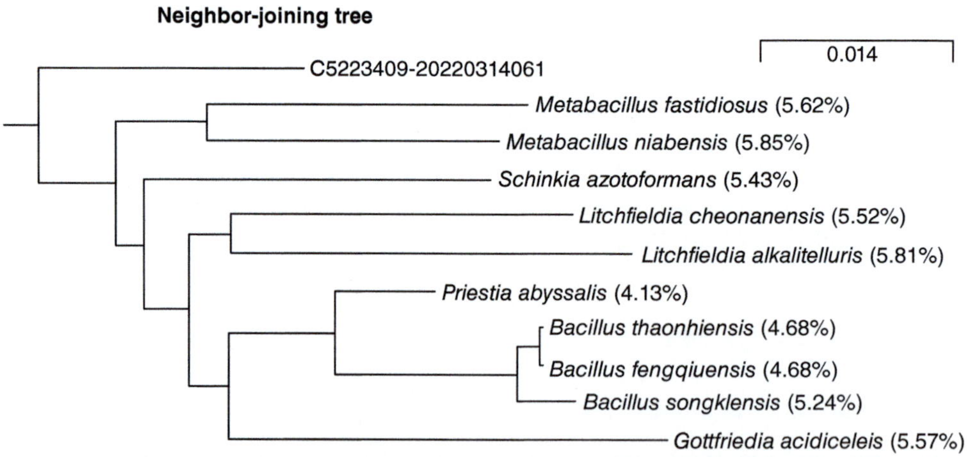

Figure 19.5 Neighbor-joining phylogenetic tree based on the first 500 bp of the 16S ribosomal gene of the unknown isolate (C5223409-20220314061) and its closest matches to the type strain sequences in the library. This isolate did not cluster closely with any known species in the library or public databases, which implies it could be a novel species or genus. The best possible identification of this isolate was to the family level (family *Bacillaceae*).

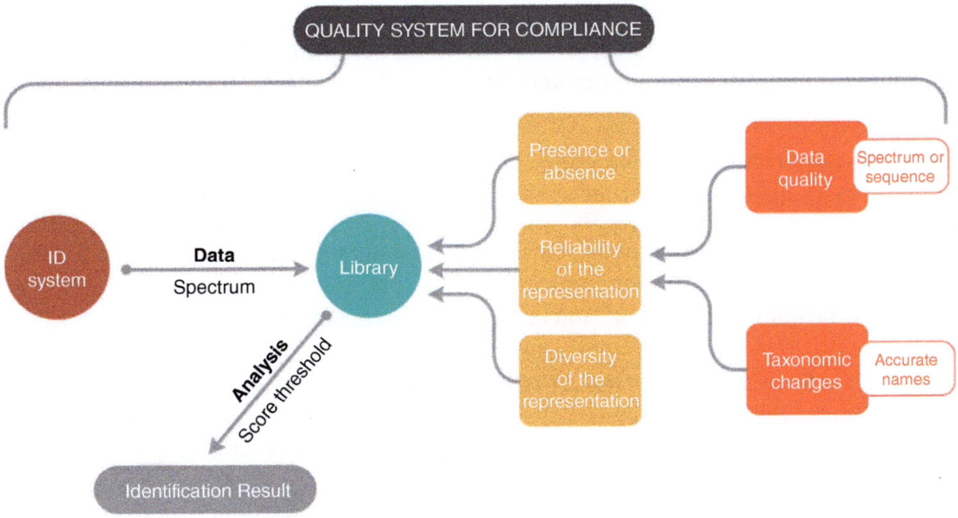

Figure 19.6 Key characteristics of a MALDI-TOF MS library impacting an identification result. A well-designed quality system ensures all features of the MALDI-TOF MS library are maintained in a state of control, so that the identification results produced are accurate and consistent.

present (or absent) in the library. Second, if present, then the reliability of this representation in terms of data quality and taxonomic accuracy will have an impact. And third, it is important for a library to have MALDI-TOF MS entries consisting of the type, the reference strain for a species, and non-type strains that represent the normal variation and diversity seen within species. Therefore, the final identification is significantly influenced by the library and the data interpretation criteria.

Having a MALDI-TOF MS processing workflow as shown in Figure 19.2 enables library entries to be made from novel and unique strains when they are encountered and have not resulted in an identification by MALDI-TOF MS. These fall-through isolates are likely to be identified by sequencing, which paves the way to justify adding them to the MALDI-TOF MS library if all the technical criteria are met. For reference libraries specifically, it is not acceptable for regulated industries to rely on non-curated public databases, clinically focused databases, or taxonomically outdated libraries. Many commercial systems that are clinically focused are typically not designed for identifying environmental isolates found in manufacturing facilities. Having some knowledge of the organisms that are routinely isolated from pharmaceutical, biotechnology, medical device, and other manufacturing facilities worldwide is necessary to determine what organisms should be added to the MALDI-TOF MS library.

Furthermore, it is important to have a compliant quality system in place to oversee all library development-related operations. A quality system should be purposely designed to ensure that laboratory processes, computerized systems, and equipment, as well as the reference libraries that are used to generate test results, are all adequately maintained in a state of control. The maintenance of validated compliant libraries includes the following features: (i) updating the names of entries to reflect taxonomic changes and accuracy of names; (ii) expanding the library with new species or strains to increase the relevancy and accuracy; (iii) maintaining a validated state of the library using a quality system that meets industry regulations and compliance; and (iv) routinely performing library re-qualifications to ensure accuracy and consistency of identifications.

19.4.2 Targeted Library Development for Gram-positive Bacteria and Water Organisms

From an EM standpoint, two groups of organisms that are under-represented in microbial identification systems focused on clinical microbiology are Gram-positive bacteria (e.g. members of the family *Bacillaceae*) and organisms recovered from water systems.

19.4.2.1 Case Study 1: Impact of MALDI-TOF MS Library Coverage for Organisms of the Family *Bacillaceae*

The family *Bacillaceae* comprises a wide phylogenetic diversity of ~130 genera. Members are commonly found in pharmaceutical manufacturing facility microbiota. Most species form spores and thus they are prevalent contaminants and resistant to disinfection. Over seven years between 2011 and 2018, Charles River expanded the MALDI-TOF MS library with nearly 400 entries that represented 67.9% of the total entries for the family *Bacillaceae*. In this retrospective case study, the identifications of ~14 200 isolates spanning 20 genera from the *Bacillaceae* that were submitted for identification by MALDI-TOF MS were analyzed. If MALDI-TOF failed to provide an identification, the isolates were identified by 16S rRNA gene sequencing. The operational impact of using MALDI-TOF MS with sequencing as a backup system for identification was calculated in terms of an identification rate. Figure 19.7 shows the impact of two libraries on the identification rate, illustrating the significant value in targeted library development of isolates pertinent to manufacturing environments.

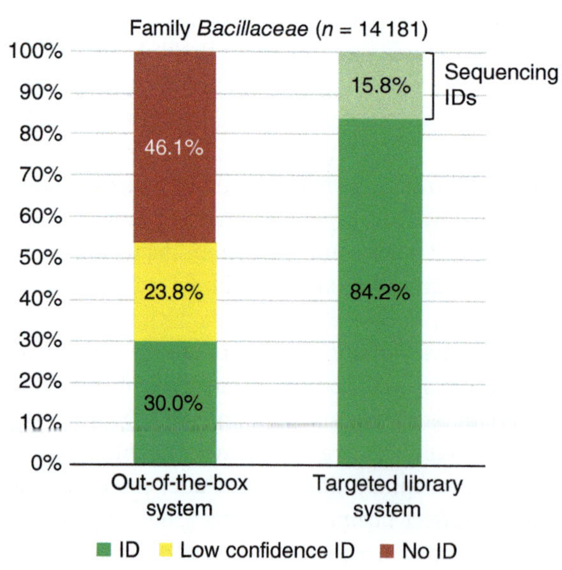

Figure 19.7 Impact of targeted library development and optimized threshold for species from the family *Bacillaceae*. Identification rates of a cohort of 14 181 isolates with a library of an out-of-the-box MALDI-TOF MS system compared with a MALDI-TOF MS system backed up by sequencing and a feedback loop of targeted library development.

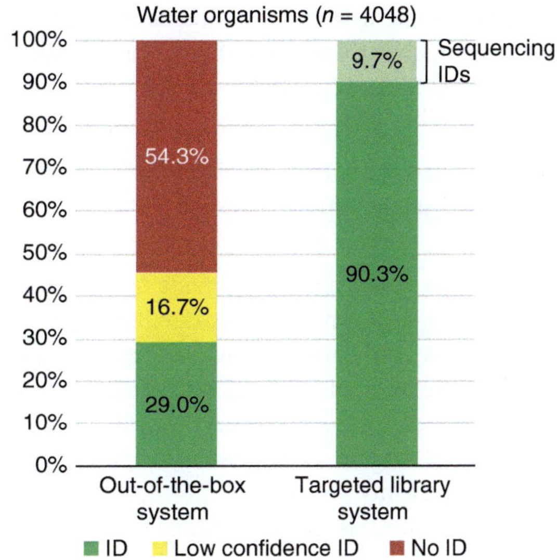

Figure 19.8 Impact of targeted library development and optimized threshold for species known to be associated with water systems. Identification rates of a cohort of 4048 isolates with a library of an out-of-the-box MALDI-TOF MS system compared with a MALDI-TOF MS system backed up by sequencing and a feedback loop of targeted library development.

19.4.2.2 Case Study 2: Impact of MALDI-TOF MS Library Coverage for Organisms Recovered from Water Systems

Fifty-five unique species described as organisms recovered from water systems were shortlisted for this retrospective case study [38–40]. These 55 species represented 50 Gram-negative rods, four Gram-positive rods, and one Gram-positive cocci. Charles River analyzed the identifications of ~4000 isolates submitted for identification by MALDI-TOF MS. If MALDI-TOF MS failed to provide an identification, the isolates were identified by 16S rRNA gene sequencing. The operational impact of using MALDI-TOF MS with sequencing as a backup system for identification was calculated in terms of an identification rate. Figure 19.8 shows impact of two libraries on the identification rate, illustrating the significant value in targeted library development of isolates pertinent to the source of contamination within the manufacturing pipeline, in this case water systems.

19.4.3 Supplemental and Custom MALDI-TOF MS Libraries

Supplemental libraries could augment the main MALDI-TOF MS library used for identifications in an EM program. Examples of supplemental libraries include those that provide extensive coverage for organism types such as filamentous fungi and *Mycobacterium* species [32, 41–44]. The species and strain diversity required for these organism types could be significant, and therefore augmentation with a supplemental library adds the required coverage to improve identification rates within a relatively short time-frame.

In some cases, there is a need to build custom libraries that are partitioned from the larger MALDI-TOF MS library. Specific applications where custom libraries may play a role are to confirm identity of quality control or product-related strains within probiotics or biotherapeutics [45]. Additionally, the evaluation, implementation, and drawbacks of a custom library for screening BSL-3 biosecurity pathogens have been reported. For example, Bruker's Security-Relevant Library covering 12 highly pathogenic species from 10 genera.

The appropriate workflows should be implemented to prevent false positives and negatives due to overlapping spectra with non-pathogenic species within the larger MALDI-TOF MS library used for EM [46–48].

19.5 Comparison of MALDI-TOF MS with Other Microbial Identification Methods

Comparison of identifications made by MALDI-TOF MS with those by phenotypic identification platforms have demonstrated that the MALDI-TOF MS is very fast and more accurate for the identification of clinical organisms [49–54]. When there were discordant results between the MALDI-TOF MS and the phenotypic system, the 16S rRNA gene sequence most often confirmed the MALDI-TOF MS result [55]. Incorrect identifications with MALDI-TOF MS are rarer than with phenotypic systems and usually occur with closely related organisms [49, 56–58]. Resolution of certain taxonomic groups of closely related species complexes is still a challenge for MALDI-TOF MS, and most of the time that is also the case with rRNA sequence analysis for these groups; examples include the *Bacillus cereus* group, *Escherichia coli* and *Shigella*, and the *Enterobacter cloacae* complex [59–61]. However, for most isolates, species resolution can be attained with MALDI-TOF MS or rRNA gene sequencing.

19.6 Future Perspectives

The use of MALDI-TOF MS for the identification of bacteria recovered from manufacturing facilities in numerous industries is continuing to expand. Phenotypic systems that have been in use for decades are being replaced by more modern methods for microbial identification. The use of MALDI-TOF MS for the identification of fungal isolates is only just emerging as a valid method in the industrial space. Additionally, as the global regulatory agencies focus on harmonization, compliance, and data integrity, the modern platforms such as MALDI-TOF MS are becoming more attractive for use. For example, compliance to FDA 21 CFR Part 11 with features such as an audit trail, user management, data security, and electronic signatures will become standard practice. This trend will continue with the implementation of automation for sample preparation and target plate spotting.

MALDI-TOF MS instruments with both positive and negative ion modes are making it possible to detect and profile a wider spectrum of analytes related to the characteristics of a microbe, e.g. lipids as markers for antibiotic resistance detection [62–65] and microbial identification [66]. The application of MALDI-TOF MS to the differentiation of closely related species has the potential to grow with the development of new data analysis techniques targeting protein profiles, or with assays targeting new analytes. So far, the use of MALDI-TOF MS for the application of strain typing has had mixed success [67]. Furthermore, MALDI-TOF MS for the identification of organism types beyond microbes such as algae for industrial microbiology has not yet become mainstream but has been shown to work in proof-of-concept studies [68, 69].

The broad category of microbial testing, and specifically microbial identification, is driven by many factors. These factors include global population growth and the increased use of manufactured goods with increased public concerns for product quality. These concerns arise because of pathogen and contamination issues in consumer goods and infections caused by contaminated pharmaceuticals. Additionally, microbial testing and identifications are driven by increased global regulatory scrutiny and requirements emphasizing safety and quality. The regulatory drivers that impact each manufacturing sector vary, but some of the key drivers having the largest impact on these markets include the need for product conformance, regulatory inspection requirements, and the proactive efforts by industry to improve quality.

References

1 Sutton, S.V.W. (2010). The environmental monitoring program in a GMP environment. *J. GXP Compliance* 14: 22–30.

2 Thomas, M.K., Vriezen, R., Farber, J.M. et al. (2015). Economic cost of a *Listeria* monocytogenes outbreak in Canada, 2008. *Foodborne Pathog. Dis.* 12: 966–971.

3 European Commission (2008). Annex 1 Manufacture of sterile medicinal products. In: *The Rules Governing Medicinal Products in the European Union, Volume 4. EU Guidelines to Good Manufacturing Practice Medicinal Products for Human and Veterinary Use*. Brussels: Belgium.

4 Food and Drug Administration (FDA) 2004. *Sterile Drug Products Produced by Aseptic Processing – Current Good Manufacturing Practice*. Guidance for Industry. US Department of Health and Human Services.

5 Food and Drug Administration (FDA) 2011. FDA Regulation 21 CFR Part 211 – *Current Good Manufacturing Practice for Finished Pharmaceuticals*. US FDA.

6 Parenteral Drug Association (PDA) 2001. Fundamentals of an environmental monitoring program. *Technical Report No. 13*. Bethesda, MD: PDA.

7 Parenteral Drug Association (PDA) 2020. Fundamentals of an environmental monitoring program Annex 1: Environmental monitoring of facilities manufacturing low bioburden products. *Technical Report No. 13-2*. Bethesda, MD: PDA.

8 Huynh-Ba, K. (2020). Analytical method validation for quality assurance and process validation professionals. *American Pharmaceutical Review*. https://www.americanpharmaceuticalreview.com/Featured-Articles/569734-Analytical-Method-Validation-for-Quality-Assurance-and-Process-Validation-Professionals/ (accessed 8 November 2022).

9 Miller, M. J. 2021. *Validating Rapid Microbiological Methods*. https://www.gcplearning.com/webinar/validating-rapid-microbiological-methods/600417REC (accessed 8 November 2022).

10 United States Pharmacopeia (USP) 2012. *Microbial Characterization, Identification, and Strain Typing*. USP 35, Chapter <1113>. Rockville, MD: The United States Pharmacopeial Convention/National Formulary.

11 GAMP5 2008. *The Good Automated Manufacturing Practice (GAMP) Guide for Validation of Laboratory Computerized Systems*. International Society for Pharmaceutical Engineering.

12 Food and Drug Administration (FDA) 2003. FDA Regulation 21 CFR Part 11 – *Electronic Records; Electronic Signatures — Scope and Application*. US FDA.

13 United States Pharmacopeia (USP) 2012. *Analytical Instrument Qualification*. USP 35, Chapter <1058>. Rockville, MD: The United States Pharmacopeial Convention/National Formulary.

14 Parenteral Drug Association (PDA) 2014. Exclusion of objectionable microorganisms from nonsterile pharmaceuticals, medical devices, and cosmetics. *Technical Report No. 67*. Bethesda, MD: PDA.

15 Sutton, S.V.W. and Cundell, A.M. (2004). Microbial identification in the pharmaceutical industry. *Pharmacop. Forum* 30: 1884–1894.

16 Drancourt, M., Bollet, C., Carlioz, A. et al. (2000). 16S ribosomal DNA sequence analysis of a large collection of environmental and clinical unidentifiable bacterial isolates. *J. Clin. Microbiol.* 38: 3623–3630.

17 Fontana, C., Favaro, M., Pelliccioni, M. et al. (2005). Use of the MicroSeq 500 16S rRNA gene-based sequencing for identification of bacterial isolates that commercial automated systems failed to identify correctly. *J. Clin. Microbiol.* 43: 615–619.

18 Petti, C.A., Polage, C.R., and Schreckenberger, P. (2005). The role of 16S rRNA gene sequencing in identification of microorganisms misidentified by conventional methods. *J. Clin. Microbiol.* 43: 6123–6125.

19 Zbinden, A., Böttger, E.C., Bosshard, P.P., and Zbinden, R. (2007). Evaluation of the colorimetric VITEK 2 card for identification of gram-negative nonfermentative rods: comparison to 16S rRNA gene sequencing. *J. Clin. Microbiol.* 45: 2270–2273.

20 Siqueira, J.F. Jr. and Rocas, I.N. (2005). Exploiting molecular methods to explore endodontic infections: part 1–current molecular technologies for microbiological diagnosis. *J. Endod.* 31: 411–423.

21 Tan, K.E., Ellis, B.C., Lee, R. et al. (2012). Prospective evaluation of a matrix-assisted laser desorption ionization-time of flight mass spectrometry system in a hospital clinical microbiology laboratory for identification of bacteria and yeasts: a bench-by-bench study for assessing the impact on time to identification and cost-effectiveness. *J. Clin. Microbiol.* 50: 3301–3308.

22 Clarridge, J.E. 3rd (2004). Impact of 16S rRNA gene sequence analysis for identification of bacteria on clinical microbiology and infectious diseases. *Clin. Microbiol. Rev.* 17: 840–862.

23 Janda, J.M. and Abbott, S.L. (2007). 16S rRNA gene sequencing for bacterial identification in the diagnostic laboratory: pluses, perils, and pitfalls. *J. Clin. Microbiol.* 45: 2761–2764.

24 Stager, C.E. and Davis, J.R. (1992). Automated systems for identification of microorganisms. *Clin. Microbiol. Rev.* 5: 302–327.

25 Griffin, M. and Reber, D. (2012). *Microbial Identification: The Keys to a Successful Program*. River Grove, IL: Davis Healthcare International Publishing, LLC.

26 Cundell, T. (2020). Transition to modern microbiological methods in the pharmaceutical industry. *Am. Pharm. Rev.* https://www.americanpharmaceuticalreview.com/Featured-Articles/571418-Transition-to-Modern-Microbiological-Methods-in-the-Pharmaceutical-Industry/ (accessed 5 November 2022).

27 Maier, T., Klepel, S., Renner, U., and Kostrzewa, M. (2006). Fast and reliable MALDI-TOF MS–based microorganism identification. *Nat. Methods* 3: i–ii.

28 Ryzhov, V. and Fenselau, C. (2001). Characterization of the protein subset desorbed by MALDI from whole bacterial cells. *Anal. Chem.* 73: 746–750.

29 Ford, B.A. and Burnham, C.A. (2013). Optimization of routine identification of clinically relevant Gram-negative bacteria by use of matrix-assisted laser desorption ionization-time of flight mass spectrometry and the Bruker Biotyper. *J. Clin. Microbiol.* 51: 1412–1420.

30 Pinar-Méndez, A., Fernández, S., Baquero, D. et al. (2021). Rapid and improved identification of drinking water bacteria using the drinking water library, a dedicated MALDI-TOF MS database. *Water Res.* 203: 117543.

31 Tekippe, M.E., Shuey, S., Winkler, D.W. et al. (2013). Optimizing identification of clinically relevant Gram-positive organisms by use of the Bruker Biotyper matrix-assisted laser desorption ionization-time of flight mass spectrometry system. *J. Clin. Microbiol.* 51: 1421–1427.

32 Heireman, L., Patteet, S., and Steyaert, S. (2020). Performance of the new ID-fungi plate using two types of reference libraries (Bruker and MSI) to identify fungi with the Bruker MALDI Biotyper. *Med. Mycol.* 58: 946–957.

33 Robert, M.G., Romero, C., Dard, C. et al. (2020). Evaluation of ID fungi plates medium for identification of molds by MALDI Biotyper. *J. Clin. Microbiol.* 58.

34 Sacheli, R., Henri, A.S., Seidel, L. et al. (2020). Evaluation of the new Id-Fungi plates from Conidia for MALDI-TOF MS identification of filamentous fungi and comparison with conventional methods as identification tool for dermatophytes from nails, hair and skin samples. *Mycoses* 63: 1115–1127.

35 Alatoom, A.A., Cunningham, S.A., Ihde, S.M. et al. (2011). Comparison of direct colony method versus extraction method for identification of gram-positive cocci by use of Bruker Biotyper matrix-assisted laser desorption ionization-time of flight mass spectrometry. *J. Clin. Microbiol.* 49: 2868–2873.

36 Khot, P.D., Couturier, M.R., Wilson, A. et al. (2012). Optimization of matrix-assisted laser desorption ionization-time of flight mass spectrometry analysis for bacterial identification. *J. Clin. Microbiol.* 50: 3845–3852.

37 Patel, R. (2019). A moldy application of MALDI: MALDI-ToF mass spectrometry for fungal identification. *J. Fungi* 5: 4.

38 Kulakov, L.A., McAlister, M.B., Ogden, K.L. et al. (2002). Analysis of bacteria contaminating ultrapure water in industrial systems. *Appl. Environ. Microbiol.* 68: 1548–1555.

39 Popović, N.T., Kazazić, S.P., Strunjak-Perović, I., and Čož-Rakovac, R. (2017). Differentiation of environmental aquatic bacterial isolates by MALDI-TOF MS. *Environ. Res.* 152: 7–16.

40 Sandle, T. (2016). Microbiological challenges to the pharmaceuticals and healthcare. In: *Pharmaceutical Microbiology Essentials for Quality Assurance and Quality Control.* Cambridge, UK: Elsevier.

41 Akyar, I., Çavuşoğlu, C., Ayaş, M. et al. (2018). Evaluation of the performance of MALDI-TOF MS and DNA sequence analysis in the identification of mycobacteria species. *Turk. J. Med. Sci.* 48: 1351–1357.

42 Brown-Elliott, B.A., Fritsche, T.R., Olson, B.J. et al. (2019). Comparison of two commercial matrix-assisted laser desorption/ionization-time of flight mass spectrometry (MALDI-TOF MS) systems for identification of nontuberculous mycobacteria. *Am. J. Clin. Pathol.* 152: 527–536.

43 Stein, M., Tran, V., Nichol, K.A. et al. (2018). Evaluation of three MALDI-TOF mass spectrometry libraries for the identification of filamentous fungi in three clinical microbiology laboratories in Manitoba, Canada. *Mycoses* 61: 743–753.

44 Wilkendorf, L.S., Bowles, E., Buil, J.B. et al. (2020). Update on matrix-assisted laser desorption ionization-time of flight mass spectrometry identification of filamentous fungi. *J. Clin. Microbiol.* 58: e01263–e01220.

45 Mohar Lorbeg, P., Golob, M., Kramer, M. et al. (2021). Evaluation of dietary supplements containing viable bacteria by cultivation/MALDI-TOF mass spectrometry and PCR identification. *Front. Microbiol.* 12: 700138.

46 Antonation, K., Tracz, D., Dreger, A., and Corbett, C. (2020). Matrix-assisted laser desorption/ ionization time-of-flight mass spectrometry-based identification of security-sensitive bacteria: considerations for Canadian Bruker users. *Can. Commun. Dis. Rep.* 46: 333–338.

47 Pomerleau-Normandin, D., Heisz, M., and Su, M. (2018). Misidentification of risk group 3/ security sensitive biological agents by MALDI-TOF MS in Canada: November 2015-October 2017. *Can. Commun. Dis. Rep.* 44: 110–115.

48 Rudrik, J.T., Soehnlen, M.K., Perry, M.J. et al. (2017). Safety and accuracy of matrix-assisted laser desorption ionization-time of flight mass spectrometry for identification of highly pathogenic organisms. *J. Clin. Microbiol.* 55: 3513–3529.

49 Bessede, E., Angla-Gre, M., Delagarde, Y. et al. (2011). Matrix-assisted laser-desorption/ ionization biotyper: experience in the routine of a university hospital. *Clin. Microbiol. Infect.* 17: 533–538.

50 Bizzini, A., Jaton, K., Romo, D. et al. (2011). Matrix-assisted laser desorption ionization-time of flight mass spectrometry as an alternative to 16S rRNA gene sequencing for identification of difficult-to-identify bacterial strains. *J. Clin. Microbiol.* 49: 693–696.

51 Carbonnelle, E., Grohs, P., Jacquier, H. et al. (2012). Robustness of two MALDI-TOF mass spectrometry systems for bacterial identification. *J. Microbiol. Methods* 89: 133–136.

52 Cherkaoui, A., Hibbs, J., Emonet, S. et al. (2010). Comparison of two matrix-assisted laser desorption ionization-time of flight mass spectrometry methods with conventional phenotypic identification for routine identification of bacteria to the species level. *J. Clin. Microbiol.* 48: 1169–1175.

53 Seng, P., Drancourt, M., Gouriet, F. et al. (2009). Ongoing revolution in bacteriology: routine identification of bacteria by matrix-assisted laser desorption ionization time-of-flight mass spectrometry. *Clin. Infect. Dis.* 49: 543–551.

54 Van Veen, S.Q., Claas, E.C., and Kuijper, E.J. (2010). High-throughput identification of bacteria and yeast by matrix-assisted laser desorption ionization-time of flight mass spectrometry in conventional medical microbiology laboratories. *J. Clin. Microbiol.* 48: 900–907.

55 Benagli, C., Rossi, V., Dolina, M. et al. (2011). Matrix-assisted laser desorption ionization-time of flight mass spectrometry for the identification of clinically relevant bacteria. *PLoS One* 6: e16424.

56 De Bel, A., Wybo, I., Vandoorslaer, K. et al. (2011). Acceptance criteria for identification results of Gram-negative rods by mass spectrometry. *J. Med. Microbiol.* 60: 684–686.

57 Dubois, D., Grare, M., Prere, M.F. et al. (2012). Performances of the Vitek MS matrix-assisted laser desorption ionization-time of flight mass spectrometry system for rapid identification of bacteria in routine clinical microbiology. *J. Clin. Microbiol.* 50: 2568–2576.

58 Neville, S.A., Lecordier, A., Ziochos, H. et al. (2011). Utility of matrix-assisted laser desorption ionization-time of flight mass spectrometry following introduction for routine laboratory bacterial identification. *J. Clin. Microbiol.* 49: 2980–2984.

59 Almuzara, M., Barberis, C., Traglia, G. et al. (2015). Evaluation of matrix-assisted laser desorption ionization-time-of-flight mass spectrometry for species identification of nonfermenting Gram-negative bacilli. *J. Microbiol. Methods* 112: 24–27.

60 Khot, P.D. and Fisher, M.A. (2013). Novel approach for differentiating Shigella species and Escherichia coli by matrix-assisted laser desorption ionization-time of flight mass spectrometry. *J. Clin. Microbiol.* 51: 3711–3716.

61 Pavlovic, M., Konrad, R., Iwobi, A.N. et al. (2012). A dual approach employing MALDI-TOF MS and real-time PCR for fast species identification within the Enterobacter cloacae complex. *FEMS Microbiol. Lett.* 328: 46–53.

62 Furniss, R.C.D., Dortet, L., Bolland, W. et al. (2019). Detection of colistin resistance in *Escherichia coli* by use of the MALDI biotyper sirius mass spectrometry system. *J. Clin. Microbiol.* 57: e01427–e01419.

63 Furniss, R.C.D., Kostrzewa, M., Mavridou, D.A.I., and Larrouy-Maumus, G. (2020). The clue is in the lipid a: rapid detection of colistin resistance. *PLoS Pathog.* 16: e1008331.

64 Smith, R.D., McElheny, C.L., Izac, J.R. et al. (2022). A novel lipid-based MALDI-TOF assay for the rapid detection of colistin-resistant *Enterobacter* species. *Microbiol. Spectrum* 10: e01445–e01421.

65 Sorensen, M., Chandler, C.E., Gardner, F.M. et al. (2020). Rapid microbial identification and colistin resistance detection via MALDI-TOF MS using a novel on-target extraction of membrane lipids. *Sci. Rep.* 10: 21536.

66 Liang, T., Leung, L.M., Opene, B. et al. (2019). Rapid microbial identification and antibiotic resistance detection by mass spectrometric analysis of membrane lipids. *Anal. Chem.* 91: 1286–1294.

67 Sauget, M., Valot, B., Bertrand, X., and Hocquet, D. (2017). Can MALDI-TOF mass spectrometry reasonably type bacteria? *Trends Microbiol.* 25: 447–455.

68 Barbano, D., Diaz, R., Zhang, L. et al. (2015). Rapid characterization of microalgae and microalgae mixtures using matrix-assisted laser desorption ionization time-of-flight mass spectrometry (MALDI-TOF MS). *PLoS One* 10: e0135337.

69 Šebela, M., Jahodářová, E., Raus, M. et al. (2018). Intact cell MALDI-TOF mass spectrometric analysis of Chroococcidiopsis cyanobacteria for classification purposes and identification of possible marker proteins. *PLoS One* 13: e0208275.

20

Identification of Mold Species and Species Complex from the Food Environment Using MALDI-TOF MS

Victoria Girard[1], Valérie Monnin[1], Nolwenn Rolland[1], Jérôme Mounier[2], and Jean-Luc Jany[2]

[1] *BioMérieux, R&D Microbiologie, La Balme Les Grottes, France*
[2] *Université de Brest, INRAE, Laboratoire Universitaire de Biodiversité et Ecologie Microbienne, France*

20.1 Fungal Taxonomy

20.1.1 Defining What Is a Fungal Species

Fungal taxonomy is based on a classification which, in accordance with the International Code of Nomenclature for algae, fungi, and plants [1], includes seven primary ranks, i.e. from the higher to the lower hierarchical rank – kingdom, phylum, class, order, family, genus, and species – as well as secondary ranks (e.g. variety and form at the subspecies level) and further ranks (e.g. subforma). Among these different hierarchical ranks, species is considered as the fundamental unit to which evolutionary changes take place [2]. Therefore, species will evolve unique characteristics over time, and although intraspecific variability does occur, global functional properties such as growth parameters or metabolite production are usually assigned to the fundamental unit, i.e. at the species level.

Several definitions of a species have been proposed, from John Ray (1686, in [3]) to Linnaeus (1797), Darwin [4], and Mayr [5]. However, most of these definitions require the existence of sexual reproduction to define species and do not apply, for instance, to asexual fungal groups. More recent definitions, such as that proposed by de Queiroz [6], are more widely accepted. For this author, species correspond to segments of evolutionary lineages that evolve independently from one another. The process that leads two lineages to evolve independently from an initial ancestral population is speciation. Speciation takes place when reproductive isolation occurs between populations, i.e. when the gene flow between populations is precluded (for reviews, see [7–11]). Most often, speciation is the result of ecologically based divergent selection. This may arise from differences in ecological niches which include abiotic and biotic factors, intra- and interspecific interactions, artificial selection (domestication), and/or sexual selection. Many forms of reproductive isolation can preclude gene flow between populations, including both pre- and post-zygotic barriers.

Microbiological Identification using MALDI-TOF and Tandem Mass Spectrometry: Industrial and Environmental Applications, First edition. Edited by Haroun N. Shah, Saheer E. Gharbia, Ajit J. Shah, Erika Y. Tranfield, and K. Clive Thompson.

When the genes under divergent selection and the genes causing reproductive isolation differ, reproductive isolation in sympatry evolves by indirect selection, i.e. linkage disequilibrium of pleiotropy [12].

20.1.2 Fungal Speciation within a Food Context

Compared with other habitats, food matrices may represent richer, less competitive, and more stable environments [13] in which retained allelic combinations in the fungal population may differ drastically from the ones of other habitats. Species mainly encountered in food environments or even only encountered in food matrices might correspond to food-adapted species. For example, the cheese-colonizing but not inoculated species *Mucor lanceolatus*, which plays a role in Tomme or Saint-Nectaire ripening [14] and exhibits traits beneficial for cheese colonization [15], might be considered as a cheese-adapted species [16].

A specific case of speciation through adaptive divergence in response to artificial selection is domestication [17]. This refers to the genetic modification of a species by breeding it in isolation from its ancestral population in an effort to enhance its utility to humans [13]; for example, *Aspergillus oryzae*, *Monascus purpureus*, *Penicillium camemberti*, *Penicillium nalgiovense* or *Saccharomyces cerevisiae* inoculated as adjunct cultures for food fermentations [17] are considered as domesticated species. One expects that domesticated fungi have been selected (by humans) for beneficial traits conferred by certain alleles (implying a global relaxation of selective constraints but a positive selection on the genes corresponding to alleles conferring beneficial traits in a food production context) and have undergone several rounds of population bottlenecks (genetic drift) [13].

20.1.3 Delimiting Species

Beyond the definition of a fungal species and the understanding of the evolutionary forces that drive speciation, it is important to consider species criteria that allow for species recognition. Indeed, what is called species concept often corresponds to species criteria to detect a species [9]. The species definition describes entities, whereas species concept corresponds to criteria to delimit species boundaries [8] and, as proposed by Taylor et al. [18], the term species concept should be replaced by operational species criteria. These operational species criteria are: the biological species concept (BSC), which emphasizes reproductive isolation; the morphological species concept (MSC), which emphasizes morphological divergences; the ecological species concept (ESC), which emphasizes adaptation to a specific niche; and the phylogenetic species concept (PSC), based on nucleotide divergence analysis [9]. These different species criteria may be discordant and importantly species recognition using these criteria eventually lead to a species hypothesis: species recognition is putative. Indeed, as highlighted by Giraud et al. [9], speciation is a temporally extended process during which the phenomena used for species recognition by BSC, MSC, ESC, or PSC, i.e. reproductive isolation, morphological divergence, niche adaptation and nucleotide divergence, do not necessarily occur and do not necessarily appear in a conserved chronology during the course of the speciation process.

More often than not, polyphasic approaches are conducted and recent descriptions of new species usually rely on the simultaneous use of the different operational species criteria. If morphological observation, mating test, or ecological niche assignment approaches rely on conserved methods, genetic divergence assessment through the PSC can be achieved using different approaches. All of them are based on sequence alignments pertaining to multiple fungal strains using different alignment algorithms followed by phylogenetic reconstructions. Sequence alignment and edition is of particular importance as phylogenetic reconstructions are often based on character-based methods (each nucleotide position being considered as a character) such as maximum likelihood or Bayesian inference [19]. The output of such pipelines is a consensus phylogenetic tree that may be accompanied by multiple single locus trees used to delineate the species boundaries within the consensus tree. The most frequently used phylogenetic approach for PSC, the Genealogical Concordance Phylogenetic Species Recognition (GCPSR), has been described by Taylor et al. [18]. It uses multiple unlinked genes to detect the absence of genetic recombination between lineages and, thus, evolutionary independence of these lineages. The coalescent-based species delimitation which uses a more probabilistic approach can also be used. In the omic era, phylogenetic studies tend to be replaced by phylogenomic studies using not handfuls of genes but rather thousands of orthologous genes.

As highlighted earlier, species are hypothetical as the operational species criteria may fail in correctly detecting segments of evolutionary lineages that evolve independently from one another [6]. Sometimes, species delimitation is difficult, not possible, or leads to incongruent data when considering different criteria. Therefore, some taxonomic entities may encompass different species with species boundaries that are not well defined. These groups are unofficially called species complex. They may correspond, for example, to entities that actually contain a number of "cryptic species", which are genetically distinct species erroneously classified as a single species due to at least superficial morphological similarity observed using the MSC. This is why the use of PSC has recently allowed mycologists to amend the fungal taxonomy by detecting species boundaries within several species complexes.

20.1.4 Foodborne Fungi within the Fungal Tree of Life

The fungal kingdom is estimated to encompass between 2.2 and 3.8 million species, among which less than 10% are named [20], and harbors considerable functional diversity, playing major roles as ecological actors as well as in industrial applications, in particular for food and feed productions. Fungal species are clustered within 12 phyla according to James *et al.* [21] or alternatively within 19 phyla according to Wijayawardene *et al.* [22]. Foodborne fungal species mainly belong to the Ascomycota and Mucoromycota and, more particularly, to the Mucorales order within the Mucoromycotina subphylum. Some food contaminants (e.g. *Wallemia sebi*) are also encountered within the Basidiomycota. As illustrated by the fungal tree of life [23], the Mucoromycota phylum diverged earlier than the dikarya, which encompass both Ascomycota and Basidiomycota phyla. Species pertaining to these taxonomic groups harbor morphological and ecological differences. Basically, most fungi within the Mucorales (Mucoromycota, Mucoromycotina, Mucorales) are characterized by coenocytic hyphae, sporangial asexual reproduction and zygospores as sexual spores [23]. Their ecological strategy has often been described as ruderal or R-selected [24]. This ruderal strategy is associated with rapid growth, rapid commitment to reproduction, utilization of easily assimilable resources, and non-persistence. Importantly, there are no mycotoxin producers among the Mucorales. Dikarya is the only sub-kingdom of Fungi and comprises the phyla Ascomycota and Basidiomycota. Their hyphae are regularly septated and, at some point in their life cycle, the hyphal compartments delimited by the septa possess two genotypically distinct nuclei [23]. Most foodborne fungi (either among the contaminants or among the adjunct cultures) belong to the Ascomycetes. The latter are characterized by a stress-tolerant ecological strategy [24] with usually high levels of conidia (asexual spore) production and strong capabilities in facing biotic and abiotic constraints. As such, numerous species within the Ascomycota can produce mycotoxins, which, among other roles, may confer an advantage against competitors.

20.2 Impact of Molds in Food

Depending on the food context, molds can have desirable or undesirable impacts on food quality and safety. Indeed, filamentous fungi can colonize a very large diversity of raw food materials (fruits, vegetables, etc.) and processed foods, and produce a large panel of enzymes which will lead to different metabolites and reaction products that allow them to increase their biomass. In most cases, fungal colonization of a given food substrate is considered as undesirable because of its negative impact on food texture/structure, visual aspect (e.g. surface color) and flavor through, for example, production of volatile compounds and sometimes because of the presence of mycotoxins which are secondary metabolites toxic for human and animal health. During human history, the colonization by specific molds of certain food substrates has led to a favorable effect on food quality attributes by increasing the shelf-life and by improving or contributing to the development of (new) organoleptic properties of food products, which are termed fermented foods [25].

20.2.1 Filamentous Fungi in Fermented Foods

Filamentous fungi encountered in fermented foods are not, strictly speaking, fermentative organisms, fermentation being defined as an anaerobic process by which organisms generate energy, but they are tightly associated with a large number of fermented foods in which they make a significant contribution, in association or not with other microorganisms (i.e. yeasts, bacteria), to the improvement of their organoleptic properties. Originally born in the East, fermentation processes using filamentous fungi have progressively spread to Western countries, mainly in food products of animal origin (i.e. fermented dairy and meat products). The main filamentous fungi associated with fermented foods and their technological functions are shown in Table 20.1.

20.2.2 Filamentous Fungi with Undesirable Impacts on Food Quality and Safety

As mentioned earlier, molds can also have undesirable impacts on food quality and safety. First, fungi are important contributors to the massive worldwide food losses and waste occurring each year, which represent in total about one-third of the food produced [44]. Concerning food losses and waste caused by fungi, Pitt and Hocking [45] estimated that fungi are responsible for 5–10% of total food losses and waste. However, surveys specifically

Table 20.1 The main filamentous fungi used to manufacture fermented foods and their functional role.

	Species	Food	Functions	References
Mucoromycota				
Actinomucor	*A. elegans, A. taiwanensis*	Sufu, fu-tu (China)	Texture, aroma	Chou et al. [26], Han et al. [27]
Amylomyces	*A. rouxii*	Ragi (Indonesia), marcha (India)	Glucose biosynthesis from starch	Tsuyoshi et al. [28]
Mucor	*M. circinelloides, M. rouxii, M. indicus*	Ragi, marcha, tempe (Indonesia), pehtze (China)	Enzymatic transformation, aroma compound production	Pedraza-Reyes and Lopez-Romero [29], Agranoff and Markham [30]
	M. lanceolatus, M. fuscus, M. racemosus, M. brunneogriseus, M. spinosus	Cheese (France), fermented meat (*M. racemosus*)	Color, texture, aroma compound production	Hermet et al. [14], Bruna et al. [31]
Rhizopus	*R. microsporus, R. oligosporus, R. oryzae*	Koji (Japan), nuruk (Korea), chu (China), marcha, tempe	Enzymatic transformation, texture, vitamins	Ginting and Arcot [32], Nout and Kiers [33]
Ascomycota				
Aspergillus	*A. sojae, A. oryzae, A. niger*	Soya sauce (eastern Asia)	Glucidolytic enzymes, proteases, other lytic enzymes	Nout and Aidoo [34], Hanya and Nakadai [35]
	A. glaucus, A. repens, A. melleus, A. candidus	Katsuobushi (Japan)	Enzymatic transformation, aroma compound production	Campbell-Platt [36]
Bisifusarium, Longinectria	*B. domesticum, B. allantoides, B. penicilloides, L. lagenoides, L. verticilliforme*	Cheese (France, Switzerland)	Color, texture, aroma compound production	Bachmann et al. [37], Ropars et al. [38], Savary et al. [39]
Monascus	*M. pilosus, M. purpureus, M. ruber*	Angkak, red rice (China)	Color, aroma compound production, secondary metabolites	Juzlova et al. [40], Akihisa et al. [41]

(*Continued*)

Table 20.1 (Continued)

	Species	Food	Functions	References
Neurospora	*N. sitophila, N. intermedia*	Oncom (Indonesia)	Texture, aroma compound production, enzymatic transformation	Beuchat [42]
Penicillium	*P. glaucum*	katsuobushi	Texture, aroma compound production, enzymatic transformation	Campbell-Platt [36]
	P. camemberti	Mold-ripened cheese (France)	Color, texture, aroma compound production	Ropars et al. [38]
	P. nalgiovense, P. camemberti	Fermented meat (Europe)	Color, aroma compound production, antioxydant activity, homogenous dehydration of product	Spotti and Berni [43]
	P. roqueforti	Blue cheeses (Europe)	Color, texture, aroma compound production	Ropars et al. [38]
Scopulariopsis	*S. fusca, S. flava, S. candida*	Cheese (France)	Color, texture, aroma compound production	Ropars et al. [38]
Sphaerosporium	*S. equinum*	Cheese (France)	Color, texture, aroma compound production	Ropars et al. [38]
Sporendonema	*S. casei*	Cheese (France)	Color, texture, aroma compound production	Ropars et al. [38]

Source: Adapted and modified from [25].

dedicated to the quantification of food product loss caused by fungi are scarce in the literature, but earlier studies from Malkki and Ravha [46] and, more recently, Kim and Xiao [47], reported that fungi were responsible for 1–3% of bakery product losses and for 1.9% (fungicide-treated) and 5.1% (untreated) of apple post-harvest losses.

Fungal spoilage can occur at all stages of the food chain, including in the field, post-harvest, during processing and storage by food producers but also during the different food handling steps at the wholesaler, retailer, and consumer's home. In addition, fungi can spoil a very diverse range of foods, from raw materials to processed foods, as they are able to grow over a wide activity water (a_w), pH, and temperature range and can use many different substrates, including sugars, organic acids, proteins, and lipids. The major fungal groups which are involved in food spoilage can be categorized as follows: xerophilic fungi able to grow at $a_w < 0.8$, thermoresistant fungi which can resist heat treatments such as pasteurization, fungi resistant to preservatives (e.g. natamycin, sorbic acid, propionic acid, benzoic acid), fungi able to develop under low O_2 levels and/or in the presence of modified atmosphere and, finally, psychrotolerant or psychrophilic fungi which can grow in refrigerated foods [48]. Given the extent of foods that can be spoiled by filamentous fungi, it is therefore not surprising that from a taxonomical perspective, a wide diversity of fungi are involved in food spoilage. This diversity is illustrated in Figure 20.1, which shows a heatmap of the different genera involved in the spoilage of the main food categories. This figure was created based on a meta-analysis of the different genera, species, and species complexes (83 genera comprising 938 species or species complex) cited in Pitt and Hocking [45] which is one of the reference books for food mycologists. Based on this meta-analysis, four major conclusions can be drawn. First, many fungal genera and species are involved in food spoilage and a single food category can be spoiled by different fungal genera represented by

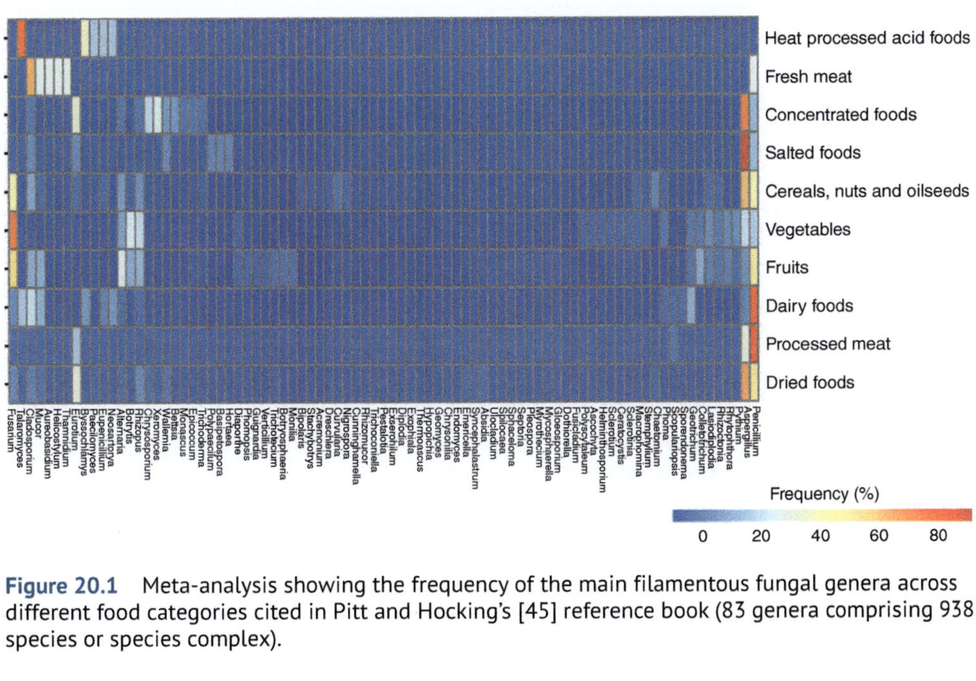

Figure 20.1 Meta-analysis showing the frequency of the main filamentous fungal genera across different food categories cited in Pitt and Hocking's [45] reference book (83 genera comprising 938 species or species complex).

different species. Second, a single genera can be encountered in different food categories. Third, *Aspergillus*, *Penicillium* and *Fusarium* represent the main genera involved in food spoilage, with *Aspergillus* and *Penicillium* spp. being encountered at relatively high frequency in most, if not all, food categories.

Besides food losses and waste linked to fungal spoilage, growth of certain fungi in certain food commodities, especially agricultural crops, may pose a risk to human and animal health because of their ability to produce mycotoxins. These mycotoxins, which are mainly produced by species of the *Alternaria*, *Aspergillus*, *Fusarium*, and *Penicillium* genera, can have toxic, carcinogenic, and mutagenic reactions in humans at low concentrations after acute, short-term, and/or long-term exposure. Aflatoxins, citrinin, trichothecenes (deoxynivalenol, T-2 and HT-2 toxins), ergot alkaloids, fumonisins, ochratoxin A, patulin, and zearalenone are the main mycotoxins of concern from a safety and regulatory point of view (Table 20.2.). Several of these fungal secondary metabolites, including aflatoxins, fumonisins, deoxynivalenol, T-2 and HT-2 toxins, zearalenone and ochratoxin A, have been estimated to occur at levels above the Codex and EU limits in 20–25% of cereals and nuts validating the FAO estimate, while their occurrence above detectable levels can reach up to 60–80% [49]. Notably, mycotoxins present in raw materials can remain in processed foods as they are not very susceptible to technological treatments.

Given the important impact of molds in food, their accurate identification is a key step for managing food safety and quality. Indeed, species-level identification can be a prerequisite for risk assessment with the aim of implementing adequate corrective actions.

Table 20.2 Examples of mycotoxins, producing fungi and host plants.

Mycotoxin	Producing fungi	Host
Aflatoxins (B1, B2, G1, G2)	*Aspergillus flavus*, *Aspergillus parasiticus*	Wheat, barley, corn, peanut
Citrinin	*Monascus purpureus*	Food supplements based on fermented rice
Trichothecenes (deoxynivalenol, T-2, HT-2)	*Fusarium* spp.	Wheat, barley, oat, corn, rice, rye
Ergot alkaloids	*Claviceps purpurea*	Rye and other grains
Zearalenone	*Fusarium* spp.	Corn, wheat, barley, sorgho, rye
Fumonisins (B1, B2, B3)	*Fusarium proliferatum*, *Fusarium verticilloides*	Corn, sorgho, rice
Ochratoxine A	*Aspergillus ochraceus*, *Penicillium viridicatum*, *Penicillium cyclopium*	Corn, grape, barley, soya, coffee, and cocoa beans
Patulin	*Penicillium expansum* and other Penicillium spp., *Aspergillus clavatus*, *A. giganteus* and *A. longivesica*, *Byssochlamys nivea*	Pome and stone fruits

20.3 Identification of Fungi

For years, fungal species identification has only been achieved through morphological approaches. Nevertheless, accurate morphological identification of fungi is tedious and challenging; compared with plants and animals, fungi have simple body plans, often with morphologically obscure structures [50]. Morphological diagnostic features are usually seen, after cultivation on an agar medium, via examination of their microscopic asexual and/or sexual spore-producing structures as well as the observation of their growth behavior under different abiotic conditions (temperature, a_w, etc.). Based on these observations, fungal specimens can be identified with the help of dichotomous identification keys such as the widely used *Laboratory Guide to Common Penicillium Species* [51] or those available in *Food and Indoor Fungi* [52].

Over the last two decades, DNA barcoding has considerably modified our ability to identify species by relying on a single feature: a DNA sequence nucleotide composition. Indeed, DNA barcoding uses a short section of DNA from a specific gene (a DNA barcode). The DNA barcode is amplified by polymerase chain reaction using generic primers (i.e. amplifying the barcode in all species within a group of organisms such as a kingdom) and the resulting amplicons are sequenced to reveal their nucleotide composition. By comparison with a curated reference library of such a DNA barcode, an individual sequence can be used to uniquely identify a specimen. The nucleotide sequence of the selected barcode must be conserved at the species level but can be variable among species.

In fungi, the universal DNA barcode is the nuclear ribosomal DNA (rDNA) internal transcribed spacer (ITS) [53]. The ITS is located between the 18S rRNA and 28S rRNA genes and includes the 5.8S rRNA gene that splits the ITS into two parts: ITS1 and ITS2. The ITS region undergoes a faster rate of evolution than rRNA genes but, fulfilling significant functions during rRNA maturation [54, 55], they are under selective pressure and remain homogeneous within a species [56]. This intraspecies conservation and interspecies variability of its nucleotide composition and its huge representation in reference libraries (e.g. >1 500 000 fungal ITS sequences available in GenBank) make ITS an ideal fungal DNA barcode. Besides ITS, other DNA barcodes may be used in fungi because they offer a better resolution for certain genera, and protein coding genes, such as the β-tubulin encoding gene [57] for *Penicillium* spp. or the elongation factor encoding gene for *Fusarium* spp. [58], are commonly used.

Beyond species identification of fungal isolates, the use of DNA barcodes, when associated with next-generation sequencing (NGS), which allow massive and parallel sequencing, enables culture-independent analyses of fungal communities in complex samples (e.g. food product) and their spatial and temporal dynamics [59].

Although DNA barcoding allows for accurate identification, when used routinely, it can be costly and time-consuming and still requires expertise [60]. That is the reason why techniques such as matrix-assisted laser desorption/ionization time-of-flight (MALDI-TOF) mass spectrometry (MS), which is more rapid, easier to handle but still reliable in identifying filamentous fungi, has emerged in the last decade as a valuable asset in the context of routine identification of fungal isolates in the food industry.

20.4 Identification of Foodborne Molds Using MALDI-TOF MS

Rapid bacterial and yeast identification by MALDI-TOF MS has revolutionized microbiology [61] and is now widespread around the world in many different types of laboratories, including food microbiology laboratories. MALDI-TOF MS for mold identification is now well established [62, 63] but it took more time for laboratories to adopt this technology in routine practice due to limited database availability and lack of standardized processes [64]. Identification of molds in general, including foodborne molds, is highly dependent on the sample preparation [65].

20.4.1 Sample Preparation

Several protocols are described in the literature, either from liquid culture or from solid medium. When grown on a solid medium, the molds can also be either spotted directly on the target slide used for analysis or can undergo a tube extraction before deposit [65].

Some initial studies used the direct transfer method [66]. A portion of a grown colony on solid media is directly spotted on the MALDI target slide or suspended in 200 μl of distilled water before spotting. Each sample is overlaid with 1 μl of absolute ethanol or formic acid and air-dried at room temperature. One microliter of a saturated solution of α-cyano-4-hydroxycinnamic acid (CHCA) in 50% acetonitrile/2.5% trifluoroacetic acid is then added and allowed to co-crystallize with the sample at room temperature. This sample preparation has many limitations and can also lead to staff and instrument contamination, especially in the case of sporulating fungi. Therefore, studies using direct transfers are outnumbered by studies using a cell lysis/extraction approach, in which an ethanol-formic acid procedure, starting mostly from agar, allows complete protein extraction [65]. The protocol, which involves formic acid and acetonitrile, is safe for the staff and for the instrument and leads to reproducible results. A standardized protocol is described in Quéro et al. [67]. Mold culture (hyphae and/or conidia) are collected from the agar plate using a moisturized cotton swab, and fungal material is then transferred into a microcentrifuge tube containing 900 μl of 70% ethanol. After homogenization, a first centrifugation step at 14 000 g for two minutes is performed, the supernatant is discarded, suspended in 40 μl of 70% formic acid and mixed. Acetonitrile (40 μl) is then added, before the second step centrifugation at 14 000 g for two minutes. Proteins in suspension in the supernatant are spotted on the target slide and covered with matrix before MALDI-TOF analysis.

This protocol is reproducible and allows spore inactivation. It is easy to perform and does not require specific expertise. The spectra can be different depending on the age of culture and solid media used for extraction (Figure 20.2). Those parameters should be taken into account for database building.

The sample preparation from liquid culture is not widely used because of its complexity. The strains are grown in liquid medium and incubated at 30°C for 24–48 hours [68]. One milliliter of the liquid fungal culture is placed into a 1.5 ml Eppendorf tube and centrifuged at 15 000 g for two minutes. The surpernatant is then discarded and submitted to several washing/centrifugation steps in order to eliminate protein from media. The pellet is resuspended in 300 μl distilled water and 900 μl of ethanol is added. After a centrifugation step at 15 000 g for two minutes, the pellet is resuspended in 70% formic acid, incubated for

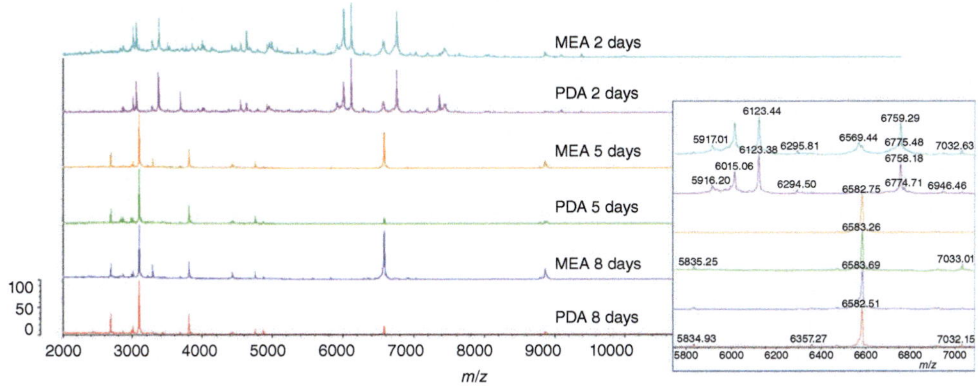

Figure 20.2 *Aspergillus brasiliensis* ATCC 16404 spectra obtained at two, five, and eight days on both solid media malt extract agar (MEA) and potato dextrose agar (PDA).

5–10 minutes at room temperature, before acetonitrile addition and centrifugation at 15 000 g for one minute. Supernatant is deposited on the target, covered by CHCA matrix and analyzed with MALDI-TOF. This technique is slow, requires specific equipment, and mix cultures are not detectable.

Overall, mold identification using MALDI-TOF MS needs a standardized protocol which includes an extraction step and preferably from solid culture.

20.4.2 Database Building and Performance of MALDI-TOF for Identification of Foodborne Molds

As previously mentioned, the adoption rate of MALDI-TOF MS for molds was slow, mainly due to poor performance of manufacturer's databases, leading to a heavy dependency on laboratory-developed databases. The species coverage of MALDI reference databases was also the Achilles' heel of MALDI-TOF MS systems when applicable to foodborne molds. These two hurdles have been overcome recently with the recent release of an FDA-cleared mold database with reliable sample preparation and performance [69–71] and the work done to expand it to foodborne microorganisms, including molds [67, 72, 73].

20.4.2.1 Database Building

To build a robust spectral database, three factors need to be taken into account. The first one is a standardized sample preparation, as explained earlier. The second factor is the microbial variability. To take into account this variability, several strains of the same species should be tested. The recommended number is 10. Those strains should ideally have different geographical origins and should come from different sample types (dairy products, beverages, pastries, etc.). This diversity is needed to capture the species characteristics, as different profiles can exist depending on the origin of the strains. The third factor is the technical variability. As shown in Figure 20.2, the mold spectra can be different depending on the growth phase. At least two different times of growth should be used to build the database in order to get hyphae and conidia. For example, for rapid growers like *Aspergillus*,

Table 20.3 Example of an experimental plan for database-building for one species.

Strain	Incubation time	Medium	Vendor
Strain 1	2 days	Sabouraud dextrose agar	bioMérieux
Strain 1	8 days	Potato dextrose agar	Becton Dickinson
Strain 2	2 days	Malt extract agar	Oxoid
Strain 2	8 days	Yeast glucose chloramphenicol agar	Becton Dickinson
Strain 3	2 days	Sabouraud dextrose agar	Becton Dickinson
Strain 3	8 days	Potato dextrose agar	bioMérieux
Strain 4	2 days	Malt extract agar	bioMérieux
Strain 4	8 days	Yeast glucose chloramphenicol agar	Oxoid

Each strain is tested after two and eight days of incubation on two different media from two different suppliers.

spectra should be acquired after two and eight days of growth. Different media and vendors should also be used to represent the diversity of media used in industrial mycology. An example of an experimental plan is shown in Table 20.3. Once spectra are acquired, they are several ways to build databases and estimate performance. The manufacturer bioMérieux developed its own algorithm [74] for their MALDI-TOF identification solution, called VITEK MS. Peak lists are binned and a log base scaling of the peak intensities is applied followed by a L1-normalization. A predictive model is established for each species using the Advanced Spectra Classifier (ASC) algorithm. This procedure provided a specific weighed bin matrix for each species, and for identification, new spectra are compared to the bin weight matrix and the sum of matching bin weights is calculated and considered as an intermediate score. All species-specific scores are then transformed into multiclass probability estimates using a Gaussian-based calibration procedure that takes into account the score distributions within species compared with all other species. To estimate the performance, a five-fold cross-validation is used to assess how accurately it would perform on independent new spectra. The spectral data are randomly split into five subsets. One round of cross-validation involved a learning phase on four subsets and validated the identification performance on the remaining subset. Five rounds of cross-validation are performed by permutation of the subsets. The results are combined across rounds to get an estimation of identification performance. Once the spectral database is built and its performance estimated, it should be challenged with external isolates as described in Quéro et al. [67].

20.4.2.2 Performance of Foodborne Mold Database

Only two studies on foodborne molds and MALDI-TOF MS have been published. The aim of the study by Quero et al. [67] was to implement a spectral database representative of food spoilage molds. To this end, 6477 spectra were acquired from 618 fungal strains belonging to 136 foodborne species and integrated in a database. The performance of this database was then evaluated by cross-validation and ~95% of correct identification to the species level was achieved, independently of the cultivation medium and incubation time. The database was also challenged with external isolates belonging to 52 species claimed in the database, and 90% were correctly identified to the species level. This comprehensive study shows that MALDI-TOF MS can be applied to the identification of foodborne molds with reliable performance when a standardized sample preparation is used. This study also showed that some phylogenetically close species such as *Penicillium camemberti* and *Penicillium commune* could not be separated. By contrast, the species of the *Cladosporium sphaerospermum* complex integrated into the database could be differentiated and correctly identified. Another 2020 study by the same author (Quero et al., [73]) showed that the differentiation of species to the complex level can be challenging. The aim of this study was to evaluate the use of MALDI-TOF to identify species of the *Aspergillus* section Flavi. Spectra were acquired from 23 *Aspergillus* species and integrated into a database for which cross-validation led to more than 99% of correctly attributed spectra. Even closely related species which cannot be differentiated from a morphological point of view and require a polyphasic analysis, such as *Aspergillus parasiticus* and *Aspergillus sojae*, could be differentiated using MALDI-TOF MS. In the second part of the study, the authors tried to differentiate *Penicillium roqueforti* isolates from three distinct genetic populations. For this study, spectra were acquired from 63 strains and a two-step calibration procedure was applied before database construction. Cross-validation and external validation led to 94% and 95% of spectra, respectively, attributed to the right population. The results suggested very good agreement between spectral and genetic data analysis, demonstrating MALDI-TOF's applicability as a fast and easy alternative to molecular techniques for strain typing of filamentous fungi in combination with the use of alternative algorithms.

Another interesting perspective for the identification of foodborne molds using MALDI-TOF MS is the direct identification from food matrices. Preliminary studies suggest that the identification performance is very variable depending on the food matrix and the fungal species studied. These results also indicate that it is necessary to wait for a significant development of the fungal contaminant to obtain a reliable identification (J. Mounier, personal communication).

Overall, MALDI-TOF MS is a promising tool for the identification of foodborne molds now that standardized protocols and robust databases exist. Published studies are scarce and additional work remains to be done to assess the performance and capabilities of such a technique to identify and type the major mold species encountered in food. One challenge will also to keep the database taxonomy updated as fungal taxonomy is constantly evolving. New species and new strains for species already encountered in the database will also need to be added on a regular basis to ensure that the databases provide the most accurate identification results.

References

1 Turland, N.J., Wiersema, J.H., Barrie, F.R. et al. (2018). *International Code of Nomenclature for Algae, Fungi, and Plants (Shenzhen Code) Adopted by the Nineteenth International Botanical Congress Shenzhen, China, July 2017*. Koeltz Botanical Books.

2 Huxley, J. (1942). *Evolution. The Modern Synthesis*. London: Allen & Unwin.

3 Cain, A.J. (1999). John Ray on the species. *Arch. Nat. His.* 26: 223–238.

4 Darwin, C. (1859). *The Origin of Species and the Descent of Man.* New York: The Modern Library.

5 Mayr, E. (1942). *Systematics and the Origin of Species*, 334. New York: Columbia University Press.

6 De Queiroz, K. (1998). The general lineage concept of species, species criteria, and the process of speciation. In: *Endless Forms: Species and Speciation*, 57–75. Oxford University Press.

7 Burnett, J.H. (2003). *Fungal Populations and Species.* USA: Oxford University Press.

8 Chethana, K.W., Manawasinghe, I.S., Hurdeal, V.G. et al. (2021). What are fungal species and how to delineate them? *Fungal Divers.* 109: 1–25.

9 Giraud, T., Refrégier, G., Le Gac, M. et al. (2008). Speciation in fungi. *Fungal Genet. and Biol.* 45: 791–802.

10 Kohn, L.M. (2005). Mechanisms of fungal speciation. *Annu. Rev. Phytopathol.* 43: 279–308.

11 Natvig, D.O. and May, G. (1996). Fungal evolution and speciation. *J. Genet.* 75: 441–452.

12 Rundle, H.D. and Nosil, P. (2005). Ecological speciation. *Ecol. Lett.* 8: 336–352.

13 Gibbons, J.G. and Rinker, D.C. (2015). The genomics of microbial domestication in the fermented food environment. *Curr. Opin. Genet. Dev.* 35: 1–8.

14 Hermet, A., Méheust, D., Mounier, J. et al. (2012). Molecular systematics in the genus mucor with special regards to species encountered in cheese. *Fun. Biol.* 116: 692–705.

15 Morin-Sardin, S., Rigalma, K., Coroller, L. et al. (2016). Effect of temperature, pH, and water activity on mucor spp. growth on synthetic medium, cheese analog and cheese. *Food Microbiol.* 56: 69–79.

16 Lebreton, A., Corre, E., Jany, J.-L. et al. (2020). Comparative genomics applied to *Mucor* species with different lifestyles. *BMC Genom.* 21: 1–21.

17 Gladieux, P., Ropars, J., Badouin, H. et al. (2014). Fungal evolutionary genomics provides insight into the mechanisms of adaptive divergence in eukaryotes. *Mol. Ecol.* 23: 753–773.

18 Taylor, J.W., Jacobson, D.J., Kroken, S. et al. (2000). Phylogenetic species recognition and species concepts in fungi. *Fun. Genet. and Biol.* 31: 21–32.

19 Felsenstein, J. and Felenstein, J. (2004). *Inferring Phylogenies.* Sunderland, MA: Sinauer Associates.

20 Hawksworth, D.L. and Lücking, R. (2017). Fungal diversity revisited: 2.2 to 3.8 million species. *Microbiol. Spect.* 5: 5–4.

21 James, T.Y., Stajich, J.E., Hittinger, C.T., and Rokas, A. (2020). Toward a fully resolved fungal tree of life. *Ann. Rev. Microbiol.* 74: 291–313.

22 Wijayawardene, N.N., Hyde, K.D., Al-Ani, L.K.T. et al. (2020). Outline of fungi and fungus-like taxa. *Mycosphere* 11: 1060–1456.

23 Spatafora, J.W., Aime, M.C., Grigoriev, I.V. et al. (2017). The fungal tree of life: from molecular systematics to genome-scale phylogenies. *Microbiol. Spect.* 5 (5): https://doi.org/10.1128/microbiolspec.FUNK-0053-2016.

24 Cooke, R.C. and Rayner, A.D. (1984). *Ecology of Saprotrophic Fungi.* Longman.

25 Nout, R.M.J. (2007). The colonizing fungus as a food provider. In: *Food Mycology: A Multifaceted Approach to Fungi and Food* (ed. J. Dijksterhuis and R.A. Samson), 335–352. Boca Raton.Food Mycology: CRC Press.

26 Chou, C.-C., Ho, F.-M., and Tsai, C.-S. (1988). Effects of temperature and relative humidity on the growth of and enzyme production by *Actinomucor taiwanensis* during Sufu Pehtze preparation. *Appl. Environ. Microbiol.* 54: 688–692.

27 Han, B.Z., Rombouts, F.M., and Nout, M.J. (2001). A Chinese fermented soybean food. *Int. J. Food Microbiol.* 65: 1–10.

28 Tsuyoshi, N., Fudou, R., Yamanaka, S. et al. (2005). Identification of yeast strains isolated from marcha in Sikkim, a microbial starter for amylolytic fermentation. *Int. J. Food Microbiol.* 99: 135–146.

29 Pedraza-Reyes, M. and Lopez-Romero, E. (1991). Chitinase activity in germinating cells of *Mucor rouxii*. *Antonie. Van. Leeuwenhoek.* 59: 183–189.

30 Agranoff, J. and Markham, P. (1997). Fatty acid components of tempe (and tapeh). In: *Proceedings or the International Tempe Symposium*, Bali, Indonesia (13–15 July 1997) (ed. S. Sudarmadji, Suparmo and S. Raharjo), 205–210. Jakarta, Indonesia: Indonesian Tempe Foundation.

31 Bruna, J.M., Ordóñez, J.A., Fernández, M. et al. (2001). Microbial and physico-chemical changes during the ripening of dry fermented sausages superficially inoculated with or having added an intracellular cell-free extract of *Penicillium aurantiogriseum*. *Meat Sci.* 59: 87–96.

32 Ginting, E. and Arcot, J. (2004). High-performance liquid chromatographic determination of naturally occurring folates during tempe preparation. *J. Agric. Food. Chem.* 52: 7752–7758.

33 Nout, R.M.J. and Kiers, J.L. (2005). Tempe fermentation, innovation and functionality: update into the third millenium. *J. Appl. Microbiol.* 98: 789–805.

34 Nout, R.M.J. and Aidoo, K.E. (2002). Asian fungal fermented food. In: *Industrial Applications* (ed. H.D. Osiewacz), 23–47. Berlin-Heidelberg-New York: Springer-Verlag.

35 Hanya, Y. and Nakadai, T. (2003). Yeasts and soy products. In: *Yeasts in Food: Beneficial and Detrimental Aspects* (ed. T. Boekhout and V. Robert), 413–428. Hamburg, Germany: Behr's Verlag GmbH & Co KG.

36 Campbell-Platt, G. (1987). *Fermented Foods of the World: A Dictionary and Guide.* London: Butterworths.

37 Bachmann, H.P., Bobst, C., Bütikofer, U. et al. (2005). Occurrence and significance of *Fusarium domesticum* alias Anticollanti on smear-ripened cheeses. *LWT Food Sci. Technol.* 38: 399–407.

38 Ropars, J., Cruaud, C., Lacoste, S., and Dupont, J. (2012). A taxonomic and ecological overview of cheese fungi. *Int. J. Food Microbiol.* 155: 199–210.

39 Savary, O., Coton, M., Frisvad, J.C. et al. (2021). Unexpected Nectriaceae species diversity in cheese, description of *Bisifusarium allantoides* sp. nov., *Bisifusarium penicilloides* sp. nov., *Longinectria* gen. nov. *lagenoides* sp. nov. and *Longinectria verticilliforme* sp. nov. *Mycosphere* 12: 1077–1100.

40 Juzlova, P., Martinikova, L., and Kren, V. (1996). Secondary metabolites of the fungus *Monascus:* a review. *J. Ind. Microbiol.* 16: 163–170.

41 Akihisa, T., Tokuda, H., Yasukawa, K. et al. (2005). Azaphilones, furanoisophthalides, and amino acids from the extracts of *Monascus pilosus*-fermented rice (red-mold rice) and their chemopreventive effects. *J. Agric. Food. Chem.* 53: 562–565.

42 Beuchat, L.R. (1986). Oncom (fermented peanut press cake). In: *Legume-Based Fermented Foods* (ed. N.R. Reddy, M.D. Pierson and D.K. Salunkhe), 135–140. Boca Raton, USA: CRC Press Inc.

43 Spotti, E. and Berni, E. (2007). Starter cultures: Molds. In: *Handbook of Fermented Meat and Poultry* (ed. F. Toldrá), 171–176. Ames, USA: Blackwell Publishing.

44 Food and Agriculture Organization of the United Nations (FAO) (2011). *Global Food Losses and Food Waste: Extent, Causes and Prevention*. Rome: FAO.

45 Pitt, J. and Hocking, A.D. (2009). Spoilage of stored, processed and preserved foods. In: *Fungi and Food Spoilage*, 3e (ed. J. Pitt and A.D. Hocking), 401–423. NY: Springer.

46 Malkki, Y. and Ravha, O. (1978). Mold inhibition by aerosols. *Baker's Digest* 52: 47–50.

47 Kim, Y.K. and Xiao, C.L. (2008). Distribution and incidence of *Sphaeropsis* rot in apple in Washington state. *Plant. Dis.* 92: 940–946.

48 Rico-Munoz, E., Samson, R.A., and Houbraken, J. (2019). Mould spoilage of foods and beverages: using the right methodology. *Food Microbiol.* 81: 51–62.

49 Eskola, M., Kos, G., Elliott, C.T. et al. (2020). Worldwide contamination of food-crops with mycotoxins: validity of the widely cited 'FAO estimate' of 25%. *Crit. Rev. Food Sci. Nutr.* 60: 2773–2789.

50 Lücking, R., Aime, M.C., Robbertse, B. et al. (2020). Unambiguous identification of fungi: where do we stand and how accurate and precise is fungal DNA barcoding? *IMA fungus.* 11: 1–32.

51 Pitt, J.I. (1988). *A Laboratory Guide to Common Penicillium Species*. Food Processing: CSI Res. Org. Div.

52 Samson, R.A., Houbraken, J., Thrane, U. et al. (2019). *Food and Indoor Fungi*. Westerdijk Fungal Biodiversity Institute.

53 Schoch, C.L. et al. (2012). Nuclear ribosomal internal transcribed spacer (ITS) region as a universal DNA barcode marker for fungi. *Proc. Nat. Acad. Sci.* 109: 6241–6246.

54 Joseph, N., Krauskopf, E., Vera, M.I., and Michot, B. (1999). Ribosomal internal transcribed spacer 2 (ITS2) exhibits a common core of secondary structure in vertebrates and yeast. *Nucleic Acids Res.* 27: 4533–4540.

55 Liu, J.S. and Schardl, C.L. (1994). A conserved sequence in internal transcribed spacer 1 of plant nuclear rRNA genes. *Plant Molecul. Biol.* 26: 775–778.

56 Liao, D. (2000). Gene conversion drives within genic sequences: concerted evolution of ribosomal RNA genes in bacteria and archaea. *J. Molecul. Evolut.* 51: 305–317.

57 Visagie, C.M., Houbraken, J., Frisvad, J.C. et al. (2005). Identification and nomenclature of the genus penicillium. *Stud. in Mycol.* 53: 53–62.

58 Geiser, D.M., del Mar, J.-G.M., Kang, S. et al. (2004). FUSARIUM-ID v. 1.0: a DNA sequence database for identifying Fusarium. *Eur. J. Plant Pathol.* 110: 4731–4479.

59 Grant, D.M., Brodnicke, O.B., Evankow, A.M. et al. (2021). The future of DNA barcoding: reflections from early career researchers. *Diversity* 13: 313.

60 Chalupová, J., Raus, M., Sedlářová, M., and Šebela, M. (2014). Identification of fungal microorganisms by MALDI-TOF mass spectrometry. *Biotechnol. Adv.* 32: 230–241.

61 Patel, R. (2015). MALDI-TOF MS for the diagnosis of infectious diseases. *Clin. Chem.* 61 (1): 100–111.

62 Gautier, M., Ranque, S., Normand, A.C. et al. (2014). Matrix-assisted laser desorption ionization time-of-flight mass spectrometry: revolutionizing clinical laboratory diagnosis of mould infections. *Clin Microbiol Infect.* 20 (12): 1366–1371.

63 L'Ollivier, C. and Ranque, S. (2017). MALDI-TOF-based dermatophyte identification. *Mycopathologia.* 182 (1-2): 183–192.

64 Lau, A.F., Walchak, R.C., Miller, H.B. et al. (2019). Multicenter study demonstrates standardization requirements for Mold identification by MALDI-TOF MS. *Front. Microbiol.* https://doi.org/10.3389/fmicb.2019.02098.

65 Sanguinetti, M. and Posteraro, B. (2017). Identification of Molds by matrix-assisted laser desorption ionization-time of flight mass spectrometry. *J. Clin. Microbiol.* 55 (2): 369–379.

66 De Carolis, E., Posteraro, B., Lass-Florl, C. et al. (2012). Species identification of *Aspergillus*, *Fusarium* and *Mucorales* with direct surface analysis by matrix-assisted laser desorption ionization time-of-flight mass spectrometry. *Clin. Microbiol. Infect.* 18: 475–484.

67 Quéro, L., Girard, V., Pawtowski, A. et al. (2019). Development and application of MALDI-TOF MS for identification of food spoilage fungi. *Food Microbiol.* 81: 76–88.

68 Vidal-Acuña, M.R., Ruiz-Pérez de Pipaón, M., Torres-Sánchez, M.J., and Aznar, J. (2018). Identification of clinical isolates of aspergillus, including cryptic species, by matrix assisted laser desorption ionization time-of-flight mass spectrometry (MALDI-TOF MS). *Med. Mycol.* 56 (7): 838–846.

69 Pinheiro, D., Monteiro, C., Faria, M.A., and Pinto, E. (2019 Oct). Vitek® MS v3.0 system in the identification of filamentous fungi. *Mycopathologia.* 184 (5): 645–651.

70 Rychert, J., Slechta, E.S., Barker, A.P. et al. (2018). Multicenter evaluation of the Vitek MS v3.0 system for the identification of filamentous fungi. *J. Clin. Microbiol.* 56 (2): e01353–e01317.

71 Sun, Y., Guo, J., Chen, R. et al. (2021). Multicenter evaluation of three different MALDI-TOF MS systems for identification of clinically relevant filamentous fungi. *Med. Mycol.* 59 (1): 81–86.

72 Akimowicz, M. and Bucka-Kolendo, J. (2020). MALDI-TOF MS – application in food microbiology. *Acta Biochim. Pol.* 67 (3): 327–332.

73 Quéro, L., Courault, P., Cellière, B. et al. (2020). Application of MALDI-TOF MS to species complex differentiation and strain typing of food related fungi: case studies with *Aspergillus* section FLAVI species and *Penicillium roqueforti* isolates. *Food Microbiol.* 86: 103311.

74 Girard, V., Mailler, S., Welker, M. et al. (2016). Identification of *Mycobacterium* spp. and *Nocardia* spp. from solid and liquid cultures by matrix-assisted laser desorption ionization-time of flight mass spectrometry (MALDI-TOF MS). *Diagn. Microbiol. Infect. Dis.* 86 (3): 277–283.

Index

Note: Page numbers in *italics* indicate figures, those in **bold** indicate tables.

Microbiological Identification using MALDI-TOF and Tandem Mass Spectrometry: Industrial and Environmental Applications, First edition. Edited by Haroun N. Shah, Saheer E. Gharbia, Ajit J. Shah, Erika Y. Tranfield, and K. Clive Thompson.

c

d

f

q

t